ENZYME ENGINEERING XI

ANNALS OF THE NEW YORK ACADEMY OF SCIENCES
Volume 672

ENZYME ENGINEERING XI

Edited by Douglas S. Clark and David A. Estell

The New York Academy of Sciences
New York, New York
1992

♾ *The paper used in this publication meets the minimum requirements of American National Standard for Information Sciences–Permanence of Paper for Printed Library Materials, ANSI Z39.48-1984.*

SP
Printed in the United States of America
ISBN 0-89766-763-8 (cloth)
ISBN 0-89766-764-6 (paper)
ISSN 0077-8923

ANNALS OF THE NEW YORK ACADEMY OF SCIENCES

Volume 672
November 30, 1992

ENZYME ENGINEERING XI[a]

Editors and Conference Organizers
DOUGLAS S. CLARK and DAVID A. ESTELL

CONTENTS

[a]The papers in this volume were presented at the Eleventh International Enzyme Engineering Conference, held in Keauhou-Kona, Hawaii, on September 22–27, 1991, sponsored by the Engineering Foundation (USA).

Part II. Improved Biocatalysts through Genetic Engineering

Part III. Enzyme Stability and Stabilization

Part IV. Novel Applications of Enzymes and Enzymelike Molecules

Part V. Enzymes under Unusual Conditions

Part VI. Applications of Enzymes in Synthesis

Part VIII. Innovations in Biocatalyst Preparation and Bioreactor Design

Introduction

DOUGLAS S. CLARK[a] AND DAVID A. ESTELL[b]

[a]*Department of Chemical Engineering*
University of California
Berkeley, California 94720

[b]*Genencor International*
180 Kimball Way
South San Francisco, California 94080

This volume contains the proceedings of the Eleventh International Enzyme Engineering Conference, which was held in Keauhou-Kona, Hawaii, on September 22–27, 1991. The conference was sponsored by the Engineering Foundation, New York. Enzyme Engineering XI focused on scientific as well as engineering aspects of enzyme technology, including recent advances in protein engineering and new applications of biocatalysts. The main themes of papers presented at the meeting were:

- enzyme structure and function
- improved biocatalysts through genetic engineering
- enzyme stability and stabilization
- novel applications of enzymes and enzymelike molecules
- enzymes under unusual conditions
- applications of enzymes in synthesis
- innovations in biocatalyst preparation and bioreactor design
- industrial enzyme processes: case studies.

Poster presentations covered a wide range of topics, including properties and applications of enzymes in organic solvents, immobilization of biocatalysts, enzyme kinetics, and the biocatalytic production of chemicals.

The Executive Committee in charge of organizing the conference was composed as follows:

Executive Chairman—Douglas S. Clark; Conference Chairman—David Estell; Members—Frances Arnold, Harvey Blanch, Isao Karube, Klaus Mosbach, Donald Renn, Frieder Schiller, Tetsuya Tosa, Johannes Tramper, Christian Wandrey, and Allen Laskin (ex officio).

We wish to acknowledge gratefully the many companies whose financial support made possible the attendance of selected plenary speakers, session chairs, and individuals presenting posters. The funds were also used to support the attendance of several graduate students. Contributions were made by Amgen Center, Archer Daniels Midland, Bio-Technical Resources, Bristol-Myers Squibb, Chemap AG, Coors Biotech, Dow Chemical, FMC Corporation, Japanese Society of Enzyme Engineering, Eli Lilly & Company, Genentech, New York Academy of Sciences, Novo Nordisk A/S, Procter & Gamble, and Sandoz Research Institute. Genencor's sponsorship of the Enzyme Engineering Award is also gratefully acknowledged. A grant from the National Science Foundation contributed to the support of United States participants from academic institutions. We also wish to thank Don Renn for his invaluable assistance in raising industrial support for the meeting, and Jonathan Dordick for his help in organizing the poster sessions.

The Twelfth Enzyme Engineering Meeting will be held in Deauville, France, in 1993.

The Enzyme Engineering Award

The Enzyme Engineering Award was established in 1982 to recognize exceptional achievements in enzyme engineering research and applications. It recognizes new discoveries, research, process, or device developments in enzyme engineering, and outstanding contributions of a scientific or engineering nature in the design, operation, and management of facilities, processes, or devices based primarily on enzyme engineering. The previous recipients have been Ichiro Chibata, Japan (1983), Klaus Mosbach, Sweden (1985), Ephraim Katchalski-Katzir, Israel (1987), and Saburo Fukui, Japan (1989). The 1991 award was sponsored by Genencor International.

THE 1991 ENZYME ENGINEERING AWARD—ALEXANDER M. KLIBANOV

Alexander Klibanov received his M.S. in Chemistry in 1971 and Ph.D. in Chemical Enzymology in 1974 from Moscow University in Russia. Following his immigration to the United States in 1977, he spent two years as a postdoctoral associate at the Chemistry Department at the University of California at San Diego. In 1979, Dr. Klibanov joined the faculty at the Massachusetts Institute of Technology, where he is currently Professor of Chemistry and a staff member of the Biotechnology Process Engineering Center.

Alexander Klibanov has made numerous pioneering contributions to enzyme technology. His seminal work with enzymes in organic solvents has served to establish the broad utility of enzymes in nonaqueous media. His research on nonaqueous enzymology has also opened new avenues in the study of enzyme selectivity and stereospecificity. In addition, he has provided an understanding of the mechanisms of biorecognition and of protein inactivation at high temperatures. His continuing research interests include stability and stabilization of proteins, enzymes as catalysts in organic chemistry, immobilized enzymes and cells, and enzymes in extreme environments. He has published more than 150 scientific papers and several patents. Professor Klibanov has received several other prestigious awards including the 1986 Leo Friend Award, the 1989 Ipatieff Prize, and the 1991 Marvin J. Johnson Award from the American Chemical Society.

The Influence of Chaperonins on Protein Folding

A Mechanism for Increasing the Yield of the Native Form

S. G. BURSTON, R. SLEIGH, D. J. HALSALL, C. J. SMITH,
J. J. HOLBROOK, AND A. R. CLARKE

Molecular Recognition Center and
Department of Biochemistry
University of Bristol
School of Medical Sciences
Bristol BS8 1TD, England

Molecular chaperones are proteins that promote the correct folding, assembly, and transport of other protein molecules.[1] The most widely studied of the molecular chaperones are the chaperonins; these form a subgroup found in prokaryotic cells, in mitochondria, and in plastids.[2–5] They are oligomeric proteins of high molecular weight, which work in conjunction with a smaller, also oligomeric, coprotein. One such chaperonin is cpn60 from *Escherichia coli,* the product of the *gro-EL* gene locus. This has a subunit mass of approximately 60,000 daltons, and the whole protein exists as an assembly of 14 subunits in a "double-doughnut" structure, each ring comprising 7 subunits. The coprotein is produced by the *gro-ES* locus and is termed cpn10 by virtue of its 10,000-dalton subunit molecular weight. Cpn10 exists as a "single-doughnut" structure of 7 subunits.[6] In conjunction, these proteins have been shown to aid the folding and/or assembly of bacteriophages, multimeric and monomeric protein molecules.[7,8]

They function by binding to unfolded or partially folded protein molecules and, by transducing the energy of ATP hydrolysis, increasing the yield of the natively assembled form.[9] The mechanistic detail of this process remains undefined.

Here we report on preliminary kinetic experiments that explore the interactions of cpn60 with (a) the folding intermediates of *Bacillus stearothermophilus* lactate dehydrogenase (LDH), (b) with ATP and an unreactive analogue (AMP-PNP), and (c) with the coprotein cpn10. From these results with LDH we suggest a mechanism that explains the ability of chaperonins to improve the efficiency of protein folding in general.

THE BINDING OF cpn60 TO PROTEIN-FOLDING INTERMEDIATES

When bacterial LDH is unfolded to equilibrium in solutions of the denaturant guanidinium chloride (Gdm-Cl), four structural states of the protein can be isolated.[10] The native state (NN) is a dimer which dissociates to give, at 1.0 M Gdm-Cl, an inactive monomer (N). This unfolds to give, at 2.2 M Gdm-Cl, an expanded monomer (G) which retains 60% of its alpha-helix content but is freely penetrated by solvent. At 4.0 M Gdm-Cl, the protein is fully unfolded (U).

Previous results have shown that when LDH is allowed to renature from increasingly unfolded states, cpn60 can retard refolding when initiated from the unfolded state (U) but not from the "molten globule" (G), or the monomer (N).[11] A kinetically identifiable, but structurally undefined, intermediate (INT) which occurs on the conversion from U to G is also bound by cpn60. A summary of the folding intermediates of LDH is presented in FIGURE 1.

From the above observations we conclude that cpn60 associates most strongly with the least-folded forms of the protein and not with its molten globule state.

To seek a fuller description of this process, we measured the return of LDH activity over a continuous time course when the fully denatured LDH refolded in the presence and absence of cpn60 (see FIGURE 2a). LDH folding alone proceeds by two, slow, unimolecular steps ($U \rightarrow INT \rightarrow G$) and an apparent bimolecular step ($2G \rightarrow NN$). To describe refolding in the presence of cpn60, further steps must be introduced to account for the interactions of these folding intermediates with the

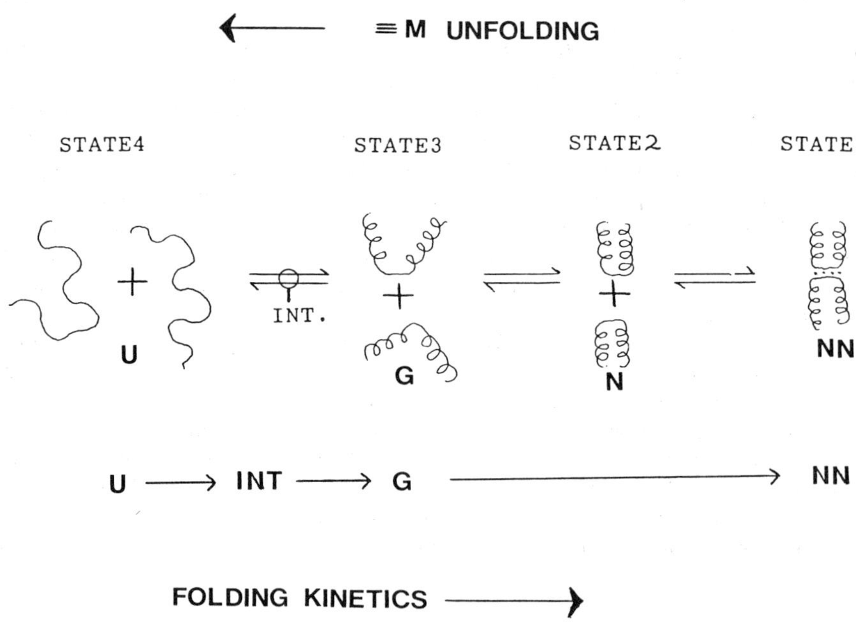

FIGURE 1. The folding mechanism of lactate dehydrogenase. The equilibrium (≡M) and kinetic states identifiable in LDH folding are summarized. In the former experiments,[10] state 1 (NN) is native, state 2 (N) predominates at 1.0 M Gdm-Cl, state 3 (G) at 2.2 M Gdm-Cl, and state 4 (U) at and above 4.0 M Gdm-Cl. In the kinetics of refolding,[11] the protein passes from U to G in two, slow unimolecular steps (3×10^{-3} and 15×10^{-3} seconds^{-1}, see FIGURE 2b). As protein refolding is performed at low protein concentration, the conversion of G to NN is measured as a single bimolecular step with an apparent rate constant of 5×10^5 M^{-1} seconds^{-1} (see FIGURE 2b).

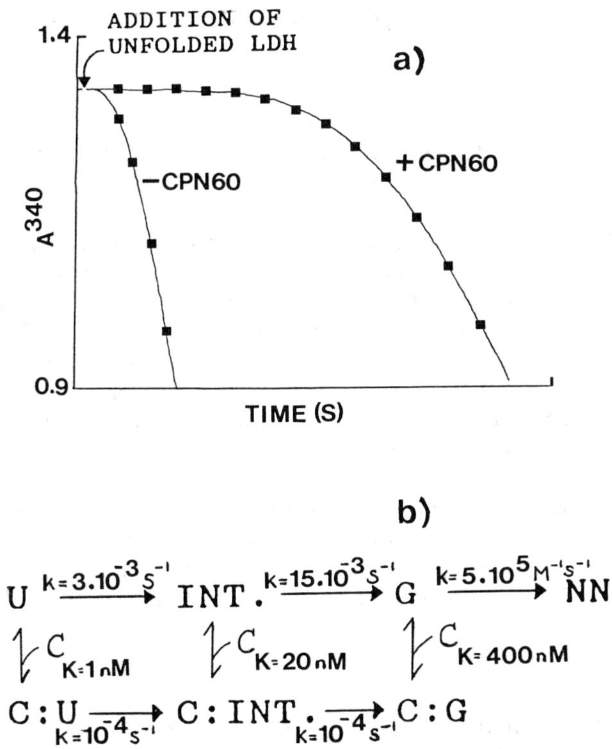

FIGURE 2. The association of cpn60 with LDH during refolding. (**a**) Aliquots of unfolded LDH (in 4.0 M Gdm-Cl/2 mM dithiothreitol) were added to an enzyme assay mix (0.2 mM NADH/10 mM pyruvate/2 mM dithiothreitol/50 mM triethanolamine-HCl pH 7.0). As the enzyme (final subunit concentration 20 nM) refolded, the turnover of NADH was measured by continuously recording absorbance at 340 nm. The experiment was performed in the presence and absence of 200 nM cpn60 (14-mer). The points on the curve represent the best fit of the model described below to the experimental data. (**b**) Model summarizing the results of fitting these curves by numerical integration using the FACSIMILE program.[12] The folding kinetics in the absence of cpn60 (C) are represented on the top line and those steps necessary to describe the interactions of intermediates with the chaperonin are included underneath (binding is described by a dissociation constant K). The analysis demonstrates that the more folded the LDH becomes, the less is its propensity to bind to cpn60.

chaperonin; this goes beyond what is feasible with standard analytical methods, so numerical integration was used to fit these continuous time courses of refolding.

The model shown in FIGURE 2b is the simplest that will fit the experimental observations; it confirms that cpn60 binds most tightly to the least structurally organized forms of LDH and shows that folding can occur on the surface of the chaperonin, but at a much retarded rate. This latter point is supported by the experimental observation that the presence of cpn60 increases the yield of active LDH by two- to threefold when allowed to fold from the fully denatured state.[11] The implication is that, by virtue of its slow folding on the chaperonin surface, the early-folding intermediates of LDH are sequestered from the bulk phase and are therefore protected from irreversible aggregation.

THE INFLUENCE OF NUCLEOTIDE BINDING IN THE ABSENCE OF HYDROLYSIS

Cpn60 is known to bind and hydrolyze ATP, this process mediating release of bound protein substrates.[9] To begin to investigate this aspect of chaperonin function, we have recorded continuous LDH-folding curves in the presence of cpn60 and at increasing concentrations of the nonhydrolyzable ATP analogue AMP-PNP. The results are shown in FIGURE 3, together with the analysis by numerical integration. This requires the introduction of new binding steps describing the interaction of the nucleotide with cpn60. The result is striking in that the cpn60/LDH complex gets weaker as the protein folds, but the binding of nucleotide to cpn60 gets stronger. The apo-cpn60 has a K_d for nucleotide of 2 μM (the same as that predicted for the unstable cpn60/molten globule complex), whereas in the stable cpn60/unfolded LDH complex the K_d rises to 200 μM. We conclude that cpn60 can adopt at least two structural forms: one associates preferentially with nucleoside-triphosphates, one with unfolded proteins; nucleotide hydrolysis is not a prerequisite of protein release.[11]

ANALYSIS OF FOLDING TIMES AND EFFICIENCY—THE EFFECT OF ATP HYDROLYSIS AND THE BINDING OF THE COPROTEIN CPN10

A simplified analysis of the rate of refolding can be made by measuring the lag time for the regaining of LDH activity (defined in FIGURE 4b) in the presence of ligands and cofactors. The influence of AMP-PNP and ATP on the lag time for refolding in the presence of cpn60 is shown in FIGURE 4a. In both cases their effects on LDH release are half-saturated at a concentration of 100–200 μM; this is in accord with the model presented in FIGURE 3. However, at saturating concentrations, ATP is able to reduce the lag time further than is AMP-PNP. This shows that either ATP itself or its turnover products, ADP · P_i, are better at displacing LDH from cpn60 than is AMP-PNP. In view of the data presented in FIGURE 4, the latter seems more likely.

FIGURE 4a also shows the effect of cpn10 on LDH displacement by AMP-PNP. The result demonstrates that cpn10 alone promotes displacement, but inhibits the binding/displacement effect of AMP-PNP. This latter effect then appears to be cooperative. This observation is, at first sight, puzzling, but can be explained if we accept that the hydrophobic cpn10[13] binds to the same form of cpn60 as unfolded LDH, thus further reducing its nucleoside-triphosphate affinity. This interaction may lead to a degree of cpn10/LDH competition for hydrophobic binding faces on cpn60.

The bar chart in FIGURE 5 summarizes these effects on lag times for folding and illustrates the important point that even in optimal conditions for LDH release from

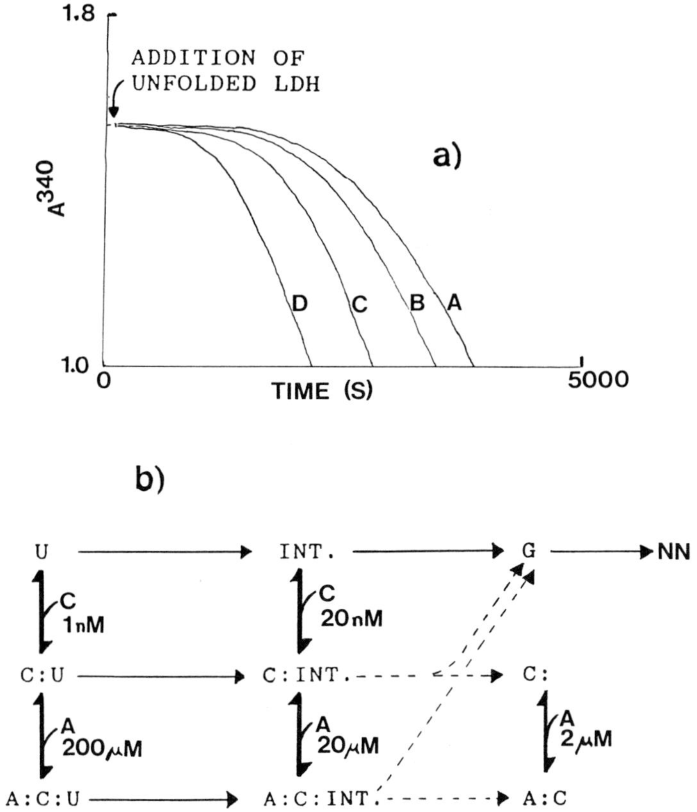

FIGURE 3. The interaction of cpn60 with AMP-PNP. (**a**) Unfolded LDH (20 nM) was added to an assay medium containing 200 nM cpn60 as described for FIGURE 2. AMP-PNP was omitted for curve A and present at 0.05 mM for B, 1.25 mM for C, and 3.75 mM for D. (**b**) This scheme summarizes the apparent binding affinities of AMP-PNP (A) for the chaperonin: protein complexes (see also FIGURE 2). The numerical values represent dissociation constants. The result emphasizes that the nucleotide binds most tightly to apo-cpn60 and most weakly to the most stable cpn60-protein complex (C:U). In these conditions the interaction of G with the chaperonin is sufficiently weak that the C:G and A:C:G complexes never accumulate. This is indicated by the dashed lines showing C:INT and A:C:INT decaying to products at a rate limited by the folding rather than the dissociation rate.

cpn60 (high [ATP] plus cpn10), the *rate* of refolding is not enhanced. In these conditions, however, the *yield* is improved dramatically and over a short time period. This result is presented in full in FIGURE 4b and shows that at high ATP concentrations cpn60 can increase the yield of active LDH over a time period approaching that of "unaided" folding. The inclusion of cpn10 accelerates this effect still further. In the natural cellular environment the chaperonin system comprises cpn60, cpn10, and ATP; we must therefore extend the model shown in FIGURE 3 to account for this.

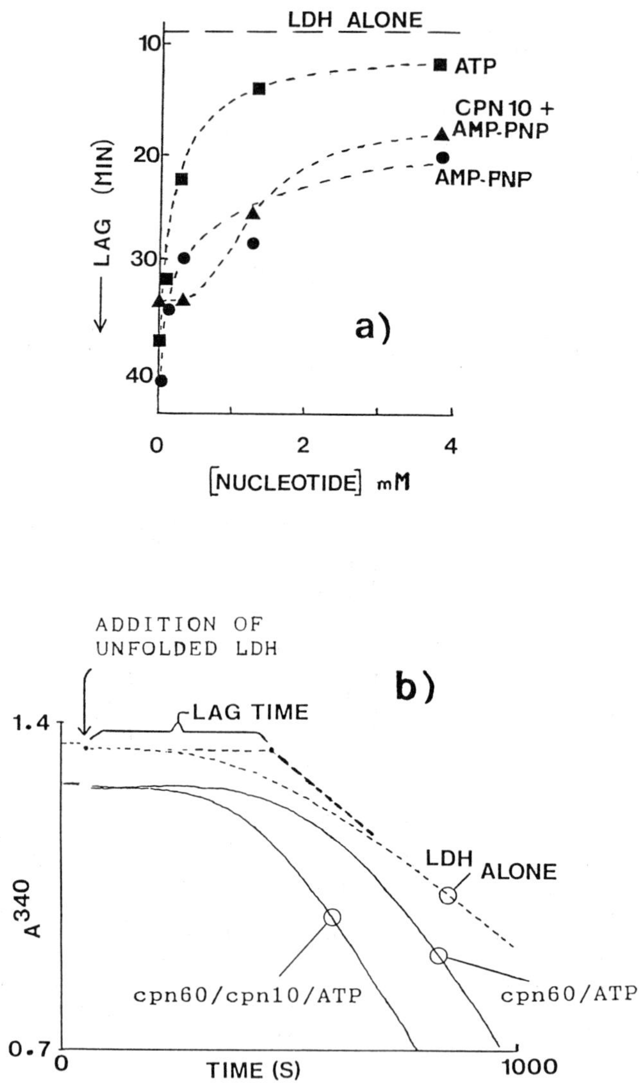

FIGURE 4. Lag times for folding; the concentration dependence of nucleotides and the effect of cpn10. Assays measuring the regaining of LDH activity were performed as described in FIGURES 2 and 3, and the lag time for renaturation recorded. **Plot a** shows the dependence of lag times on nucleotide concentration. In the case of experiments in the presence of cpn10, the concentration of 7-mer was 300 nM. Three progress curves are shown in **b**: LDH refolding in the absence of chaperone (giving the definition of the lag time) and in the presence of cpn60/3.75 mM ATP and cpn60/cpn10/3.75 mM ATP. The latter are included to show the improvement of yield in these conditions which occurs over a period comparable to that of unchaperoned folding.

A PRELIMINARY MECHANISM TO EXPLAIN THE ACTION OF CHAPERONINS IN IMPROVING THE EFFICIENCY OF PROTEIN FOLDING

Several properties of cpn60 are shown by the results presented here. They are:

1. The protein binds most tightly to the most unfolded forms of LDH, thereby stabilizing these states preferentially. We propose that the affinity is dictated by the degree of exposure of hydrophobic residues.
2. Release of the substrate protein is facilitated by the binding of AMP-PNP. This effect is more pronounced in the case of ATP; possibly due to the formation of a cpn60/ADP · P_i complex through hydrolysis.

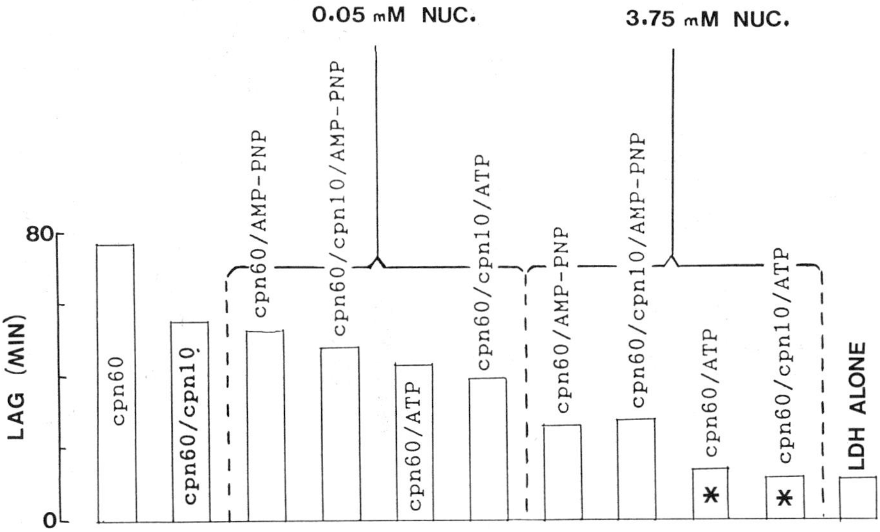

FIGURE 5. Summary of folding times; the influence of cpn60, cpn10, and nucleotides. The lag times for the regaining of LDH activity were measured as described for FIGURES 2, 3, and 4 and are represented here as a histogram. The asterisks mark conditions in which the yield is improved.

3. The binding of cpn10 to cpn60 stabilizes the form that associates strongly with unfolded proteins, not the one that binds AMP-PNP. In doing this, it competes for the hydrophobic binding surface.
4. In optimal conditions (i.e., with cpn10 and ATP), the chaperonin is able to increase the yield of folded protein over a short time period. We see no increase in the rate of folding. These observations demonstrate the ability of the chaperonin either to prevent misfolding or to convert misfolded protein to the native state. We suggest that the latter is more likely.

The previous results of Chandrasekhar *et al.*[13] show that:

5. Cpn10 associates with cpn60 after prolonged incubation with ATP, suggesting

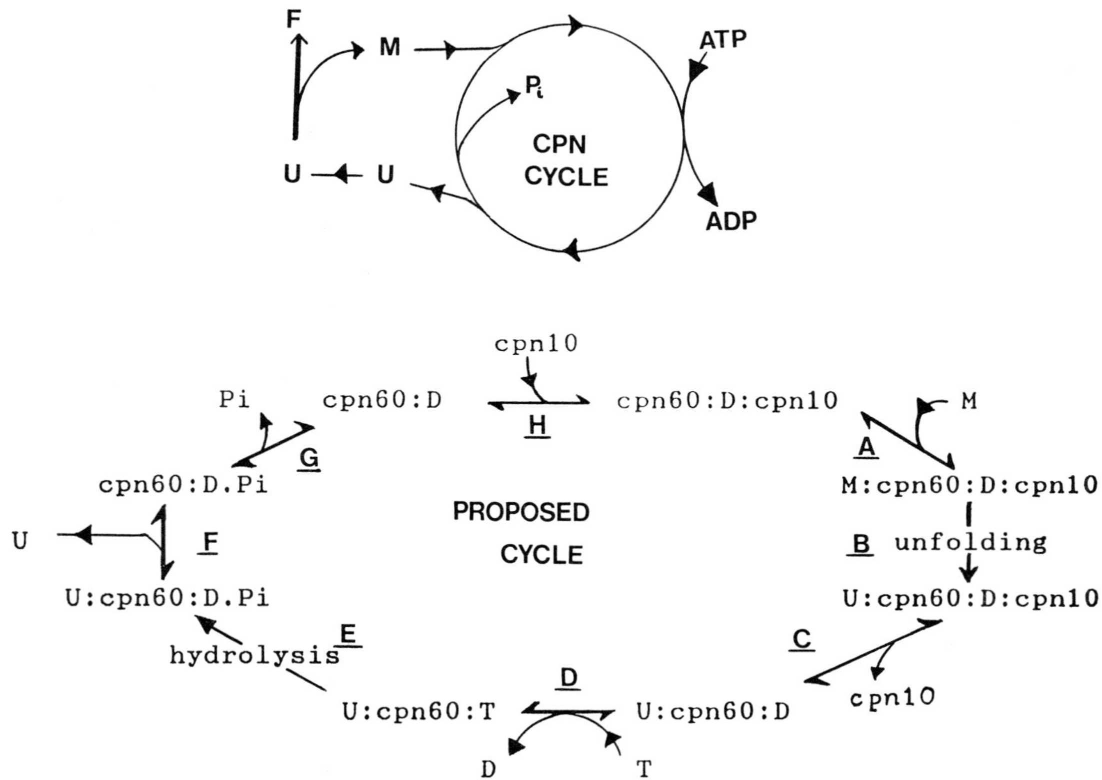

FIGURE 6. Mechanism for the enhancement of protein folding efficiency by chaperonins. Two schemes are shown, the first representing the coupling between folding and ATP hydrolysis where U can fold to its native state F or to a misfolded state M. In net terms, the energy of hydrolysis is coupled to protein unfolding. The second shows the chaperonin cycle in more detail, the steps of which are described in the text. The abbreviation T represents ATP and D represents ADP.

that hydrolysis is required. It does not associate in the presence of nonhydrolyzable analogues. This accords with number 3 above.

6. The rate of ATP hydrolysis by cpn60 is enhanced by association with unfolded protein and inhibited by cpn10.[9,13] We suggest that cpn10 stabilizes a product of ATP hydrolysis in the cpn60 nucleotide site, preventing constant turnover. This product is unlikely to be ADP · P_i as this would lock the chaperonin in a state that has a low affinity for protein substrates. For the mechanistic model, we tentatively suggest that this product is ADP.

The experimental observations are consistent with the mechanism presented in FIGURE 6. When LDH folds from the completely denatured state, a minority (10–15%) folds correctly. Misfolded structures, by virtue of their exposed hydrophobic surfaces, bind to the cpn60/cpn10/ADP complex (step A), this induces the substrate protein to unfold (step B). Cpn10 is then competitively displaced (step C), allowing release of ADP and association of ATP (step D). This reduces the affinity of cpn60 for the unfolded protein, and the hydrolysis of ATP to give ADP · P_i (step E) amplifies this effect. The unfolded protein is then released (step F), followed by P_i (step G), and the reassociation of cpn10 (step H) returns the system to its original state.

The net result of this cycle is to use the energy of hydrolysis of ATP to drive the unfolding of misfolded protein. Chaperonins therefore prevent irreversible aggregation of misfolded structures by recycling them. This allows proteins to refold to their native conformation through self-assembly and requires no specific recognition event.

REFERENCES

1. ELLIS, R. J. & S. M. VAN DER VIES. 1991. Annu. Rev. Biochem. **60:** 321–347.
2. GEORGOPOULOS, C. P. & B. HOHN. 1978. Proc. Natl. Acad. Sci. USA **75**(1): 131–135.
3. HENDRIX, R. W. 1979. J. Mol. Biol. **129:** 375–392.
4. HEMMINGSEN, S. M., C. A. WOOLFORD, S. M. VAN DER VIES, K. TILLY, D. T. DENNIS, C. P. GEORGOPOULOS, R. W. HENDRIX & R. J. ELLIS. 1988. Nature **333:** 330–334.
5. GOLOUBINOFF, P., J. T. CHRISTELLER, A. A. GATENBY & G. H. LORIMER. 1989. Nature **342:** 884–889.
6. SAIBIL, H., Z. DONG, S. WOOD & A. AUF DER MAUER. 1991. Nature **353:** 25–26.
7. ELLIS, R. J. & S. M. HEMMINGSEN. 1989. Trends Biochem. Sci. **14:** 339–342.
8. WICKNER, W. 1989. Trends Biochem. Sci. **14:** 280–283.
9. MARTIN, J., T. LANGER, R. BOTEVA, A. SCHRAMEL, A. L. HORWICH & F. ULRICH-HARTL. 1991. Nature **352:** 36–42.
10. SMITH, C. J., A. R. CLARKE, W. N. CHIA, L. I. IRONS, T. ATKINSON & J. J. HOLBROOK. 1991. Biochemistry **30:** 1028–1036.
11. BADCOE, I. G., C. J. SMITH, S. WOOD, D. J. HALSALL, J. J. HOLBROOK, P. LUND & A. R. CLARKE. 1991. Biochemistry **30:** 9195–9200.
12. CURTIS, A. R. & W. P. SWEETENHAM. 1985. United Kingdom Atomic Energy Authority (Unclassified) HMSO Report AERE-R 11771. Harwell, Oxon.
13. CHANDRASEKHAR, G. N., K. TILLEY, C. A. WOOLFORD, R. HENDRIX & C. P. GEORGOPOULOS. 1986. J. Biol. Chem. **261:** 12414–12419.

Using Structural Comparison as a Guide in Protein Engineering

RICHARD BOTT, JUDY DAUBERMAN,
ROBERT CALDWELL, COLIN MITCHINSON,
LORI WILSON, BRIAN SCHMIDT, CURRAN SIMPSON,
SCOTT POWER, PUSHKARAJ LAD, I. HELENA SAGAR,
THOMAS GRAYCAR, AND DAVID ESTELL

Genencor International
180 Kimball Way, South
San Francisco, California 94080

INTRODUCTION

The subtilisin family of enzymes is being extensively studied as a model system for protein engineering. There are several reasons for selecting these enzymes. All occur as single polypeptide chains ranging from 269–279 residues. All share the catalytic triad formed by aspartic acid, histidine, and serine. There are many primary sequences of subtilisinlike enzymes from prokaryotic and eukaryotic sources for comparison. There are also a number of three-dimensional structures available. Finally, subtilisins are used commercially, primarily as detergent additives.

The wealth of data on the structure and function of different subtilisins provide ample evidence that it would be possible to "engineer" subtilisin. The amino acid sequences of subtilisin BPN′ from *Bacillus amyloliquefaciens* and subtilisin Carlsberg from *B. licheniformis* differ at 88 out of 275 positions including one deletion in subtilisin Carlsberg,[1] and yet the three-dimensional structures of these enzymes are quite similar.[2–4] Given these results, it is not surprising that the single amino acid substitutions at a number of different sites on the surface of the enzyme have resulted in only localized changes in the overall tertiary folds of these enzymes.[5]

These findings gave impetus to an experiment to "recruit" the up to 10 times higher k_{cat} of subtilisin Carlsberg into subtilisin BPN′.[6] Only a few of the differences in the amino acid sequences of subtilisins Carlsberg and BPN′ are found in the substrate binding region, at positions 156, 169, and 217 (numbering is with respect to subtilisin BPN′). In subtilisin Carlsberg, serine, alanine, and leucine residues are found at positions 156, 169, and 217 respectively, while in subtilisin BPN′, glutamic acid, glycine, and tyrosine are present at the same positions. Substituting at positions 156, 169, and 217 in subtilisin BPN′ with the corresponding amino acids found in the sequence of subtilisin Carlsberg resulted in a variant enzyme (E156S/G169A/Y217L) that has a fivefold increase in k_{cat} relative to the native subtilisin BPN′.[6] Subsequently, we determined the three-dimensional structure of a quadruple variant of subtilisin BPN′ (M50F/E156S/G169A/Y217L) having the additional substitution of phenylalanine for methionine at position 50; Phe50 is found in the amino acid sequence of subtilisin Carlsberg. The three-dimensional structure of this quadruple variant showed that all of the substituted side chains adopted the same conformations seen in the three-dimensional structure of subtilisin Carlsberg.[7] These results are entirely consistent with the assumption that substitutions of amino acids found in homologous enzymes would individually and collectively cause similar structural and functional changes in the variant enzyme.

In this paper, we present results of additional experiments to exchange structural features of two enzymes, subtilisin BPN′ and the subtilisin from *B. lentus.* In one case, we attempted to introduce a deletion found in *B. lentus* into subtilisin BPN′ and, in a second, show an instance where substitutions of side chains found in subtilisin BPN′ and the related enzyme proteinase K from *Tritirachium album limber* resulted in at least one significant difference that would not have been predicted when the same side chains were introduced into *B. lentus* subtilisin.

RESULTS

B. lentus subtilisin differs from subtilisin BPN′ at 103 of 275 positions. These differences include six deletions in *B. lentus* subtilisin: positions 37 and 58 where

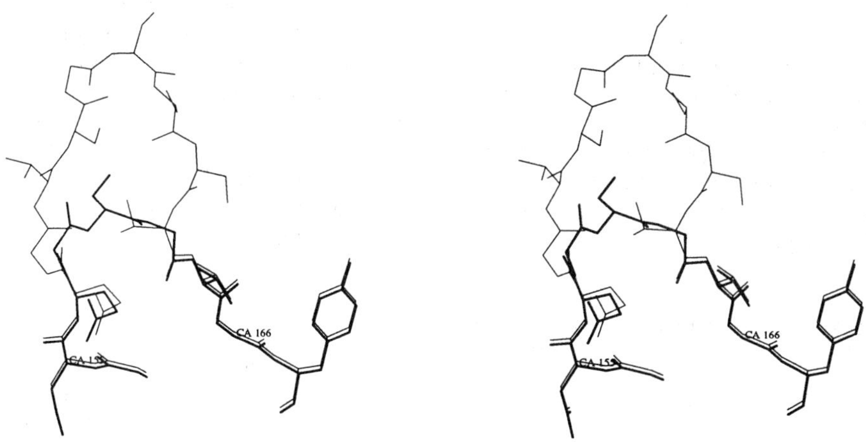

FIGURE 1. Stereographic view comparing residues 156–167 of subtilisin BPN′ (thin lines) with a prediction based on molecular modeling for the subtilisin BPN′ variant T158S/S159G/delta 160–164 (bold lines).

single amino acid residues are deleted, and residues 160–163, a four-residue deletion. To explore what effect the large four-residue deletion has on the activity of the enzyme, we attempted to model a four-residue deletion into subtilisin BPN′, in which residues 156–166 form an extended loop. Modeling studies suggested that a five-residue deletion, residues 160–164, could be easily modeled while the four-residue deletion of *B. lentus* could not. The modeled five-residue deletion superposed on the loop formed by residues 156–166 in subtilisin BPN′ is shown in FIGURE 1. In this model, residues 158 and 159 were also changed to Ser and Gly respectively to facilitate any conformation adjustments that might be required in the structure to

TABLE 1. Kinetic Parameters for the Substrate Succinyl-alanyl-alanyl-prolyl-phenylalanyl-*p*-nitroanilide at 25°C, pH 8.6

	k_{cat} (seconds^{-1})	K_m (mM)	k_{cat}/K_m (seconds^{-1}M^{-1})
Subtilisin BPM′	50	0.14	3.6×10^5
Subtilisin BPN′ variant			
(T158S/S159G del 160–164)	12	0.13	9.2×10^4
B. lentus subtilisin	110	0.78	1.4×10^5

accommodate the deletion of residues 160–164. The values of k_{cat}, K_m, and k_{cat}/K_m all decrease rather than approach the values seen for *B. lentus* subtilisin *lentus* (TABLE 1). In order to determine whether some unexpected structural changes might account for these changes, we determined the structure of the variant enzyme using x-ray crystallography. The three-dimensional structure of this variant has been determined to 2.5Å resolution. The variant crystallized in a nonisomorphous space group P4$_3$, a crystalline lattice that is unique to this variant. The unique crystallization lattice may well be a consequence of the removal of the 160–164 loop which participates in a lattice interaction in the $P2_12_12_1$ space group obtained with native subtilisin BPN′ and all other subtilisin BPN′ variants. The statistics of the refinement of this structure are presented in TABLE 2. A comparison of the three-dimensional structure for the variant compares very well with the modeling prediction as shown in FIGURE 2. The major difference between the modeled prediction and the crystal structure involves a 180-degree flip in the peptide between residues 158 and 159. In this case, we have obtained a *B. amyloliquefaciens* variant whose structure was predicted by modeling but did not result in an enzyme recruiting the improvement in kinetic parameters of *B. lentus subtilisin* for this substrate.

TABLE 2. Crystallographic and Refinement Statistics

	Protein			
	Subtilisin BPN′ T158S S169G del 160–164	Subtilisin *B. lentus* S87N	Subtilisin *B. lentus* K27R V104Y N123S T274A form 1	Subtilisin *B. lentus* K27R V104Y N123S T274A form 2
Space group	$P4_3$	$P2_12_12_1$	$P2_12_12_1$	$P2_12_12_1$
Unit cell				
a (Å)	63.35	53.30	53.00	48.15
b (Å)	63.35	61.50	61.25	54.10
c (Å)	73.60	75.10	75.10	84.55
Resolution	10–2.5 Å	10–1.8 Å	10–1.7 Å	10–2.0 Å
Number of reflections	7997	16480	19999	12204
Crystallographic R factor	0.161	0.148	0.148	0.154
Rms deviation from ideality				
Bond length	0.008 Å	0.015 Å	0.015 Å	0.014 Å
Bond angle	2.5°	2.8°	2.6°	3.1°

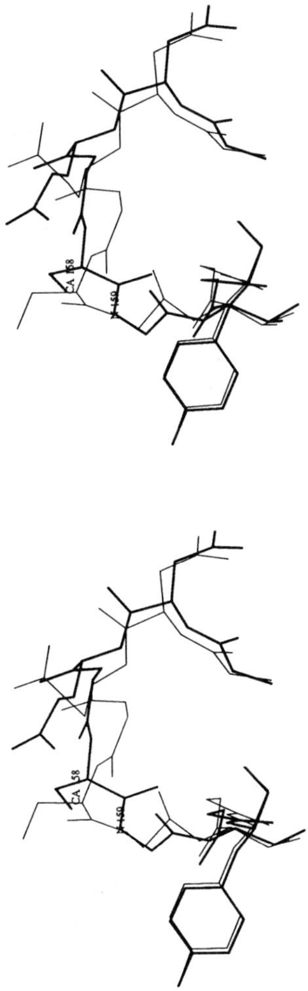

FIGURE 2. Stereographic view of residues 156–167 predicted by molecular modeling for the subtilisin BPN′ variant T158S/S159G/delta 160–164 (thin lines) with the same region from the structure determined by x-ray crystallography at 2.5 Å resolution (bold lines).

Our second experiment brought together substitutions at four positions in *B. lentus* subtilisin—positions 27, 104, 123, and 274 in subtilisin BPN′ numbering. At these positions, Val104 is replaced with tyrosine and Thr274 is replaced with alanine. Both of these residues occur at corresponding positions in the amino acid sequence of subtilisin BPN′. In addition, arginine replaces lysine at position 27 and serine replaces asparagine at position 123. The amino acid sequence of proteinase K also has tyrosine and serine at positions corresponding to 104 and 123 in subtilisin BPN′.

The kinetic data comparing the variant with the native *B. lentus* subtilisin are presented in TABLE 3. The variant shows an increase in the overall catalytic efficiency as measured by k_{cat}/K_m on the substrate succinyl-alanyl-alanyl-prolyl-phenalanyl-*para*-nitoranilide relative to the native enzyme. We have also found that this variant shows improved performance in simulated detergent applications. We have determined the three-dimensional structures of a *B. lentus* subtilisin and this variant at high resolution. The crystallographic refinement statistics are presented in TABLE 2. The variant was crystallized in two crystal forms under the same crystal growth conditions, and the three-dimensional structure has been determined in both forms. The two variant structures are highly conserved in the two crystal forms.

TABLE 3. Kinetic Parameters for the Substrate Succinyl-alanyl-alanyl-prolyl-phenylalanyl-*p*-nitroanilide at 25°C, pH 8.6

	k_{cat} (seconds^{-1})	K_m (mM)	k_{cat}/K_m (seconds^{-1}M^{-1})
Subtilisin BPN′	50	0.14	3.6×10^5
B. lentus subtilisin	110	0.78	1.4×10^5
B. lentus subtilisin variant (K27R/V104Y/N123S/T274A)	520	1.3	4.3×10^5

We have used method of Bott and Frane[8] to determine the mean error and the variance about this mean for equivalent atoms of two related crystallographic structures. This method exploits the linear relationship between the logarithm of the distance between equivalent atoms in the two structures and their crystallographic temperature factors. FIGURE 3 illustrates that there is a linear relationship between the logarithm of the distance between equivalent atoms as a function of the crystallographic temperature factor between *B. lentus* subtilisin and the variant. Linear regression is used to determine the mean error in their coordinates as a function of their mean temperature factor. In addition, it is possible to determine the variance of the distances from the mean as measured by the root-mean-squared error. This has been shown to be useful in identifying potentially significant difference in related crystallographic structures. Similar results were obtained comparing the equivalent atoms of different crystal forms of the variant. TABLE 4 lists the potentially significant residues that were detected in pairwise comparisons of either variant structure with the native *B. lentus* subtilisin or with each other. The potentially significant differences are found to occur in the vicinity of lattice contacts in one or both crystal forms involving residues 175–181. Our present interpretation is that these differences represent the distortion introduced by different lattice packing contacts. These differences are in the range of 0.7–1.0 Å. There are no significant

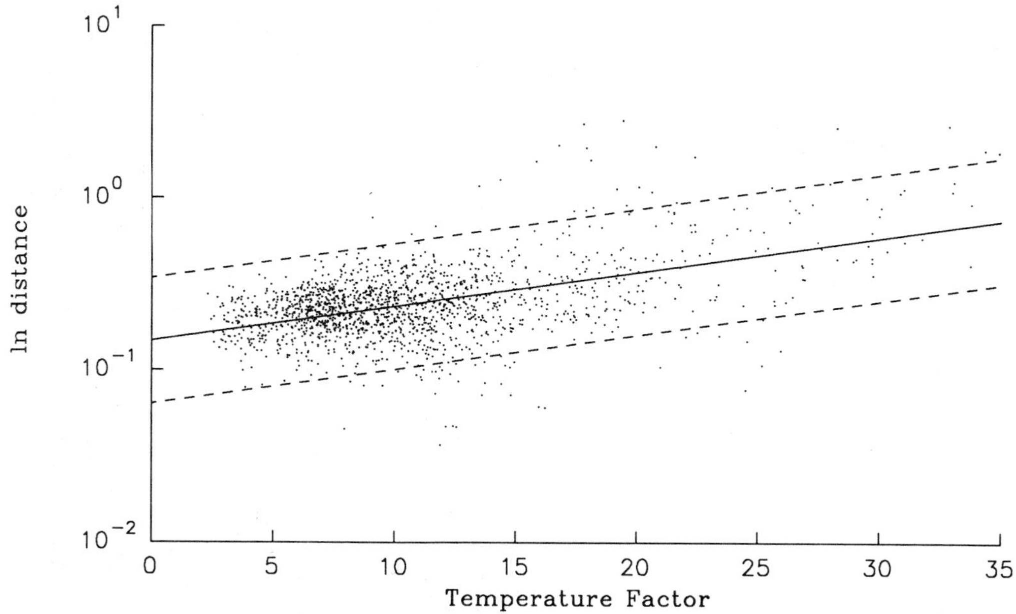

FIGURE 3. Plot illustrating the relationship between the distance between two equivalent atoms in the x-ray crystallographic structure of native *B. lentus* subtilisin and a variant K27R/V104Y/N123S/T274A as a function of their temperature factor. As has been previously reported,[5] there is a linear relationship between the natural logarithm of the distance and the crystallographic temperature factor B. This allows a linear regression fitting of the expected mean distance between equivalent atoms and the variance about this mean distance as a function of B.

TABLE 4. Potentially Significant Difference between Variant and Native *B. lentus* Subtilisin

	Residue Z Score[a] in Pairwise Comparison			
Residue[b]	Native/ Form 1	Native/ Form 2	Form 1/ Form 2	Probable Cause[c]
Ile79		3.5		
Ser125	7.5	4.8		N123S and/or V104Y
Ser166	3.0			
Tyr167	4.0		4.0	
Arg170	4.2	6.3	8.2	
Asp181		3.3	3.4	Different Crystal Form
Gln182		4.5	4.8	Different Crystal Form
Asn183		3.7	4.3	Different Crystal Form
Asn184		3.2	3.6	Different Crystal Form
Arg186		4.1	3.2	Different Crystal Form
Ala187		3.3	3.3	Different Crystal Form
Ser188		3.1		
Phe189		3.4		
Ala194	3.6			
Gly195	3.2			
Gly219			3.1	
Thr220			3.3	
Lys237			3.0	
Glu271	3.8	4.1		T274A
Ala274	4.1	3.0		T274A
Arg275	5.7	3.2		T274A

[a]As defined in Reference 5.

[b]Residue included if in any pairwise comparison, a residue Z score ≥ 3.0 was encountered. The numbering used corresponds to subtilisin BPN′.

[c]If the pattern of difference could be isolated to variant vs. native or form 1 vs. form 2, it is noted.

differences between the variant structures determined in different crystal lattices involving residues 27, 104, 123, or 274 where substitutions have been introduced. The comparison between *B. lentus* subtilisin and the *B. lentus* variant in the same crystal form has potentially significant differences at nine residues indicated in TABLE 4. The comparison between *B. lentus* subtilisin and the variant in the second crystal form is complicated by the appearance of many comparable or larger differences arising from lattice contacts seen between the two variant structures. However, some potentially significant differences are common to both variants when compared with the native *B. lentus* subtilisin structure. These occur at position 125 and at positions 269, 274, and 275 at or near two of the sites of substitution and are likely to represent the structural consequences of the site-specific substitutions made in the *B. lentus* variant.

The structures of native *B. lentus* subtilisin and the variant in the vicinity of the active site are shown in FIGURE 4. In this figure, the main chain of residues 125–126 adopts different conformations in native and variant *B. lentus* subtilisin. A hydrogen bond between the Ser125 OG and the carbonyl O of Ser221 is lost in the variant enzyme. The quality of the electron density map also indicates that Ser125 is less ordered in the variant enzyme. The cause of this change is somewhat uncertain. In the native structure, the Ser125 OG also forms a hydrogen bond with the side chain of Asn123 which has been replaced by Ser123 in the variant enzyme.

This region is part of the substrate binding region that is highly conserved in the tertiary structures of all subtilisins determined to date. It is also conserved in the tertiary structure of proteinase K which shares a common active site with *B. Lentus* subtilisin and also has a serine at position 123. In the structure of proteinase K, an additional solvent atom is seen to link the Ser125 OG with the Ser123 OG without disrupting the common hydrogen bond linking the Ser125 OG to the carbonyl of Ser221. Thus a serine at 123 does not necessarily disrupt the main chain folding of residues 125–126.

The main interest in the substitution of serine for asparagine at position 123 is that the enzyme's proteolytic activity is increased when serine is introduced at position 123[9,10] and any structural change may be related to this increase. There is another substituted site in the vicinity of the active site, namely, position 104, where tyrosine replaces valine. This single change alone results in a variant with increased activity for the substrate succinyl-alanyl-alanyl-prolyl-phenylalanyl-*p*-nitroanilide. Three subtilisins whose structures have been determined have tyrosine at position 104: subtilisin BPN′, subtilisin Carlsberg, and proteinase K. All three enzymes have active sites that are identical (i.e., no significant differences). *B. lentus* subtilisin differs from all of the above subtilisins and proteinase K by having serine at position 128. In the variant structure, tyrosine 104 forms a hydrogen bond between Ser128 OG and Tyr104 OH. This has the effect of spacing the loop formed by residues 96–107 and the segment formed by residues 126–129. The side chain of Leu126 is seen to be within van der Waals contact with both Leu96 and Ile107. The interaction between the side chain of Leu126 and the side chains of Leu96 and Ile107 may be affected by the formation of the hydrogen bond between Tyr104 and Ser128. These changes in the interactions between the side chain of Leu126 may give rise to the changes seen in the variant that are not seen in proteinase K. Residues 98–102 in the 96–107 loop as well as residues 126–129 are known to form an antiparallel beta sheet with the main chain of substrate a position P4-P1 (using the nomenclature suggested by Berger and Schecter).[11] The shifts we observe, along with the evidence of

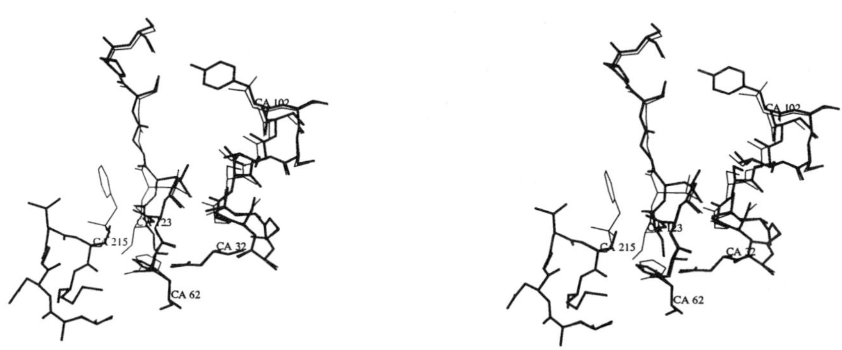

FIGURE 4. Stereographic view of the active site of the variant K27R/V104Y/N123S/T274A variant (bold lines) and the native *B. lentus* subtilisin (thin lines). The shift in residues 125–126 can be seen to stand out compared to the other segments forming the active site.

increased disorder, suggest that the increased activity may be correlated with the increased flexibility in the segments of the active site. Additional experiments are needed to confirm this.

Another significant change is found at the carboxyl terminus of the variant and native *B. lentus* subtilisin. Here, threonine is replaced by alanine, leading to a contraction in the segment formed by residues 269–275. This, in turn, might be expected to stabilize this region by reducing the volume of the enzyme, freeing up more volume to be occupied by solvent. Instead, the reverse is true. Replacing the threonine removes two hydrogen bonds between the OG1 of Thr274 and the carbonyl O of Ala270, and the NH2 of Arg19 and the same OG1 of Thr274. This is somewhat destabilizing. Replacing Thr274 with serine is also destabilizing but to a much lesser degree.[9] Alanine is found at position 274 in subtilisin BPN′. However, unlike the active sites of subtilisin BPN′ and *B. lentus* subtilisin which are highly conserved, the carboxyl termini of these two enzymes are much less conserved. In subtilisin BPN′, there are a number of compensating substitutions in the vicinity of Ala274.

In summary, the presence of the different amino acids in the vicinity of the sites selected for recruitment, in this case the presence of serine 128, can result in structural changes not found in the related enzymes. Modeling studies exploring the specific changes would have highlighted this possibility and will be used in future attempts to exchange structural features. The significance of rigidity versus flexibility has been noted previously in our laboratory[12] and also by Petsko and coworkers[13] to be a function of the particular site substituted.

CONCLUSIONS

Our experience with these substitutions, as well as earlier experience, illustrates that homology between related structure can be a powerful asset in protein engineering. In addition, molecular modeling can be utilized to preview the potential enzyme-specific consequences arising from amino acid sequence variation in neighboring residues. These results also stress the need to improve prediction algorithms and to develop methods to extrapolate the functional consequences of the predicted structural changes that a substitution may cause.

REFERENCES

1. SMITH, E. L., R. J. DELANGE, W. H. EVANS, M. LANDON & F. S. MARKLAND. 1968. Subtilisin Carlsberg. V. The complete sequence comparison with subtilisin BPN′; evolutionary relationships. J. Biol. Chem. **243:** 2184–2191.
2. MCPHALEN, C. A. & M. N. G. JAMES. 1988. Structural comparison of two serine proteinase-inhibitor complexes: eglin-C-subtilisin Carlsberg and CI-2-subtilisin novo. Biochemistry **27:** 6582–6598.
3. NEIDHART, D. J. & G. A. PETSKO. 1988. The refined structure of subtilisin Carlsberg at 2.5 Å. Protein Eng. **2:** 271–276.
4. BOTT, R., M. ULTSCH, A. KOSSIAKOFF, T. GRAYCAR, B. KATZ & S. POWER. 1988. The three-dimensional structure of *Bacillus amyloliquefaciens* subtilisin at 1.8 Å and an analysis of the consequences of peroxide inactivation. J. Biol. Chem. **263:** 7895–7906.
5. BOTT, R. & M. ULTSCH. 1987. The structural consequences of site-specific mutagenesis. *In*

Fifth International Symposium on the Genetics of Microorganisms. M. Alacevic, D. Hranueli & Z. Toman, Eds.: 375–385. Pliva Press. Zagreb, Croatia.

6. WELLS, J. A., B. C. CUNNINGHAM, T. P. GRAYCAR & D. A. ESTELL. 1987. Recruitment of substrate-specificity properties from one enzyme into a related one by protein engineering. Proc. Natl. Acad. Sci. USA **84:** 5167–5171.
7. BOTT, R., T. GRAYCAR & D. ESTELL. 1991. Engineering altered substrate specificity into subtilisin BPN′. *In* Biotechnology: Bridging Research and Applications. D. Kamely, A. M. Chakarabarty & S. E. Kornguth, Eds.: 429–440. Kluwer Academic Publishers. Boston, Mass.
8. BOTT, R. & J. FRANE. 1990. Incorporation of crystallographic temperature factors in the statistical analysis of protein tertiary structures. Protein Eng. **3:** 649–657.
9. CALDWELL, R., D. ESTELL & T. GRAYCAR. European Patent Office Application no. 90915958:4.
10. GRAYCAR, T. P., R. R. BOTT, R. M. CADWELL, J. L. DAUBERMAN, P. J. LAD, S. D. POWER, I. H. SAGAR, R. A. SILVA, G. WEISS, L. R. WOODHOUSE & D. ESTELL. Altering the proteolytic activity of subtilisin through protein engineering. Ann. N.Y. Acad. Sci. (This volume.)
11. BERGER, A. & I. SCHECTER. 1970. Mapping the active site of papain with the aid of peptide substrates and inhibitors. Phil. Trans. R. Soc. Lond. B. **257:** 249–264.
12. BOTT, R., M. ULTSCH, J. WELLS, D. POWERS, D. BURDICK, M. STRUBLE, J. BURNIER, D. ESTELL, J. MILLER, T. GRAYCAR, R. ADAMS & S. POWER. 1987. Importance of conformational variability in protein engineering of subtilisin. ACS Symp. Ser. **334:** 139–147.
13. RASMUSSEN, B. F., A. M. STOCK. D. RINGE & G. A. PETSKO. 1992. Nature **357:** 423–424.

Isoenzymes of Lipase from *Candida cylindracea*

Studies Related to Carbohydrate Composition[a]

MARIA L. RUA, TERESA DIAZ-MAURIÑO,[b] CRISTINA OTERO, AND ANTONIO BALLESTEROS[c]

Instituto de Catálisis, CSIC
Universidad Autónoma
28049 Madrid, Spain

[b]*Instituto Rocasolano, CSIC*
28006 Madrid, Spain

INTRODUCTION

Lipases are unique among enzymes in that their activity increases when the substrate is present at an oil-water interphase. Although these enzymes are generally water soluble, they act near the lipid membrane. The three-dimensional structure of two neutral lipases, *Mucor miehei*[1] and human pancreatic lipase,[2] has been recently reported: the active site has homology with the serine proteases.

Lipases have received increased attention after they were shown to be active even in nearly anhydrous, water-immiscible organic solvents. The lipase from *Candida cylindracea* has possibly been one of the most used owing to its high activity in hydrolisis as well as in synthesis.

Multiple forms of exocellular lipases have been described for fungi. The heterogeneity can be due to posttranslational processing like partial proteolysis and deglycosylation, and to the biosynthesis of really different lipases. It has been reported by Kawaguchi *et al.* that *C. cylindracea* lipase produces at least two kinds of extracellular lipases (they called them lipase I and lipase II).[3]

In the present work, the *C. cylindracea* lipase has been purified yielding two main isoenzymes which differ in the carbohydrate content. The nature of the sugar moieties has been analyzed by hydrolysis with endoglycosidase (PNGase F) and by interaction with different lectins.

MATERIALS AND METHODS

Lipase from *Candida cylindracea* (type VII) was purchased from Sigma Chemical Co. The enzyme peptide-*N*-(*N*-acetyl-β-glucosaminyl) asparaginase amidase F

[a] This research was supported by the EEC BRIDGE Program (No. BIOT-CT90-0176) and by the Spanish CICYT (No. BI088-0241).

[c] Author to whom correspondence should be addressed.

("PNGase F") and the glycan differentiation kit were obtained from Boehringer Mannhein.

N-terminal sequence analysis was carried out by Edman degradation in a model 890C Beckmann sequenator. Deglycosylation of lipases with PNGase F was effected at 30°C during 20 hours; the extent of deglycosylation was followed by sodium dodecyl sulfate-polyacrylamide gel electrophoresis (SDS-PAGE) and by interaction with lectins.

RESULTS AND DISCUSSION

C. cylindracea lipase was purified from a commercial preparation (Sigma). On DEAE-Sephacel, the enzyme activity was separated into two peaks eluted at 60 and 100 mM NaCl, isoforms that were named lipase A and lipase B, respectively. Both isoenzymes were further purified by hydrophobic chromatography and gel filtration.[4]

TABLE 1. Amino Acid Composition of Lipases A and B[a]

Amino Acid	Lipase A	Lipase B	Lipase I
Asp	13.1	12.7	12.4
Thr	5.5	6.5	6.9
Ser	8.2	8.8	8.4
Glu	6.6	7.3	6.4
Pro	7.2	7.0	5.8
Gly	11.2	10.8	10.9
Ala	9.7	9.1	8.8
Val	4.7	4.6	5.1
Met	2.7	2.5	2.6
Ile	4.5	4.3	5.4
Leu	8.9	8.9	8.9
Tyr	3.9	4.7	3.7
Phe	5.5	5.7	5.8
Lys	4.3	3.6	3.7
His	1.3	0.8	0.9
Arg	2.6	2.6	2.6
Trp	ND	ND	0.9
Cys	ND	ND	0.6
TOTAL No. of amino acids	566	563	534

[a]ND is not determined. The amino acid composition of lipase I from *C. cylindracea* found by Kawaguchi *et al.*[3] is also shown for comparative purposes.

Lipases A and B are glycoproteins that hydrolyze both short and long *p*-nitrophenyl esters as well as tributyrin, triolein, and olive oil emulsions. However the two isoenzymes differ on specificity: lipase B is a better enzyme for high chain substrates whereas lipase A has a more esterasic character.[4]

Amino acid analyses of the two isoenzymes showed (see TABLE 1) no significant differences (being also similar to those deduced for lipase I by Kawaguchi *et al.*[3] from

TABLE 2. *N*-Terminal Sequence of Lipases A and B

Lipase A	<APTAKLANGDTITGLNAIINEAFLGIPFA
Lipase B	<APTATLANGD
Lipase I[a]	APTATLANGDTITGLNAIINEAFLGIPFA

[a]As reported by Kawaguchi *et al.*[3]

cDNA sequence determinations). Also the molecuiar weight of lipases A and B was very similar [aprox. 60 kilodaltons (kDa)].

Amino terminal sequences are very similar (cf. TABLE 2), and only the fifth amino acid was different: −Thr (T) in lipase B is Lys (K) in lipase A−. The amino terminal sequence of lipase B is exactly the same that found by Kawaguchi *et al.* for lipase I.[3]

By isoelectric focusing, lipase A gave one major band (pI 5.5), and lipase B gave 3 more acidic bands (pI 4.98, 4.84, and 4.95) all having the same molecular weight (found by two-dimensional PAGE, not shown). All lipase B isoforms have similar hydrophobicity (coelute on hydrophobic cromatography), the same amino-terminal sequence, and apparently the same oligosaccharide chain (see below). On the other hand, lipase A is a more hydrophobic protein.

The carbohydrate composition of the two isoenzymes was studied by interaction with lectins following the affinoblotting technique. The two lipases interacted (see TABLE 3) with Con A and GNA, which suggests the presence of terminal mannose.

Next, enzymic deglycosylation was probed. PNGase F completely removed the oligosaccharide from denatured lipases A and B. TABLE 4 presents the lipases' sugar-content values, estimated by measuring in SDS-PAGE the decrease in molecular weight of the protein after deglycosylation. (The values are not very different from those already determined by us following the phenol-sulfuric method: 10% for

TABLE 3. Interaction of Lipase A and B with Lectins[a]

Lectin	Specificity	Interaction: Lipase A	Interaction: Lipase B
Datura stramonium (DSA) (*)	Galβ(1–4)GlcNAc GlcNac–Ser/Thr	±	±
Galanthus nivalis (GNA) (*)	Manα(1–3)Man (α1–3 > α1–6 > α1–2)	+++	+++
Maackia amurensis (MAA)(*)	NeuNAcα(2–3)Gal	−	−
Arachis hypogaeaa (PNA)(*)	Galβ(1–3)GalNAc	−	−
Sambucus nigra (SNA)(*)	NeuNAcα(2–6)Gal/ GalNAc	−	−
Aleutia arantia (AAA)(*)	Fucα(1–6)GlcNAc	−	−
Ricinus comunis (RCA_{60})(**)	Gal	−	−
Cannavalis ensiformis (Con A) (**)	Man,Glc	+++	+++

[a]Lipases were subjected to SDS-PAGE, transferred to nitrocellulose, and incubated with either digoxigenin-labeled (*) or ^{125}I-labeled (**) lectins.

TABLE 4. Deglycosylation of Lipases A and B with Peptide:*N*-Glycosidase F

Enzyme	Sugar Content
Lipase A	6.7%
Lipase B	2.7%

lipase A and 3% for lipase B−). These investigations suggest that lipases A and B have N-linked, high mannose oligosaccharide chains.

REFERENCES

1. BRADY, L., A. M. BRZOZOWSKI, Z. S. DEREWENDA, E. DODSON, G. DODSON, S. TOLLEY, J. P. TURKENBURG, L. CHRISTIANSEN, B. HUGE-HENSEN, L. NORSKOV, L. THIM & U. MENGE. 1990. A serine protease triad forms the catalytic centre of a triacyglycerol lipase. Nature **343:** 767–770.
2. WINKLER, F. K., A. D'ARCY & W. HUNZIKER. 1990. Structure of human pancreatic lipase. Nature **343:** 771–774.
3. KAWAGUCHI, Y., H. HONDA, J. TANIGUCHI-MORIMURA & S. IWASADI. 1989. The codon CUG is read as serine in an asporogenic yeast *Candida cylindracea.* Nature **341:** 164–166.
4. RUA, M. L., T. DIAZ-MAURIÑO, V. M. FERNANDEZ, C. OTERO & A. BALLESTEROS. Biochim. Biophys. Acta. (Accepted.)

Functional Roles of Active Site Residues of *Bacillus polymyxa* β-Amylase

NOBUYUKI UOZUMI

Department of Biotechnology
Faculty of Engineering
Nagoya University
Nagoya 464-01, Japan

INTRODUCTION

β-Amylase (α-1,4-glucan maltohydrolase, EC 3.2.1.2) catalyzes the liberation of β-anomeric maltose from the nonreducing ends of α-1,4-glucan and is present in certain bacteria as well as in higher plants. Five genes encoding β-amylase have been cloned and sequenced from both prokaryotes and eukaryotes (FIGURE 1). Three highly conserved sequences are recognized among them, and are suggested to comprise the active site. β-Amylases characterized to date are sensitive to various sulfhydryl-modifying reagents and are considered to contain a sulfhydryl (SH) group essential for the activity. The exact role of active site residues in the enzymatic action, however, remains uncertain. This report describes the effects of the mutation of cysteine, histidine, and glutamic acid in the conserved regions on the catalytic properties of β-amylase.

RESULTS AND DISCUSSION

Construction and Expression of the Mutant β-Amylase

Bacillus polymyxa produces multiform β-amylases with approximate molecular masses of 70, 56, and 42 kilodaltons (KDa) and a 48-KDa α-amylase.[1] The amylase gene contains in-phase β- and α-amylase-coding sequences in the 5′ and 3′ regions, respectively[2] (FIGURE 2). A precursor protein, a 130-KDa amylase, has both β- and α-amylase activities and is proteolytically cleaved to produce multiform β-amylases and a 48-KDa α-amylase after secretion.[3] The three major β-amylases contain three cysteine residues and have essentially the same enzymatic properties.

Identification of Disulfide Bond

To assess the role of cysteine residues in the catalytic reaction, cysteine residues were replaced by site-directed mutagenesis to construct four mutant enzymes, C83S, C91V, C323S, and C-free. Wild-type and mutant forms of the enzyme were expressed in and purified to homogeneity from *Bacillus subtilis.*[4] A disulfide bond between Cys83 and Cys91 was identified by isolation of tryptic peptides bearing a fluorescent label, IAEDANS, from wild-type and C91V enzymes followed by amino acid

```
                                      I                         II                  III

                                 83      91                                        323
                                 ▽       ▽                                          ▽
B.polymyxa              77 IISTHKCGGNVGDDCNIPLPSW  98 159 GPSGELRYPSYYP 172 319 LTFTCLEMSDS 329
                           * *** *******  *** ***         *********** *          ******* **
C.thermosulfurogenes    77 IMSTHACGGNVGDTVNIPIPSW  98 159 GPSGELRYPSYNP 172 320 MTFTCLEMDDS 330
                           *** * ******* ****** *         ** ******** *           ****** **
Soybean                 89 IMSFHQCGGNVGDIVNIPIPQW 110 182 GPAGELRYPSY-P 193 339 LNFTCLEMRDS 349
                           ************* ********         ***** ***** *          **** ****
Barley                  87 IMSFHQCGGNVGDAVNIPIPQW 108 180 GPAGEMRYPSY-P 191 337 INFTCAEMRDL 347
                           *********** *** ******         * *** ***** *          **** ****
Sweet Potato            90 IMSFHQCGGNVADAVFIPIPQW 111 184 GAAGELRYPSY-P 195 341 LNFTCLEMRDS 351

CONSERVED                  I-S-H-CGGNV-D---IP-P-W         G--GE-RYPSY-P         --FTC-EM-D-
```

FIGURE 1. Conserved amino acid sequences in three homologous regions (I, II, and III) of β-amylases. Identical residues are denoted by asterisks between the sequences. The amino acids are numbered from the N terminus of each mature enzyme. Closed triangles indicate positions of cysteine residues of the *B. polymyxa* β-amylase. Amino acid sequences of the following amylases were aligned to maximize the homologies: *B. polymyxa*, *C. thermosulfurogenes*, soybean, barley, sweet potato. Conserved amino acids among five β-amylases are shown at the bottom.

sequencing. This, in turn, indicates that Cys323 contains a free SH group. No disulfide bonds have been identified in β-amylases from other origins.

Effect of Cysteine Replacement on Structure

Mutations to prevent disulfide formation are expected to induce some perturbations in tertiary structure. The mutation of Cys83 to Ser83 resulted in a gross structural alteration as judged by the circular dichroism (CD) spectrum, while the mutation of Cys91 to Val91 had no apparent effect on the structure. The enzyme activity of the former mutant at 50°C was inactivated at almost twice the rate of that of the latter mutant. Val91. A well-conserved residue among β-amylases might be

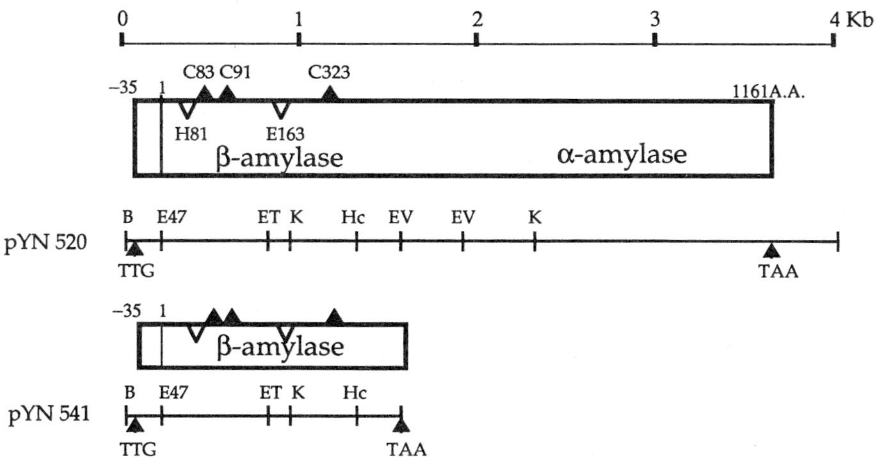

FIGURE 2. Diagramatic presentation of the β-amylase gene from *B. polymyxa.* The upper half represents the correlation of the β- and α-amylase coding regions on plasmid pYN520, while the lower half shows the amylase gene on plasmid pYN541 used in this study. Closed triangles indicate the positions of cysteine residues. The start codon (TTG) and stop codon (TAA) are also indicated. Abbreviations for restriction enzyme sites: B, *BamHI*; E47, *Eco47III*; ET, *EcoT221*; EV, *EcoRV*; Hc, *HincII*; K, *KpnI*.

able to restore, to some extent, the structural perturbations induced by disruption of a disulfide bond, which might result in the simultaneous enhancement of the thermostability.

Kinetic Parameters of Wild-Type and Mutant Enzymes

Kinetic constants K_m and k_{cat}, were determined from initial rate measurements for hydrolysis of soluble starch (TABLE 1). The mutants C91V and C323S had K_m values for this substrate comparable with that of Wt, while the other mutants, C83S and C-free, exhibited a marked increase in the K_m, approximately 7- and 8-fold increases, respectively. For all the mutants, the k_{cat}/K_m ratio, which reflects the

TABLE 1. Kinetic Parameters of Various β-Amylases with Soluble Starch as a Substrate[a]

	K_m (mg/ml)	k_{cat} (/second)	k_{cat}/K_m (ml/mg seconds)
Wt	0.071 ± 0.014	128 ± 2	1820
C83S	0.460 ± 0.007	14.5 ± 0.1	31.5
C91V	0.099 ± 0.021	9.5 ± 0.1	96.6
C323S	0.118 ± 0.012	40.5 ± 4.0	343
C-free	0.549 ± 0.005	16.2 ± 0.2	29.5

[a]Initial reaction rates were determined at 25°C. Kinetic data were fitted to the Michaelis-Menten relationship to evaluate K_m and V_{max}. Turnover number (k_{cat}) was calculated based on the amount of enzyme used.

specific activity for soluble starch of the enzyme, was reduced drastically compared to that for Wt; approximately 5-fold for C323S, 20-fold for C91V, and 60-fold for both C83S and C-free.

Inactivation of Wild-Type and Mutant Enzymes by SH Reagents

To determine whether substitution of cysteine residues has any effect on inactivation, the mutant enzymes as well as the Wt enzyme were subjected to modification by SH reagents such as pCMB, NEM, and DTNB and then assayed for enzyme activity (TABLE 2). Reaction of pCMB with enzymes resulted in the complete inactivation of Wt, C83S, and C91V. Both C323S and C-free retained full activity under the same condition. Of the mutant enzymes, only C91V was inactivated by all three SH reagents, although DNTB modification caused only a partial inactivation of its activity. Assuming that a loss of the enzyme activity is due to the chemical modification of cysteine residues, all these data clearly indicate that alkylation of Cys323 by pCMB causes a loss of the activity as a consequence of introduction of steric bulk. Therefore, none of the cysteine residues participate in the catalytic reaction.

TABLE 2. Effects of Various Sulfhydryl Group Reagents on the Activity of β-Amylases[a]

	pCMB	NEM		DTNB	
	40 μM	2 mM	10 mM	1 mM	5 mM
Wt	<1	112	95	94	95
C83S	<1	90	98	103	100
C91V	<1	25	<1	46	53
C323S	115	86	97	100	100
C-free	103	92	97	102	99

[a]Enzyme (200 nM) was incubated at 4°C for 60 minutes in 50 mM phosphate buffer (pH 7.0) was various sulfhydyl group reagents at concentrations indicated, followed by measurement of the activity at 25°C with soluble starch as a substrate.

TABLE 3. Glu 163 is Essential for β-Amylase Catalytic Reaction

	Specific Activity (U/mg)	Activity Ratio
Wt	142	1
H81L	1.79	1.26×10^{-2}
E163Q	0.0151	1.06×10^{-4}

Glu163 Involves the β-Amylase Catalytic Center

Furthermore, the Glu163 residue might participate in the catalytic reaction, since affinity labeling of the enzyme with α-EPG (2,3-epoxypropyl α-D-glucopyranoside) causes the inactivation of the activity as a consequence of esterification of Glu163 with the affinity label. The mutant E163Q exhibited barely detectable activity for soluble starch, approximately 10^{-4} of the specific activity of the wild-type enzyme (TABLE 3). The catalytic triad of serine protease consists of Ser-His-Glu(Asp). We predicted that the β-amylase catalytic reaction involves histidine residue. The His (H) in the homologous region is located at position 81. Other His residues are not observed from 1 to 248 amino acids. The truncated β-amylase, which consists of 248 AA, retains β-amylase activity to some extent. Thus His was replaced with Leu (L). However, the activity of the H81L was at a detectable level. We conclude that Glu163 was essential for the β-amylase catalytic reaction.

REFERENCES

1. KAWAZU, T., Y. NAKANISHI, N. UOZUMI, T. SASAKI, H. YAMAGATA, N. TSUKAGOSHI & S. UDAKA. 1987. Cloning and nucleotide sequence of the gene coding for enzymatically active fragments of the *Bacillus polymyxa* β-amylase. J. Bacteriol. **169:** 1564–1570.
2. UOZUMI, N., S. SAKURI, S. TAKEKAWA, T. SASAKI, H. YAMAGATA, N. TSUKAGOSHI & S. UDAKA. 1989. A single gene directs synthesis of a precursor protein with β- and α-amylase activities in *Bacillus polymyxa.* J. Bacteriol. **171:** 375–382.
3. TAKEKAWA, S., N. UOZUMI, N. TSUKAGOSHI & S. UDAKA. 1991. Proteases involved in generation of β- and α-amylases from a large amylase precursor in *Bacillus polymyxa.* J. Bacteriol. **173:** 6820–6825.
4. UOZUMI, N., T. MATSUDA, N. TSUKAGOSHI & S. UDAKA. 1991. Structural and functional roles of cysteine residues of *Bacillus polymyxa* β-amylase. Biochemistry **30:** 4594–4599.

Photosensitive Nitrile Hydratase from *Rhodococcus* sp. N-771

Structure and Function of the Enzyme[a]

JUN HONDA,[b] TERUYUKI NAGAMUNE,[b,c,e] YOSHITAKA TERATANI,[d] AKIRA HIRATA,[d] HIROYUKI SASABE,[b] AND ISAO ENDO[c]

[b]*Frontier Research Program*
[c]*Chemical Engineering Laboratory*
RIKEN Institute
2-1 Hirosawa, Wako-shi
Saitama 351-01, Japan

[d]*School of Science and Engineering*
Waseda University
3-4-1 Ohkubo, Shinjuku-ku
Tokyo 169, Japan

INTRODUCTION

Nitrile hydratase (NHase) hydrates various nitrile compounds to the corresponding amides, viz., R-CN + $H_2O \rightarrow$ R-$CONH_2$. The bacteria containing this enzyme were used for the production of acrylamide from acrylonitrile in Nitto Chemical Industry Co., Ltd., Japan. This marked the first industrial-scale production of commodity chemicals by an enzymatic method in the history of the chemical industry.[1] There are several bacterial species known to contain NHase, but the NHase from *Rhodococcus* (*R.*) sp. N-771 shows an extraordinary reaction to light. Exposure of this enzyme to near-ultraviolet light induces an enhancement of its enzymatic activity, whereas dark aerobic incubation of this enzyme *in vivo* induces a diminution of its enzymatic activity.[2] The activity of the enzyme before and after light exposure can differ by a factor of more than 30. This property in the beginning hampered the development of the bioprocess in Nitto Chemical. In the scientific sense, however, this photosensitive property is an interesting phenomenon worthy of investigation, since not very much has been reported about this type of enzyme activity regulation by light. Elucidation of this photosensitive mechanism would probably lead to a characterization of a novel photosensitive reaction center of an enzyme. So far, we have used various approaches to the investigation of this photosensitive mechanism, one of the main ones being structural analysis by x-ray crystallography. This approach enables us to see the structure of this enzyme on an atomic scale. Once we have this information, it will be much easier to interpret

[a]This research was partially supported by a grant for the "Biodesign Research Program" from RIKEN Institute.
[e]Author to whom correspondence should be addressed.

TABLE 1. Some of the Properties of the Photosensitive NHase from *Rhodococcus* sp. N-771

Molecular weight:	60 kDa by high-performance liquid chromatography
Subunits:	2 heterogeneous, each approx. 27.5 and 28 kDa (by sodium dodecyl sulfate-polyacrylamide gel electrophoresis)
Temperature	Unstable above 20°C, optimum 30°C
pH:	Stable between 6–8, optimum 7.8
Cofactors:	2 Fe atoms/enzyme, pyrroloquinoline quinone (PQQ) or its derivative present
Substrate specificity:	Broad

results obtained from various spectroscopic analyses, another of our main approaches. In this paper, we shall report some of the latest results from our investigation.

PURIFICATION AND CHARACTERIZATION OF NHase

Purification procedures have been reported previously.[3] Briefly, the bacteria is cultivated with yeast-malt medium, aerobically incubated in the dark for about 30 hours, disrupted with lysozyme treatment, and after ammonium sulfate precipitation, purified with 6 successive column chromatographies. All the purification

```
α subunit
SVTIDHTTEN AAPAQAPVSD RAWALFRALD GKGLVPDGYV EGWKKTFEED
FSPRRGADLV ARAWTDPEFR QLLLTDGTAA VAQYGYLGPQ GEYIVAVEDT
PTLKNVIVCS LCSCTAWPIL GLPPTWYKSF EYRARVVREP RKVLSEMGTE
IASDIEIRVY DTTAETRYMV LPQRPAGTEG WSQEQLQEIV TKDCLIGVAI
PQVPTV

β subunit
MDGVHDLAGV QGFGKVPHTV NADIGPTFHA EWEHLPYSLM FAGVAELGAF
SVDEVRYVVE RMEPRHYMMT PYYERYVIGV ATLMVEKGIL TQDELESLAG
GPFPLSRPSE SEGRPAPVET TTFEVGQRVR VRDEYVPGHI RMPAYCRGRV
GTISHRTTEK WPFPDAIGHG RNDAGEEPTY HVKFAAEELF GSDTDGGSVV
VDLFEGYLEP AA

α subunit: 206 residues  Mr=22,787
β subunit: 212 residues  Mr=23,428
    NHase: 418 residues  Mr=46,215
```

FIGURE 1. Amino acid sequence of photosensitive NHase from *Rhodococcus* sp. N-771 and N-774 deduced from the DNA sequence coding the enzyme.

procedures are performed in the dark to prevent irreversible activation of the enzyme. The purified enzyme was characterized, and some of its properties are shown in TABLE 1.

DETERMINATION OF PRIMARY STRUCTURE

For amino acid sequence determination, NHase and protease-treated NHase (using lysil endopeptidase) were each set onto a protein sequencer (ABI 477A/120A) and a total of 81 amino acid residues were determined from the amino terminal. These sequences were found to be identical to the amino acid sequence of NHase deduced from the DNA sequence of *R.* sp. N-774.[4] This led us to believe that

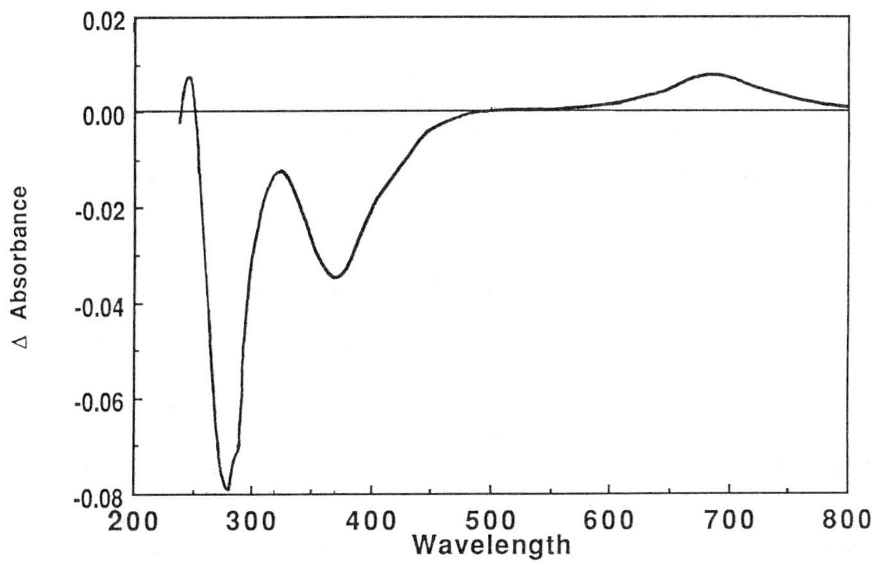

FIGURE 2. Difference absorption spectrum of active NHase–inactive NHase.

the DNA sequence coding this enzyme from *R.* sp. N-771 must also be identical, or at least very similar. The DNA sequence of *R.* sp. N-771 that codes the amino acid sequence of NHase was determined using the polymerase chain reaction (PCR) method. Briefly, chromosomal DNA of *R.* sp. N-771 was isolated, primers for PCR were synthesized, the DNA region coding NHase was amplified with a thermal cycling reactor, and the DNA fragments were sequenced with a DNA sequencer (ABI 373A). As a result, it was found that the amino acid sequences of NHases from *R.* sp. N-771 and N-774 are identical (FIGURE 1). Also, it has recently been found that the amino acid sequence of NHase from *R.* species is identical to that from *Brevibacterium* sp. R312.[5] Therefore, it is highly probable that the NHases from the two bacteria are same enzyme.

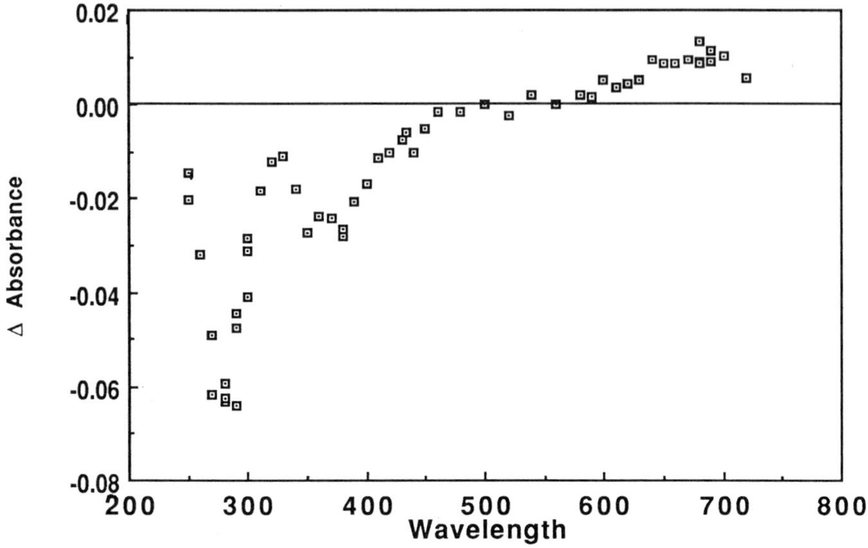

FIGURE 3. Flash-induced absorption change of inactive NHase 1 microsecond after 355 nm flash (laser irradiation).

X-RAY CRYSTALLOGRAPHIC ANALYSIS

If you want to see the tertiary structure of NHase at the atomic scale, x-ray crystallographic analysis is the only method available for a protein of this molecular size. Thus, crystallization of NHase was attempted. NHase crystals were grown by the vapor diffusion method as previously described.[6] Many trials were performed by varying the crystallization parameters such as enzyme concentration, pH, temperature, type of buffer and its concentration, precipitating reagents, type of metals added, etc. Among them, two types of crystals were formed which were characterized by x-ray precession photography. One crystal form, belonging to an orthorhombic crystal system and space group $P2_12_12$, has two molecules per asymmetric unit and diffracts beyond 2.5 Å, whereas the other crystal form, belonging to a hexagonal crystal system and space group $P6_322$, has three molecules per asymmetric unit and diffracts to 2.8 Å. Thus the orthorhombic crystals are more appropriate for high-resolution data collection, and they are used for the actual analysis which is in progress.

FLASH SPECTROPHOTOMETRY

NHase exhibits drastic changes in absorption spectra before and after photoirradiation.[2] A shoulder at 370 nm of the inactive NHase can probably be attributed to a nonheme iron center. After photoirradiation, this shoulder diminishes and a new absorption band with a peak at 680 nm shows up. The difference spectrum of the two absorption spectra (inactive and active) is shown in FIGURE 2. In order to investigate and identify the reaction center(s) involved in the photoactivation process of this enzyme, time-resolved measurement (in a very short time scale) of the changes in absorption spectrum was attempted. The third harmonics of an Nd:YAG laser (355 nm) were irradiated at a 5-nanosecond pulse to the dilute solution (5 mg/ml) of inactive NHase, and the differences in absorption from 250 nm to 720 nm were measured. The result is shown in FIGURE 3. With the present apparatus, it was only possible to trace the change after 1 microsecond, by which time the photoactivation reaction is complete (cf. FIGURE 2). This result suggests that the photoactivation process of NHase does not accompany any major change of protein conformation, which is a much slower reaction, but a small change around the photosensitive and catalytic reaction center which immediately leads to the expression of catalytic activity.

BEHAVIOR OF IRON ATOMS

Judging from the change in absorption spectra, the change in the state of iron atoms seemed to play an important role in the expression of catalytic activity. In order to examine the behavior of iron atoms before and after enzyme activation,

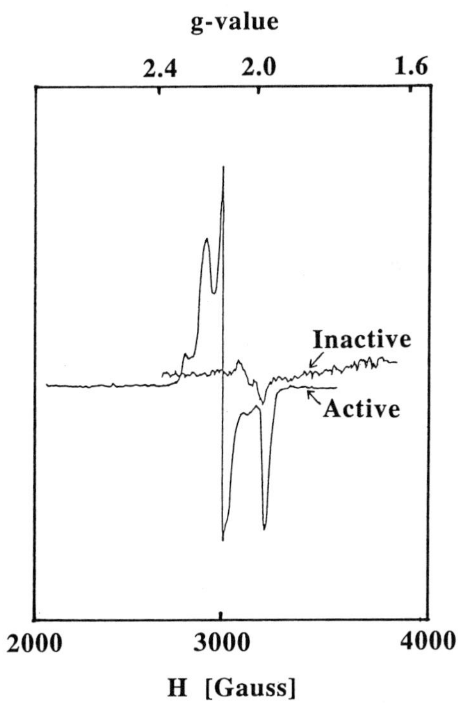

FIGURE 4. ESR spectra of inactive and active NHases taken at 4.3 K.

electron spin resonance (ESR), Mössbauer spectrum, and magnetic susceptibility studies were undertaken. The results from ESR (FIGURE 4) showed that the active NHase was ESR active while the inactive NHase was ESR silent. This indicated the formation of an unpaired electron upon photoactivation, and the active spectrum indicated the presence of an Fe-S rhombic center in the active state. In order to measure Mössbauer spectra, the iron atom in the NHase must be replaced by a stable isotope, ^{57}Fe. This was achieved by cultivation of *R.* sp. N-771 with metal-free medium and a suitable amount of ^{57}Fe, and then purification. Mössbauer spectra of both active and inactive NHase were measured at 77 K (FIGURE 5). The Mössbauer spectrum of the inactive NHase showed the presence of two species, A and B, whereas that of the active NHase showed the presence of one species, equal to species B of the inactive NHase. Thus the Mössbauer spectra clearly demonstrate the change in the state of one of the two iron atoms in the NHase molecule. Magnetic susceptibility studies using a superconducting quantum interference device (SQUID) gave values of 5.68 per mole inactive enzyme and 7.00 per mole active enzyme. These values gave unambiguous interpretations of the Mössbauer spectra. Species A is attributed to low-spin Fe(II) having no unpaired electron, and species B is attributed to low-spin Fe(III) having an unpaired electron. Thus, the photoactivation process of NHase is found to be accompanied by the following oxidation reaction for iron atoms

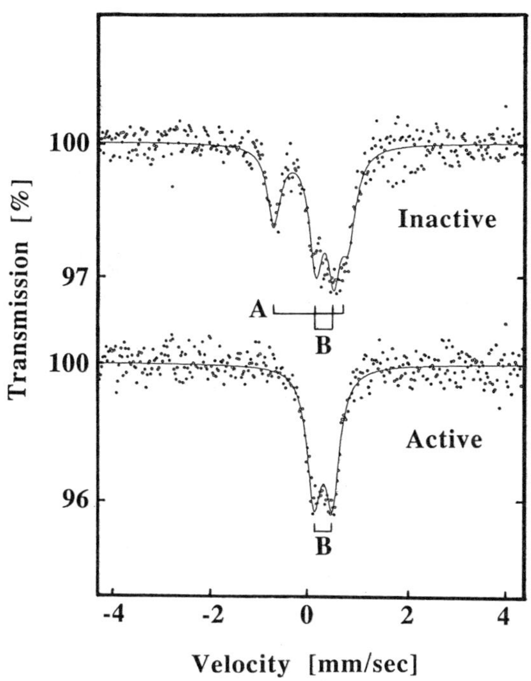

FIGURE 5. Mössbauer spectra of inactive and active NHases taken at 77 K. A and B indicate type of species in the spectra.

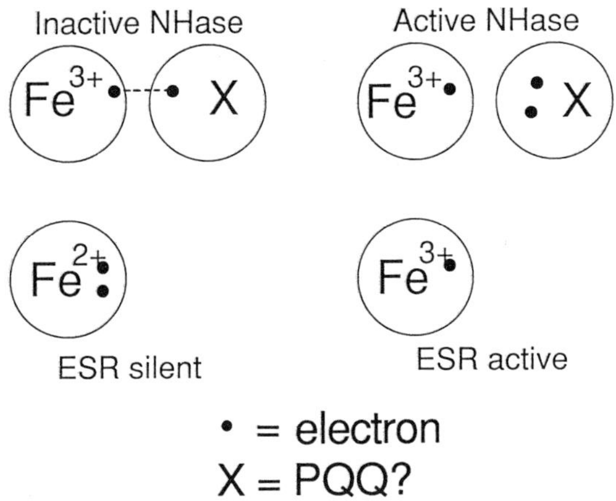

FIGURE 6. A hypothetical model for the photoactivation reaction involving the cofactors in the NHase.

(LS = low spin):

$$\underset{LS}{Fe(II)}\underset{LS}{Fe(III)} \xrightarrow{h\nu} \underset{LS}{Fe(III)}\underset{LS}{Fe(III)}$$

If the iron atom is oxidized, a question remains as to the destination of the released electron. At this point, we can only speculate, but the hypothetical model shown in FIGURE 6 can explain the electron transfer mechanism of this enzyme and simultaneously satisfy the results of Mössbauer spectra and ESR spectra. We have evidence (fluorescence spectra and bioassay, unpublished data) to believe that a pyrroloquinoline quinone (PQQ) or its derivative is incorporated in this enzyme as a cofactor, and in this model, the quinone is acting as an electron acceptor. In the inactive state, an unpaired electron of this quinone forms an apparent electron pair with the unpaired electron of the iron atom, which explains why Fe(III) in the inactive NHase is not detected by ESR spectroscopy.

An electron transfer similar to this type is well observed in protein complexes such as the photosynthetic reaction center,[7] but this type of intramolecular photooxidation involving iron in a single protein molecule is, to our knowledge, the first to be reported.

ACKNOWLEDGMENTS

We are grateful to our coworkers in RIKEN Institute for their contribution: Drs. N. Kamiya, W. Cho, Y. Kobayashi, and Y. Sakaguchi

REFERENCES

1. WATANABE, I., Y. SATOH & K. ENOMOTO. 1987. Agric. Biol. Chem. **51:** 3193–3199.
2. NAGAMUNE, T., H. KURATA, M. HIRATA, J. HONDA, A. HIRATA & I. ENDO. 1990. Photochem. Photobiol. **51:** 87–90.

3. NAGAMUNE, T., H. KURATA, M. HIRATA, J. HONDA, H. KOIKE, M. IKEUCHI, Y. INOUE, A. HIRATA & I. ENDO. 1990. Biochem. Biophys. Res. Commun. **168:** 437–442.
4. IKEHATA, O., M. NISHIYAMA, S. HORINOUCHI & T. BEPPU. 1989. Eur. J. Biochem. **181:** 563–570.
5. MAYAUX, J., E. CERBELAUD, F. SOUBRIER, D. FAUCHER & D. PETRE. 1990. J. Bacteriol. **172:** 6764–6773.
6. NAGAMUNE, T., J. HONDA, W. CHO, Y. TERATANI, N. KAMIYA, A. HIRATA, H. SASABE & I. ENDO. 1991. J. Mol. Biol. **220:** 221–222.
7. HUBER, R. 1989. EMBO J. **8:** 2125–2147.

Lipase Specificity against Some Fatty Acids?

F. ERGAN, S. LAMARE,[a] AND M. TRANI

Biotechnology Research Institute
National Research Council Canada
6100 Royalmount
Montréal, Québec H4P 2R2, Canada

PRINCIPLE

Lipases hydrolyze triglycerides according to their specificity, which has been defined by Jensen *et al.* as a comparative difference in rates of catalysis of certain reactions.[1] For example, if the substrate contains three different fatty acids, A, B, and C, they can be hydrolyzed at three different rates. If one is able to follow the rate of appearance of A, B, and C, it will be possible to take advantage of this fact to discriminate between the fatty acids. If A is hydrolyzed faster than B, and B is hydrolyzed faster than C, then the product of the hydrolytic reaction will be a mixture of a free fatty acid fraction, mainly A and B with very little C, and a glyceride fraction, enriched in C. One can then separate the two fractions in order to produce the fatty acid C which is of interest.

A number of oil crops contain in their triglycerides a fatty acid of interest: crambe and high erucic acid rapeseed (HEAR) both contain erucic acid (C22:1), whereas evening primrose oil (EPO) and borage oil each contain γ-linolenic acid, or GLA (C18:3). The use of lipases to hydrolyze with discrimination the triglycerides of these oils allows for the separation of the fatty acid of interest from all the other fatty acids contained in the oil.

MATERIALS AND METHODS

Hydrolysis of HEAR Oil

Forty grams of oil was stirred together with 16 ml of water and 6.4 ml of a lipase solution (0.01 g/ml) in a beaker maintained at 25°C. The lipase used was *Candida rugosa* type VII from Sigma.

Hydrolysis of Borage or Evening Primrose Oil

Half a gram of the oil was stirred together with 0.24 ml of water and 0.04 ml of the same lipase solution as above, in an Eppendorf tube placed on a vortex shaker. (The ratio between the water phase and the oil phase is the same for both reactions.)

[a]Laboratoire de Technologie Enzymatique, Université de Technologie de Compiègne, B.P. 649, 60206 Compiègne Cedex, France.

Analyses

The high-performance liquid chromatography (HPLC) method described previously[2] was used, with minor modifications. In order to follow both hydrolyses, 10-μl aliquots are sampled at preset time intervals. The aliquots are diluted in 800 μl of acetone and 15 μl are injected in an ODS-2 Reverse Phase C18 column (from CSC, Montréal, Canada), 25 cm × 4.6 mm, 5 μm, where the mobile phase (acetone: acetonitrile, 50/50) is run using a flow gradient:

- For HEAR oil, the flow is 0.8 ml/minute for 6 minutes, increases up to 4 ml/minute in 4 minutes, remains at 4 ml/minute for 30 minutes, and then drops back to 0.8 ml/minute in 4 minutes. The peaks observed have the following retention times: monoerucin, 6.8 minutes; erucic acid, 7.4 minutes; and dierucin, 12.5 minutes.
- For GLA-containing oils, the flow follows the same pattern as above for the first 21 minutes, where it is decreased in 4 minutes back to 0.8 ml/minute. The following peaks are detected, with the corresponding retention times: GLA, 4.2 minutes; linoleic acid, 4.6 minutes; and oleic acid, 5.2 minutes.

RESULTS

Production of Erucic Acid from HEAR Oil

In the case of erucic acid, the lipase of *Candida rugosa* shows almost complete specificity with negligible amounts of erucic acid released during hydrolysis.

HEAR oil contains 40–50% of erucic acid, the other fatty acids being mainly linolenic, linoleic, oleic, and eicosenoic acids. Erucic acid can only be at positions 1 and 3 of the glycerol backbone, but does not occupy all of them. For this reason, it is not possible to use a 1,3-specific lipase to separate erucic acid from the other fatty acids. FIGURE 1 shows the progress curve of the hydrolysis of HEAR oil where very little erucic acid is released with time while the analysis of the reaction medium shows the accumulation of dierucin. When calculating from this curve the percentage of erucic acid under the form of dierucin, one reaches a maximum value of 70% after 16 hours of hydrolysis. Seventy percent of the erucic acid originally present in the oil is under the form of dierucin. It is then easy to separate dierucin from the hydrolyzed fatty acids. The free fatty acid fraction contains 10% erucic acid. The yield of recovery of dierucin is 85%, and its purity 75%. Further hydrolysis of dierucin to erucic acid with another lipase (*Rhizopus arrhizus* or *Mucor miehei*) showing no specificity against erucic acid gives a yield of 100%.

Production of γ-Linolenic Acid from Borage Oil or Evening Primrose Oil

GLA does not have a specific position on the glycerol backbone. In borage oil, the total content of GLA is 21%, with 41.5% on positions 1 or 3 and 58.5% in position 2. Evening primrose oil contains 8% of GLA, 57.6% of it on 1 or 3 positions and the rest

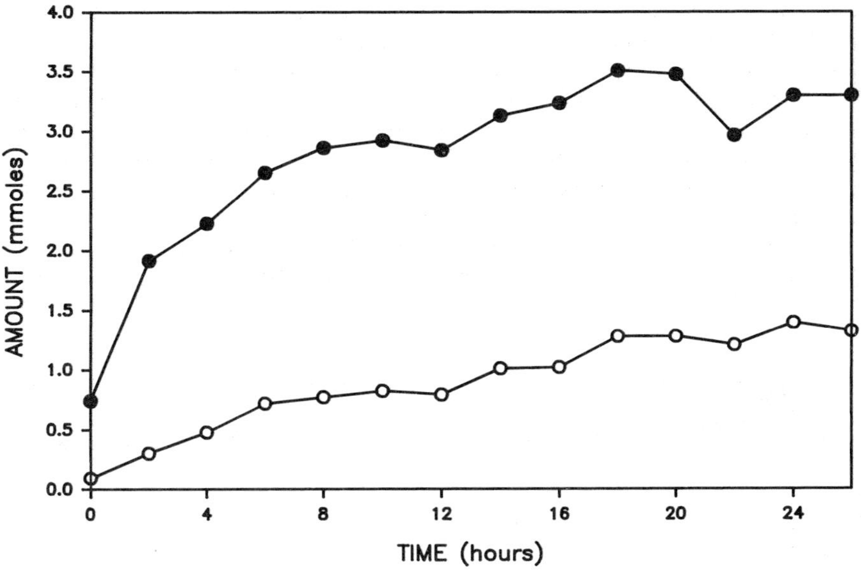

FIGURE 1. Progress curve of HEAR oil hydrolysis (conditions described in text). ●, dierucin; ○, erucic acid.

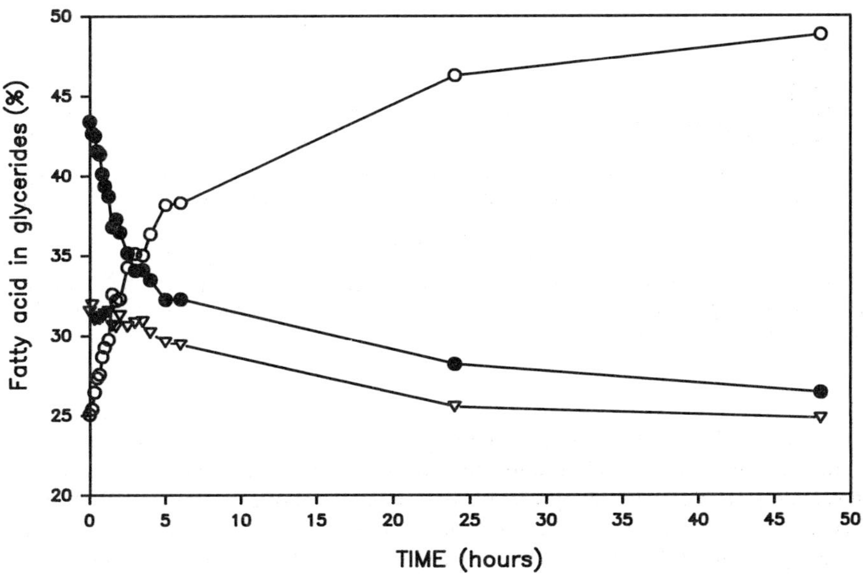

FIGURE 2. Progress curve of borage oil hydrolysis (conditions described in text). The results are expressed in terms of percentages of each fatty acid remaining in the glyceride fraction. ○, GLA; ●, linoleic acid; ▽, oleic acid.

(42.4%) at position 2. The use of a position-specific lipase will not be efficient to make the difference between GLA and the other fatty acids. In the case of GLA, the lipase shows only a marked preference against this fatty acid. GLA is released during the hydrolysis, albeit at a much slower rate than the rest of the fatty acids. In this instance, kinetic control of the reaction is necessary. The ultimate purity of the GLA obtained is governed by the end point of the reaction. FIGURES 2 and 3 show the percentage of the main fatty acids in glycerides as a function of time for borage oil and evening primrose oil. The use of borage oil as a substrate allows for the enrichment of the glyceride fraction in GLA from 25% up to 50% in 48 hours. (The initial value of 25% of GLA is due to the fact that our HPLC analyses detect only

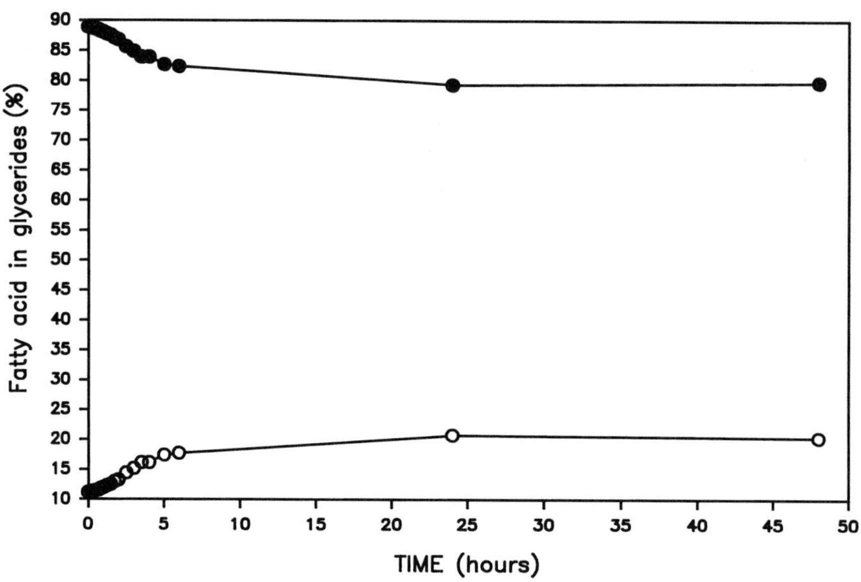

FIGURE 3. Progress curve of evening primrose oil hydrolysis (conditions described in text). The results are expressed in terms of percentages of each fatty acid remaining in the glyceride fraction. ○, GLA; ●, linoleic acid.

GLA and linoleic and oleic acids, accounting for about 85% of the total fatty acids of the oil.) A similar pattern is followed when using evening primrose oil as a substrate. It has been possible to enrich by a factor of 2 the content of the glyceride fraction in GLA, going from 10% up to 20% in approximately 24 hours. In both cases, the recovery yield of GLA is around 90%. FIGURE 4 shows the use of 8 different lipases on borage oil. The enrichment in GLA of the glyceride fraction is plotted versus time for all lipases tested: only *Candida rugosa* yields a glyceride fraction of 50%. *Mucor miehei, Rhizopus arrhizus,* and Lipozyme™ lead to a certain enrichment, but only up

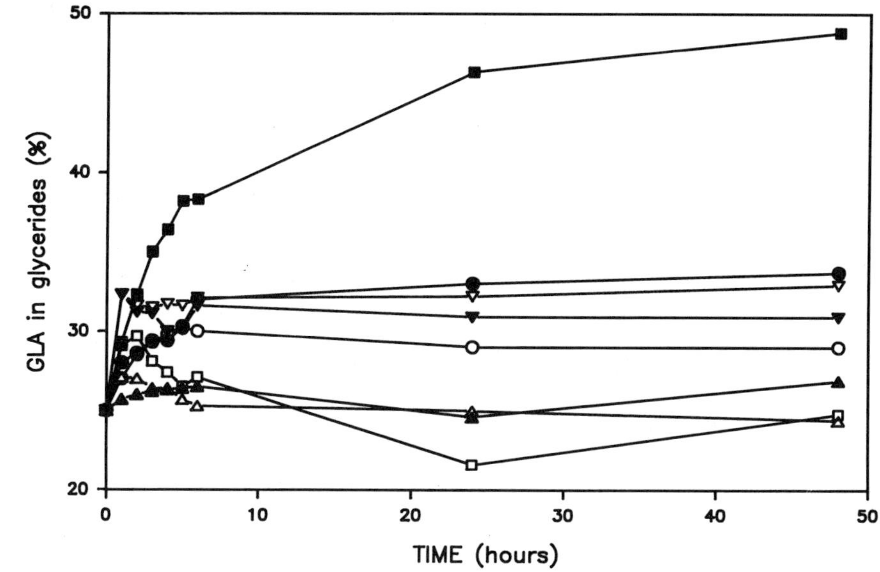

FIGURE 4. Enrichment in GLA in the glyceride fraction obtained from borage oil hydrolysis with different lipases (conditions described in text). ■, *Candida rugosa;* ●, *Mucor miehei;* ▽, *Rhizopus arrhizus;* ▼, *Rhizopus delemar;* ○, Lipozyme(TM); ▲, porcine pancreas; □, *Pseudomonas fluorescens;* △, *Chromobacterium viscosum.*

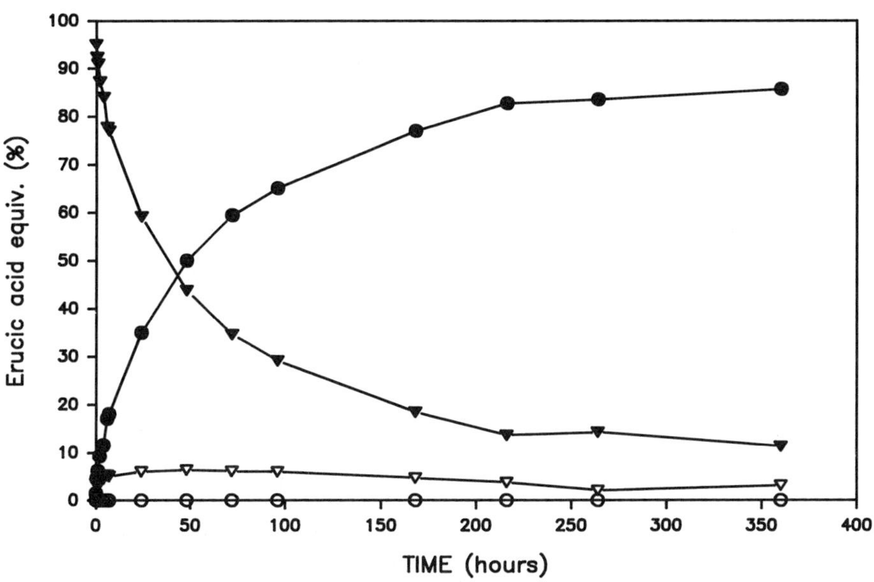

FIGURE 5. Hydrolysis of trierucin with *Candida rugosa* lipase as a function of time. Results are expressed in erucic acid equivalents. ▼, trierucin; ▽, dierucin; ○, monoerucin; ●, erucic acid.

to 30–35%. Porcine pancreas, *Pseudomonas fluorescens,* and *Chromobacterium viscosum* show no enrichment at all in GLA.

DISCUSSION

We have taken advantage of an unusual "specificity" against some fatty acids: erucic acid and γ-linolenic acid. Erucic acid, or *cis*-13-docosenoic acid, is a C22 fatty acid with one double bond. GLA, or *cis*-6,9,12-octadecatrienoic acid, is a C18 fatty acid with three double bonds. We observed that these two fatty acids were not hydrolyzed, or at least at a much slower rate than all other fatty acids involved. Do these two fatty acids have something in common that none of the other fatty acids involved has? The other fatty acids are palmitic (hexadecanoic C16:0), stearic (octadecanoic C18:0), oleic (*cis*-9-octadecenoic C18:1), linoleic (*cis*-9,12-octadecadienoic C18:2), linolenic (*cis*-9,12,15-octadecatrienoic C18:3), and eicosenoic (*cis*-11-eicosenoic acid C20:1). No positive answer could be given to that question. We are also concerned with a possible neighboring fatty acid effect on lipase specificity. More work with defined triglycerides was carried out to better understand the results.

Using trierucin in the same conditions as HEAR oil in the hydrolysis leads to complete transformation in erucic acid (FIGURE 5). The behavior of pure dierucin in the same reaction conditions in the presence of *Candida rugosa* lipase was checked, and confirmation that no hydrolysis occurs is shown in FIGURE 6. More work is in progress to understand this behavior concerning erucic acid.

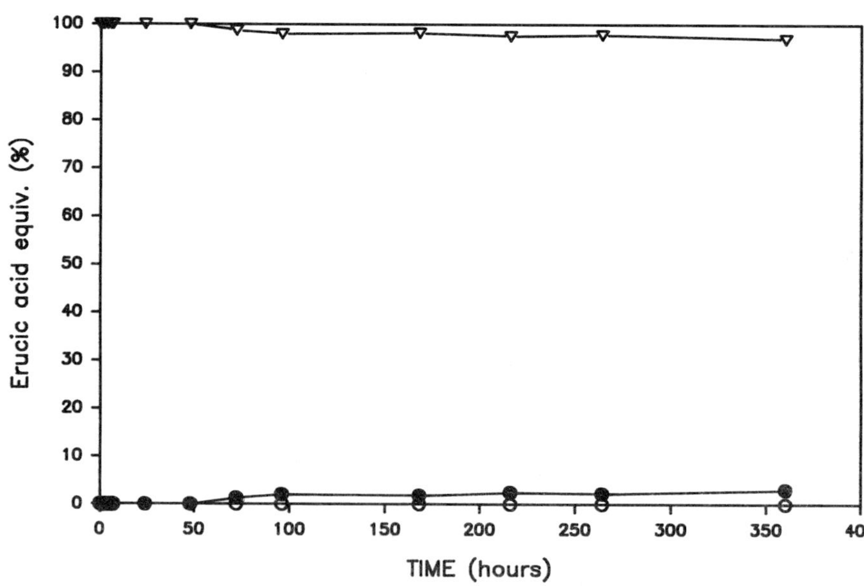

FIGURE 6. Hydrolysis of dierucin with *Candida rugosa* lipase as a function of time. Results are expressed in erucic acid equivalents. ▽, dierucin; ○, monoerucin; ●, erucic acid.

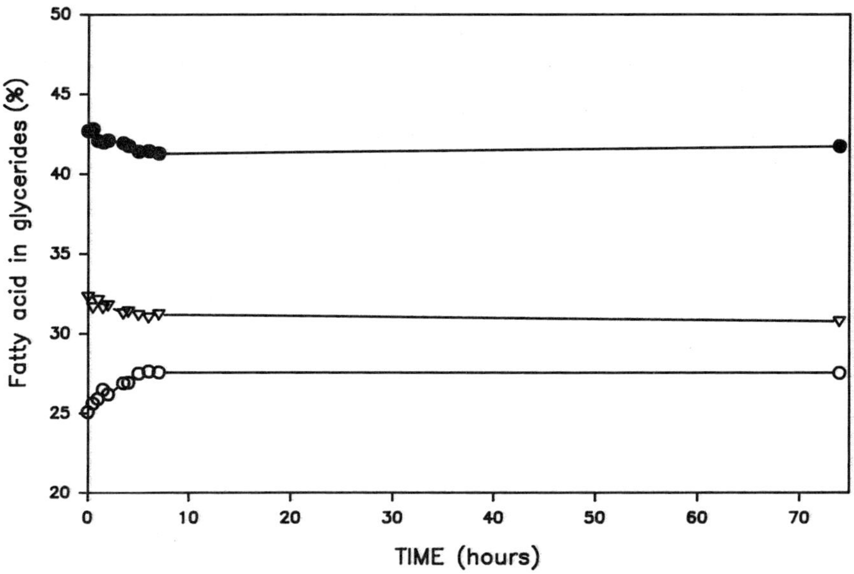

FIGURE 7. Progress curve of hydrolysis of synthetic oil resembling borage oil (conditions described in text). The results are expressed in terms of percentages of each fatty acid remaining in the glyceride fraction. ○, GLA; ●, linoleic acid; ▽, oleic acid.

We also used a synthetic oil resembling borage oil, made in-house according to our previously published method.[3] This synthetic oil contains only the three main fatty acids of the oil, that is to say, GLA (25%), linoleic acid (43%), and oleic acid (32%). All measurable fatty acids account for 100% of the fatty acids of the oil. The distribution of the fatty acids on the glycerol backbone is random, and does not necessarily reflect the natural oil arrangement. When using the synthetic oil in the conditions used for borage oil, no discrimination was observed between the fatty acids present. The results are plotted in FIGURE 7 for a 48-hour reaction. The GLA content varies from 25% up to only 28% in 8 hours and stays at this level for at least 40 more hours. The same reaction on natural oil led to a level of GLA of 50% in the glyceride fraction.

CONCLUSION

Experiments were carried out in our laboratory to focus and elaborate on the mechanism of enzyme "specificity" as pertaining to the efficient separation of a fatty acid of interest from other contaminating fatty acids. Experiments were also performed to investigate the feasibility of improving the yields of the enzymatic reaction to enrich even more the fatty acid of interest. We are currently investigating the use of the chosen lipase, free and immobilized, to enzymatically produce erucic acid from HEAR and crambe oils, and GLA from evening primrose and borage oils.

REFERENCES

1. JENSEN, R. G., F. A. DEJONG & R. M. CLARK. 1983. Determination of lipase specificity. Lipids **18**(3): 239–252.
2. ERGAN, F. & G. ANDRÉ. 1989. Simple high performance liquid chromatography methods for monitoring lipase reactions. Lipids **24**(1): 76–78.
3. ERGAN, F., M. TRANI & G. ANDRÉ. 1990. Production of glycerides from glycerol and fatty acid by immobilized lipases in non-aqueous media. Biotechnol. Bioeng. **35:** 195–200.

Identification of Residues Involved in Active-Site Formation in *Aspergillus ficuum* Phytase

ABUL H. J. ULLAH AND H. CHARLES DISCHINGER, JR.

Southern Regional Research Center
Agricultural Research Service
United States Department of Agriculture
1100 Robert E. Lee Boulevard
New Orleans, Louisiana 70124

INTRODUCTION

Aspergillus ficuum produces an extracellular phytase, myo-inositol hexakisphosphate phosphohydrolase, capable of hydrolyzing inorganic orthophosphate from phytate. The protein has been purified to near homogeneity, and its physicochemical and catalytic properties determined.[1] Recently, we have deduced the total covalent structure of the protein by chemical means. The enzyme is a special form of acid phosphatase (EC 3.1.3.8) exhibiting a K_{cat} of 220 seconds^{-1} and a K_m of 40 μM at pH 5.0 and 58°C.[1,2] It is also able to hydrolyze phytate at pH 2.5. Because of these desirable features, phytase has drawn interest from enzyme-producing companies. Two potential uses of the enzyme are (1) as a food supplement to hydrolyze phytate in feed for monogastric animals, enhancing mineral bioavailability and reducing the level of phytate in animal waste, and (2) in the manufacture of myo-inositol tetra- and triphosphate in immobilized enzyme-based packed-bed bioreactors. Based on the catalytic properties of both the soluble and immobilized forms of the enzyme, it is apparent that improvement in the turnover number of the protein is needed to justify use of the enzyme as a feed supplement and biocatalyst. As a prelude to performing any protein engineering (PE) work, it is imperative that the active catalytic center of the enzyme be mapped. In this presentation, we report the identification of residues involved in "active-site" formation. Using chemical probes and examining the sequence of phytase for homology with other acid phosphatases, we conclude that Arg-58 and His-59 are involved in "active-site" formation. The identity of an essential Trp residue is proposed.

MATERIALS AND METHODS

Purified *Aspergillus ficuum* phytase was used throughout the study. For arginine modification, phytase was modified under conditions similar to those of Patthy and Smith.[3] Upon addition of 1,2-cyclohexanedione (CHD) solution to the dialyzed protein (sodium borate, pH 9.5), the final concentration of CHD was 2 mM in 37.5 mM borate buffer. The mixture was incubated at 37°C for 2 hours. For histidine modification, phytase was treated with 5 to 50 mM diethyl pyrocarbonate (DEP) at pH 6.0 and 25°C.[4] For tryptophan modification, phytase was incubated with 1 mM *N*-bromosuccinimide (NBS) at pH 4.0 (100 mM sodium acetate) at 25°C.[5] The phytase sequence was analyzed for hydrophobic regions using the IBI Pustell Sequence Analysis Programs (Ver. 2.03).

RESULTS

Arg Modification

When *A. ficuum* phytase was treated with varying concentrations of CHD (2 to 15 mM) for 1.0 hour, different degrees of inactivation were observed (FIGURE 1). The

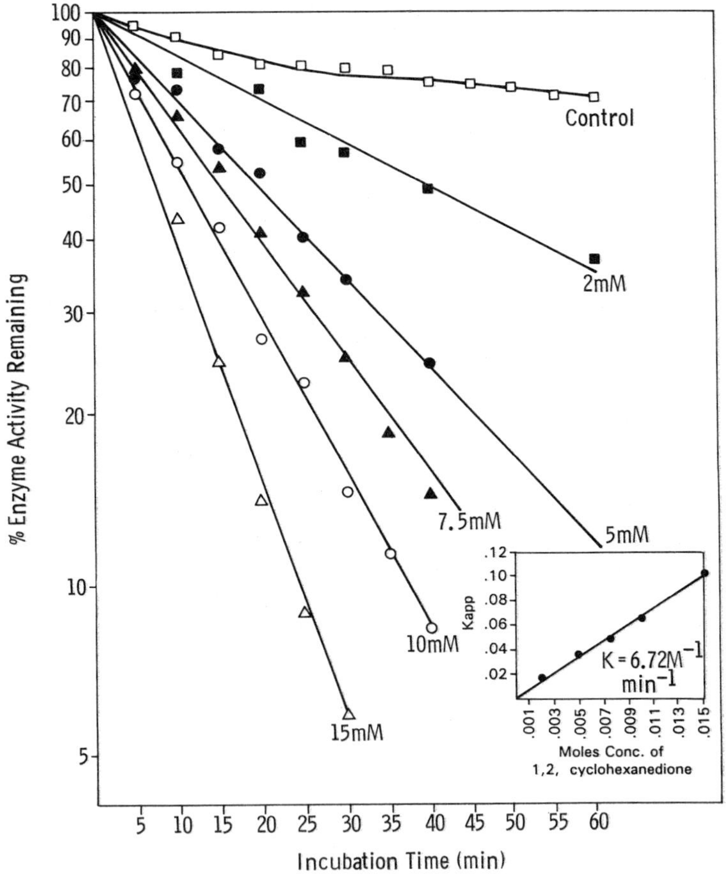

FIGURE 1. Inactivation of phytase by varying concentrations of 1,2-cyclohexanedione as a function of time. The inset shows the first-order rate constant derived from these curves.

AfPhytase	(48)	C	R	V	T	F	A	Q	V	L	S	**R**	**H**	**G**	A	**R**	Y	**P**	T	D	S	K	G	K
EcACP	(6)	L	K	L	E	S	V	V	I	V	S	**R**	**H**	**G**	V	**R**	A	**P**	T	K	A	T	Q	L
HuPACP	(1)	K	E	L	K	F	V	T	L	V	F	**R**	**H**	**G**	D	**R**	S	**P**	I	D	T	F	P	T
HuLACP	(1)	R	S	L	R	F	V	T	L	L	Y	**R**	**H**	**G**	D	**R**	S	**P**	V	K	T	Y	P	T
YspACP1	(29)	C	K	I	K	Q	V	H	T	L	Q	**R**	**H**	**G**	S	**R**	N	**P**	T	G	G	N	A	A
YscACP3	(46)	C	E	M	K	Q	L	Q	M	L	A	**R**	**H**	**G**	E	**R**	Y	**P**	T	Y	S	K	G	A
YscACP5	(46)	C	E	M	K	Q	L	Q	M	V	G	**R**	**H**	**G**	E	**R**	Y	**P**	T	V	S	L	A	K
AnACP	(129)	P	G	I	V	Q	R	R	R	A	G	**R**	**H**	**G**	L	H	Q	R	S	R	N	T	Q	A
RtF2,6BP	(250)	.	.	.	P	R	S	I	Y	L	C	**R**	**H**	**G**	E	S	E	L	N	L	R	G	R	I
YePGM	(1)	.	.	.	.	P	K	L	V	L	V	**R**	**H**	**G**	Q	S	E	W	N	E	K	N	L	F
HuPGM-M	(1)	.	M	A	T	H	R	L	V	M	V	**R**	**H**	**G**	E	T	T	W	N	Q	E	N	R	F
HuBPGM	(1)	.	.	S	K	Y	K	L	I	M	L	**R**	**H**	**G**	E	G	A	W	N	K	E	N	R	F

FIGURE 2. Alignment of *A. ficuum* phytase sequence with published acid phosphatase, fructose-2,6-biphosphatase, and phosphoglycerate mutase sequences. EcACP = *E. coli* acid phosphatase,[9] HuPACP = human prostatic acid phosphatase,[10] HuLACP = human lysosomal acid phosphatase,[11] YspACP1 = *Sacchromyces pombe* acid phosphatase,[12] YscACP3 = *S. cerevisae* acid phosphatase,[13] YscACP5 = *S. cerevisiae* acid phosphatase,[14] AnACP = *A. niger* acid phosphatase,[15] RtF2,6BP = rat fructose-2,6-bisphosphatase,[16] YePGM = yeast phosphoglycerate mutase,[17] HuPGM-M = human muscle phosphoglycerate mutase,[18] HuBPGM = human brain phosphoglycerate mutase.[19,20]

extent of inactivation is concentration dependent and exhibits a pseudo-first-order kinetics. The reaction order for the inactivation was estimated to be 0.87, and the second-order rate constant ($K = 6.72\text{M}^{-1}\text{minute}^{-1}$) is derived from the inset of FIGURE 1. This inactivation can be partially reversed by the addition of hydroxylamine, which is characteristic of Arg modification. Upon prolonged inactivation (2.0 hours), the enzyme loses its ability to cleave phosphate from a variety of phosphomonoesters, including phytate. Since Arg has been shown to be essential in a mammalian acid phosphatase,[6] we compared sequences among various acid phosphatases. The results are summarized in FIGURE 2. One septapeptide containing two Arg residues showed strong homology in seven acid phosphatases. The sequence of this peptide is **RHGXRXP.**

His Modification

Incubation of fungal phytase with DEP in 0.1 M sodium acetate buffer, pH 6.0, resulted in a time-dependent loss of enzyme activity using phytate as a substrate, and a concomitant increase in absorbance at 240 nm (FIGURE 3). The latter indicates the modification of the imidazole ring of histidine. Incubation of DEP-treated enzyme with other substrates shows similar inactivation. The inactivation curve depicted in FIGURE 3, showing rapid initial loss of activity followed by a slower phase, is characteristic of the modification of His by DEP.[4] The sequence homology study of the putative active site shows a His residue to be present in the septapeptide adjacent to an Arg.

Trp Modification

Incubation of fungal phytase with NBS in 0.1 M sodium acetate buffer, pH 4.0, led to inability of the enzyme to hydrolyze phosphate from phytase. Examination of the sequence for both hydrophobicity and hydrophilicity using the IBI Pustell protein secondary structure prediction program (Ver. 2.03) revealed two of the four Trp (no. 25 and no. 267) to be in hydrophilic regions. The remaining two (residues 376 and 399) were found to be in extremly hydrophobic environments (FIGURE 4).

DISCUSSION

Fungal phytase can be readily inactivated by CHD, implicating the involvement of Arg in the catalysis of the substrate. The partial reversal of the inactivation by 0.2 M hydroxylamine further confirms the identity of the modified residue. The enzyme

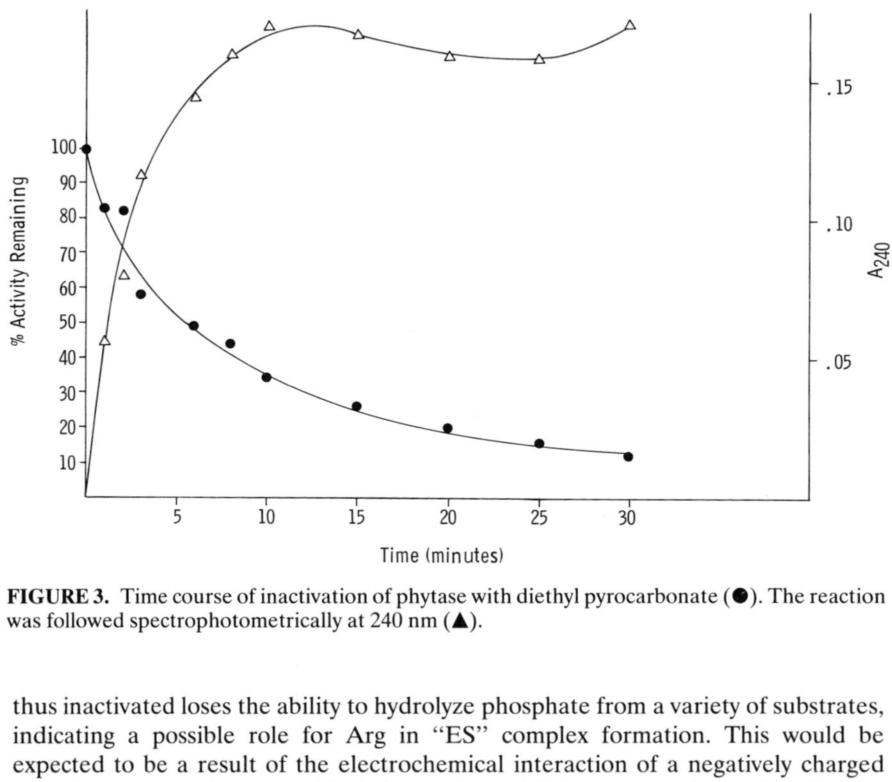

FIGURE 3. Time course of inactivation of phytase with diethyl pyrocarbonate (●). The reaction was followed spectrophotometrically at 240 nm (▲).

thus inactivated loses the ability to hydrolyze phosphate from a variety of substrates, indicating a possible role for Arg in "ES" complex formation. This would be expected to be a result of the electrochemical interaction of a negatively charged phosphate group with the positively charged guanidinium group of Arg. Examination of sequence homology among phosphohydrolases shows a highly conserved tripeptide containing an Arg residue (FIGURE 2). This is a part of a septapeptide which shows homology with diverse acid phosphatases. This larger peptide contains a second Arg residue. Since the first Arg is more widely conserved, it is likely that this residue is the one that causes enzyme inactivation upon treatment with CHD. This could be further substantiated by site-directed mutagenesis experiments. Similarly, modification of His by DEP leads to loss of catalytic activity in phytase. This indicates

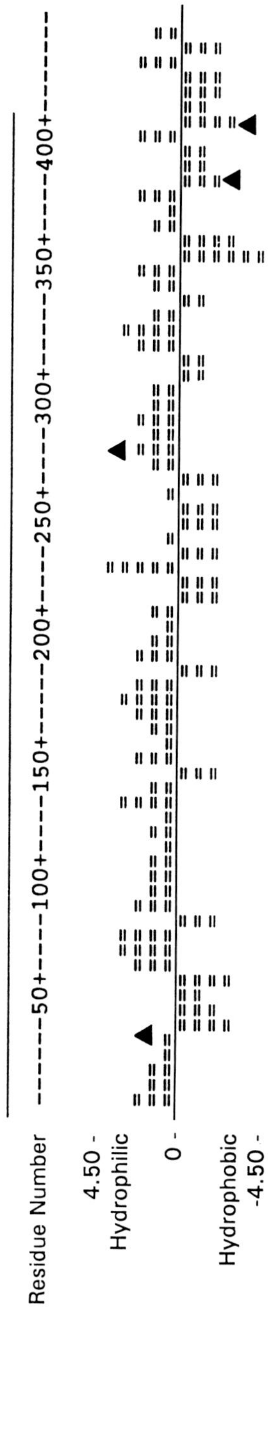

FIGURE 4. IBI Pustell Sequence Analysis of fungal phytase; relative plot of hydrophobicity. ▲ denotes positions of Trp residues.

involvement of His in the formation of the active site in fungal acid phosphatases. The proposed active site derived from sequence alignment data (FIGURE 2) contains a conserved His residue adjacent to the most conserved Arg residue. It is likely that this is the labile His. The formation of a covalent phosphoenzyme via a His residue in a mammalian acid phosphatase was proposed by Van Etten,[7] and the involvement of phospho-histidine in *Mycoplasma* phospho-carrier protein HPr has been established.[8] The proximity of these two labile residues fits with the hydrolysis scheme proposed by Van Etten.[7] We propose that the positively charged guanidinium group of Arg aligns the susceptible phosphate group to the active center. This allows the formation of the "phospho-histidine" transition complex. In this scheme, the role of the labile Trp residue is not clear, and is the subject of ongoing research on this molecule. An x-ray-deduced structure and its refinement by molecular modeling for any acid phosphatase would be helpful in clarifying the roles of each of these labile residues in phytase. We have not yet identified which of the four Trp residues are essential to activity. TRP-25 is a likely candidate, since it is in close proximity to the proposed active site and is in a hydrophilic region. The other three Trp residues, by comparison, are located at the C-terminal end of the molecule (positions 267, 376, 399), and two of them are in extremely hydrophobic regions, rendering them inaccessible for participation for catalysis.

CONCLUSIONS

1. By chemical probing of fungal phytase Arg, His, and Trp have all been shown to be involved in active-site formation. Modification of each led to inactivation of the enzyme.
2. Examination of sequence homology in acid phosphatases and other related proteins revealed a conserved septapeptide.
3. Arg and His residues were included in this peptide, which we propose to be involved in the catalysis of phosphomonoesters in this class of enzyme.
4. While not found in the septapeptide, one of the four Trp residues is likely to be involved in "active-site" formation.

REFERENCES

1. ULLAH, A. H. J. & D. M. GIBSON. 1987. Extracellular phytase (E.C.3.1.3.8) from *Aspergillus ficuum* NRRL 3135: purification and characterization. Prep. Biochem. **17:** 63–91.
2. ULLAH, A. H. J. 1988. *Aspergillus ficuum* phytase: partial primary structure, substrate selectivity, and kinetic characterization. Prep. Biochem. **18:** 459–471.
3. PATTHY, I. & E. L. SMITH. 1975. Reversible modification of arginine residues. Application to sequence studies by restriction of tryptic hydrolysis to lysine residues. J. Biol. Chem. **250:** 557–564.
4. ABDULWAJID, A. W. & F. Y-H. WU. 1986. Chemical modification of *Escherichia coli* RNA polymerase by diethyl pyrocarbonate: evidence of histidine requirement for enzyme activity and intrinsic zinc binding. Biochemistry **25:** 8167–8172.
5. SARTIN, J. L., T. E. HUGLI & T-H. LIAO. 1980. Reactivity of the tryptophan residues in

bovine pancreatic deoxyribonuclease with *N*-bromosuccinimide. J. Biol. Chem. **255:** 8633–8637.

6. McTigue, J. J. & R. L. Van Etten. 1978. An essential arginine residue in human prostatic acid phosphatase. Biochim. Biophys. Acta **523:** 422–429.
7. Van Etten, R. L. 1982. Human prostatic acid phosphatase: a histidine phosphatase. Ann. N.Y. Acad. Sci. **390:** 27–51.
8. Ullah, A. H. J. & V. P. Cirillo. 1976. Mycoplasma phosphoenolpyruvate dependent sugar phosphotransferase: purification and characterization of the phosphocarrier protein. J. Bacteriol. **127:** 1298–1306.
9. Touati, E. & A. Danchin. 1987. The structure of the promoter and amino terminal region of the pH 2.5 acid phosphatase structural gene app-a of *Escherichia coli* a negative control of transcription mediated by cyclic AMP. Biochimie 10 **69:** 215–221.
10. Vihko, P., P. Virkunnen, P. Henttu, K. Roiko, T. Solin & M.-L. Huhtala. 1988. Molecular cloning and sequence analysis of cDNA encoding human prostatic acid phosphatase. FEBS Lett. **236:** 275–281.
11. Pohlmann, R., C. Krentler, B. Schmidt, W. Schroder, G. Lorkowski, J. Culley, G. Mersmann, C. Geier, A. Waheed, S. Gottschalk, K.-H. Grzeschik, A. Hasilik & K. Von Figura. 1988. Human lysosomal acid phosphatase cloning, expression, and chromosomal assignment. EMBO J. **7:** 2343–2350.
12. Elliot, S., C-W. Chang, M. E. Schweingruber, J. Schaller, E. E. Rickli & J. Carbon. 1986. Isolation and characterization of the structural gene for secreted acid phosphatase from *Schizosaccharomyces pombe*. J. Biol. Chem. **261:** 2936–2941.
13. Bajwa, W., B. Meyhack, H. Rudolph, A.-M. Schweingruber & A. Hinnen. 1984. Structural analysis of the two tandemly repeated acid phosphatase genes in yeast. Nucleic Acids Res. **20:** 7721–7739.
14. Arima, K., T. Oshima, I. Kubota, N. Nakamura, T. Mizunaga & A. Toh-E. 1983. The nucleotide sequence of the yeast PHO5 gene: a putative presursor of repressible acid phosphatase contains a signal peptide. Nucleic Acids Res. **11:** 1657–1672.
15. Macrae, W. D., F. P. Buxton, S. Sibley, S. Garven, D. I. Gwynne, R. W. Davies & H. N. Arst, Jr. 1988. A phosphate-repressible acid phosphatase gene from *Aspergillus niger:* its cloning, sequencing and transcriptional analysis. Gene **71:** 339–348.
16. Bazan, J. F., R. J. Fletterick & S. J. Pilkis. 1989. Evolution of a bifunctional enzyme: 6-phosphofructo-2-kinase/fructose-2,6-biphosphatase. Proc. Natl. Acad. Sci. USA **86:** 9642–9646.
17. Fothergill, L. A. & R. N. Harkins. 1982. The amino acid sequence of yeast *Saccharomyces cerevisiae* phosphoglycerate mutase EC 2.7.5.3. Proc. R. Soc. Lond. B **215:** 19–44.
18. Shanske, S., S. Sakoda, M. A. Hermodson, S. Dimauro & E. A. Schon. 1987. Isolation of a cDNA encoding the muscle-specific subunit of human phosphoglycerate mutase. J. Biol. Chem. **262:** 14612–14617.
19. Sakoda, S., S. Shanske, S. Dimauro & E. A. Schon. 1988. Isolation of a cDNA encoding the B isozyme of human phosphoglycerate mutase (PGAM) and characterization of the PGAM family. J. Biol. Chem. **263:** 16899–16905.
20. Blouquit, Y., M-C. Calvin, R. Rosa, D. Prome, J-C. Prome, F. Pratbernou, M. Cohen-Solal & J. Rosa. 1988. Sequence of the human erythrocyte phosphoglycerate mutase by microsequencer and mass spectrometry. J. Biol. Chem. **263:** 16906–16910.

Studies on the Structure and Function of Subtilisin E by Protein Engineering

HIROSHI TAKAGI,[a] HIROSHI MATSUZAWA,[b]
TAKAHISA OHTA,[b] MAKARI YAMASAKI,[b]
AND MASAYORI INOUYE[c]

[a] *Central Research Laboratories*
Ajinomoto Company, Inc.
1-1 Suzuki-cho, Kawasaki-ku
Kawasaki 210, Japan

[b] *Department of Agricultural Chemistry*
The University of Tokyo
1-1-1 Yayoi, Bunkyo-ku
Tokyo 113, Japan

[c] *Department of Biochemistry*
Robert Wood Johnson Medical School at Rutgers
University of Medicine and Dentistry of New Jersey
675 Hoes Lane
Piscataway, New Jersey 08854

Subtilisin is an alkaline serine protease produced by a wide variety of *Bacillus* species. In view of the industrial application for detergent and food processing, subtilisin has been extensively investigated as a promising target for protein engineering.[1]

We have previously cloned and expressed the gene for *B. subtilis* subtilisin E in *Escherichia coli*,[2,3] and in addition, using this system, we isolated mutant subtilisin E with higher activity by optimizing the amino acid adjacent to the catalytic triad.[4] In this study, on the basis of structural comparison with other serine proteases belonging to the subtilisin family, we attempted to enhance the thermostability[5] and to alter the substrate specificity of the enzyme.[6]

First of all, sites for Cys substitutions for formation of a disulfide bond were chosen in subtilisin E based on the structure of aqualysin I.[5] Aqualysin I, an alkaline serine protease belonging to the subtilisin family, is secreted into culture medium by *Thermus aquaticus* YT-1, an extreme thermophile.[7] The optimum temperature for its caseinolytic activity is 80°C. Aqualysin I contains four Cys residues, which form two disulfide bonds, Cys67/Cys99 and Cys163/Cys194.[8] These disulfide bonds seem to be responsible for the thermostability of aqualysin I.

The amino acid residues of subtilisin BPN′, Asn61, Ala98, Ser161, and Glu195, correspond to the Cys residues at position 67, 99, 163, and 194 of aqualysin I, respectively, upon alignment of their sequences. Crystallographic data for subtilisin BPN′ (FIGURE 1) have shown that the distance between the α-carbons of the residues is estimated to be 5.8 Å for Asn61 and Asn98, and 12.7 Å for Ser161 and Glu195. As the sequence of subtilisin E is 86% identical with that of subtilisin BPN′, positions 61(Gly) and 98(Ser) in subtilisin E were chosen as candidates for substitution with Cys residues.

Three mutant plasmids, pHTC61, pHTC98 and pHTC61C98, were constructed by site-directed mutagenesis in order to replace Gly61 with Cys, Ser98 with Cys, and both Gly61 and Ser98 with Cys, respectively. The wild-type and mutant subtilisin E genes were expressed in *E. coli* in the presence of isopropyl β-thiogalactopyranoside

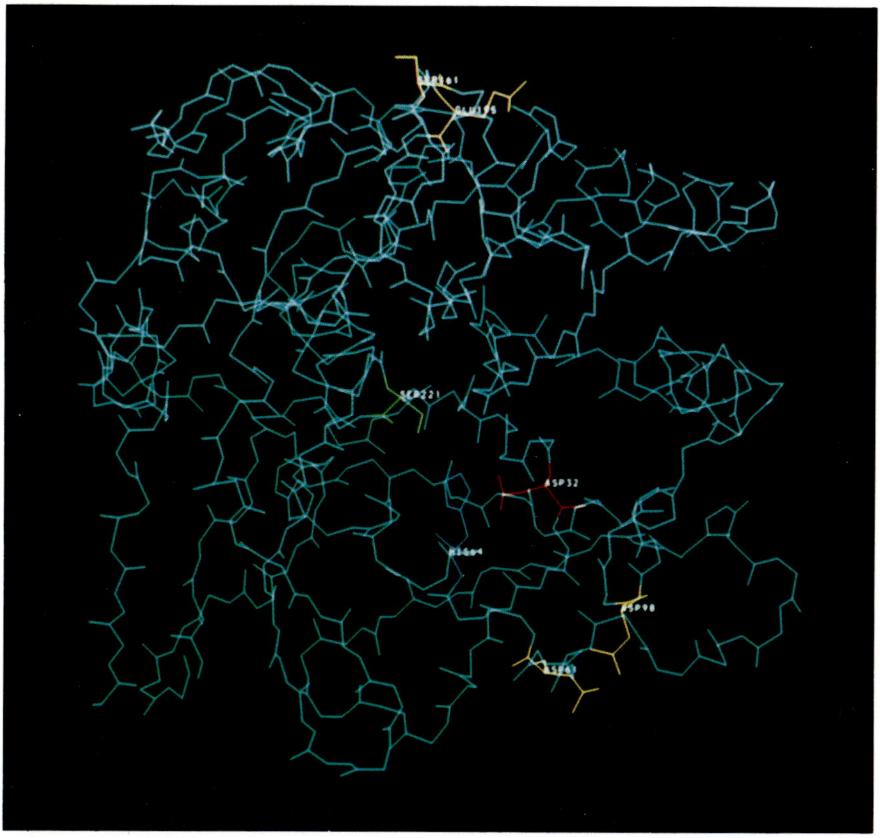

FIGURE 1. α-Carbon backbone structure of subtilisin BPN′ showing the catalytic triad. (Ser221, His64, and Asp32), Asn61, Ala98, Ser161, and Glu195, which were substituted by Cys residues using site-directed mutagenesis.

(IPTG), and the periplasmic fraction of the induced cells was prepared. All enzymes were purified from the periplasmic fraction to give a single band upon sodium dodecyl sulfate-polyacrylamide gel electrophoresis (SDS-PAGE). Disulfide bond formation can be detected by SDS-PAGE under nonreducing conditions because a protein that contains a disulfide bond has a smaller radius of gyration and therefore migrates further down the gel.[9] Under nonreducing conditions, Cys61/Cys98 mutant subtilisin indeed migrated faster than the wild-type enzyme, whereas under reducing conditions, the two bands were virtually indistinguishable. This result suggested that the disulfide bond was formed spontaneously between Cys61 and Cys98 in the present expression system. The evidence for disulfide formation in the mutant was supported by the experiment according to the method of Ellman[10] using 5,5′-dithiobis(2-nitrobenzoic acid).

To investigate the kinetic constants, enzyme assay was carried out at 37°C using a 0.13-mM solution of the synthethic peptide succinyl-Ala-Ala-Pro-Phe-*p*-nitroanilide, a typical substrate for subtilisin. The single-Cys mutants at position 61 or 98 showed a slight decrease in catalytic efficiency relative to the wild-type enzyme. On the other hand, the catalytic efficiency of the disulfide mutant (Cys61/Cys98) was essentially equivalent to that of the wild-type enzyme. Besides, in the case of casein protein, the specific activity of the disulfide mutant was virtually unchanged relative to that of the wild-type enzyme.

It has been shown for some serine protease that increased thermal stability reduces autolysis.[11] Therefore, we investigated the thermodynamic characteristics of these enzymes in terms of thermal stability and autolysis. First, to determine the autolytic stability, the rate of thermal inactivation was measured at 45–60°C in the presence of Ca^{2+}. The half-life ($t_{1/2}$) of the Cys61/Cys98 mutant was found to be 2–3 times longer than that of the wild-type enzyme. When enzymes were treated with 10 mM dithiothreitol (DTT) for reduction, the Cys61/Cys98 mutant had a half-life nearly equivalent to that of the wild-type enzyme (TABLE 1). Similar results were obtained by differential scanning calorimetry analysis for evaluating the change in thermodynamic stability of the molecules. The thermal stability of these enzymes was estimated in the presence of Ca^{2+} after treatment with phenylmethanesulfonyl fluoride (PMSF) to prevent autolysis accompanying the unfolding of the enzyme (FIGURE 2). The disulfide mutant showed a T_m of 63.0°C, which was 4.5°C higher than that observed for the wild-type enzyme. Under reducing conditions, however, the thermal stability of the mutant enzyme was found to be decreased to a level lower than that of the wild-type enzyme. These results strongly suggest that the introduc-

TABLE 1. Effect of DTT on the Half-Life Due to Autolytic Inactivation of Wild-Type and Disulfide Mutant Subtilisin E[a]

	Half-life at:					
	45°C			55°C		
Enzyme	−DTT (hours)	+DTT (hours)	−DTT/+DTT	−DTT (minutes)	+DTT (minutes)	−DTT/+DTT
Wild-type	4.4	3.3	1.3	31	31	1.0
Cys61/Cys98	10.2	4.4	2.3	81	32	2.5

[a]Purified enzymes (30 μg/ml) were either treated or not treated with 25 mM DTT at room temperature for 20 minutes. Remaining activity after heating for various times at indicated temperatures in the presence of 1 mM $CaCl_2$ was determined at 37°C using succinyl-Ala-Ala-Pro-Phe-*p*-nitroanilide as the substrate. Half-life due to autolytic inactivation was determined from semilog plots, $\log_{10}$ (residual activity) versus time.

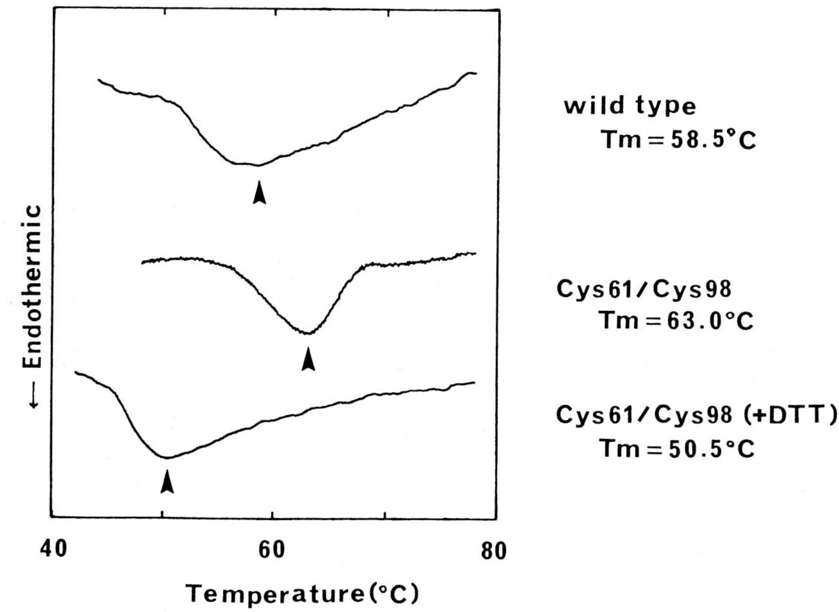

FIGURE 2. Differential scanning calorimetry data for wild-type and Cys61/Cys98 mutant subtilisin E. These results were obtained in 10 mM sodium phosphate buffer (pH 6.2) containing 1 mM $CaCl_2$ and 1 mM PMSF. Data for the reduced forms of these proteins were obtained in the same buffer plus 100 mM DTT. The concentration of protein samples was 25 mg/ml, and 60 μl of each sample was used for the experiment. All the samples were scanned at a heating rate 0.6°C minute^{-1} from 20°C to 90°C.

tion of a disulfide bond by site-directed mutagenesis enhanced the thermostability of subtilisin E without changing the catalytic efficiency of the enzyme.

Secondly, we gave attention to a new alkaline elastase secreted by the alkalophilic *Bacillus* strain Ya-B.[6] Although this enzyme showed extensive homology with subtilisins (54–58% identical), it possesses the three main characteristics of high elastin-hydrolyzing activity, high elastin-binding ability, and substrate preference for aliphatic amino acid residues.[12] Despite the high homology of the sequences, a characteristic deletion of Ser161-Thr162-Ser163-Thr164 of subtilisin, which might influence P1 specificity, was discovered in the corresponding sequence of alkaline elastase.[13] Crystallographic data for subtilisin BPN′ (FIGURE 3) have shown that the segment lies close to the P1 substrate binding cleft where Gly 166 is located, and the deletion of the four amino acids in alkaline elastase might affect the conformation of the cleft. We postulated that the conformational change around the P1 cleft results in a difference of the substrate specificity between the two enzymes. To investigate the influence of the deletion on the substrate specificity, we attempted to delete the four amino acid residues of subtilisin E in the corresponding sequence of alkaline elastase. Considering the distance between the α-carbons of the residues, three segments, Glu156-Gly157-Ser158-Ser159(EGSS), Gly160-Ser161-Thr162-Ser163(GSTS), and Ser161-Thr162-Ser163-Thr164(STST), in subtilisin E were selected for deletion.

Three mutant plasmids were constructed by site-directed mutagenesis to delete the four amino acid residues. Wild-type and mutant subtilisin E genes were expressed in *E. coli* in the presence of IPTG, and all enzymes in the periplasmic

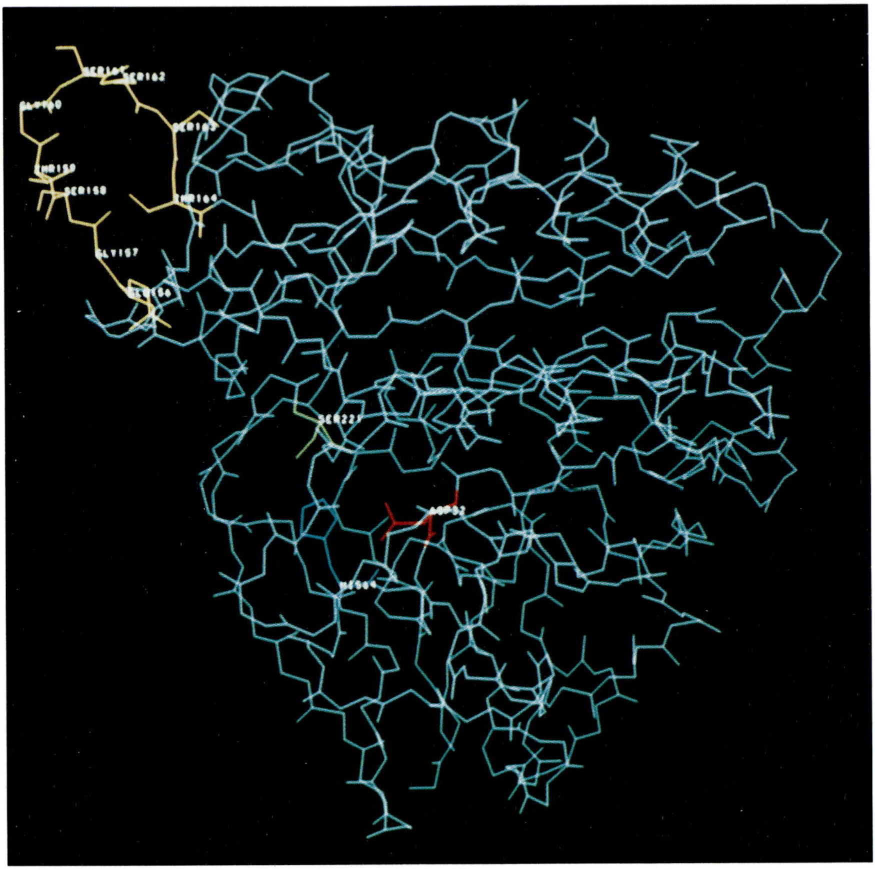

FIGURE 3. α-Carbon backbone structure of subtilisin BPN′ showing the catalytic triad. (Ser221, His64, and Asp32) and Glu156-Gly157-Ser158-Ser159-Gly160-Ser161-Thr162-Ser163-Thr164 (shown in yellow).

fraction without further purification were assayed with succinyl-Ala-Ala-Pro-Phe-*p*-nitroanilide (AAPF), Ala-Ala-Val-Ala-*p*-nitroanilide (AAVA), and Ala-Ala-Ala-*p*-nitroanilide (AAA). The activity of the STST deletion mutant was essentially equivalent to that of the wild-type enzyme, whereas a significant decrease in enzymatic activity occurred when EGSS or GSTS deletion was introduced.

TABLE 2. Hydrolysis of Peptide Substrates by the Wild-Type Subtilisin E, STST Mutant Subtilisin E, and Alkaline Elastase[a]

Enzyme	Specific Activity (U/mg)[b] for			Relative Activity	
	AAA[c]	AAVA[d]	AAPF[e]	AAA/AAPF	AAVA/AAPF
Wild-type	10	1400	2800	0.0036	0.51
STST mutant	5.9	940	1100	0.0052	0.83
Alkaline elastase	64	3600	2500	0.026	1.5

[a]Enzyme assays were performed in 50 mM Tris-HCl (pH 8.5) and 1 mM $CaCl_2$ at 37°C.
[b]Variations in the values were below 5%.
[c]*N*-succinyl-L-Ala-L-Ala-L-Ala-*p*-nitroanilide.
[d]*N*-succinyl-L-Ala-L-Ala-L-Val-L-Ala-*p*-nitroanilide.
[e]*N*-siccinyl-L-Ala-L-Ala-L-Pro-L-Phe-*p*-nitroanilide.

Mutant subtilisin E deleting Ser161-Thr162-Ser163-Thr164 (STST deletion mutant) was purified from the periplasmic fraction to give a single band upon SDS-PAGE, and its enzymatic properties were evaluated. Wild-type subtilisin E and alkaline elastase were examined as controls. AAPF has been used as an authenthic substrate for subtilisin, and also AAVA and AAA have been developed as specific elastase substrates based on the fact that the sequence of Ala-Ala-(Ala)- and Ala residues are abundant in elastin.[14]

TABLE 2 shows the specific activity toward AAPF as the substrate for subtilisin and AAVA and AAA as the substrates for elastase, and the relative activity of AAVA/AAPF and AAA/AAPF of the enzymes. It is worth noting that the relative activity of AAVA/AAPF and AAA/AAPF of the mutant enzyme was approximately 1.5-fold higher than that of the wild-type subtilisin E, despite the fact that the mutant had 40% of the specific activity of the wild-type enzyme for AAPF. The AAVA/

TABLE 3. K_m and k_{cat} Values of the Wild-Type Subtilisin E, STST Mutant Subtilisin E, and Alkaline Elastase[a]

Enzyme	Kinetic Constants[b] for	
	AAVA[c]	AAPF[d]
Wild-type	3.0 (1.0, 0.33)	9.5 (19, 2.0)
STST mutant	2.8 (0.62, 0.22)	4.1 (7.3, 1.8)
Alkaline elastase	5.0 (0.95, 0.19)	5.2 (17, 3.3)

[a]Assays were performed in 50 mM Tris-HCl (pH 8.5) and 1 mM $CaCl_2$ at 37°C for subtilisin, and 50 mM Carbonate (pH 10.5) and 1 mM $CaCl_2$ at 37°C for elastase. Values represent k_{cat}/K_m (seconds^{-1} mM^{-1}), and values in parentheses represent k_{cat} (seconds^{-1}) and K_m (mM), respectively.
[b]Variations in the values were below 5%.
[c]*N*-succinyl-L-Ala-L-Ala-L-Val-L-Ala-*p*-nitroanilide.
[d]*N*-succinyl-L-Ala-L-Ala-L-Pro-L-Phe-*p*-nitroanilide.

AAPF and AAA/AAPF ratios for the mutant are slightly closer to those for alkaline elastase.

To facilitate comparisons, kinetic constants were determined from initial rate measurements of AAPF and AAVA hydrolysis. As shown in TABLE 3, the higher relative activity of the deletion mutant for AAVA is most likely due to the lowered

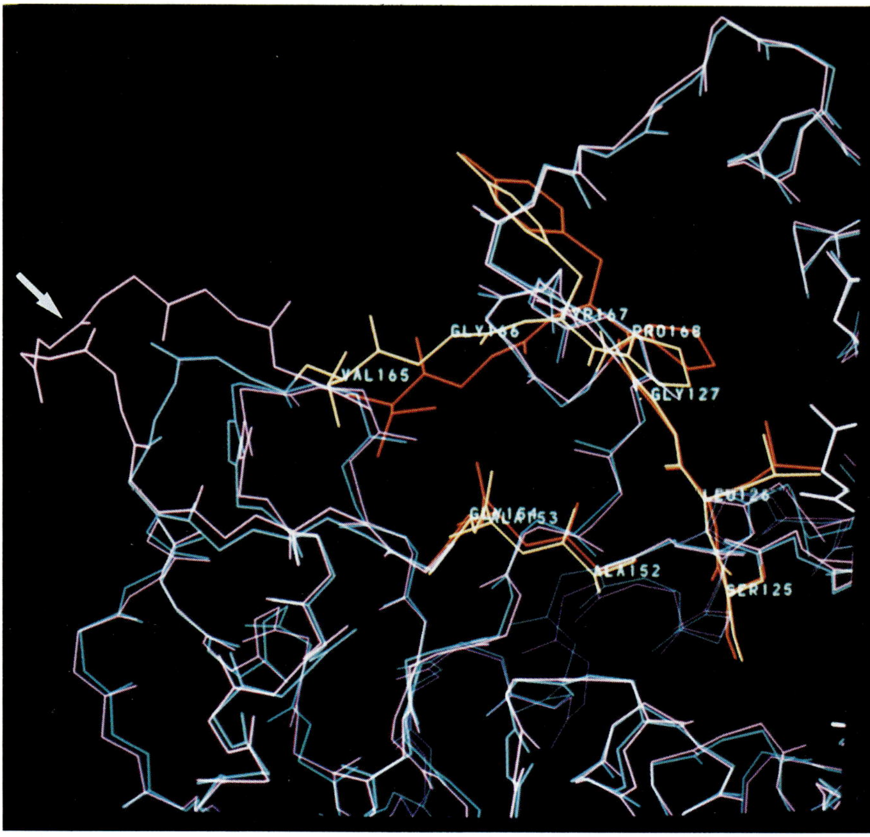

FIGURE 4. Computer modeling of the P1 substrate binding cleft in the crystal (shown in violet red) and the predicted (shown in blue yellow) structure of the α-carbon backbone structure of subtilisin BPN′ using the program SYBYL. The mutant enzyme with deleted Ser161-Thr162-Ser163-Thr164 is drawn as the predicted structure. The white arrow indicates the area of Glu156-Gly157-Ser158-Ser159-Gly160-Ser161-Thr162-Ser163-Thr164 that lies in the vicinity of the P1 binding cleft.

K_m, indicating that the affinity of the enzyme for AAVA is increasing. The mutant enzyme had a K_m value for AAVA that was close to that of alkaline elastase, although the value for AAPF was nearly equal to that of the wild-type subtilisin E. However, the k_{cat}/K_m ratio, which reflects the catalytic efficiency for the deletion mutant, was lower than that of the wild-type enzyme, apparently because of the reduction in the

k_{cat} value. It is interesting that the changes in substrate preference between wild-type subtilisin E and alkaline elastase are predominantly caused by an alteration in K_m and not in k_{cat}.

When elastin and casein were the natural substrates, the ratio of elastin/casein of the deletion mutant slightly shifted to the alkaline elastase, while the hydrolyzing activity of the mutant enzyme for these substrates was considerably lower than those of the wild-type enzyme in the same manner as the peptide substrates. In general, the specificity profile of the STST deletion mutant somewhat approached that of alkaline elastase as compared with its parent, subtilisin E.

In order to predict the structure around the substrate-binding pocket in the mutant enzyme, models were built using the program SYBYL (Tripos Software) to display the three-dimensional structure of subtilisin BPN′ (FIGURE 4). The computational modeling of the mutant suggested that the structural change might occur at the back of the P1 binding cleft which is composed of Val165-Gly166-Tyr167-Pro168, not at the entrance of the cleft, which is formed by the segments Ser125-Leu126-Gly127 and Ala152-Ala153-Gly154. It is possible that such a difference around the P1 binding site due to the four amino acids being deleted would be responsible for the different specificities of the two enzymes.

REFERENCES

1. WELLS, J. A. & D. A. ESTELL. 1988. Trends Biochem. Sci. **13:** 291–297.
2. IKEMURA, H., H. TAKAGI & M. INOUYE. 1987. J. Biol. Chem. **262:** 7859–7864.
3. TAKAGI, H., Y. MORINAGA, M. TSUCHIYA & M. INOUYE. 1988. Bio/Technology **6:** 948–950.
4. TAKAGI, H., Y. MORINAGA, H. IKEMURA & M. INOUYE. 1988. J. Biol. Chem. **263:** 19592–19596.
5. TAKAGI, H., T. TAKAHASHI, H. MOMOSE, M. INOUYE, Y. MAEDA, H. MATSUZAWA & T. OHTA. 1990. J. Biol. Chem. **265:** 6874–6878.
6. TAKAGI, H., S. ARAFUKA, M. INOUYE & M. YAMASAKI. J. Biochem. (In press.)
7. MATSUZAWA, H., K. TOKUGAWA, M. HAMAOKI, M. MIZOGUCHI, H. TAGUCHI, I. TERADA, S.-T. KWON & T. OHTA. 1988. Eur. J. Biochem. **171:** 441–447.
8. KWON, S.-T., H. MATSUZAWA & T. OHTA. 1988. J. Biochem. **104:** 557–559.
9. POLITT, S. & H. ZALKIN. 1983. J. Bacteriol. **153:** 27–32.
10. ELLMAN, G. L. 1959. Arch. Biochem. Biophys. **82:** 70–77.
11. DANIEL, R. M., D. A. COWAN, H. W. MORGAN & M. P. CURRAN. 1982. Biochem. J. **207:** 641–644.
12. TSAI, Y.-C., S. F. LIN, Y.-F. LI, M. YAMASAKI & G. TAMURA. 1986. Biochim. Biophys. Acta **883:** 439–447.
13. KANEKO, R., N. KOYAMA, Y.-C. TSAI, R.-Y. JUANG, K. YODA & M. YAMASAKI. 1989. J. Bacteriol. **171:** 5232–5236.
14. FOSTER, J. A. 1982. Methods Enzymol. **82:** 559–570.

Induced Rebuilding of Aspartase Conformation

LIN MA, SHU-GUI CAO, BAI-XU YAN, HAN-TIAN ZUO, ZHONG-TIAN DING, QING-FAN MENG, AND YU-HUA CHENG

The National Laboratory of Enzyme Engineering
Jilin University
Changchun, 130023, China

INTRODUCTION

There have been many papers on denaturation and renaturation of proteins. It is one of *the* theoretical problems in biochemistry. As we all know, not all denatured proteins can be renatured when the denaturing agents are removed from the system. It is very interesting from the standpoint of enzyme engineering whether or not the enzyme with changed conformation gains some new characteristics and functions.

In this article, we describe studies of the denaturation of aspartase by 3 mol/L guanidine-HCL, the renaturation of aspartase by inducer, and the UV-differential and fluorescence spectra of the native and renatured aspartases. The substrate specifities of the native and renatured aspartases were determined separately.

MATERIALS AND METHODS

L-Aspartic acid is a product of Chinese Peiking Chemicals; guanidine-HCL is a product of Chinese Huzhou Biochemicals; L-aspartyl-phenylalanine methyl ester, L-aspartic acid di-*t*-butyl ester, L-aspartic acid-benzyl ester, L-malic acid, D-aspartic acid, and L-asparagine are from Sigma. All other chemicals were of reagent grade.

Purification of L-Aspartase and Assay of Its Activity

L-Aspartase was extracted and purified electrophoretically to homogeneous preparations from *Escherichia coli* AS 1.881 in our laboratory; its specific activity was 50 units/mg protein.[1] The activity was routinely determined spectrophotometrically by measuring the formation of fumarate or its derivatives by the increase in absorbance at 240 nm and 30°C.[2] Protein concentration was determined according to the method of Bradford.[3]

Denaturation and Renaturation of Aspartase

The denaturation and renaturation of aspartase were carried out as follows: to 1 ml of 0.1 mol/L Tris-HCl buffer, pH 6.8, containing 0.1 mol/L KCl, 5 mmol/L

2-mercaptoethanol, 3mol/L guanidine-HCl, we added 200 μg purified aspartase and then incubated 30 minutes at 25°C. When the activity of aspartase was completely gone, 1 ml of 0.1 mol/L inducer was added to the solution above (the inducers consisted of L-Asp, L-malic acid, and so on), the solution containing inducer dialyzed against 1 L of 0.1 mol/L Tris-HCL buffer, pH 6.8, containing 0.1 mol/L KCl, 5 mmol/L 2-mercaptoethanol for 8 hours at 4°C, and then incubated at 25°C for 4 hours. After incubation the solution was dialyzed against the same buffer two times at 4°C for 4 hours each time.

RESULTS

Substrate Specifity of Aspartases

We used L-Asp, L-Asp-L-phe methyl ester, L-malic acid, L-Asn, and D-Asp as the inducers to obtain induced renatured aspartase, and we used these inducers as substrates, and determined the activity of the native and renatured aspartase. The

TABLE 1. Substrate Specificity of Aspartases[a]

Inducer and Enzyme	Relative Activity						
	L-Asp	D-Asp	L-Malic Acid	L-Asn	L-Asp-Phe Methyl Ester	L-Asp-di-*t*-butyl	L- - Benzyl -Asp
L-Asp	20.0	0.0	3.4	8.3	5.0	13.75	4.0
D-Asp	0.0	0.0	0.0	0.0	0.0	0.0	0.0
L-Malic acid	0.0	0.0	0.0	0.0	0.0	0.0	0.0
L-Asn	0.0	0.0	0.0	0.0	0.0	0.0	0.0
L-Asp-Phe-ome	0.0	0.0	0.0	0.0	0.0	0.0	0.0
Native enzyme	100.0	0.0	1.6	4.6	0.05	0.07	0.0
Enzyme renatured by dilution	51.0	0.0	1.5	4.5	0.0	0.0	0.0

[a]Enzyme activity: ΔOD 240/minute mg protein, and the native enzyme activity is 100%.

results are shown in TABLE 1. TABLE 1 shows that the native aspartase can catalyze the ammonia lysis of L-Asp-phe methyl ester, L-Asp-di-*t*-butyl ester, and so on. Its activity is very low, but the L-Asp-induced renatured enzyme has higher activity than the native enzyme. This indicates that the induced renatured aspartase is different from the native enzyme and that renatured by dilution.

Optimum pH of Aspartase

The effect of pH on the enzyme reaction was examined under assay conditions with different pH, and the pH was varied over a range of 6.0 to 9.5. The buffers (0.05 mol/L) used were phosphate buffers of pH 6.0–7.5 and Tris-HCL buffers of pH 7.5–9.5. The results are shown in FIGURE 1, which shows that the optimum pH is 8.5 for the native aspartase, and 8.0 for the aspartase renatured by L-Asp induction.

Thermostability of Aspartases

The thermostability of the native and induced renatured aspartases was examined by incubation in enzyme solutions (100 μg/ml, 0.05 mol/L Tris-HCl buffer, containing 2 mmol/L $MgCl_2$ pH 7.5) at 30°C to 55°C for 30 minutes and then determination of the residual enzymatic activities. The results are shown in FIGURE 2, and demonstrate that the thermostability of the renatured aspartase is much more unstable than that of the native aspartase.

Spectroscopy of Aspartases

The UV-differential and fluorescence spectra of the native and renatured aspartase have shown that the native, renatured aspartases by inducer and by

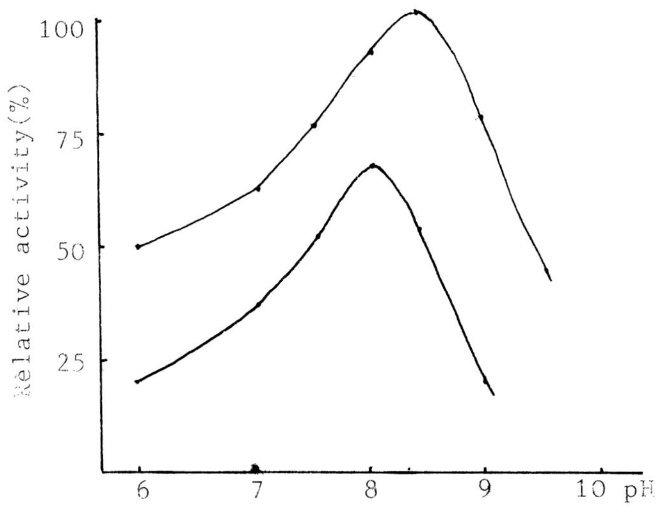

FIGURE 1. Optimum pH of aspartases.

dilution have different conformations and the conformation of the aspartase renatured by L-Asp may be relaxed. The results are shown in FIGURES 3 and 4.

Storage and Stability of Aspartases

The enzyme renatured by L-Asp induction was stored in Tris-HCl buffer for 72 hours at 4°C, and then we determined the residual enzymatic activity using L-Asp, L-Asp-phe methyl ester, L-Asp di-*t*-butyl ester, L- -benzyl-Asp as substrate respectively. The results are shown in TABLE 2. When L-Asp was used as substrate, the activity of aspartase increased gradually, but when other L-Asp derivatives were used as substrate, the activity of aspartase decreased gradually. The results indicate that

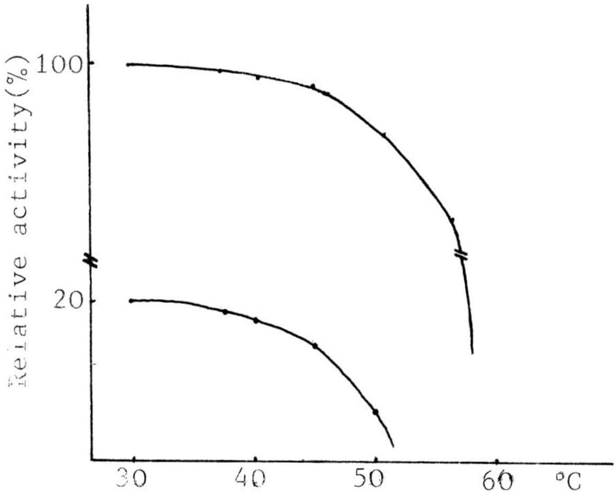

FIGURE 2. Thermostability of aspartases.

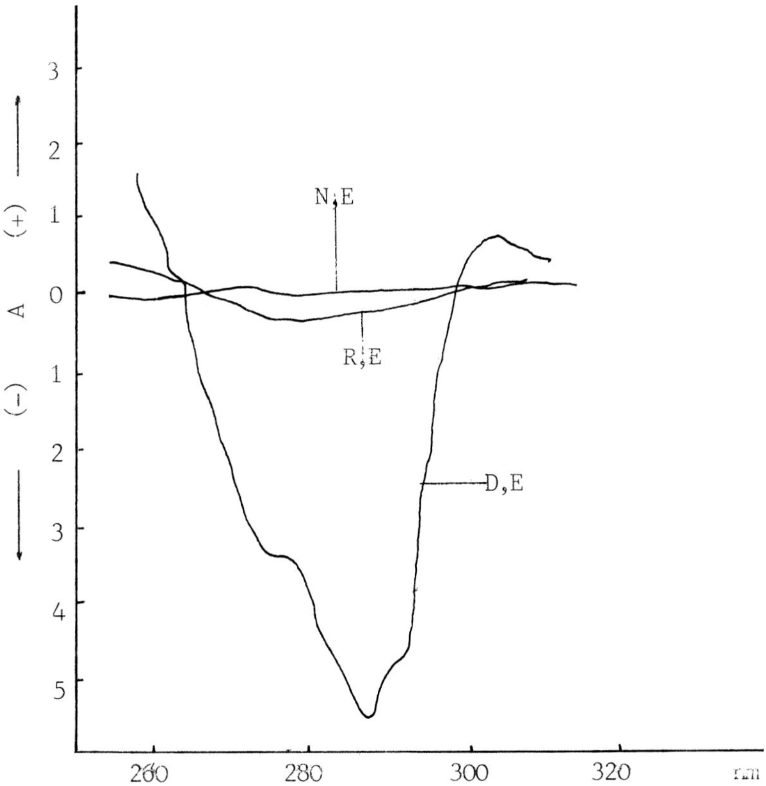

FIGURE 3. UV-differential spectra of aspartase.

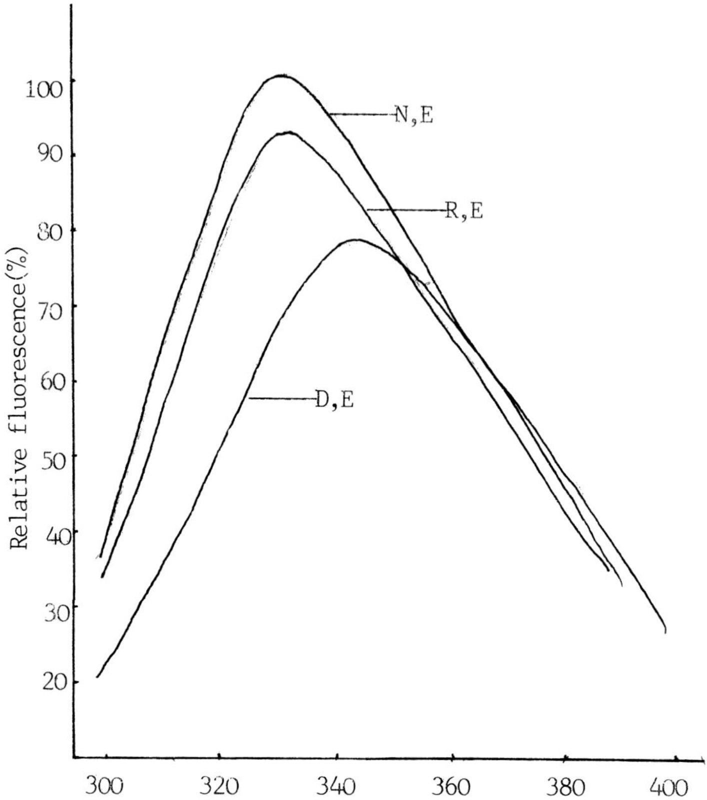

FIGURE 4. Fluorescence spectra of aspartases.

the new conformation of aspartase, which demonstrates new properties, is not very stable, and the native conformation of aspartase is renewed gradually.

DISCUSSION

The above results show that a modified conformation (M) can be obtained by the method of "induced rebuilding of conformation of the enzyme." It has been shown

TABLE 2. The Stability of the Aspartase Renatured by Induction[a]

	Relative Activity			
Substrate	L-Asp	L-Asp-Phe-methyl Ester	L-Asp-di-*t*-butyl Ester	L- -benzyl -Asp
Aspartase renatured by L-Asp induction	30	2.4	6 .35	2.0

[a]The native enzyme activity is 100%.

by spectra and enzyme activity that the M conformation is relatively stable for several days and that some new activity appears.

The theoretical pattern of denaturation and renaturation of the enzyme is shown as follows:

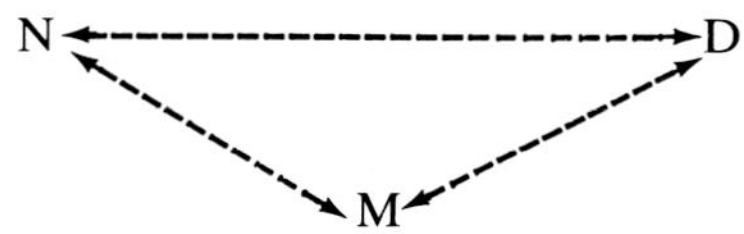

where N = the native enzyme; D = the denatured enzyme; and M = the renatured enzyme.

We think that the changed conformation of the enzyme must change its substrate specifity and other properties, and there is a stable intermediate conformation during the renaturation of the enzyme. Our results indicate that the aspartase with the M conformation is different from the native enzyme.

REFERENCES

1. ZENG, C. B., *et al.* 1990. Acta Sci. Nat. Univ. JiLinensis **3:** 105.
2. WATANABE, Y., *et al.* 1981. Biochim. Biophys. Acta **661:** 261.
3. BRADFORD, M. M. 1976. Anal. Biochem. **72:** 248.

Effects of Sila Substitution of Substrates on Activity of Alcohol Dehydrogenase[a]

KENJI SONOMOTO,[b] HIROKAZU OIKI,[c]
AND YASUHIKO KATO[c]

[b]*Department of Biochemical Engineering and Science*
Faculty of Computer Science and Systems Engineering
Kyushu Institute of Technology
Iizuka 820, Japan

[c]*Department of Applied Chemistry*
Faculty of Engineering
Kyushu Institute of Technology
Tobata-ku
Kitakyushu 804, Japan

INTRODUCTION

The silicon atom belongs to group IVb in the periodic table similarly to carbon-one. This atom has a lower electronegativity and a larger atomic radius than carbon. The 3d orbitals are of low enough energy to be available for bonding to silicon but not to carbon. Owing to these fundamental differences, organosilicon compounds have a specific character in comprehensive organic chemistry. They are of considerable use as new functionality materials, their precursors, and versatile synthetic equivalents of active species in highly selective organic syntheses. The production of organosilicon compounds with desired chemical and physical properties by employing biocatalysts is of great importance and interest from an industrial viewpoint. Thus far, however, only a few results have been reported on the bioconversion of organosilicon compounds.[1–3] There have been no fundamental studies on the mode of recognition and reactions of enzymes toward organosilicon compounds as nonnatural substrates in comparison with conventional ones, except that Kawamoto *et al.* reported a fast and highly stereoselective reaction with yeast lipase.[4]

In this paper, we describe enzymatic conversion of artificial organosilicon compounds, sila-substituted alcohols, by alcohol dehydrogenase. Our study has placed a special emphasis on the interaction of biocatalysts with structurally analogous carbon and silicon compounds in order to see whether biochemical sila-substitution effects are detectable or not.

MATERIALS AND METHODS

Unless stated otherwise, reactions were carried out at 25°C as follows: 0.02 ml of equine liver alcohol dehydrogenase (HLADH) (3 U/ml) was added to a mixture

[a]This work was supported in part by the Iwatani Naoji Foundation's Research Grant.

TABLE 1. Effects of the Trimethylsilyl Group in Organosilicon Compounds on Activity of Equine Liver Alcohol Dehydrogenase

Alcohol	Chemical Formula	Relative Reaction Rate (%)[a]
Trimethylsilylmethanol	$(CH_3)_3SiCH_2OH$	27
Trimethylsilylethanol	$(CH_3)_3SiCH_2CH_2OH$	212
Trimethylsilylpropanol	$(CH_3)_3SiCH_2CH_2CH_2OH$	65
2,2-Dimethylpropanol	$(CH_3)_3CCH_2OH$	133
3,3-Dimethylbutanol	$(CH_3)_3CCH_2CH_2OH$	108
3-Methoxy-3-methylbutanol	$(CH_3)_2C(OCH_3)CH_2CH_2OH$	114
Ethanol	CH_3CH_2OH	100

[a]The rate of dehydrogenation reaction with ethanol was expressed as 100.

composed of 2.75 ml of 10.9 mM Tris-HCl buffer (pH 8.8) and 0.25 ml of 10 mM NAD^+ solution containing 130 mM alcohol. The increase in absorbance at 340 nm was measured.

Three kinds of alcohol dehydrogenase preparations from the following sources were used: yeast (Oriental Yeast Co., Ltd., Tokyo, Japan); equine liver (Sigma Chemical Company, St. Louis, Mo.); *Thermoanaerobium brockii* (Sigma Chemical Company, St. Louis, Mo.). 3,3-Dimethylbutanol (DMB), trimethylsilylethanol (TMSE), trimethylsilylmethanol (TMSM), and trimethylsilylpropanol (TMSP) were obtained from Aldrich Chemical Co. Milwaukee, Wis. 2,2-Dimethylpropanol was purchased from Wako Pure Chemical Industries, Osaka, Japan. Other organosilicon compounds having hydroxyl groups were gifts from Shin-Etsu Chemical Co., Ltd., Tokyo, Japan. 3-Methoxy-3-methylbutanol was a product of Tokyo Kasei Kogyo Co., Ltd., Tokyo, Japan. All other chemicals were obtained from commercial sources.

RESULTS AND DISCUSSION

Among the commercially available alcohol dehydrogenases described above, only HLADH catalyzed the oxidative reaction not only of TMSM, TMSE, and TMSP but also of their derivatives (butyldimethylhydroxymethylsilane, butyldimethyl-3-hydroxypropylsilane, and heptamethyl-3-(3-hydroxypropyl)trisiloxane). No activity

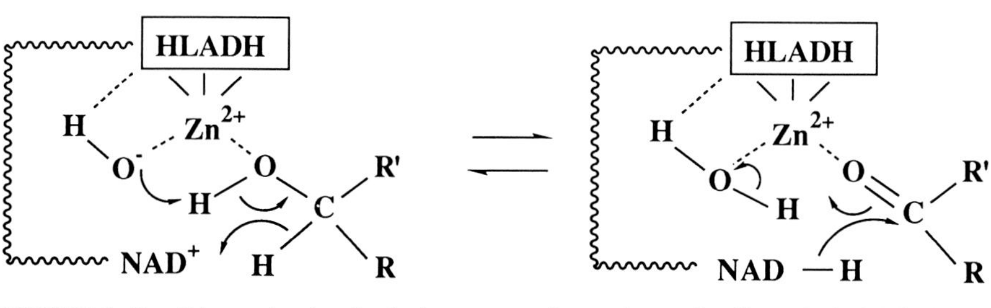

FIGURE 1. Possible mechanism for hydrogen transfer at the equine liver alcohol dehydrogenase active site. HLADH, equine liver alcohol dehydrogenase.

TABLE 2. Kinetic Constants for Dehydrogenation of Trimethylsilyl Alcohols and the Corresponding Carbon Compounds by Equine Liver Alcohol Dehydrogenase

Alcohol	K_m (mM)	V_{max} (μmol · minute^{-1} · mg protein^{-1})
Trimethylsilylmethanol	6.28	0.255
Trimethylsilylethanol	0.567	2.14
Trimethylsilylpropanol	1.01	0.713
2,2-Dimethylpropanol	20.4	0.974
3,3-Dimethylbutanol	1.80	0.966
Ethanol	4.51	1.12

was, however, observed with the corresponding analogues where the methyl group in the trimethylsilyl group was replaced by phenyl-one. Thus, recognition of several sila-substituted alcohols by HLADH may be attributed to its broad substrate specificity.

In order to examine the effect of the silicon atom in sila-substituted alcohols on HLADH activity, kinetic investigations with trimethylsilyl alcohols and their carbon counterparts were performed. HLADH showed greatly different activities with organosilicon compounds having differing chain lengths between the trimethylsilyl group and the hydroxyl group (TABLE 1). Considering the possible mechanisms for hydrogen transfer at the HLADH-active site,[5] in general, the alcohol oxidation rate increases with decreasing electron density at the α-carbon to the hydroxyl group in alcohols (FIGURE 1). The trimethylsilyl group shows a β-effect, that is, it stabilizes a negative charge on the α-carbon and a positive charge on the β-carbon to the silicon.

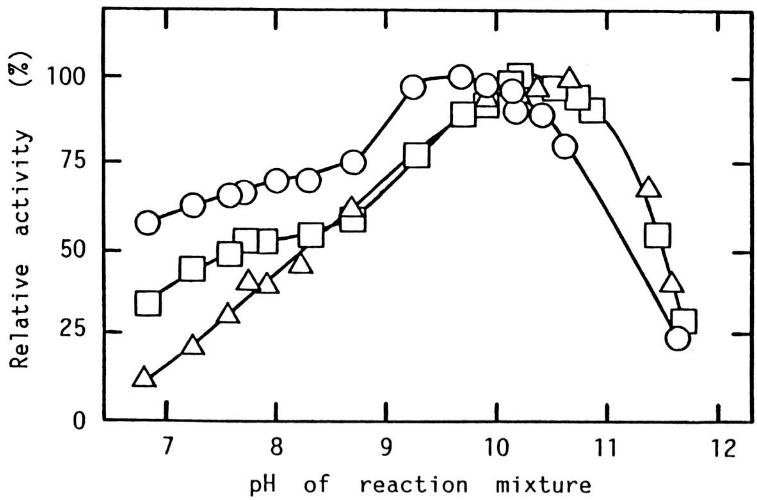

FIGURE 2. Effect of pH on equine liver alcohol dehydrogenase activity. Reactions were carried out at an indicated pH as described in Materials and Methods: ○, trimethylsilylethanol; △, 3,3-dimethylbutanol; and □, ethanol as alcohol.

Of trimethylsilyl alcohols tested, thus, TMSE enables a fast dehydrogenation reaction, whereas TMSM causes a difficult hydrogen transfer at the HLADH active site. On the other hand, the carbon analogues having a *tert*-butyl group did not exhibit the unique phenomena mentioned above (TABLE 1). HLADH activity decreased with an increase in chain length of the carbon compounds as in the case of that with straight-chain primary alcohols. Furthermore, the K_m value (mM) of HLADH for TMSE was smaller than that for DMB (TMSE, 0.567; DMB, 1.80) and the V_{max} value (μmol · minute^{-1} · mg protein^{-1}) was increased (TMSE, 2.14; DMB, 0.966). TMSE was also found to be an excellent substrate of highest affinity to HLADH among the other alcohols tested (TABLE 2).

We tried an increase in HLADH activity with the carbon analogues having a *tert*-butyl group by substitution of the methyl group for electron-attractive groups

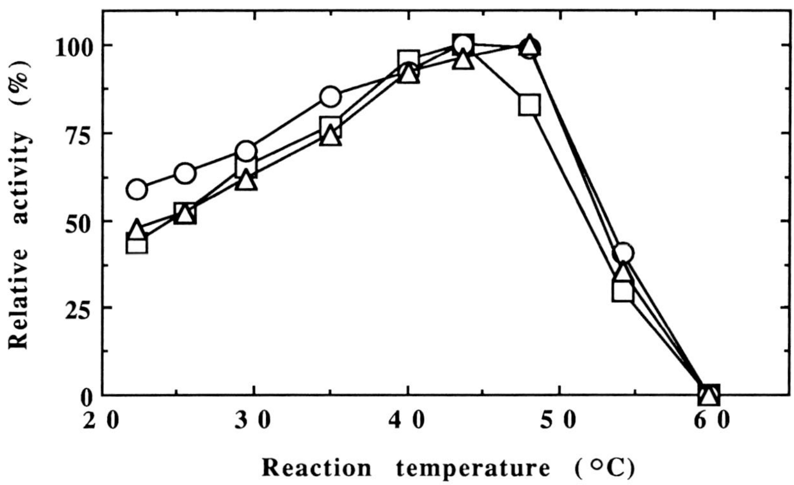

FIGURE 3. Effect of reaction temperature on equine liver alcohol dehydrogenase activity. Reactions were carried out in 25 mM $NaHCO_3$-NaOH buffer (pH 10.4) containing ○, trimethylsilylethanol; △, 3,3-dimethylbutanol; and □, ethanol at an indicated reaction temperature. Other conditions were described in Materials and Methods.

such as the methoxy group and the amino group. However strong the substituents used may be, they could not give a high oxidative activity of HLADH as obtained with TMSE (for example, 3-methoxy-3-methylbutanol as shown in TABLE 1).

With respect to the dependencies of pH and temperature for the HLADH reaction, TMSE was almost identical with ethanol and DMB (FIGURES 2 and 3). These facts suggested that TMSE was recognized as a nonnatural substrate by HLADH.

CONCLUSIONS

HLADH recognized several organosilicon compounds, sila-substituted alcohols, as a substrate and exhibited oxidative activity. Especially, large differences in catalytic rates were observed between trimethylsilyl alcohols [$Me_3Si(CH_2)_nOH$] but not between the carbon analogues. The activity of HLADH was extremely and peculiarly affected by the trimethylsilyl group in the organosilicon compounds. These differences might result from a β effect of the trimethylsilyl group (stabilization of a negative charge on the α-carbon and a positive charge on the β-carbon to the silicon) when considering the possible mechanisms for hydrogen transfer at the HLADH active site. As a result of kinetic investigations with HLADH, TMSE (n = 2) was found to show high affinity compared with the others. Furthermore, enhancement of HLADH activity derived from sila substitution was not easily elicited even by the use of the corresponding carbon analogues having electron-attractive groups.

This research might not only develop the basic and practical aspects of organosilicon biochemistry but also suggests latent abilities of biocatalysts toward many artificial organoelement species.

ACKNOWLEDGMENTS

The authors are indebted to Shin-Etsu Chemical Co., Ltd., for its generous supply of several kinds of organosilicon compounds.

REFERENCES

1. Syldatk, C., H. Andree, A. Stoffregen, F. Wagner, B. Stumpf, L. Ernst, H. Zilch & R. Tacke. 1987. Enantioselective reduction of acetyldimethylphenylsilane by *Trigonopsis variabilis* (DSM 70714). Appl. Microbiol. Biotechnol. **27**(2): 152–158.
2. Syldatk, C., A. Stoffregen, F. Wuttke & R. Tacke. 1988. Enantioselective reduction of acetyldimethylphenylsilane: a screening with thirty strains of microorganisms. Biotechnol. Lett. **10**(10): 731–736.
3. Fritsche, K., C. Syldatk, F. Wagner, H. Hengelsberg & R. Tacke. 1989. Enzymatic resolution of *rac*-1,1-dimethyl-1-sila-cyclohexan-2-ol by ester hydrolysis or transesterification using a crude lipase of *Candida cylindracea.* Appl. Microbiol. Biotechnol. **31**(2): 107–111.
4. Kawamoto, T., K. Sonomoto & A. Tanaka. 1991. Efficient optical resolution of 2-(4-chlorophenoxy)propanoic acid with lipase by the use of organosilicon compounds as substrate: the role of silicon atom in enzymatic recognition. J. Biotechnol. **18**(2): 85–92.
5. Brändén, C.-I., H. Jörnvall, H. Eklund & B. Furugren. 1975. Alcohol dehydrogenases. *In* The Enzymes. P. D. Boyer, Ed. 3rd edit. **11:** 103–190. Academic Press. New York, N.Y.

Altering the Proteolytic Activity of Subtilisin through Protein Engineering

THOMAS P. GRAYCAR, RICHARD R. BOTT, ROBERT M. CALDWELL, JUDY L. DAUBERMAN, PUSHKARAJ J. LAD, SCOTT D. POWER, I. HELENA SAGAR, ROBIN A. SILVA, GEOFFREY L. WEISS, LESLIE R. WOODHOUSE,[a] AND DAVID A. ESTELL

Department of Research and Development
Genencor International, Inc.
180 Kimball Way
South San Francisco, California 94080

The utility of protein engineering in redesigning the structure of a protein to tailor its functional properties has been firmly established. In particular, the *Bacillus* serine protease subtilisin has proven to be a useful model protein for examining the use of systematic structural modification to incorporate novel functional properties into an enzyme.[1,2] The list of properties that have been altered in subtilisin via such modification includes oxidative stability,[3,4] thermal stability,[5] alkaline pH stability,[6] stability in organic solvent,[7] substrate specificity in aqueous buffer,[8–11] nucleophile specificity,[12,13] and pH activity profile.[14] In addition to demonstrating the versatility of protein engineering, these studies have also provided valuable insight into the expected consequences of protein structure modification. For example, it is now recognized that while amino acid substitutions generally lead to only slight structural perturbations, these minor changes in structure can cause significant changes in protein function. Furthermore, it is apparent from several studies with subtilisin that multiple amino acid substitutions may additively affect a particular functional property.

Provided with this extensive data base of structure-function relationships in subtilisin, the engineering of subtilisin for altered proteolytic activity is now being attempted. Increasing the proteolytic activity of subtilisin could boost the enzyme's effectiveness as an additive to household laundry detergents. Subtilisin sold for use in laundry detergents accounts for the largest share of the worldwide industrial enzyme market with sales estimated for 1991 at $200 million. Furthermore, the utility of subtilisin for peptide synthesis in aqueous systems can be enhanced by decreasing the enzyme's proteolytic activity. This would alleviate the problem of low synthesis yields obtained due to proteolysis of the peptide product.

[a]Department of Nutritional Sciences, University of California, Berkeley, California 94720.

IMPROVING THE PROTEOLYTIC ACTIVITY OF SUBTILISIN

Subtilisin from several different species of *Bacillus* is currently used in commercial laundry products. Although the active sites of these enzymes display a high degree of structural homology, considerable difference in wash performance at different pH levels has been demonstrated. Consequently, subtilisin expressed by *B. licheniformis* and *B. amyloliquefaciens* is used primarily in liquid detergents which typically have a wash water pH below 9. Powdered detergents, which command the largest market share, have a wash water pH above 9 and use primarily the subtilisin from *B. lentus.* The subtilisin from *B. amyloliquefaciens* has previously been engineered to have improved activity and wash performance.[15] Here, we report the engineering of *B. lentus* subtilisin for improved proteolytic activity.

The first approach to improving the proteolytic activity of *B. lentus* subtilisin consists of incorporating amino acid changes in the active site at locations believed to be associated with this property. Using numbering based on the *B. amyloliquefaciens* subtilisin sequence, one of these is residue 104.[b] This residue makes contact with the side chain of a bound substrate molecule's P4 amino acid.[c] Substitution of the tyrosine at position 104 in *B. amyloliquefaciens* subtilisin with valine, the amino acid found at this site in *B. lentus,* resulted in a significant decrease in catalytic efficiency for a tetrapeptide substrate (TABLE 1). Because of the structural homology of the active sites of these two enzymes, it was expected that making the corresponding substitution of the valine (V) at position 104 in *B. lentus* subtilisin with tyrosine (Y) would increase enzyme activity. The tetrapeptide kinetics listed in TABLE 1 for the V104Y variant of *B. lentus* indicate that catalytic efficiency is increased substantially by this substitution.

TABLE 1. Tetrapeptide Substrate Kinetics[a] and Proteolytic Activity[b] Measured for Native and Variant Subtilisins

Enzyme	k_{cat} (seconds^{-1})	K_m (M)	k_{cat}/K_m (seconds^{-1} M^{-1})	Proteolytic Activity
B. amyloliquefaciens	50	1.4×10^{-4}	3.6×10^{5}	ND[c]
Y104V	3.5	9.2×10^{-4}	3.8×10^{3}	ND
B. lentus	170	7.8×10^{-4}	2.2×10^{5}	27
V104Y	470	3.1×10^{-4}	1.5×10^{6}	34
K27R/V104Y/T274A	480	3.3×10^{-4}	1.5×10^{6}	ND
K27R/V104Y/N123S/T274A	520	1.2×10^{-3}	4.3×10^{5}	50

[a]Kinetic measurements were performed at 25°C in 0.1 M Tris, pH 8.6, using the substrate *N*-succinyl-L-Ala-L-Ala-L-Pro-L-Phe-*p*-nitroanilide.

[b]Proteolytic activity data are percent hydrolysis at pH 10 of a casein solution following a one-hour incubation at 37°C.

[c]ND denotes data not determined.

[b]The amino acid numbering used for the different subtilisins discussed in this paper corresponds to that of mature *B. amyloliquefaciens* subtilisin (commonly referred to as subtilisin BPN′). The complete subtilisin BPN′ amino acid sequence can be found in Reference 22.

[c]The nomenclature used in referring to the binding of a polypeptide substrate to the active site of proteases follows a numbering system initially proposed by Schechter and Berger.[16]

Although the V104Y variant's tenfold increase in activity for the tetrapeptide substrate is impressive, proteolytic activity as measured by casein hydrolysis showed only a 26% increase. This result is not surprising since the residue 104 change was based upon improved hydrolysis of a specific tetrapeptide substrate rather than a protein substrate. Because improved proteolytic activity is the objective, however, a different approach was employed involving the screening of random mutation variants for increased proteolytic activity. The gene used for random mutagenesis was that of the *B. lentus* V104Y variant with two other site-specific mutations. Both of these mutations, a lysine (K) replaced with arginine (R) at position 27, and a threonine (T) replaced with alanine (A) at position 274, are located far away from the enzyme's active site. The residue 27 and 274 substitutions do not affect the enzyme's tetrapeptide substrate activity which is a sensitive assay of alterations to the active site structure (TABLE 1). Random mutations were generated via poisoned oligonucleotide synthesis and subsequent replacement of corresponding wild-type gene segments coding for various regions of the enzyme's active site. Culture broths of the transformants were then screened for caseinolytic activity.

Several enzymes were identified as having improved proteolytic activity after screening more than 40,000 random variants of the *B. lentus* K27R/V104Y/T274A subtilisin. One enzyme in particular appeared to have considerably more proteolytic activity than the parent enzyme. Subsequent DNA sequencing revealed a single mutation at position 123 where an asparagine (N) is replaced by serine (S). Kinetic analysis of the K27R/V104Y/N123S/T274A variant with the tetrapeptide substrate (TABLE 1) shows a small increase in substrate turnover rate, k_{cat}, but a significant increase in K_m with the net result that catalytic efficiency (k_{cat}/K_m) is reduced. In contrast, the caseinolytic activity of the new variant is almost twice that of the *B. lentus* subtilisin and approximately 50% higher than that of the *B. lentus* V104Y variant. Clearly, the tetrapeptide substrate serves as a good probe for alteration in the active site structure, but not as a good predictor for proteolytic activity. Binding of the tetrapeptide substrate is directed primarily by the phenylalanine at the P1 position. Hence, it is possible that secondary binding sites play more of a role in the hydrolysis of protein substrates.

Comparison of the three-dimensional structures of the native *B. lentus* subtilisin and the *B. lentus* K27R/V104Y/N123S/T274A variant provides some insight into the structural alteration that could affect the improvement in proteolytic activity.[17] Notable differences in the active site conformation of these two enzymes include an increase in the disorder at residue 125 in the variant enzyme as well as a widening of the spacing between the loop formed by residues 96–107 and the segment comprising residues 126–129. x-Ray crystal structures of subtilisin with bound inhibitors have shown that residues in the 96–107 loop and the 126–129 segment form a three-stranded antiparallel β-sheet with the main chain of substrate residues P4–P1.[18] The structural alterations observed in the variant enzyme suggest an increase in the flexibility of this region in the active site. This increased flexibility could account for the improvement in proteolytic activity observed in the variant subtilisin by allowing the active site conformation to adapt more readily to reactive sites on a protein surface. This rationale is also consistent with the tetrapeptide substrate being a poor indicator of proteolytic activity since increased flexibility in the enzyme's active site would not be expected to improve catalysis for a small, homogenous, and specific substrate.

ENGINEERING REACTION SPECIFICITY

Protease-catalyzed peptide synthesis in aqueous systems can suffer poor yields due to proteolytic product degradation. This problem can often be circumvented by trapping the reaction product in an organic solvent phase or by product precipitation. Alternatively, proteolytic product degradation can be avoided through modification of the enzyme's reaction specificity. Decreasing the amidase activity of a protease independent of its esterase activity would allow kinetically controlled peptide synthesis to proceed with minimal product degradation. This may be accomplished in a number of ways including chemical modification of catalytic residues[19–21] and protein engineering of the active site. We have utilized this latter approach to modify the ratio of amidase to esterase activity in subtilisin.

TABLE 2. *p*-Nitroanilide and Thiobenzyl Ester Tetrapeptide Substrate Kinetics Measured for Native and Variant Subtilisins[a]

Enzyme	k_{cat} (seconds^{-1})	K_m (M)	k_{cat}/K_M (seconds^{-1} M^{-1})
N-Succinyl-L-Ala-L-Ala-L-Pro-L-Phe-*p*-nitroanilide			
B. amyloliquefaciens	50	1.4×10^{-4}	3.6×10^{5}
P225A	4	6.2×10^{-4}	6.4×10^{3}
M222K	0.24	9.1×10^{-4}	2.6×10^{2}
N-Succinyl-L-Ala-L-Ala-L-Pro-L-Phe-thiobenzyl ester			
B. amyloliquefaciens	2000	1.1×10^{-4}	1.8×10^{7}
P225A	3000	3.1×10^{-4}	9.7×10^{6}
M222K	2600	4.4×10^{-4}	5.9×10^{6}

[a]Kinetic measurements were performed at 25°C in 0.1 M Tris, pH 8.6.

The subtilisin family of serine proteases has a conserved proline at position 225.[22] Analysis of a highly refined x-ray crystal structure of *B. amyloliquefaciens* subtilisin reveals that this proline residue introduces a discontinuity in an extended α-helix comprising residues 224–237. Preceding this α-helix is a turn of 3_{10}-helix formed by residues 220–222 which includes the catalytic serine at position 221. Computer modeling suggests that substitution of the proline at position 225 with alanine could remove the discontinuity from the α-helix and extend the helix by one turn to include residues 220–223. The replacement of proline (P) at position 225 with alanine in the *B. amyloliquefaciens* subtilisin was made by site-specific mutagenesis. The x-ray crystal structure determined for the P225A variant shows substantial movement of residues 220–223 and that α-helix hydrogen bonds are now formed by this segment while the 3_{10}-helix hydrogen bonds are weakened.[23]

The kinetics for native *B. amyloliquefaciens* subtilisin and the P225A variant on amide and ester substrate hydrolysis show a dramatic change in reaction specificity (TABLE 2). Amidase activity of the variant, as measured by the *p*-nitroanilide tetrapeptide substrate, is decreased 60-fold in catalytic efficiency compared to native enzyme. In contrast, catalytic efficiency for the variant using a thiobenzyl ester derivative of the same tetrapeptide is decreased only 2-fold. The alanine substitution for proline at position 225 effectively decreases amidase activity in subtilisin independent of esterase activity. Amide bond hydrolysis by serine proteases is known to be acylation rate limited while ester bond hydrolysis is usually deacylation rate limited. In addition, the *p*-nitroanilide and the thiobenzyl ester substrates have the same acyl group so that the deacylation step is identical for both reactions. Hence, it is apparent from these results that the acylation rate for the *p*-nitroanilide substrate but not the deacylation rate is substantially decreased by the alanine substitution.

How might the extension of the α-helix seen in the x-ray crystal structure of the P225A variant relate to the observed decrease in acylation rate? It has been suggested that a partial positive charge at the N-terminus of an active site α-helix, resulting from the helix's inherent dipole, may play a functional role in enzyme catalysis.[24] However, as noted above, the catalytic serine in native subtilisin is outside the α-helix on a turn displaced from the helix axis by the kink caused by the proline at position 225. Inclusion of the catalytic serine into the α-helix by removing the kink in the P225A variant might locate the partial positive charge from the helix dipole at a specific location in the active site that is detrimental for acylation. Indeed, this might explain why a proline at position 225 is strictly conserved in the subtilisin-type serine proteases. Alternatively, a slight disruption of the geometry of the catalytic triad in the P225A variant may also be responsible for the decrease in acylation rate. Some movement of the catalytic serine 221 is observed in the P225A structure as the main chain of residues 220–223 adopts a new conformation.

To more directly examine the functional consequences of introducing a positive charge close to the catalytic site of *B. amyloliquefaciens* subtilsin, a site-specific variant having the methionine (M) at position 222 replaced with lysine was characterized kinetically using the same amide and ester substrates. The M222K variant was constructed as part of a previous study aimed at removing the sensitivity of subtilisin to peroxide oxidation.[3] Because the methionine at position 222 is conserved in the known *Bacillus* subtilisin sequences, all 19 of the coded amino acids were introduced singly at this site. The lysine variant proved the least active of all the position 222 variant enzymes on the *p*-nitroanilide tetrapeptide substrate. Although the x-ray crystal structure of the M222K variant has yet to be determined, it is apparent from the native enzyme structure that the positively charged amino group at the end of the lysine side chain could be located within 5 Å of the site of amide bond hydrolysis.

Although more dramatic, the effect of the lysine substitution at position 222 on reaction specificity is similar to that of the alanine substitution at position 225. Catalytic efficiency for hydrolysis of the acylation rate limited *p*-nitroanilide tetrapeptide substrate by the M222K variant is reduced by over three orders of magnitude compared to the native enzyme (TABLE 2). The dominant effect of the lysine substitution is on the substrate turnover rate k_{cat} which is decreased 200-fold indicating that the acylation step in the reaction is severely destabilized. In comparison, hydrolysis of the thiobenzyl ester substrate is only 3-fold lower in catalytic efficiency for the M222K variant. Since the side chain volumes of methionine and lysine are comparable, the decreased acylation rate in the M222K variant is most likely due to the positively charged amino group at the end of the lysine's side chain.

Introduction of a positive charge at the catalytic site of subtilisin would be expected to alter substantially the enzyme's pH activity profile. The pH dependence for catalysis by serine proteases follows the protonation state of the catalytic histidine residue. *B. amyloliquefaciens* subtilisin is most active at alkaline pH where the catalytic histidine having a pK_a of 7 is deprotonated. Making the electrostatic environment of the catalytic histidine more positive should lower it's pK_a by destabilizing the protonated form. In FIGURE 1, however, it is evident that the M222K variant's apparent pK_a for k_{cat}/K_m follows the ionization of an amino acid more basic than histidine. With an apparent pK_a of 9 it appears that catalytic activity in the alkaline pH range is dependent upon the protonation state of the lysine at position 222. Determination of the pK_a for the catalytic histidine in the M222K variant has yet to be determined, although it is expected to be significantly lower than 7.

The mechanism of catalysis in serine proteases involves a general base catalyzed nucleophilic attack by the catalytic serine residue on the substrate carbonyl carbon to form a tetrahedral intermediate. Breakdown of the tetrahedral intermediate is general acid catalyzed and results in a covalently bonded acyl-enzyme intermediate. To a large extent, deacylation is believed to follow the same mechanism as acylation except that water (or some other exogenous nucleophile) bound at the site previously occupied by the first leaving group is the attacking nucleophile. In acylation and deacylation, one of the side chain nitrogen atoms of the catalytic histidine acts as both the general base and the general acid catalyst. The kinetic results obtained for the P225A and the M222K variants indicate that the presence of a full or partial positive charge in close proximity to the catalytic site of subtilisin significantly impairs acylation for amide and anilide substrates, possibly through destabilizing the formation of a positive charge on the catalytic histidine as it transfers a proton from the serine hydroxyl group to the leaving group. If the positive charge has any effect on acylation for the thiobenzyl ester substrate, it is apparently not significant enough to cause this step to become rate limiting for the reaction.

Although acylation and deacylation are believed to proceed along a similar reaction pathway, the observation that acylation for the M222K and the P225A variants is decreased independent of deacylation suggests that the activation barrier for these two reactions is different. Indeed, k_{cat} for both variants on the thiobenzyl ester substrate is actually increased compared to the native enzyme (TABLE 2). The added positive charge from either substitution appears to improve the stabilization of the oxyanion formed on the transition state intermediate in deacylation. A mechanism most consistent with these results would be that the activation barrier in acylation is proton transfer to the catalytic histidine, whereas stabilization of the tetrahedral oxyanion intermediate is rate determining in deacylation. In a separate study by Carter *et al.*, kinetic analysis of a *B. amyloliquefaciens* variant in which the catalytic histidine is replaced by alanine has also revealed a difference in the activation barrier for acylation compared to deacylation.[25]

CONCLUSION

Alteration of the proteolytic activity of subtilisin has been achieved through modification of the structure and charge distribution in the enzyme's active site.

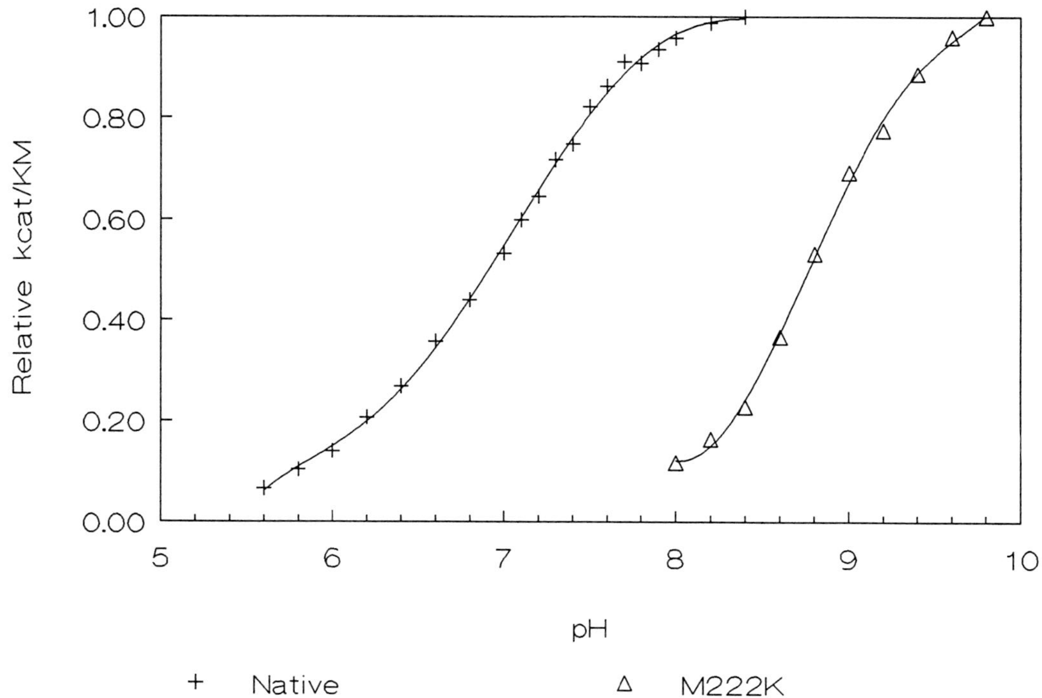

FIGURE 1. Comparison of the pH dependence of k_{cat}/K_m at 25°C, ionic strength 0.1 M, for the hydrolysis of succinyl-L-Ala-L-Ala-L-Pro-L-Phe-*p*-nitroanilide by native *Bacillus amyloliquefaciens* subtilisin and a variant enzyme (M222K) having methionine substituted with lysine at position 222.

Improvement in proteolytic activity was achieved with two substitutions that appear to increase flexibility in the substrate binding region. Modification of reaction specificity in subtilisin via a decrease in proteolytic activity while leaving esterase activity intact was obtained by introducing a positive charge close to the catalytic site. Both studies have provided useful mechanistic insight into the proteolytic action of subtilisin and, perhaps, serine proteases in general.

Acquiring mechanistic information, however, is not the only application of these subtilisin variants. In a laundry detergent wash test, the *B. lentus* K27R/V104Y/N123S/T274A variant has demonstrated a twofold improvement in protein stain removal compared to native *B. lentus* subtilisin.[26] The *B. amyloliquefaciens* M222K variant has shown considerably reduced proteolytic product degradation in peptide synthesis reactions.[27] Hence, the value of protein engineering can extend beyond the study of protein structure-function relationships to include the development of improved commercial biocatalysts as well.

REFERENCES

1. GRAYCAR, T. P. 1991. Protein engineering of subtilisin. *In* Biocatalysts For Industry. J. S. Dordick, Ed. Plenum Publishing Corporation. New York, N.Y.
2. WELLS, J. A. & D. A. ESTELL. 1988. Subtilisin—an enzyme designed to be engineered. Trends Biochem. Sci. **13:** 291–297.
3. ESTELL, D. A., T. P. GRAYCAR & J. A. WELLS. 1985. Engineering an enzyme to be resistant to chemical oxidation. J. Biol. Chem. **260**(11)**:** 6518–6521.
4. BOTT, R., M. ULTSCH, A. KOSSIAKOFF, T. GRAYCAR, B. KATZ & S. POWER. 1988. The three-dimensional structure of *Bacillus amyloliquefaciens* subtilisin at 1.8 Å and an analysis of the structural consequences of peroxide inactivation. J. Biol. Chem. **263**(16)**:** 7895–7906.
5. PANTOLIANO, M. W., M. WHITLOW, J. F. WOOD, S. W. DODD, K. D. HARDMAN, M. L. ROLLENCE & P. N. BRYAN. 1989. Large increases in general stability for subtilisin BPN′ through incremental changes in the free energy of unfolding. Biochemistry **28**(18)**:** 7205–7213.
6. CUNNINGHAM, B. C. & J. A. WELLS. 1987. Improvement in the alkaline stability of subtilisin using an efficient random mutagenesis and screening procedure. Protein Eng. **1**(4)**:** 319–325.
7. CHEN, K., A. ROBINSON, M. VAN DAM, P. MARTINEZ, C. ECONOMOU & F. H. ARNOLD. 1991. Enzyme engineering for nonaqueous solvents II. Additive effects of mutations on the stability and activity of subtilisin E in polar organic media. Biotechnol. Prog. **7:** 125–129.
8. ESTELL, D. A., T. P. GRAYCAR, J. V. MILLER, D. B. POWERS, J. P. BURNIER, P. G. NG & J. A. WELLS. 1986. Probing steric and hydrophobic effects on enzyme-substrate interactions by protein engineering. Science **233:** 659–663.
9. WELLS, J. A., D. B. POWERS, R. R. BOTT, T. P. GRAYCAR & D. A. ESTELL. 1987. Designing substrate specificity by protein engineering of electrostatic interactions. Proc. Natl. Acad. Sci. USA **84:** 1219–1223.
10. WELLS, J. A., B. C. CUNNINGHAM, T. P. GRAYCAR & D. A. ESTELL. 1987. Recruitment of substrate-specificity properties from one enzyme into a related one by protein engineering. Proc. Natl. Acad. Sci. USA **84:** 5167–5171.
11. BONNEAU, P. R., T. P. GRAYCAR, D. A. ESTELL & J. B. JONES. 1990. Alteration of the specificity of subtilisin BPN′ by site-directed mutagenesis in its S_1 and S'_1 Binding Sites. J. Am. Chem. Soc. **112:** 1026–1030.
12. ARBIGE, M., D. ESTELL, M. PEPSIN & A. J. POULOSE. Preparation of enzymes having altered activity. European Patent Office Application No. 0260105.

13. XU, Z.-F., K. E. FOCHT & J. S. DORDICK. Engineering enzymes for use in organic solvents. Ann. N.Y. Acad. Sci. (This volume.)
14. RUSSEL, A. J. & A. R. FERSHT. 1987. Rational modification of enzyme catalysis by engineering surface charge. Nature **328:** 496–500.
15. VENEGAS, M. G., M. S. SHOWELL, R. D. LAURIE & K. SARLO. 1991. Protein engineered detergent proteases. Abstr. Enzyme Engineering XI conference; Keauhou-Kona, Hawaii, September 22–27.
16. SCHECHTER, I. & A. BERGER. 1967. Biochem. Biophys. Res. Commun. **27:** 157–162.
17. BOTT, R., J. DAUBERMAN, R. CALDWELL, C. MITCHINSON, L. WILSON, B. SCHMIDT, C. SIMPSON, S. POWER, P. LAD, I. H. SAGAR, T. GRAYCAR & D. ESTELL. Using structural comparison as a guide in protein engineering. Ann. N.Y. Acad. Sci. (This volume.)
18. JAMES, M. N. G. & C. A. MCPHALEN. 1988. Structural comparison of two serine proteinase–protein inhibitor complexes: Eglin-C-subtilisin Carlsberg and CI-2-subtilisin novo. Biochemistry **27**(17)**:** 6582–6598.
19. NAKATSUKA, T., T. SASAKI & E. T. KAISER. 1987. Peptide segment coupling catalyzed by the semisynthetic enzyme thiolsubtilisin. J. Am. Chem. Soc. **109:** 3808–3810.
20. WEST, J. B., J. SCHOLTEN, N. J. STOLOWICH, J. L. HOGG, A. I. SCOTT & C.-H. WONG. 1988. Modification of proteases to esterases for peptide synthesis: methylchymotrypsin. J. Am. Chem. Soc. **110:** 3709–3710.
21. WU, Z.-P. & D. HILVERT. 1989. Conversion of a protease into an acyl transferase: selenosubtilisin. J. Am. Chem. Soc. **111:** 4513–4514.
22. SIEZEN, R. J., W. M. DE VOS, J. A. M. LEUNISSEN & B. W. DIJKSTRA. 1991. Homology modelling and protein engineering strategy of subtilases, the family of subtilisin-like serine proteases. Protein Eng. **4**(7)**:** 719–737.
23. CALDWELL, R., T. GRAYCAR, M. ULTSCH, F. GRAHAM, D. ESTELL & R. BOTT. 1989. Analysis of the catalytic importance of a hydrogen bonding network in subtilisin using site-directed Mutagenesis. J. Cell. Biochem. **13A** (suppl.): 51.
24. HOL, W. G. J. 1985. The role of the α-helix dipole in protein function and structure. Prog. Biophys. Mol. Biol. **45:** 149–195.
25. CARTER, P., L. ABRAHMSEN & J. A. WELLS. 1991. Probing the mechanism and improving the rate of substrate-assisted catalysis in subtilisin BPN′. Biochemistry **30**(25)**:** 6142–6148.
26. CALDWELL, R., D. ESTELL & T. GRAYCAR. European Patent Office Application No. 90915958:4.
27. GRAYCAR, T. Unpublished results.

Opportunities and Limits in Creating New Enzymes

Experiences with the NAD-Dependent Lactate Dehydrogenase Frameworks of Humans and Bacteria[a]

H. M. WILKS, A. CORTES,[b] D. C. EMERY, D. J. HALSALL, A. R. CLARKE, AND J. J. HOLBROOK

Molecular Recognition Center and
Department of Biochemistry
University of Bristol
School of Medical Sciences
University Walk
Bristol BS8 1TD, United Kingdom

[b]*Department of Biochemistry and Physiology*
University of Barcelona
Marti y Franquès 1
08028 Barcelona, Spain

In one sense, opportunity for enzyme engineering is truly limitless: for a 300-residue protein of 20 amino acids, there are 20^{300} (10^{390}) possible sequences. Over biological time (3000 million years) and with a rate of cell division of 3 hours^{-1}, only a maximum of 10^{55} sequences can have been evaluated naturally (assuming the earth's oceans were *Escherichia coli* cells filled with phage). Thus there are still $10^{390} - 10^{55} \approx 10^{390}$ sequences that have not been explored in nature and that the opportunistic protein engineer could choose to create new biological function. Even if the volume of creation is increased to that of the known universe, the number 10^{390} is not significantly approached (FIGURE 1).

The number 10^{55} places a limit on the length (N) of DNA and derived protein that can have been optimized for function over evolutionary time. Thus from solving $4^N = 10^{55}$ we conclude $N = 91$ bases or 30 amino acids. A typical large protein of 300 amino acid residues cannot then represent a global optimum for any biological function. The best it could be is an assembly of empirically optimized 25–35 amino acid domains, and thus if the modern enzyme engineer defines a target function, there is considerable hope that the large framework could be further improved towards that goal (as seen in this present and other work). The lengths of coding regions in eukaryotic DNA are highly variable but average 110 bases, i.e., 36 amino acids. In the discussion of this paper, G. Petsko (Brandeis) reported that the concept that DNA coding for proteins has evolved by concatenating a series of small optimized domains is supported by the recent observation that the cDNA transcript of a *Tetrahymena* gene that codes for the enzyme phosphoenolpyruvate mutase has

[a]Supported by the SERC, Genzyme (United Kingdom) Ltd, the British Council (Acciones Integradas), and SmithKline-Beecham Research (United Kingdom) Ltd.

two in-frame amber stop codons (TAG) and each of these codons is precisely followed by an intron [J. R. Knowles (Harvard), personal communication].

Over the past two years, methods have been developed to express random sequences of DNA as protein fused to phage M13 coat protein, and it has been suggested that it will be possible to mimic the process of evolution by suitable affinity chromatography to isolate both the required protein sequence and its gene.[1] However, just as evolution has been unable to sample all of sequence space, so too the protein engineer will be limited to the number of M13 phage that can be screened (10^{15} plaque-forming units are produced per liter culture of *E. coli* cells containing the phage M13). With 10^{15} variants, the length of DNA that can be optimized is obtained from $4^N = 10^{15}$, i.e., $N = 24$ bases or 8 amino acids.

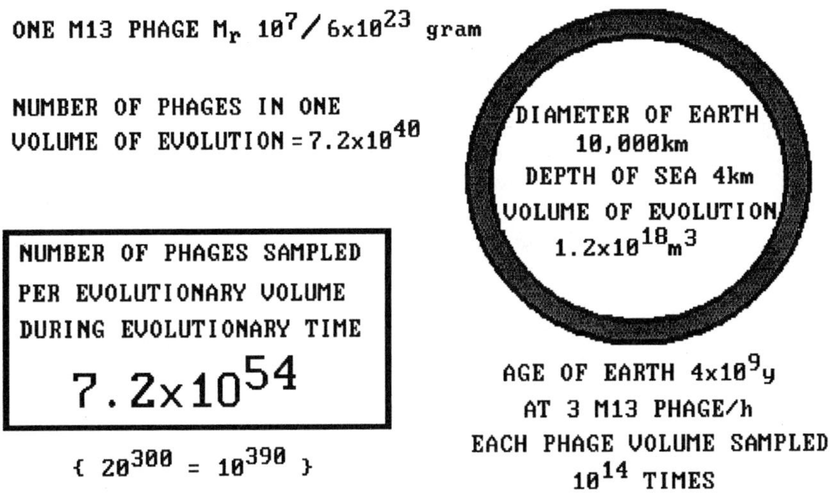

FIGURE 1. Limits on the variability of natural proteins. Evolution is assumed to take place in the earth's oceans completely filled with M13 phage replicating in E. coli each 20 minutes. The 7.2×10^{54} sequence variations that can thus be tested for advantage are infinitesimal compared to the 10^{390} possible variants of a 300-residue protein composed of 20 amino acids.

Random mutagenesis of existing proteins is also limited in its ability to produce radically altered proteins by problems of sampling all the possible variants. In addition, the genetic code is very resistant to change. Not only are codons redundant at the third position, but also amino acid residues with similar properties are coded for by similar sequences and thus resistant to sparse mutagenesis. For example (1) a codon with a T at the second position always codes for an amino acid residue with a hydrophobic side chain, (2) the codons for aspartate and glutamate differ only at the third position. Therefore, strategies, such as use of thioate nucleotides,[2] that create randomly dispersed mutations (in which only one mutation is likely to be present in any codon) are unlikely to yield new proteins with dramatically different properties to those of the parent proteins. Thus protein engineering by man based on mimicking random evolution is not likely to significantly increase the limited suite of protein frameworks we inherit from nature. Only when the forces and derived rules that enable fold to be predicted from sequence are fully understood will the enzyme engineer be able to grasp the unlimited opportunity that will be presented by the more unnatural of the 10^{390} variants.

NATURE IS NOT LIMITED IN THE FOLD IT USES FOR FUNCTION

After comparing the structures and properties of the NAD(P)-dependent dehydrogenases we concluded:

a. Protein thermal stability is not directly dependent on the protein fold.
b. The same redox chemistry is catalyzed by different protein folds in nature.
c. Different redox chemistries can be catalyzed on the same protein fold.
d. It is possible to redesign enzymes that are not under evolutionary pressure to have considerably improved catalytic properties.

These conclusions are applicable to (but not limited to) the lactate dehydrogenase framework. However, although it should be possible to engineer any desired property into any protein framework, it is also apparent that only enzyme frameworks that have been well-characterized are likely to be successfully rationally redesigned. This point is expanded in Reference 3.

ANALYSIS OF THE NATURAL LACTATE DEHYDROGENASE MECHANISM

To obtain the fundamental knowledge required for rational redesign, we first used a combination of crystallography, site-directed mutagenesis, and transient kinetic techniques to relate function to structure in the NAD-dependent lactate dehydrogenases from both prokaryotes and eukaryotes (FIGURE 2). That knowledge not only revealed those amino acids required for the catalytic pathway, but it also mapped those amino acids that are part of a major rearrangement of shape that is induced when the negatively charged ketoacid enters the active site and causes the protein to sequester the substrate in an internal vacuole that is sensitive to the size of the substrate (FIGURE 3) and that contains exactly balanced charge. Using this knowledge we have been able to rationally design specific new enzymic properties and so avoid the low statistical probabilities associated with random mutagenesis. Some of the residues that determine substrate specificity are carried on the under surface of the "upper jaw" of FIGURE 3.[22] The rate-limiting step in lactate dehydrogenase catalysis is the rate at which this loop can sweep through a viscous solvent to close onto the upper surface of helix α2G. The rate-limiting step is, within a factor of 3, independent of the sequence of amino acids on the "upper jaw" (FIGURE 4), and since the chemical step is much faster than the shape change, the lactate dehydrogenase (LDH) system has the advantage that the loop sequence can be easily varied to achieve different substrate specificities without much danger that the chemical step will become rate limiting. The disadvantage of the vacuole mechanism is that the natural vacuole, without alteration, will only accept small, singly charged substrates.

CHEAP NADPH REGENERATION FROM A DUAL COENZYME SPECIFICITY ENZYME

Bacillus stearothermophilus LDH is an NADH-requiring enzyme. Replacement of a conserved aspartic acid residue at position 52 with a serine residue allows the 2′ phosphate of NADPH to be accommodated without charge repulsion.[4] The enzyme with this mutation has a 20-fold shift in coenzyme specificity towards NADPH (mainly as a result of weakened NADH binding due to the loss of a hydrogen bond, but also as a result of a 2.5-fold increase in the k_{cat} for NADPH). The new thermotolerant enzyme with activity with both NADH and NADPH can be used to feed reducing equivalents from formic acid into the reactions of reductive biosynthesis for chiral chemical production (FIGURE 5). The reverse change, to convert the NADPH-dependent glutathione reductase to an NADH-dependent one has also been made by Perham's group.[5]

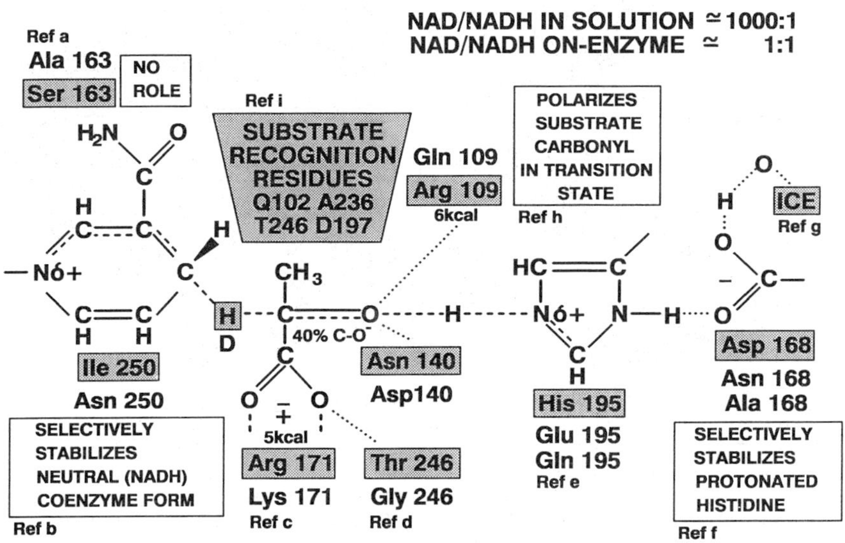

FIGURE 2. A map showing the roles of amino acids around the extended catalytic pathway of *B. stearothermophilus* lactate dehydrogenase. (Reproduced from Reference 24 with permission.) Original results are for Reference a,[25] Reference b,[26] Reference c,[27–29]. Reference d,[6,30] Reference e,[15,17,31,32] Reference f,[33] Reference g,[24] Reference h,[23] Reference i.[6,13] The residues of the wild-type enzyme are in shaded boxes next to their replacements.

A SEMISYNTHETIC MALATE DEHYDROGENASE MADE ON THE LDH FRAMEWORK

To insert the large and more negatively charged oxaloacetate substrate into the vacuole that normally balances the charge of pyruvate, we (1) removed negative

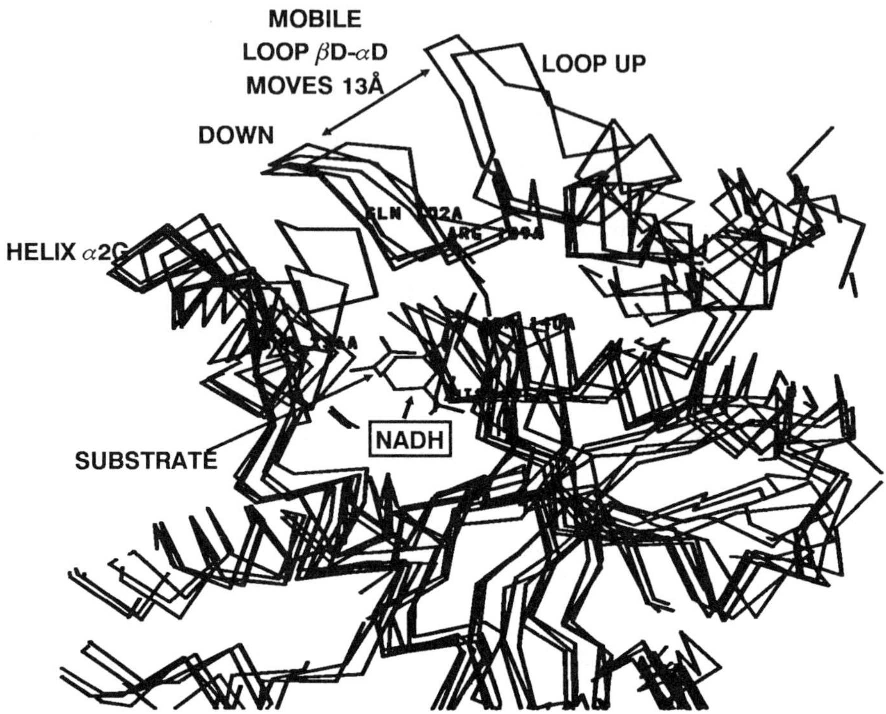

FIGURE 3. The substrate-induced movement of the upper "jaw" loop (βD-αD) down onto helix α2G. In the free lactate dehydrogenase or in the "binary" complex with NADH, the position of the loop can be up or down, depending on species. In the ternary complexes with NADH and the competitive inhibitor oxamate, the loop is always tightly closed down onto helix α2G. The structures are pig M_4 binary (NADH) and ternary (NADH + oxamate),[25] mouse X_4 binary, binary and ternary dogfish M_4 LDH,[34] and *B. stearothermophilus* LDH quaternary (NADH + oxamate + FBP).[35]

charge from the vacuole (D197N, E107Q), (2) enlarged the vacuole by the T246G mutation, and (3) introduced a positively charged amino acid residue on the inside surface of the catalytic vacuole by mutation of glutamine-102 to arginine (FIGURE 6). All of these changes altered the specificity of lactate dehydrogenase from *B. stearothermophilus*[6,7] towards a malate dehydrogenase (MDH) but it was the Q102R mutation that gave the most dramatic result (FIGURE 7). Note that as predicted by the charge balance rule,[8] constructions where the framework is made more positive by $+2(-(-1)+1)$ such as D197N, Q102R are less effective than that (Q102R) where exact charge balance is maintained. In this new construction, oxaloacetate provides a counter ion for the new arginine residue at 102 whereas pyruvate does not. The result is a switch in specificity of nearly 10^7-fold in favor of oxaloacetate and away from pyruvate—the largest achieved by enzyme engineering of any framework. Not only is the specificity switch very large, but the new enzyme is almost twice as active (k_{cat}) with the new substrate as is the natural malate dehydrogenase from *B. stearothermophilus.* Thus we succeeded in creating a perfect malate dehydrogenase which had the added, but not obviously useful, property of being activated by fructose-1,6-biphasphate (FBP) (unlike any known natural MDH).

ENZYME REDESIGNED NOT TO NEED THE ALLOSTERIC ACTIVATOR FBP

Many bacterial lactate dehydrogenases including that from *B. stearothermophilus* are activated by FBP.[9] Binding of 2 FBPs across the dimer-dimer Q-axis interface activates the enzyme in part by causing the formation of tetramers. *B. stearothermophilus* lactate dehydrogenase was a favorable enzyme in which to reveal the mechanism of regulation since, in the absence of FBP, it dissociates at physiological protein concentration to dimers with K_d = 10 μM. Many other bacterial LDHs simply open the contacts at the P-axis interface, but remain as tetramers. Positive charge repulsion by arginines-173[10] and histidines-188[11] across the P-axis prevents assembly of the tetramer, and neutralization of that charge by the negatively charged FBP normally allosterically activates the enzyme (K_m for pyruvate reduced from 3 mM to 50 μM). Thus the rational replacement of arginine-173 at the regulatory site with neutral glutamine removed charge repulsion across the P-axis and enabled the active (low K_m) tetramer to form spontaneously. Indeed very high concentrations of FBP then became an inhibitor because the extra negative charge then destabilized the

Secondary structures.
------βD------| (92–99) |-binary------αD- (110–119)
------βD---| (92–98) |ternary-αD- (113–119)

	92	93	94	95	96	97	98	99	100	101	102	103	105	106	107	108	109	110	111	112	113	114	115	116	117	118	119
NATURAL MDHs	**D**	**I**	**V**	**V**	**I**	**T**	**A**	**G**	**A**	**R**	**Q**	**K**	**P**	**G**	**E**	**T**	**R**	**L**	**D**	**L**	**V**	**Q**	**R**	**N**	**V**	**N**	**I**
AND LDHs	K	W	A	I	L	C	G	A	V	P	R	Q	E	T	M	S		D	N	E	L	D	K	T	G	A	V
All 150-400s^{-1}	R	Y	I	L	V	P	L	L	M	N	M	R	S		Q	E		A	A		F	N	T		I	T	M
k_{cat}s.		M		C	S	S	S		K	A	K	V	A		G	D		F	E			E	A		A	Q	
Lowered 10x		A				I			P	K		G						S	Q			K	I		S	K	
by viscosity.						V			N			A						K				M	V			R	
												D						R					F			G	
SYNTHETIC												I														S	
VARIANTS:																										M	
200% active MDH[6]											R																
100% active[21]						T																					
50% active[22]														W													
100% active[7]															Q												
TS up by 6 kcal[23]																	Q										
50% k_{cat}[13]											M	V	S														

FIGURE 4. Variation of the Sequence of the βD-αD specificity loop in hydroxy acid dehydrogenases. Sequences for the alignment were taken from the GenEMBL database and aligned initially with the multiple alignment program CLUSTAL and then manually with reference to the known 3-D structures. "Viscosity" is the ratio of k_{cat} with and without 44% w/v glycerol.[24]

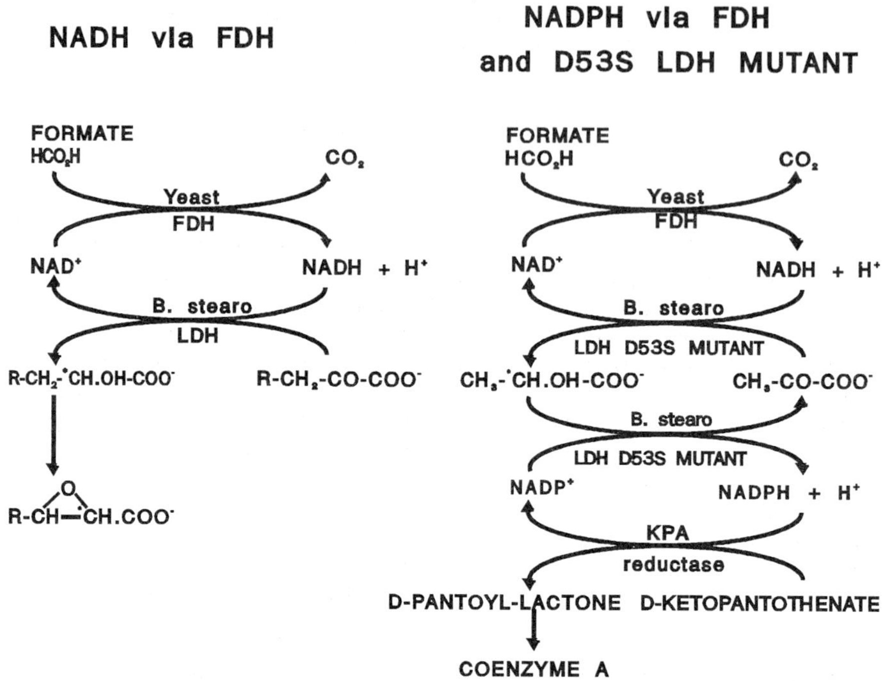

FIGURE 5. The ability to regenerate NADPH from NADP and the NAD-dependent formate dehydrogenase (FDH) by coupling through the dual coenzyme specificity D53S mutant of LDH and a trace of lactate.[4] This reaction makes it possible to carry through bulk biotransformations that would otherwise require the very expensive cofactor NADPH. The left-hand diagram is the conventional Wandrey-Kuhla route for the bulk synthesis with reducing equivalents from formic acid. The right-hand scheme illustrates the use of the dual specificity D53S and a trace of lactic acid to link reducing equivalents derived from formate to the enzymes of reductive biosynthesis.

charge balance across the P-axis of the tetramer and promoted inactive dimer formation (see Reference 12 for a short review).

BROAD SUBSTRATE SPECIFICITY: COOPERATIVE MUTATIONS AT TWO MATING SURFACES

Introduction of a set of five mutations on complementary surfaces (α2G and the mobile active site loop βD-αD) that border the catalytic vacuole (FIGURE 3) produced a broad substrate specificity α-hydroxy acid dehydrogenase.[13] The design goal here was to obtain a "chemists enzyme" that had a broad specificity and that would reduce almost any substituted pyruvate to give 100% ee *S*-lactate as a generic technology in the synthesis chiral pharmaceutical intermediates.[14] The three mutations in the loop region (Q102M, K103V, P105S) increased the hydrophobicity and flexibility of the loop and gave rise to lower K_ms for substrates larger than pyruvate but only marginal improvements in k_{cat} for the two largest substrates tested. The mutations in helix α2G (A235G, A236G) resulted in eightfold increased k_{cat}s for long straight-chain substrates such as ketocaproate. It was only when these two sets of mutations were combined in one enzyme molecule that a useful broad substrate specificity α-hydroxy acid dehydrogenase was obtained (FIGURE 8). The K_m values for most of these substrates was in the ideal 1–10 mM range. Low-K_m substrates, such as pyruvate with the wild-type enzyme (K_m = 60 μM), are very hard to transform on a preparative scale because all tight-binding pyruvates show very high substrate inhibition at the >0.2 M substrate concentrations used in enzyme reactors.

These experiments illustrate the importance of distinguishing two classes of mutations at two sites within a protein: those where the number of possibilities increases multiplicatively, and those where it increases additively. For example, were we to randomly vary the 4 (n) amino acids on the under surface of the loop (where the number of variants is 20^n) to obtain the best sequence for a given target, and also vary a further 5 (m) residues on the surface of helix α2G and obtain another local

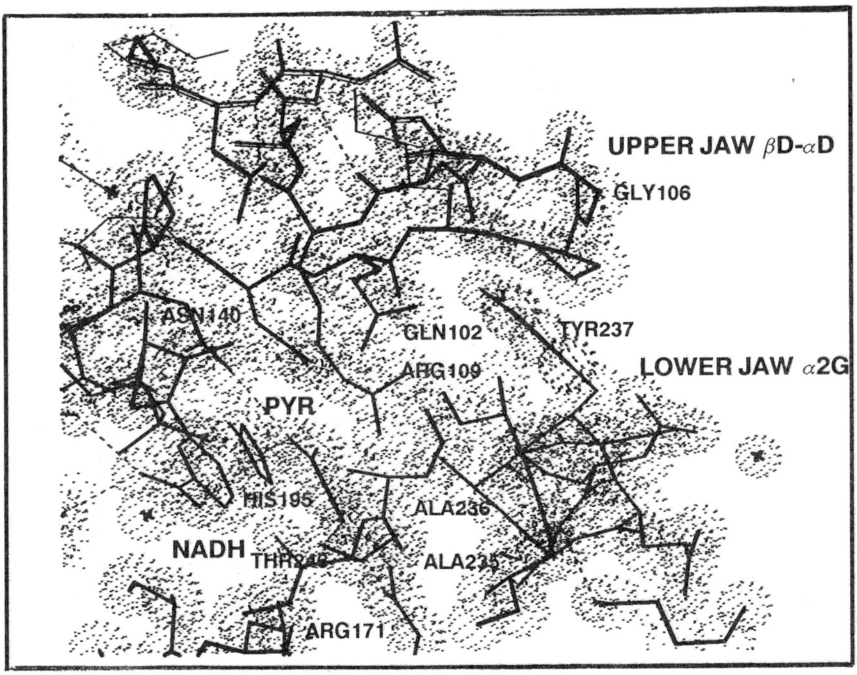

FIGURE 6. A section through a space-filling model of the active center of *B. stearothermophilus* lactate dehydrogenase with the loop in the down position. The dotted spheres are drawn at 60% of the true Van der Waals radii so that the boundary between the "upper" and "lower" jaw region is more easily discerned. Notice the free volume next to the —CH_3 of pyruvate which enables undistorted pyruvate and ketobutyrate to bind in the active center. Larger side chains cannot fit into the closed active site (FIGURE 8). × = low B-value waters. This view is from the other side of the molecule than showed in FIGURE 3. The fully closed vacuole will not allow even protons to equilibrate with bulk solvent.

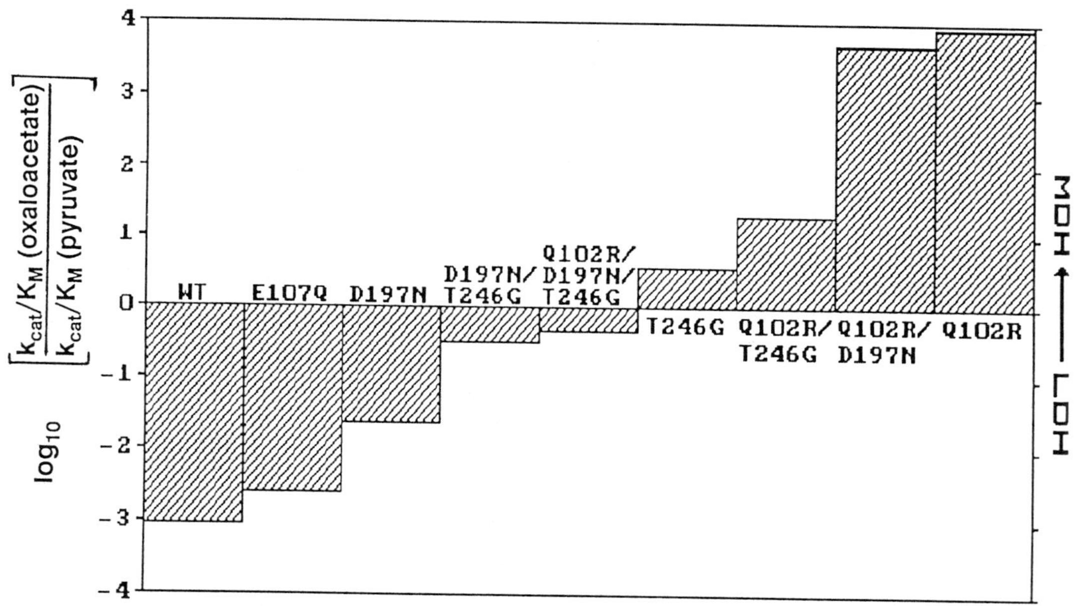

FIGURE 7. Progress in the design and construction of a malate dehydrogenase on the lactate dehydrogenase framework.[36] The wild-type enzyme is 1000-fold more active with pyruvate than with oxaloacetate. The first design idea was to remove negative charge from the periphery of the vacuole to compensate for the increased negative charge of the new target (E107Q, D197N). The second design idea was to increase the vacuole size by reducing the size of the protein side chains (T246G). This gave a protein that was more active with the new target than with pyruvate. The best design involved the insertion of a positive charge (Q102R) to provide a counterion for the negative charge of the new substrate. This is the greatest switch (10^7-fold) in substrate specificity achieved by mutagenesis and gave a fully active (k_{cat} = 250secondsseconds^{-1}) new malate dehydrogenase. Results are plotted from Reference 6.

optimum sequence, the results of Reference 13 suggest we should not have obtained a global optimum sequence at both surfaces. This again points to the limitations to be expected from the random mutagenesis approach: were the mutational affects additive (as they are for example when the allosteric FBP-binding sites at remote positions 188 and 173 are combined with the active site specificity mutations), then a global optimum is obtained by the very reasonable task of screening $20^n + 20^m$ colonies (3.10^7). Where the affects are clearly cooperative (as they are here), it would be necessary to screen an impossibly large number of mutants ($20^{(m+n)} = 10^{11}$ if $n = 4$ and $m = 5$) to obtain the optimal sequence at both surfaces for any given target substrate.

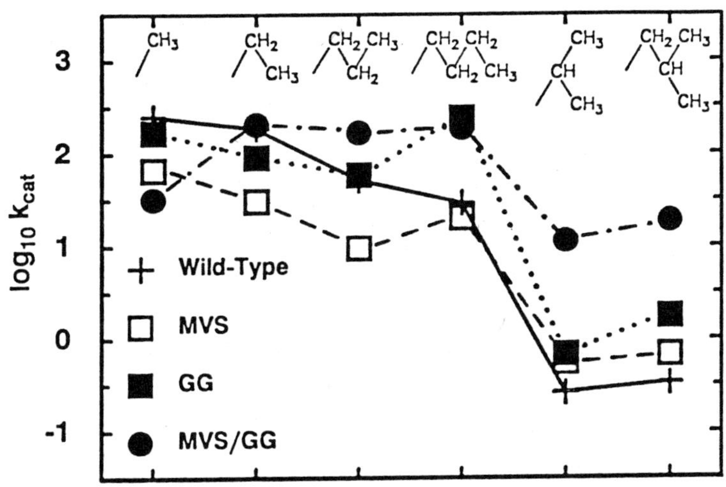

FIGURE 8. The successful design of a broad specificity hydroxyacid dehydrogenase for use in chiral synthesis. A plot of $\log_{10}(k_{cat})$ for a series pyruvates (R.CO.COO$^-$) for the wild-type enzyme (+), a mutant with $^{235\text{-}6}$AA → GG (lower jaw, ■), a mutant with $^{102\text{-}105}$QKP → MVS (upper jaw, □), and all five mutations at once (●). Only when both surfaces are changed together is a broad specificity enzyme formed. (Replotted from Reference 13 and reproduced from Reference 24 with permission.) The fixed size vacuole of the wild-type will not accommodate side chains larger than —CH_2—CH_3.

CHARGE BALANCE IN THE HYDROXY ACID DEHYDROGENASE VACUOLE

These experiments have shown how sensitive the active site is to the size and the charge of the substrate, and this provides discrimination between very similar small substrates which otherwise have too few functional groups to be distinguished at a solvated protein surface.

Originally the concept that the functional vacuole has an overall charge balance of zero (summed over the substrate, coenzyme ring, and histidine-195) arose because (a) only complexes of zero overall charge were bound tightly[8,15,16] and (b) high concentrations of either the competitive inhibitor oxamate or of a slow substrate (nitrophenylpyruvate) could drive the uptake of a proton from bulk solvent at pH 8 where histidine-195 in the complex with NADH is normally unprotonated.[17]

With the advent of site-directed mutagenesis, it has been possible to directly show that the active site of LDH is sensitive to charge imbalance by alteration of Asn140 to the negatively charged anion of Asp140. The lactate dehydrogenase from *B. stearothermophilus* requires FBP binding at histidine-188 for full activity, and the pH dependence of K_m for substrate is complicated by the changed phosphate ester ionization in FBP and histidine-188 ionization in a range above pH 6. Therefore, mutagenesis was carried out on cloned, overexpressed human muscle M_4 lactate dehydrogenase,[18] which is not regulated by FBP and which has simple binding versus pH curves.[19] The mutation of asparagine-140 to aspartate resulted in an enzyme with the pH optimum shifted from 7 to 4.5 due to the fact that only at acid pH where the aspartate group is protonated is the essential charge neutrality achieved which enables catalysis and substrate binding in the closed vacuole. With an aspartic acid residue at position 140, no charge-imbalanced complexes formed at above 1/1000th of the concentration of those of the charge-balanced complexes. The maximum rate of catalysis of LDH is limited by the rate of formation of the vacuole which is limited by solvent friction. Experimental results and molecular modeling studies indicate that the pK_a of the active site histidine-195 in the enzyme-NADH complex is raised to greater than 10 by the presence of the anion at position 140.

ENLARGING THE CATALYTIC VACUOLE: SPECIFICITY FOR LARGE SUBSTRATES

Using conventional site-directed mutagenesis, unique DNA restriction sites have been inserted at either end of the mobile active site loop. This enables DNA coding for new loops to be inserted into the gene to easily test how the size and charge of the loop affect the substrate specificity of the enzyme. The restriction sites were inserted on a template that had the mutations A235G and A236G, which increase the size of the catalytic vacuole so that larger substrates can be accommodated.[13]

Using DISCOVER and INSIGHT version 2.5 (Biosym Technologies, San Diego, Calif.) a search was made for loops that could be used to shorten the mobile active site loop of LDH without distorting the adjacent tertiary structure. It was found that the loop could be shortened by three amino acid residues by insertion of a tripeptide loop between asparagine-101 and arginine-109 (which we decided to maintain in all mutant enzymes, as it is essential for polarizing the substrate carbonyl bond). The best loops were modeled into the original LDH structure on the graphics to optimize the match of the new side chains to the new environments. A serine residue was chosen for the first position of the tripeptide, as it can exist in either a hydrophobic or a hydrophilic environment. A lysine residue was chosen for the tip of the loop which will be in a solvent-exposed environment. A glycine residue was inserted at old position 108 in order not to constrain the conformation of the arginine at 109. (FIGURE 9). The resulting mutant LDH (with the sequence SerLysGly between positions 101 and 109) was found to have significantly different substrate specificity to the wild-type enzyme. The mutant enzyme still binds pyruvate tightly (K_m was

0.065 mM in the presence of FBP and 0.5 mM in the absence of the allosteric activator), but catalyzes the reaction at a very low rate (0.07 seconds^{-1} in the presence of FBP compared with 250 seconds^{-1} for the wild-type enzyme). The mutant enzyme is, however, better at catalyzing the reduction of phenyl pyruvate than is the wild-type enzyme. For phenylpyruvate the mutant enzyme has a k_{cat} of 100 seconds^{-1} in the presence of FBP (compared with 33 seconds^{-1} for the wild-type) and a K_m of 3 mM which is not much worse than that of the wild-type at 1.8 mM. Modeling of the tripeptide loop suggests that the loop-down LDH structure is stabilized, as three residues is too short to bridge the apo or binary (loop-up)

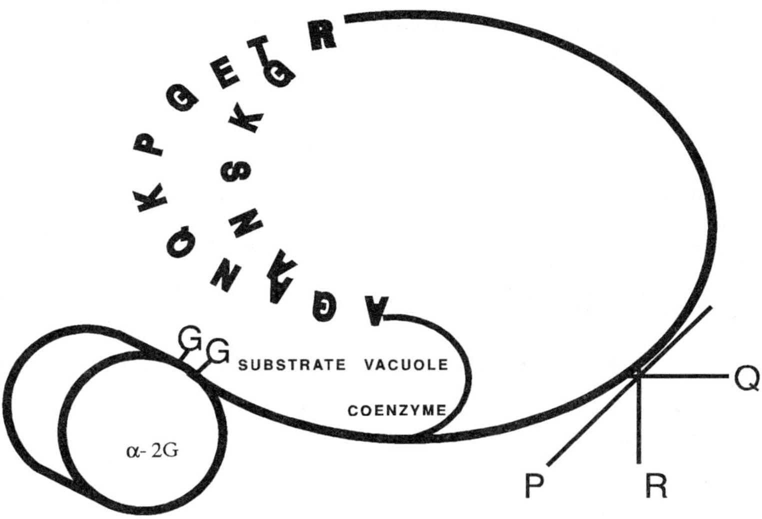

FIGURE 9. Design of a specific phenyl-lactate dehydrogenase by shortening the mobile loop that normally caps the enzyme internal vacuole. By using a short tripeptide to bridge between Arg109 and Asn101, the active site is made more exposed to the solvent. The short loop is constructed on a framework where the vacuole is already enlarged by the A235G, A236G mutations. Small substrates do not have enough bulk to exclude the solvent, and the mutant is a poorer catalyst for these. Large aromatic substrates such as phenylpyruvate have enough bulk to seal the enlarged opening from external solvent and are transformed more rapidly than by the wild-type enzyme. Arg109 is retained as it polarizes the substrate carbonyl in the transition state.

structures. The tight binding of pyruvate by this mutant enzyme supports the prediction that the loop is always in the high-affinity loop down but is exposed to the solvent and so turns it over only very slowly. In accordance with the design idea, the bulky aromatic side chain of phenylpyruvate is able to seal the more exposed vacuole isolating the catalytic pathway from the solvent so that this substrate is turned over very well by the mutant enzyme.[21]

In the future, it should be possible to use the construct with the restriction sites at either end of the loop region to produce a series of dehydrogenases with loops of variable length and variable sequence. Thus, by restricting random mutagenesis to the region of lactate dehydrogenase that we have already identified as being important for substrate recognition, it is possible to isolate enzymes that can carry out any desired chiral reduction. However, this method does requires a suitable screen for clones that express mutant enzymes of the desired specificity. For the dehydrogenases, this is simply provided by coupling NADH production with phenazine methosulfate to formation of insoluble blue formazan dye.

One of the conclusions of the 10th Enzyme Engineering meeting was that, from a practical point of view, random mutagenesis seems to be the most powerful method that can be used (to obtain enzymes with improved properties) at present for all the enzymes cloned.[20] However, although improvements in the stability of enzymes have been obtained by random mutagenesis, large alterations in the properties of enzymes have not. Localized random mutagenesis of well-understood enzyme frameworks is a method that is more likely to succeed.

REFERENCES

1. KANG, A. S., C. F. BARBAS, K. D. JANDA, S. J. BENKOVIC & R. A. LERNER. 1991. Proc. Natl. Acad. Sci. USA **88:** 4363–4366.
2. HOLM, L., A. K. KOIVULA, P. M. LEHTOVAARA, A. HEMMINKI & J. K. C. KNOWLES. 1990. Protein Eng. **3:** 181–191.
3. CLARKE, A. R., S. COLEBROOK, A. CORTES, D. C. EMERY, D. J. HALSALL, K. W. HART, R. M. JACKSON, H. M. WILKS & J. J. HOLBROOK. 1991. Biochem. Soc. Trans. **19:** 576–581.
4. FEENEY, R., A. R. CLARKE & J. J. HOLBROOK. 1989. Biochem. Biophys. Res. Commun. **166:** 667–672.
5. SCRUTTON, N. S., A. BERRY & R. N. PERHAM. 1990. Nature **343:** 38–43.
6. WILKS, H. M., K. W. HART, R. FEENEY, C. R. DUNN, H. MUIRHEAD, W. N. CHIA, D. A. BARSTOW, T. ATKINSON, A. R. CLARKE & J. J. HOLBROOK. 1988. Science **242:** 1541–1544.
7. CLARKE, A. R. & J. J. HOLBROOK. 1989. Science Prog. Oxford **73:** 117–131.
8. PARKER, D. M. & J. J. HOLBROOK. 1977. An oil-water:histidine mechanism for the activation of coenzyme in the α-hydroxyacid dehydrogenases. *In* Pyridine Nucleotide–Dependent dehydrogenases. H. Sund, Ed.: 485–495. Walter de Gruyter. New York & Berlin.
9. GARVIE, E. I. 1980. Microbiol. Rev. **44:** 106–139.
10. CLARKE, A. R., D. B. WIGLEY, D. A. BARSTOW, W. N. CHIA, T. ATKINSON & J. J. HOLBROOK. 1987. Biochim. Biophys. Acta **913:** 72–80.
11. MATZUZAWA, H., M. MICHIDA, K. KUNAI, Y. ITOH & T. OHTA. 1988. FEBS Lett. **233:** 375–378.
12. CLARKE, A. R., T. ATKINSON & J. J. HOLBROOK. 1989. Trends Biochem. Sci. **14:** 101–105, 145–148.
13. WILKS, H. M., D. J. HALSALL, T. ATKINSON, W. N. CHIA, A. R. CLARKE & J. J. HOLBROOK. 1990. Biochemistry **29:** 8587–8591.
14. HOLBROOK, J. J., H. M. WILKS, A. R. CLARKE & T. ATKINSON. 1989. Modified NAD-dependent dehydrogenases specific for long chain α-keto carboxylic acids. United Kingdom. Patent Application No. 8905269.0.
15. LODOLA, A., D. M. PARKER, R. JECK & J. J. HOLBROOK. 1978. Biochem. J. **173:** 597–605.
16. PARKER, D. M., A. LODOLA & J. J. HOLBROOK. 1978. Biochem. J. **173:** 597–605.
17. HOLBROOK, J. J. 1973. Biochem. J. **133:** 847–849.
18. BARSTOW, D. A., G. W. BLACK, A. F. SHARMAN, M. D. SCAWEN, T. ATKINSON, S.-S. LI, W. N. CHIA, A. R. CLARKE & J. J. HOLBROOK. 1990. Biochim. Biophys. Acta **1087:** 73–79.
19. HOLBROOK, J. J. & R. A. STINSON. 1973. Biochem. J. **131:** 737–748.

20. URABE, I. 1990. Annals. N.Y. Acad. Sci. **613:** xviii–xix.
21. WILKS, H. M., K. M. MORETON, D. J. HALSALL, K. W. HART, R. M. SESSIONS, A. R. CLARKE & J. J. HOLBROOK. Biochemistry. (In press.)
22. WALDMAN, A. D. B., K. W. HART, A. R. CLARKE, D. B. WIGLEY, D. A. BARSTOW, T. ATKINSON & J. J. HOLBROOK. 1988. Biochem. Biophys. Res. Commun. **150:** 752–759.
23. CLARKE, A. R., D. B. WIGLEY, W. N. CHIA, D. A. BARSTOW, T. ATKINSON & J. J. HOLBROOK. 1986. Nature. **324:** 699–702.
24. DUNN, C. R., H. M. WILKS, D. J. HALSALL, T. ATKINSON, A. R. CLARKE, H. MUIRHEAD & J. J. HOLBROOK. 1991. Phil. Trans R. Soc. Lond. **332:** 177–185.
25. WIGLEY, D. B., A. R. CLARKE & J. J. HOLBROOK. 1987. Protein Eng. **1:** 260–260.
26. WIGLEY, D. B., A. R. CLARKE, C. R. DUNN, D. A. BARSTOW, T. ATKINSON, W. N. CHIA, H. MUIRHEAD & J. J. HOLBROOK. 1987. Biochim. Biophys. Acta **916:** 145–148.
27. HART, K. W., A. R. CLARKE, D. B. WIGLEY, W. N. CHIA, D. A. BARSTOW, T. ATKINSON & J. J. HOLBROOK. 1987. Biochem. Biophys. Res. Commun. **146:** 346–353.
28. HART, K. W., A. R. CLARKE, D. B. WIGLEY, A. D. B. WALDMAN, W. N. CHIA, D. A. BARSTOW, T. ATKINSON, J. B. JONES & J. J. HOLBROOK. 1987. Biochim. Biophys. Acta **914:** 294–298.
29. WIGLEY, D. B., A. LYALL, K. W. HART & J. J. HOLBROOK. 1987. Biochem. Biophys. Res. Commun. **149:** 927–929.
30. BUR, D., A. R. CLARKE, J. D. FRIESEN, M. GOLD, K. W. HART, J. J. HOLBROOK, J. B. JONES, M. A. LUYTEN & H. M. WILKS. 1989. Biochem. Biophys. Res. Commun. **161:** 59–63.
31. PORTER, J. P. & J. J. HOLBROOK. 1991. Unpublished results.
32. HOLBROOK, J. J. & V. A. INGRAM. 1973. Biochem. J. **131:** 729–738.
33. CLARKE, A. R., H. M. WILKS, D. A. BARSTOW, T. ATKINSON, W. N. CHIA & J. J. HOLBROOK. 1988. Biochemistry **27:** 1617–1622.
34. HOLBROOK, J. J., A. LILJAS, S. J. STEINDEL & M. G. ROSSMANN. 1975. Enzymes **11a:** 191–293.
35. WIGLEY, D. B., S. J. GAMBLIN, J. P. TURKENBURG, E. DODSON, K. PIONTEK, H. MUIRHEAD & J. J. HOLBROOK. 1992. J. Mol. Biol. **223:** 317–335.
36. HOLBROOK, J. J. & T. ATKINSON. 1988, 1989. Recombinant modified thermostable NAD-dependent dehydrogenases and processes for their production. European Patent Convention 90904396.0-. U.K. Priority 8806358 and International Patent Application No. GB 8903099.

Engineering Subtilisin for Use in Organic Solvents[a]

ZU-FENG XU,[b] KATHLEEN FOCHT, AND JONATHAN S. DORDICK

Department of Chemical and Biochemical Engineering
125 CB
The University of Iowa
Iowa City, Iowa 52242

INTRODUCTION

Protein engineering has been used successfully in aqueous biocatalysis to increase enzyme thermostability[1,2] and alter substrate specificity.[3–6] There is no fundamental reason why enzyme function in nonaqueous media cannot be altered or improved by using the same technique. Our previous study with subtilisin Carlsberg indicated that increased polarity in the active site of the enzyme, via water addition, improves catalytic efficiency in tetrahyrofuran (THF).[7] Instead of adding water, active site polarity can be increased by protein engineering. To this end, we have studied the effect of polar mutations in the related subtilisin BPN′, namely, G166N and M222Q, in which glycine and methionine are replaced by asparagine and glutamine at positions 166 and 222, respectively, in both aqueous and organic media with specific ester substrates. The altered polarity of the enzyme's active site may affect its substrate specificity and provide an opportunity to improve enzymatic activity towards unfavorable substrates in nonaqueous media. The goal of this ongoing research is to rationally design subtilisin BPN's active site for optimal function in nonaqueous media.

EXPERIMENTAL

Subtilisin BPN′ wild-type (WT, from *Bacillus amyloliquefaciens*) and its mutants G166N and M222Q were generous gifts from Genencor International, Inc. They were dialyzed (MW cutoff: 1000) against phosphate buffer (20 mM, pH 7.8) for 24 hours. Following dialysis, the enzyme solutions (which contained 25 mg/mL protein) were freeze-dried.

[a]This work has been supported by grants from the National Science Foundation (PYI award to JSD), Army Research Office, Mead Corporation, and Olin Corporation.

[b]Present address, Department of Chemistry, Massachusetts Institute of Technology, Cambridge, Massachusetts 02139.

The water content of the freeze-dried proteins ranged from 8 to 10% w/w, as measured via Karl-Fischer titration.[7] The concentration of enzyme active centers in aqueous buffer (as determined by *N-trans*-cinnamoylimidazole titration)[8] was BPN′ WT, 54.2 ± 1.0%; G166N, 36.2 ± 0.5%; and M222Q, 35.8 ± 2.5%. An indirect titration method showed that about 15% of active protein could be titrated in THF with 1 M *n*-propanol and 0–2% v/v water added to the solvent.[7] Above this water content, the fraction of active centers increased dramatically.

N-Acetyl-L-phenylalanine ethyl ester (NAPheEE) is from Sigma. *n*-Propanol and all the solvents used in this work were from Aldrich, distilled twice, and stored over molecular sieves. Initial rates of enzyme-catalyzed transesterification between the

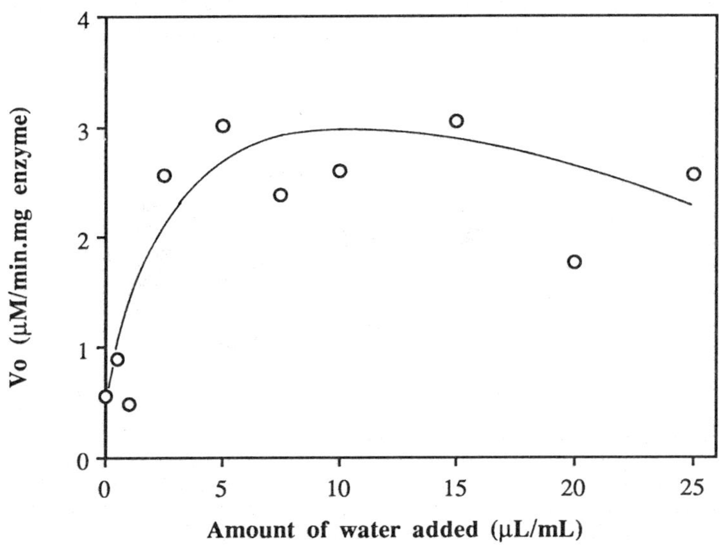

FIGURE 1. Catalytic activity of subtilisin BPN′ on the transesterification in tetrahydrofuran between NAPheEE and *n*-propanol as a function of the amount of water added to the reaction system. The water content on the dry enzyme and in THF was 9.8% (w/w) and 0.087% (v/v), respectively. No ester hydrolysis was observed during the measurement of transesterification below 25 μL/mL water added.

ester substrate and *n*-propanol in solvents were measured by gas chromatography (GC) by following *N*-acetyl-L-phenylalanine propyl ester formation. Conditions were $[\text{enzyme}]_{\text{in solvent}}$, 1 mg/mL; $[\text{enzyme}]_{\text{in buffer}}$, 2 to 50 μg/mL; [NAPheEE], 20.0 mM; [*n*-propanol], 1.0 M in 1 mL solvent. Each transesterification was carried out in a 4-mL sealed vial, shaken at 250 rpm, and 30°C. All the data were normalized by using 0.1-mM dodecane as internal GC standard, and the initial rates were corrected by the amount of active enzyme. The rates of the hydrolysis of NAPheEE with different enzymes were carried out in phosphate buffer, and measured on high-performance liquid chromatography (HPLC) with a C_{18} reversed-phase column.

RESULTS AND DISCUSSION

Effect of Mutation on Catalytic Efficiency

Addition of water improved subtilisin BPN′ catalysis in nonaqueous media much as the Carlsberg enzyme[7] (FIGURE 1). Improved activity is due to increased polarity at the enzymic active site. Therefore, we studied polar mutation in BPN′ to determine whether similar effects could be obtained.

The active site of the subtilisin BPN′ consists of two subsites (FIGURE 2). The S_1 pocket is the acyl binding site where, in the wild-type enzyme, the Gly_{166} is located. The S'_1 site is the region in which the leaving group of the ester substrate is located. Met_{222} is located at the S'_1 site adjacent to the catalytic Ser_{221} in the wild-type enzyme. We have hypothesized from our previous study that water may be replaced by constructing polar mutations in the enzyme's active site and result in an improved

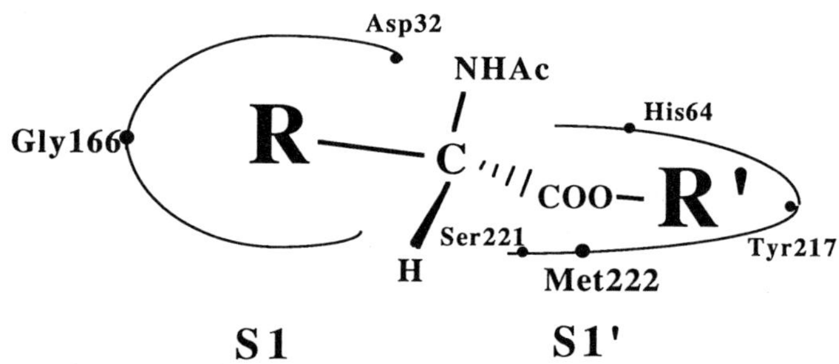

FIGURE 2. Simplified schematic representation of the active site of subtilisin BPN′ with an *N*-acetyl-L-amino acid alkylester substrate (modified from Reference 3, p. 1027). R is the amino acid residue of the substrate; R′ represents the alkyl chain of the leaving group of the substrate. The S_1 pocket is the acyl binding site where, in the wild-type enzyme, the hydrophobic Gly_{166} is located. The S'_1 site is the region in which the leaving group (R′) of the ester substrate is located. Met_{222} is located at the S'_1 site adjacent to the catalytic Ser_{221} in the wild-type enzyme.

biocatalysis in nonaqueous media. Using genetically engineered subtilisin BPN′, namely, G166N (at the S_1 site) and M222Q (at the S'_1 site), we examined this hypothesis.

In aqueous buffer, the two mutants showed lower catalytic efficiency in the hydrolysis of NAPheEE; as much as 3-fold for G166N and 25-fold for M222Q (FIGURE 3a). However, in dry THF, the G166N became twice as catalytically efficient as the wild-type enzyme, and 3-fold more efficient in hexane (FIGURE 3b and c). Moreover, the catalytic efficiency gap between M222Q and WT was significantly reduced in both THF and hexane. The M222Q mutant had roughly 50% of the WT activity in THF, and 40% in hexane as compared to only 4% of WT activity in aqueous buffer.

All enzymes underwent a dramatic loss in catalytic efficiency (k_{cat}/K_m) when placed in organic media as compared to aqueous buffer (TABLE 1). This is primarily because of the drop in catalytic turnover number (k_{cat}). While catalytic specificity

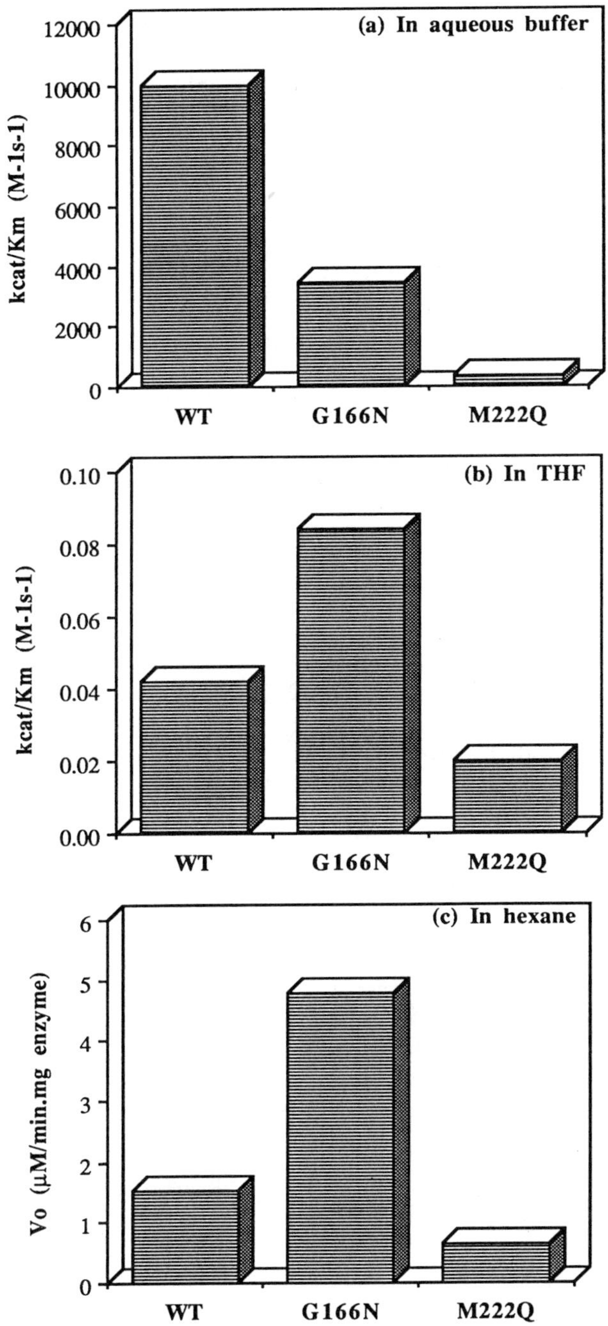

FIGURE 3. The performance of wild-type and engineered subtilisin BPN′ in aqueous buffer (a), dry THF (b), and hexane (c). The reactions studied are (a) hydrolysis of NAPheEE in sodium phosphate buffer, and (b and c) transesterification between NAPheEE and *n*-propanol in THF and hexane, respectively.

TABLE 1. The Catalytic Parameters of Wild-Type and Engineered Subtilisin BPN′ in Aqueous Buffer and in Dry THF[a]

	Enzymes	k_{cat} (seconds^{-1})	K_m (mM)	k_{cat}/K_m (M^{-1} seconds^{-1})
In buffer	WT	149.4	14.6	1.03×10^4
	G166N	81.2	23.8	3.42×10^3
	M222Q	5.2	13.3	393.8
In THF	WT	1.56×10^{-3}	37.1	0.042
	G166N	8.86×10^{-3}	105.9	0.084
	M222Q	1.18×10^{-3}	60.3	0.020

[a]The reactions studied are the same as those in FIGURE 3.

changed between aqueous buffer and dry solvents, all enzymes were dramatically less active in nonaqueous media as compared to aqueous buffer. However, as showed in TABLE 2, both mutants recover some lost k_{cat} in organic media.

Combined Effect of Solvent and Protein Engineering

FIGURE 4 shows the combined effect of water addition and protein engineering on enzymatic catalysis in nonaqueous media. The positive effects of mutation and water addition are additive. The increased polarity at the S_1 binding site (the G166N) due to the addition of water and the polar mutation substantially increased the catalytic activity of subtilisin BPN′ in dry THF. The increase of the catalytic activity was about 40-fold for G166N compared to WT when 10 μL/mL (or 1% v/v) of water was added to dry THF.

CONCLUSION

Flexibility and polarity of the active site of enzymes appear to represent important factors that govern enzymatic catalysis in organic solvents. Elucidation of these factors should be considered as a starting point for the rational design and optimization of enzyme function in nonaqueous media. In the case of subtilisin BPN′, these factors can be optimized by adding a certain amount of water (1% v/v is optimum) to the reaction system so as to increase the flexibility and polarity of the active site

TABLE 2. Effect of Mutation and Solvent on the Catalytic Parameters of Subtilisin BPN′ in Aqueous Buffer and in Dry THF[a]

	In Buffer	In THF	THF/Buffer[b]
$k_{cat}/K_m(N)/k_{cat}/K_m(G)$	0.33	2.0	6.0
$k_{cat}/K_m(Q)/k_{cat}/K_m(M)$	0.11	0.47	4.3
$k_{cat}(N)/k_{cat}(G)$	0.54	5.8	10.7
$k_{cat}(Q)/k_{cat}(M)$	0.04	0.77	22.0
$K_m(N)/K_m(G)$	1.63	2.9	1.8
$K_m(Q)/K_m(M)$	0.91	1.6	1.8

[a]The reactions studied are the same as those in FIGURE 3.
[b]This ratio represents $[k_{cat}/K_m(N)/k_{cat}/K_m(G)]_{in\ THF}/[k_{cat}/K_m(N)/k_{cat}/K_m(G)]_{in\ buffer}$.

region or having a polar mutation at the S_1 site. These techniques, combined with appropriate solvent selection, can substantially improve enzymatic catalysis in non-aqueous media. Future research in this field will enable us to rationally design better enzymes for specific applications as well as to further understand enzyme function in dehydrated environments.

ACKNOWLEDGMENTS

We are grateful to Thomas Graycar and Dr. David Estell at Genencor International for supplying the subtilisin BPN′ wild-type and mutant enzymes.

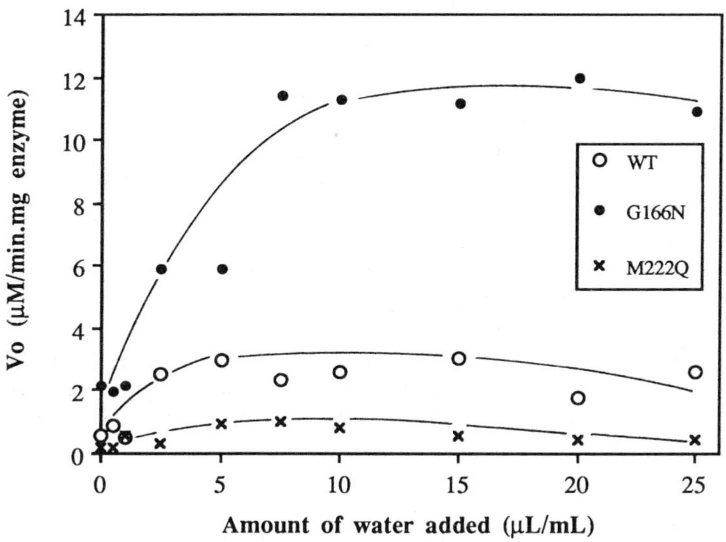

FIGURE 4. The activity of subtilisin BPN′ and its mutants on the transesterification of NAPheEE with *n*-propanol in THF as a function of the amount of water added to the reaction medium.

REFERENCES

1. PERRY, L. J. & R. WETZEL. 1984. Science **226:** 555–557.
2. VILLAFRANCA, J. E., E. E. HOWELL, D. M. VOET, M. S. STROBEL, R. C. OGDEN, J. N. ABELSON & J. KRAUT. 1983. Science **222:** 782–788.
3. BONNEAU, P. R., T. P. GRAYCAR, D. A. ESTELL & J. B. JONES. 1991. J. Am. Chem. Soc. **113:** 1026–1030.
4. ESTELL, D. A., T. P. GRAYCAR, J. V. MILLER, D. B. POWERS, J. P. BURNIER, P. G. NG & J. A. WELLS. 1986. Science **233:** 659–663.
5. CRAIK, C. S., C. LARGMAN, T. FLETCHER, S. ROCZNIAK, P. J. BARR, R. FLETTERICK & R. W. J. RUTTER. 1985. Science **228:** 291–297.
6. FERSHT, A. R., J. P. SHI, J. KNILL-JONES, D. M. LOWE, A. J. WILLKINSON, D. M. BLOW, P. BRICK, M. M. Y. WAYE & G. WINTER. 1985. Nature **314:** 325.
7. AFFLECK, R., Z.-F. XU, V. SUZAWA, K. FOCHT, D. S. CLARK & J. S. DORDICK. 1992. Proc. Natl. Acad. Sci. USA. **89:** 1100–1104.
8. SCHONBAUM, G. R., B. ZERNER & M. BENDER. 1961. J. Biol. Chem. **236:** 2930–2935.

Hyperproduction of Thermostable Lipase by Genetically Engineered *Pseudomonas* Species

KOICHI NAKAMURA, TARO IIZUMI, AND TETSURO FUKASE

Kurita Central Laboratory
Kurita Water Industries Ltd.
7-1 Wakamiya Morinosato
Atsugi, Kanagawa 243-01, Japan

INTRODUCTION

Lipase has been applied to industrial uses because of its unique character. However, an improvement in productivity is required to expand the field of application.

One *Pseudomonas* species was isolated, and named KWI-56, that produced an extracellular thermostable lipase. More than 90% of the enzyme activity remained after 24 hours of heat treatment at 60°C.[1]

In order to improve lipase productivity, a multicopy plasmid harboring the lipase genes including the structure gene, *lip,* and its activator gene, *act,* which was found necessary to express the enzyme activity in the process of genetic analysis,[2] was introduced to the parental *Pseudomonas* strain. We report here the effect of gene dosage on lipase productivity.

MATERIALS AND METHODS

The *lip* and *act* genes located on a 2.9 kilobase (kb) fragment were inserted to a multicopy plasmid pSUP104 resulting pSUP-lip64. The linker-ligation technique using *Bam*HI-*Hin*dIII adapter was employed, in order to insert lipase genes to pSUP104 in tandem. As a result, pSUP-lip642 harboring two sets of lipase genes and pSUP-lip643 harboring three sets were constructed (FIGURE 1). All the plasmids were introduced into the parental *Pseudomonas* strain by mobilization using *Escherichia coli* S17-1.

The plasmid copy number was determined by Southern hybridization followed by densitometric analysis. On agar gel electrophoresis, the *Eco*RI-*Bgl*II fragments derived from the plasmid DNA and the chromosomal DNA give bands of 4.4 kb and of 2.9 kb, respectively. Plasmid stability was evaluated by measuring the marker retention in test tube culture. Generation was calculated from a ratio of increase in optical density at 660 nm.

Three transconjugants and the parental strain without a plasmid were cultivated

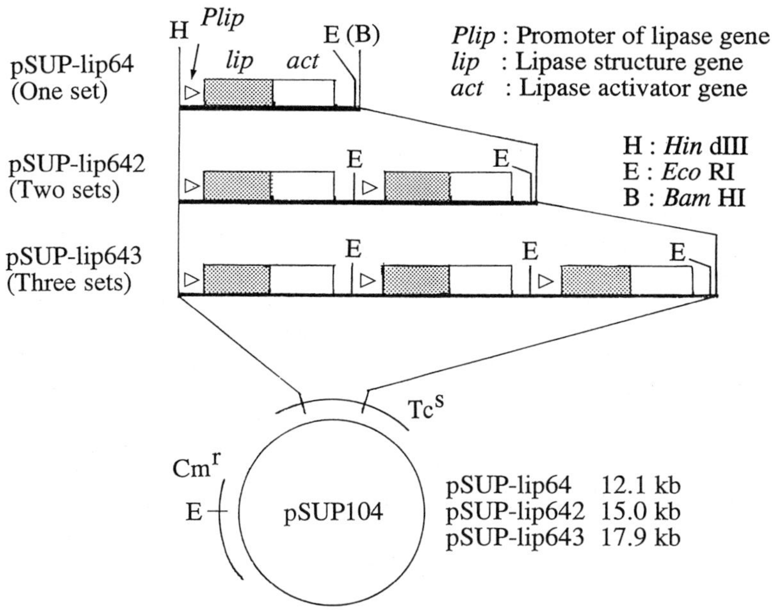

FIGURE 1. Structure of plasmids.

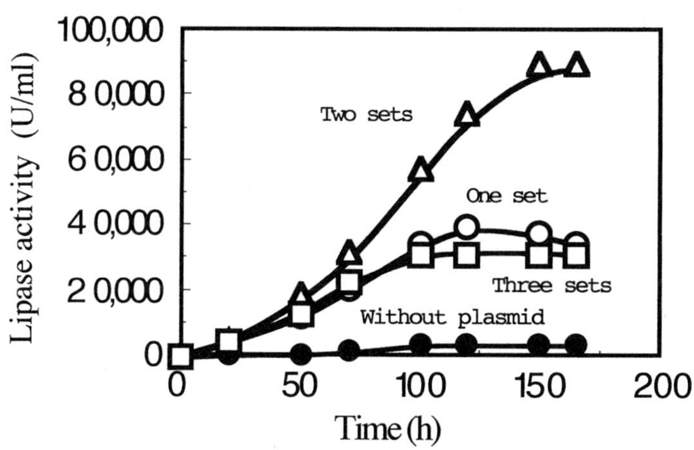

FIGURE 2. Lipase production in jar fermenter.

in 2-L jar fermenters. A peptone medium was used, and the culture conditions were at 30°C and neutral pH. Oleic acid was fed into the fermenter at a constant rate to carry out the fed-batch culture.

RESULTS AND DISCUSSION

The plasmid copy number was measured as 15 to 20 by densitometric analysis. The plasmids harboring one and two sets of lipase genes were stable in the parental *Pseudomonas* strain for 10 generations, and more than 90% of the plasmid remained at generation 50. On the other hand, when harboring three sets of lipase genes in tandem, the plasmid was unstable and 25% was lost at 10 generations.

The results of lipase production by the parental strain and the transconjugants in jar fermenters are summarized in FIGURE 2. The parental strain produced 2500 U/ml lipase, whereas the productivity of the transconjugants was 40,000 and 90,000 U/ml for one and two sets of lipase genes, respectively, indicating direct proportionality to the number of lipase genes. The gene dosage, i.e., the number of lipase genes and the plasmid copy number, was found to contribute to increasing lipase productivity.

For the plasmid harboring three sets of lipase genes in tandem, lipase production decreased to 33000 U/ml. Plasmid instability greatly affected lipase productivity in this case.

As a result, the lipase productivity of *Pseudomonas* KWI-56 was successfully increased about 30 times by increasing the number of lipase genes. Further improvement may be possible if more genes can be dosed without instability.

REFERENCES

1. IIZUMI, T., K. NAKAMURA & T. FUKASE. 1990. Purification and characterization of a thermostable lipase from newly isolated *Pseudomonas* sp. KWI-56. Agric. Biol. Chem. **54**(5): 1253–1258.
2. IIZUMI, T., K. NAKAMURA, Y. SHIMADA, A. SUGIHARA, Y. TOMINAGA & T. FUKASE. 1991. Cloning, nucleotide sequencing, and expression in *Escherichia coli* of a lipase and its activator genes from *Pseudomonas* sp. KWI-56. Agric. Biol. Chem. **55**(9): 2349–2357.

Promenade in the Sequence Space of Bacterial Catalase by Random Mutagenesis

TETSUYA YOMO, SAVITR TRAKULNALEAMSAI, AND ITARU URABE[a]

Department of Biotechnology
Faculty of Engineering
Osaka University
2-1 Yamada-oka, Suita-shi
Osaka 565, Japan

INTRODUCTION

One of the most striking characteristics of proteins is that they occupy a huge sequence space; for example, even a small protein with 100 amino acid residues makes a space containing 20^{100} different sequences. This number is much larger than the number of proteins in the world. This means that there is enough room for the existence of new enzymes. Therefore, we planned to explore the sequence space by random mutagenesis. The sequence space can be divided into subspaces based on such categories as stability and function. If we intend to change the specific activity of an enzyme, all the sequences in the subspace, including the wild-type enzyme, are the candidates. Such an activity subspace is generally surrounded by the sequence space of inactive proteins. Therefore, if we intend to change the function of the enzyme, we have to traverse the inactive field.

RANDOM MUTAGENESIS

We started with *Bacillus stearothermophilus* catalase (formerly peroxidase).[1,2] This enzyme has weak peroxidase activity, and seems to be in a fused subspace having two functions, catalase and peroxidase. In this case, it may be possible to pass from catalase to peroxidase without crossing the inactive field. The catalase gene (*perA*) was randomly mutated by the method of Myers *et al.*,[3] *Escherichia coli* cells were transformed with the mutated gene, and 2648 transformants were obtained (FIGURE 1). The catalase and peroxidase activities of all the transformants were examined by colony assay after heat treatment (70°C, 10 minutes), and the transformants were classified into the following three groups: S (both activities are as strong as those of the wild type), M (at least one of the activities is not detected or is weaker than that of the wild type), and B (both activities are not detected). The numbers of transformants that belonged to the S, M, and B groups were 1323, 191, and 1134, respectively. Sodium dodecyl sulfate–polyacrylamide gel electrophoresis (SDS-

[a] To whom correspondence should be addressed.

PAGE) showed that 6 out of 328 transformants in the B group and 82 out of 191 transformants in the M group produced the mutant proteins before the heat treatment. It is noteworthy that all the mutant enzymes that were produced at detectable levels (checked by SDS-PAGE) were active before the heat treatment. This means that, for this enzyme, the activity subspace includes the stability subspace.

TOPOGRAPHY OF THE SEQUENCE SPACE OF THE CATALASE

We selected 108 transformants producing a detectable amount of the mutant protein, 20 from the S group, 82 from the M group, and 2 from the B group, and their

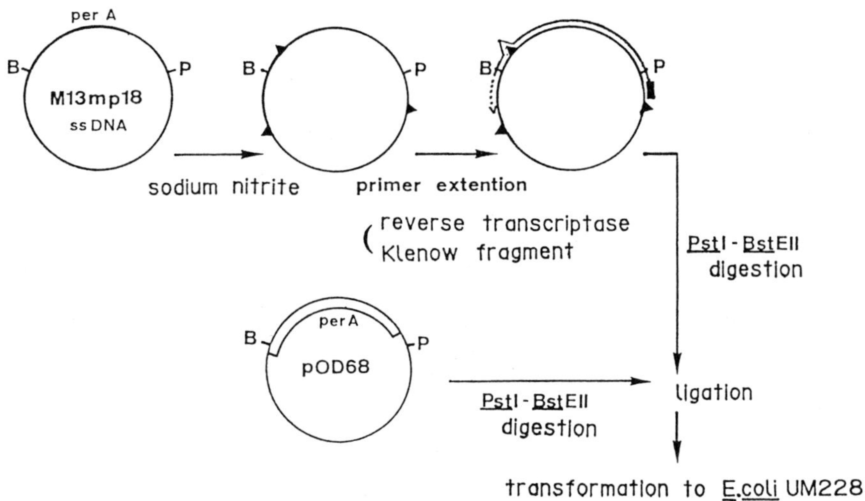

FIGURE 1. Procedure for the *in vitro* random mutagenesis of the catalase gene (*perA*).

enzyme activities were measured spectrophotometrically before and after heat treatment. The amounts of enzyme proteins were also determined by SDS-PAGE and densitometry before and after heat treatment. Statistical analysis of these data showed the topography of the wild-type enzyme near the space point. As for the stability subspace, the wild-type enzyme is in a higher region; but the activity subspace is at a middle level, and the region of high activity is different from that of high stability. Similarly, the wild-type enzyme is in a higher region of the catalase subspace and in a middle region of the peroxidase subspace, and the region of high peroxidase activity is different from that of high catalase activity.

Therefore, it seems possible to find the way from catalase to peroxidase. Thus, we have improved the ratio of the two activities (P/C ratio) to 220 times greater than that of the wild-type enzyme.

REFERENCES

1. LOPRASERT, S., S. NEGORO & H. OKADA. 1988. Thermostable peroxidase from *Bacillus stearothermophilus*. J. Gen. Microbiol. **134:** 1971–1976.
2. LOPRASERT, S., S. NEGORO & H. OKADA. 1989. Cloning, nucleotide sequence, and expression in *Escherichia coli* of the *Bacillus stearothermophilus* peroxidase gene (*perA*). J. Bacteriol. **171:** 4871–4875.
3. MYERS, R. M., L. S. LERMAN & T. MANIATIS. 1985. A general method for saturation mutagenesis of cloned DNA fragments. Science **229:** 242–247.

Use of Charged Tails in Protein-Purification Fusions

Polyaspartic Acid Tails in *Aspergillus* Glucoamylase Expressed in Yeast

ILARI SUOMINEN,[a] HELI NURMELA,[a] HEIKKI HEIMO,[a] CLARK FORD,[b] DEBORAH STACHON,[b] AND CHARLES GLATZ[c]

[a]*Department of Biochemistry*
University of Turku
20500 Turku, Finland

[b]*Department of Food Science and Human Nutrition*
[c]*Department of Chemical Engineering*
Iowa State University
Ames, Iowa 50011

INTRODUCTION

Protein recovery and purification from large-scale fermentations can be very expensive, often exceeding the cost of fermentation itself. In order to improve the cost-efficiency of protein purification, several systems have been developed recently based on purification fusion technology.[1] A major drawback in most of these systems is that they require chromatographic techniques, making them impractical for recovery of low-cost products from large-scale fermentations.

We are studying the use of charged fusion tails to enhance selectivity of economical, charge-based recovery methods such as polyelectrolyte precipitation.[2,3] Polyelectrolytes (PEs) like cationic polyethylene imine (PEI) and anionic polyacrylic acid (PAA) can be used at low concentrations, and they have the potential to be recycled. PE precipitation is essentially volume independent and can be carried out under mild conditions, minimizing protein denaturation and side-chain modifications. Nonchromatographic methods such as PE precipitation allow lower production costs for large-scale products. They can also be employed as the first step in the purification of high-cost pharmaceutical proteins, since full benefit from high-specificity techniques such as affinity chromatography can be obtained only if they are preceded by a low-specificity recovery step.

In this paper, we describe construction of amino terminal aspartic acid fusions to *Aspergillus* glucoamylase. The modified enzymes were expressed in yeast, purified for biochemical characterization, and the behavior of unpurified crude enzymes in PEI precipitation was studied. Fusion of negatively charged tails of moderate length was shown to enhance the efficiency of precipitation.

RESULTS AND DISCUSSION

Tail Design

To study the recovery of an industrial enzyme, we chose to engineer negatively charged polyaspartic acid tails to *Aspergillus awamori* glucoamylase,[4] normally produced by large-scale fermentation of the fungus. The enzyme consists of an N-terminal catalytic domain and a C-terminal starch-binding domain separated by a stretch having heavy O-glycosylation (FIGURE 1). The nascent enzyme has a transient signal sequence, which is removed by leader peptidase upon entry into the endoplasmic reticulum and a hexapeptide removed by the yeast KEX2 peptidase,[5] when the enzyme is secreted into the growth medium.

We have earlier constructed negatively charged aspartic acid tails in the carboxy terminus of *Escherichia coli* β-galactosidase.[2] While nucleic acids interfere with the separation of negatively charged intracellular proteins by polyelectrolyte precipitation,[3] this obviously should not be a problem with extracellular proteins. With aspartate tails, it is also easy to introduce in the protein a recognition sequence for the protease enterokinase for removal of the tail after purification.

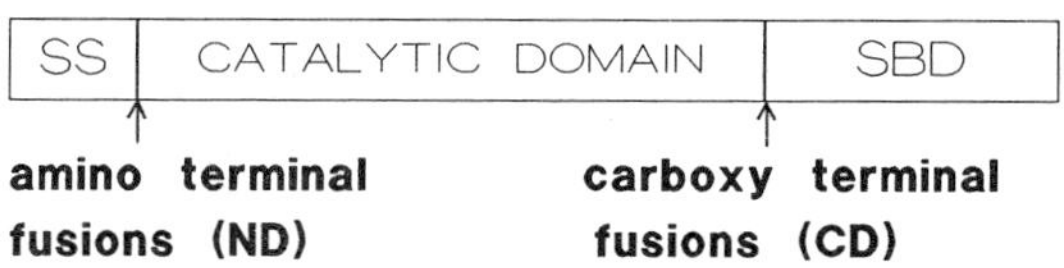

FIGURE 1. Schematic presentation of *Aspergillus* glucoamylase. The mature enzyme has an N-terminal catalytic domain and a C-terminal starch-binding domain (SBD) separated with a heavily O-glycosylated stretch of amino acids. Amino terminal (ND) poly-Asp fusions were created in the N-terminus between the transient signal sequence and the mature end of the catalytic domain. C-terminal Asp fusions (CD) will be described elsewhere.

The enzyme was produced in *Saccharomyces cerevisiae* using a yeast expression vector, pGAC9[4] (FIGURE 2), kindly provided by Cetus Corporation (Emeryville, Calif.). Expression of the glucoamylase gene in this plasmid is controlled by yeast enolase promoter and terminator. Three pairs of oligonucleotides (FIGURE 3A) were designed and synthesized to create tails of different lengths. The annealed oligonucleotides were inserted into the BssHII site residing in 3′ end of the KEX2 hexapeptide coding region. The resulting amino terminal oligo aspartic acid fusions are outlined in FIGURE 3B. Four aspartate residues followed by a lysinyl residue comprise a recognition sequence for the protease enterokinase,[6] which can thus be employed in analysis of tail integrity as well as in removal of the tail after purification. A tyrosinyl residue was included at position +2 in the fusions to enable monitoring of the tail removal after purification and to provide a less ambiguous sequence analysis of the amino terminus.

For construction of the fusion tails, the small 5′-terminal PstI-HindIII fragment of the glucoamylase gene was cloned from pGAC9 into pBS+ vector (Stratagene, La Jolla, Calif.). The oligonucleotides were then inserted in the BssHII site, and the PstI-HindIII fragments were transferred into M13mp18 for sequencing. After con-

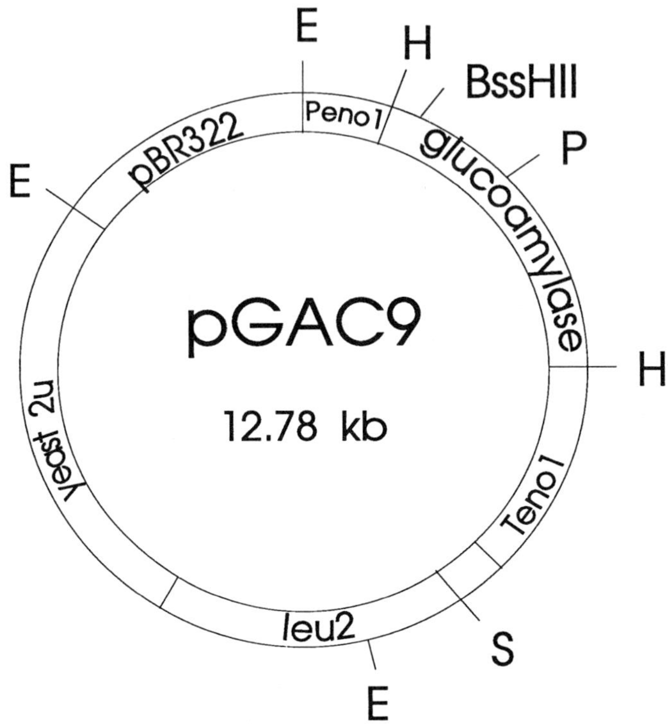

FIGURE 2. Map of pGAC9.[4] pGAC9 is an *E. coli–S. cerevisiae* shuttle plasmid having pBR322 and yeast 2 μ replicons, respectively. Glucoamylase is inserted in the HindIII site of pAC1[4] (pAC1 is pGAC9 without the glucoamylase gene). Peno1 and Teno1 stand for yeast enolase gene promoter and terminator, respectively. The BssHII site is used for inserting the charged-tail fusions in the glucoamylase gene. E = EcoRI, H = HindIII, P = PstI, S = SalI.

firming the validity of the fusions, the genes containing the fusions were reconstructed in the *E. coli*–yeast shuttle plasmid pAC1.[4] The plasmids isolated from *E. coli* were then transformed into yeast *S. cerevisiae* C468.

Expression and Characterization of the Modified Enzymes

Yeast strains transformed with the fusion constructions as well as with pAC1 and pGAC9 were streaked on SD-His plates containing 1% (w/v) soluble starch and grown for 4 days at 30°C. The plates were then stained with iodine vapor to detect glucoamylase activity as haloes (clearing zones) around the colonies. The haloes around C468/pGAND5 colonies were of the same size as in the control strain carrying pGAC9, whereas C468/pGAND7 produced slightly smaller haloes. C468/pGAND10 produced comparably smaller haloes, and C468/pAC1 colonies were totally haloless as expected.

To analyze whether decreasing halo size with increasing tail length is due to decreased production or lowered specific activity of the fusion enzymes, the strains were grown in SD-His liquid cultures for 5 days and the enzyme activities in culture supernatants assayed. The expression levels obtained for pGAND5 and pGAND7 (44 U/ml and 36 U/ml, respectively) suggest no decrease in specific activity, but for pGAND10 the expression level was considerably smaller (4 U/ml).

To confirm that the specific activities were not affected by the fusions and to further analyze the produced fusion proteins, they were each purified from 4 liters of 5-day SD-His cultures. The supernatants were concentrated approximately 20-fold and diafiltered (0.1 M NaOAc, pH 4.4, 0.5 M NaCl) with an Amicon S-1 circular flow ultrafiltrator (NMWCO 10,000). The retentate was then loaded into an Acarbose-Sepharose column (kindly provided by Dr. Birte Svenson, Carlsberg Laboratories, Copenhagen, Denmark), the unbound proteins were washed with the diafiltration

```
A.

GAND5:
5' -CGC GAC TAC GAC GAC GAT GAC AAG GCT- 3'            and
     3' -TG ATG CTG CTG CTA CTG TTC CGA GCG C- 5'

GAND7:
5' -CGC GAC TAC GAC GAC GAC GAT GAC GAC AAG GCT- 3'        and
     3' -TG ATG CTG CTG CTG CTA CYG CTG CTT CGT GCG C- 5'

GAND10:
5' -CGC GAC TAC GAC GAC GAT GAC GAC GAC GAT GAC GAC AAG GCT- 3'   and
     3' -TG ATG CTG CTG CTA CTG CTG CTG CTA CTG CTG TTC CGA GCG C- 5'

B.

Wild type:

    Lep                 KEX2
     |                   |
ss-Ala-Asn-Val-Ile-Ser-Lys-Arg-Ala-Thr...
3'-... AAT GTC ATT TCC AAG CGC GCG ACC...
                          |
                        BssHII
GAND5:
    Lep                 KEX2                                EK
     |                   |                                  |
ss-Ala-Asn-Val-Ile-Ser-Lys-Arg-Asp-Tyr-Asp-Asp-Asp-Asp-Lys-Ala-Arg-Ala-Thr...
3'-... AAT GTC ATT TCC AAG CGC GAC TAC GAC GAC GAT GAC AAG GCT CGC GCG ACC...

GAND7:
    Lep                 KEX2                                        EK
     |                   |                                          |
ss-Ala-Asn-Val-Ile-Ser-Lys-Arg-Asp-Tyr-Asp-Asp-Asp-Asp-Asp-Asp-Lys-Ala-Arg-Ala-Thr...
3'-... AAT GTC ATT TCC AAG CGC GAC TAC GAC GAC GAC GAT GAC GAC AAG GCT CGC GCG ACC...

GAND10:
    Lep                 KEX2                                                    EK
     |                   |                                                      |
ss-Ala-Asn-Val-Ile-Ser-Lys-Arg-Asp-Tyr-Asp-Asp-Asp-Asp-Asp-Asp-Asp-Asp-Asp-Lys-Ala-Arg-Ala...
3'-... AAT GTC ATT TCC AAG CGC GAC TAC GAC GAC GCT GAC GAC GAC GAT GAC GAC AAG GCT CGC GCG...
```

FIGURE 3. A: Three pairs of oligonucleotides used for creating GAND5, GAND7, and GAND10 derivatives of glucoamylase. The pairs are shown annealed with each pair having BssHII cohesive overhangs at each end. **B:** N-terminal sequences of wild-type and ND derivatives of glucoamylase. ss denotes the transient signal sequence, which is removed by leader peptidase cleavage (Lep) during secretion. The subsequent hexapeptide (-NVISKR-) is also removed during secretion by yeast KEX2 gene encoded peptidase (denoted as KEX2 above the sequence). The BssHII restriction site used in cloning is shown below the wild-type sequence. The C-terminal DDDDK sequence in each tail comprises the recognition sequence for enterokinase (EK) that can be used to remove the tails after recovery and purification, if desired. Due to duplication of the R residue coded by the cloning site and the presence of a K residue in the EK site, the actual changes of the net charges for GAND5, GAND7, and GAND10 compared to the wild-type enzyme are −3, −5, and −8, respectively.

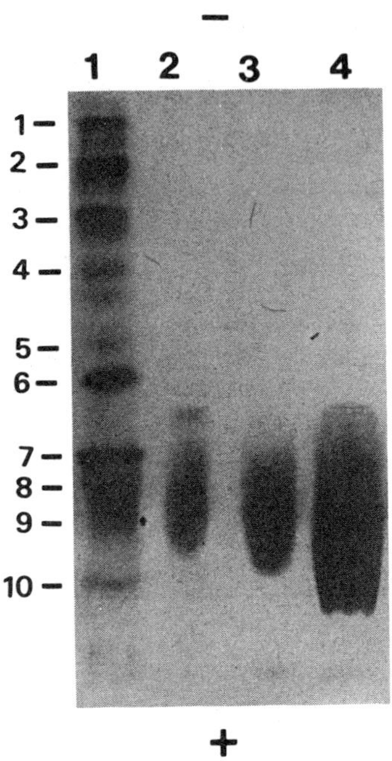

FIGURE 4. Isoelectric focusing of affinity-purified glucoamylases. The enzymes were purified from ultrafiltrated yeast culture supernatants with acarbose-Sepharose affinity chromatography and applied to a pH 3–9 PhastSystem IEF gel (Pharmacia, Uppsala, Sweden). Lane 2, GAC9 (wild type); Lane 3, GAND5; Lane 4, GAND7. The standard proteins (Pharmacia) and their pI's from cathode to anode in lane 1 were (1) lentil lectin, basic, 8.65; (2) lentil lectin, middle, 8.45; (3) lentil lectin, acidic, 8.15; (4) myoglobin, basic, 7.35; (5) myoglobin, acidic, 6.85; (6) human carbonic anhydrase B, 6.55; (7) bovine carbonic anhydrase B, 5.85; (8) β-lactoglobulin A, 5.20; (9) soybean trypsin inhibitor, 4.55; (10) amyloglucosidase, 3.50.

buffer, and glucoamylase was eluted with 1.7 M Tris-HCl, pH 7.6. The fractions containing more than 10% compared to the most active fraction were pooled, dialyzed against water, and lyophilized. The specific activities of the affinity-purified proteins were 42 U/μg, 71 U/μg, and 57 U/μg for GAND5, GAND7, and GAND10, respectively. Sodium dodecyl sulfate-polyacrylamide gel electrophoresis (SDS-PAGE) showed roughly 70% purity for each of the modified proteins (not shown), so the measured specific activities may not be accurate enough to make final conclusions, but there does not seem to be any decrease in specific activity with increasing tail length. Isoelectric focusing of the purified wild-type enzyme (GAC9), GAND5, and GAND7 (FIGURE 4) in a pH gradient from 3 to 9 clearly showed decrease of isoelectric point with increasing tail length, suggesting that the fusion tails were intact. The observed band spreading is most probably caused by differential glycosylation of the enzyme in yeast resulting in closely spaced multiple band patterns.

The purified enzymes were also subjected to amino terminal amino acid sequence analysis to verify that the tails had remained intact and to study whether processing by both leader peptidase and KEX2 peptidase had occurred faithfully. After two independent sequencing runs of 5 to 10 cycles for each enzyme, it was evident that both processing steps had been correct and that the fusion tails were exactly as designed. Thus, ineffective leader peptidase processing cannot be the cause for lower production level for the loger-tailed version. Also, depletion of tRNA-Asp due to many Asp residues in the tail can be ruled out, since similar tails at the C-terminus do not cause lower production levels (Ford *et al.,* results to be described elsewhere). Most probably the cause is to be found earlier in the secretion process. However, a tail length can be found that optimizes both secretion and separation behavior (see below).

Precipitation of the Charged-Tail Glucoamylases with Polyethylene Imine

To test the amenability of PEI precipitation for glucoamylase, the effective range of PEI/protein ratio was first determined. The wild-type enzyme, GAND5, and GAND7 derivatives were produced in 250 ml SD-His shake-flask cultures, and the culture supernatants were concentrated and diafiltered to 10 mM NaOAc, pH 4.5. Equal amounts of protein from each preparation were precipitated with increasing doses of PEI. The results (FIGURE 5) suggested that, in this case, less than half of the PEI dose (<0.6 g PEI/1 g protein) is required for optimal precipitation of the tailed glucoamylases compared to the wild-type enzyme and that excessive doses of PEI cause resolubilization of the complex.

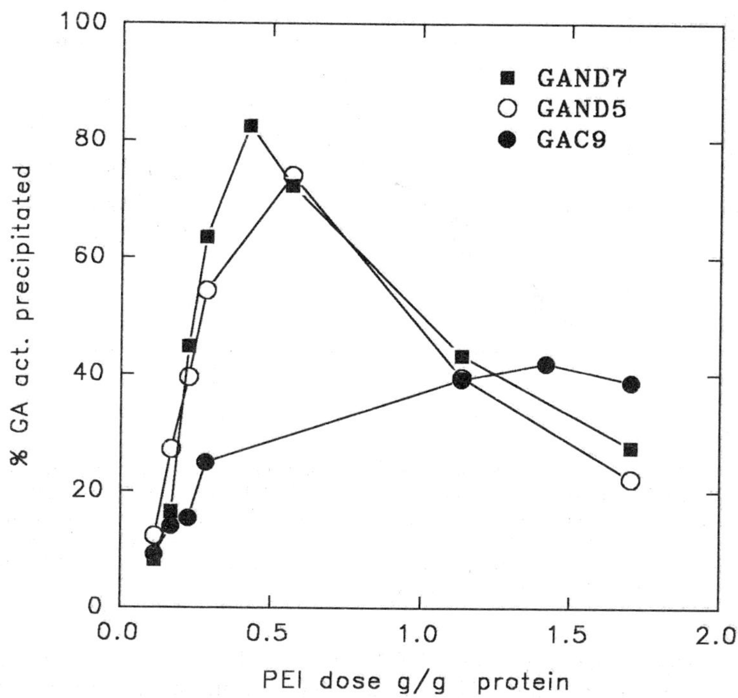

FIGURE 5. Effective range of PEI doses for precipitating glucoamylase. *S. cerevisiae* strains carrying pGAC9, pGAND5, and pGAND7 were grown for 6 days in shake-flask cultures, and the culture supernatants were concentrated and diafiltered to 10 mM NaOAc, pH 4.5 by ultrafiltration (Centricon, NMWCO 10,000 Da). Same amounts (4.4 μg) for each protein were precipitated in a fixed volume by varying doses of PEI (PEI solution was adjusted to pH 4.5 prior to precipitation). Glucoamylase activities were determined in the supernatants after centrifugation and correlated to the 0 PEI control (= % GA activity precipitated).

Each enzyme was then produced in a 10-liter lab fermenter, and the enzymes were harvested by adsorption to raw starch followed by desorption in a smaller volume and final concentration with a Centriprep concentrator (NMWCO 10,000). These crude enzyme preparations were then subjected to PEI precipitation. Equal amounts of protein were precipitated with a narrow range of PEI doses. The results (FIGURE 6) show that GAND10 can be recovered with 90% efficiency with 0.3 g of PEI per 1 g of total protein while only 50% of the wild-type enzyme precipitates with

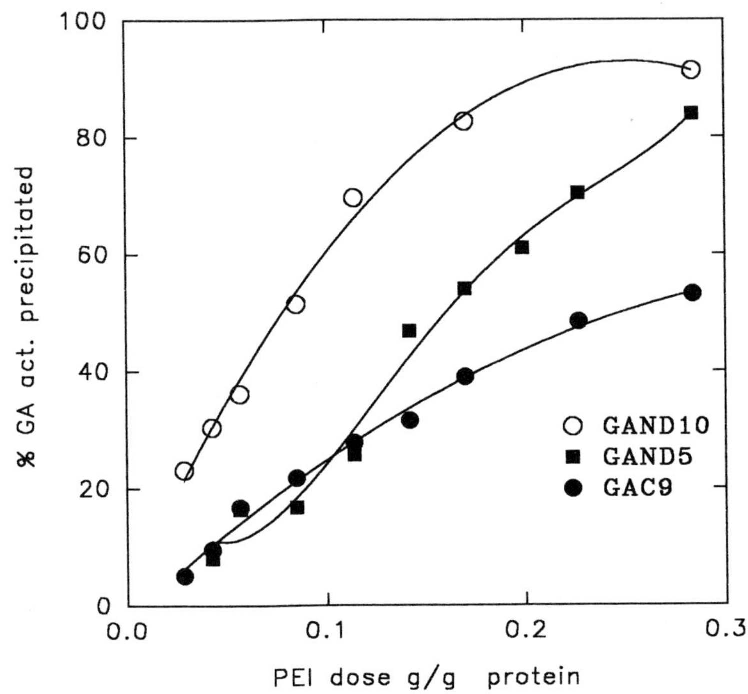

FIGURE 6. PEI precipitation of charged-tail glucoamylases. *S. cerevisiae* strains carrying pGAC9, pGAND5, pGAND7 (not shown), and pGAND 10 were separately cultured in a 10-liter fermenter for 5 days. Glucoamylases were harvested by adsorption to insoluble starch followed by desorption with 1.7M Tris-HCl, pH 7.6, dialysis against 10 mM NaOAc, pH 4.5, and finally concentrated by ultrafiltration. PEI precipitation was as in FIGURE 5. GAND7 behaved similarly to GAND5 and is omitted for clarity.

this dose. Approximately 80% of GAND5 (and GAND7, not shown) was precipitated with the same PEI dose. These results demonstrate that by fusing negatively charged tails of only moderate length (5 to 7 Asp) to the N-terminus of glucoamylase, a significant increase in the efficiency of PEI precipitation can be obtained. This suggests that a charged-tail glucoamylase could be effectively and selectively recovered directly from culture supernatants of large-scale fermentations. This possibility is currently under investigation.

CONCLUSIONS

In this paper, we have shown that the charged-tail purification fusion approach can be extended also to a secreted protein of yeast in addition to our previous model enzyme, the *E. coli* β-galactosidase.[2,3] This is significant, since with secreted proteins removal of nucleic acids prior to precipitation is avoided. Also, most of the commercial large-scale proteins are extracellular products. We have here demonstrated that keeping the tail length comparably short (5 to 7 Asp, in this case), it is possible to avoid secretion defects caused by longer tails, and at the same time achieve considerable enhancement of precipitation. It was also shown that addition of extra amino acid residues to the N-terminus does not interfere with leader peptidase and KEX2 peptidase processing during secretion of glucoamylase. The enhancement of PEI precipitation of glucoamylase from crude culture concentrates also suggests that this approach can be used to efficiently and selectively recover glucoamylase directly from large-scale industrial fermentations.

REFERENCES

1. FORD, C. F., I. SUOMINEN & C. E. GLATZ. 1991. Protein Exp. Purif. **2:** 95–107.
2. ZHAO, J., C. F. FORD, C. E. GLATZ, M. A. ROUGVIE & S. M. GENDEL. 1990. J. Biotechnol. **14:** 273–284.
3. PARKER, D. E., C. E. GLATZ, C. F. FORD, S. M. GENDEL, I. SUOMINEN & M. A. ROUGVIE. 1990. Biotechnol. Bioeng. **36:** 467–475.
4. INNIS, M. A., M. J. HOLLAND, P. C. MCCABE, G. E. COLE, V. P. WITTMAN, R. TAL, K. W. K. WATT, D. H. GELFAND, J. P. HOLLAND & J. H. MEADE. 1985. Science **228:** 21–26.
5. FULLER, R. S., A. BRAKE & J. THORNER. 1989. Proc. Natl. Acad. Sci. USA **86:** 1434–1438.
6. MAROUX, S., J. BARATTI & P. DESNUELLE. 1971. J. Biol. Chem. **246:** 5031–5039.

Improvement of Germoplasms of Enzyme-Producing Microbial Strains by Genetic Engineering

GUANG-ZHEN MENG

Department of Biosciences and Biotechnology
Chinese Academy of Sciences
Beijing 100864, China

INTRODUCTION

Successful application of enzymes depends to a great degree on the selection of excellent microorganisms. Although conventional technology is still useful in breeding and screening of high-activity microbial strains, it has been proven that recombinant DNA technology is a more powerful method for the same purpose. In recent years, Chinese scientists have been making attempts to improve the germoplasm of some microbial strains through genetic engineering. The genes coding for penicillin G acylase, glutaral-7-aminocephalosporanic acid acylase, thermostable alpha-amylase, prochymosin, human prourokinase. L-asparaginase, D-xylose isomerase, and aspartase, etc., have been cloned and expressed efficiently in various host microorganisms. Some overproducing mutants have been obtained and used for production of enzymes on a pilot scale. This paper concentrates on reviewing the current status of this research field in China.

PENICILLIN G ACYLASE

Penicillin G acylase is an enzyme that hydrolyzes penicillin G to 6-aminopenicillanic acid (6-APA), which is the intermediate for production of various semisynthetic penicillins. The acylase gene has been cloned using EcoR1-Pst1 digested pBR322 as vector from the chromosome of *Escherichia coli* D816. The gene has been expressed in hybrid strain *E. coli* A 56 (pPA22). The expression was sensitive to temperature. When cells were incubated at 37°C, there was no accumulation of penicillin acylase. However, cells incubated below 28°C synthesized a large amount of the enzyme. The yield of the penicillin acylase from engineered *E. coli* was about 1288 units/100 ml. The enzymatic properties of the acylase from the hybrid strain are identical with those of *E. coli* D816. Phenylacetic acid induces and glucose represses the synthesis of the penicillin acylase in both hybrid and parent strains. Immobilized engineered bacteria cells in a 500-L column have been used for pilot production of 6-APA.[1-3]

GLUTARAL-7-AMINOCEPHALOSPORANIC ACID ACYLASE

In two-step enzymatic conversion of cephalosporin C, the enzymes D-amino acid oxidase and glutaral-7-aminocephalosporanic acid (GL-7-ACA) acylase catalyze the formation of GL-7-ACA and 7-ACA sequentially. Using BamH1-digested and dephosphorylated pBR322 as vector, a GL-7-ACA acylase gene from *Pseudomonas* sp. 130 chromosomal DNA digested with Sau3A was cloned and expressed in *E. coli* C600. Seven positive clones were detected from 3205 recombinant plasmids with the ^{32}P-labeled oligonucleotide *in situ* hybridization. Out of those, three clones that produce active GL-7-ACA acylase were identified by radioimmunological assay, chemical test, and chromatographic analysis of the reaction mixture. Analysis of gel electrophoresis indicated that pMR5 and pMR7 contained the same 6.8 kilobase (kb) fragment and the size of the insert in pMR6 was 5.7 kb. The effect of various *E. coli* hosts on the expression of cloned GL-7-ACA acylase gene was studied. The enzyme yield of the engineered strain Cu334 was about 540 units per liter.[4,5]

THERMOSTABLE ALPHA-AMYLASE

Thermostable alpha-amylase is an enzyme widely used in starch processing. With a strain of *Bacillus stearothermophilus* 658 as donor and plasmid pBR322 as vector, two different clones of the alpha-amylase gene were obtained. The properties of the alpha-amylases produced by the two clones are quite different. Hybridization experiments indicated that the two different genes code for isozymes of alpha-amylase of the same strain. One gene has been subcloned and expressed in *Bacillus subtilis.* In the presence of stabilizer, more than 90% of its activity still remained after heat treatment at 90°C for 5 minutes. The recombinant plasmid held stable during the fermentation process. The enzyme yield reached 150–300 units per milliliter.[6,7]

PROCHYMOSIN

With the constant growth of cheese production, the shortage of calf stomachs, as the most important source of milk-clotting enzyme, has become more serious. To close the gap between the demand for chymosin and its traditional source, a promising route lies in recombinant DNA technology. At first, the mRNA coding for prochymosin was isolated from the gastric mucosa of calf fourth stomach by a guanidine isothiocyanate procedure. Using mRNA as a template, double-strained cDNA can be synthesized by standard reverse transcription techniques and introduced into *E. coli* plasmid pBR 322. The two positive clones pCT 5 and pCT 15 were selected from the constructed cDNA liberary using the C-500 fragment of human pepsinogen gene as a probe. The results obtained from both restriction mapping and DNA sequence indicated that DNA inserts cover the entire coding sequence for prochymosin. Using Tac as promotor, the expression of the gene was induced by ZPTG and also regulated by temperature. The yield of chymosin was about 80–100 mg per liter. After downstream treatment, about 20% active enzyme was recovered.[8–10]

HUMAN PROUROKINASE

mRNA coding for prourokinase was isolated from human Detroit 562 cells after they were induced by PMA. Reverse transcription was carried out by using 18mer synthetic oligodeoxyribonucleotides as primer, and the cDNA library was established by annealing the ds-cDNA to the Pst1 site of pBR322 by the dC-dG tailing method. One positive clone, UK-1, was obtained when ^{32}P-labeled pHUK-8-Pst1 DNA was used as a probe in screening the transformants. Expression vectors containing the pro-UK cDNA (pSV$_2$-proUK) and a dihydrofololate reductase cDNA (pSV$_2$-dhfr or MMTV-dhfr) were cotransfected into CHO-dhfr$^-$ cells by the calcium phosphate precipitation technique. The dhfr$^+$ transformants were selected by fibrinolytic agarose plate assay. Two colonies, named CLF-14 and CLF-8, exhibited significantly higher expression levels of u-PA activity. They reached more than 24 and 16 IU/10^6 cell per 48 hours, respectively.[11,12]

L-ASPARAGINASE

L-Asparaginase from *E. coli* has been produced in China as an antileukemia agent.[13] Repeated injection of the enzyme results in formation of specific antibody and immunological side effects. It has been proved that *Vibrio succinogenes* is a new source of L-asparaginase with different antigenicity. A group of Chinese scientists are making an attempt to produce the enzyme with the aid of a recombinant DNA technique. L-Asparaginase gene was cloned from *Vibrio succinogenes* chromosomal DNA using the expression vector λgt11. One positive recombinant coding for L-asparaginase was cloned. Nessler reaction analysis performed on the recombinant phage showed positive color reaction, which indicated that the expressed protein possessed L-asparaginase activity.[14]

D-XYLOSE ISOMERASE

D-Xylose isomerase catalyzes the conversion of both D-xylose to xylulose and D-glucose to D-fructose. By the first reaction, D-xylose, the main component of hemicellulose from wood and agricultural wastes, can be converted to D-xylulose, which is fermentable to alcohol by yeast. The second reaction is used to produce high-fructose corn syrup in large tonnage in many countries. The *E. coli* D-xylose operon was obtained by virtue of DNA recombination *in vitro,* and selection by complementating D-xylose isomerase gene was isolated through subcloning. The cloned DNA containing the D-xylose operon and the isomerase gene were further cloned into various high-copy plasmids, respectively, in an effort to increase activity of the enzyme in the host cells. Meanwhile, effects of different host strains on the expression of the gene were examined. The results show that both the D-xylose operon and the isomerase gene are expressed in different *E. coli* strains.[15]

ASPARTASE

Aspartase catalyzes the formation of aspartic acid from ammonium fumarate. The aspartic acid is widely used as a raw material directly or indirectly in the food, the pharmacy, cosmetics, and chemical industries. A high-yield strain, *E. coli* AS

1.881, has been screened and used for production of aspartic acid in large scale in China.[16–17] It was proven that removing a small peptide fragment from the C-terminal of the enzyme molecule could result in a significant decrease in its activity. In order to obtain the modified enzyme molecules, the gene coding for the enzyme was cloned, sequenced, and exressed. The site-directed mutagenesis of the enzyme is under way.[18]

REFERENCES

1. YANG, S. L. 1985. Chin. J. Biotechnol. **1**(1): 29–35.
2. WU, R. P. 1985. Chin. J. Biotechnol. **1**(3): 12–19.
3. YANG, S. L. 1988. Chin. J. Biotechnol. (English Edit.) **4**(1): 21–28.
4. YANG, Y. L. 1991. Chin. J. Biotechnol. **7**(2): 99–107.
5. JIAO, R. S. 1990. *In* Proceedings of the Seventh Five-Year Plan Key Project Biotechnology, Beijing: 55–56.
6. REN, D. M. 1987. Chin. J. Biotechnol. **3**(3): 197–202.
7. REN, D. M. 1990. *In* Proceedings of the Seventh Five-Year Plan Key Project Biotechnology, Beijing: 189.
8. TAN, S. Y. 1988. Chin. J. Biotechnol. **4**(2): 91–97.
9. TAN, S. Y. 1988. Chin. J. Biotechnol. (English Edit.) **4**(1): 29–37.
10. LIU, N. J. 1990. *In* Proceedings of the Seventh Five-Year Plan Key Project Biotechnology, Beijing: 199–200.
11. LI, X. Z. 1990. Chin. J. Biotechnol. **6**(3): 252–255.
12. LI, F. Z. 1991. Chin. J. Biotechnol. **7**(2): 114–119.
13. MENG, G. Z. 1985. Molecular Architecture of Proteins and Enzymes. R. A. Bradshaw & J. Tang, Eds.: 135–154. Academic Press. New York, N.Y.
14. GWAN, Y. C. 1988. Chin. J. Biotechnol. (English Edit.) **4**(1): 1–12.
15. ZHANG, Q. J. 1988. Chin. J. Biotechnol. (English Edit.) **4**(1): 38–44.
16. MENG, G. Z. 1978. Acta Microbiol. Sin. **18**(1): 39–44.
17. MENG, G. Z. 1988. *In* Biotechnology and Food Industry, Proceedings of the International Symposium, Budapest, Hungary: 19–25.
18. ZHANG, H. Y. 1990. Annual Report of the Laboratory of Enzyme Engineering: 45–58. Jilin University. Changchun, China.

Effect of Native and Modified Forms of Superoxide Dismutase and Catalase on Experimental Silicosis in Rats

ALEXANDER V. MAKSIMENKO, LJUDMILA M. BEZRUKAVNIKOVA,[a] EKATERINA L. GRIGORIEVA, VLADIMIR V. YAGLOV,[a] AND VLADIMIR P. TORCHILIN

Institute of Experimental Cardiology
Cardiology Research Center
Academy of Medical Sciences of Russia
3rd Cherepkovskaya str. 15A
Moscow, Russia

[a]Institute of Trade Hygiene and Occupational Diseases
Academy of Medical Sciences of Russia
Moscow, Russia

INTRODUCTION

Silicosis is a rather widespread occupational disease which is induced by industrial quartz-containing aerosols affecting the lungs. According to published data and our data, modified forms of enzymes with depolymerizing and antiradical action (hyaluronidase, collagenase, superoxide dismutase) are capable of inhibiting the progress of fibrosis.[1] The efficacy of superoxide dismutase (a scavenger for superoxide radicals) is likely to be increased with its simultaneous administration with catalase, which converts toxic hydrogen peroxide to harmless water and oxygen.[2] Preparations of native and modified catalase (CAT) were used to study the possibility of enhancing the effects. Native CAT and experimentally obtained aldehyde dextran–modified catalase (CAT-AD) and catalase–aldehyde dextran–superoxide dismutase covalent conjugate (CAT-AD-SOD) were used for the investigation.

The aim of the present study was to perform chemical modification and to carry out a comparative study on the properties of native and modified forms of SOD and CAT, as well as their covalent conjugate, and to test them on the model of experimental silicosis in rats.

EXPERIMENTAL WORK

Carrier

Dextran [m.m. 70 kilodaltons (kDa)] modified by partial periodate oxidation was used as a carrier for SOD modification. This aldehyde dextran contained 20 aldehyde groups per 100 glycoside units of polymer chain.

Preparation of SOD-AD

Reaction mixture containing 10 μM SOD from rat liver and 0.3 mM AD was incubated for 18 hours (0.05 M K,Na-phosphate buffer) at room temperature (pH 8.5). Then $NaBH_4$ reduction was performed and the conjugate obtained was isolated by ultrafiltration (Amicon, XM-50) and subsequently lyophilized.

Enzyme activity of SOD was defined by a spectrophotometry method (560 nm) following inhibition of nitrotetrasolium blue reduction during generation of superoxide radical in a xanthine–xanthine oxidase system at pH 7.8.

Carrier and Cross-linking Agent

Aldehyde dextran obtained as the result of partial periodate oxidation of dextran (m.m. 150 kDa) was used for CAT modification and covalent cross-linking of SOD and CAT. The dextran mentioned exceeded dextran of another molecular mass in the following: catalase amount coupled to carrier, its residual catalytic activity, simplicity of the isolation of modification products. All these factors indicated the choice of this dextran (m.m. 150 kDa) as an optimal carrier or cross-linking agent (for SOD coupling) for chemical modification of CAT. The aldehyde dextran obtained contained 20–22 aldehyde groups per 100 glycoside units of polymer chain.

Preparation of CAT-AD

Catalase (10 μM) from bovine liver was incubated with AD (0.02–30 μM) in 0.05 M buffer saline (pH 8.5) for 18 hours at 4°C. Then the mixture was carefully reduced by $NaBH_4$ and the product obtained was isolated by ultracentrifugation (Amicon, XM-100) and rinsed with 0.05 M of KH_2PO_4 (pH 4.0). Such acid washing provides dissociation of enzyme not coupled to subunits, and it can be easily separated from the CAT-AD covalent complex by ultracentrifugation. The product isolated was lyophilized.

Catalase enzyme activity was determined by spectrophotometry following disappearance of hydrogen peroxide at 240 nm.

Preparation of CAT-AD-SOD

AD (18 mg) was dissolved in 30 ml of 0.05 M buffer saline (pH 8.5) and supplemented with 7.2 mg of CAT (243 μl), and then the mixture was incubated with careful agitation for 3 hours at room temperature. After that 1.92 mg of SOD was added to the reaction mixture and incubation was carried out for 18 hours at 4°C. The incubation mixture was reduced with $NaBH_4$, the conjugate was isolated by ultrafiltration, and then lyophilized (Amicon, XM-300).

Model of Experimental Silicosis in Rats

Investigation was done on Wistar male rats. Intensive progression of silicosis (morphology similar to that observed in humans) was noted upon administration of quartz dust into the lungs. Experimental silicosis was induced by a single intratracheal administration of 20 mg of quartz dust (dispersity less than 5 μm). Respirable fractions of dust were isolated by sedimentation in water. Quartz dust accumulation in the lungs induced development of fibrosis, which results in an increase of lung mass and an elevation of connective tissue proteins (determined by oxyproline content). Therefore, fibrosis progression was monitored by mass of wet lungs and lymph nodes, mass of dry lungs and lymph nodes, as well as oxyproline level in dry tissue. Rats were killed at different steps of the experimental procedure and the indices mentioned above were determined calculating lung and lymph node mass coefficients (i.e., mass of the latter per 100 g of body weight). Stable fibrosis was observed during a three-week administration of quartz dust into rat lungs.

Experimental Therapy

Inhalation was applied to animals three weeks after quartz dust administration. SOD in native and modified (SOD-AD) form was administered in a dose of 200 U per animal. Inhalation continued for two months (three times a week), each duration being 1 hour.

Native and modified (CAT-AD) catalase preparations were inhaled into rats in a similar way. The dose of native or modified CAT was 200 U per animal, and the dose of the CAT-SOD mixture of native enzymes was 200 U of each enzyme per animal.

CAT-AD-SOD bienzyme conjugate and the mixture of modified enzymes SOD-AD and CAT-AD were injected into rats intraperitoneally to study the antifibrosis effect of these preparations. CAT-AD-SOD was administered in a dose of 400 U SOD and 750 U CAT, and modified enzyme mixture was administered in a dose of 400 U of each enzyme per animal. Duration of injection course was two months (three times a week).

Animals were killed after termination of experimental therapy, and mass coefficients of dry lungs and lymph nodes were determined, as well as oxyproline content in the lungs and lymph nodes, which allowed us to assess the degree of fibrosis progression.

Statistical analysis of the data obtained was performed using Student's t-test ($p < 0.05$).

Morphological changes were determined by lung sections stained with hematoxyline and eosine, and using oxyproline-pikrofuxine staining to reveal connective tissue proteins (according to Van-Gieson).

RESULTS AND DISCUSSION

Modified Enzyme Preparations

The present method using CAT-AD-SOD affords the highest yield according to protein and residual catalytic activity. A three-step procedure of obtaining conjugate (CAT modification by aldehyde dextran; SOD covalent coupling; reduction of

$NaBH_4$ products) proved to be preferable as compared with a one-step synthesis procedure or preliminary modification of SOD.

CAT modification increases its thermostability in CAT-AD complex and in CAT-AD-SOD conjugate. SOD also exhibits enhanced thermostability in this conjugate, as well as in SOD-AD complex.

Some data on comparative properties of enzyme preparations are presented in TABLE 1.

TABLE 1. Indices of Native and Modified Forms of CAT and SOD Preparations

Parameter	Preparation	Native SOD	SOD-AD	Native CAT	CAT-AD	CAT-AD-SOD
Protein content (%)		80	30	57	22	29
Retained activity of superoxide dismutase	U/mg protein (%)	100	86			< 70[a]
	U/mg preparation	1400[b]	450			354
Retained catalase activity	U/mg protein (%)			100	83	< 60[a]
	U/mg preparation			306[c]	98	93
Amino group content per 1 M protein (± 2%)		100	32	100	16	17
Residual catalytic activity after 3 hours of incubation of enzyme preparations at 50°C in 0.1 M K,Na-phosphate buffer pH 7.4 (% from initial value, ± 3%)	SOD activity	75	90			90
	Catalase activity			0	40	40

[a]Specific conjugate activity (per mg of protein) according to CAT and SOD was tentatively calculated assuming CAT and SOD amount to be equal to the total amount of protein of the preparation indicated in the table.

[b]Enzyme quantity that is necessary to inhibit reduction of nitrotetrasole blue by 50% under conditions of the experiment (pH 7.8, 25°C, 560 nm) was assumed as SOD activity subunit.

[c]Enzyme quantity that is required for decomposition of 1 μM hydrogen peroxide within 1 minute at 25°C, pH 7.0, 240 nm was assumed as a unit of catalase activity.

Biomedical Study

Data on experimental therapy of silicosis-induced rats using SOD preparations are shown in TABLE 2. We did not observe any significant difference in mass coefficients of the liver, kidney, spleen, and heart in any animal group. Lung mass coefficients of those rats that received SOD-AD were significantly lower than those of silicosis-induced animals. SOD-AD therapy appeared to significantly decrease oxyproline content as compared with the control. However, mass coefficients of dry lymph nodes significantly exceeded this value in silicosis groups. Native SOD, as well as dextran, appeared to exert no inhibitory effect on fibrosis progression under these conditions.[1] Thus, SOD-AD complex exhibits a pronounced antifibrosis effect, lowering lung mass coefficients by 18–22% due to SOD chemical modification.

Results of experimental therapy using inhalation of CAT preparations are presented in TABLE 3. Administration of native or modified CAT preparations failed to demonstrate any marked antifibrosis effect. We noted a significant decrease in dry lung mass and mass coefficients in the rat group that received a mixture of native CAT and SOD as compared with the group of untreated animals, thus pointing to an

TABLE 2. Indices of Fibrosis Progression in Rat Lungs after Inhalation of Native and Modified SOD Preparations

Index	Control	Silicosis (untreated)	Native SOD	SOD-AD Complex
Body weight (g)	423 ± 16	461 ± 59	475 ± 21	445 ± 15
(%)		100	103.2	96.5
Dry lung mass (g)	0.5147 ± 0.1686	1.1449 ± 0.468	1.1678 ± 0.0513	0.9386 ± 0.0801[a]
(%)		100	102.4	82.0
Mass coefficient of dry lungs (g/100 g)	0.1237 ± 0.0429	0.2718 ± 0.297	0.2696 ± 0.231	0.2119 ± 0.0169[a]
(%)		100	99.2	78.0
Mass of dry lymph nodes (g)	0.0233 ± 0.0019	0.0939 ± 0.0191	0.1113 ± 0.0203	0.1261 ± 0.0187
(%)		100	118.5	134.3
Mass coefficient of dry lymph nodes (g/100 g)	0.0055 ± 0.0004	0.0207 ± 0.0034	0.0233 ± 0.0035	0.0283 ± 0.0036[a]
(%)		100	112.4	136.6
Oxyproline level in the lungs (mg)	5.6055 ± 0.2795	12.0191 ± 1.2843	12.6802 ± 1.2371	9.9103 ± 0.6800[a]
(%)		100	105.5	82.5
Oxyproline level in lymph nodes (mg)	0.0323 ± 0.0029	1.1479 ± 0.1186	1.1628 ± 0.01342	1.6521 ± 0.2593[a]
(%)		100	101.3	143.9

[a]Significant difference from the group of untreated rats with silicosis ($p < 0.05$).

inhibitory effect on fibrosis progression. However, mass coefficients of dry lymph nodes significantly exceeded values obtained with silicosis-induced rats. Administration of native SOD and CAT mixtures significantly decreases oxyproline content in the animal lungs. In all groups of animals, there was no pronounced difference between peripheral blood indices and the control data (hemoglobin content, number of erythrocytes and leukocytes).

Comparative data allow one to assume that the treatment of experimental silicosis with native SOD and CAT mixtures has an impact on lung stroma and significantly inhibits development of collagen fibers in the stroma and granulomas.

TABLE 3. Indices of Fibrosis Progression in Rat Lungs after Inhalation of Native and Modified Catalase Preparations

Index	Control	Silicosis (untreated)	Native Catalase	CAT-AD Complex	Combination of Native CAT and SOD
Body weight (g)	464 ± 14	470 ± 12	471± 13	490 ± 27	458 ± 15
(%)		100	100.2	100.3	97.4
Dry lung mass (g)	0.4050 ± 0.212	0.9350 ± 0.0415	0.9732 ± 0.0296	0.9467 ± 0.0610	0.7265 ± 0.0426[a]
(%)		100	104.1	101.2	77.7
Mass coefficient of dry lungs (g/100 g)	0.0879 ± 0.0055	0.2008 ± 0.0091	0.2077 ± 0.0062	0.1973 ± 0.0146	0.1470 ± 0.0121[a]
(%)		100	103.5	98.3	73.2
Mass of dry lymph nodes (g)	0.0155 ± 0.0025	0.1004 ± 0.0085	0.1146 ± 0.0104	0.1097 ± 0.0072	0.1275 ± 0.0093
(%)		100	114.1	109.3	127.0
Mass coefficient of dry lymph nodes (g/100 g)	0.0055 ± 0.0017	0.0217 ± 0.0022	0.0245 ± 0.0023	0.0231 ± 0.0020	0.0282± 0.0024[a]
(%)		100	112.0	106.3	129.9
Oxyproline level in the lungs (g)	7.8989 ± 0.8156	11.6279 ± 1.6078	13.7251 ± 1.3427	12.0421 ± 1.5206	8.1046 ± 1.519[a]
(%)		100	118.0	104.0	69.7
Oxyproline level in lymph nodes (mg)	0.0043 ± 0.0017	0.6348 ± 0.0994	0.6364 ± 0.0856	0.5881 ± 0.0828	0.7186 ± 0.0863
(%)		100	100.2	92.6	113.2

[a]Significant difference between these indices and the group of silicosis untreated rats ($p < 0.05$).

The mixture of native enzymes has a marked antifibrosis effect, lowering lung mass coefficients by 23–27%.

Data on experimental therapy with CAT-AD-SOD injections are presented in TABLE 4. We observed a significant decrease in dry lung mass coefficient (by 15%) in those rats that received CAT-AD-SOD, as well as a drop in oxyproline content as compared with the indices of untreated animals. Thus, comparative study has

TABLE 4. Indices of Fibrosis Progression in Rat Lungs after Intraperitoneal Injection of Bienzyme Preparations of Modified Catalase and Superoxide Dismutase

Index	Control	Silicosis (untreated)	Combination of Modified Enzymes CAT-AD and SOD-AD	CAT-AD-SOD Bienzyme Covalent Conjugate
Body weight (g)	514 ± 15	466 ± 12	473 ± 13	470 ± 15
Dry lung mass (g)	0.4919 ± 0.0210	0.8743 ± 0.0511	0.9438 ± 0.0346	0.7511 ± 0.0426
Mass coefficient of dry lungs (g/100 g)	0.0946 ± 0.0047	0.1876 ± 0.0080	0.1999 ± 0.0132	0.1599 ± 0.0088[a]
(%)		100	106.5	85.2
Mass of dry lymph nodes (g)	0.0153 ± 0.0260	0.0647 ± 0.0097	0.0722 ± 0.0078	0.0808 ± 0.0123
(%)		100	111.4	124.7
Mass coefficient of dry lymph nodes (g/100 g)	0.0033 ± 0.0120	0.0139 ± 0.0018	0.0152 ± 0.0026	0.0170 ± 0.0018[a]
(%)		100	109.3	122.3
Oxyproline level in the lungs (mg)	5.1011 ± 0.6987	7.2241 ± 0.7422	6.6316 ± 0.5700	5.7109 ± 0.6014[a]
(%)		100	91.8	74.1
Oxyproline level per 100 mg dry lung tissue (g/100 g)	0.9924 ± 0.1548	1.0310 ± 0.1710	0.6950 ± 0.0709[a]	0.7109 ± 0.0358[a]
(%)		100	67.4	68.9

[a]Significant difference between indices and group of silicosis untreated rats ($p < 0.05$).

demonstrated that treatment with CAT-AD-SOD inhibits fusion of miliary granulomas to larger ones and impedes formation of collagen fibers in them.

On the whole, a marked efficacy of native SOD and CAT compositions is of particular interest. Neither native SOD nor native CAT exhibited an antifibrosis effect when used separately. A combined application of native CAT and SOD allows one to achieve an effect comparable with one using SOD-AD. This not only emphasizes the importance of the application of native enzymes for silicosis treatment, but also demonstrates an additional protective effect of catalase, which is clearly shown under these conditions. As a part of the CAT-AD-SOD covalent conjugate, catalase also produced an antifibrosis effect, but a less pronounced one. Application of CAT-AD and SOD-AD mixtures failed to exert any noticeable effect. It is rather likely that the difference in efficacy of native and modified enzyme mixtures is explained either by different localization zones of CAT-AD and SOD-AD in the lung tissue, or by a different rate of their delivery to injury areas, since CAT-AD-SOD exerts antifibrosis effect under these conditions.

CONCLUSION

Thus, chemical modification of catalase by aldehyde dextran leads to an increase of enzyme thermostability both in CAT-AD covalent complex and in the composition of CAT-AD-SOD covalent bienzyme conjugate, pronounced catalytic activity being preserved. A three-step procedure of obtaining bienzyme conjugate is regarded to be an optimal one.

Inhalation of native CAT and SOD mixture into the rats with experimental silicosis produces effects on the lung stroma and significantly lowers development of collagen fibers in the stroma and granulomas. The mixture of these enzymes exhibits a marked antifibrosis effect, while separate use of these components fails to produce such an effect. Application of native SOD and CAT mixtures, as compared to SOD-AD, demonstrates a greater effect, thus speaking in favor of an additional effect of catalase. Such an effect is also observed, though to a less degree, in the case of intraperitoneal injections of CAT-AD-SOD covalent bienzyme conjugate into silicosis-induced rats. This results in slowing down of fusion of miliary granulomas to larger ones, as well as deceleration of the formation of collagen fibers in them. The fact that a rise in mass coefficients of lymph nodes is noted during such a therapy points to the necessity of varying dosages as well as the composition of the mixtures applied.

REFERENCES

1. Maksimenko, A. V., O. G. Arkhipova, V. V. Yaglov & V. P. Torchilin. 1988. Aldehyde dextran modified enzymes for medical application. Prog. Biotechnol. **4:** 509–522.
2. Opie, L. H. 1989. Reperfusion injury and its pharmacologic modification. Circulation **80**(4): 1049–1062.

Operational Stability of Enzymes

Acylase-Catalyzed Resolution of *N*-Acetyl Amino Acids to Enantiomerically Pure L-Amino Acids

ANDREAS S. BOMMARIUS, KARLHEINZ DRAUZ, HERBERT KLENK, AND CHRISTIAN WANDREY[a]

Degussa AG
Organic and Biological Chemistry R&D
Post Office Box 1345
W-6450 Hanau 1, Germany

[a]*Institute for Biotechnology 2*
Research Center Jülich
Post Office Box 1913
W-5170 Jülich, Germany

INTRODUCTION

Enantiomerically pure L-amino acids are interesting compounds in parenteral nutrition (infusion solutions), as feed and food additives, as intermediates for pharmaceuticals, cosmetics, and pesticides, and as chiral synthons for organic synthesis. L-Amino acids can be obtained from extraction of protein hydrolyzates, from enantioselective crystallization of racemates, fermentation, and enzymatic synthesis.

Production of L-amino acids by enzymatic synthesis has a number of advantages:

- Enzymes, in contrast to microorganisms, often accept a wide range of substrates. The precursors to natural, both proteinogenic and nonproteinogenic, and unnatural amino acids are accepted as substrates.
- Enzymes in most cases are very enantioselective catalysts; for many enzymes, the enantiomeric excess of a range of substrates is higher than 98% and thus surpasses enantiospecificity of enantioselective crystallizations or asymmetric synthesis via organometallic intermediates.

One of the methods for enzymatic L-amino acid synthesis is the resolution of racemates of *N*-acetyl-DL-amino acids by the enzyme acylase I (aminoacylase; *N*-acetylamino-acid amidohydrolase, E.C. 3.5.1.14.). There are other enzymatic methods employing L-amidases[1–3] or L-hydantoinases.[4–6] As shown in FIGURE 1, in the acylase process the L-isomer of the *N*-acetyl-amino acid is cleaved to yield L-amino acid whereas the *N*-acetyl-D-amino acid is not reacted. After separation of the L-amino acid, the remaining *N*-acetyl-D-amino acid can be racemized by acetic anhydride in alkali or, since recently, by using a racemase.[7] This way, high conversions to the *L*-amino acid are achieved. The *N*-acetyl-DL-amino acid precursors are conveniently accessible on laboratory as well as pilot scale through acetylation of

DL-amino acids with acetyl chloride or acetic anhydride in alkali in a Schotten-Baumann reaction (for a review, see Reference 8).

The first publication on the use of aminoacylase for splitting of acetyl amino acid racemates in a large-scale continuous process appeared in 1966.[9] Three years later, the first industrial process was started by Tanabe Seiyaku with an immobilized acylase.[10,11] Based on the work by Wandrey *et al.*,[12] Degussa introduced an acylase process based on an enzyme-membrane-reactor configuration in 1981.[13,14] The process has since been scaled up to a multi-100-ton level of annual production.

The substrate specificity of aminoacylase I is unusually wide; it has been investigated by Chenault *et al.*[15] FIGURE 2 lists both proteinogenic and nonproteinogenic amino acids prepared at Degussa in bulk quantities by the resolution of the respective *N*-acetyl amino acids.

Acylase I is available from two sources: porcine kidney and *Aspergillus oryzae* fungus. The properties of both enzymes have been compared:[16] both are dimeric

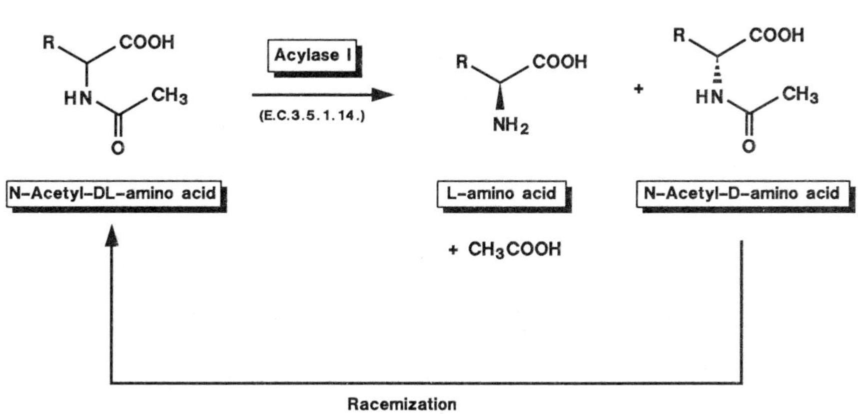

FIGURE 1. Resolution of *N*-acetyl-DL-amino acids to L-amino acids with acylase I.

zinc-containing enzymes (kidney, MW 85.5 kDa, 1 Zn/subunit; fungus, MW 73 kDa, 3 Zn/subunit). The specific activity of porcine kidney preparations is much higher than those from fungus; however, the price per unit activity is comparable. Both enzymes depend on divalent metal ions for activation. Different metal ions have been investigated for both enzymes, but much more is known for porcine kidney[17] than for fungal[18] acylase. Commonly, Co^{2+} ions are used for both.[16,19] Similarly, stability of porcine kidney enzyme has been studied in more detail[12,18] than that of fungal acylase (but see Reference 15).

In this paper, a method of measuring operational stability of enzymes based on monitoring deactivation at constant conversion in a recycle reactor is described. Results of such stability measurements with acylase I from both porcine kidney and *Aspergillus oryzae* sources are presented, and enzyme stability from both sources is compared. Data on activation of both enzymes by divalent metal ions are discussed and supplemented by data on the impact of metal ions on acylase stability.

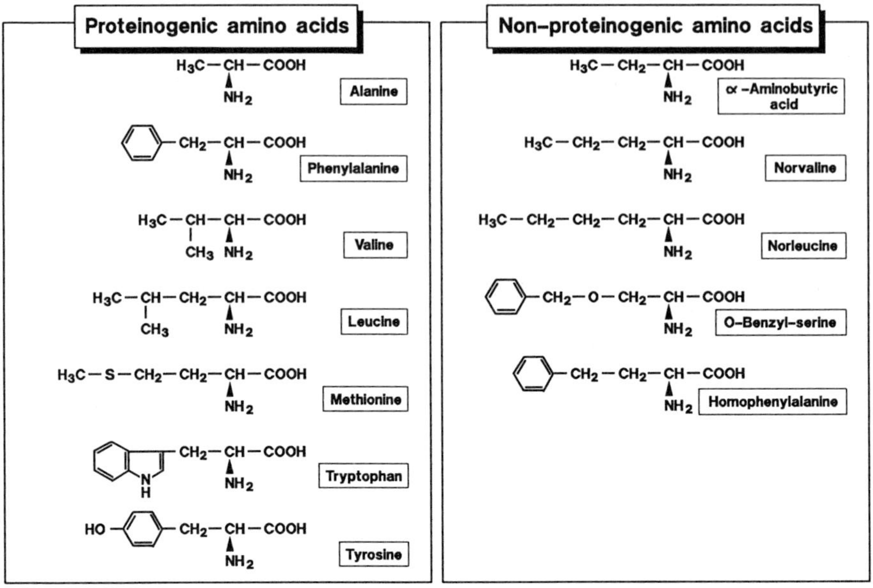

FIGURE 2. L-Amino acids prepared in bulk quantities by acylase I resolution of *N*-acetyl-DL-amino acids.

MEASUREMENT OF OPERATIONAL ENZYME STABILITY

Often, enzyme stability is expressed in terms of half-life, which is the time after which enzyme activity has declined to 50% of initial activity. Determinations of half-life are conducted either by measuring residual activity at standard conditions after resting the enzyme for various lengths of time at a preset temperature, or by conducting repeated batches at set conditions and measuring total reaction time until the enzyme activity has declined to half the initial value.

Both methods are inadequate for characterization of enzyme stability under operating conditions. The first method emphasizes the stability of the enzyme protein molecule; the deactivation process occurs in the absence of reaction and thus is usually not a measure of operational stability. The second method determines enzyme stability at varying conversions; the results are only reproducible with the same conversion profile. Defined measurements of operational enzyme stability are carried out at constant conversion of substrate, which implies constant turnover number in case of stable enzymes. Keeping conversion constant necessitates the use of a continuous reactor. With immobilized enzymes, a fixed-bed arrangement in a plug-flow reactor is most advantageous, while a continuously stirred tank reactor (CSTR) is more appropriate for soluble enzymes, so it was used in this work (FIGURE 3).[12]

The relevant parameter for operational stability studies of enzymes is the product of active enzyme concentration [E] and residence time τ, $[E] \times \tau$. Its derivation starts with the definition of reaction rate:

$$-\,d[S]/dt = r'[\text{mol}/(1 \times \text{minute})] = r[\text{mol}/(g \times \text{minute})] \times [E][g/l], \quad (1)$$

with volume-based reaction rate r' and mass-based reaction rate r. Introducing the degree of conversion x:

$$[S_0] \times dx/dt = r \times [E] \quad (2)$$

Rearranged and integrated over the degree of conversion from time 0 to residence time τ:

$$\int_0^x (1/r)dx = [E]/[S_0] \int_0^\tau dt = ([E] \times \tau)/[S_o] \quad (3)$$

In a CSTR, the integral $\int (1/r)\,dx$ reduces to an algebraic equation because r is constant throughout the reactor:

$$([E] \times \tau)/[S_0] = x/r(x) \quad (4)$$

A very useful feature of Equation 4 is its independence of the enzyme kinetics; however, r still depends on x. At constant mean residence time τ and substrate concentration $[S_0]$ at the reactor inlet, a constant degree of conversion x results as long as the concentration of active enzyme [E] is constant. If it is assumed that no washout of enzyme occurs through the membrane, [E] can only decrease through enzyme deactivation. A lower value of [E] from enzyme deactivation results in a decreased degree of conversion x. To reach the previous value of $[E] \times \tau$ and thus the

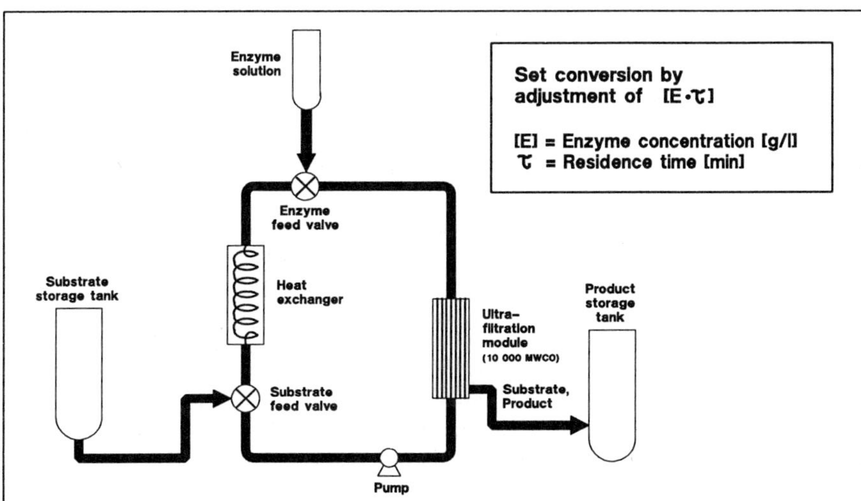

FIGURE 3. Recycle reactor for measurements of operational stability of enzymes (conditions, see Materials and Methods).

preset level of conversion, two strategies are possible:

- Either the residence time τ is increased by lowering the flow-rate F to compensate for the decrease in [E], or
- Fresh enzyme is added to bring the concentration of active enzyme [E] back to previous levels.

While the first option is the convenient one to use for immobilized enzyme, soluble enzymes are most easily investigated with the second method in a CSTR. The amount of enzyme added over time can be scaled to the amount of product formation ($[P] = x[S]$) to yield an accurate measure of operational stability.

MATERIALS AND METHODS

Enzymes

Acylase from porcine kidney was obtained from Reanal (Budapest, Hungary); its specific activity was 2000 U/mg. Acylase from *Aspergillus* was from Amano Corp. (Nagoya, Japan); its specific activity was 31,000 E/g, i.e., 1067 U/g enzyme preparation.[b] Renal acylase was used as received. Fungal acylase was dissolved in the 10-fold volume of 0.6 M L-methionine (L-Met) solution (pH 7) and retained with an ultrafiltration membrane while L-Met solution was flushed through the setup until the degree of transmission of the acylase solution was larger than 80%.

Chemicals

L-Methionine (pure) was from Rexim (Ham, France) and DL-methionine was from Degussa (Frankfurt, Germany); both were used as received. Acetic anhydride, acetic acid (both synthesis grade), HCl, NaOH, $ZnCl_2$, $CoCl_2 \cdot 6H_2O$, $MgCl_2$, and $CaCl_2$ were from E. Merck (Darmstadt, Germany).

Acetylation of DL-Methionine

DL-Met (149.12 g, 1 mol) is added to a well-stirred 500-ml flask containing 65 ml water and 25 ml 50% NaOH solution. The pH is adjusted to 9 with the NaOH solution and 113.5 ml (1.2 mol) acetic anhydride, and enough NaOH is added so that the pH is kept between 8 and 9 and the temperature does not rise above 30°C. After addition of acetic anhydride, the pH is left to decrease to 7 and the suspension is stirred at room temperature for another 1.5 hours. The suspension is then acidified with conc. HCl to pH 1.5 and stirred another 2 hours to complete crystallization of the product. Subsequently, the crystals are filtered off, washed three times with 30 ml cold water, and dried under vacuum at 50°C. The isolated yield is 140 g *N*-Ac-DL-Met (80% theor. yield).

[b]The acylase activity from Amano is expressed in E/g [≡ μmol/(g enzyme × 30 minutes)]. Since units "U" are based on 1 minute instead of 30, 31,000 E/g correspond to 1067 U/g [≡ μmol/(g enzyme × minute)].

Recycle Reactor

N-Ac-DL-Met substrate solution (0.6 M) was carefully sterile filtered (0.22 μm) before entering the reactor. The reactor volume *V* was 1100 ml; the residence time τ was 2.9 hours. The reactor was operated as a CSTR with a recycle ratio of 250. Reaction temperature was 37°C; the pH was kept at 7.0. The enzyme was retained in the recycle reactor by a Romicon ultrafiltration hollow-fiber membrane module PM 10 (MWCO 10,000; membrane area 0.5 m^2). Conversion of *N*-Ac-DL-Met was set at 80% and monitored by change of optical rotation with a Perkin-Elmer polarimeter 241. If conversion fell below 78% owing to biocatalyst deactivation, enzyme solution was automatically added into the system until the degree of conversion was back to more than 80%.

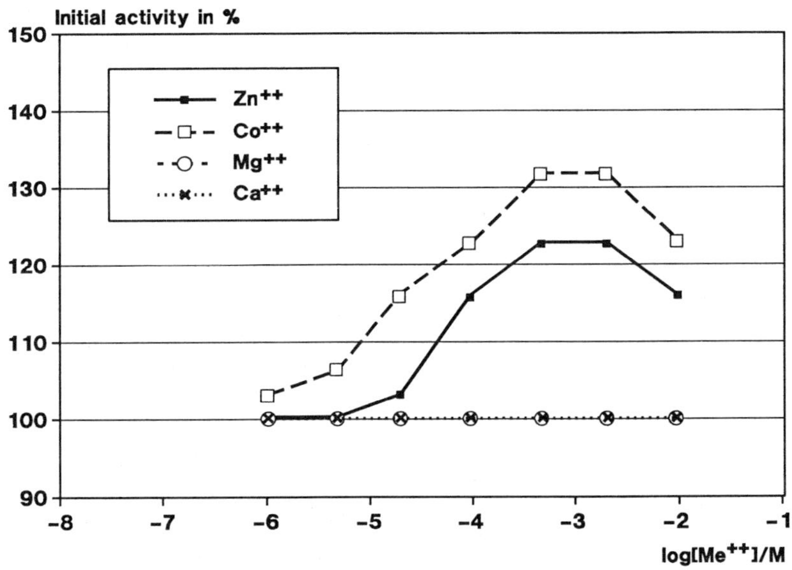

FIGURE 4. Activation of fungal acylase I by divalent metal ions (conditions, batch reactor; otherwise, see Materials and Methods).

RESULTS AND DISCUSSION

Acylase Activity: Activation by Divalent Metal Ions

In contrast to the case for porcine kidney enzyme, the scope of activation by metal ions has not been fully explored for *Aspergillus* acylase. FIGURE 4 illustrates the results for different divalent metal ions.

The alkaline earth ions Mg^{2+} and Ca^{2+} do not activate at all. Both Co^{2+} and Zn^{2+} are activators with a maximum from 5×10^{-4} to 2×10^{-3} M; however, the degree of activation is smaller than a factor of two. Fe^{2+} ions were found to activate the enzyme

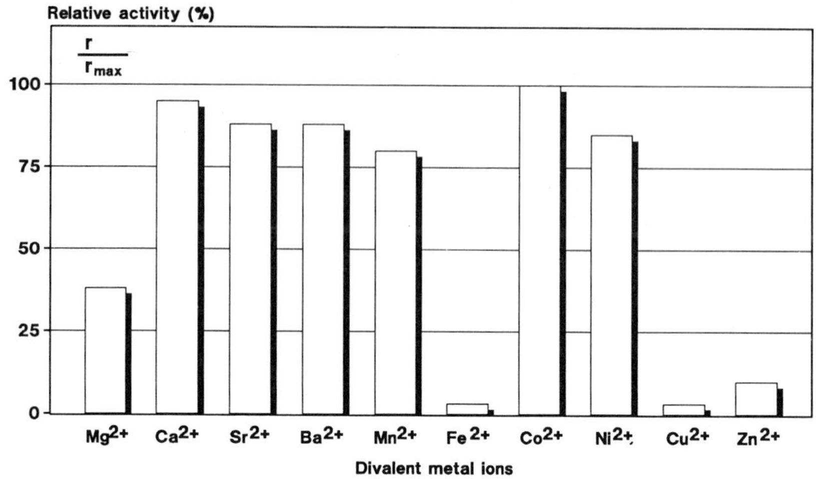

FIGURE 5. Activation of hog kidney acylase I by divalent metal ions ($[Me^{2+}] = 5 \times 10^{-4}$ M; recycle reactor, soluble enzyme. (Data from Reference 17.)

too; however, even under N_2, oxidation to Fe^{3+} with subsequent precipitation of Fe(O)OH could not be prevented, so this option was not pursued further. The findings in FIGURE 4 contrast markedly with those for activation of porcine kidney acylase I in FIGURE 5.[17]

Here, Co^{2+}, most often used for acylase activation, gave the best results followed by Ca^{2+}, which did not activate the *Aspergillus* enzyme at all. Mg^{2+}, Fe^{2+}, and even Zn^{2+} were found to be weak activators. The Co^{2+} dissociation constants of fungal and kidney enzyme are very similar, $10^{-7.5}$ and 10^{-7}, respectively;[20,21] the value for the Zn^{2+} constant is the same, 10^{-10}.[20,22] The numbers for Co^{2+} help to explain activation by Co^{2+}: loose binding of indigenous metal ions is compensated by external addition. The activation patterns by Zn^{2+} or Ca^{2+} (no dissociation constant known) do not agree with this picture; however, since both acylases are zinc enzymes,[22,23] externally added Zn^{2+} ions should be able to replace dissociated zinc atoms.

Operational Stability of Acylase I: Continuous Addition of Activator

The study of operational stability was focused on Co^{2+} and Zn^{2+} as the two most significant activators. In the first part of the investigation, the metals were added to the substrate solution and thus were present at constant concentration in the recycle reactor throughout the run. TABLE 1 lists the measured operational stabilities for both porcine kidney and *Aspergillus* acylase.

The most obvious result is the markedly different stability of the two enzymes: fungal acylase is more stable than kidney enzyme by more than an order of magnitude. Although both forms of acylase sell at a comparable price per unit of activity, the lower operational stability eliminates porcine kidney as a source of acylase I for a large-scale process. For fungal acylase, the measured stability with both Co^{2+} and Zn^{2+} ions was very good at the optimal activation concentration of 5×10^{-4} M. Operational stability decreases at lower [Me^{2+}] values; however, Zn^{2+} seems to be slightly less affected than Co^{2+}. As TABLE 1 also demonstrates, running a continuous acylase process without activating metal ions is not a good choice: acylase stability decreases, so enzyme consumption increases by a factor of four. Since addition of certain metal ions to the reaction medium leads to a decrease of deactivation by about 75%, it has been hypothesized[17] that deactivation is largely caused by loss of metal ions, shown to be essential for activity,[20] unless they are externally resupplied. These metal ions should also be able to reactivate partially inactivated acylase.

Operational Stability of Acylase I: Reactivation Experiments by Pulsed Addition of Activator

To corroborate the hypotheses of metal leaching from acylase and reactivation by externally added metal ions, fungal enzyme was reacted without activator and intermittently exposed to several metal ions in a continuous recycle reactor run; the results are plotted in FIGURE 6. Running enzymatic *N*-Ac-DL-Met hydrolysis without activator led to expected continual decrease in conversion. Whereas additional enzyme did not increase conversion by much (point 1 in FIGURE 6, addition of 10% of initial enzyme; point 2, 20% of initial activity), complete restoration of activity was achieved by addition of Zn^{2+} (2×10^{-3} M in reactor after addition; point 3, 470 hours); this effect has also been observed for Co^{2+} with renal acylase.[17] According to the conversion profile after Zn^{2+} addition, the ion is incorporated into the enzyme within a few hours or less whereas the time constant of Zn^{2+} leaching is on the order of two days. Ca^{2+} (point 4) and Mg^{2+} (point 5) were found to have no stabilizing effect whereas Zn^{2+} (point 6, 2×10^{-3} M) again caused reactivation by about 30 percentage points of conversion. Addition of less Zn^{2+} after 24 hours with 7% points decrease in conversion (point 7, 5×10^{-4} M) caused conversion to again reach almost previous levels. It was decided to check enzyme deactivation with pulsed addition of Zn^{2+} every 24 hours ([Zn^{2+}] = 4×10^{-4} M after addition) with proportional enzyme addition to keep the degree of conversion constant at 80% (point 8).

TABLE 1. Measured Operational Stabilities of Acylase I[a]

Enzyme Source		Metal Ion (M)	Consumption (U/mol L-Met)
Hog kidney			
	Co^{2+}	5×10^{-4}	6000
Fungus			
	Co^{2+}	5×10^{-4}	308
		5×10^{-5}	616
	Zn^{2+}	5×10^{-4}	308
		5×10^{-5}	370
	—	0	~1321

[a]0.6 M *N*-Ac-DL-Met; τ = 3 hours; length of test, $\geq 100\,\tau$; degree of conversion = 80%.

After 250 hours, the Zn^{2+} pulse was decreased to 8×10^{-5} M after addition (point 9), and to 1.6×10^{-5} M after an additional 350 hours (point 10). Peak Zn^{2+} concentrations just after addition of metal ions were considerably higher than average ones owing to washout of free Zn^{2+} from the recycle reactor. Since the reactor was operated as a CSTR, the Zn^{2+} concentration profile over time can be calculated by Equation 5:

$$[Zn^{2+}]_t = [Zn^{2+}]_0 \times \exp(-t/\tau) \tag{5}$$

To determine the average concentration of Zn^{2+} in the reactor, Equation 5 has to be integrated over the time period between Zn^{2+} pulses:

$$\frac{[Zn^{2+}]_{avg}}{[Zn^{2+}]_0} = \frac{\int_0^\theta ([Zn^{2+}]_t/[Zn^{2+}]_0)dt}{\int_0^\theta dt} = \frac{\int_0^\theta \exp(-t/\tau)dt}{\int_0^\theta dt} \tag{6}$$

θ is the time between Zn^{2+} pulses. If θ is much higher than τ, as is the case here (θ = 24 hours, τ = 2.9 hours), the solution to Equation 6 is simply $[Zn^{2+}]_{avg}/[Zn^{2+}]_0 = \tau/\theta$. Peak as well as average concentrations during periods 8, 9, and 10 are listed together with measured values of enzyme consumption in TABLE 2.

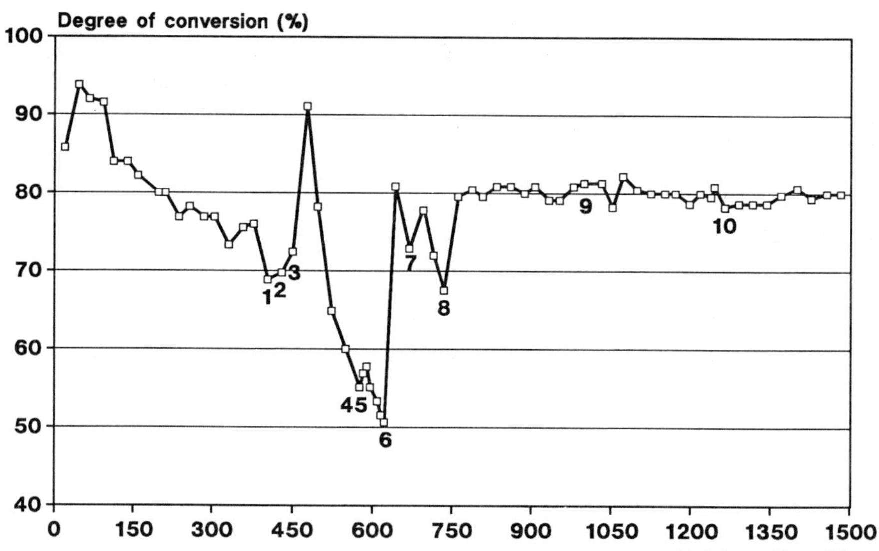

FIGURE 6. Test of operational stability of fungal acylase in a CSTR with various additions of metal ions (numbers, see text; reactor and conditions, see Materials and Methods).

TABLE 2. Measured Operational Stability of Fungal Acylase with Pulsed Addition of Activator[a]

Zn^{2+} Concentration (M)		Consumption
Average	Peak	(U/mol L-Met)
4.5×10^{-5}	4×10^{-4}	477
1.0×10^{-5}	8×10^{-5}	525
2.2×10^{-6}	1.6×10^{-5}	572

[a]Activator, Zn^{2+}; acylase source, fungus.

It seems that a pulse of 4×10^{-4} M of Zn^{2+} is sufficient to stabilize the enzyme. At lower Zn^{2+} concentrations, the enzyme is somewhat less stable. However, a lower average Me^{2+} concentration leaving the reactor might have advantages, such as in the situation where overall metal ion discharge from a processing plant has to be minimized.

SUMMARY

The method of measuring enzyme deactivation by monitoring necessary addition of fresh enzyme to keep a constant degree of conversion in a CSTR at constant $[E] \times \tau$, the product of concentration of active enzyme [E] and residence time τ, was successfully applied to acylase I from porcine kidney and *Aspergillus oryzae* fungus. Fungal enzyme was found to be more stable than kidney enzyme. Activation by both Co^{2+} and Zn^{2+} ions also yielded increased operational enzyme stability: Co^{2+} and Zn^{2+} are better stabilizers than activators. Mg^{2+} and Ca^{2+} are found to be neither activators nor stabilizers. Fungal acylase partially deactivated by exposition to a metal-free medium in the CSTR was reactivated by addition of Zn^{2+}, demonstrating that loss of Zn^{2+} from the enzyme molecule is mainly responsible for deactivation in a continuous reactor.

ACKNOWLEDGMENTS

The authors thank Wolfgang Leuchtenberger for preparing some of the results, and Heinz Both, Helge Werner, and Dittmar Orzewski for excellent technical assistance.

REFERENCES

1. DSM/Stamicarbon. 1975, US patent 4 172 846.
2. DSM/Stamicarbon. 1976, US patent 3 971 000.
3. DSM/Stamicarbon, 1990, European patent 0 199 407.
4. SYLDATK, C., V. MACKOWIAK, H. HÖKE, C. GROSS, G. DOMBACH & F. WAGNER. 1990. J. Biotechnol. **14:** 345–362.
5. GROSS, C., C. SYLDATK, V. MACKOWIAK & F. WAGNER. 1990. J. Biotechnol. **14:** 363–376.
6. WAGNER, F., C. SYLDATK, V. LEHMENSIEK, K. KROHN, H. HÖKE & A. LÄUFER. 1990. European patent, 0 377 083.

7. TAKAHASHI, T., O. IZUMI & K. HATANO. 1989. Takeda Chemical Industries, Ltd. European patent EP 0 304 021.
8. SONNTAG, N. O. V. 1953. Chem. Rev. **52:** 237–416.
9. TOSA, T., T. MORI, N. FUSE & I. CHIBATA. 1966. Enzymologia **31:** 214.
10. TOSA, T., T. MORI, N. FUSE & I. CHIBATA. 1969. Agric. Biol. Chem. **33:** 1047–1052.
11. CHIBATA, I., T. TOSA, T. SATO & T. MORI. 1976. Methods Enzymol. **44:** 746–759.
12. WANDREY, C. & E. FLASCHEL. 1979. Adv. Biochem. Eng. **12:** 147–218.
13. WANDREY, C., R. WICHMANN, W. LEUCHTENBERGER & M. R. KULA. 1981. Degussa AG/GBF. US 4 304 858.
14. LEUCHTENBERGER, W., M. KARRENBAUER & U. PLÖCKER. 1984. Enzyme Engineering 7. Ann. N.Y. Acad. Sci. **434:** 78.
15. CHENAULT, K., J. DAHMER & G. M. WHITESIDES. 1989. J. Am. Chem. Soc. **111:** 6354–6364.
16. GENTZEN, I., H.-G. LÖFFLER & F. SCHNEIDER. 1980. Z. Naturforsch. **35c:** 544–550.
17. WANDREY, C. 1977. Habilitationsschrift. TU Hannover, Germany.
18. GENTZEN, I., H.-G. LÖFFLER & F. SCHNEIDER. 1979. *In* Metalloproteins. U. Eser, Ed.: 270–274. Thieme. Stuttgart, Germany.
19. MARSHALL, R., S. M. BIRNBAUM & J. P. GREENSTEIN. 1956. J. Am. Chem. Soc. **78:** 4636–4642.
20. GILLES, I., H.-G. LÖFFLER & F. SCHNEIDER. 1981. Z. Naturforsch. **36c:** 751–754.
21. KUMPE, E., H.-G. LÖFFLER & F. SCHNEIDER. 1981. Z. Naturforsch. **36c:** 951–955.
22. KÖRDEL, W. & F. SCHNEIDER. 1977. Z. Naturforsch. **32c:** 337–341.
23. GILLES, I., H.-G. LÖFFLER & F. SCHNEIDER. 1984. Z. Naturforsch. **39c:** 1017–1020.

Cellulolytic and Hemicellulolytic Enzymes Functional above 100°C[a]

L. D. RUTTERSMITH, R. M. DANIEL, AND H. D. SIMPSON

Thermophile Research Unit
School of Science and Technology
The University of Waikato
Hamilton, New Zealand

INTRODUCTION

Cellulases and hemicellulases have a variety of potential industrial applications, including the production of fermentable sugars from biomass and enzyme-assisted pulp bleaching. There are several advantages, in industrial terms, to be gained from employing thermostable enzymes in processes operating at elevated temperatures. For example, the lignin component of hemicellulose, which often blocks enzyme access, softens and melts over the temperature range 90–180°C.

Thermotoga sp. strain FjSS3-B.1 is a member of the most thermophilic order of eubacteria, the Thermotogales. It has an optimum growth temperature of 85°C.[1] The genus *Thermotoga* is distinguished by its sheathlike structure which balloons over the ends of the rod-shaped cells forming the so-called togas. FjSS3-B.1 was previously shown to possess cellulolytic and hemicellulolytic enzymes.[2] We are currently studying several of these: *endo* – β-1,4-xylanase (EC 3.2.1.8), arabinofuranosidase (EC 3.2.1.55) with β-1,4-xylosidase activity, *exo*-acting cellobiohydrolase (EC 3.2.1.91), and β-glucosidase (EC 3.2.1.2). We report here on some characteristics of these enzymes, in particular their remarkable thermostability.

METHODS

The enzymes were all purified to homogeneity using the Pharmacia fplc (Pharmacia, Uppsala, Sweden) system in conjunction with three columns: the Mono Q (10/10) anion exchange column, the Mono P (5/20) chromatofocusing column, and the Superdex 200 (26/60) gel filtration column. Purity was judged by Coomassie staining of sodium dodecyl sulfate-polyacrylamide gel electrophoresis (SDS-PAGE) gels using the Pharmacia Phast system.

The release of reducing sugars from polysaccharides was measured using the method of Lever.[3] Sugars produced were identified by high-performance liquid chromatography (HPLC) using a Bio Rad Aminex HPX.42A oligosaccharide column. Action on cellobiose was quantified by measuring the release of glucose using a Glucose GOD-Perid test kit (Boehringer Mannheim, Mannheim, Germany). Other

[a]This work was partially supported by Pacific Enzymes Limited and by the award of a postdoctoral fellowship to one of us (LDR) from the New Zealand University Grants Committee.

assays were carried out by measuring the absorbance increase of synthetic substrates such as pNP-mono or disaccharides or methylumbelliferyl cellobioside.

For experiments involving temperatures above 98°C, samples were incubated either in sealed Mininert vials (Pierce Chemical Co.) fully immersed in an oil bath, or in sealed glass capillary tubes. A thermistor probe (Omega) was sealed into a Mininert vial under experimental conditions to measure the temperature inside.

RESULTS

All of the cellulolytic and hemicellulolytic enzymes in FjSS3-B.1 were >95% cell associated and may well have been located within the toga of the organism. A combination of freeze-thawing and homogenization released >70% of cell-associated enzyme activity from the cells. This left the cells apparently intact (microscopic examination), although the "togas" appeared to be somewhat disrupted. Thus cells could be removed by centrifugation.

All the enzymes were purified to homogeneity (a second β-glucosidase has been identified but not yet purified), and more detailed results have been published on two of the enzymes.[4,5] The enzymes all have a pH optimum around neutral except the xylanase with an optimum of pH 5.3.

Substrate Specificities

FjSS3-B.1 cellobiohydrolase was active against amorphous cellulosic substrates, releasing cellobiose as its sole product. The xylanase was shown to be *endo*-acting by HPLC analysis of end products (see TABLE 1).

The β-glucosidase described here has a high (60 mM) K_m for cellobiose and is active against pNP-xylose as well as pNP-glucose (TABLE 2). However, it is not active against xylobiose.

On the other hand, despite showing its highest activity against the synthetic substrate pNP-α-L-arabinofuranose and having quite low activity against pNP-xylose, the "arabinofuranosidase" shows good activity against, and has a low K_m for,

TABLE 1. Enzyme Properties

Enzyme	Substrate Specificity	V_{max} (U/mg)	K_m	pH_{opt}
Cellobiohydrolase ($M_r \approx$ 36,000)	CMC (38), amorphous cellulose (8), MUC (4), filter paper (3), pNP cellobiose (2), all hydrolyzed. Cellobiose is sole product of amorphous cellulose degradation.		0.04% (CMC) 0.2 mM (MUC) 0.9mM (pNPx)	7.3
Xylanase ($M_r \approx$ 31,000)	Hydrolyzes oatspelts xylan to xylobiose and xylotriose	126	0.007%	5.3
β-glucosidase ($M_r \approx$ 75,000)	See TABLE 2	134	See TABLE 2	7.0
Arabinofuranosidase ($M_r \approx$ 92,000)	See TABLE 2	67	See TABLE 2	7.0

TABLE 2. β-Xylosidase/Arabinofuranosidase and β-Gucosidase I Substrate Specificity

Substrate	β-Glucosidase I [% activity (K_m)]	Arabinofuranosidase/ β-Xylosidase [% activity (K_m)]
pNP-β-D-glucopyranoside	148 (0.1 mM)	7 (2.75 mM)
pNP-β-D-xylopyranoside	100 (8 mM)	7 (S inhibition)
oNP-β-D-xylopyranoside	23 (2.5 mM)	2
pNP-β-D-fucopyranoside	34	0
pNP-α-L-arabinofuranoside	30 (80 mM)	100 (3 mM)
pNP-α-L-arabinopyranoside	0	67 (4.5 mM)
Xylobiose	0	30 (2 mM)
Cellobiose (β-1,4)	28 (59 mM)	0
Sophorose (β-1,2)	21	0
Gentiobiose (β-1,6)	8	0
Laminaribiose (β-1,3)	4	0
pNP-β-D-mannopyranoside	0	0
pNP-α-D-mannopyranoside	0	0
pNP-α-L-rhamnopyranoside	0	0
pNP-β-D-galactopyranoside	0	0
pNP-α-D-glucopyranoside	0	0
pNP-α-D-galactopyranoside	0	0
pNP-β-L-fucopyranoside	0	0
pNP-α-L-fucopyranoside	0	0
Sucrose	0	0
Lactose	0	0
Maltose	0	0
	(zero means <0.1%)	
	Xylose—no inhibition at 40 mM	**Arabinose**—no inhibition at 40 mM
	Glucose—K_i = 0.42 mM	**Xylose**—K_i = 3.3 mM

xylobiose, such that it could well be termed a β-D-xylosidase. This result shows that data obtained from activity with synthetic substrates can be misleading. Although β-D-xylosidase (EX 3.2.1.37) and α-L-arabinofuranosidase (EC 3.2.1.55) are usually considered to be two separate enzymes, substrate specificities are often determined using synthetic substrates alone. Utt *et al.* recently reported on the cloning of a novel bifunctional protein with β-D-xylosidase and α-L-arabinofuranosidase activities,[6] although this was on the basis of activities with synthetic substrates only. The data reported here show that activity on pNP-xylose does not necessarily mean an enzyme is active on xylobiose, and vice versa.

Thermal Stability

All of the enzymes had significant half-lives above 100°C (TABLE 3). Certain additives increased the thermal stability still further, depending on the enzyme concerned (TABLE 3). Sodium chloride was effective in stabilizing the cellobiohydrolase, sorbitol was effective for the xylanase, and trehalose was effective for the β-xylosidase/arabinofuranosidase. Polyols often have a stabilizing effect on enzymes, due to a reduction in water activity. A major cause of activity loss at these temperatures is likely to be hydrolysis.[7] However, it is interesting to note that glycerol was not as effective a stabilizing agent as sorbitol for the xylanase, and sorbitol actually destabilized the β-xylosidase/arabinofuranosidase. Using an assay of 20–40 minutes duration, the cellobiohydrolase and xylanase both show maximum substrate hydrolysis above 100°C.

DISCUSSION

When grown on xylan or cellulosic substrates, *Thermotoga* FjSS3-B.1 cells grow attached to the surface of the substrate (unpublished observation). The localization of cellulolytic and hemicellulolytic enzymes in the "togas" of this organism might indicate that this is the site of hydrolytic processing of substrate prior to uptake of carbohydrate into the cell itself. The finding that amylases are associated with "togas" in *Thermotoga maritima*[8] seems to support this assertion.

TABLE 3. Thermal Stabilities above 100°C

Enzyme	Half Life
Cellobiohydrolase	7 minutes at 113°C
+ 0.8 M NaCl	14 minutes at 113°C
Xylanase[a]	2 minutes at 105°C
+ xylan	9 minutes at 105°C
Xylanase (immobilized)	10 minutes at 105°C
+ xylan	4 minutes at 115°C
+30% sorbitol	5 minutes at 115°C
+100% sorbitol	10 minutes at 130°C
β-Glucosidase[a]	150 minutes at 98°C
+12% trehalose	7 minutes at 111°C
Arabinofuranosidase[a]	240 minutes at 98°C
+15% trehalose	5 minutes at 106°C
+60% trehalose	10 minutes at 111°C

[a]Bovine serum albumin at 40 μg/ml added.

The four enzymes reported on here are all the most thermostable examples of their type described to date. The organism from which these enzymes have been isolated is the most extremely thermophilic of the eubacteria, with an upper growth temperature of 90°C.[1] However, this is 15–20°C lower than the upper growth temperature for several archaebacterial species, from which one might expect to isolate even more thermostable enzymes. Some reports of these more thermostable enzymes have recently been published.[9,10] It is therefore clear that upper temperature limits for industrial enzyme reactions may well be over 100°C.

Future work in this area will involve the cloning of the genes for these enzymes to overcome the problem of the poor biomass yields of this thermophile. N-terminal amino acid sequences have already been obtained to facilitate the construction of oligonucleotide probes.

ACKNOWLEDGMENTS

The authors would like to thank U. Haufler for preliminary work on the xylanase and M. Broughton for technical assistance.

REFERENCES

1. HUSER, B. A., B. K. C. PATEL, R. M. DANIEL & H. W. MORGAN. 1986. Isolation and characterisation of a novel extremely thermophilic anaerobic chemo-organotrophic eubacterium. FEMS Microbiol. Lett. **37:** 121–127.
2. BRAGGER, J. M., R. M. DANIEL, T. COOLBEAR & H. W. MORGAN. 1989. Very stable enzymes from extremely thermophilic archaebacteria and eubacteria. Appl. Microbiol. Biotechnol. **31:** 556–561.
3. LEVER, M. 1973. Colorimetric and fluorometric carbohydrate determination with *p*-hydroxybenzoic acid hydrazide. Biochem. Med. **7:** 274–281.
4. SIMPSON, H. D., U. R. HAUFLER & R. M. DANIEL. 1991. An extremely thermostable xylanase from the thermophilic eubacterium *Thermotoga.* Biochem. J. **277:** 413–417.
5. RUTTERSMITH, L. D. & R. M. DANIEL. 1991. Thermostable cellobiohydrolase from the thermophilic eubacterium *Thermotoga* strain FjSS3-B.1. Biochem. J. **277:** 887–890.
6. UTT, E. A., C. K. EDDY, K. F. KESHAV & L. O. INGRAM. 1991. Sequencing and expression of the *Butyrivibrio fibrisolvens* xyl B gene encoding a novel bifunctional protein with β-D-xylosidase and α-L-arabinofuranosidase activities. Appl. Environ. Microbiol. **57:** 1227–1234.
7. AHERN, T. J. & A. M. KLIBANOV. 1985. The mechanism of irreversible enzyme inactivation at 100°C. Science **228:** 1280–1284.
8. SCHUMANN, J., A. WRBA, R. JAENICKE & K. O. STETTER. 1991. Topographical and enzymatic characterization of amylases from the extremely thermophilic eubacterium *Thermotoga maritima.* FEBS Lett. **282:** 122–126.
9. KOCH, R., A. SPREINAT, K. LEMKE & G. ANTRANIKIAN. 1991. Purification and properties of a hyperthermoactive α-amylase from the archaebacterium *Pyrococcus woesei.* Arch. Microbiol. **155:** 572–578.
10. BLUMENTALS, I. I., A. S. ROBINSON & R. M. KELLY. 1990. Characterisation of SDS-resistant proteolytic activity in the hyperthermophilic archaebacterium *Pyrococcus furiosus.* Appl. Environ. Microbiol. **56:** 1992–1998.

Thermostable Enzymes from Thermophiles

SHU-ZHENG ZHANG, SHOU-JUN YANG, YI ZHOU, AND ZI-ZHENG YAN

Institute of Microbiology
Academia Sinica
Beijing, 100080, China

We have undertaken research work on thermophiles and their enzymes. First, we purified and characterized an α-glucosidase from *Thermus thermophilus* HB 8.[1] Then we isolated many thermophiles from hot springs in South China and screened them for useful enzymes. A maltooligosaccharide-forming amylase from *Bacillus* sp. HY-1 has been purified and characterized.[2] Purification and characterization of a highly thermostable neutral protease from *Bacillus* sp. HY-69 are now under way. The optimum was at 90–95°C, pH 7.5–8.6. The half-life ($t_{1/2}$) at 95°C was about 108 minutes; at 100°C, 20 minutes.

This report deals with an extracellular catalase from *Thermostreptomyces* sp. T485, an L-lactate dehydrogenase from *Thermus nonliquefaciens* HG-25, and a pullulanase from *Becillus* sp. HY-1.

The catalase was purified to homogeneity by ammonium sulfate fractionation and successive column chromatography on Sephadex G-100, DEAE-Sephadex A-50, and hydroxyapatite with 954-fold purification and 7% recovery. Some properties of the enzyme were determined. The results are summarized in TABLE 1. This catalase is an extracellular enzyme, different from all catalases previously reported.[3–5] In addition, the M_r value of the total molecule is about 2.7 times that of the subunit; whether it contains 3 subunits needs further confirmation. The absorption spectra of catalase showed a sharp peak at 405 nm, similar to that of the heme-containing catalase. The amino acid composition of the catalase has been determined as in TABLE 2. Phenyalanine is the most abundant, next are glycine and alanine, while tyrosine has not been detected.

A thermophilic bacteria, strain HG-25, was found to produce thermostable intracellular L-lactate dehydrogenase (LDH). It has characteristics of *Thermus* sp. The cells were gram-negative, nonsporulating, nonmotile, aerobic rods containing yellow pigment. The optimum temperature for growth was between 65°C and 75°C; the maximum was 85°C and the minimum 40°C. The generation time at the optimum was about 80 minutes. The G + C content in DNA was 62–65 mol%. But some properties of strain HG-25 are different from those of *Thermus aquaticus* and T. *thermophilus* HB 8. Starch was not hydrolized. Acid was not produced from glucose. It is designated as a new species of *Thermus, Thermus nonliquefaciens* sp. nov.

The thermostable LDH was partially purified by ammonium sulfate fractionation

TABLE 1. Some Properties of Catalase

Optimum temperature	60°C
Optimum pH	7.0
K_m for H_2O_2	50 mmol/L
V_{max}	6.0 mmol × min^{-1} × mg^{-1}
pI	pH 4.25 by PAG-IEF
Molecular weight	152,000 by concentration gradient PAGE 57,000 bt SDS-PAGE
Inhibitors (activity %)	Hg^{2+} 0.1 mM (23) 1.0 mM (0) NaN_3 1.0 mM (23) 10.0 mM (0) EDTA 10.0 mM (94) 37.5 mM (12)

and DEAE-cellulose column chromatography. For pyruvate reduction, the optimum temperature of the enzyme was 60°C at pH 8.0. The thermostability of the LDH is as in TABLE 1; the half-life at 85°C was 10 minutes.

Microbial amylases, such as α-amylase, β-amylase, glucoamylase, and pullulanase, have been widely used in starch-processing industries. The thermostable pullulanase from *Bacillus* sp. HY-1 was an intracellular enzyme. The enzyme was most active at 60–65°C and pH 6.2. It was stable in the pH range from 5.8 to 7.5. After incubation at 70°C for 30 minutes in 0.1 M acetate buffer pH 6.2, the enzyme retained about 90% of its original activity. The half-life at 70°C was 60 minutes as in FIGURE 1.

TABLE 2. Amino Acid Composition of Catalase

Amino Acid	Number/100 Residues
Asx	7.30
Thr	2.88
Ser	1.29
Glx	8.28
Pro	3.54
Gly	10.47
Ala	10.55
1/2Cys	1.50
Val	7.64
Met	0.25
Ile	5.04
Leu	6.79
Tyr	Undetected
Phe	20.83
His	1.94
Lys	4.01
Arg	4.36
Hypro	1.24

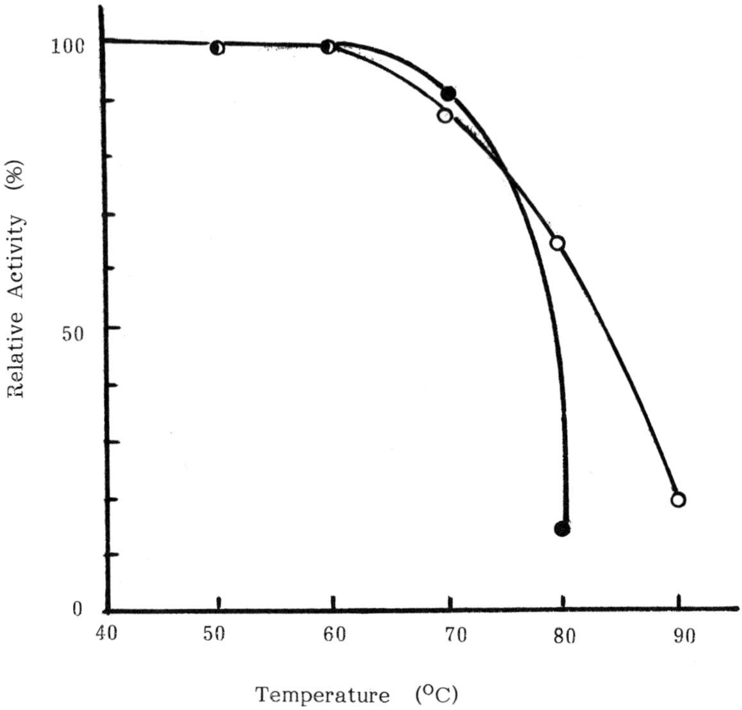

FIGURE 1. Thermostability of the LDH (○) and pullulanase (●). After the LDH was incubated at various temperature for 10 minutes in 0.1 M phosphate buffer pH 7.4, the relative activity was determined.

REFERENCES

1. YANG, SHOU-JUN & SHU-ZHENG ZHANG. 1988. Purification and characterization of thermostable α-glucosidase from an extreme thermophile, *Thermus thermophilus* HB 8. Ann. N.Y. Acad. Sci. **542:** 210–213.
2. YANG, SHOU-JUN, YOU-HOUG LI, YI-ZHI LIU, ZHENG-YI WANG & SHU-ZHENG ZHANG. 1990. Characterization of a new thermostable maltooligosaccharide-forming amylase. Ann. N.Y. Acad. Sci. **613:** 568–570.
3. JACOB, G. S. & W. H. ORME-JOHNSON. 1979. Catalase of *Neurospora cressa.* 1. Induction, purification and physical properties. Biochemistry **18:** 2967–2975.
4. KONO, Y. & I. FRIDOVICH. 1983. Isolation and characterization of the pseudocatalase of *Lactobacillus plantarum.* J. Biol. Chem. **258:** 6015–6019.
5. ALLGOOD, G. S. & J. J. PERRY. 1986. Characterization of a manganese-containing catalase from the obligate thermophile *Thermoliphilum album.* J. Bacteriol. **168:** 563–566.

Secondary Alcohol Dehydrogenases from Extremely Thermophilic Bacteria

H. D. SIMPSON[a] AND D. A. COWAN[b]

[a] *UCLbe*
Department of Chemical and Biochemical Engineering
University College London
Torrington Place
London WC1E 7JE, England

[b] *Department of Biochemistry*
University College London
Gower Street
London WC1E 6BT, England

INTRODUCTION

Alcohol:$NADP^+$ oxidoreductases (EC1.1.1.2) or secondary alcohol dehydrogenases (sADH) are useful chiral synthesizing reagents.[1] The potential biotechnological applications of these enzymes include the conversion of prochiral diols to lactols and lactones, the resolution of racemic alcohols by enantioselective oxidation, and the conversion of prochiral ketones to alcohols. There may be several advantages of exploiting enzymes from thermophilic bacteria for organic syntheses, for example, the robustness of the enzymes to heat and organic solvents[2,3] and the ease of purification after the enzymes are cloned into mesophilic host organisms.

In the literature, there have been few reports on alcohol dehydrogenases from thermophilic bacteria. $NADP^+$-linked alcohol dehydrogenase activity has been detected during the spore-forming phase of *Clostridium thermosaccharolyticum.*[4,5] Both NAD^+-linked and $NADP^+$-linked alcohol dehydrogenase activities were detected in cell extracts of *Thermoanaerobium brockii*[6,7] and *Clostridium thermohydrosulfuricum.*[8,9] However, in extracts of *Clostridium thermocellum,* the alcohol dehydrogenase activity required NAD^+ as cofactor[7] and no activity was detected with $NADP^+$. *Thermoanaerobacter ethanolicus* has primary and secondary alcohol dehydrogenases, which are both $NADP^+$-linked.[10] Conversely, an NAD^+-dependent alcohol dehydrogenase was detected in cell extracts of the extremely thermophilic archaebacterium *Sulfolobus solfataricus.*[11]

METHODS

Growth of Organisms

The organisms were isolated from samples obtained from thermal areas in New Zealand. The isolates were extremely thermophilic glycolytic anaerobic bacteria which grew optimally at 60°C in glucose complex medium.[12]

Enzyme Assay

Alcohol dehydrogenase activity was routinely assayed at 60°C in 20 mM glycine buffer, pH9, containing 150 mM 2-butanol and 0.5 mM $NADP^+$. One unit of enzyme activity represents the amount of enzyme that catalyzed the oxidation of 1 μmol of 2-butanol/minute at 60°C.

Purification Procedure

Alcohol dehydrogenase was isolated from the three strains selected and purified to homogeneity by a four-step purification protocol. Washed cell pellet was sonicated and centrifuged to remove cell debris. The crude cell extract was loaded onto a Hi-Load Q Sepharose anion exchange column (2.6 × 10 cm), and alcohol dehydrogenase activity was eluted with 0.3 M sodium chloride in ethanolamine buffer, pH 9. The peak of activity was concentrated and the salt removed before the Reactive Red Sepharose dye-ligand chromatography (2 × 13 cm) step. Activity was eluted with 1 mM $NADP^+$ in phosphate buffer, pH 7. Ultrospherogel size exclusion chromatography (0.75 × 30 cm) was used to further purify the enzyme. As a final purification to enable N-terminal sequence determination, semipreparative gel electrophoresis was performed.[13] The band corresponding to alcohol dehydrogenase activity was detected in polyacrylamide gels using an activity stain[14] based on the reaction of phenazine methosulfate and methyl thiazolyl tetrazolium.

TABLE 1. Summary of Purification of Ket4B1 Alcohol Dehydrogenase

Column Step	Total Protein (mg)	Total Activity (U)	Specific Activity (U/mg)	Yield (%)	Purification (−fold)
Crude extract	790	1214	1.5	100	1
1. Hi-Load Q Sepharose	230	850	3.7	73	2
2. Red Sepharose	18	612	34	58	21
3. Ultrospherogel	6	413	68	34	42

RESULTS AND DISCUSSION

Fifteen isolates of thermophilic bacteria were obtained from samples taken from New Zealand thermal sites. These organisms grew anaerobically on Zeikus medium[12] with the production of quantities of hydrogen sulfide. All isolates were nonmotile, nonsporulating rods which grew optimally at 60°C. The isolates were tentatively characterized as *Thermoanaerobium*-like, but further characterization is necessary for taxonomy. The crude extract preparations of the 15 isolates were screened for alcohol dehydrogenase activity, and 3 exhibited significant secondary alcohol dehydrogenase activity. NAD^+-linked alcohol dehydrogenase was not detected in the extracts when assays employed 0.5 μM or 0.5 mM NAD^+. This could

TABLE 2. Summary of Substrate Specificity Results

Substrate	Isolate		
	Tau13B1	Tok18B4	Ket4B1
2-Butanol	100	100	100
2-Propanol	70	87	97
Benzyl alcohol	0	1	0
Cyclohexanol	5	40	17
Ethanol	3	7	3
Hexyl alcohol	0	2	1
2-Hexanol	1	28	20
3-Hexanol	2	31	20
±1,3-Butanediol	8	15	17

reflect the fact that extracts were assayed under aerobic conditions since it has been reported that the *Thermoanaerobium brockii* NAD-linked ethanol dehydrogenase was not stable towards oxygen.[6]

The NADP-linked secondary alcohol dehydrogenase was purified 42-fold from Ket4B1 cells, 38-fold from Tau13B1, and 29-fold from Tok18B4 cells. Recovery of enzyme activity was 34%, 49%, and 61% for Ket4B1, Tau13B1, and Tok18B4, respectively. The purification protocol for isolate Ket4B1 is summarized in TABLE 1. The characterization studies were performed on semipurified enzyme preparations obtained after the size exclusion column chromatography step of the purification protocol. Sodium dodecyl sulfate-polyacrylamide gel electrophoresis (SDS-PAGE) revealed that the preparations were at least 90% homogeneous. Protein staining revealed the presence of one major band corresponding to the band obtained by activity staining.

The three isolates exhibited a preference for secondary alcohols, and only very low activity was detected with primary alcohols (TABLE 2). The activity with cyclic alcohols varied between the different isolates, but only very low activity was measured with aromatic alcohols as the substrate. Assays performed in parallel with *Thermoanaerobium brockii* secondary alcohol dehydrogenase obtained from Sigma Chemical Company suggested that the substrate specificity of the Tok18B4 isolate was very similar to that of the *T. brockii* enzyme.[8] The K_m values for 2-butanol (TABLE 3) were slightly higher than the value of 0.31 mM reported for *Thermoanaerobium brockii.*[8] However, the secondary alcohol dehydrogenase from *Themoanaerobacter ethanolicus* exhibited a considerably higher K_m value of 16.4 mM with 2-propanol.[10]

The pH optima for secondary alcohol dehydrogenase activity of the isolates Tau13B1, Tok18B4, and Ket4B1 were 8.9, 8.85, and 9.0, respectively. All three

TABLE 3. Michaelis Constants with 2-Butanol

Isolate	K_m (mM)	V_{max} (μmol/minute per mg protein)
Tau13B1	0.95	38
Tok18B4	0.7	44
Ket4B1	1.0	28

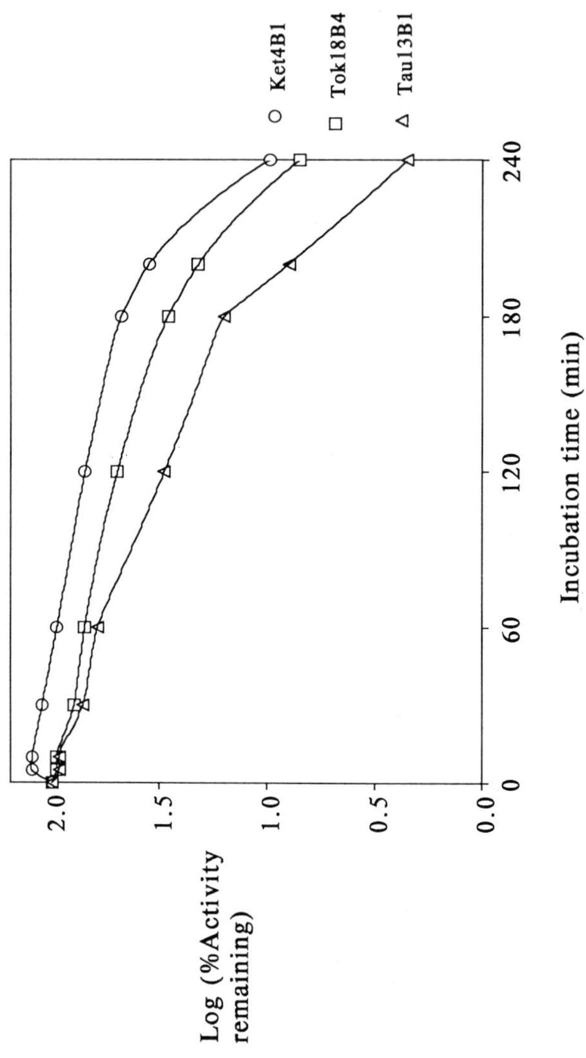

FIGURE 1. Thermostability of Ket4B1, Tok18B4, and Tau13B1 secondary alcohol dehydrogenases at 70°C.

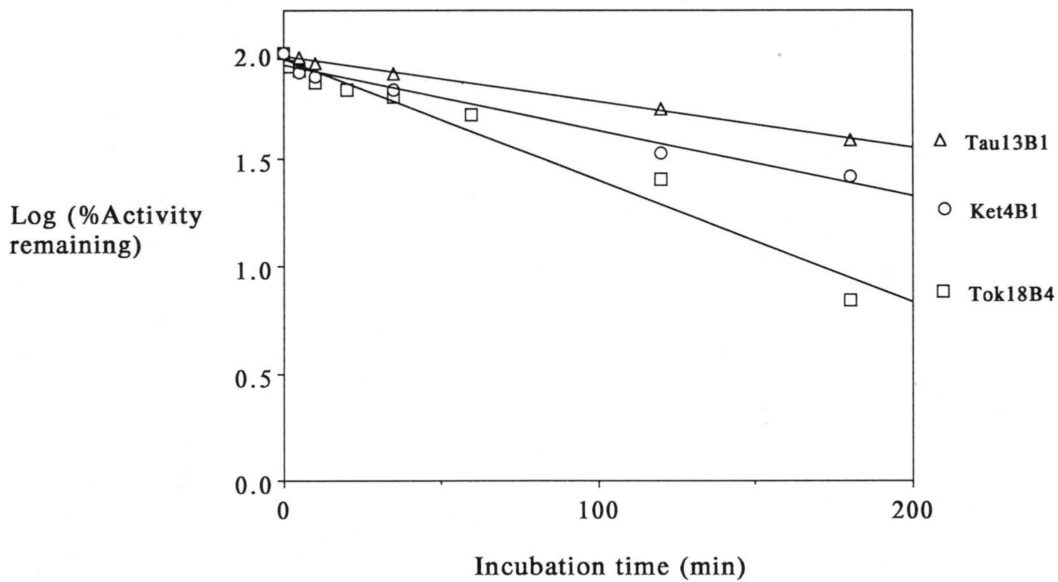

FIGURE 2. Thermostability at 70°C in the presence of 20% ethanol.

isolates exhibited only 20% of the optimal activity when assayed at pH 7.5 and approximately 40% activity at pH 10.

The molecular weight of the Ket4B1 enzyme, as indicated by size-exclusion chromatography and PAGE, was 158,000. Under denaturing conditions, the molecular weight was 41,000, which suggested that the enzyme exists as a homotetramer. Similarly, the subunit molecular weights for the yeast and horse liver alcohol dehydrogenases[15,16] and the *T. brockii* enzyme were reported to be 40,000. A molecular weight of 172,000 was observed for *T. ethanolicus.*[10]

For the thermostability experiments, the alcohol dehydrogenase samples were preincubated at low protein concentrations (65 μg/ml) at 80°C in 50 mM glycine buffer, pH 9, containing 5 mM dithiothreitol. After the required incubation time, the samples were removed to 4°C and subsequently assayed at 60°C. The secondary alcohol dehydrogenase isolated from Ket4B1 appeared to be the most thermostable enzyme, exhibiting a half-life at 80°C of 196 minutes, whereas the Tok18B4 and Tau13B1 enzymes exhibited half-lives of 126 minutes and 92 minutes, respectively (FIGURE 1). However, when incubated with organic solvent at 70°C, the Tau13B1 sADH appeared to be more stable than the Ket4B1 and Tok18B4 enzymes (FIGURE 2).

REFERENCES

1. LAMED, R. J., E. KEINAN & J. G. ZEIKUS. 1981. Potential applications of an alcohol-aldehyde/ketone oxidoreductase from thermophilic bacteria. Enzyme Microb. Technol. **3:** 144–148.
2. TAGUCHI, H., M. HAMAOKI, H. MATSUZAWA & T. OHTA. 1983. Heat-stable extracellular proteolytic enzyme produced by *Thermus caldophilus* strain GK24, an extremely thermophilic bacterium. J. Biochem. **93:** 7–13.
3. OWUSU, R. K. & D. A. COWAN. 1989. Correlation between microbial protein thermostability and resistance to denaturation in aqueous:organic solvent two-phase systems. Enzyme Microb. Technol. **11:** 568–574.
4. HSU, E. J. & Z. J. ORDAL. 1970. Comparative metabolism of vegetative and sporulating cultures of *Clostridium thermosaccharolyticum.* J. Bacteriol. **102:** 369–376.
5. VANCANNEYT, M., P. DE VOS, L. VENNENS & J. DE LEY. 1990. Lactate and ethanol dehydrogenase activities in continuous cultures of *Clostridium thermosaccharolyticum* LMG 6564. J. Gen. Microbiol. **136:** 1945–1951.
6. LAMED, R. J. & J. G. ZEIKUS. 1980. Glucose fementation pathway of *Thermoanaerobium brockii.* J. Bacteriol. **141:** 1251–1257.
7. LAMED, R. & J. G. ZEIKUS. 1980. Ethanol production by thermophilic bacteria: relationship between fermentation product yields of and catabolic enzyme activities in *Clostridium thermocellum* and *Thermoanaerobium brockii.* J. Bacteriol. **144:** 569–578.
8. LAMED, R. J. & J. G. ZEIKUS. 1981. Novel NADP-linked alcohol-aldehyde/ketone oxidoreductase in thermophilic ethanologenic bacteria. Biochem. J. **195:** 183–190.
9. TURUNEN, M., E. PARKKINEN, J. LONDESBOROUGH & M. KORHOLA. 1987. Distinct forms of lactate dehydrogenase purified from ethanol- and lactate-producing cells of *Clostridium thermohydrosulfuricum.* J. Gen. Microbiol. **133:** 2865–2873.
10. BRYANT, F. O., J. WIEGEL & L. R. LJUNGDAHL. 1988. Purification and properties of primary and secondary alcohol dehydrogenases from *Thermoanaerobacter ethanolicus.* Appl. Environ. Microbiol. **54:** 460–465.
11. RELLA, R., C. A. RAIA, M. PENSA, F. M. PISANI, A. GAMBACORTA, M. DE ROSA & M. ROSSI. 1987. A novel archaebacterial NAD^+-dependent alcohol dehydrogenase. Eur. J. Biochem. **167:** 475–479.
12. ZEIKUS, J. G., P. W. HEGGE & M. A. ANDERSON. 1979. *Thermoanaerobium brockii* gen.

nov. and sp. nov. A new chemoorganotrophic, caldoactive, anaerobic bacterium. Arch. Microbiol. **122:** 41–48.

13. HAGER, D. A. & R. R. BURGESS. 1980. Elution of proteins from sodium dodecyl sulphate-polyacrylamide gels, removal of sodium dodecyl sulphate, and renaturation of enzymatic activity. Anal. Biochem. **109:** 76–86.
14. FOWLER, P. W., A. J. S. BALL & D. E. GRIFFITHS. 1972. The control of alcohol dehydrogenase isoenzyme synthesis in *Saccharomyces cerevisiae.* Can. J. Biochem. **50:** 35–43.
15. JORNVALL, H. 1973. Partial similarities between yeast and liver alcohol dehydrogenases. Proc. Natl. Acad. Sci. USA **70:** 2295–2298.
16. JORNVALL, H. 1970. Horse liver alcohol dehydrogenase. The primary structure of the ethanol-active isoenzyme. Eur. J. Biochem. **13:** 565–576.

A Study on Increasing Enzymatic Stability and Activity in Organic Solvents

CAO SHUEGUI, FENG YAN, MA LIN, YANG HONG, DING ZHONG-TIAN, SUN TIAN-XIAO, AND CHENG YU HUA

National Laboratory of Enzyme Engineering
Jilin University
Changchun, 130023, China

INTRODUCTION

Enzymes often exhibit their catalytic function in an aqueous solution. With the development of biological technology, more and more organic solvents might be used in reaction systems. Due to the difference between water and organic solvents in polarity, an organic solvent often affects the stability and activity of the enzyme. Different organic solvents have different effects on enzymes. In this paper, we report some positive results on improving enzymatic properties and stability in the presence of organic solvent. We found that adding amphopathics to the reaction system and using immobilized enzyme are useful methods.

MATERIALS AND METHODS

Enzyme Assays

Cellulase (Onozuka R-10, Yakult Honsha Co. Ltd., Japan) was partially purified by column chromatography. Activity was measured according to the dinitrosalicylic acid (DNS) method.[1] The effect of amphopathics on activity determinations was canceled by control experiments. Lipase from *Penicillium expansum* was extracted with phosphate buffer (pH 8.8). The supernatants were lyophilyzed into enzyme powder. Lipase activity was determined by incubating the appropriate enzyme with 0.8 mol/L lauric acid and laurinol at 40°C for 24 hours under agitation. Activity was calculated from the increase of the product peak area which was determined by high performance liquid chromatography (HPLC).

The Preparation of Immobilized Enzyme

Lipase was absorbed for 2 hours at 4°C in the carrier which was prepared in a mixture containing Ca-alginate, PEI (MW, 10,000), and glutaraldehyde. Immobilized enzyme was washed with benzene and filtered.

RESULTS AND DISCUSSION

The Properties of Cellulase in the Presence of Ethanol

The activity of cellulase (C_x) was inhibited by ethanol (FIGURE 1). The higher the concentration of ethanol, the greater the inhibition C_x. From CD (FIGURE 2), we know the ellipticity of C_x in 208 nm increased in the presence of ethanol. By $X^H = ([\theta]_{208} - 4000)/(33000 - 4000)$, it is obvious that ethanol had caused the enzyme α-helix content to increase.

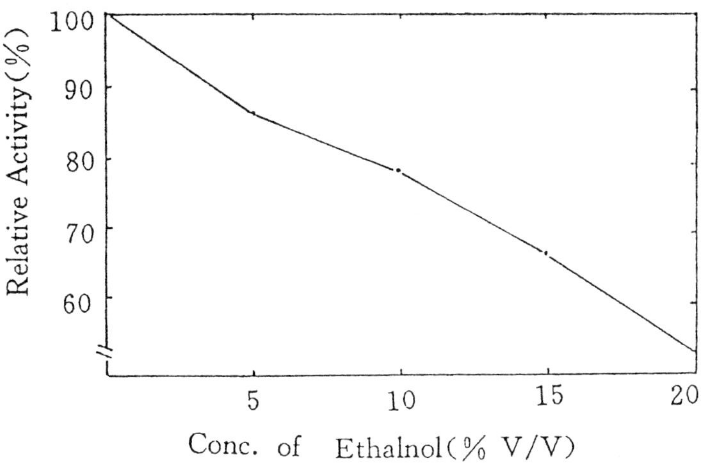

FIGURE 1. The effect of ethanol on the activity of cellulase (C_x).

The Properties of Cellulase in the Presence of Ethanol and Amphopathics

When we added Tween 80, Tween 20, and Triton X-100 in the cellulase reaction system in the presence of 5% (V/V) ethanol, the activity of C_x was changed (FIGURE 3). The C_x activity was increased by 10% when Tween 80 (0.3% V/V) was added to the enzyme solution with 5% ethanol, and the half-life of C_x was prolonged about threefold at 50°C. As shown by UV and CD spectra, the actions of ethanol and Tween 80 are opposite (FIGURE 4, FIGURE 5). When the concentration of Tween 80 was 0.3% (V/V), C_x's structure was close to the native one but looser.

The Effect of Amphopathics on Lipase Activity

Lipase as an unsoluble powder in benzene catalyzed the esterification of lauriol and lauric acid. When amphopathics (see A, B, and C in FIGURE 6) were added into the reaction system, the activity of lipase increased about threefold (FIGURE 6). It may be that (1) the polar heads of the amphopathics were close to the enzyme and

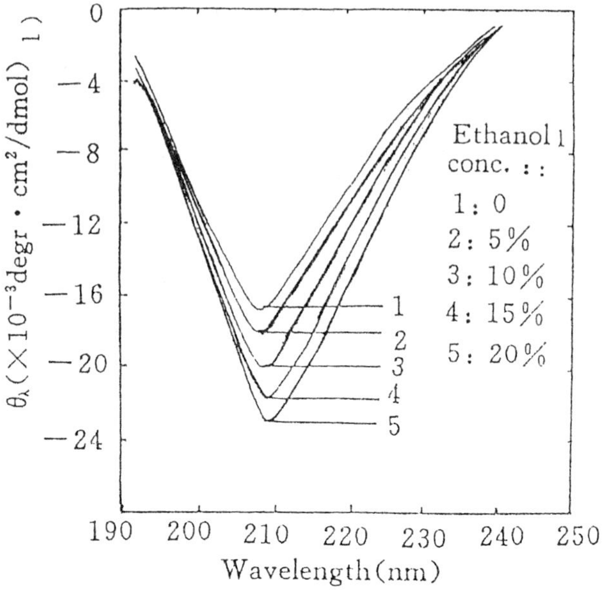

FIGURE 2. The effect of ethanol on CD spectra of cellulase (C_x).

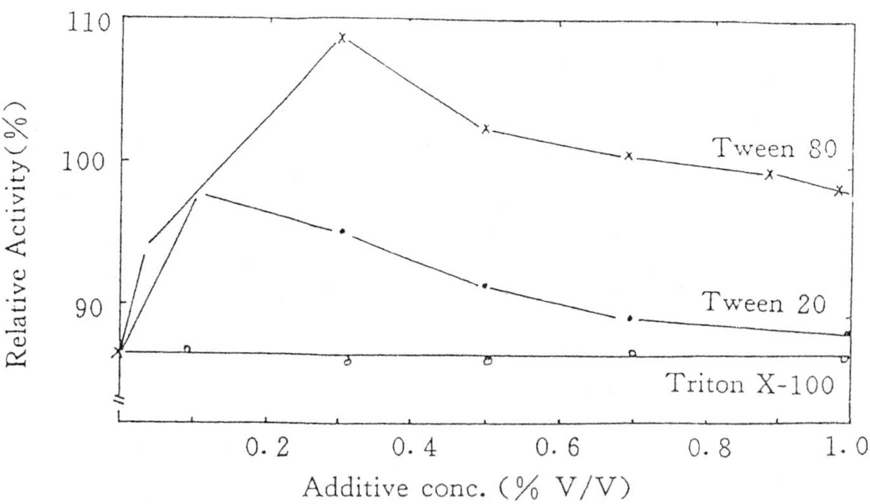

FIGURE 3. Effect of additives on C_x activity in the presence of ethanol.

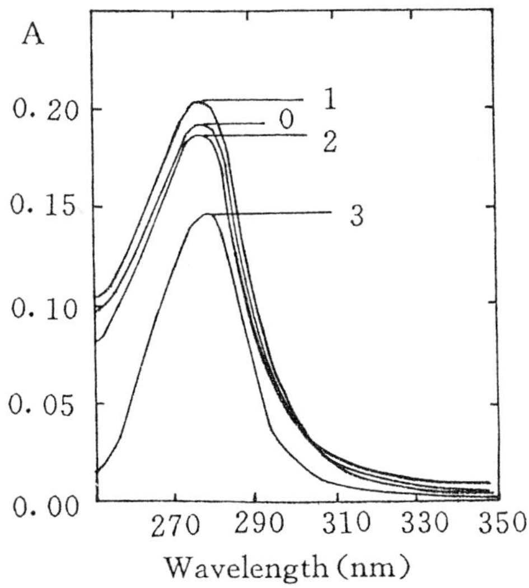

FIGURE 4. Effect of Tween 80 and ethanol on UV spectra of C_x.

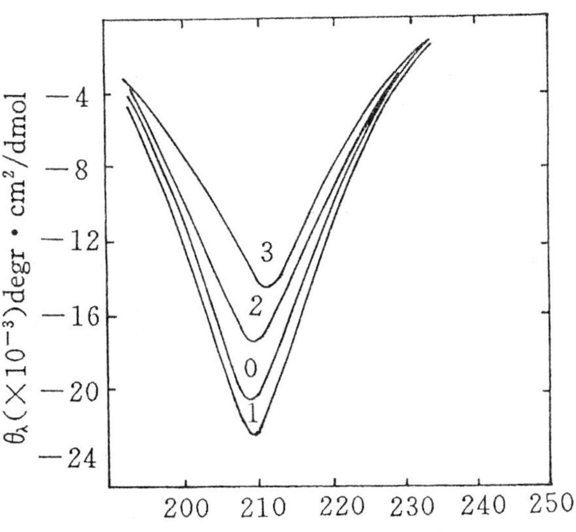

FIGURE 5. Effect of Tween 80 and ethanol on CD spectra of C_x. [C_x, 0.1 mg/mL, pH, 4.8; 0, C_x; 1, C_x + ethanol (5% V/V); 2, C_x + ethanol (5% V/V) + Tween 80 (0.3% V/V); 3, C_x + ethanol (5% V/V) + Tween 80 (0.6% V/V)].

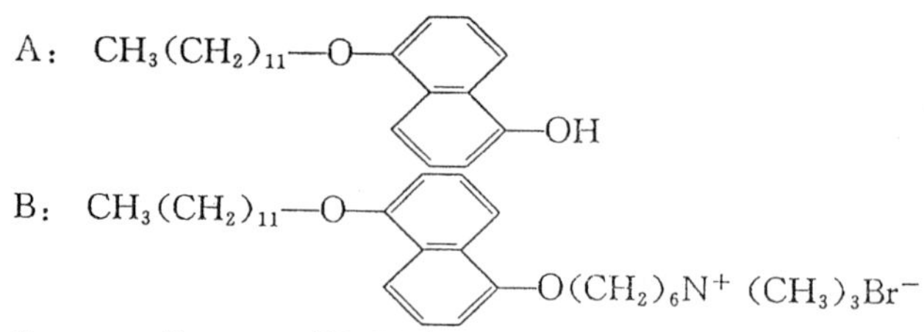

C: meso-Zn tetra-(N,N-dimethylhexadecylaminophenyl) porphyrin bromide.

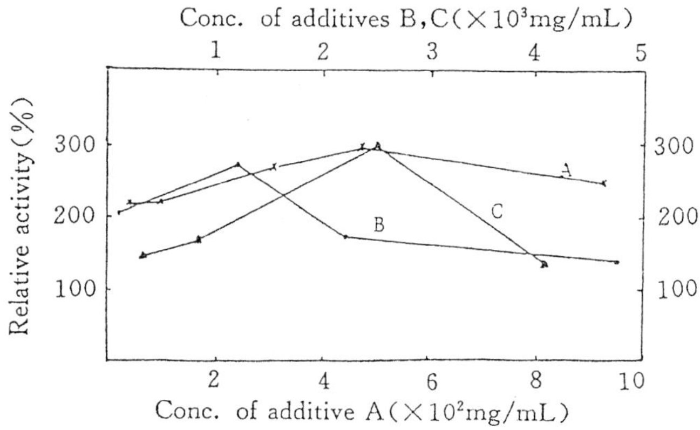

FIGURE 6. The effect of additives A, B, and C on the activity of lipase.

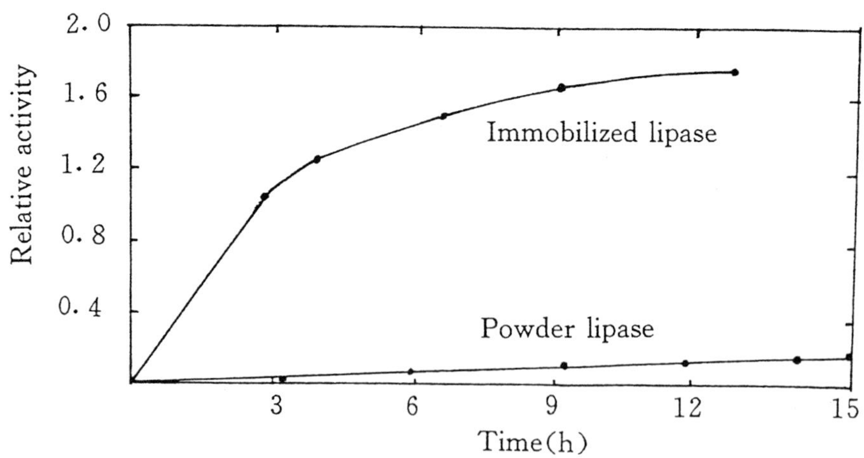

FIGURE 7. The process curve of lipase.

the hydrophobic tails pointed toward the organic solvent, thus forming reverse micelles. The structure increased the dispersion of enzyme in benzene and the likelihood of encounters between enzyme and substrate. (2) The hydrophobic tail of the amphopathics attracted the hydrophobic substrate from the medium to the enzyme.

The Properties of Immobilized Lipase

We found that the immobilized enzyme not only exhibited higher activity but also came to equilibrium sooner (FIGURE 7). The initial rate of immobilized enzyme was 20-fold greater than that of native enzyme.

REFERENCES

1. MILLER, G. H. 1959. Anal. Chem. **31:** 4426.

Fully Dispersed and Covalently Attached Chymotrypsin Derivatives as Industrial Catalysts in Biphasic Systems

Activity, Stability, and Reactivation[a]

JOSÉ M. GUISÁN, ROSA M. BLANCO, AGATHA BASTIDA, ROBERTO FERNANDEZ-LAFUENTE, CARMEN CUESTA, AND PETER J. HALLING[b]

Instituto de Catálisis
CSIC
Universidad Autónoma
Cantoblanco
28049 Madrid, Spain

[b]*Department of Bioscience and Biotechnology*
University of Strathclyde
Glasgow, Scotland

INTRODUCTION

Fully dispersed insolubilized enzymes (e.g., covalently attached on preexisting supports) may be a very interesting option for developing industrial processes under nonconventional conditions (e.g., in biphasic systems). Such multiphase systems may be used both to dissolve hydrophobic substrates and to shift equilibria away from hydrolysis. At an academic level, newer types of enzyme derivatives (e.g., "enzyme aggregates" or "reverse micelles") are now becoming more popular as biocatalysts in organic solvents. However, in our opinion, the use of more classical enzyme derivatives (e.g., enzymes fully dispersed and covalently attached on porous supports) presents a number of important advantages from a practical as well as from a more basic point of view:

1. There may be very high volumetric activity because of full dispersion of enzyme and the high loading capacity of porous supports.
2. The selection of suitable supports allows us to prepare enzyme derivatives able to work in different reactor systems.
3. When, as usual, crude or semipurified enzymatic extracts are used to prepare insolubilized derivatives, the intrinsic properties (e.g., activity/stability) of the fully dispersed ones do not depend on the exact composition of the extracts.
4. These fully dispersed derivatives may be used as catalyst in any type of biphasic system [e.g., low water content–low water activity (LWC/LWA-BS),

[a]This work was supported by the Spanish CICYT (Projects No. BIO88-276 and BIO91-485) and by the Commission of the European Communities, Biotechnology Action Program, Contract No. 398.E.

low water content–high water activity (LWC/HWA-BS), or high water content (HWC-BS)].

5. The insolubilized derivatives may be periodically washed, even inside the industrial reactor system, with any type of medium (e.g., water, urea solutions, cosolvents, and so on) in order to develop enzyme reactivation strategies.

On the other hand, full dispersion of these insolubilized enzyme molecules enables more direct enzyme-solvent interactions which may promote some important distortions and/or deleterious effects: (1) interaction of the enzyme molecule in the aqueous phase with saturating concentration of organic solvents; (2) interaction of the enzyme molecule with the solvent interphase; (3) effect of reduction of water activity on the 3-D structure of the enzyme. However, these interactions may be fairly correlated to simple molecular processes, so we do not have to consider other very complex and overlapping effects (e.g., protein aggregations, enzyme-detergent interactions). The simpler physicochemical description of such a system, isolated enzyme molecule–organic solvent, permits an easier and more "rationale" optimization of the enzyme derivatives to be used in biphasic systems (e.g., by site-directed mutagenesis, by physicochemical techniques).

In spite of these excellent prospects, it is evident that a number of integrated and multidisciplinary studies are still necessary to develop the industrial application of such biphasic systems. It is necessary to prepare nondistorted and high loaded insolubilized enzyme derivatives. It may be necessary to develop additional strategies for enzyme stabilization. We must understand the mechanism of enzyme inactivation in biphasic systems. And it may be interesting to develop reactivation strategies, and so on. These types of studies have been and are being developed in our laboratory by using chymotrypsin as the model enzyme. In addition to very interesting practical prospects (e.g., catalyst of the synthesis of peptide bonds), chymotrypsin also offers a number of advantages to studies in biphasic systems: e.g., this simple and well-known enzyme is able to catalyze very different processes (hydrolysis, synthesis of peptide bonds, esterifications, transesterifications) involving a number of very different substrates (acyl donors and nucleophiles).

In this paper, we first comment on different aspects of the preparation of multipoint covalently attached chymotrypsin derivatives. Further, we discuss the activity-stability and reactivation behavior of these enzyme derivatives in biphasic systems. Finally, we describe an interesting practical application of these fully dispersed chymotrypsin derivatives—their use as catalysts for the "kinetically controlled synthesis" of hydrophobic dipeptides in high water content–biphasic systems (HWC-BS).

PREPARATION OF MULTIPOINT COVALENTLY ATTACHED CHYMOTRYPSIN DERIVATIVES

The fact that enzyme-support multipoint attachment may exert an important stabilizing effect on insolubilized enzyme derivatives is commonly stated.[1] These "more rigid" enzyme molecules "must" be more resistant to conformational changes induced by heat, organic solvents, or any other distorting agent. However, the practical development of this idea has only succeeded in the case of complex enzyme copolymerization, it has been poorly developed in the simpler and more interesting enzyme insolubilization by covalent attachment on preexisting porous supports.

It is evident that a very intense multiinteraction process between two very complex structures (the enzyme and the porous support) is not very easy to accomplish. However, we have been able to develop such strategies for enzyme stabilization through (1) a careful selection of the "*immobilization-stabilization system*" (the attachment of enzymes, through their amine groups, to very dense monolayers of simple aldehyde groups moderately secluded from support surfaces)[1] and (2) a detailed study of the variables that control the intensity of this enzyme-support multiinteraction (e.g., we have found that this intensity increases with the surface density of active groups, with pH, with temperature, with the contact time between the immobilized enzyme and the activated support, and so on).[2] Using this immobilization-stabilization strategy, we have been able to prepare a number of interesting derivatives of several industrial enzymes (trypsin, lipase, penicillin G acylase). In every case we have been able to find "compromise solutions" that allow us to prepare nondistorted, high loaded, and highly stabilized enzyme derivatives.

By using this same stabilization strategy, we have also been able to prepare very active and very stable chymotrypsin-agarose derivatives.[3] The intensity of the multiinteraction process between the amine groups of chymotrypsin and the activated supports followed the same trend commented on above. Our best derivatives preserved 70% of intrinsic catalytic activity corresponding to soluble enzyme, and they had 100 mg of pure chymotrypsin insolubilized per milliliter of catalyst. In addition, these derivatives were more than 60,000-fold more stable than native chymotrypsin in experiments of inactivation against heat in the absence of autolysis phenomena, and similar dramatic stabilizations were also found against other inactivating agents (urea, organic cosolvents). These industrial parameters, individually and as a whole (activity, loading, and stabilization), are much better than any other ones reported in literature for insolubilized chymotrypsin derivatives. Therefore, our multipoint-attached chymotrypsin derivatives offer important industrial prospects for use as industrial catalysts in any type of reaction media including biphasic systems.

STABILITY OF FULLY DISPERSED CHYMOTRYPSIN IN WATER SATURATED WITH ORGANIC SOLVENTS

We have tested the stability of our multipoint attached chymotrypsin derivative in the presence of saturating concentrations of two different organic solvents, ethyl acetate ($\log P = 0.65$) and dichloromethane ($\log P = 1.26$) (FIGURE 1). For comparative purposes, we have also tested the behavior of soluble chymotrypsin. In these experiments of "short-term apparently irreversible inactivation," both chymotrypsin structures behave very differently. The multipoint attached enzyme molecule, much more stable than native enzyme against heat, also is much more stable against the deleterious effect of organic solvents. In fact, this enzyme derivative retains 100% activity after incubation with saturating concentration of both solvents for 40 days. In contrast, soluble enzyme is poorly stable against mainly the more apolar solvents (e.g., the half-life in dichloromethane is 10 days, and it is still reduced to 6 days in chloroform.[4] Since we have not observed aggregation phenomena during incubation of soluble enzyme in these saturating solutions of organic solvents, we can assume its behavior is like that of every nonstabilized and nondistorted chymotrypsin derivative.

Hence, we can conclude that additional stabilization of chymotrypsin derivatives (e.g., by multipoint covalent attachment) is necessary if we wish to use this enzyme as an industrial catalyst in biphasic systems with apolar solvents.

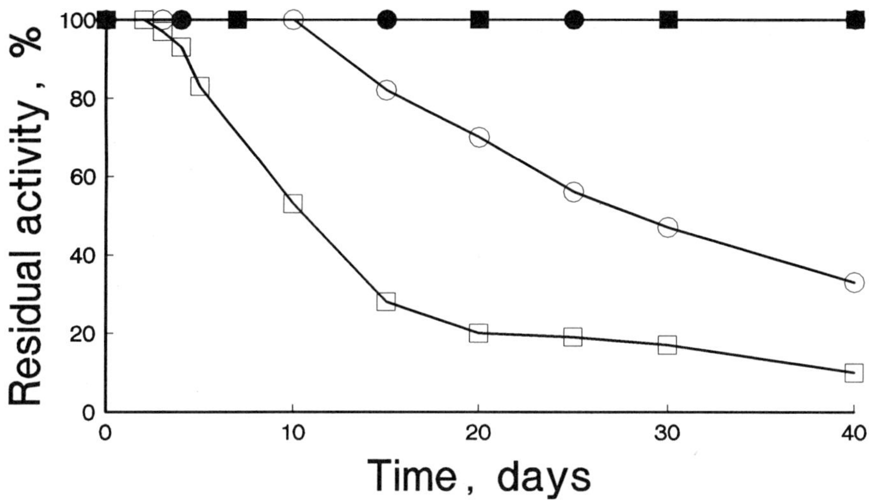

FIGURE 1. Stability of fully dispersed chymotrypsin in water saturated with organic solvents. Short-term apparently irreversible inactivation of soluble chymotrypsin (empty figures) and multipoint derivative (solid figures). Soluble and immobilized chymotrypsin were suspended in 0.1 M phosphate aqueous solution saturated with a slight excess of organic solvents: ethyl acetate (circles) and dichloromethane (squares). The solutions and suspensions were incubated at 25°C without stirring. Catalytic activities were measured by spectrophotometric assay of the hydrolysis of BTEE (benzoyl tyrosyne ethyl ester) in fully aqueous medium (0.1 M phosphate buffer pH 7.0). Activity is expressed as percentage, taking as 100% the activity of the aqueous solution of enzyme before saturating it with organic solvent.

STABILITY OF FULLY DISPERSED CHYMOTRYPSIN IN BIPHASIC SYSTEMS UNDER STIRRING

Immobilization inside a porous structure seems to protect insolubilized enzyme molecules from dramatic interactions with solvent interfaces. In FIGURE 2 we show that our multipoint attached chymotrypsin derivative remains fully stable after long-term incubation in biphasic systems under stirring. Hence, even in very low water content systems, immobilized chymotrypsin only suffers the effect of the small-concentration organic solvents that saturate the aqueous phase. As commented on above, this effect is almost negligible for our multipoint attached chymotrypsin derivatives.

In contrast, in FIGURE 2 we also show the dramatic effect of hydrophobic interfaces on the stability of nonimmobilized chymotrypsin. In fact, the half-life of soluble enzyme is reduced from 10 days (in saturating solutions of dichloromethane) down to 15 minutes in a 1:10 water/dichloromethane biphasic mixture under stirring. Interestingly and similarly, the more polar solvents (e.g., ethyl acetate) exert a less deleterious effect on the structure of native chymotrypsin. This behavior is exactly opposite to the trend usually reported in the literature[5] and allows us to open a new discussion on the general validity of such a trend.[4]

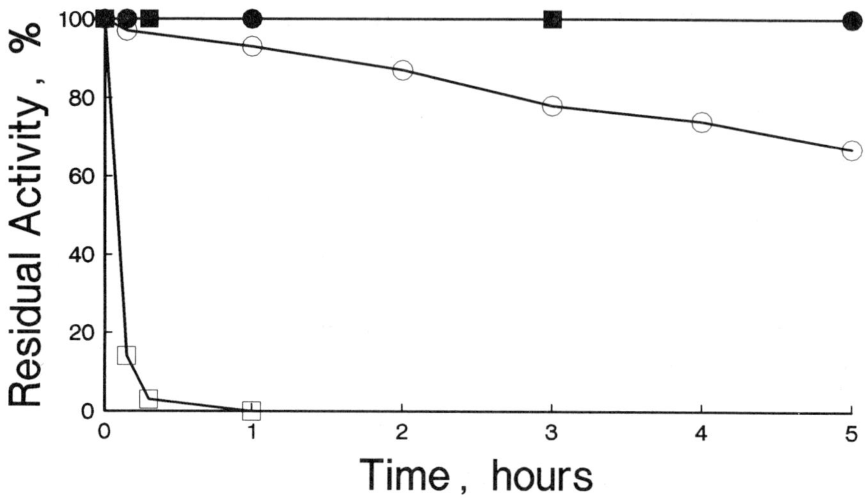

FIGURE 2. Stability of fully dispersed chymotrypsin under stirring. Short-term, apparently irreversible inactivation of soluble enzyme (empty figures) and multipoint derivative (solid figures) under vigorous stirring. Soluble and immobilized chymotrypsin was suspended in biphasic systems containing 0.1 M phosphate buffer and organic solvent (volume ratio 1:10), ethyl acetate (circles), and dichloromethane (squares). These biphasic mixtures were incubated at 25°C with magnetic stirring at 1000 rpm; the stirrer bar length was 90% of the vessel diameter. Enzymatic assays were performed as described in FIGURE 1.

REACTIVATION STRATEGIES

In addition to these interesting stability results, we have proven the possibility of reactivating both stabilized and nonstabilized chymotrypsin derivatives after a partial or total inactivation in the presence of organic solvents (cosolvents or immiscible solvents).[6] When the incubation in organic solvents is carried out under "chemically inert" conditions (e.g., room temperature, neutral pH), these fully dispersed enzyme derivatives are strongly protected from irreversible inactivation even under "extreme conditions" (e.g., very high concentrations of cosolvents, very low water activity–biphasic systems and so on). Under these conditions, enzyme aggregation is impossible (enzyme molecules are covalently fixed on the support and they cannot suffer any intermolecular process) and chemical modifications are almost negligible. Hence, inactivation by organic solvents only promotes distortions on the 3-D structure of these enzyme molecules and these distortions must be completely reversible. However, we have also tested whether simple washings with fully aqueous media were enough to get complete reversibility.[6] We have found that some kinetic hindrances may prevent the distorted enzyme from complete reactivation simply by long-term reincubation in fully aqueous media. However, we have demonstrated that unfolding of those distorted enzymes with high urea concentrations permits a further, more complete and reproducible reactivation in fully aqueous media. In this way we have been able to get excellent reactivations of both stabilized and nonstabilized chymotrypsin derivatives after complete inactivations in low water activity–biphasic systems (e.g., $a_w = 0.4$) or in high concentrations of organic cosolvents (e.g., 90% dioxane).[6] In addition, we have observed that reactivation is more complete (e.g., 90% of initial activity of the enzyme derivative) and faster for multipoint covalently attached chymotrypsin-agarose. It seems that the existence of a number of "reference points" (different lysine groups belonging to distinct protein chains but exactly fixed on the support at a distance corresponding to the native structure) may promote an easier and more reproducible refolding process.

ACTIVITY OF CHYMOTRYPSIN DERIVATIVES IN BIPHASIC SYSTEMS

Hydrolytic Reactions

First of all, we have used a model reaction, hydrolysis of benzoyl tyrosine ethyl ester (BTEE), in order to test the overall possibilities of our multipoint attached derivatives as catalysts in the presence of saturating concentrations of very different organic solvents (from ethyl acetate, $\log P = 0.65$, to trichloroethane, $\log P = 2.8$). In addition, this model reaction allowed us to compare the behavior of our derivatives in organic solvents with that in fully aqueous media, which is not possible for synthetic processes.

In general, activity values were fairly good—from 35 to 75% of activity corresponding to native chymotrypsin in fully aqueous media.[4] That is, the combined effect of immobilization, multipoint attachment, and presence of organic solvents only promotes a very slight lost of catalytic activity.

Even at the nonoptimal conditions in which the model tests were carried out (e.g., very low substrate concentration), the combined effect of the possible distortions of 3-D structure of the enzyme promoted by the solvents plus their possible direct inhibitory effects (e.g., hydrophobic adsorption on the hydrophobic subsites of the active center) only promotes a very slight decrease of catalytic activity. Thus, we can conclude that the exact structure/properties of the chymotrypsin active center are almost fully preserved in the presence of very different organic solvents. Therefore, we can also assume that the catalytic activity of our fully dispersed and multipoint-attached derivatives in any type of biphasic systems (at least at high water activity values) will be practically the highest obtainable of any type of chymotrypsin derivatives for any type of enzymatic process (hydrolysis, synthesis, transesterifications, and so on).

Synthetic Processes

In addition we have also directly tested the activity of our derivatives in different synthetic processes in low water content biphasic systems and different water activity values.[7–9] In every case, we have obtained very interesting synthetic activity values, which compare fairly with the scarce reports in the literature for similar synthetic processes catalyzed by other fully dispersed chymotrypsin derivatives.[10] For instance, the direct condensation of acetyl tryptophane and ethanol in 3-pentanone to yield acetyl tryptophane ethyl ester was 2.5 μmol of synthetic product per minute per milliliter of catalyst.

In addition, synthetic activity values obtained for our fully dispersed and multipoint attached derivatives are much higher than those reported for "aggregate chymotrypsin," as will be discussed below.

PEPTIDE SYNTHESIS IN HIGH WATER CONTENT–BIPHASIC SYSTEMS BY CHYMOTRYPSIN-AGAROSE DERIVATIVES

Although the use of very low water content and/or anhydrous systems is now becoming very popular in enzyme technology, we think that most of the practical advantages of such systems can also be achieved by using simpler high water content–biphasic systems, e.g., handling of insoluble substrates, recovery of hydrophobic products, and so on. In addition, these HWC-BSs, present additional and very interesting practical advantages: (1) We can perform condensation processes involving both very high concentrations of hydrophilic and hydrophobic substrates. (2) Hydrophilic substrates can be used in a very large excess (e.g., to improve synthetic yields) but the real consumption of substrate is only the stoichiometric one (the excess of reagent may be easily reused because it becomes "immobilized" in the aqueous phase with the enzyme derivative). (3) Large amounts of buffer can be used. (4) When necessary, pH can be very easily controlled and the reaction mixture can be accurately titrated (e.g., we have directly measured pH values in 1:1 water-solvent mixtures under vigorous stirring and these values are exactly the same as those shown in fully aqueous solutions).

In TABLE 1 we show interesting parameters corresponding to the synthesis of a hydrophobic peptide AcPhe-LeuNH_2 from a hydrophobic acyl donor AcPhe-OMe and an excess of hydrophilic nucleophile LeuNH_2 catalyzed by our multipoint attached chymotrypsin derivatives in HWC-BSs. Synthetic yields obtained with different organic phases are very good (higher than 90%) and similar to ones reported for the same process catalyzed by "aggregate chymotrypsin" in anhydrous media.[11] In contrast, synthetic activities obtained for our fully dispersed derivatives in HWC-BSs (4.8 μmol of peptide/minute per mg of immobilized chymotrypsin and 480 μmol of peptide/minute per ml of catalyst) are dramatically higher than those reported for a very similar process (synthesis of Z-Phe-LeuNH_2) catalized by aggregate chymotrypsin in anhydrous media (approx. 0.03 μmol/minute per mg of chymotrypsin precipitated on celite or 1 μmol/minute per ml of catalyst).[11] In addition, by using different organic phases [organic solvents with different polarity (log P)], different nucleophile concentrations (e.g., up to 500 mM), and different composition of the aqueous phase (e.g., presence of cosolvents), we can modulate very different and important industrial parameters corresponding to these kinetically controlled synthetic processes (synthetic yields, synthetic activities, solubilities and partitions of substrates and products, and so on.[12] These results clearly indicate that some popular assumptions about the use of enzymes in organic solvents may not be valid. For example, there may be a number of cases in which the use of "enzyme aggregates" in anhydrous solvents may be clearly disadvantageous compared to other, simpler industrial reaction systems. In these cases, e.g., "kinetically" controlled synthesis of hydrophobic peptides, the use of anhydrous conditions is not strictly necessary and the reaction may also be carried out in HWC-BSs. Hence the use of "enzyme aggregates" as biocatalysts may be a limitation (they are poorly active and require the use of these unnecessary anhydrous conditions) rather than an additional industrial advantage.

TABLE 1. Kinetically Controlled Synthesis of AcPhe-LeuNH_2 Catalyzed by Chymotrypsin-Agarose in High Water Content Biphasic Systems[a]

	Maximum Yield	V_s	V_h	V_s/V_h	P[b]	log P[c]
Ethyl acetate	90	4.8	0.5	9	11.5	0.65
Dichloromethane	91	1.2	0.12	10	15	1.25
Chloroform	92	0.27	0.02	13	49	2.01

[a]Experimental: reaction mixture consisted of aqueous phase (0.1 M phosphate buffer) and organic solvent in 1:1 volume ratio. Total concentration of substrates was: 50 mM acyl donor (AcPheOMe) and 150 mM nucleophile (leucinamide). At different times, aliquots from both phases were withdrawn and analyzed by high-performance liquid chromatography. Maximum yield given in percent conversion of the acyl donor. Rates given in micromoles of dipeptide per minute and milligram of immobilized chymotrypsin.

[b]Partition of AcPheOMe between organic and aqueous phase.

[c]log P values of the solvents taken from literature.

CONCLUDING REMARKS

In this paper, we have briefly commented on a number of studies directly focused on improving and pointing out the very broad possibilities of fully dispersed enzyme derivatives as industrial catalysts in biphasic systems. Our multipoint covalently attached chymotrypsin derivatives appear to be very useful industrial catalysts in any type of biphasic system. These excellent practical prospects arise as a consequence of the combined effect of:

1. The establishment of an original strategy for immobilization-stabilization of chymotrypsin by multipoint covalent attachment on preexisting porous supports.
2. The subsequent stabilization of multipoint covalently attached chymotrypsin derivatives against the presence of organic solvents.
3. The role of immobilization inside porous structures on the prevention of enzyme molecules from very deleterious enzyme-solvent interphase interactions.
4. The development of enzyme reactivation strategies that are only applicable to fully dispersed immobilized enzyme derivatives.

The industrial prospects of our chymotrypsin derivatives have also been demonstrated in terms of a set of excellent industrial parameters for the synthesis of an interesting dipeptide bond (phenylalanine-leucinamide). We have been able to show that these synthetic processes may be much better designed by using simple reaction systems (e.g., our fully dispersed immobilized enzymes as catalysts in high water content–biphasic media) than by using much more sophisticated ones (e.g., "enzyme aggregates" in anhydrous media).

ACKNOWLEDGMENTS

Agarose gels were obtained from HISPANAGAR S:A. (Burgos, Spain), courtesy of Rafael Armisen. The authors are grateful to Ms. Maria del Carmen Ceinos for skillfull technical assistance.

REFERENCES

1. Guisan, J. M. 1988. Agarose-aldehyde gels as supports for immobilization-stabilization of enzymes. Enzyme Microb. Technol. **10:** 375–382.
2. Blanco, R. M., J. J. Calvete & J. M. Guisan. 1989. Immobilization-stabilization of enzymes: variables that control the intensity of the trypsin (amine)-agarose (aldehyde) multipoint attachment. Enzyme Microb. Technol. **11:** 353–359.
3. Guisan, J. M., A. Bastida, C. Cuesta, R. Fernandez-Lafuente & C. M. Rosell. Immobilization-stabilization of α-chymotrypsin by covalent attachment to aldehyde-agarose gels. Biotechnol. Bioeng. **38**. (In press.)
4. Blanco, R. M., P. J. Halling, A. Bastida, C. Cuesta & J. M. Guisan. Effect of immiscible organic solvents on activity/stability of native chymotrypsin and immobilized-stabilized derivatives. Biotechnol. Bioeng. **38**. (In press.)
5. Reslow, M., P. Adlercreutz & B. Mattiasson. 1987. Organic solvents for bioorganic synthesis. 1. Optimization of parameters for a chymotrypsin catalyzed process. Appl. Microbiol. Biotechnol. **26:** 1–8.
6. Guisan, J. M., A. Bastida, R. M. Blanco, C. Cuesta & R. Fernandez-Lafuente. 1992. Insolubilized enzyme derivatives in organic solvents: mechanisms of inactivation and strategies for reactivation. *In* Proceedings "Fundamentals of Biocatalysis in Nonconventional Media," Noordwijkerhout, the Netherlands, April.
7. Blanco, R. M., J. M. Guisan & P. J. Halling. 1989. Agarose-chymotrypsin as a catalyst for peptide and amino acid ester synthesis in organic media. Biotechnol. Lett. **11:** 811–816.
8. Blanco, R. M., J. L. L. Rakels, J. M. Guisan & P. J. Halling. Effect of thermodynamic water activity on amino-acid ester synthesis catalyzed by agarose-chymotrypsin in 3-pentanone. (Submitted.)
9. Blanco, R. M., J. M. Guisan & P. J. Halling. The equilibrium and kinetics of *N*-acetyl-tryptophan phenylethyl ester synthesis by agarose-chymotrypsin in organic media. (Submitted.)
10. Turkova, J., T. Vanek, R. Turkova, B. Veruovic & V. Kubanec. 1982. Chymotrypsin adsorbed to a polyethylene terephtalate support (Sorsilen) as a suitable model for the investigation of enzymatically catalyzed organic synthesis in nonaqueous media. Biotechnol. Lett. **4:** 165–170.
11. Clapes, P., P. Adlercreutz & B. Mattiasson. 1990. Enzymatic peptide synthesis in organic media: nucleophile specificity and medium engineering in α-chymotrypsin-catalyzed reactions. Biotechnol. Appl. Biochem. **12:** 376–386.
12. Bastida, A., R. M. Blanco & J. M. Guisan. Peptide synthesis in high water content–biphasic systems by chymotrypsin-agarose derivative. (In preparation.)

Purification and Characterization of a Pepstatin-Insensitive, Thermostable, Extracellular Acid Proteinase from a Bacterium[a]

MARK PRESCOTT, KEITH PEEK,[b] ELIZABETH PRENDERGAST, AND ROY M. DANIEL

Thermophile Research Unit
University of Waikato
Private Bag 3105
Hamilton, New Zealand

[b]*Pacific Enzymes Limited*
Post Office Box 1496
Hamilton, New Zealand

INTRODUCTION

Aspartate proteinases (also referred to as acid or carboxyl) represent one of the four major classes of proteinase. Work on members of this group has centered largely around those enzymes isolated from mammalian, fungal, or plant sources. Some of these have proven to be of great economic importance, for example, the renneting enzyme, chymosin.

The first report on the isolation of an acid proteinase from a bacterium was reported as recently as 1987.[1] There have been only three further descriptions,[2,3] one from an archaebacterial source.[4] Hence, the available information on bacterial acid proteinases is somewhat limited.

It is against this background that we decided to investigate the properties of acid proteinases from bacterial sources. In particular, we were interested in studying those that possessed some degree of thermostability. In this preliminary report, we describe some characteristics of an enzyme isolated as the result of a screen for such sources of acid proteinase.

MATERIALS AND METHODS

Proteinase assays were performed at 60°C in 50 mM sodium formate, pH 3.0 and 5 mM $CaCl_2$ using bovine hemoglobin (0.2% w/v) as substrate. Enzyme activity was monitored by determination of trichloroacetic acid–soluble products at 280 nm.

Samples of hot pool water were enriched for acid proteinase–producing organisms in liquid medium adjusted to pH 3.0 containing 0.2 g/l $(NH_4)_2SO_4$, 0.5 g/l

[a]This work was made possible with the financial support of Pacific Enzymes Limited.

$MgSO_4$, 0.1 g/l $CaCl_2 \cdot H_2O$, 3.0 g/l KH_2PO_4, 0.01 g/l yeast extract, and 2.0 g/l casein, and isolated on agar plates containing the same medium. Trace elements and vitamins were initially included but later found not to be required for growth.

Preparative quantities of isolate Wai 21a were grown in the same medium without casein (600 l) supplemented with 2 g/l glucose and aeration (200 l/minute) to the stationary phase, where proteinase accumulation was at its greatest. Cells were harvested by centrifugation, and the cell-bound proteinase removed by soaking the cells in 30 l fresh medium overnight at 4°C followed by the removal of cells by centrifugation.

The released proteinase was purified to homogeneity by the following chromatography steps. Fast Flow S-Sepharose at pH 3.0 and stepwise elution with 0.2 M NaCl. Phenyl Sepharose at pH 3.0 and elution with a gradient of ethoxyethanol. Elution from Mono Q at pH 4.7 with a gradient of 0–350 mM NaCl.

RESULTS AND DISCUSSION

Isolation of Acid Proteinase–Producing Organisms

New Zealand hot pools in the range pH 1.5–5.0 and temperature range 47°C–97°C were sampled and screened for organisms with the capability of producing extracellular acid proteinases. Enrichment at pH 3.0 on casein-containing medium yielded a total of 14 separate isolates possessing a range of growth temperature optima between 55°C and 70°C. All of the isolates were spore-forming bacilli.

In general, the isolates obtained produced only small amounts of extracellular proteinase when grown in liquid culture. In addition, the activity was not freely available and remained cell bound until released by extraction at 4°C. Our attempts at obtaining reproducible release during the growth cycle were unsuccessful. However, up to 45% of acid proteinase could be retrieved when the harvested cells were soaked in fresh medium overnight at 4°C.

Characterization of Wai 21a Acid Proteinase

We chose to pursue further the properties of an acid proteinase produced by an isolate enriched at 60°C, referred to here as Wai 21a. Homogeneous Wai 21a enzyme (0.6 mg) was purified with a yield of 50% from 1 kg (wet weight) of cells.

As might be expected for an extracellular enzyme isolated from an organism growing at acid pH, Wai 21a proteinase displays optimal activity at pH 3.0 when assayed using a hemoglobin substrate. Fifty percent of the activity is retained at pH 1.8 and 4.1. Although some cysteine proteinases express activity under acidic conditions and some aspartate proteinases are active at near neutral pHs, the low pH optimum for the proteinase suggests that it belongs to the aspartate class of proteinases. This conclusion was supported by the finding that Wai 21a proteinase activity was not inhibited by a range of serine, metallo, and cysteine inhibitors and inactivators (results not shown).

The aspartate class of proteinases are by definition susceptible to inhibition by the microbial proteinase inhibitor pepstatin and the active site-directed affinity

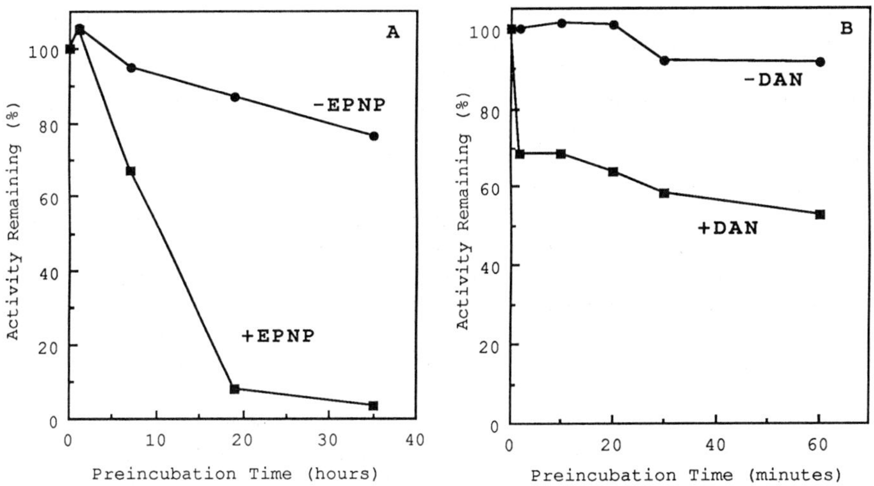

FIGURE 1. The inhibition of Wai 21a acid proteinase by EPNP and DAN. Wai 21a proteinase activity was preincubated at room temperature with (+) or without (−) 20 mg/ml EPNP at pH 3.5 with vigorous shaking[7] (panel A) or 12.7 mM, 0.01 mM $CuCl_2$[9] (panel B) for the times before assaying for residual activity.

labels DAN (diazoacetyl-DL-norleucine methyl ester) and EPNP (1,2-epoxy-3-(*p*-nitrophenoxy)propane).[5] The latter inhibit by inactivating either one or the other of the catalytic aspartate residues. Although aspartate proteinases are particularly susceptible to pepstatin[6] (e.g., pepsin, K_i = 1 nM), Wai 21a activity was found to be resistant to a concentration (5 μM) that produced complete inhibition of pepsin. Such behavior towards pepstatin is rare and likens Wai 21a to the *Scytalidium* subclass of acid proteinases from the Basidiomycetes.[10]

The presence of an aspartate residue at the active site of Wai 21a is implied by the inhibition of activity by EPNP (FIGURE 1, panel A).[7] This property distinguishes

TABLE 1. A Comparison of Wai 21a Acid Proteinase with Those from Other Bacterial Sources[a]

	Enzyme Source				
Property	Wai 21a	*Xanthomonas*[1]	*Pseudomonas*[2]	*Bacillus*[3] MN-32	*Sulfolobus*[4]
pH optima	3.0	2.7	3.0	3.5	2.0
Isoelectric point	3.8	4.0	3.2	3.2	ND
Molecular weight (SDS-PAGE)	50 kDa	41 kDa	43 kDa	41 kDa	51 kDa
Pepstatin inhibition	−	−	−	−	+
DAN inhibition	+[b]	−	−	−	−
EPNP inhibition	+[b]	+	+	−	−
Thermostable	Yes	ND	ND	Yes	Yes

[a]A + indicates inhibition; a − indicates lack of inhibition. Superscript numbers refer to references. ND is not determined. SDS-PAGE is sodium dodecyl sulfate-polyacrylamide gel electrophoresis.

[b]See FIGURE 1.

it from all members of the *Scytalidium* subclass with the sole exception of the EPNP-sensitive proteinase B from *S.lignocolum.*[8]

This pattern of inhibition by pepstatin and EPNP is similar to that reported for two of the three other acid proteinases so far isolated from bacteria (TABLE 1). The thermostable proteinase from bacillus MN-32 was found to be resistant to EPNP (TABLE 1). However, the partial inhibition of Wai 21a proteinase by DAN distinguishes it from other acid proteinases so far described (FIGURE 1, panel B). This observation requires further investigation but could be rationalized if the inhibitor were labeling an aspartate residue, not at the active site but at a subsite in the long binding cleft characteristic of this class of enzymes.[10] The determination of the kinetic parameters for the inhibited enzyme will confirm this.

TABLE 1 indicates that the physical properties of Wai 21a proteinase with the exception of a slightly higher relative molecular weight of 50 kilodaltons (kDa) are similar to those previously described.

Wai 21a displays thermostable properties. In the presence of 5 mM Ca^{++} and at its optimal pH for activity, the half-life of the homogeneous enzyme at 75°C and 60°C was 34 minutes and 9.65 hours respectively. Although pretreatment of the enzyme with EDTA reduced the half-life of the enzyme activity at 75°C to 15 minutes, it appears that the enzyme retains a basal level of thermostability that can be enhanced in the presence of Ca^{++} ions. Interestingly, the activity is most thermostable between pH 4 and 5 and not at its optimal pH for enzyme activity (results not shown). In addition, Ca^{++} produces a more marked enhancement in thermostability at this higher pH. On the basis of the limited data available for the thermostable acid proteinases described by Murao (TABLE 1), Wai 21a would appear to possess a similar degree of thermostability.

In conclusion, this study indicates that bacterial acid proteinases are more common than the sparsity of literature would suggest. In addition, although the physical properties of Wai 21a are similar to those of enzymes already described from bacterial sources, its EPNP sensitivity distinguishes it from the thermostable bacillus MN-32, and its partial sensitivity to DAN sets it aside from those other described. Work continues to determine the implications of these findings.

REFERENCES

1. ODA, K., T. NAKAZIMA, T. TERASHITA, K. SUZUKI & S. MURAO. 1987. Agric. Biol. Chem. **51**(11): 3073–3080.
2. ODA, K., M. SUGITANI, K. FUKUHARA & S. MURAO. 1987. Biochim. Biophys. Acta **923:** 463–469.
3. MURAO, S., K. OHKUNI, M. NAGAO, K. ODA & T. SHIN. 1988. Agric. Biol. Chem. **52**(6): 1629–1631.
4. LIN, X & J. TANG. 1990. J. Biol. Chem. **265**(3): 1490–1495.
5. KAY, J. 1985. *In* Aspartic Proteinases and Their Inhibitors. V. Kostka, Ed.: 1–19. Walter de Gruyter. Berlin, Germany.
6. TANG, J. 1976. Trends Biochem. Sci. **1**(9): 205–208.
7. TANG, J. 1971. J. Biol. Chem. **246**(14): 4510–4517.
8. MURAO, S. & K. ODA. 1985. *In* Aspartic Proteinases and Their Inhibitors. V. Kostka, Ed.: 379–399. Walter de Gruyter. Berlin, Germany.
9. RAJAGOPALAN, T. G., W. H. STEIN & M. STANFORD. 1966. J. Biol. Chem. **241:** 4295–4297.
10. TANG, J., M. N. G. JAMES, I. N. HSU, J. A. JENKINS & T. L. BLUNDELL. 1978 Nature **271:** 618–621.

Modulation of Lipolytic Activity and Stability by Using Sugars

SYLVAIN LAMARE, JOSE-MARIA SANCHEZ-MONTERO, AND MARIE-DOMINIQUE LEGOY

URA 1442 du CNRS
Laboratoire de Technologie Enzymatique
Université de Compiègne,
Compiègne, France

Lipases have been extensively studied in recent years because of their value in synthesis and their potential applications.[1] Their prime function is to catalyze the hydrolysis of esters and especially triacylglycerols. The catalytic process is heterogeneous since it involves a water/lipid interface.[2]

Candida rugosa lipase is known to remain active in organic phases where water availability is restricted.[3] The degree of water molecule association can be modified by the presence of additives such as sugars. These compounds are able to modulate and structure the microenvironment of the biocatalyst, and their influence on the stability has been studied for different enzymes.[4–8]

In the present paper, the effect of hydration state of the biocatalyst on the activity and stability of *Candida rugosa* lipase is investigated for hydrolysis reactions.

MATERIALS AND METHODS

Lipase (E.C. 3.1.1.3.) from *Candida rugosa* (formerly *C. cylindracea*) (type VII) and pNP acetate were obtained from Sigma (St. Louis, Mo.). Tributyrin was purchased from Aldrich-Chimie.

Lactose and ethanol were obtained from Carlo Erba, glucose, sorbitol, and sucrose from Prolabo, and fructose from Sigma.

Buffer salts were of high purity.

Dialysis of the Enzymatic Preparations

The commercial preparation of the lipase was dialyzed in order to remove lactose and small molecules from the enzymatic preparation. Concerning the experiments on the modulation of lipase activity, powdered lipase was mixed with double-distilled water and transferred to a dialysis bag. The bag was submerged in a vessel containing double-distilled water and mechanically stirred at 4°C. After 6, 24, or 52 hours of dialysis, the enzymatic preparation was lyophilized for 48 hours using a SERAIL RP2V freeze-dryer.

Concerning the study of the modulation of lipase stability, dialysis was performed using a hollow fiber cartridge AMICON (H1P10-43 type, nominal cutoff 10,000 MW, the same as the dialysis bag). The enzymatic preparation was used after 70 hours of dialysis.

Addition of Sugars to Dialyzed Enzymatic Preparations

Different sugars (lactose, sucrose, fructose, glucose, and sorbitol) were added to the enzymatic preparation after 24 hours of dialysis followed by lyophilization. The sugars were added to the lipase preparation dialyzed. In a typical experiment, 0.08 g of "purified" lipase was mixed with 15 mL of double-distillated water and placed in a dialysis bag. The bag was submerged in a vessel containing a solution of sugar (5 g/mL) and mechanically stirred at 4°C for 5 or 24 hours for lactose, or 24 hours for the other sugars. The lipase mixture was then lyophilized as described above.

Sugar Determination by High-Performance Liquid Chromatography

Concentrations of lactose and other sugars were monitored by HPLC using a Waters system (Milford, Mass.) composed of a 510 pump, a R 410 refractive index detector, and a carbohydrate column (3.9 mm × 30.0 cm) (Waters). The mobile phase was 75:25 (v/v) acetonitrile:water (flow rate, 1.5 mL/minute).

For each lipase preparation, a double-distilled water solution of the lyophilized enzyme was prepared and analyzed to determine the sugar concentration.

Concentrations were calculated from peak areas after calibration with standard solutions.

Protein Determination

The protein content of each lipase preparation was determined using the Bradford method.[9] In a typical experiment, 0.1 ml of a 4-mg/mL lipase solution was mixed with 1 mL Bradford reagent and the protein concentration was determined spectrophotometrically at a wavelength of 595 nm by using a calibration curve.

Assay of Lipase Hydrolytic Activity

Enzyme activity was assayed using a Metrohm pH-stat (Dosimat E 285 and Titrator E 526) at 30°C and pH = 7.0 with tributyrin as substrate. In a typical experiment, 10 mL of 0.2-M phosphate buffer containing 0.15-M tributyrin and 0.154-M NaCl, pH = 7.0, was placed in the 30°C thermostated cuvette of the pH-stat (tributyrin emulsion was previously sonicated for 2 minutes at 50 W with a Branson B-12 sonifier). The reaction mixture was shaken vigorously for 20 minutes at 30°C, then 2 mL of lipase solution (4 mg/mL, 0.2-M phosphate buffer, pH 7.0) was added. The butyric acid produced by the reaction was automatically titrated by addition of 0.1-M NaOH and the initial hydrolytic activity was calculated. Experiments were performed in duplicate or triplicate and the maximum deviations from the mean were less than 5%. One unit of lipase activity is defined as the amount of enzyme needed to produce 1 μmol of butyric acid per hour at 30°C.

For the stability studies, pNPacetate was used as substrate (modified from Reference 10). In a typical experiment, 40 μl of an enzymatic solution (0.525 mg of protein) was added to 960 μl of a 0.5-mM pNPacetate solution in 50-mM pH 7.0 phosphate buffer containing 1% (v/v) ethanol for the solubilization of the substrate. Hydrolytic activity was followed at 25°C with spectrophotometer (HITACHI U 2000) at 400 nm. The slopes of the resulting curves were determined every 30 seconds, and the mean of the values was taken into account for the calculation of the reaction rates.

RESULTS AND DISCUSSION

Hydrolytic Lipase Activity

The lipases are characterized by a common catalytic feature: the involvement of a lipid-water interface in the catalytic process.[11] This unique feature of interfacial catalysis provides the lipases an inherent affinity for hydrophobic environments and distinguishes them from other hydrolytic enzymes.

TABLE 1. Hydrolytic Activity of the Lipase Preparations

Enzyme Preparation	Hydrolytic Activity (μmol/hour · mg of enzyme preparation)	Percent of Sugar (w/w)	Percent of Protein (w/w)	Hydrolytic Activity (μmol/hour · mg of protein)
L	326	38.4% lactose	15	2170
D6	384	11.3% lactose	59	650
D24	406	0	65	624
D52	438	0	77	569
D24-L5	256	69.6% lactose	18	1422
D24-L24	250	73.8% lactose	15	1667
D24-Sorb	226	75.7% sorbitol	17	1330
D24-Gluc	192	78.7% glucose	11	1745
D24-Suc	242	75.7% lactose	14	1728
D24-Fruc	244	78% fructose	13	1877

Tributyrin is practically immiscible in water and therefore forms an emulsion in aqueous media. However it is hydrolyzed at a higher rate than solid or poorly water soluble lipids. A variety of lipase preparations were studied: commercial lipase (L) without treatment; lipase dialysed for 6 (D 6), 24 (D 24), or 52 hours (D 52); lipase dialyzed for 24 hours with addition of lactose for 5 (D 24-L 5) or 24 hours (D 24-L 24); and lipase dialyzed for 24 hours with sorbitol (D 24-Sorb), glucose (D 24-Gluc), sucrose (D 24-Suc), or fructose (D 24-Fruc) added for 24 hours.

The results obtained for tributyrin hydrolysis are shown in TABLE 1. Longer dialysis times resulted in increased activity per milligram of the enzyme preparation. Dialysis acts as a purification step, eliminating impurities from the commercial preparation such as lactose and small peptides which can be detected by electrophoresis.[12]

However, the dialyzed enzymes show a significant decrease of hydrolytic activity per milligram of protein.

When sugars are added, the enzymatic activity per milligram of enzyme preparation is lowered because 70 or 80% of the enzyme mass is sugar. As an example, the results for 24-hour dialyzed lipase (D 24) show a hydrolytic activity of 406 μmol/hour · mg of enzyme preparation or 624 μmol/hour · mg protein. The same lipase, posttreated for 5 hours with lactose (D 24-L 5), shows a hydrolytic activity of 256 μmol/hour · mg of enzyme containing 70% lactose (w/w) and 1422 μmol/hour · mg of protein.

Addition of different sugars gives the same result: the hydrolytic activity per milligram of protein for the lipase preparations containing sugar is greater than for the lipase preparations that are free of additive.

The loss of hydrolytic activity for the D 6, D 24, and D 52 enzyme preparations is not due to denaturation since much of the activity can be restored by adding carbohydrate.

We have also tested the hydrolytic abilities of the lipase dialyzed for 24 hours (D 24) as a function of the contact time between this D 24 enzyme and lactose.

For this experiment, D 24 lipase was used. At the initial time, 2.95 mg lactose was added to 1.05 mg D 24 lipase, giving 73.8% (w/w) lactose in the lipase preparation. The hydrolytic activity was tested after 20 minutes, 1 hour, 2 hours and 3 hours of contact between the D 24 lipase and lactose. The measured activities after 20 minutes, 1 hour, 2 hours, and 3 hours of incubation time were 1760, 1620, 1830, and 1970 μmol/hour · mg of protein, respectively.

It appears, therefore, that the observed enhancement of hydrolytic capacity does not depend on the contact time between the enzyme and the sugar.

Hydrolysis was also studied with D 24 lipase tested in 10 mL reaction medium containing tributyrin and lactose (2.95 mg), glucose (3.15 mg), fructose (3.12 mg), sucrose (3.03 mg), or sorbitol (3.03 mg). The results are 1575, 1620, 1555, 1440, and 1350 μmol/hour · mg of protein, respectively. Thus, the increase in enzyme activity can be obtained to the same order of magnitude if the sugars are added directly to the reaction medium. These results cannot be attributed to the effect of sugars on the water activity, representing the availability of water, because the concentration of sugar in the reaction medium is very small.

Hydrolytic Lipase Stability

Using different preparations (L and D 70 dialyzed during 70 hours), we tested the effect of lactose on the thermal stability for four temperatures (40, 50, 60, and 70°C). FIGURES 1, 2, 3, and 4 show the evolution of the residual activity versus the time of incubation.

At 40°C an activation effect was noted for both preparations L (40% lactose w/w) and D 70 (0% lactose). The activity gain for the L preparation after two hours of incubation is significant and represents 2.25 times the initial value. For the D 70 preparation, a similar phenomenon is observed but the activity after two hours is only 1.4 times greater. At 50°C, the phenomenon of activation is still present for the L preparation, but after 45 minutes, the thermal inactivation counteracts. For the D 70 preparation, a decrease of activity is observed, corresponding to a loss of 30% after two hours. The results are totally different for higher temperatures. In all cases, the lactose seems to have a negative effect upon the activity.

At 60°C, the different curves are quite similar, but at 70°C, the denaturation is

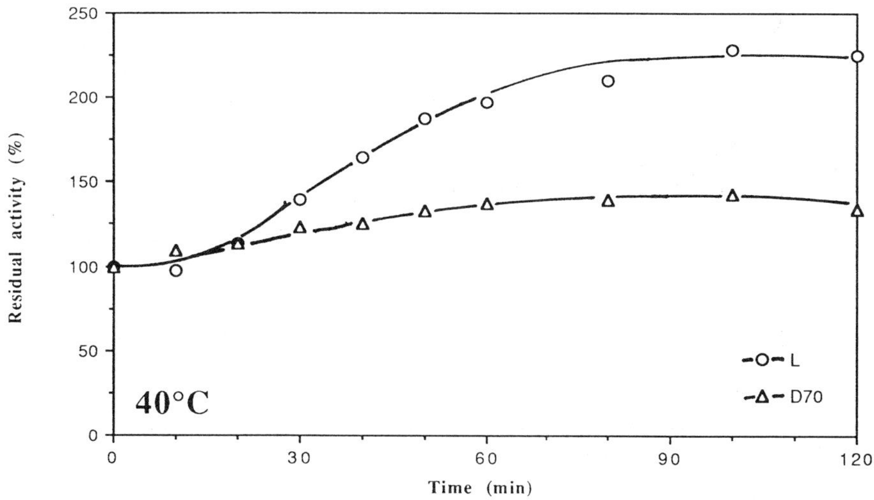

FIGURE 1. Stability of different lipase preparations for pNP acetate hydrolysis at 40°C. Lipase L (○) and lipase D 70 (△).

more important for the preparation containing lactose. The half-lives at 70°C are 5 and 15 minutes for the L and D 70 preparations respectively.

Higher temperatures (80°C, data not shown) lead to a rapid loss of activity in all cases. The half-lives are 1 and 2 minutes for L and D 70 preparations respectively.

The same experiments were done using a preparation containing 23% lactose (w/w). In all cases, the results were intermediate in comparison with the two extremes.

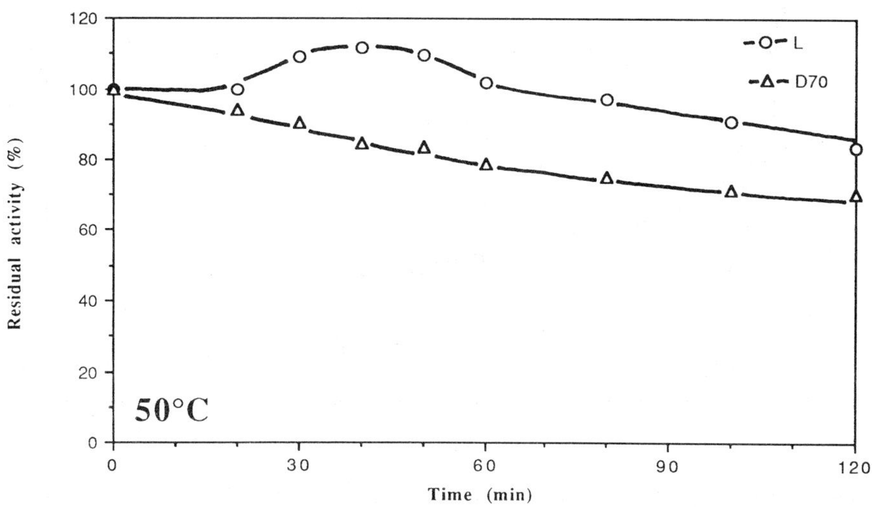

FIGURE 2. Stability of different lipase preparations for pNP acetate hydrolysis at 50°C. Lipase L (○) and lipase D 70 (△).

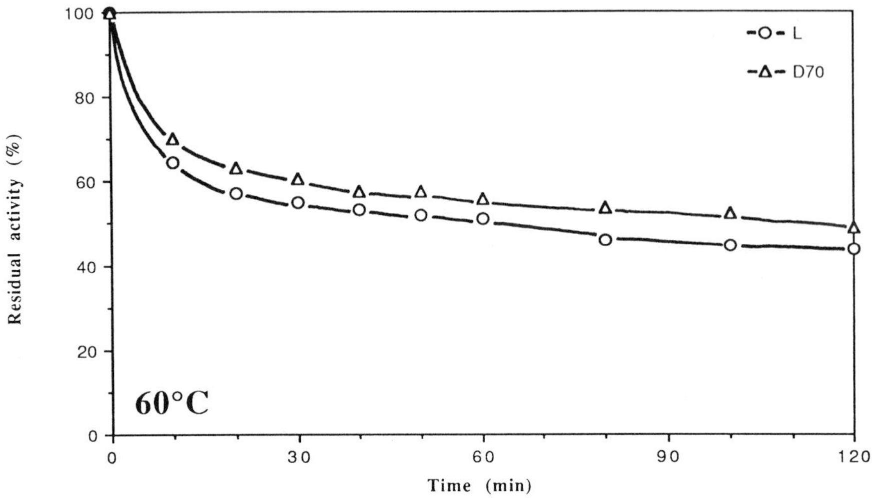

FIGURE 3. Stability of different lipase preparations for pNP acetate hydrolysis at 60°C. Lipase L (○) and lipase D 70 (△).

Other experiments were done adding carbohydrates directly to the incubation medium. Lactose, glucose, and sorbitol were tested. The added quantity was calculated to obtain an enzymatic preparation containing 75% (w/w) carbohydrates (same conditions as in TABLE 1). Surprisingly, the direct addition of glucose, sorbitol, or lactose has no positive effect on stability even at 40°C.

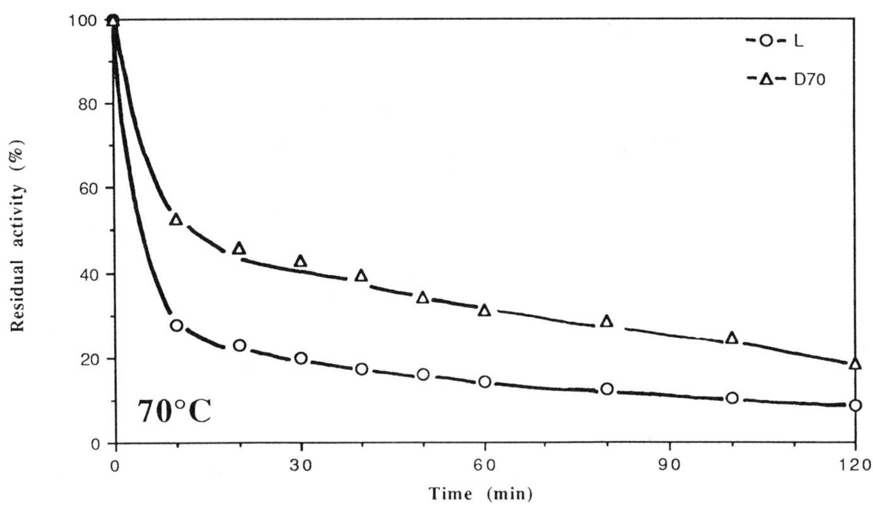

FIGURE 4. Stability of different lipase preparations for pNP acetate hydrolysis at 70°C. Lipase L (○) and lipase D 70 (△).

CONCLUSION

Carbohydrates in the microenvironment of the *Candida rugosa* lipase are able to modulate hydrolytic activity and stability. Addition of carbohydrates to the enzyme preparation or to the reaction medium produces a significant increase in hydrolytic activity for all carbohydrates tested, regardless of whether they are water activity depressors or not. Since these carbohydrates are not chemically linked to the enzyme molecules, evidently they are solubilized in the reaction medium.

The presence of lactose in the enzymatic preparation has a protective effect against thermal denaturation for temperatures in the range of 40–50°C. For higher temperatures (60°C and more), the effect is negative, leading to an important decrease of stability. Lactose increases the denaturation process. Nevertheless, the direct addition of carbohydrates (glucose, sorbitol, or lactose) to the incubation medium has no effect on stability.

Further studies are now in progress to provide a better understanding of the effect of carbohydrates on the hydrolytic activity of lipases, and the preparations listed in TABLE 1 are being studied for stability experiments.

REFERENCES

1. POSORKE, L. H. 1984. Industrial scale application of enzymes to the fats and oil industry. J. Am. Oil Chem. Soc. **61:** 1758–1760.
2. DESNUELLE, P. 1972. The lipases. *In* the Enzymes. P. D. Boyer, Ed. 3rd edit.: 575–616. New York & London.
3. GOLDBERG, M., F. PARVARESH, D. THOMAS & M. D. LEGOY. 1988. Enzymatic ester synthesis with continuous measurement of water activity. Biochim. Biophys. Acta **957:** 359–362.
4. GEKKO, K. & S. TIMASHEFF. 1981. Mechanism of protein stabilization by glycerol: preferential hydration in glycerol-water mixture. Biochemistry **20:** 4667–4676.
5. MONSAN, P. & D. COMBES. 1984. Effect of water activity on enzyme action and stability. Ann. N.Y. Acad. Sci. **434:** 48–60.
6. ANDERSON, E. & B. HAHN-HAGERDAL. 1987. Enzyme action in polymer and salt solutions. I. Stability of penicillin acylase in poly(ethylene glycol) and phosphate solutions in relation to water activity. Biochim. Biophys. Acta **912:** 317–324.
7. SANCHEZ-MONTERO, J-M., V. HAMON, D. THOMAS & M. D. LEGOY. 1991. Modulation of lipase hydrolysis and synthesis reactions using carbohydrates. Biochim. Biophys. Acta **1078**(3): 345–350.
8. GUARDIA, E. & G. J. HAAS. 1967. Influence of water binders on the activity and thermal inactivation of lipases. J. Agric. Food Chem. **15** (3): 412–416.
9. BRADFORD, M. M. 1976. A rapid and sensitive method for the quantitation of microgram quantities of protein utilizing the principle of protein-dye binding. Anal. Biochem. **72:** 248–254.
10. SHIRAI, K. & R. L. JACKSON. 1982. Lipoprotein lipase-catalyzed hydrolysis of *p*-nitrophenyl butyrate. J. Biol. Chem. **257** (3): 1253–1258.
11. BROCKMAN, H. L. 1984. General features of lipolysis: relation scheme, interfacial structure and experimental approaches. *In* Lipases B. Borgstrom & H. L. Brockman, Eds. 3–46. Elsevier. Amsterdam, The Netherlands.
12. GOLDBERG, M., D. THOMAS & M. D. LEGOY. 1990. The control of lipase-catalyzed transesterification and esterification reaction rates. Eur. J. Biochem. **190:** 603–609.

Continuous Culture of *Bacillus stearothermophilus* and Enzyme Preparation from Its Cell

KOSUKE TOMITA,[a] KAZUHIKO NAGATA,[b]
AND KENJI OKADA[b]

[a]*Faculty of Engineering*
Kanto Gakuin University
4834 Mutsuura, Kanazawa-ku
Yokohama 236, Japan

[b]*Department of Biochemistry*
Research and Development Center
Unitika Limited
23 Kozakura, Uji
Kyoto 611, Japan

INTRODUCTION

Bacillus stearothermophilus is a useful thermophile, as we have previously shown. *B. stearothermophilus* produces many kinds of thermostable enzymes. The properties and analytical applications of one of these, glucokinase, were presented by us in *Enzyme Engineering 10.*[1]

Atkinson *et al.* have described the batch culture of *B. stearothermophilus* NCA 1503 and the large-scale preparation of 15 types of enzymes.[2] In this paper, we will describe the continuous culture of a strain of *B. stearothermophilus* and the preparation of other thermostable enzymes from its cell.

CONTINUOUS CULTURE OF *B. STEAROTHERMOPHILUS*

A thermophilic strain of bacteria was isolated from the soil of the suburbs of Uji. The strain was identified as *B. stearothermophilus.* The maximum specific growth rate was high, 1.69 hour^{-1}, at 60°C. The strain was found to yield useful enzymes in quantity. We decided to use the strain for the production of thermostable enzymes.

Seed culture was carried out in a 30-L fermentor. Culture medium contained the following components (%): glucose, 0.35; yeast extract, 0.30; $(NH_4)_2SO_4$, 0.40; KH_2PO_4, 0.05; $Na_2HPO_4 \cdot 12H_2O$, 0.05; $MgSO_4 \cdot 7H_2O$, 0.015; trace minerals, 10 mL $\times$ L^{-1}. The pH was maintained at 6.8–7.2, and the temperature was 58.5–60.5°C. After 2.5 hours of culture, the broth was inoculated to a 500-L fermentor containing the medium (370 L working volume) of the same composition as above. The culture

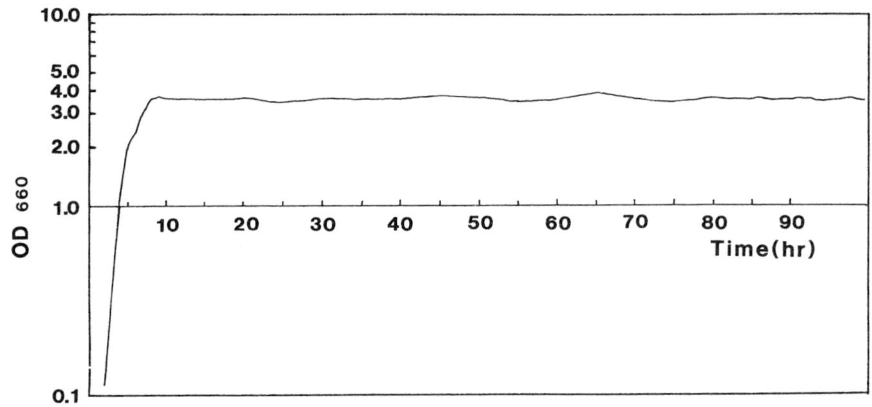

FIGURE 1. Growth profile of continuous culture of *B. stearothermophilus.*

was carried out under the same conditions. After glucose was consumed, the continuous culture was started at a dilution rate of 1.25 $hour^{-1}$. As shown in FIGURE 1, the culture continued stably over 100 hours of constant cell growth. Cells were harvested by centrifugation at 8000 *g* for 10 minutes. The productivity of cells was about 5 g × L^{-1} × $hour^{-1}$.

The high growth rate and temperature contribute to the high productivity, and the contamination problem can be left out of consideration.

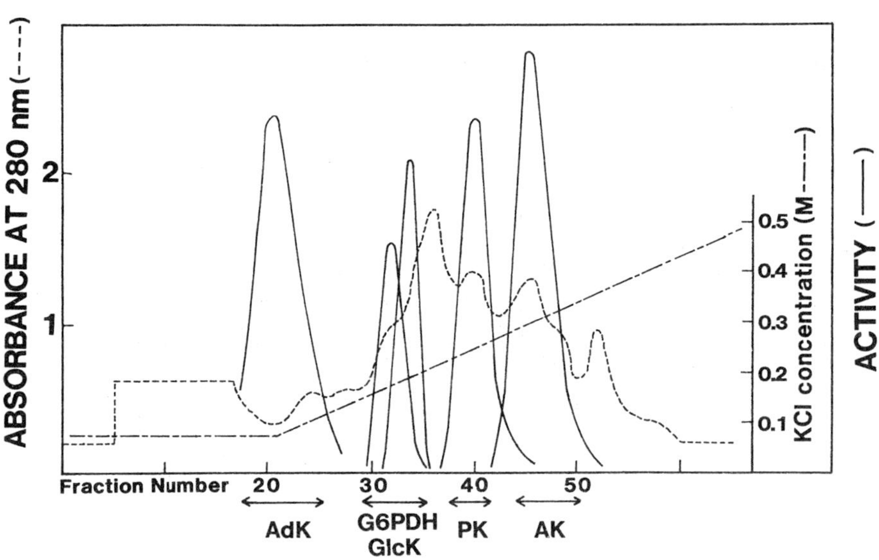

FIGURE 2. Chromatography of *B. stearothermophilus* enzymes on DEAE-Sepharose.

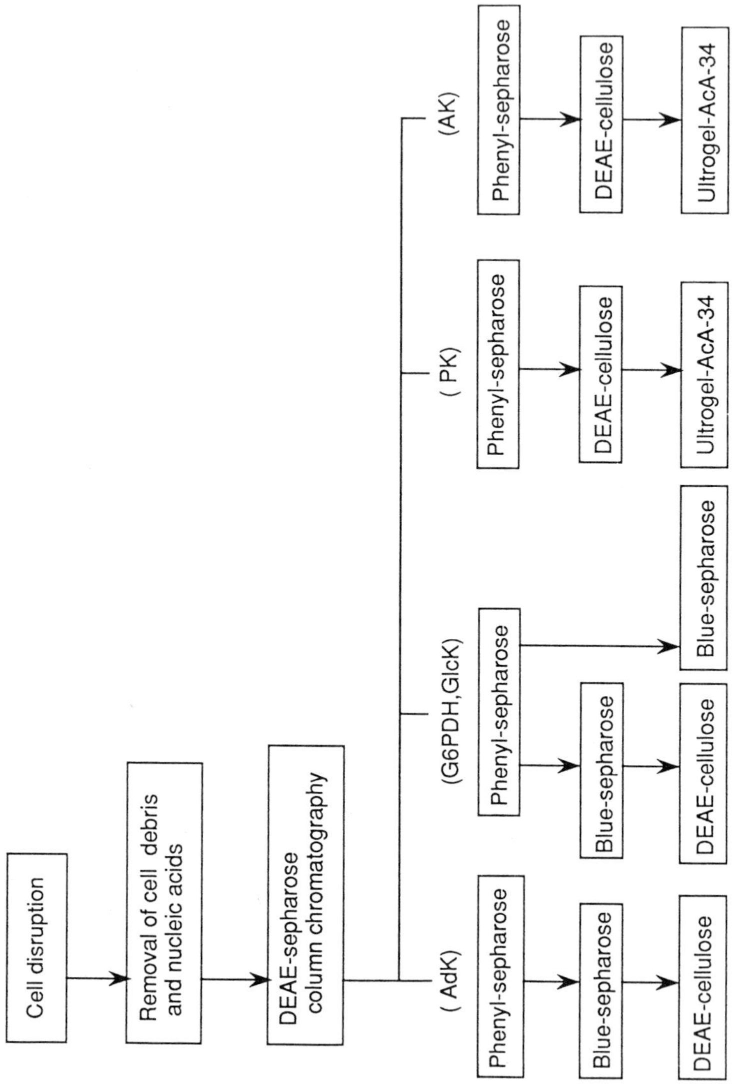

FIGURE 3. Flow diagram of the preparation of five enzymes.

PURIFICATION OF ENZYMES

The harvested cells were suspended in 0.1-M potassium phosphate buffer, pH 7.0, and disrupted with a homogenizer. Nucleic acids were precipitated with cationic polymer coagulant, and removed by centrifugation.

Extract from 10 kg of cells was loaded onto a DEAE-Sepharose column (35 by 20 cm) previously equilibrated with 50-mM potassium phosphate buffer (pH 7.5). The column was eluted with a linear gradient consisting of 80 L of each of several concentrations of 0.5-M potassium chloride added to the buffer. Fractions were collected and assayed for adenylate kinase (AdK), glucose 6-phosphate dehydrogenase (G6PDH), glucokinase (GlcK), pyruvate kinase (PK), and acetate kinase (AK) activities. The elution pattern is illustrated in FIGURE 2 with protein peaks. The peaks of AdK, PK, and AK appear separately, and the peaks of G6PDH and GlcK appear overlapping each other.

Further purification with a combination of different column chromatography techniques gave each purified enzyme—AdK, G6PDH, GlcK, PK, and AK. A summary of the purification procedures is given in FIGURE 3. All the procedures were carried out at 4°C.

The procedures resulted in a 2700-fold purification with a 35% recovery of total activity for AdK, 3100-fold with 31% for G6PDH, 2100-fold with 48% for GlcK, 550-fold with 40% for PK, and 4000-fold with 42% for AK. Polyacrylamide gel electrophoresis showed a single band for every enzyme.

The results show that five kinds of highly purified enzymes were simultaneously isolated in high yield from the one lysate.

CONCLUSIONS

Continuous culture of *B. stearothermophilus,* a thermophile, and enzyme preparation from its cell went on successfully. The high growth rate and the high growth temperature were advantageous to the continuous culture. The productivity of cells was sufficiently high, and the contamination problem was not serious.

The five enzymes were quite well isolated from the one lysate, and purified to homogeneity. Such simultaneous preparation is favorable for saving production cost.

REFERENCES

1. TOMITA, K., K. NAGATA, H. KONDO, T. SHIRAISHI, H. TSUBOTA, H. SUZUKI & H. OCHI. 1990. Thermostable glucokinase from *Bacillus stearothermophilus* and its analytical application. Ann. N.Y. Acad. Sci. **613:** 421–425.
2. ATKINSON, T., G. T. BANKS, C. J. BRUTON, M. J. COMER, R. JAKES, T. KAMALAGHARAN, A. R. WHITAKER & G. P. WINTER. 1979. Large-scale isolation of enzymes from *Bacillus stearothermophilus.* J. Appl. Biochem. **1:** 247–258.

New Directions for Bioselective Sensors and Probes

GARRY A. RECHNITZ

Hawaii Biosensor Laboratory
Department of Chemistry
University of Hawaii
Honolulu, Hawaii 96822

The biosensing abilities of natural organisms still far exceed the capabilities of laboratory devices with respect to sensitivity, information content, stability, and reliability. Our biosensor research has aimed to mimic natural processes through the use of animal and plant materials as molecular recognition elements. Particularly promising results have been obtained with neuronal structures containing highly selective chemoreceptors and with cell cultures of certain plant species; the latter offer especially attractive uniformity and time stability as biosensor components. Future prospects include the use of transgenic plant materials and of drug-selective receptors from brain tissue.

Pineapple-Tissue-Based Bioelectrode for the Determination of Hydrogen Peroxide

A flow-injection analysis (FIA) system for the determination of hydrogen peroxide was developed by utilizing a pineapple-tissue-based modified carbon paste electrode. Pineapple stem core tissue contains peroxidase, which is not available commercially in the pure isolated form. Hydrogen peroxide concentrations were determined in the presence of *O*-phenylenediamine as a hydrogen donor at a potential −200 mV. The detection limit (S/N = 3) for hydrogen peroxide using this tissue electrode is 2.1 μM with a relative standard deviation of 0.44% ($n = 12$). The tissue-modified carbon paste electrode is stable for up to 20 days with no significant activity changes.

Mixed Carbon Paste–Pea Seedling Electrochemical Sensor for Measuring Plant-Growth-Regulating Activity of Amines

The use of pea seedlings as a source of diamine oxidase for the determination of enzymatic oxidation rates, which are important in evaluating the plant-growth-regulating activity of various amines, is reported. A mixed carbon paste–plant tissue amperometric sensor was constructed and characterized using speramidine as a substrate ($K_m = 5 \times 10^{-6}$ M). Generation of hydrogen peroxide due to enzymatic oxidation of amine was monitored at 0.9 V vs. Ag/AgCl under steady-state conditions using this sensor. Easy construction, reliability of the data, reusability, shorter response times (of the order of 4 seconds), and a detection limit of 7.1 nmol l^{-1} are some of the advantages.

Improved Plant-Tissue-Based Biosensor Using In Vitro *Cultured Tobacco Callus Tissue*

In vitro cultured tobacco callus tissue, which contains high activity of peroxidase, is used to construct an amperometric biosensor for the detection of hydrogen peroxide. The tissue is incorporated into a carbon paste matrix along with ferrocene, an electron mediator. The detection of hydrogen peroxide was accomplished through its enzymatic reduction mediated by ferrocene. Comparison studies conducted with normally grown types of tobacco tissues indicate that the callus tissue has the highest biocatalytic activity and the least variance between tissue preparations with respect to sensor response toward hydrogen peroxide. Additionally, the sensor exhibits a remarkably long lifetime of more than three months. Other important characteristics of the sensor include fast response times (less than 2 seconds), wide dynamic range (5×10^{-6} M–1.10×10^{-4} M), and a lower detection limit of 7.5×10^{-7} M hydrogen peroxide.

Intact Chemoreceptor Biosensors Based on Hawaiian Aquatic Species

A receptrode biosensor is presented that uses intact-chemoreceptor-based molecular recognition from antennular structures of the Hawaiian swimming crab species *Portunis sanguinolentus.* The sensor is coupled to a learning, pattern-recognition calculation for performing analytical chemistry. Action potential waveforms are used to establish the identiy of individual action potential types that can be associated to particular analytes. The pattern recognition calculations used are referred to as cluster analysis (CA) and principal component analysis (PCA). Action potential similarities are determined by using a dendrogram plot of the cluster analysis results and further elucidated by using principal component score plots. Quantitative analysis was performed after classification of analyte and background responses. Chemoresponses to salinity and trimethylamine *N*-oxide, two chemical constituents that are found in the crustacean living environment, were investigated and gave analytic responses over several orders of magnitude.

Construction and Properties of a Pyrazinamide-Selective Biosensor Using Chemoreceptor Structures from Crayfish

Receptor-based biosensors using structures from freshwater species (crayfish) are shown to give selective analytical responses to the antitubercular drug pyrazinamide. The construction of such biosensors, in general, is greatly facilitated through the use of a video imaging inverted biological microscope during the manipulation and assembly of the delicate structures involved and points the way toward microscale receptrodes with enhanced analytical properties in the future. The novel sensor is characterized in terms of selectivity, sensitivity, response time, dose/response relationship, and lifetime. Advantages of using freshwater species in constructing neuronal biosensors, e.g., extended lifetime and fewer interferences, are explored.

Industrial Application of Microbial Sensors for Bioprocess Control

TETSUYA TOSA, MASARU SENUMA, YOSHIMI NAKAGAWA, AND TOMOAKI MORIMOTO

Research Laboratory of Applied Biochemistry
Tanabe Seiyaku Company, Limited
16-89, Kashima-3-chome, Yodogawa-ku
Osaka 532, Japan

INTRODUCTION

Since the beginning of the 1960s, we have been studying immobilized biocatalysts.[1] In 1969, we succeeded in the industrial application of the continuous optical resolution of DL-amino acids using immobilized aminoacylase.[2] This is the first industrial application of immobilized enzymes in the world. After this, we developed industrial applications of immobilized microbial cells for the continuous production of L-aspartic acid, L-malic acid, L-alanine, and D-aspartic acid.[3] On the basis of these, we expanded the techniques to immobilized biomaterials for bioseparation such as immobilized tannin[4] and immobilized histidine.[5]

On the other hand, since the latter half of the 1950s, by applying basic knowledge of molecular biology and genetic engineering, we have succeeded in constructing new microbial strains for amino acid fermentation. At present, we are producing L-amino acids such as L-isoleucine, L-proline, and L-threonine by fermentation.[6–8] Further, we have techniques to produce L-histidine, L-arginine, etc.[9,10] With these technologies, Tanabe Seiyaku commercialized an amino acid infusion in 1959 for the first time in the world.

In order to improve the productivity of the bioreactor systems, we have investigated biosensors for bioprocess control and quality control. As a result, we developed both L-aspartic acid and urea sensors by coupling immobilized microbial cells to a sensing electrode. In this paper, we will present the industrial applications of both microbial sensors.

ASPARTIC ACID SENSOR

Production of L-Aspartic Acid and L-Alanine

L-Alanine is a useful amino acid as a medicine and a food additive, and D-aspartic acid is used in Tanabe Seiyaku as one of the important components of semisynthetic penicillin, Aspoxicillin. FIGURE 1 shows the simultaneous production of D-aspartic acid and L-alanine, which was commercialized in Tanabe Seiyaku in 1989. DL-Aspartic acid is chemically synthesized from fumaric acid and ammonia, and the resulting DL-aspartic acid is converted to L-alanine and D-aspartic acid by the action of L-aspartate β-decarboxylase of immobilized *Pseudomonas dacunhae,* because this enzyme has higher optical specificity, namely, it acts only on L-aspartic acid.

After detailed engineering analysis, we designed a simultaneous production system of D-aspartic acid and L-alanine from DL-aspartic acid.[11] In this bioreactor system, the determination of L-aspartic acid is important for the calculation of the reaction ratio and for process and quality control.

Construction of an Aspartic Acid Sensor

The conventional method for the determination of L-aspartic acid is the UV method using glutamate-oxaloacetate transaminase (GOT) and malate dehydrogenase (MDH).[12] This method is tedious and expensive due to the use of expensive

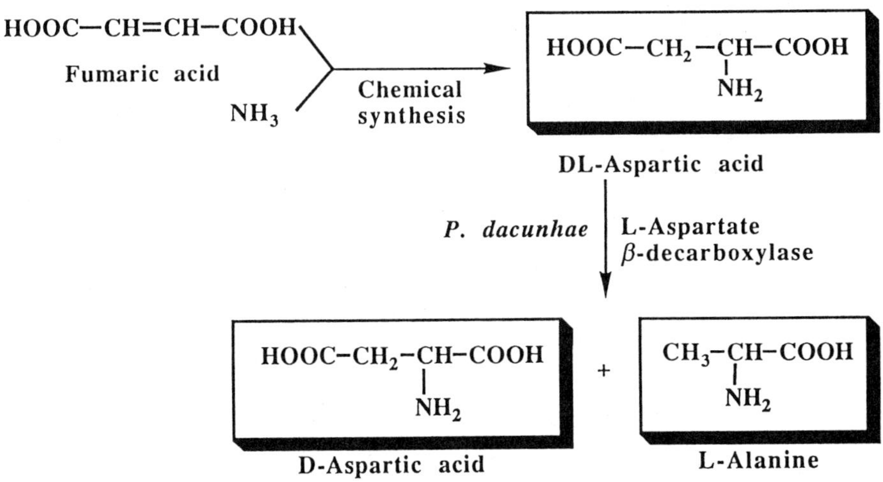

FIGURE 1. Simultaneous production of D-aspartic acid and L-alanine.

enzymes. In order to overcome these disadvantages, we developed an L-aspartic acid sensor by coupling a membrane coated with *Pseudomonas dacunhae* immobilized with κ-carrageenan to a carbon dioxide electrode.

FIGURE 2 shows the outline of the preparation method. *Pseudomonas dacunhae* (8 g wet weight) cells are suspended in physiological saline (8 ml) at 45°C, and κ-carrageenan (2.1 g) is also dissolved in physiological saline (45 ml) at 50°C. Both are mixed, and the mixture is coated onto the acetyl cellulose membrane (Millipore Co., 0.45 μm). Then the membrane is soaked in 0.3 M KCl solution for gelation at 5°C for 10 minutes. The resulting membrane is treated with 80-mM hexamethylenediamine (HMDA) and 80-mM glutaraldehyde (GA) at 5°C for 20 minutes in order to increase its strength. After this treatment, the membrane is soaked in 1 M ammonium L-aspartate acid and 0.1-mM pyridoxal phosphate (PLP) to activate it.

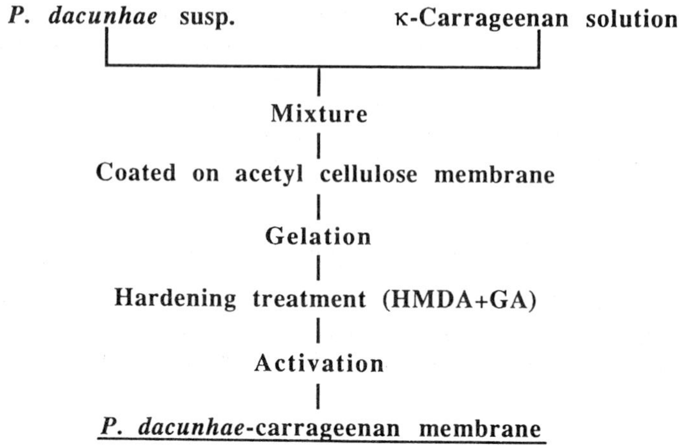

FIGURE 2. Preparation of *Pseudomonas dacunhae* κ-carrageenan membrane.

FIGURE 3 is an illustration of the microbial electrode for aspartic acid. A commercially available carbon dioxide electrode (Orion Research, Model 95-02) is used. It consists of a carbon dioxide–permeable membrane and a combined glass electrode. The immobilized *Pseudomonas dacunhae*–holding acetyl cellulose membrane is cut into a circle and fixed on the surface of this carbon dioxide–permeable membrane. Both membranes are fastened with a bottom cap. When the sample is introduced into the *Pseudomonas dacunhae* membrane, L-aspartic acid is decarboxylated by the action of the L-aspartate β-decarboxylase of the microorganisms to produce carbon dioxide. The liberated carbon dioxide gas permeates through the

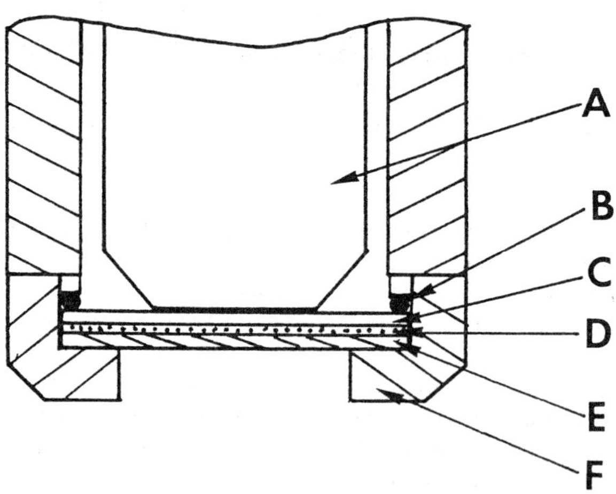

FIGURE 3. Cross-sectional view of the microbial electrode for L-aspartic acid: (A) combined glass electrode; (B) O-ring; (C) carbon dioxide–permeable membrane; (D) *P. dacunhae* κ-carrageenan membrane; (E) acetyl cellulose membrane; (F) cap.

gas-permeable membrane, and is detected by the electrode. FIGURE 4 shows the fundamental equipment of the aspartic acid sensor. This consists of a circulating thermostat, a microbial electrode, a jacketed cell (10 ml), a magnetic stirrer, an ion analyzer (Orion Research, Model 901), and a recorder.

Determination of L-Aspartic Acid

A potentiometric measurement for a calibration curve is carried out by pipetting a sample solution containing L-aspartic acid into a cell containing 0.1-M sodium citrate and 0.1-mM PLP (pH 6.0) at 25°C. The sample solution is stirred, and the steady-state potential is recorded.

FIGURE 5 shows a typical calibration curve of L-aspartic acid. The response was linear from 0.2 to 5 mM L-aspartic acid, with a slope of 50 mV/decade in sodium

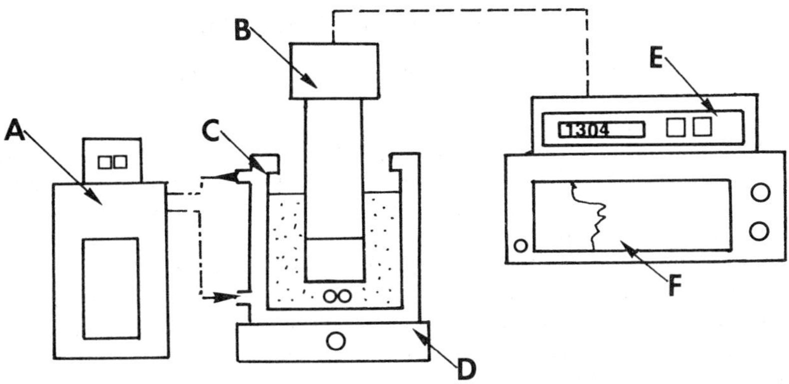

FIGURE 4. Schematic diagram of fundamental equipment of aspartic acid sensor: (A) circulating thermostate; (B) microbial electrode; (C) jacketed cell; (D) magnetic stirrer; (E) ion analyzer; (F) recorder.

citrate buffer, and the response time was about 10 minutes. TABLE 1 shows selectivity or specificity of this sensor. It responded slightly to some amino acids such as L-methionine, L-tryptophan, and L-isoleucine, but the response is considered to be within the margin of experimental error. Therefore, we think that the selectivity of the aspartic acid sensor is satisfactory for the desired process control. The stability of this sensor was tested. As shown in FIGURE 6, it was stable for at least 70 days during intermittent use consisting of over 140 tests. The stability was extremely higher than that of sensor consisting of *Bacterium cadaveris* and an ammonia gas electrode reported by Kobos and Rechnitz.[13]

Determination of L-Aspartic Acid in Sample Solution

As this sensor responds to carbon dioxide gas, it is necessary to remove any such gas initially contained in the sample solution. Carbon dioxide gas can be easily

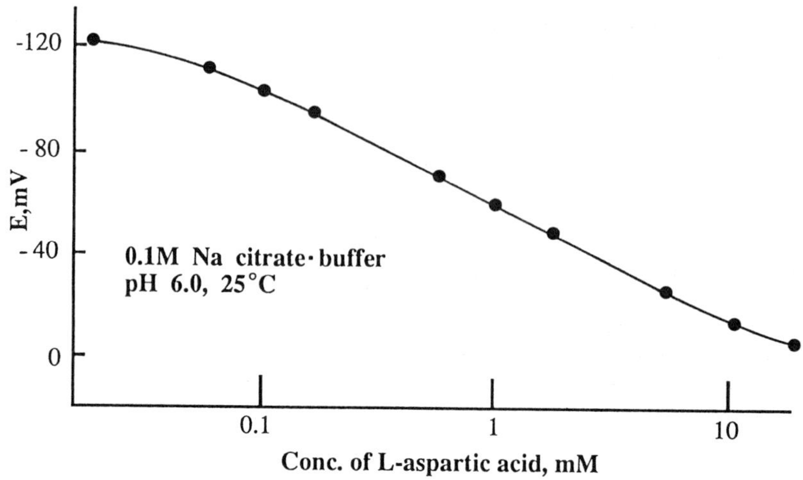

FIGURE 5. Calibration curve of L-aspartic acid.

removed by aerating the sample solution, which is first given low-pH treatment. Thus, we designed the analytical system as shown in FIGURE 6, and applied this to the determination of L-aspartic acid in the effluent from the bioreactor for production of D-aspartic acid and L-alanine.

The sample (1 ml) was diluted with buffer (9 ml), aerated (200 ml/minute, 10 minutes) for removal of carbon dioxide gas, and introduced (10 ml) into the sensor. The concentration of L-aspartic acid was determined from the standard addition method, and the results were compared with those obtained by the conventional UV method using GOT and MDH.[12]

As shown in TABLE 2, there are no difference between the results by both methods. The reproducibility expressed as a coefficient of variation is within 4.3% in this sensor. The total determination time for each sample was about 25 minutes,

TABLE 1. Selectivity of the Aspartic Acid Sensor

Amino Acid[a]	Selectivity[b] (%)	Amino Acid[a]	Selectivity[b] (%)
L-Asp	100	L-Lys	3
D-Asp	0	L-Met	6
L-Ala	1	L-Phe	0
L-Arg	3	L-Pro	0
L-Cys	0	L-Ser	1
L-Glu	0	L-Thr	1
Gly	0	L-Tyr	3
L-His	1	L-Trp	5
L-Ile	4	L-Val	2
L-Leu	0		

[a]Determination was carried out in 1-mM solution.
[b]Relative potential.

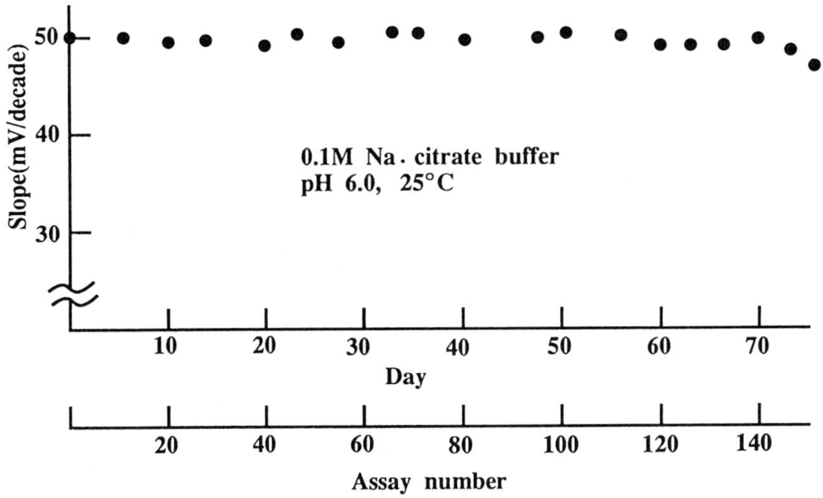

FIGURE 6. Stability of aspartic acid sensor. The slope was calculated by determination of 0.5 and 5 mM L-aspartic acid.

including dilution, rinsing, and carbon dioxide gas removal. This method can be used for the determination of L-aspartic acid in a sample containing a large excess of carbon dioxide gas, L-alanine, and D-aspartic acid.

On the basis of previous results, we designed and commercialized a microbial L-aspartic acid sensor system. This system mainly consists of the removal unit for carbon dioxide, the sensor, and a computer. The system is connected with the simultaneous production system of D-aspartic acid and L-alanine as shown in FIGURE 7. The concentration of L-aspartic acid in the effluent at the center and outlet of the bioreactor is automatically determined. A few rates of substrate solution and the temperature of the reactor are controlled according to the assay value. That is, when the concentration of L-aspartic acid is high, the flow rate of the substrate is reduced and/or the reactor temperature is raised automatically.

TABLE 2. Comparative Analysis of L-Aspartic Acid in Effluent of Bioreactor

	Concentration of L-Aspartic Acid (mM)	
Sample[a]	Microbial Sensor	Conventional Enzyme Method
A	7 (4.3)[b]	6
B	22 (3.2)	20
C	28 (2.8)	27
D	19 (4.0)	20
E	30 (3.6)	31

[a]Effluent of bioreactor for production of D-aspartic acid and L-alanine from DL-aspartic acid.
[b]Coefficient of variation, $n = 5$, is given in parentheses.

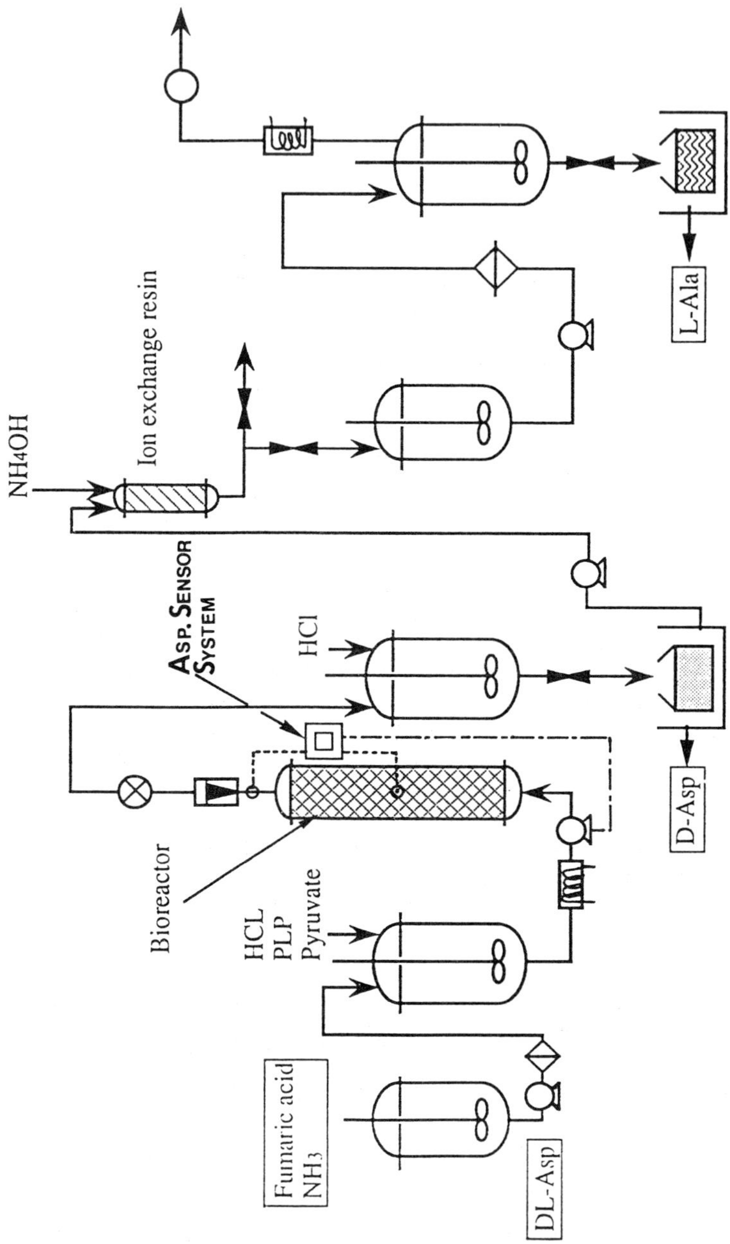

FIGURE 7. Bioreactor system for simultaneous production of D-aspartic acid and L-alanine.

UREA SENSOR

Purpose of a Urea Sensor

In amino acid fermentation, carbohydrates, urea, vitamins, salts, and other compounds are used as nutrients in the culture medium. Among them, urea is an important nutrient as a nitrogen source, and has an advantage over ammonium salts. The ammonium ion in large concentrations is toxic for the growth of microorganisms; but urea is not itself toxic; and the ammonium ion is slowly liberated by the action of urease in the organism. So urea is more conveniently used. Thus, it is important to determine the urea concentration in the fermentation broth for the effective production of amino acids.

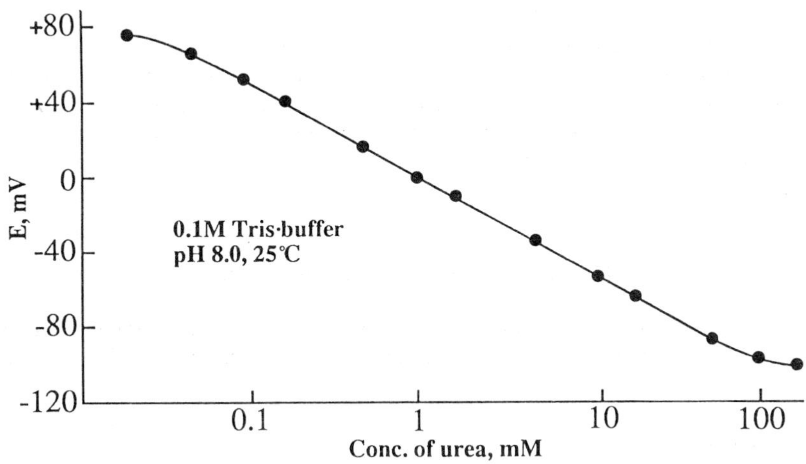

FIGURE 8. Calibration curve of urea.

Determination of Urea

At present, urea is determined by the diacetyl monooxime method[14] based on color reaction or by the urease method.[15] The diacetyl monoxime method is tedious, has a long determination time, and smells bad. The urease method's disadvantage is that the enzyme is expensive. Therefore, we tried to develop a microbial sensor for urea by coupling a membrane coated with *Sporosarcina ureae* immobilized with κ-carrageenan to an ammonia gas electrode (Orion Research Model 95-12). This bacterium has high urease activity. The preparation method is fundamentally the same as that of the *P. dacunhae* membrane for the L-aspartic acid sensor.

FIGURE 8 shows a typical calibration curve of urea. The response was linear from 0.1 to 50 mM urea, with a slope of 55 mV/decade in 0.1-M Tris buffer, containing 0.1-mM edetic acid, pH 8.0 at 25°C, and the response time was about 7 minutes. TABLE 3 shows selectivity or specificity of this sensor. It responded slightly to some amino acids, but the response is considered to be within the margin of experimental

TABLE 3. Selectivity of the Urea Sensor

Amino Acid[a]	Selectivity[b] (%)	Amino Acid[a]	Selectivity[b] (%)
Urea	100	L-Lys	3
D-Asp	3	L-Met	1
L-Ala	0	L-Phe	0
L-Arg	1	L-Pro	4
L-Cys	2	L-Ser	0
L-Glu	1	L-Thr	0
Gly	0	L-Tyr	0
L-His	0	L-Trp	2
L-Ile	0	L-Val	1
L-Leu	0		

[a]Determination was carried out in 0.5-mM solution.
[b]Relative potential.

error. Further, this sensor did not respond to organic compounds such as glucose, sucrose, and acetic acid. Therefore, we think that the selectivity of this urea sensor is satisfactory for the control of amino acid fermentation. The stability of this sensor was tested. As shown in FIGURE 9, it was stable for at least 30 days during intermittent use consisting of over 50 tests. Thus, the stability is satisfactory for industrial use.

Determination of Urea in Sample Solution

As this urea sensor uses an ammonia gas electrode, ammonia and volatile amines in the culture broth influence the determination of urea. Therefore, it is necessary to

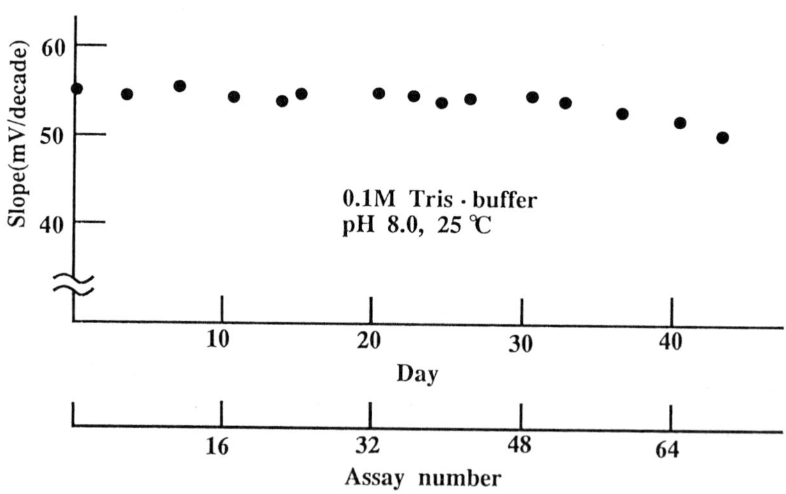

FIGURE 9. Stability of the urea sensor. The slope was calculated by determination of 0.5 and 5 mM urea.

remove these chemicals if they are present initially in the sample solution. These compounds can be easily removed by a column containing cation exchange resin. Thus, an analytical system was developed for the determination of urea in amino acid fermentation broth.

The broth (0.1 ml) was directly charged into the column (Dowex 50W-X8, 1 ml) without removal of contaminants such as microbial cells and $CaCO^3$. The column was washed with water (5 ml), and the column effluent was diluted with buffer solution (5 ml). The concentration of urea in the diluent was determined from the standard addition method. As a control experiment, urea was determined by the diacetyl monoxime method.[14] As shown in TABLE 4, the determination values in the case of the microbial sensor were constant regardless of dilution rate. The reproducibility expressed as a coefficient of variation is within 3.7% in this urea sensor. On the other hand, the values in the case of the diacetyl monoxime method[14] decreased slightly according to the decrease of dilution rate. These results suggest that inhibitors of color reaction existed in the fermentation broth. Therefore, it is considered that this sensor can be successfully applied to the determination of urea in colored and turbid samples such as fermentation broth.

TABLE 4. Comparative Analysis of Urea in Fermentation Broths

Sample[a]	Dilution Rate	Concentration of Urea (mM)	
		Microbial Sensor	Diacetyl Monoxime Method
Fermentation broth	100	36 (3.0)[b]	39
	80	35 (3.7)	36
	60	34 (2.8)	33
	30	36 (3.4)	31

[a]Fermentation broth containing CaCO3, urea, ammonia, and other compounds.
[b]Coefficient of variation, $n = 5$, is given in parentheses.

CONCLUSION

A microbial sensor for L-aspartic acid consists of immobilized *Pseudomonas dacunhae* membrane and a carbon dioxide electrode. The response of this sensor was linear between 0.2 and 5 mM L-aspartic acid with a slope of 50 mV/decade. The response time was about 10 minutes. The stability was about 70 days at 25°C, and selectivity was satisfactory.

Further, a microbial sensor for urea consists of immobilized *Sporosarcina ureae* membrane and an ammonia gas electrode. The response of the sensor was linear between 0.1 and 50 mM urea with a slope of 55 mV/decade. The response time was about 7 minutes. The stability and selectivity were also satisfactory for industrial purposes.

Both sensors were found to be useful as a simple, economical, and selective method for the determination of L-aspartic acid and urea in bioprocess analysis, and since 1988 they have been successfully used in our plants for process and quality control.

It is very difficult to quantitatively evaluate the economical efficiency, but it has become possible to produce a high and constant quality of amino acids at all times after introduction of these biosensors into our bioreactor systems. There have been many reports on the biosensors, but practical application is rather limited. In our case, a clear target was made of the improvement of the productivity and quality of conventional processes. This philosophy is one reason why we succeeded in the industrial application of biosensors.

REFERENCES

1. Tosa, T., T. Mori, N. Fuse & I. Chibata. 1966. Studies on continuous enzyme reactions—screening of carriers for preparation of water-insoluble aminoacylase. Enzymologia **31:** 214–224.
2. Chibata, T., T. Tosa, T. Sato, T. Mori & Y. Matuo. 1972. Preparation and industrial application of immobilized aminoacylases. *In* Proceedings IV IFS: Fermentation Technology Today: P383.
3. Chibata, I. & T. Tosa. 1984. Production of optically active compounds using immobilized biocatalysts. *In* IUPAC Chemistry for the Future. H. Grunewald, Ed.: 343–346. Pergamon Press. Oxford, England.
4. Chibata, I., T. Tosa, T. Mori, T. Watanabe & N. Sakata. 1986. Immobilized tannin—a novel adsorbent for protein and metal ion. Enzyme Microb. Technol. **8:** 130–136.
5. Matsumae, H., S. Minobe, K. Kindan, T. Watanabe, T. Sato & T. Tosa. 1990. Specific removal of endotoxin from protein solutions by immobilized histidine. Biotechnol. Bioeng. **12:** 129–140.
6. Komatsubara, S., M. Kisumi & I. Chibata. 1980. Transductional construction of an isoleucine-producing strain of *Serratia marcescens.* J. Gen. Microbial. **119:** 50–61.
7. Sugiura, M., Y. Imai, T. Takagi & M. Kisumi. 1985. Improvement of a proline-producing strain of *Serratia marcescens* by subcloning of a mutant allele of the proline gene. J. Biotechnol. **3:** 47–58.
8. Sugita, T. & S. Komatsubara. 1989. Construction of a threonine-hyperproducing strain of *Serratia marcescens* by amplifying the phosphoenolpyruvate carboxylase gene. Appl. Microbiol. Biotechnol. **30:** 290–293.
9. Takagi, T., M. Kisumi & I. Chibata. 1985. Construction of an arginine-producing strain of *Serratia marcescens.* Appl. Microbiol. Biotechnol. **21:** 378–382.
10. Sugiura, M. & M. Kisumi. 1984. Stabilization of a histidine-producing strain of *Serratia marcescens.* Appl. Environ. Microbiol. **48:** 43–47.
11. Masaru, S., O. Ohtuki, N. Sakata, M. Furui & T. Tosa. 1989. Industrial production of D-aspartic acid and L -alanine from DL-aspartic acid using a pressured column reactor containing immobilized *Pseudomonas dacunhae* cells. J. Ferment. Bioeng. **67:** 233–237.
12. Bergmeyer, H. U., M. Grasl & H. E. Walter. 1983. Reagents for enzymatic analysis. *In* Method Enzymatic Analysis. H. U. Bergmeyer, Ed. 3rd edit. **2:** 160–247. Verlag chemie. Weinheim, Germany.
13. Kobos, R. K. & G. A. Rechnitz. 1977. Regenerable bacterial membrane electrode for L-aspartate. Anal. Lett. **10:** 751–758.
14. Mar, R. L. & D. Bootzin. 1957. Modification of the diacetyl determination of urea. Anal. Chem. **29:** 1233–1236.
15. Fawcett, J. K. & J. F. Scott. 1960. A rapid and precise method for the determination of urea. J. Clin. Pathol. **13:** 156–159.

Enzymatic Electrocatalysis

Spatial and Kinetic Coupling between Electrochemical Reaction and Enzymatic Catalysis[a]

CHRISTIAN BOURDILLON

Laboratoire de Technologie Enzymatique
URA 1442 du CNRS
Université de Compiègne, BP 649
60206 Compiègne, France

INTRODUCTION

Enzymes from the class of oxydoreductases catalyze a wide variety of specific redox reactions. All of these enzymes need an electron carrier: the cosubstrate (or cofactor or mediator) whose concentration is the means to introduce energy in Reaction 1:

$$\text{substrate} + \text{cosubstrate} \overset{\text{enzyme}}{\longleftrightarrow} \text{product} + \text{modified cosubstrate} \quad (1)$$

Numerous cosubstrates—physiological redox couples like NAD(P)H, O_2, cytochrome c or artificial like viologens, quinones, ferrocenes—are able to exchange electrons directly on an electrode surface at a significant rate (Reaction 2):

$$\text{modified cosubstrate} \pm e^- \pm H^+ \overset{\text{electrode}}{\longleftrightarrow} \text{cosubstrate} \quad (2)$$

Since so much industrial fine chemical synthesis involves oxidation/reduction processes, the use of oxidoreductases to carry out specific transformations is a major area of interest. Of particular significance in this regard are regiospecific or asymmetric syntheses of amino acids, steroids, carbohydrates, and other pharmaceuticals (see for example References 1–3 for review).

The coupling of Reactions 1 and 2 in an electroenzymatic reactor is a good solution to bring energy into the enzymatic catalysis without any introduction of by-product. But the basically heterogeneous nature of electrochemical reactions implies some constraints, i.e. diffusional, to optimize the coupling. FIGURE 1 gives a schematic view of the problem.

The simplest answer should be an efficient electron transfer directly between the enzyme and the electrode (no cosubstrate and thus no electron transport), but despite some success with small redox proteins like cytochrome c,[4,5] ferredoxins,[6] or a small enzyme like cytochrome c peroxydase,[7] the thick protein or carbohydrate shells

[a]We gratefully acknowledge le Ministère de la Recherche et de la Technologie, le Ministère de l'Agriculture et de la Forêt, and Electricité de France for their financial support.

that surround the redox centers of enzymes forbid, in general, direct electrochemistry with the active and folded enzyme.[8] Additionally, such a transfer cannot occur with NAD(P) dehydrogenases since they do not bear a built-in redox prosthetic group.

Thus, during the last decade, different strategies were developed to optimize the coupling and to organize the relative spatial position of the three required components—the electrochemical interface, the enzyme, and the cosubstrate:

1. The direct grafting of the enzyme onto the electrode surface, the cosubstrate being free in solution.[9–12] The electrochemical site and the enzymatic site are very close at the molecular level, and the electron transport by the cosubstrate is very efficient.
2. The direct grafting of both the cosubstrate and the enzyme onto the electrode surface: NAD and a dehydrogenase,[13] FAD and glucose oxidase,[14] or viologen and an NAD oxidoreductase.[15] This method was not very successful since the

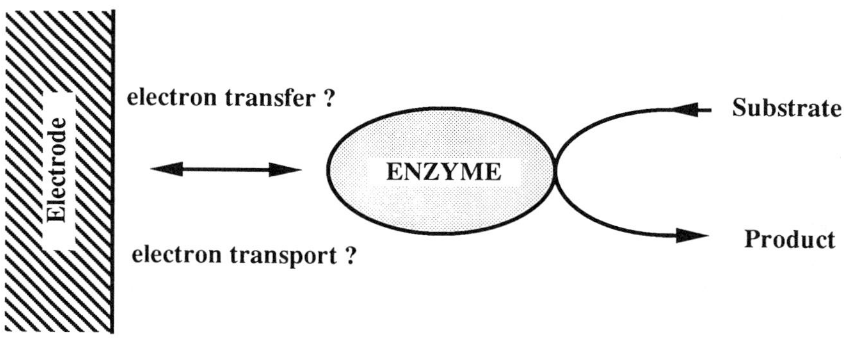

FIGURE 1. How to optimize the electron transfers and the electron transport between an enzyme and an electrode?

mobilities of the components were not sufficient to produce high catalytic current (less than 1 $\mu A\ cm^{-2}$).
3. The entrapment of enzymes into an organic matrix exhibiting an electron collector capacity:
   - In "organic metal" or conductive salts like TCNQ complexes.[16–18]
   - In conducting polymers like polypyrrole[19,20] or substituted polypyrrole.[21]
   - In redox polymers like substituted poly(vinylpyridine) or polysiloxanes.[22–24]

   The immobilized cosubstrate-mediator works as an electron relay[25] between the enzyme active site and the electrode.
4. Recently, highly organized two-dimensional structures like self-assembled hydrophobic films on electrodes[26,27] or Langmuir-Blodgett films on electrodes[28] were used as templates for immobilization of oxydoreductases. These structures mimic some properties of biological membranes and particularly the 2-D mobility of the components.

Whatever the strategy used, a good level of catalytic current (at least 50 $\mu A\ cm^{-2}$ of actual surface area) is obtained only if the rigidity between each component is not too high. For example, when the cosubstrate is covalently linked, it has to be sufficiently mobile to perform an efficient electron transfer alternatively with the enzyme and the electron collector.

There is no doubt that this high efficiency of coupling is already important for analytical applications like biosensors, but it is actually crucial for preparative applications in electroenzymatic reactors. The results presented here are focused on the optimization of this coupling with the help of two structures:

- Enzyme immobilized on carbon electrode (free cosubstrate).
- Enzyme and cosubstrate both immobilized on a hydrophobic modified electrode.

Some recent examples of applications of these concepts in electrosynthesis and analytical chemistry are presented.

IMMOBILIZED ENZYMES ON CARBON ELECTRODES

Enzyme Immobilization

Carbonaceous material has been used extensively as the most suitable conducting matrix in the construction of enzyme-modified electrodes. This is due to the presence of various surface functional groups, the concentration and nature of which can be easely modified. Twelve years ago, we developed a very simple method of protein grafting using surface carboxylic as anchoring groups.[29,9] This method has been largely used[30,11,12] and recently revisited.[31]

The result of the carbodiimide activation of the carboxylic surface groups followed by the reaction of amines from the proteins is a monolayer of enzymes directly linked to the carbon surface. This layer, even in its most compact form, is always open for electrochemistry of small molecules, without internal diffusion (FIGURE 2).

The maximum distance between the two sites of electron exchange, the enzyme active site and the electrochemical surface, is, at most, on the order of magnitude of the protein size. The electron transport controlled by the cosubstrate diffusion coefficient is thus very efficient (with a diffusion coefficient of 10^{-5} cm^2 $seconds^{-1}$, a molecule of cosubstrate spends an average of $5\ 10^{-8}$ seconds to cover a distance of 10 nm).

The maximal theoretical activity of this immobilized layer can be evaluated with the following assumptions:

- All immobilized enzyme molecules are active.
- The surface area is supposedly geometrical (no roughness).
- The enzyme molecule is a rigid sphere.
- The molecules are immobilized one after another.

In this case, the steric hindrances between the first immobilized molecules and the next to be immobilized allow a statistical coverage of 60% of the surface area.[32]

For example, with a Stokes radius of 4.3 nm (glucose oxidase), the calculation gives a surface concentration $\Gamma_{max} = 1.7\ 10^{-12}$ mole cm^{-2}. This value may be used as the basis for immobilization yield calculations:

$$\text{immobilization yield} = \frac{\Gamma\text{:measured amount of immobilized active enzyme}}{\Gamma_{max}} \times 100$$

With robust enzymes like glucose oxidase or hydrogenases, the immobilization yields easily reach 50%.[31,33] This is not as easy with dehydrogenases and their unstable tetrameric structures.[10]

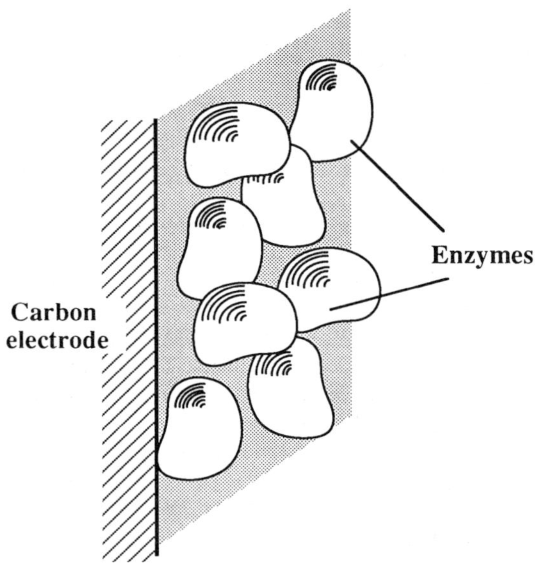

FIGURE 2. Schematic view of the modified electrode bearing a monolayer of enzyme molecules.

Kinetic Coupling Model

From the electrochemical point of view, it is simple to model the coupling in a steady-state condition using experimentally hydrodynamic methods. For example, the rotating disk method involves the control of the convective mass transport of substrate, product, and cosubstrates at the electrode interface.

With the immobilized glucose oxidase/free benzoquinone system, the enzymatic reaction is:

$$\text{Glucose (G)} + \text{Benzoquinone (BQ)} \rightarrow \text{Gluconolactone} + \text{Hydroquinone (HQ)} \quad \textbf{(3)}$$

and the electrochemical coupled reaction:

$$HQ \rightarrow BQ + 2H^+ + 2e^- \tag{4}$$

The steady-state catalytic current (I_{cat}) at a controlled potential is given by calculation of the flux balance[9,12] at the electrode interface with the following assumptions:

- There is no diffusional constraint inside the enzyme monolayer.
- The thickness of the diffusion layer, δ, for all species, is controlled by the rotation speed and given by the Levich equation.[34]
- The electrochemical reduction rate of BQ is negligible compared to the electrochemical oxidation rate of HQ.

$$I_{cat} = \frac{nFAJ_v k_{elec}}{k_{elec} + \dfrac{D_{HQ}}{\delta_{HQ}}} \tag{5}$$

with J_v, the enzymatic flux (in mole cm^{-2} seconds^{-1}):

$$J_V = \frac{\Gamma k_2}{1 + \dfrac{K_G}{G_0} + \dfrac{K_{BQ}}{BQ_0}} \tag{6}$$

n, number of electrons exchanged per mole of glucose; F, Faraday; *A*, electrode surface area (cm^2); k_{elec}, electrochemical rate constant at a given potential (cm seconds^{-1}); D_{HQ}, diffusion coefficient for HQ (cm^2 seconds^{-1}); k_2, molecular catalytic activity of the enzyme (seconds $^{-1}$); K_G and K_{BQ}, glucose oxidase kinetic constants respectively for G and BQ (M); and G_0 and BQ_0, interfacial concentrations for G and BQ (M).

Equation 5 can be simplified if the electrochemical reaction is not limiting. For a high overpotential, D/δ is negligible behind k_{elec}:

$$I_{cat} = nFAJ_V = \frac{nFA\Gamma K_2}{1 + \dfrac{K_G}{G_0} + \dfrac{K_{BQ}}{BQ_0}} \tag{7}$$

In this situation BQ_0 is known = HQ + BQ in the bulk solution since the electrochemical reaction is very efficient. If additionally the glucose concentration in the bulk solution is high, G_0 is very close to G in the bulk solution and the catalytic current can be evaluated knowing the enzymatic kinetic constants.

For example with glucose oxidase:

$$\Gamma_{max} = 1.7\ 10^{-12} \text{ mole cm}^{-2}$$

$$k_2 = 1000 \text{ seconds}^{-1}$$

$$K_G = 7\ 10^{-2}\text{ M}$$

$$K_{BQ} = 2\ 10^{-3}\text{ M.}$$

This simple calculation gives I_{cat} = 156 μA per cm^2 for G = 0.1 M. and BQ = 5 10^{-3} M.

Although only one layer of enzyme is immobilized, it is important to note the high level of this catalytic current density. Until now, only a few configurations built with macroscopic layers of enzymes molecules entrapped in various conductive matrices were able to reach this level.[17,18,24]

Measurement of the Immobilized Enzyme Activity

The typical measurement technique is a steady-state voltametric experiment at controlled potential. The catalytic current is measured when the introduction of the substrate in the solution starts the catalysis (FIGURE 3). The response rate is limited only by the mixing time of the viscous glucose solution inside the electrochemical cell, since from the theoretical point of view, the catalytic current reaches its value as soon as the concentration profiles inside the diffusion layer are stationary (less than 1 second from the Cottrell equation).

If the experimental conditions are adequate as in FIGURE 3 (high overpotential and high k_{elec}), the current is controlled only by the enzymatic catalysis and Equation 7 can be used to calculate Γ, the immobilized active enzyme concentration. Here, for example I_{cat} = 5.1 μA ⇒ Γ = 7.8 10^{-13} mole cm^{-2} and the immobilization yield is 46%.

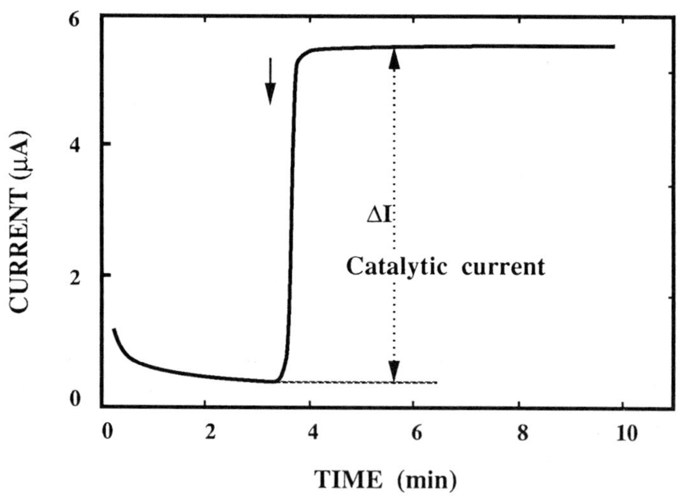

FIGURE 3. Typical measurement of the amount of immobilized enzyme (glucose oxidase here) on a carbon disk electrode (diameter 3 mm, rotation speed 600 rpm). After stabilization of the ground current at 0.7 V/SCE in a benzoquinone solution (5 10^{-3} M in 0.1 M acetic acid buffer, pH 5.6), a concentrate solution of glucose (final 0.1 M) was added (arrow). For enzyme immobilization, the glassy carbon disk was anodized 20 seconds at 2.2 V/SCE and the produced carboxylic groups were activated with a carbodiimide 0.1 M. After rinsing, the enzymatic molecules reacted with the activated surface overnight at 4°C.[31]

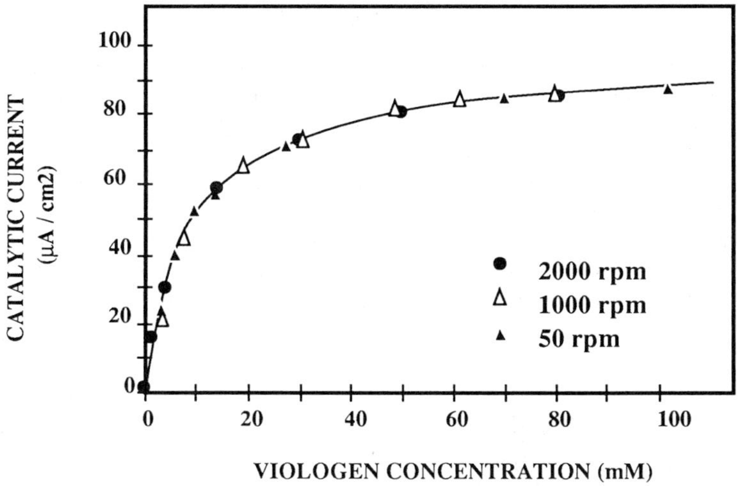

FIGURE 4. Kinetic behavior of the hydrogenase electrode submitted at various diffusional constraints with a saturating concentration of hydrogen (H_2 bubbling pressure: 1 atm). The enzyme was immobilized on a carbon disk electrode with the procedure described in FIGURE 3. Glycine buffer 0.1 M, pH 9. Hydrogenase from *Desulfovibrio gigas* was activated 3 hours before kinetic measurements. Working potential: −0.2 V/SCE (KCl-saturated calomel electrode).

The global measurement of the protein immobilization yield (catalytically active and inactive) is not very simple due to the very low amount of immobilized protein. We have developed a fluorimetric assay at the picomole level, convenient for electrode surface area greater than 1 cm^2,[10] but this delicate method is not suitable for routine analysis on carbon disk electrodes.

Typical Kinetic Behavior

As theoretically shown above, the main characteristic of the coupling is the local recycling of the cosubstrate without diffusional constraints. This means that the convective layer has no effect on the cosubstrate concentration in the enzyme microenvironment, whatever the cosubstrate concentration.

The catalytic current is thus independent of the electrode rotation speed if, on the other hand, the interfacial substrate concentration is itself in excess, for example at 5 or 10 times the respective Michaelis constant.

FIGURE 4 shows such a behavior with the hydrogenase/free viologen system. At pH 9 and at −0.2 V/SCE, the reactions are driven in the following direction:

$$1/2\, H_2 + MV^{++} \rightarrow H^+ + MV^{+\circ} \tag{8}$$

$$MV^{+\circ} \rightarrow MV^{++} + e^- \tag{9}$$

The catalytic currents were measured by the jump method described in FIGURE 3, the substrate feeding and elimination being very simple by way of alternative bubbling of H_2 and N_2.

Analytical Applications

The currents obtained from this enzymatic electrocatalysis configuration are logically dependant on substrate concentrations in solutions correctly buffered and supplemented with a convenient electroactive cosubstrate. Thus an "enzyme electrode," in the analytical meaning, could be made using this principle.

But, unfortunately, despite some potential qualities like fast response time and high current density, this principle is not a good way to produce a highly selective and easy to use biosensor. The microscopic structure is too open to the electroactive interfering species eventually present in crude biological or industrial fluids. More sophisticated solutions will have to be found to increase the overall selectivity of the modified electrode.

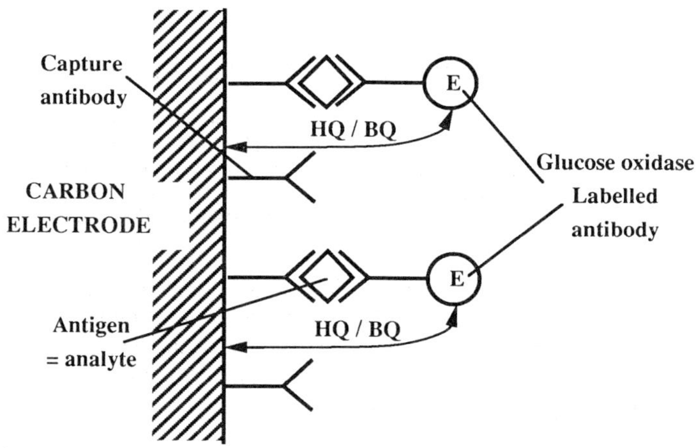

FIGURE 5. Molecular stacking at the end of an ELISA (sandwich variant) conducted directly on a carbon electrode. The labeled antibody, immobilized in an enzymatic electrocatalysis configuration, can be detected by the catalytic current due to HQ/BQ regeneration. The amount of linked antigen is calculated from this current.

The situation is completely different if the goal is to measure not the substrate concentration but the amount of immobilized enzyme at the end of a heterogeneous immunoassay. A synthetic solution of substrate and cosubstrate can be used for the final quantification of the enzymatic activity.

In ELISA procedure, the enzyme-labeled antibody is used as detector and amplifier of the antigen molecule, captured (sandwich assay) or adsorbed (competitive assay) on the solid phase. If the solid phase is a carbon electrode (FIGURE 5), the electrocatalytic detection gives the amount of immobilized labeled antibody directly.[35,36]

This pure heterogeneous measurement is faster and more sensitive compared to the homogeneous spectrophotometric or fluorimetric detections performed in the bulk solution. For example, with a labeled antibody built by coupling with glucose oxidase (stoichiometry 1/1), the molecular activity easily reaches 500 $second^{-1}$ in the experimental conditions of FIGURE 3. For a ground current of about 2 $\mu A\ cm^{-2}$ on a glassy carbon electrode, it is easy to read a catalytic jump of 0.1 $\mu A\ cm^{-2}$. Such a catalytic current is equivalent to the detection of $0.1\ 10^{-6}/2F \times 500 = 10^{-15}$ mole cm^{-2} of labeled antibody—a very good sensitivity; better, for example, than direct methods like ellipsometry.[37]

Another point of interest is the possibility of selection of any shape and size for the surface of the solid phase. This is particularly convenient for the design of a flow system monitored by an automatic apparatus.

We are working on the feasibility and industrial development of such a system.

Applications in Electrosynthesis

Preparative scale applications in electrochemistry need high current density to reduce investment cost. As our best catalytic currents are at the level of 0.1 $mA\ cm^{-2}$ of actual surface area, the use of electrodes exhibiting a high specific surface area is a key point in the development of electroenzymatic conversions.

Fortunately, carbon materials are produced in very different shapes and sizes. For example, carbon felts made with fibers of about 10 μm in diameter have a high specific surface area of 2600 cm^2/g or a geometric area of about 300 cm^2/cm^3.[38] This material is already used as flow-through electrodes for metal recovery in wastewaters.

We have developed such a reactor, at the laboratory scale, to study two kinds of electroenzymatic conversions:

- By the coupling of a dehydrogenase and the electrochemical regeneration of NAD^+.
- By the coupling of an oxidase with the electrochemical regeneration of an artificial cosubstrate.

We do not speak here about the first point since the electrochemical regeneration of NAD^+ was recently treated in detail in other papers[39–41] and an example of coupling with a dehydrogenase is given in this volume (Laval *et al.*).

For the coupling between oxidase and artificial cosubstrate, our model is the selective oxidation of glucose into gluconic acid with quinone regeneration. The problem was, in fact, the very low operational stability of glucose oxidase functioning with oxygen as cosubstrate. But if a quinone is used instead of oxygen, the operational stability is increased dramatically (at least 50 times).[42] An electroenzymatic reactor has been developed with a carbon felt anode bearing immobilized glucose oxidase (FIGURE 6).

A summary of the running parameters and of recent results is given in TABLE 1. It is important to note that the high volumetric productivity is only slightly reduced (20%) when concentrated and viscous sugars solutions as high as 400 g/l are used.

We are now working on the scale-up of this reactor, to reach a level of 10 A with a simple amperostatic control.

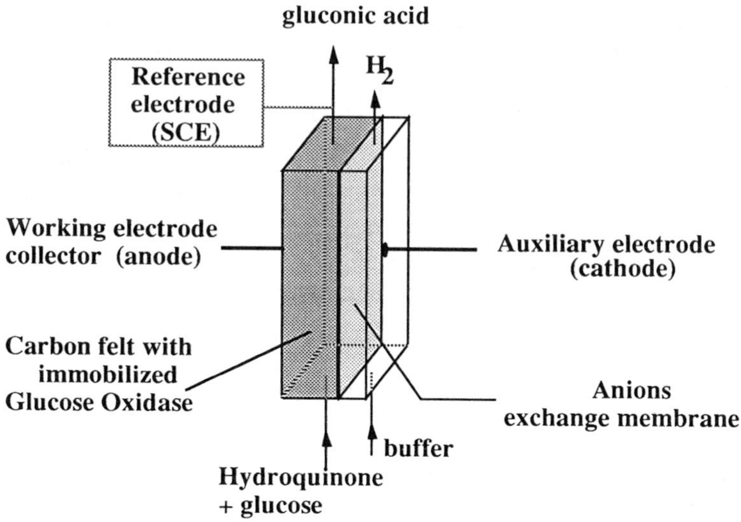

FIGURE 6. Schematic view of the electroenzymatic reactor used for the conversion of glucose into gluconic acid. The electrical section area is 6 × 2.5 cm.

ENZYME AND COSUBSTRATE IMMOBILIZED ON A HYDROPHOBIC ELECTRODE

The results briefly presented above demonstrate that a free cosubstrate is a very efficient shuttle to carry electrons between the electrode and the enzymatic site. This solution is undoubtedly the best since it is simpler for preparative scale applications.

But for analytical applications like enzyme electrodes, it would be interesting to produce a more integrated device with both the enzyme and the cosubstrate immobilized at the electrochemical interface. As cited in the Introduction, some of the attempts in this direction were not very successful due to lack of mobility between the components. Taking this constraint into account, we have recently shown that the use of a supported hydrophobic bilayer in interaction with enzyme and cosubstrate molecules could solve both the immobilization and the mobility problems.[26,27]

TABLE 1. Typical Running Parameters for the Flow-Through Reactor

Carbon felt volume	15 cm^3
Immobilized glucose oxidase	0.6 mg
Working potential	0.7 V/SCE
Flow rate	400 cm^3/hour
Glucose concentration	100 g/l
Hydroquinone concentration	0.55 g/l
Catalytic current	0.7 A at 30°C (500 A/m^2)
Volumic productivity (glucose transformed)	3800 g/l per day
Operational stability	Half-life: 300 hours

Structure of the Hydrophobic Modified Electrode: Cosubstrate Immobilization

The modified electrode used for enzyme immobilization was described previously.[43] It is an organized bilayer assembly in a microporous aluminium oxide template at a gold electrode (FIGURE 7).

The bilayer consisted of octadecyltrichlorosilane (OTS) and a positively charged electroactive amphiphile like $C_{18}Fc^+$ (ferrocenyl-methyl-dimethyl-octadecylammonium chloride) or $C_{18}MV^{2+}$ (*N*-Methyl-*N′*-octadecyl-4,4′bipyridinium chloride). While OTS molecules were covalently attached to the alumina surface, the electroactive amphiphile, self-assembled to form the second half-bilayer structure (FIGURE 7B), was mobile along OTS-coated surfaces.[43]

The template's internal structure provided for the perpendicular orientation of

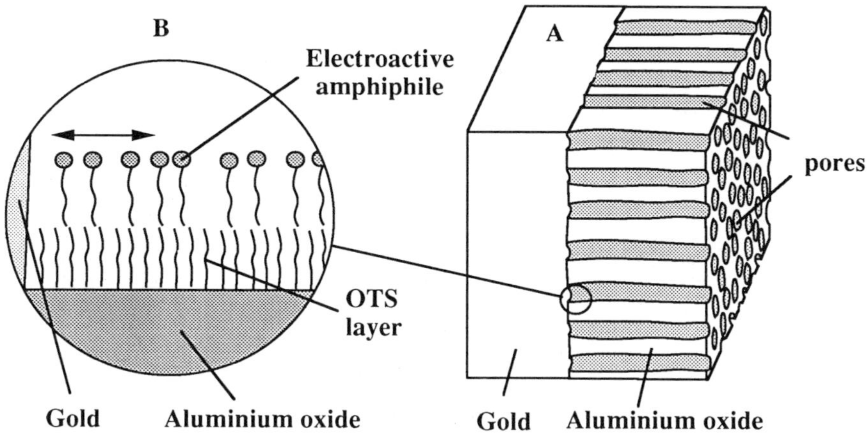

FIGURE 7. A drawing showing the microscopic structure of a porous aluminium oxide film at a gold electrode. **A**: The oxide films were about 2.2 μm thick and the thickness of the vapor-deposited gold layer was 1.5 μm, pores about 100 nm in diameter. This structure was glued on the top of a glass tube to give the final electrode of 3 mm in diameter. **B**: Enlarged cross section of a pore showing the bilayer structure and the lateral mobility of the electroactive amphiphile.

the bilayer assembly with respect to the electrode surface. The lateral mobility of the electroactive amphiphile allowed electrochemistry (2-D electrochemistry) without electroactive species in the bulk solution. To sum up, this structure is especially convenient to immobilize, by hydrophobic interactions, an electroactive cosubstrate able to exhibit a high level of local mobility.

Enzyme Immobilization

Our first experiments performed for enzyme immobilization in the bilayer structure have demonstrated that the peripheral shells of the enzymes play a key role in the adsorption process. Two enzymes were used: glucose oxidase on a $C_{18}Fc^+$ layer and hydrogenase from *Desulfovibrio gigas* on a $C_{18}MV^{2+}$ layer (TABLE 2).

The very hydrophilic glucose oxidase was immobilized with a correct yield only if the $C_{18}Fc^+$ was present to protect the enzyme molecules against hydrophobic denaturation on the OTS layer. A simple analysis of the effect of glucose oxidase immobilization on the dynamics of the lateral diffusion suggests that a large number of $C_{18}Fc^+$ molecules aggregate around the enzyme molecules.[26] This is a kind of micellization of the enzyme which does not interact directly with the OTS layer.

With hydrogenase from *D. gigas,* the situation was quite different since this enzyme is known to be relatively hydrophobic. As expected, the amount of immobilized active enzyme was high, at the level of 3 picomoles per cm^2 (the equivalent of 50% of the surface area of the porous aluminium oxide). Another important observation is concerned with the effect of chloride ions which prevent hydrogenase immobilization (TABLE 2). This effect is related to the stabilizing effect of Cl^- on the

TABLE 2. Stable Adsorption of Active Enzyme onto the Supported Hydrophobic Bilayers[a]

Electrode	Enzyme	Immobilization Medium	Amount of Immobilized Active Enzyme (per cm^2 of alumina)
OTS only	Glucose oxidase	Phosphate buffer	$<0.02\ 10^{-12}$ mole
OTS + $C_{18}Fc^+$	Glucose oxidase	Phosphate buffer	$0.6\ 10^{-12}$ mole
OTS only	Hydrogenase	Phosphate buffer	$0.5\ 10^{-12}$ mole
OTS + $C_{18}MV^{2+}$	Hydrogenase	Phosphate buffer	$3.0\ 10^{-12}$ mole
OTS + $C_{18}MV^{2+}$	Hydrogenase	Phosphate buffer + KCl 0.01 M	$<0.02\ 10^{-12}$ mole

[a]Experimentally, the OTS-treated electrodes, loaded or not with the amphiphile, were dipped 30 minutes in the immobilization medium (0.5 mg/ml of enzyme). After extensive rinsing, the amount of immobilized active enzyme was measured kinetically by an external method (electrochemical for glucose oxidase, spectrophotometrical for hydrogenase) (see References 26 and 27 for details).

$C_{18}MV^{2+}$ layer by ion pairing. That means that the immobilization process involves, as for membrane enzymes in biological membranes, hydrophobic interactions between the enzyme and the hydrocarbon region of the bilayer. The enzyme molecules have to displace $C_{18}MV^{2+}$ to establish such interactions, and the final $C_{18}MV^{2+}$ layer concentration was reduced after a successful enzyme immobilization.[27] This conclusion was confirmed by an experiment in which enzyme immobilization was attempted and failed when conducted in a buffer solution without chloride but containing 50 μM $C_{18}MV^{2+}$. Under these conditions, the presence of $C_{18}MV^{2+}$ in solution again stabilizes the OTS/$C_{18}MV^{2+}$ bilayer and impedes enzyme immobilization. FIGURE 8 summarizes the main characteristics of the structure in the case of immobilization of a hydrophobic enzyme like hydrogenase.

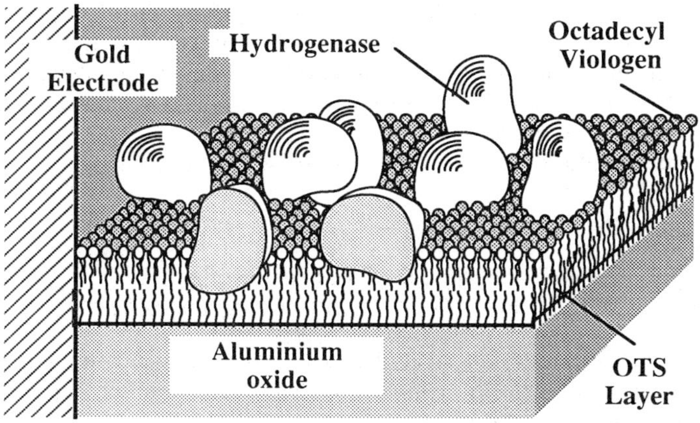

FIGURE 8. A schematic view of the supported bilayer electrode bearing immobilized hydrogenase and viologen.

Enzymatic Electrocatalysis: Amperometric And Potentiometric Results

The typical behavior of the enzymatic electrode (FIGURE 9) demonstrates that the electroactive amphiphile, here $C_{18}MV^{2+}$, the cosubstrate for hydrogenase, efficiently mediates the catalytic reaction in a two-dimensional manner.

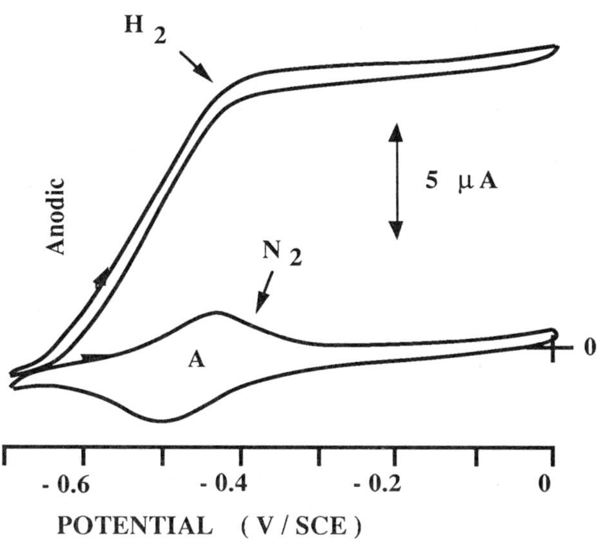

FIGURE 9. Cyclic voltammetry under N_2 or H_2 bubbling of the hydrophobic electrode bearing immobilized $C_{18}MV^{2+}$ and hydrogenase. Glycine buffer + KCl 0.01 M, pH 9. Scan rate: 10 mV/second. Electrode area: 0.07 cm^2.

Voltamogram A, which shows the electrochemistry of the $C_{18}MV^{2+}$ in the bilayer, could be obtained either in the presence of H_2 but prior to the hydrogenase immobilization, or under N_2 bubbling following enzyme immobilization. In the presence of both hydrogen and hydrogenase, one observes a catalytic current by coupling of Reactions 8 and 9.

The successive steps for the electron transport are in this case:

- Electron exchange between the hydrogenase and the $C_{18}MV^{2+}$ to produce $C_{18}MV^{\circ+}$.
- Transport of $C_{18}MV^{\circ+}$ by lateral diffusion in the bilayer.
- Electron exchange between $C_{18}MV^{\circ+}$ and the electrode to regenerate $C_{18}MV^{2+}$.

The high catalytic current produced (170 μA cm^{-2}) is not dependent on the rotation rate of the electrode, which gives an indication that the rate-limiting step is related to a process within the enzymatic assembly.

The same behavior was observed with the glucose oxidase electrode and $C_{18}Fc^+$. The catalytic currents were controlled by the enzymatic kinetics (FIGURE 10) and by the dynamics of transport inside the two-dimensional $C_{18}Fc^+$ layer. For example, addition of a small amount of octanol in the solution increased the catalytic current. As the lateral mobility of $C_{18}Fc^+$ is directly related to the intercalation of a such cosurfactant inside the layer,[43] this result demonstrated that the catalytic current is, at least, partially controlled by the overall mobility along the layer.

Amperometric measurements showed that the immobilized amphiphile redox couple is able to exchange electrons efficiently both with the enzyme and the

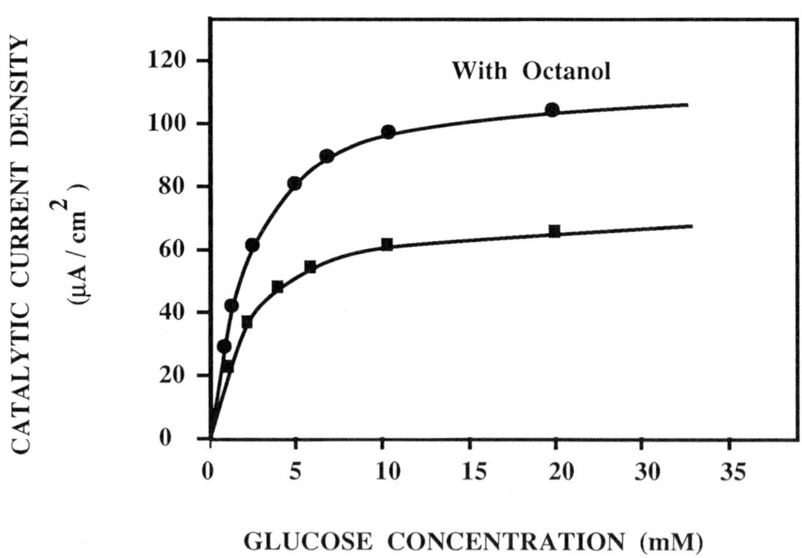

FIGURE 10. Dependence of the steady-state catalytic current on the concentration of glucose. Phosphate buffer 0.01 M at pH 6.5 + KNO_3 0.1 M. Final concentration of octanol: 3 mM.

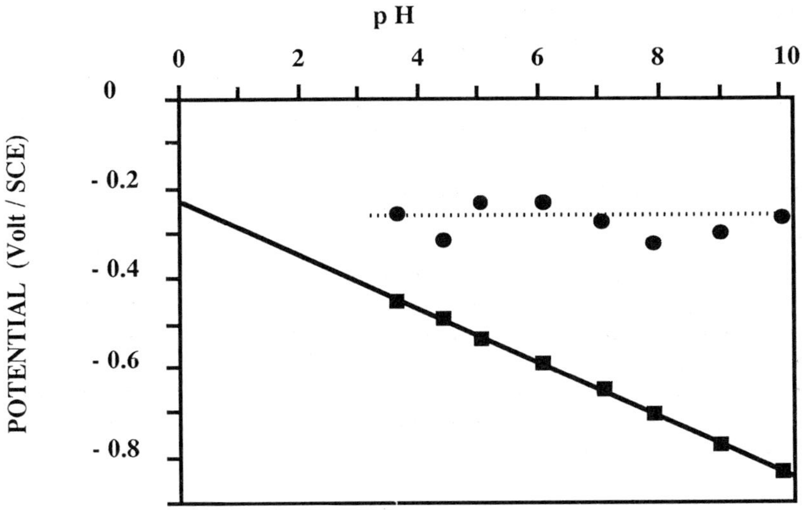

FIGURE 11. pH dependence of the potentiometric response of the bilayer electrode bearing immobilized hydrogenase (■) or not (●). The data were collected in a deaerated 0.05-M acetic buffer (pH 3.5–5.0), phosphate buffer (pH 6.0–8.0), and glycine buffer (pH 9.0–11). All buffer solutions were also 0.01 M in KCl.

electrode. If additionally these two reactions are easily reversible, a potentiometric measurement of the equilibrium could be performed without mediator in solution.

The hydogenase/$C_{18}MV^{2+}$ couple is a good candidate for this purpose since the enzymatic reaction is very easily reversible between pH 3 and 11. FIGURE 11 shows the thermodynamic response of the hydrogenase electrode to the H^+/H_2 redox couple, for a hydrogen pressure of 770 mm Hg at 24°C and without viologen in solution. The potentials, measured directly between the hydrogenase electrode and a calomel reference electrode, were stabilized very rapidly, in less than 30 seconds.[27] The blanks, without hydrogenase, gave a rather unstable potential poised at the foot of the $C_{18}MV^{2+}$ reduction wave.

The linear regression line through the data points can be expressed by $E = -0.245\text{–}0.059$ pH vs. SCE, which is in agreement with the theoretical behavior. Moreover, this response was found to be independent of the level of enzyme activity and the surface concentration of the redox amphiphile.

These features make this, and other analogous hydrophobic electrodes, attractive candidates for the development of selective potentiometric sensors.

CONCLUSION

We have described in this report some ways to optimize the kinetic coupling between electrochemical reactions and enzymatic catalysis. There is no doubt that the main parameter is the mobility of the cosubstrate-mediator in the environment of the two other components of the cycle: the electrode interface and the enzymatic site. Thus we pointed out the interest of the structural organization of the electrochemical interface at the molecular level.

Catalytic currents as high as 100 μA per cm^2 of actual surface area were obtained with only one monolayer of enzyme molecules directly grafted onto carbon electrodes. This level of current density allowed us to reach the preparative scale of application with an electroenzymatic reactor.

A more sophisticated molecular organization was produced with the help of the amphiphilic character of bilayer assemblies. We demonstrated that enzyme immobilization in such a structure is a viable strategy, not only for hydrophilic enzymes like glucose oxidase, but also for more hydrophobic enzymes which, traditionally, have been more difficult to work with. It is apparent that the success of this approach depends, in general, on the self-segregation phenomenon which is ubiquitous in biological systems.

We are hopeful that the strategy developed here will open the possibility of extending such schemes to integral membrane enzymes or receptors in order to construct efficient biomimetic systems.

REFERENCES

1. MAY, S. W. & S. R. PADGETTE. 1983. Oxidoreductase enzymes in biotechnology; current status and future potential. Bio/technology **1:** 677–686.
2. SIMON, H., H. GUNTHER & J. THANOS. 1986. On the use of viologen dyes for stereospecific bioreduction. *In* Enzymes as Catalysts in Organic Synthesis. M. P. Schneider, Ed.: 35–44. D. Reidel Publishing Company.
3. HUMMEL, W. & M. R. KULA. 1989. Dehydrogenases for the synthesis of chiral compounds. Eur. J. Biochem. **184:** 1–13.
4. ARMSTRONG, F. A., H. A. O. HILL & N. WALTON. 1988. Direct electrochemistry of redox protein. Acc. Chem. Res. **21:** 407–413.
5. ARMSTRONG, F. A., A. M. BOND, H. A. O. HILL, B. N. OLIVER & I. S. M. PSALTI. 1989. Electrochemistry of cytochrome c, plastocyanin and ferredoxin at edge and basal-plane graphite electrodes interpreted via a model based on electron transfer at electroactive sites of microscopic dimensions in size. J. Am. Chem. Soc. **111:** 9185–9189.
6. HALADJAN, J., P. BIANCO, L. ASSO, F. GUERLESQUIN & M. BRUSCHI. 1986. Direct electrochemical reduction of spinach and *Desulfovibrio desulfuricans* Norway ferredoxins. Electrochim. Acta **31:** 1513–1517.
7. ASSEFA, H. & E. F. BOWDEN. 1986. Electron transfer reaction of cytochrome C peroxydase at tin oxide electrodes. Biochem. Biophys. Res. Commun. **139:** 1003–1008.
8. SZUCS, A., G. D. HITCHENS & J. O. M. BOCKRIS. 1989. Electrochemical reactions of glucose oxidase at graphite electrodes. Bioelectrochem. Bioenerg. **21:** 133–148.
9. BOURDILLON, C., J. P. BOURGEOIS & D. THOMAS. 1980. Covalent linkage of glucose oxidase on modified glassy carbon electrodes. Kinetic phenomena. J. Am. Chem. Soc. **102:** 4231–4235.
10. LAVAL, J. M., C. BOURDILLON & J. MOIROUX. 1984. Enzymatic electrocatalysis: electrochemical regeneration of NAD with immobilized lactate dehydrogenase modified electrodes. J. Am. Chem. Soc. **106:** 4701–4706.
11. IANNIELLO, R. M. & A. M. YACYNYCH. 1981. Immobilized enzyme chemically modified electrode as an amperometric sensor. Anal. Chem. **53:** 2090–2095.
12. RAZUMAS, V. J., J. J. JASAITIS & J. J. KULYS. 1984. Electrocatalysis on enzyme modified carbon electrode. Bioelectrochem. Bioenerg. **12:** 297–322.
13. TORSTENSSON, A., G. JOHANSSON, M. O. MANSSON, P. LARSSON & K. MOSBACH. 1980. Electrochemical regeneration of NAD covalently bound to liver alcohol dehydrogenase. Anal. Lett. **13:** 837–849.
14. MIYAWAKI, O. & L. B. WINGARD. 1985. Electrochemical and glucose oxidase coenzyme

activity of FAD covalently attached to glassy carbon. Biochim. Biophys. Acta **838:** 60–68.

15. GUNTHER, H., A. S. PAXINOS, M. SCHULZ, C. VAN DIJK & H. SIMON. 1990. Direct electron transfer between carbon electrodes, immobilized mediator and an immobilized viologen-accepting pyridine nucleotide oxidoreductase. Angew. Chem. Int. Ed. Engl. **29:** 1053–1054.
16. KULYS, J. J., A. S. SAMALIUS & G. J. SVIRMICKAS. 1980. Electron exchange between enzyme active center and organic metal. FEBS Lett. **114:** 7–12.
17. YAROPOLOV, A. I., V. MALOVIC & V. A. ISUMRUDOV. 1984. Study of catalytic properties of glucose oxidase immobilized on polymeric semiconductive carriers. Biokhimiya **49:** 1703–1707.
18. ALBERY, W. J. & P. N. BARLTLETT. 1985. Amperometric enzyme electrodes: parts 1 and 2. J. Electroanal. Chem. **194:** 211–222, 223–235.
19. FOULDS, N. C. & C. R. LOWE. 1986. Enzyme entrapment in electrically conducting polymers. J. Chem. Soc. Faraday Trans. **82:** 1259–1264.
20. FORTIER, G. & D. BELANGER. 1991. Characterization of the biochemical behavior of glucose oxidase entrapped in a polypyrrole film. Biotechnol. Bioeng. **37:** 854–858.
21. FOULDS, N. C. & C. R. LOWE. 1988. Immobilization of glucose oxidase in ferrocene-modified pyrrole polymers. Anal. Chem. **60:** 2473–2478.
22. DEGANI, Y. & A. HELLER. 1989. Electrochemical communication between redox centers of glucose oxidase and electrodes via electrostatically and covalently bound redox polymers. J. Am. Chem. Soc. **111:** 2357–2358.
23. PISHKO, M. V., I. KATAKIS, S. E. LINDQUIST, A. HELLER & Y. DEGANI. 1990. Electrical communication between graphite electrodes and glucose oxidase redox polymer complexes. Mol. Cryst. Liq. Cryst. **190:** 221–249.
24. HALE, P. D., L. I. BOGUSLAVSKY, T. INAGAKI, H. I. KARAN, H. LEE, T. A. SKOTHEIM & Y. OKAMOTO. 1991. Amperometric glucose biosensors based on redox polymer-mediated electron transfer. Anal. Chem. **63:** 677–682.
25. DEGANI, Y. & A. HELLER. 1987. Direct electrical communication between chemically modified enzymes and metal electrodes. 1. Electron transfer from glucose oxidase to metal electrodes via electron relays, bound covalently to the enzyme. J. Phys. Chem. **91:** 1285–1289.
26. BOURDILLON, C. & M. MAJDA. 1990. Microporous aluminium oxide films at electrodes. Mediation of the catalytic activity of glucose oxidase via lateral diffusion of a ferrocene amphiphile in a bilayer assembly. J. Am. Chem. Soc. **112:** 1795–1799.
27. PARPALEIX, T., J. M. LAVAL, M. MAJDA & C. BOURDILLON. 1992. Potentiometric and voltammetric investigations of H_2/H^+ catalysis by periplasmic hydrogenase from *Desulfovibrio gigas* immobilized at the electrode surface in an amphiphilic bilayer assembly. Anal. Chem. **64:** 641–646.
28. FIOL, C., J. M. VALLETON, N. DELPIRE, G. BARBEY, A. BARRAUD & A. RUAUDEL-TEIXIER. 1991. Elaboration of a glucose biosensor based on Langmuir-Blodgett technology. 5th International Conference LB films, FP18, Paris.
29. BOURDILLON, C., J. P. BOURGEOIS & D. THOMAS. 1979. Chemically modified electrodes bearing grafted enzymes. Biotechnol. Bioeng. **21:** 1877–1879.
30. CASS, A. E. G., G. DAVIS, G. D. FRANCIS, H. A. O. HILL, W. J. ASTON, I. J. HIGGINS, H. PLOTKIN, L. D. L. SCOTT & A. P. F. TURNER. 1984. Ferrocene-mediated enzyme electrode for amperometric determination of glucose. Anal. Chem. **56:** 667–671.
31. BIANCO, P., J. HALADJIAN & C. BOURDILLON. 1990. Immobilization of glucose oxidase on carbon electrodes. J. Electroanal. Chem. **293:** 151–163.
32. FINEGOLD, L. & J. T. DONNELL. 1979. Maximum density of random placing of membrane particules. Nature **278:** 444–445.
33. MEGE, R. M. & C. BOURDILLON. 1985. Nickel controls the reversible anaerobic activation/inactivation of the *Desulfovibrio gigas* hydrogenase by the redox potential. J. Biol. Chem. **260:** 14701–14706.
34. LEVICH, V. G. 1962. *In* Physicochemical Hydrodynamics. N. R. Admunson, Ed.: 60–78. Prentice Hall. Englewood Cliffs, N.J.

35. GYSS, C. & C. BOURDILLON. 1987. Enzymatic electrocatalysis as a strategy for electrochemical detection in heterogeneous immunoassays. Anal. Chem. **59:** 2350–2355.
36. HUET, D., C. GYSS & C. BOURDILLON. 1990. A heterogeneous immunoassay performed on a rotating disk electrode with electrocatalytic detection. J. Immunol. Methods **135:** 33–41.
37. NYGREN, H. & M. STERNBERG. 1985. Calibration by ellipsometry of the enzyme linked immunosorbent assay. J. Immunol. Methods **80:** 15–24.
38. DELANCHE, B., S. TELLIER & M. ASTRUC. 1990. Mass transfer to a carbon or graphite felt electrode. Electrochim. Acta. **35:** 1369–1376.
39. BONNEFOY, J., J. MOIROUX, J. M. LAVAL & C. BOURDILLON. 1988. Electrochemical regeneration of NAD^+. A new evaluation of its actual yield. J. Chem. Soc. Faraday Trans. **84:** 941–950.
40. FASSOUANE, A., J. M. LAVAL, J. MOIROUX & C. BOURDILLON. 1990. Electrochemical regeneration of NAD in a plug-flow reactor. Biotechnol. Bioeng. **35:** 935–939.
41. BIADE, A., C. BOURDILLON, J. M. LAVAL, G. MAIRESSE & J. MOIROUX. 1992. Complete conversion of L-lactate into D-lactate. A generic approach involving enzymatic catalysis, electrochemical oxidation of NADH and electrochemical reduction of pyruvate. J. Am. Chem. Soc. **114:** 893–897.
42. BOURDILLON, C., R. LORTIE & J. M. LAVAL. 1988. Gluconic acid production by an immobilized glucose oxidase reactor with electrochemical regeneration of an artificial electron acceptor. Biotechnol. Bioeng. **31:** 553–558.
43. MILLER, C. J., C. A. WIDRIG, D. H. CHARYCH & M. MAJDA. 1988. Microporous aluminium oxide films at electrodes. Lateral charge transport in self organized bilayer assemblies. J. Phys. Chem. **92:** 1928–1936.

Inversion of Chirality with an Electroenzymatic Reactor

JEAN-MARC LAVAL,[a] CHRISTIAN BOURDILLON,[a] AZZ-EDDINE BIADE,[b] GILLES MAIRESSE,[b] AND JACQUES MOIROUX[b]

[a]*Laboratorie de Technologie Enzymatique*
URA 1442 du CNRS
Universite de Technologie de Compiegne, BP 649
60206 Compiegne Cedex, France

[b]*Laboratoire d'Electrochimie des Substrats Enzymatiques*
Faculte de Pharmacie
Universite de Picardie
1 rue des Louvels
80037 Amiens Cedex 1, France.

Previous work from our group[1–4] has demonstrated that electrochemical oxidation on a carbon electrode is an efficient way to regenerate NAD^+. Direct electrochemical regeneration of NAD^+ can be performed with a yield better than 99.99% without any addition of reactants, catalysts, and products into the medium. Dehydrogenases are potentially useful catalysis in chiral synthesis (reviews in References 5–8). Unfortunately the production of chiral compounds by direct enzymatic catalysis is classically NAD^+ producing and there is presently no way of achieving the direct electrochemical reduction of NAD^+ to NADH with a good yield (reviews in References 9–12).

In this work,[13] we developed a process where one enantiomer, namely, L-lactate, can be totally transformed into the other, i.e., D-lactate, by coupling the stereospecificity of the L-lactate dehydrogenase (L-LDH, E.C. 1.1.1.27.) catalyzed oxidation of L-lactate by electrochemically regenerated NAD^+ on the one hand, with electrochemical reduction of the oxidized form of the substrate, i.e., pyruvate, into the racemic reduced form of the substrate (D,L-lactate) on the other hand. As D-lactate is not transformed by any reaction, it progressively accumulates (FIGURE 1). It appears also that the electrochemical reactions, NADH oxidation and pyruvate reduction, allow the total shift of the enzymatically catalyzed equilibrium, a condition that must be fulfilled in order to achieve the selective oxidation of L-lactate.

Experimentally we used two flow-through reactors with two separated potentiostats, one for the anodic regeneration of NAD^+ on a carbon felt electrode, the other for the cathodic reduction of pyruvate on a mercury electrode, so as to control and monitor the essential parameters. The two reactors were coupled in a recycling configuration, and the experiment was allowed to progress on its own except for pH adjustment (pH 9) of the solution flowing through the working electrode compartments. The mercury electrode was held at a potential of -1.75 V vs. SCE (saturated calomel electrode) to reduce the pyruvate into the racemic D,L-lactate. Then NAD^+ was reduced into dimers $(NAD)_2$ and NADH isomers.[9,10] The carbon felt electrode

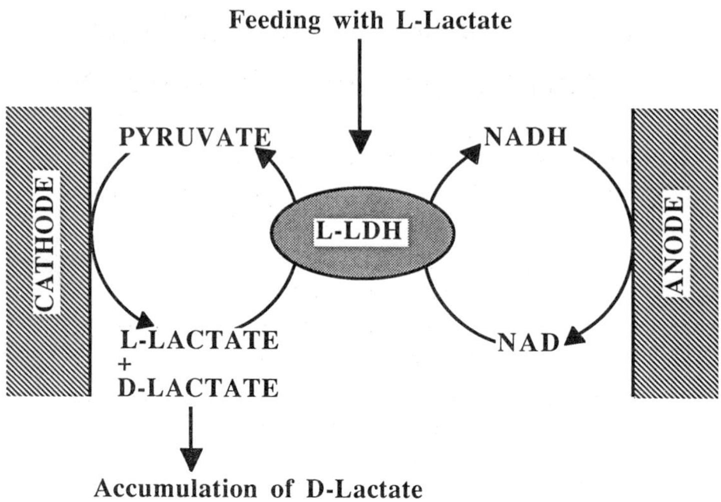

FIGURE 1. Scheme of the process.

was held at a potential of 0.5 V vs. SCE to oxidize both dimers $(NAD)_2$ and NADH isomers into enzymatically active NAD^+.[9,10,14]

The data obtained at the end of a typical experiment are shown in TABLE 1. The conversion coefficient of the L-lactate was 0.985 ± 0.015 and the production yield of the D-lactate was 0.99 ± 0.02. Therefore, within experimental uncertainty, L-lactate was completely converted into D-lactate. NAD^+ had to be regenerated more than 200 times and ca. 60% remained enzymatically active in the end. Moreover the fact that we succeeded under our conditions proves that we overcame two potential pitfalls: (1) With the electrochemical reduction, the pyruvate concentration was always maintained low enough to avoid significant side reactions.[15] (2) At pH 7, the dimers $(NAD)_2$ are known to decompose rapidly into NADH and NAD^+, but also

TABLE 1. Conversion of L-Lactate into D-Lactate: a Typical Experiment

	Working Electrode Compartments		Auxiliary Electrode Compartments	
	Initial Conditions	End of the Experiment	Initial Conditions	End of the Experiment
Volume	29 mL	47 mL	43 mL	25 mL
pH[a]	9.0	9.1	9.0	9.0
L-Lactate	2.00×10^{-2} M	$(1.5 \pm 1) \times 10^{-4}$ M[b]	0	$(1 \pm 1) \times 10^{-4}$ M[b]
D-Lactate	0	$(1.18 \pm 0.02) \times 10^{-2}$ M[b]	0	$(7 \pm 1) \times 10^{-4}$ M[b]
Pyruvate	0	$< 10^{-4}$ M	0	$< 10^{-4}$ M
NAD+	$(2.0 \pm 0.1) \times 10^{-4}$M[c]	$(0.7 \pm 0.1) \times 10^{-4}$ M	0	$< 10^{-5}$ M

[a]0.5 M carbonate buffer.
[b]Means and standard deviations (5 experiments).
[c]Initial enzyme concentration: 0.04 mg mL^{-1}.

into enzymatically inactive species.[16] The small inactivation of the coenzyme observed during experiments proves that the dimer is satisfactorily stable at pH 9.

If the configuration used in this work demonstrated the feasability of the chiral inversion scheme, the whole process was kinetically controlled by the rate of the α-keto-acid reduction. Improvement of productivity implies the use of a cathode material allowing an electroactive area per volume ratio close to the carbon felt. Then it would be possible to perform both the anodic and cathodic reactions in the compartments of the same reactor.

REFERENCES

1. LAVAL, J. M., C. BOURDILLON & J. MOIROUX. 1984. J. Am. Chem. Soc. **106:** 4701.
2. LAVAL, J. M., C. BOURDILLON & J. MOIROUX. 1987. Biotechnol. Bioeng. **30:** 157.
3. BONNEFOY, J., J. MOIROUX, J. M. LAVAL & C. BOURDILLON. 1988. J. Chem. Soc. Faraday Trans. I. **84:** 941.
4. FASSOUANE, A., J. M. LAVAL, J. MOIROUX & C. BOURDILLON. 1990. Biotechnol. Bioeng. **35:** 935.
5. JONES, J. B. 1985. Asymmetric Synthesis. J. D. Morrison, Ed. Academic Press. New York, N.Y.
6. WHITESIDES, G. M. & C. H. WONG. 1983. Aldrichim. Acta **16:** 27.
7. HIRSCHBEIN, B. L. & G. M. WHITESIDES. 1982. J. Am. Chem. Soc. **104:** 4458.
8. WANDREY, C., A. F. BUCKMANN & M. R. KULA. 1981. Biotechnol. Bioeng. **23:** 2789.
9. ELVING, P. J. 1976. *In* Topics in Bioelectrochemistry and Bioenergetics. G. Milazzo, Ed. **1:** 278. Wiley Interscience. New York, N.Y.
10. ELVING, P. J., W. T. BRESNAHAM, J. MOIROUX & Z. SAMEC. 1982. Bioelectrochem. Bioenerg. **9:** 365.
11. ALLEN, P. M. & W. R. BOWEN. 1985. Trends Biotechnol. **3:** 145.
12. BERGEL, A., H. DURLIAT & M. COMTAT. 1987. J. Chim. Phys. **84:** 593.
13. BIADE, A. E., C. BOURDILLON, J. M. LAVAL, G. MAIRESSE & M. MOIROUX. 1992. J. Am. Chem. Soc. **114:** 893.
14. DRAKESMITH, F. G. & B. GIBSON. 1988. J. Chem. Soc. Chem. Commun.: 1493.
15. CHENAULT, H. K. & G. M. WHITESIDES. 1989. Bioorg. Chem. **17:** 400.
16. BRESNAHAM, W. T. & P. J. ELVING. 1981. Biochim. Biophys. Acta **678:** 151.

Antiidiotypic Antibodies Exhibiting an Acetylcholinesterase Abzyme Activity

LAURENT JORON,[a] LADAN IZADYAR,[b] ALAIN FRIBOULET,[b,c] MARIE-HÉLÈNE REMY,[b] GIANFRANCO PANCINO,[a] ALBERTO ROSETO,[a] AND DANIEL THOMAS[b]

[a]*Division d'Immuno-Cytologie Appliquée*
[b]*Laboratoire de Technologie Enzymatique*
URA 1442 du CNRS
Université de Technologie de Compiègne, BP 649
60206 Compiègne Cedex, France

INTRODUCTION

In 1986, catalytic antibodies (abzymes) were obtained with monoclonal antibodies elicited to haptens that are analogues of the transition state for hydrolysis of esters.[1,2] Since then more sophisticated transformations have been reported, including bimolecular reactions, pericyclic rearrangements, photoreductions, redox reactions, elimination reactions, or peptide bond hydrolysis.[3] Since the network theory of immune regulation, the properties of antiidiotypic antibodies (anti-IdAbs) have been investigated widely. The combining site of an anti-IdAb may contain information regarding the three-dimensional structure of an antigen.[4] Anti-IdAbs have thus been used as probes of protein active sites and in structure-function analysis in receptor-ligand interactions. However, these remarkable properties of "internal imagery" have not been exploited to design catalytic anti-IdAbs.

The monoclonal antibody AE-2 raised against human erythrocyte acetylcholinesterase (AChE) has been previously described to inhibit the enzyme activity.[5] The epitope resides on each of the individual subunits of the enzyme, but is still not completely characterized. However, competition studies with active site-specific directed reversible inhibitors of AChE (BW 284C51, edrophonium, or 2-PAM) indicate that the AE-2 epitope is putatively one of the anionic sites.[6,7] In the present paper, the monoclonal antibody AE-2 was used to induce antiidiotypic antibodies. These anti-IdAbs are exhibiting an acetylcholinesterase abzyme activity.

MATERIALS AND METHODS

Materials

ABTS (2,2′-azino-bis[3-ethylbenzthiazoline-6-sulfonic acid]), acetylthiocholine (ATCh), BW 284 C 51 (1,5-bis(4-allyldimethylammoniumphenyl)pentan-3-one dibromide), DTNB (5,5′-dithio-bis[2-nitrobenzoic acid]), DFP (diisopropyl fluorophosphate), edrophonium, ethopropazine, gallamine, and AChE from human erythro-

[c]Author to whom correspondence should be addressed.

cytes (type XIII) were from Sigma (St. Louis, Mo.). Goat antirabbit immunoglobulin G (IgG), peroxidase-linked goat antirabbit IgG, and peroxidase-linked goat antimouse IgG were from Biosys (Compiègne, France). All other chemicals and reagents used were of analytical grade or of the highest available purity.

Methods

Production of Anti-IdAbs

Monoclonal antibody (mAb) AE-2 was prepared by Fambrough *et al.*,[5] and the cell line secreting this antibody was obtained from the American Type Culture Collection (New York, N.Y.). mAb AE-2 was amplified in mouse ascitic fluids and purified by protein A–Sepharose (Biorad kit). Antisera were obtained by immunizing rabbits with the protein A–Sepharose–purified AE-2 preparation (300 μg/rabbit) in the presence of Freund's complete adjuvant. Booster injections (100 μg) were given 7, 21, and 64 days after the primary immunization. Anti-IdAbs were purified on protein A–Sepharose column and on an affinity column prepared by covalent binding of AE-2 to cyanogen-bromide-activated Sepharose 4B. Anti-AE-2 immunoglobulins from rabbit retained on the latter column were eluted with 0.2-M glycine/HCl buffer, pH 2.5. Fractions of 2 ml were collected in test tubes containing 1 ml 1-M Tris buffer, pH 7.4, in order to neutralize the acidic elution buffer rapidly.

Solid Phase Assays

Wells of polystyrene microtiter plates from Nunc (Roskilde, Denmark) were treated overnight with protein A purified $F(ab')_2$ fragments of mAb AE-2 (10 μg/ml) in carbonate buffer pH 9.3, and preadsorbed with 1% gelatine in phosphate-buffered saline pH 7.2 (PBS) for 1 hour at room temperature to reduce the nonspecific binding. The plates were washed three times with PBS containing 0.05% Tween-20, then PBS, and either used immediately or stored at −20°C. Microtiter plates were coated following the same procedure as above with either human erythrocyte AChE (50 μg/ml), for neutralization inhibition immunoassays (NI-EIA), or with goat antirabbit IgG (20 μg/ml), for immobilization of anti-IdAbs.

Enzymatic Activity Determination

AChE activity and inhibition kinetics of both the enzyme and anti-IdAbs were monitored in solution by the method of Ellman[8] using 1-mM acetylthiocholine (ATCh) and 1-mM DTNB in 10-mM phosphate buffer pH 7.2. The catalytic activity of anti-IdAbs attached to microtiter plates was evaluated after extensive washing with PBS by a modified Ellman procedure.[9] In both cases, the increase in absorbance was followed at 405 nm. The peroxidase activity was followed at 405 nm with 1-mM ABTS and 0.003% H_2O_2 as substrates in 100-mM phosphate-citrate buffer pH 5.0.

Sucrose Density Gradient Ultracentrifugation

The sedimentation of rabbit serum AChE and anti-IdAbs was examined on 11-ml gradients of sucrose (5–20% w/v) prepared in 10 mM Tris-HCl (pH 7.2), 1 M NaCl, and 0.5% Triton X-100. Samples (75 μl) were centrifuged at 4°C for 40 hours at 38,000 rpm in a SW 41 rotor. Horse liver alcohol dehydrogenase (ADH_{HL}, 4.8 S), dimeric (G2) AChE from *Torpedo californica* (5.6 S), and yeast alcohol dehydrogenase (ADH_Y, 7.4 S) were used as sedimentation markers. Gradient solutions were recovered by collecting 315-μl fractions and 100-μl aliquots were used for AChE activity measurement or immunoglobulin determination.

RESULTS

Production and Characterization of the Polyclonal Anti-IdAbs

To induce anti-idiotypic antibodies, rabbits were immunized with protein A–Sepharose–purified AE-2. Rabbit serum was titrated on its immunoglobulin content to AE-2-$(Fab')_2$ fragments with a solid phase enzyme immunoassay. Polystyrene microtiter wells were incubated overnight with protein A–purified $(Fab')_2$ fragments of AE-2 in carbonate buffer (pH 9.3), serial dilutions of rabbit antiserum were added, and bound antibodies revealed with peroxidase-labeled goat antirabbit immunoglobulins. The titer of the adsorbed rabbit antiserum was about 1/16,000. The inhibition of binding of mAb AE-2 to AChE by the serum containing the anti-IdAbs was determined by a neutralization inhibition enzyme immunoassay (NI-EIA)[10] (FIGURE 1). AE-2 (20 μg/ml) was preincubated (1 hour, 37°C) with serial dilutions of the immune serum or of a nonimmune rabbit serum. Binding of AE-2 to coated AChE (50 $\mu g \times ml^{-1}$) was thus tested. AE-2 binding was inhibited by immune serum, but not by a nonimmune serum. The immune serum in the absence of AE-2 was used as a negative control.

Anti-IdAbs (against both combining site and non–combining site idiotopes) were isolated from rabbit serum following sequential adsorption of the serum to immunoadsorbent columns (protein A–Sepharose then AE-2-Sepharose) and subsequent elution of anti-IdAbs from the AE-2-Sepharose column. A similar inhibition of AE-2 binding to AChE was obtained when AE-2 was preincubated with protein A–Sepharose- and AE-2-Sepharose-purified antibodies instead of the immune serum (data not shown). AE-2-Sepharose-purified anti-IdAbs were adsorbed on goat antirabbit antibodies, and AChE activity was assayed. The curves (FIGURE 2) indicate linearity between adsorbed anti-AE-2 level and AChE activity. Moreover, when a normal rabbit serum was subjected to the same purification steps as anti-IdAbs (protein A–Sepharose then AE-2-Sepharose) and then adsorbed on goat antirabbit antibodies, no rabbit immunoglobulin binding or AChE activity could be detected. Thus, the measured AChE activity cannot be explained by the presence of free AChE molecules contaminating the anti-IdAb preparations, after the purification steps, but could be due to the formation of a specific active complex between the anti-IdAbs and rabbit serum AChE.

Molecular Characteristics of the Anti-IdAbs

Sedimentation velocity analysis allows the resolution of the distinct molecular entities bearing AChE activity. In normal rabbit serum, AChE is mainly present as a tetrameric (G4) enzyme,[11] but a small amount (about 5% of total AChE activity) of a 6.1-S entity, corresponding to a dimeric (G2) form, may be identified (FIGURE 3a). The formation of a specific active complex between anti-IdAbs and a monomeric or dimeric AChE form would lead to an active form sedimenting at least at 9 S or 11 S respectively.[12] The AE-2-Sepharose-purified anti-IdAbs were thus analyzed on continuous 5–20% sucrose gradients (FIGURE 3b). When anti-IdAbs were revealed by peroxidase-linked goat antirabbit immunoglobulins, a main peak was obtained

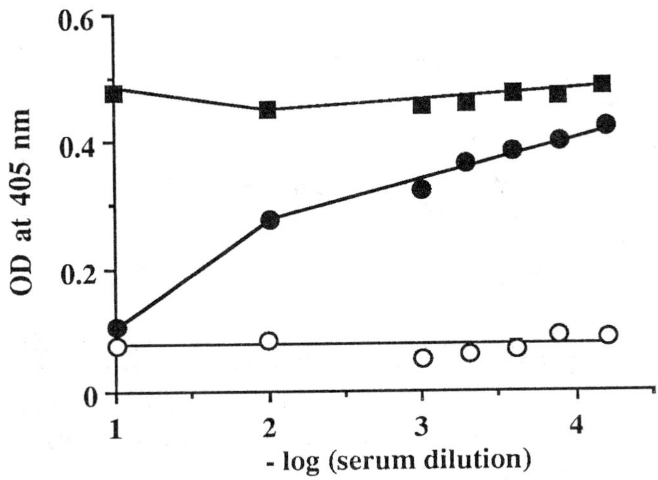

FIGURE 1. Neutralization of AE-2 binding to absorbed erythrocyte AChE by anti-IdAbs present in the immune serum. AE-2 (20μg/ml) was preincubated with different dilutions, from 1/10 to 1/16 000, of an immune serum (●) and a nonimmune serum (■) for 1 hour at 37°C and then added to microtiter plates coated with AChE (50 μg/ml). Bound AE-2 was revealed by purified peroxidase-linked goat antimouse IgG. Immune serum alone (○), without preincubation with AE-2, was used as a negative control.

with a sedimentation coefficient of 5.7 S with a marked shoulder at 6.8 S. When the fraction contents were tested for their AChE activity, a single peak (6.8 S) was obtained. Thus, the catalytic activity revealed in the anti-IdAb preparation is not due to the formation of specific active complexes between anti-IdAbs and AChE present in immunized rabbit serum.

Kinetic Properties and Inhibition Susceptibility of Anti-IdAbs in Solution

When studied in solution, the hydrolysis of ATCh by anti-IdAbs follows characteristic Michaelis-Menten kinetics. In contrast to human erythrocyte AChE (HE-

AChE) and rabbit serum AChE (RS-AChE), which exhibit the usual excess substrate inhibition above 10^{-3} M substrate, the Lineweaver-Burk plot of ATCh hydrolysis kinetics by anti-IdAbs does not deviate from linearity. The kinetic parameters determined by the method of Ellman are summarized in TABLE 1. It is important to note that if the value of the catalytic constant (k_{cat}) is evaluated from the concentration of antibodies, and due to their polyclonal nature is thus a minimal value, the Michaelis constant value (K_m) is an intrinsic property of the catalyst that does not depend on its concentration. This value largely differs from the values measured with HE-AChE and RS-AChE (more than 500-fold) and from the values reported for hydrolysis of ATCh by pseudocholinesterases (more than 80-fold). Such

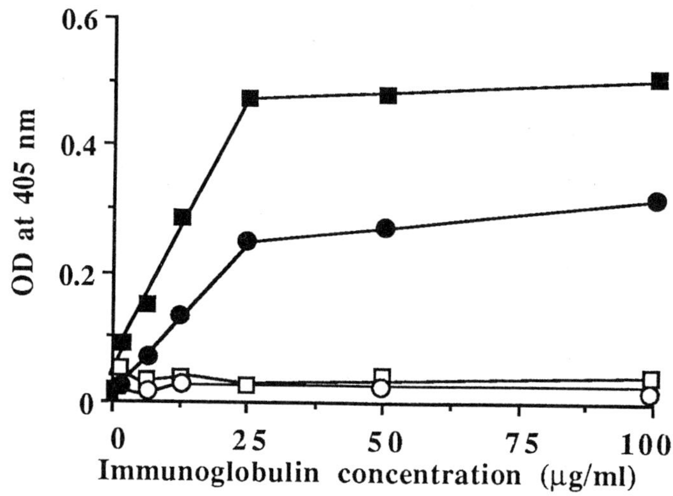

FIGURE 2. AChE activity (●) and immunoglobulin level (■) adsorbed on goat antirabbit immunoglobulins as a function of AE-2-Sepharose-purified anti-IdAbs concentration. Different dilutions of AE-2-Sepharose-purified anti-IdAbs (98 μg/ml) were added (100 μl/well) for 1 hour at 37°C in wells coated with goat antirabbit IgG. Adsorbed anti-IdAbs AChE activity was measured, and bound immunoglobulins determined with horseradish-peroxidase-labeled goat antirabbit antibodies. Nonimmune rabbit IgG submitted to the same purification steps was used at the same concentration as negative controls in both parts of the assay, (□) for peroxidase-labeled anti-rabbit and (○) acetylcholinesterase activity.

a high K_m value would be indicative of the action of a nonspecific esterase. However, when the hydrolysis of different esters of thiocholine (1 mM final concentration) by the anti-AE-2 preparation was checked, we found that the catalytic reaction is highly specific for the acetyl ester. The relative velocities for propionyl and butyryl esters when compared with the acetyl ester reach 48% and 5% respectively.

Many specific inhibitors of AChE are known. Among them, organophosphates irreversibly block the enzyme by phosphorylation of the active site serine, while many quaternary ammonium compounds are specific reversible inhibitors of the enzyme. The AChE activity of the anti-IdAbs was reduced to 47% and 2% of its original value after 1 hour incubation at 20°C with 2.5 μM and 25 μM DFP (an irreversible

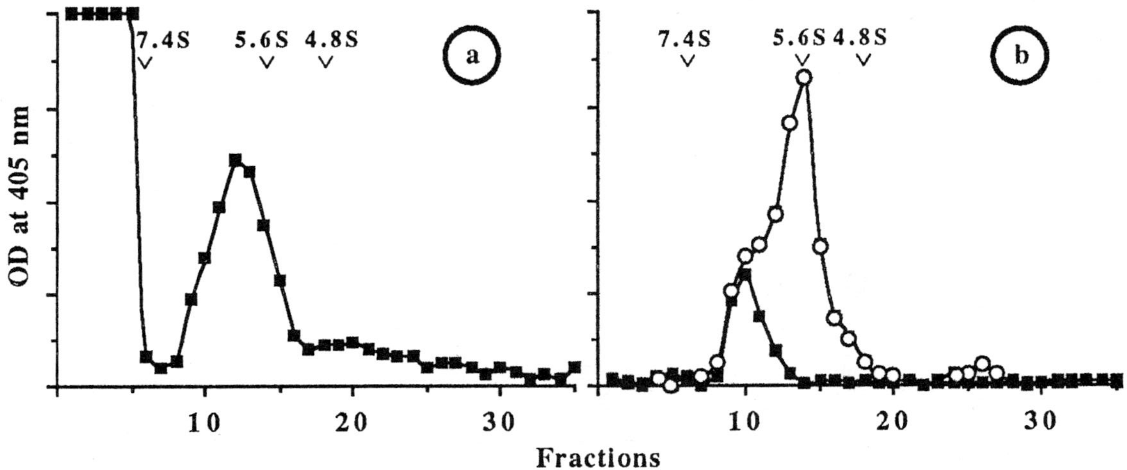

FIGURE 3. Sedimentation profiles of normal rabbit serum AChE (a) and AE-2-Sepharose-purified anti-IdAbs (b). The fractions collected from the anti-IdAbs preparation were tested both for their AChE activity (■) and their content in anti-IdAbs (○). AChE activity was measured in the presence of 10 μM ethopropazine to inhibit pseudocholinesterase in the case of rabbit serum. The presence of anti-IdAbs was evaluated by incubating 2 hours, at 37°C, the gradient fractions in 96-well microtiter plates previously coated with $(Fab')_2$ fragments of mAb AE-2. After rinsing, bound anti-IdAbs were revealed with peroxidase-linked goat antirabbit IgG. The sedimentation coefficients were calculated by comparison with sedimentation standards.

phosphorylating inhibitor of AChE) respectively. This result is strongly indicative of the involvement of a serine residue in the hydrolytic mechanism. Moreover, in the presence of 1-mM substrate, 1 μM BW 284C51 or 20 μM edrophonium, which are specific reversible inhibitors of AChE, reduced the hydrolytic activity to 5% of its original value. On the other hand, gallamine (1 mM), which is known as a peripheral site-directed inhibitor of AChE, and ethopropazine (1mM), a specific inhibitor of pseudocholinesterases, never affect the catalytic activity of the anti-IdAbs. Thus the cholinesterase activity exhibited by anti-IdAbs is a true AChE activity with respect to substrate and inhibitor specificity, but is not inhibited by excess substrate or by peripheral site-directed inhibitors in contrast to the enzyme present on red blood cells or in the serum.

TABLE 1. Kinetic and Equilibrium Data for the Hydrolysis of Acetylthiocholine by Human Erythrocyte AChE (HE-AChE), Rabbit Serum AChE (RS-AChE), and Anti-IdAbs[a]

	K_m (mM)	K_{ss} (mM)	k_{cat} (s^{-1})	k_{uncat} (s^{-1})	Enhancement Ratio (k_{cat}/k_{uncat})
HE-AChE[b]	0.13	22	8000	$1.95\ 10^{-7}$	$4.10\ 10^{10}$
RS-AChE	0.12	18.3	3470	$1.95\ 10^{-7}$	$1.78\ 10^{10}$
Anti-IdAbs	65		17.5	$1.95\ 10^{-7}$	$8.97\ 10^{7}$

[a]RS-AChE active sites were titrated with the irreversible inhibitor *O*-ethyl-*S*-(diisopropyl-aminoethyl-2) methanethiophosphonate (MPT), and kinetics were followed in the presence of 10 μM ethopropazine to inhibit pseudocholinesterases. Values are the means of three individual experiments.

[b]Calculated from References 6 and 15.

DISCUSSION

The anti-IdAbs elicited against the inhibitory monoclonal antibody AE-2 exhibit a cholinesterase activity. This activity is only present in the immune serum, and can thus not be explained by the presence of contaminating free serum cholinesterase molecules, after the purification steps on protein A–Sepharose then AE-2-Sepharose columns. The hypothesis that a specific active complex between anti-IdAbs and serum cholinesterase that could copurify with anti-IdAbs can be ruled out by sedimentation velocity analysis on sucrose gradients and inhibition experiments. On the other hand, the measured K_m value for the hydrolysis of ATCh by the anti-IdAb preparation unambiguously indicates that the measured activity cannot be due to a classical cholinesterase. However, from inhibition studies with active-site-specific inhibitors of AChE, and from substrate-specificity experiments, it can be concluded that the antiidiotypic abzyme exhibits a specific AChE activity. When compared with esterase abzyme activities previously described, the affinity is lower but the catalytic activity and enhancement ratio (10^8) are very high if we take into account the polyclonal nature of our preparation.

The described study using "internal imagery" could be a general method for studying enzyme molecules by creating similar three-dimensional structures with different amino acid sequences. The production of monoclonal catalytic anti-

IdAbs,[13] obtained from the same enzymatic active site, could be a new way for studying structure-activity relationships of a catalytic activity and also a new approach for the definition of idiotopes.

In the framework of enzyme technology, catalytic anti-IdAbs could be an alternative solution versus site-directed mutagenesis to get different specificity and affinity. For instance a higher K_m value gives rise to wider ranges of linearity when enzymes (i.e., AChE) are used as the active part of a biosensor. The described method is able to give polyclonal catalytic antibodies, and the induction of human catalytic anti-IdAbs could be a new therapeutical tool (i.e., urokinase or tPA activities). Our results could also be an explanation for the catalytic activities observed with human autoantibodies.[14]

REFERENCES

1. TRAMONTANO, A., K. D. JANDA & R. A. LERNER. 1986. Science **234:** 1566–1570.
2. POLLACK, S. J., J. W. JACOBS & P. G. SCHULTZ. 1986. Science **234:** 1570–1574.
3. LERNER, R. A., S. J. BENKOVIC & P. G. SCHULTZ. 1991. Science **252:** 659–667.
4. LUDWIG, D. S., R. A. FINKELSTEIN, A. E. KARU, W. S. DALLAS, E. R. ASHBY & G. K. SCHOOLNIK. 1987. Proc. Natl. Acad. Sci. USA **84:** 3673–3677.
5. FAMBROUGH, D. M., A. G. ENGEL & T. L. ROSENBERRY. 1982. Proc. Natl. Acad. Sci. USA **79:** 1078–1082.
6. SORENSEN, K., U. BRODBECK, A. G. RASMUSSEN & B. NORGAARD PEDERSEN. 1987. Biochim. Biophys. Acta **912:** 56–62.
7. WOLFE, A. D. 1989. Biochim. Biophys. Acta **997:** 232–235.
8. ELLMAN, G. L., K. D. COURTNEY, V. ANDRES & R. M. FEATHERSTONE. 1961. Biochem. Pharmacol. **7:** 88–95.
9. DOCTOR, B. P., L. TOKER, E. ROTH & I. SILMAN. 1987. Anal. Biochem. **166:** 399–403.
10. OOSTERLAKEN, T. A. M., M. HARMSEN, C. TANGERMAN, P. SCHIELEN, C. A. KRAAIJEVELD & H. SNIPPE. 1988. J. Immun. Methods **115:** 255–261.
11. SKAU, K. A. 1985. Comp. Biochem. Physiol. **80C:** 207–210.
12. BRIMIJOIN, S. & K. P. MINTZ. 1985. Biochim. Biophys. Acta **828:** 290–297.
13. IZADYAR, L., A. FRIBOULET, M. H. REMY, A. ROSETO & D. THOMAS. (In preparation.)
14. PAUL, S., D. J. WOLLE, C. M. BEACH, D. R. JOHNSON, M. J. POWEL & R. J. MASSEY. 1989. Science **244:** 1158–1162.
15. BRODBECK, U., R. GENTINETTA & P. OTT. 1981. *In* Membrane Proteins. A Laboratory Manual. A. Azzi, U. Brodbeck & P. Zahler, Eds.: 85–96. Springer-Verlag. Berlin, Germany.

A New Enzymatic Method for the Determination of 3-Methoxy-4-hydroxymandelic Acid

ROGER J. HINTON, J. RICHARD RAMSAY,
TONY ATKINSON, KEVIN J. BOWN,
AND PETER M. HAMMOND

Division of Biotechnology
Center for Applied Microbiology and Research
Porton Down, Salisbury
Wiltshire, SP4 0JG, United Kingdom.

INTRODUCTION

Concurrent with the discovery of 3-methoxy-4-hydroxymandelic acid (vanillylmandelic acid, VMA) in human urine was the observation that the level of excretion of this acid was highly elevated in patients with pheochromocytoma.[1] Pheochromocytoma is a tumor that arises from the chromaffin cells in the adrenal medulla. It can occur at any age from between 9 months and 90 years, but is found predominantly in the middle-aged, with an estimated incidence of at least 0.1% of the hypertensive (high blood pressure) population.[2] Clinical diagnosis is extremely difficult, with patients presenting a wide variety of symptoms. A review of a 50-year autopsy series has indicated that more tumors were missed than were diagnosed.[3] Therefore, the only reliable indicator is biochemical analysis, with the demonstration of increased urine levels of the catecholamine metabolite VMA. Over 90% of patients are reliably diagnosed by the proper analysis of a single 24-hour urine sample for VMA.[4] A 5-year survival rate of 96% is possible provided early detection is made.[5]

The urinary level of VMA is also relevant to the diagnosis of neuroblastoma,[6] a malignant tumor arising from the adrenal medulla. This is the most frequent solid tumor in children with an incidence of 1 in 5000; more than 80% of patients are under the age of 4 years.[7] The prognosis for neuroblastoma is poor and most patients die before age 7 years unless treatment begins in the initial stages.[8] If the primary tumor is detected early and removed before the child is a year old, there is a very good chance of complete recovery. There are few other conditions where the responsibility for diagnosis depends so heavily on the laboratory as pheochromocytoma and neuroblastoma,[9] and there have been many calls for the introduction of mass screening.[10] This was introduced in Japan in 1986,[11] and on a smaller scale, regional neonatal screening has been implemented in Newcastle in the United Kingdom.[12] However, the popularity of screening programs has hitherto been limited by the lack of suitable methods for the quantitative determination of VMA. A number of procedures have been published, including gas chromatography,[12] high-performance liquid chromatography,[13] gas chromatography–mass spectrometry,[14] and enzyme immunoassay,[15] but as these are time-consuming, expensive, and require technical expertise, they are not suited to mass screening.

In contrast, microbial-enzyme-based assays have found increasing application in clinical chemistry because of the advantages they offer in terms of speed, specificity, lack of toxicity, and their economical cost of production.[16] Such assays can be readily automated, further reducing the cost per test and allowing rapid analysis of a large number of samples. Through the use of "dipstick" technology, extralaboratory testing can be rapidly performed.[17] We describe the development of an enzymatic method for the analysis of urinary VMA that may provide a practical means of mass screening, thereby allowing the early diagnosis of two eminently curable tumors.

EXPERIMENTAL

A microorganism identified using the API system as *Pseudomonas putida* was isolated from soil on the ability to utilize DL-mandelate as a sole source of carbon. The use of vanillylmandelic acid as a substrate was also investigated, but the presence of a methoxy-group at the meta position was found to result in the compound showing a high degree of toxicity towards microorganisms (results not shown), and no isolates were obtained with this compound as sole carbon source. Mandelate acted as a good growth substrate and induced an enzyme that also expressed activity towards the methoxy derivative VMA.

The pathway of mandelate degradation by the organism was similar to that described by Durham,[18] although our strain produced both D- and L-mandelate dehydrogenase and did not interconvert these substrates by means of a racemase. The isolate was initially grown and maintained at pH 7.0 on a minimal salts medium comprised of (g/L) K_2HPO_4, (3.0); KH_2PO_4, (1.0); KCl, (3.0); NaCl, (5.4); NH_4Cl, (1.1); $MgCl_2$, (0.1); $NaSO_4$, (0.02); $CaCl_2$, (0.01); $FeCl_3$, (0.0001); DL-mandelate (1.0). A heavy metals supplement was added as a concentrate at 1 ml/L.[19,20] Cultures were grown in 100-mL aliquots of medium, in 250-mL Erlenmeyer flasks at 30°C in an orbital incubator and subcultured at three-day intervals. The tolerance of the organism to DL-mandelate was raised over a period of weeks from 0.1%(w/v) to 0.5%(w/v). Further improvement of growth and enzyme yield was achieved by the use of the complex medium tryptone soya broth (TSB) and the inducer phenoxyacetic acid. The incorporation of phenoxyacetic acid was particularly useful; it acted as a powerful inducer of mandelate dehydrogenase (MDH) but not as a substrate, and so its concentration did not need monitoring or control during fermentation. For pilot-scale production, a 3.6-L seed culture was initially grown in a minimal salts medium containing DL-mandelate as a sole carbon source for 24 hours. This was used to inoculate a 150-L fermenter containing 2% (w/v) TSB and 0.5% (w/v) phenoxyacetic acid. Typical cell yields of 8 g/L were obtained from these batch cultures, with enzyme yields of 160 U/g (wet weight) cells.

MDH assays were performed at 30°C in preincubated cuvettes of 100-mM potassium phosphate buffer pH 7.4 + 0.5% Tween-80 containing 0.4-mM iodonitrotetrazolium violet (INT: extinction coefficient E500 nm = 35,200) and 0.15-mM phenazine methosulfate (PMS). The reductive formation of the colored formazan was monitored spectrophotometrically at 500 nm. One unit of enzyme activity was

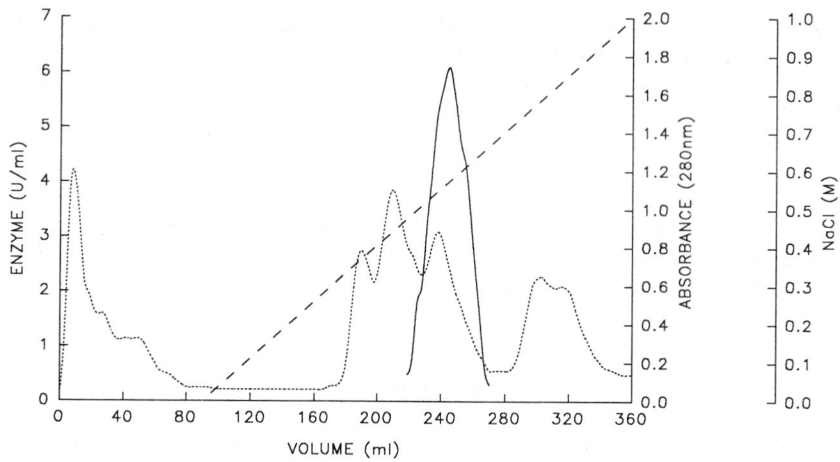

FIGURE 1. Chromatographic separation of mandelate dehydrogenase on Q-Sepharose. Dotted line: absorbance (280 nm). Solid line: mandelate dehydrogenase (U/mL). Dashed line: NaCl concentration (M).

defined as the quantity of enzyme required to reduce one μmole of substrate per minute.

Small-scale cell disruption was performed by sonication in 100-mM Tris-HCl (pH 7.4), containing 0.3% (v/v) Triton X-100 to aid the release of membrane bound MDH from the cell debris. The cleared cell lysate was loaded onto a 100 × 30 mm Q-Sepharose column equilibrated with 50-mM Tris-HCl pH 7.5 containing 0.1-mM phenylmethylsulfonyl fluoride. The column was washed with loading buffer and

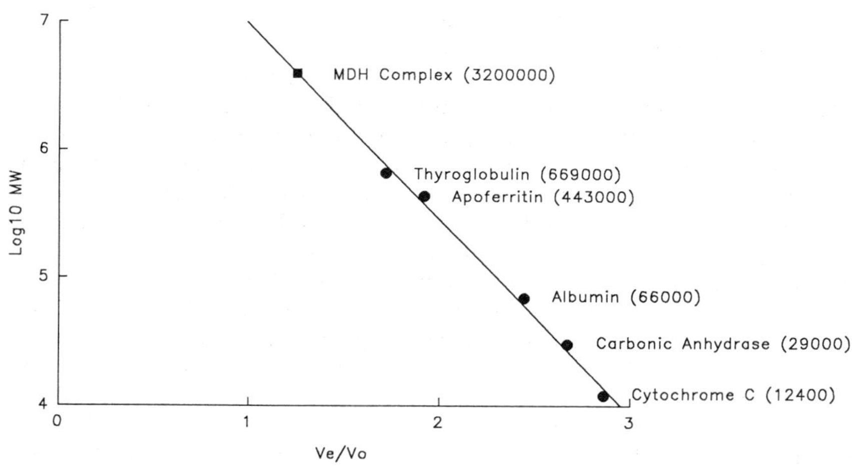

FIGURE 2. Determination of native molecular weight of the DL-mandelate dehydrogenase complex by gel filtration fast performance liquid chromatography on Superose 6 HR. (Numbers in parentheses refer to relative molecular mass.)

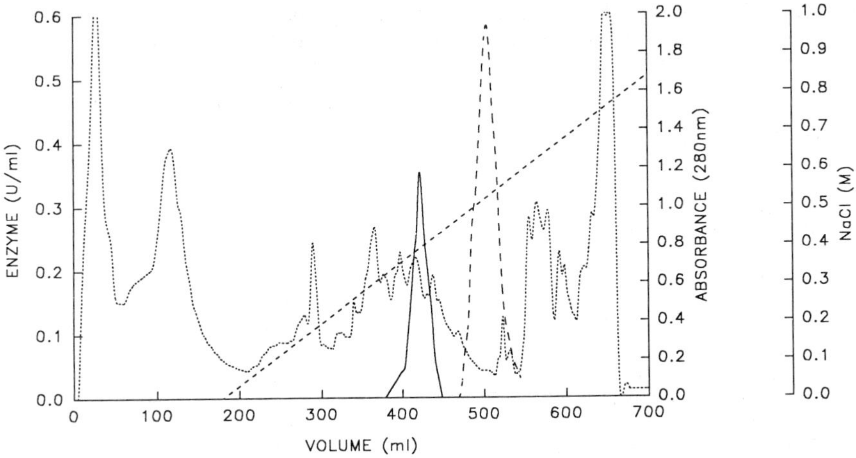

FIGURE 3. Chromatographic separation of D- and L-mandelate dehydrogenase by anion exchange fast performance liquid chromatography on Mono-Q HR. Dotted line: absorbance (280 nm). Solid line: D-mandelate dehydrogenase (U/mL). Large-dashed line: L-mandelate dehydrogenase (U/mL). Small-dashed line: NaCl concentration (M).

eluted with a 270-mL gradient from 0–1 M NaCl in the same buffer. MDH eluted at approximately 550-mM NaCl, with a recovery of 85% and a 5-fold purification (FIGURE 1). Gel-filtration in 50-mM Tris/HCl buffer at pH 8.0 using Superose-6 HR 10/30 indicated that the enzyme existed as a large-molecular-weight complex (FIGURE 2); this was confirmed by an activity stained native polyacrylamide gel electrophoresis gel. The complex contained both D- and L-MDH, which were separated by

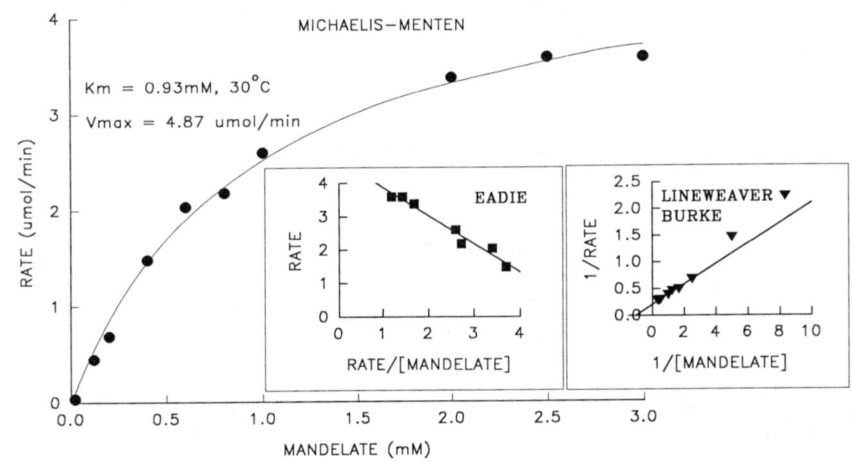

FIGURE 4. Substrate kinetics for D-mandelate dehydrogenase. Michaelis Menten, Eadie, and Lineweaver-Burke plots were calculated using the Enzfitter computer data analysis program, available from Sigma Chemicals Ltd.

TABLE 1. Substrate Specificity of D-Mandelate Dehydrogenase from *Pseudomonas putida,* Expressed as a Percentage of the Activity towards D-Mandelate

Substrate	Concentration in Assay (mM)	Percent Activity
D(−) Mandelate	1.31	100
L(+) Mandelate	1.31	0
L-Phenylalanine	1.21	0
L-Tyrosine	1.10	0
L-Dihydroxyphenlalanine	1.01	56
L-Dopamine	1.06	70
(±) Noradrenaline HCl	0.97	29
(±) Adrenaline	0.54	41
3,4-Dihydroxymandelic acid	1.09	61
DL-Metanephrine HCl	0.86	0
4-Hydroxymandelic acid	1.19	66
3-Methoxy-4-hydroxtmandelic acid (VMA)	1.00	75
3-Methoxymandelic acid	1.10	50
3-Methoxy-4-hydroxybenzyl alcohol	1.61	0
Vanillic acid	1.19	6

anion-exchange chromatography using Mono Q HR 16/10. The enzyme was adsorbed in 50-mM Tris-phosphate buffer (pH 7.0) containing 2% (v/v) Tween-80, and eluted with a 600-mL gradient from 0–1 M NaCl in the same buffer (FIGURE 3). D-MDH eluted at approximately 400-mM and L-MDH at 520-mM NaCl.

Preliminary enzyme kinetics (determined in 100-mM phosphate buffer, pH 7.4) for D-mandelate (FIGURE 4) showed that the enzyme had a K_m of 0.93 mM (30°C) and a V_{max} of 4.87 μmol/minute. TABLE 1 shows the activity of the enzyme towards related catecholamines and metabolites. The enzyme has been linked to an INT/PMS system in an assay for the quantitative determination of D-VMA (FIGURE 5). The assay, which can be performed as an end-point or kinetic determination is currently undergoing clinical trials and is expected to be the first enzymatic assay available to date.

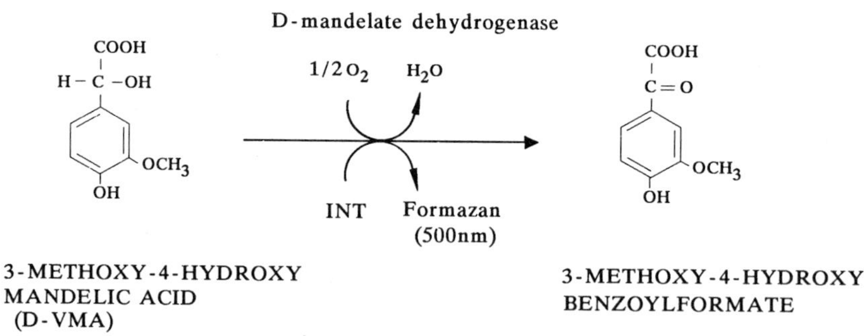

FIGURE 5. Reaction scheme for the enzyme-based determination of D-vanillylmandelic acid using D-mandelate dehydrogenase.

REFERENCES

1. Armstrong, M. D., A. McMillan & K. N. F. Shaw. 1957. 3-Methoxy-4-hydroxymandelic acid, a urinary metabolite of norepinephrine. Biochim. Biophys. Acta **25:** 422–423.
2. Manager, W. M. & R. W. Gifford, Jr. 1977. Pheochromocytoma. Springer-Verlag. New York, N.Y.
3. St. John Sutton, M. G., S. G. Sheps & J. T. Lie. 1981. Prevalence of clinically unsuspected pheo chromo cytoma review of a 50 year autopsy series. Mayo Clin. Proc. **56:** 354–360.
4. Al-Nuaim, A. & M. El-Desouki. 1989. Pheochromocytoma; diagnosis and treatment. Saudi Med. J. **10:** 446–453.
5. Remine, W. H., G. C. Chong, J. A. Van Heerden, S. G. Sheps & E. G. Harrison, Jr. 1974. Current management of pheochromocytoma. Ann. Surg. **179:** 740–748.
6. Gitlow, S. E., L. B. Dziedzic, S. M. Greenwood & S. W. Dziedzic. 1973. Biochemical and histologic determinants in the prognosis of neuroblastoma. Cancer **32:** 898–905.
7. Young, J. L., L. G. Reis, E. Silverberg, J. W. Horm & R. W. Miller. 1986. Cancer incidence, survival and mortality for children younger than age 15 years. Cancer **58:** 598–602.
8. Pillai, D. N., J. W. Earl & M. W. Duncan. 1991. Clin. Biochem. Revs. **12:** 14–21.
9. Weinkove, C. 1991. Measurement of catecholamines and their metabolites in urine. J. Clin. Pathol. **44:** 269–275.
10. Craft, A. W. & G. Dale. 1987. Screening for neuroblastoma. Lancet **ii:** 1082.
11. Sawada, T., T. Sugimoto & T. Tanaha. 1987. Number and cure rate of neuroblastoma cases detected by the mass screening programme in Japan; future aspects. Med. Pediatr. Oncol. **15:** 14–17.
12. Dale, G., A. C. McGill, J. A. Seviour & A. W. Craft. 1988. Urinary excretion of HMMA and HVA in infants. Ann. Clin. Biochem. **25:** 233–236.
13. Matsumoto, M., A. Anazawa, K. Suzuki, M. Owada, T. Kitagawa, K. Okabe & K. Morita. 1985. Urine mass screening for neuroblastoma by high-performance liquid chromatography. Pediatr. Res. **19:** 625.
14. Seviour, J. A., A. C. McGill & G. Dale. 1988. Method of measurement of urinary homovanillic acid by gas chromatography mass spectrometry suitable for neuroblastoma screening. J. Chromatogr. **432:** 273–277.
15. Yokomori, K., T. Hori, Y. Tsuchida, M. Kuroda & M. Yoshioka. 1989. A new urinary mass screening system for neuroblastoma in infancy by use of monoclonal antibodies against VMA and HVA. J. Pediatr. Surg. **24:** 391–394.
16. Hammond, P. M., R. J. Hinton & M. D. Scawen. 1987. Microbial enzymes in laboratory diagnosis. Med. Lab. Sci. **44:** 170–177.
17. Tryding, N. 1986. Clinical need for analyses nearer the patient. Upsala J. Med. Sci. **91:** 185–188.
18. Durham, D. R. 1984. Initial reactions involved in the dissimilation of mandelate by *Rhodotorula graminis.* J. Bacteriol. **160:** 778–780.
19. Cohen-Bazire, G., W. R. Siston & R. Y. Stanier. 1957. Genetic studies of pigment synthesis by non-sulphur purple bacteria. J. Cell. Comp. Physiol. **49:** 25–68.
20. Kamin, H., R. H. White-Stevens & R. P. Presswood. 1978. Salicylate hydroxylase. Methods Enzymol. **53:** 527–543.

Coupled Enzyme Reactions in Flow Injection Analysis and Liquid Chromatography[a]

E. DOMÍNGUEZ,[b,f] G. MARKO-VARGA,[c] H. L. LAN,[d]
Y. OKAMOTO,[d] P. D. HALE,[e] T. A. SKOTHEIM,[e]
B. HAHN-HÄGERDAL,[b] AND L. GORTON[c]

[b]Department of Applied Microbiology
[c]Department of Analytical Chemistry
Chemical Center
Post Office Box 124
University of Lund
22100 Lund, Sweden

[d]Department of Chemistry
Polytechnic University
Brooklyn, New York 11202

[e]Moltech Corporation
Chemistry Building
State University of New York at Stony Brook
Stony Brook, New York 11794

Immobilized enzymes are suitable analytical reagents because of their selective molecular recognition and long-term operational performance. The use of enzymes as analytical reagents has considerably increased by their introduction in flow systems, e.g., flow injection analysis (FIA) and liquid chromatography (LC).[1] However, the number of compounds that can be enzymatically analyzed is limited by the physical or chemical properties of the substrates and products. In cases where none of the substrates or products are measurable, coupled sequential reactions open the possibility to determine one of these components. By the use of coupled reactions, additional advantages can be gained: (A) compounds involved in unfavorable equilibria may also be efficiently converted by shifting the equilibrium constant with the subsequent reaction, resulting in a system thermodynamically favorable and with a linear response range; (B) the selectivity of the analysis can be improved and therefore the accuracy, diminishing the risk of interferences; (C) inhibitory products can be continuously removed; and (D) the sensitivity is increased if any of the specimen (substrates, products, or cofactors) enters in a cyclic sequence with the possibility of signal amplification and recovering of cofactors.[2]

[a]We acknowledge financial support from the Swedish National Energy Administration (STEV) and the Swedish Board for Technical Development (STU and STUF).

[f]On leave from the Department of Nutrition and Food Analysis, University of Alcalá de Henares, 28871 Alcalá de Henares (Madrid), Spain.

The pyridine nucleotide–dependent dehydrogenases constitute the largest group of redox enzymes, and a great number of them are commercially available. A reaction catalyzed by a dehydrogenase can be written schematically as follows:

$$\text{Substrate} + NAD^+ \rightleftarrows \text{Product} + NADH + H^+ \quad (1)$$

The formal potential, $E^{\circ\prime}$, of the $NAD^+/NADH$ redox couple is −560 mV vs. SCE (pH 7, 25°C). For most systems, the equilibrium of Reaction 1 favors the substrate rather than the product side because of the low oxidizing power of the cofactor, reflected by the low $E^{\circ\prime}$. For a restricted number of reactions, the subsequent addition of water in Reaction 1 or Reaction 2 drives the equilibrium towards the product side, and thus favors NADH production, Reaction 3. Examples of such systems are the oxidation of aldoses, e.g., β-D-glucose and β-D-xylose with glucose dehydrogenase, GDH,[3,4] of acetaldehyde with aldehyde dehydrogenase, AlDH,[5] and formate with formate dehydrogenase.[6]

These reactions can be summarized as follows:

$$\beta\text{-D-glucose} + NAD^+ \overset{GDH}{\rightleftarrows} \partial\text{-gluconolactone} + NADH \quad (2)$$

The lactone will spontaneously and rapidly react with water in a second irreversible reaction:

$$\partial\text{-gluconolactone} + H_2O \rightarrow \text{gluconic acid} + H^+ \quad (3)$$

The overall reaction will be an irreversible oxidation of β-D-glucose with the production of a stoichiometric amount of NADH.

$$\beta\text{-D-glucose} + NAD^+ + H_2O \overset{GDH}{\rightarrow} \text{gluconic acid} + NADH + H^+ \quad (4)$$

A similar addition of water to the product first formed is also valid for the enzymatic oxidation of acetaldehyde.[5] The decomposition of carbonic acid in the presence of water will drive the enzymatic oxidation of formate into an irreversible step.[6] Consequently, the analytical applications of these dehydrogenase-catalyzed reactions can be designed with total conversion of the analyte. It is always desirable to reach 100% conversion because the response is then unaffected by (A) minor losses of enzyme activity with time (B) slight variations in the flow rate, pH, ionic strength, and temperature; and (C) the presence of small amounts of activators and/or inhibitors.[1,7] However, in many other dehydrogenase-catalyzed reactions, the low value of the equilibrium constant, K_e, keeps the conversion of the analyte far below 100%. Moreover, if the equilibrium of the reaction is not achieved, the determination becomes easily nonlinear. For this reason many of the immobilized enzyme reactors (IMERs) incorporated into flow systems are designed with large volumes, that is to say that by increasing the amount of enzyme, the equilibrium will be achieved more easily and rapidly. But, large reactor volumes should always be avoided in FIA an especially in LC when the reactor is used in the postcolumn mode due to the additional band broadening that they originate. The problem still remains if the enzyme is immobilized on the surface of an electrode where the restricted area of the electrode limits the amount of biocatalyst.

The most general way to overcome an unfavorable equilibrium is based on the introduction of a succeeding step that will remove any of the products formed, and thus the overall reaction will favor the forward direction. This step can be chemical, enzymatic, or electrochemical.

ENZYMATIC DRIVING FORCE

Dehydrogenase-catalyzed reactions with favorable equilibrium constants, e.g., those catalyzed by GDH or AlDH, can be added to drive an equilibrium reaction and to yield NADH which is easily detectable. We have developed two analytical applications based on the addition of GDH and AlDH for the control of a fermentation process. The first one includes the selective determination of xylulose and xylose in an FIA system.[4] Xylulose is a ketopentose for which no dehydrogenase yielding NADH is available. This sugar can be isomerized to D-xylose in an equilibrium reaction catalyzed by xylose isomerase, XI (see FIGURE 1). This equilibrium can be shifted by the deliberate addition of GDH which will catalyze the oxidation of the xylose formed in the presence of NAD^+. By doing that, xylulose can be monitored through the NADH formed. GDH is only active for the β-form of xylose and because the α-anomer is produced in the isomerization step, mutarotase is added to speed up the anomeric equilibrium. An overall reaction is represented in FIGURE 1. Due to the low turnover rate of GDH with β-D-xylose, an excess of this enzyme has to be added to make the system thermodynamically favorable. Three different ratios between GDH and XI were studied considering the reaction rate for xylose and xylulose. Mutarotase was kept constant in relation to the GDH charged and in a ratio of MT/GDH equal to 3.48 to avoid any limitation in the anomeric equilibrium. The different reaction rates for xylose and xylulose with three different

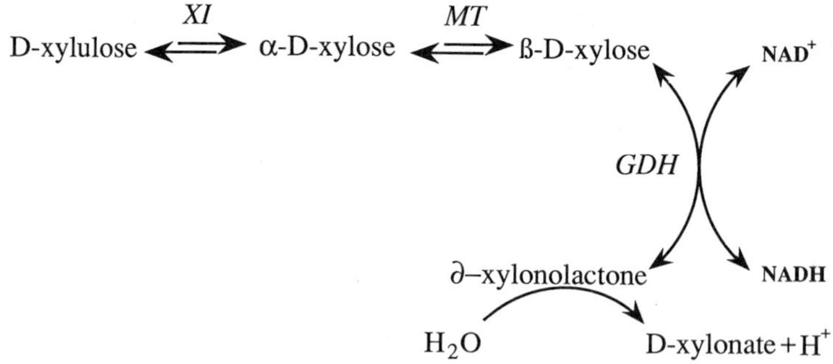

FIGURE 1. Coupled enzyme reaction sequence for the determination of D-xylulose and D-xylose. XI, xylose isomerase; MT, mutarotase; GDH, glucose dehydrogenase. NAD^+ and NADH denote the oxidized and reduced form of the nicotinamide adenine dinucleotide, respectively.

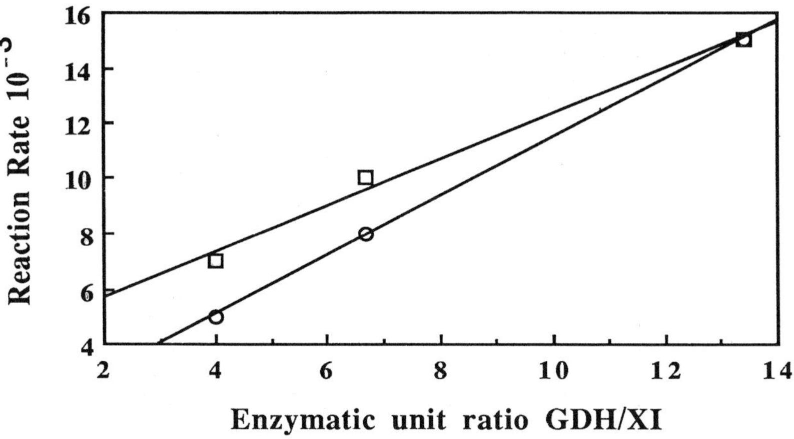

FIGURE 2. Effect of the GDH/XI unit ratio charged on the reaction rate of xylulose (○) and xylose (□). Reaction rate is expressed as the slope of the straight lines obtained when −ln (1 fractional conversion) are plotted versus 1/Flow rate; 50-μl reactors containing the coimmobilized enzymes were studied in the flow injection mode with 50-μl injections of 100-μM D-xylose and 100-μM D-xylulose at each flow rate. The carrier was a 0.1-M phosphate buffer pH 7.0 containing 3-mM NAD^+ and 8-mM magnesium (II) sulfate.

ratios are shown in FIGURE 2. According to these results, the higher the ratio of GDH/XI, the higher the reaction rate for both pentoses. This increase is larger for xylulose than for xylose due to the first equilibrium in which xylulose is involved. A large excess of GDH is required to drive this equilibrium, and the addition of water, see FIGURE 1, becomes the sole thermodynamic driving force of the whole system. Having such an excess of GDH also results in the reaction rate of xylulose equaling the reaction rate of xylose, indicating no limitation in the isomerization step.[8]

The enzymatic determination of ethanol by alcohol dehydrogenase, ADH, suffers from a very unfavorable equilibrium (K_e 8 10^{-12} M, 25°C) which results in nonlinear calibration characteristics when a 50-μl reactor containing immobilized ADH is introduced in an FIA system (FIGURE 3). By coimmobilizing ADH with AlDH, a linear calibration graph for ethanol is achieved using the same reactor volume (FIGURE 3), showing the efficiency of this enzyme in driving the unfavorable equilibrium. Moreover, a new molecule of NADH is produced which increases the sensitivity of the analysis.

AlDH can also be added to monitor the oxidation of ethanol through the NADH formed when alcohol oxidase, AOD, is used in a first catalytic reaction. The reaction sequence is presented in FIGURE 4 and includes the addition of catalase in order to remove the hydrogen peroxide formed in the oxidation of ethanol which could oxidize and inactivate AlDH.[5] By doing that, an additional advantage is obtained based on recovery of the oxygen consumed in the first reaction. The effect of this oxygen recovery is proved when comparing the conversions of acetaldehyde and ethanol at different flow rates (FIGURE 4). In this case a ratio between AOD and AlDH of 10/1 was assayed. A higher conversion for ethanol than for acetaldehyde is

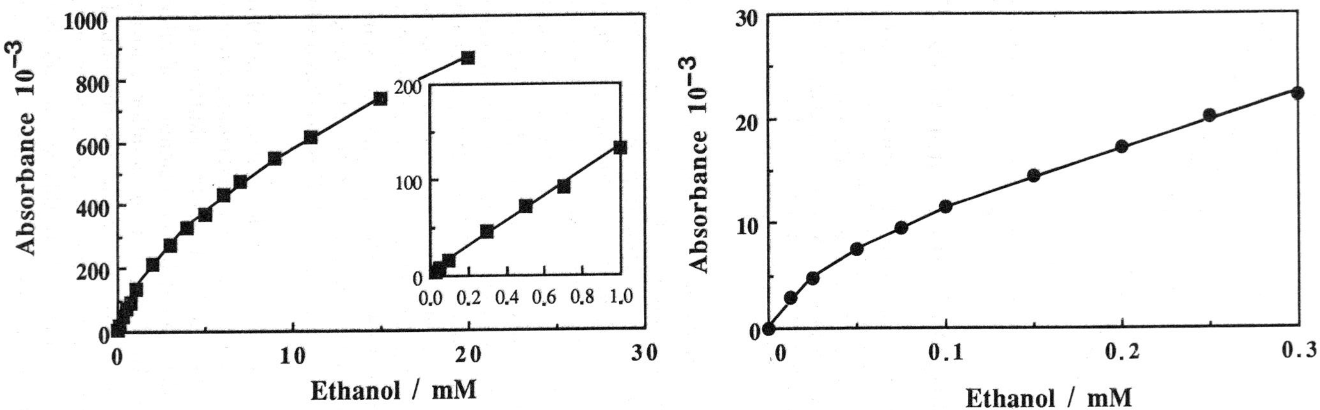

FIGURE 3. Calibration graph obtained with ethanol when a 50-μl alcohol dehydrogenase and aldehyde dehydrogenase coimmobilized reactor (■) or a 50-μl alcohol dehydrogenase immobilized reactor (●) are incorporated into an FIA system. The carrier contained 200-mM glycine, 1-mM dithiothreitol, and 4-mM NAD^+ in 0.1-M phosphate buffer pH 7.5. The injection volume was 10 μl and the flow rate was 1 ml $minute^{-1}$.

then achieved if the flow rate is lower than 0.8 ml/minute, indicating that the addition of catalase shuttles the hydrogen peroxide into a cyclic sequence with a signal amplification for ethanol.[2] Under these conditions, where a large excess of catalase is present, the amount of alcohol oxidase limits the overall reaction for ethanol. This is demonstrated by coimmobilizing AOD and AlDH with a ratio of 1/2.5; in this case, no amplification is observed due to the limited amount of AOD.[2]

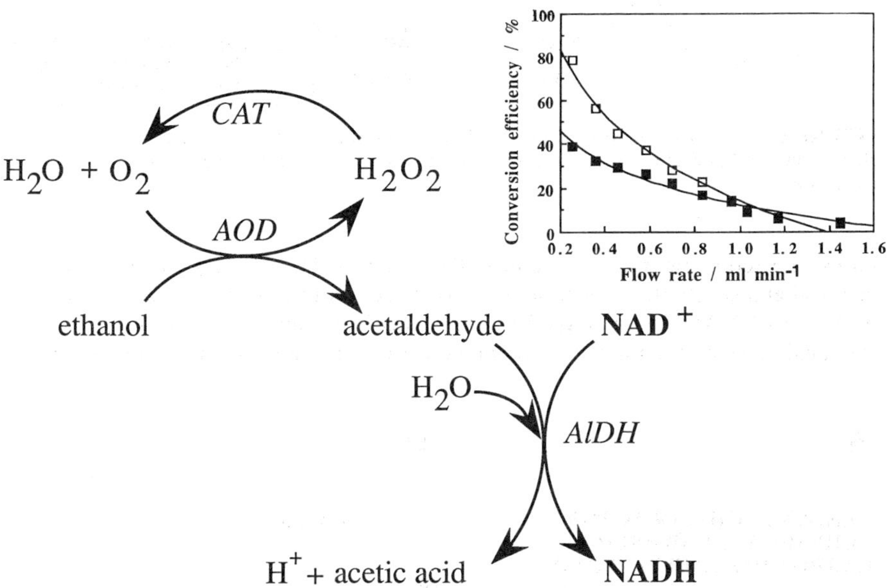

FIGURE 4. Coupled enzyme reaction sequence for the determination of ethanol with alcohol oxidase, AOD, and addition of aldehyde dehydrogenase, AlDH, for monitoring of the reaction by the NADH produced. CAT, catalase. Variation of the conversion efficiency for ethanol (□) and acetaldehyde (■) with the flow rate in an FIA system when a 50-μl coimobilized enzyme reactor containing AOD and AlDH in a unit ratio of 10/1 was used; 16 units of catalase were charged per unit of alcohol oxidase; 25-μl injections of 125-μM ethanol or 125-μM acetaldehyde were made into the carrier stream containing 2-mM NAD^+, 0.15-M KCl in 0.1-M phosphate buffer at pH 7.0. This carrier solution was saturated with O_2.

ELECTROCHEMICAL DRIVING FORCE

Dehydrogenase-catalyzed reactions call for electrochemical detection, as an electron-transfer step is involved in the natural cycle of the redox enzyme.[7] Direct amperometric detection by oxidation of NADH suffers from irreversibility, large overpotentials (consequently risk of interferences in complex samples), and electrode fouling. These limitations can be solved by the use of chemically modified electrodes (CMEs) with mediating functionalities. The introduction of mediators on the electrode material, as in the case of the carbon paste electrode, will facilitate the charge transfer between the reduced form of the cofactor and the electrode. In other

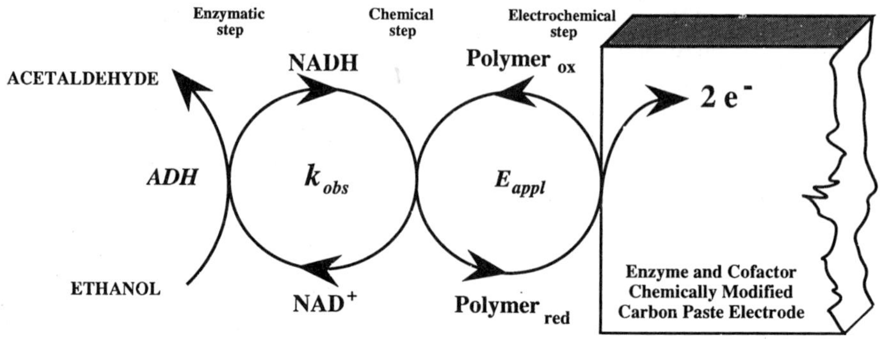

FIGURE 5. Schematic reaction sequence that occurs in an alcohol dehydrogenase, cofactor, chemically modified electrode when a sample containing ethanol reaches the surface of the electrode.

terms, the redox mediator can shuttle the electrons from NADH to the electrode material at a substantially decreased overvoltage.[7,9] In this way, the electrochemical oxidation of NADH is catalyzed by these mediators and involves two steps, one chemical between the mediator and one electrochemical between the mediator and

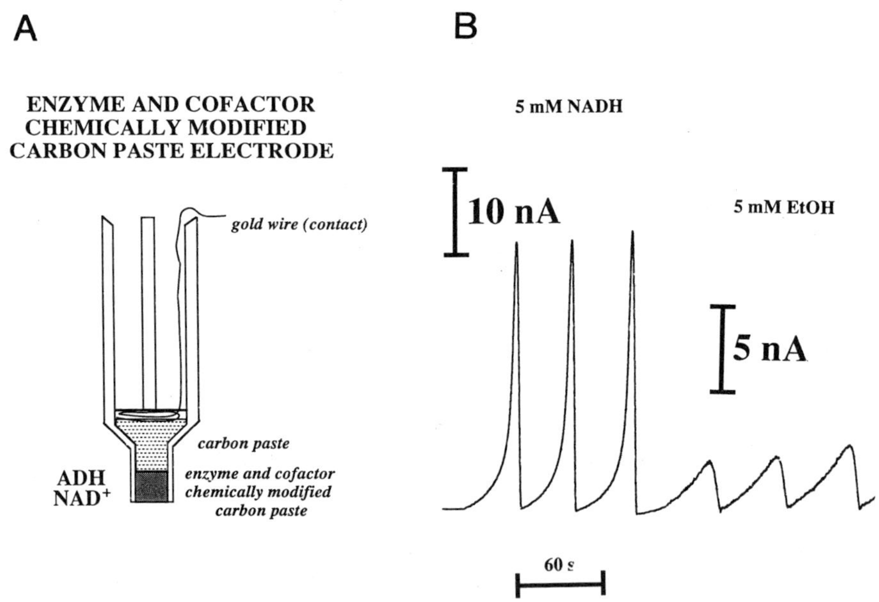

FIGURE 6. **(A)** Schematic representation of an enzyme, coenzyme, chemically modified carbon paste electrode used in this work. **(B)** FIA peaks obtained when this electrode is inserted in an FIA system and 50-μl injections of a 5-mM ethanol and 5-mM NADH are made into a 0.1-M phosphate buffer pH 7.5 carrier stream. The applied potential was 200 mV vs. SCE, and the flow rate was 0.7 ml minute^{-1}.

the electrode. This can be simplified in the following reactions:

$$\text{NADH} + \text{Med}_{ox} \rightarrow \text{NAD}^+ + \text{Med}_{red} \quad (5)$$

$$\text{Med}_{red} \xrightarrow{E_{appl} > E^{\circ\prime}} \text{Med}_{ox} + \text{H}^+ + 2e^- \quad (6)$$

Once the NADH has reached the electrode surface, it reacts with the oxidized form of the immobilized mediator (Med_{ox}) yielding NAD^+ and the reduced form of the

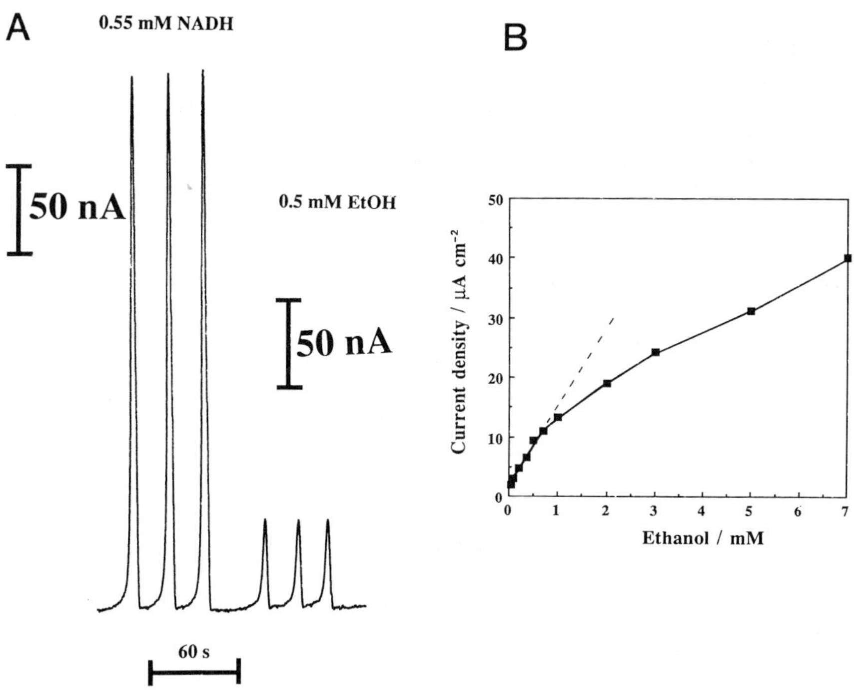

FIGURE 7. (**A**) FIA peaks obtained with an enzyme, coenzyme, chemically modified carbon paste electrode when the enzyme and coenzyme are previously dissolved in a 0.2% aqueous solution of polyethylenimine. The FIA system was the same that in FIGURE 4. The applied potential was 100 mV vs. SCE. (**B**) Calibration graph obtained for ethanol with this electrode in the FIA system.

mediator (Med_{red}), Reaction 5. The reduced form of the mediator will be electrochemically reoxidized if the applied potential is more positive than the $E^{\circ\prime}$ value of the mediator, Reaction 6. That is to say that the mediator dictates at what potential the electron transfer occurs. Consequently, the detection of NADH can now take place at a much lower potential minimizing the risk of interferences.

The electrocatalytical oxidation of NADH can be used as the driving force in the case of an unfavorable equilibrium, as mentioned above for the enzymatic oxidation of ethanol. The system described and summarized in FIGURE 5 makes use of a modified carbon paste electrode containing all the necessary reagents, yeast alcohol dehydrogenase, NAD^+, and a polymer containing covalently bound mediating molecules recently synthesized.[10] When this reagentless electrode is incorporated into an FIA system and a potential of 200 mV vs. SCE is applied, it responds to injections of ethanol and NADH. The peaks obtained with this electrode are presented in FIGURE 6, together with a schematic representation of the electrode. In this case, 1 mg of enzyme and 5.2 mg of NAD^+ were added as dry powders to 10 mg of modified graphite and mixed carefully after each addition. The paste was formed by mixing 4 μl of paraffin oil per 10 mg of the enzyme, and coenzyme modified graphite. A much better performance of the electrode is achieved if the enzyme and coenzyme are previously dissolved in a 0.2% aqueous solution of polyethylenimine, PEI, and the amount of redox polymer is increased up to 4% (w/w graphite). In this case, 1.85 mg of the enzyme and an equal amount of NAD^+ were added, dissolved in PEI, per 10 mg of modified graphite. After drying this mixture under reduced pressure, the paraffin oil was added in the same proportion as mentioned above. The peaks obtained with this electrode in the FIA system and the calibration graphs with an applied potential of 100 mV are presented in FIGURE 7. A linear calibration is obtained up to 0.7 mM in the FIA system at pH 7.5, and a much higher current is obtained with this reagentless ethanol electrode for both NADH and ethanol. These results prove the efficiency of the electrocatalytical oxidation of NADH in driving the unfavorable oxidation of ethanol with ADH. The polymeric structure of the redox polymer used in this work avoids its leakage and diffusion from the paste. Electrostatic interactions between PEI, NAD^+, and ADH may stabilize the keeping of the cofactor in the paste. Some of these electrodes were used in the FIA system for one day, and only a decrease of 10% in the signal for ethanol was observed.

ACKNOWLEDGMENTS

The authors thank Carina Peratikon for technical assistance in the alcohol dehydrogenase work.

REFERENCES

1. MARKO-VARGA, G. & E. DOMINGUEZ. 1991. Enzymes as analytical tools. Trends Anal. Chem. **10:** 290–297.
2. DOMINGUEZ, E., G. MARKO-VARGA, B. HAHN-HÄGERDAL & L. GORTON. 1991. Applications and kinetics of immobilised enzymes and coupled enzyme reactions. GBF Monogr. **14:** 225–228.
3. APPELQVIST, R., G. MARKO-VARGA, L. GORTON, A. TORSTENSSON & G. JOHANSSON. 1985. Enzymatic determination of glucose in a flow system by catalytic oxidation of the nicotinamide coenzyme at a modified electrode. Anal. Chim. Acta **169:** 237–247.
4. DOMINGUEZ, E., G. MARKO-VARGA, B. HAHN-HÄGERDAL & L. GORTON. 1988. A flow injection system for the amperometric determination of xylose and xylulose with coimmobilized enzymes and a modified electrode. Anal. Chim. Acta **213:** 139–150.
5. DOMINGUEZ, E., G. MARKO-VARGA, B. HAHN-HÄGERDAL & L. GORTON. 1991. A study of the activity of immobilised yeast aldehyde dehydrogenase in a flow injection system. Anal. Chim. Acta **249:** 145–154.

6. ORTEGA, F., M. BALLESTEROS, J. I. CENTENERA & E. DOMINGUEZ. 1990. Enzymatic flow injection method for the determination of formate. J. Pharm. Biomed. Anal. **8:** 991–994.
7. GORTON, L., E. CSÖREGI, E. DOMINGUEZ, J. EMNÉUS, G. JÖNSSON-PETTERSSON, G. MARKO-VARGA, & B. PERSSON. 1991. Selective detection in flow analysis based on the combination of immobilised enzymes and chemically modified electrodes. Anal. Chim. Acta **250:** 203–249.
8. DOMINGUEZ, E., G. MARKO-VARGA, B. HAHN-HÄGERDAL & L. GORTON. 1991. Optimization of enzyme ratios in a coimmobilised enzyme reactor for the analysis of D-xylose and D-xylulose to be used in a flow system. (In manuscript.)
9. GORTON, L. 1986. Chemically modified electrodes for the electrocatalytic oxidation of nicotinamide coenzymes. J. Chem. Soc. Faraday Trans. I **82:** 1245–1258.
10. GORTON, L., B. PERSSON, P. D. HALE, L. I. BOGUSLAVSKY, H. I. KARAN, H. S. LEE, T. A. SKOTHEIM, H. L. LAN & Y. OKAMOTO. 1992. Electrocatalytic oxidation of nicotinamide adenine dinucleotide cofactor at chemically modified electrodes. ACS Symp. Ser. **487:** 56–83.

Use of Immobilized Alkaline Phosphatase as an Analytical Tool for Flow-Injection Biosensing of Zinc(II) and Cobalt(II) Ions

IKUO SATOH

Department of Chemical Technology
Faculty of Engineering
Kanagawa Institute of Technology
1030 Shimo-Ogino, Atsugi-shi
Kanagawa-ken 243-02, Japan

INTRODUCTION

Bioanalytical methods have been proposed[1–13] based on a novel idea: "an apoenzyme reactivation microassay for heavy metal ions with use of immobilized metalloenzymes as recognition elements in flow streams." The details of the assay principle and procedure have been reported earlier.[1–13] Submicromolar levels of zinc(II) ions and submillimolar levels of cobalt(II) ions have been successfully evaluated through the reactivation of immobilized apoenzymes (in other words, cofactor-free enzymes).

In this work, alkaline phosphatase immobilized onto oxirane-acrylic beads (Eupergit C) was used as a recognition element for zinc(II) and cobalt(II) ions in combination with flow-through transducers, i.e., a thermistor and a photomultiplier for monitoring the enzyme activity.

EXPERIMENTAL WORK

Alkaline phosphatase (ALP; EC 3.1.3.1, from *Escherichia coli,* 50 U/mg of powder) was kindly supplied by Toyo Jozo Company, Limited. *p*-Nitrophenyl phosphate, the substrate for the enzyme, and other reagents were all commercially available and were used without further purification. Doubly deionized water purified with a Millipore Milli-Q system ($\geq$ 17.0 megohm-cm resistivity) was used throughout. The enzyme, immobilized onto Eupergit C (100–200 μm particle diameter, 40-nm pore diameter, 180 m^2/g surface area; Röhm Pharma, Darmstadt, Germany),[14] was loaded into a small polymer column (packed volume 0.3 ml), which was then mounted into a water-jacketed holder (303 + 0.1 K)[3] or a thermistor probe housed in a thermostated aluminum bath (303 + 0.001 K).[15] FIGURE 1 shows a schematic diagram of the proposed biosensing system. Buffer solutions (0.1 M Tris-HCl, pH 8.0, containing 1 M NaCl) were pumped continuously through the system (1.0 ml/minute). Enzyme activity was assessed by injection of *p*-nitrophenyl phosphate solution, following which each change in temperature and absorbance at 405 nm owing to the enzymatic reaction was separately detected with the thermistor probe and the photomultiplier housed in a spectrophotometer with a quartz flow-through cell (volume 18 μl, light-path length 10 mm).

2,6-Pyridine dicarboxylate solution (pH 6.0) was used as the cofactor-complexing agent for removing zinc(II) and cobalt(II) ions from the immobilized enzymes. Sample solutions were introduced via a loop-injection rotary valve.

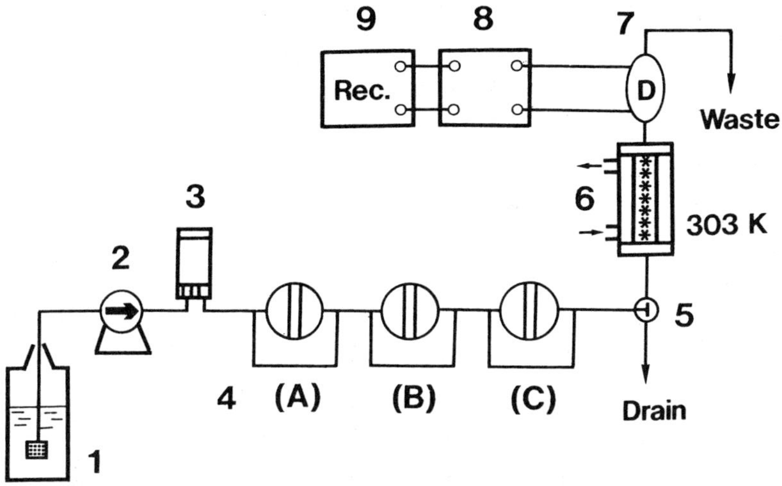

FIGURE 1. Schematic diagram of the proposed biosensing system: (1) carrier reservoir (0.1 M Tris-HCl buffer, pH 8.0, containing 1M NaCl); (2) peristaltic pump (flow rate of 1.0 ml/minute); (3) pulse damper; (4) rotary injection valves (A) for cofactor-complexing (chelating) agent; (B) for standard Zn(II)/Co(II) or water samples; (C) for substrate (*p*-nitrophenyl phosphate); (5) three-way valve; (6) the reactor (packed with immobilized enzymes beads) housed in a water-jacketed holder or attached to a temperature probe; (7/8) thermistor/Wheatstone bridge equipped with a chopper-stabilized amplifier or a quartz flow-through cell (18 μl)/spectrophotometer; (9) recorder.

RESULTS AND DISCUSSION

Biosensing of Zinc(II) Ions

Alkaline phosphatase catalyzes the hydrolytic reaction of *p*-nitrophenyl phosphate to *p*-nitrophenol and orthophosphate:

$$p\text{-Nitrophenyl phosphate} + H_2O \xrightarrow{\text{ALP}} p\text{-nitrophenol} + \text{orthophosphate}$$

Therefore, enzyme activity can be measured by monitoring not only the amount of *p*-nitrophenol formed in the above reaction, but also the hydrolytic heat evolved.

Injection of a 0.1-ml *p*-nitrophenyl phosphate solution of various concentrations caused an increase in absorbance at 405 nm owing to the *p*-nitrophenol liberated. There was a linear relationship with a slope of 2.40 × 10^3 absorbance unit/M between the concentration (0.05 to 0.50 mM) of *p*-nitrophenyl phosphate and the

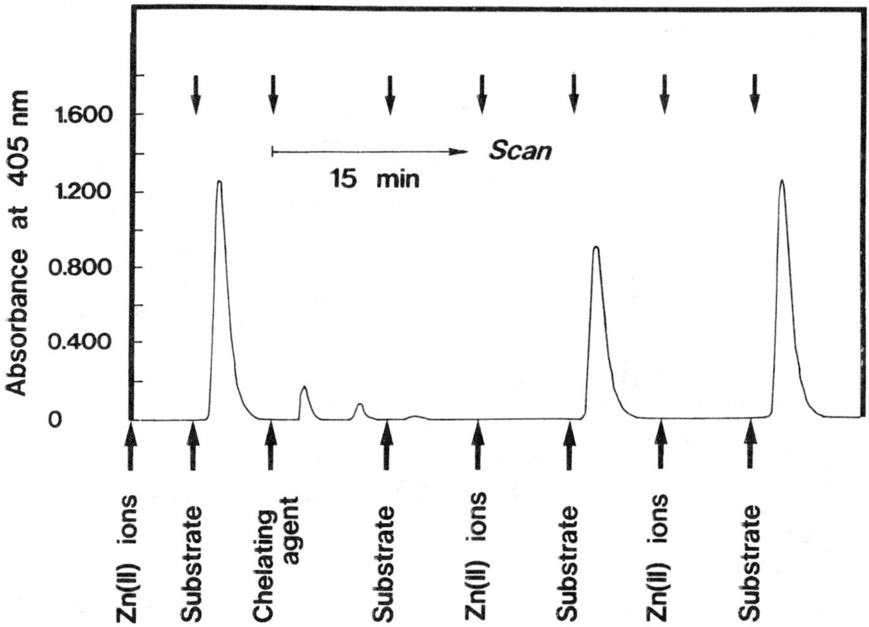

FIGURE 2. Response curves for the injections of the samples in the assay cycle for the photometric flow-biosensing of zinc(II) ions. The arrows indicate changes in the perfusing medium (flow rate 1.0 ml/minute). The cycle starts with injection of 0.5 ml of a 1.0-mM zinc(II) ion solution. At this time, the enzyme reactor is fully occupied with the cofactor, that is, zinc(II). *p*-Nitrophenyl phosphate (0.1 ml, 0.5 mM) is then injected for measuring the activity of the holoenzyme. 2,6-Pyridine dicarboxylate solution (20 mM, pH 6.0) is subsequently introduced for removing the cofactor from the catalytic site of the enzyme. The reactor is then successively washed with carrier (0.1 M Tris-HCl buffer (pH 8.0) containing 1 M NaCl). Here the cofactors coordinated in the catalytic site are virtually removed, and thereby, only a tiny residual response is observed. Now a trace amount of zinc(II) ion solution is injected so that the ions may be trapped in the active sites. The response resulting from the partially reactivated enzymes is registered as an increase in absorbance at 405 nm owing to *p*-nitrophenol formation during a 0.1-ml injection of the substrate. Finally, a sufficient amount of zinc(II) ions is added to the reactor and the holoenzyme is generated again. Thus, the system is ready for another assay.

absorbance. The catalytic activity was then spectrophotometrically monitored by injecting 0.1 ml of 0.5-mM *p*-nitrophenyl phosphate. A typical example of response curves to the sample solutions in the process of the proposed reaction cycle is shown in FIGURE 2. Exposing the cofactor-complexing agent to the reactor packed with the beads of immobilized holoenzymes dramatically decreased the absorbance at 405 nm attributable to the formation of *p*-nitrophenol. Regeneration of the reactor was achieved by introducing 5.0 ml of the chelating agent. The effect of pH in the chelating agent on the regeneration was investigated, and no difference in the level of regeneration was observed at each pH. The recovery of the enzymes was revealed by subsequent addition of a trace amount of zinc(II) ions to the chelator-exposed reactor. The response resulting from the partial reactivation of the cofactor-free enzymes was dependent not only on the concentration of zinc(II) ions injected to the reactor, but also on their volume. A sufficient reactivation of the reactor was obtained by adding smaller volumes of higher concentrations of zinc(II) ions. In practice, the reactor could be reactivated by introducing 0.5 ml of 1.0-mM zinc(II) ion solution. Thus, zinc(II) ions in a range of 0.1 to 10 μM could be determined for 0.5-ml injections. The effect of pH in the zinc(II) solution on the reactivation of the apoenzyme reactor was also examined in the weakly acidic pH region (from 4.0 to 6.0 in steps of 0.5). No appreciable difference in reactivation of the reactor was observed.

The other use of ALP as an analytical tool for microdetermination of zinc(II) ions in combination with a calorimetric monitoring method based on a versatile detection principle can be approached because most enzymatic reactions are accompanied by considerable heat evolution. The calorimetric monitoring of the catalytic activity using an enzyme-thermistor apparatus[15] also demonstrated results of the assay similar to the photometric monitoring and covered the determination range of 10 μM to 1.0 mM.[13] Repeated use of the reactor via the regeneration/reactivation process has been done successfully at least 120 times during 2 months of operation under conditions held at 303 K throughout.

Biosensing of Cobalt(II) Ions

Separate injections of several kind of divalent cations in 1-mM concentration to the apoenzyme reactor were tested. Marked reactivation attributable to addition of cobalt(II) ions was observed in both spectrophotometric and calorimetric biosensing. Substitution of cobalt(II) ions for zinc(II) ions in the enzyme molecules was successfully performed in the process of converting the holoenzyme to the apoenzyme. Thus, cobalt(II) ions could be spectrophotometrically measured in a range of 1.0 to 200 μM and also calorimetrically in a range of 0.04 to 1.0 mM. The reactor packed with the cobalt-substituted enzymes has been repeatedly used at least 120 times during 2 months of operation. In connection with long-term stability, the cobalt-substituted enzyme seemed to be somewhat less stable than the zinc enzyme. Consequently, the reactor, once packed with the enzyme beads, could be proof against the regeneration/reactivation process done over 570 times during one year of operation at ambient temperature.

Further studies are currently in progress toward establishing the generality of the proposed analytical methods. A trial using apoenzymes immobilized onto porous membrane as supporting materials instead of the porous beads to realize more rapid assay are also to be reported in the near future.

REFERENCES

1. SATOH, I., K. IKEDA & N. WATANABE. 1986. Microanalysis for zinc ion by using an apoenzyme thermistor. *In* Proceedings of the Sixth Sensor Symposium (Tsukuba, Japan: 203–206. Institute of Electrical Engineers of Japan. Tokyo, Japan.
2. SATOH, I., S. KIMURA & T. NAMBU. 1987. Biosensing of copper(II) ions with an apoenzyme thermistor containing immobilized metalloenzymes in flow system. *In* Digest of Technical Papers of the Fourth International Conference on Solid-State Sensors and Actuators. (Transducers '87), Tokyo, Japan: 789–792. Institute of Electrical Engineers of Japan. Tokyo, Japan.
3. SATOH, I., R. ABE & T. NAMBU. 1988. Bioelectrochemical sensing of copper(II) ions using an immobilized apoenzyme column. Denki Kagaku **56**(12): 1045–1049.
4. SATOH, I. 1989. Continuous biosensing of heavy metal ions with use of immobilized enzyme-reactors as recognition elements. *In* Mater. Res. Soc. Int. Symp. Proc. **14:** 45–50.
5. SATOH, I., T. KASAHARA & N. GOI. 1990. Amperometric biosensing of copper(II) ions with an immobilized apoenzyme reactor. Sensors Actuators **B1:** 499–503.
6. SATOH, I. & T. MASUMURA. 1990. Flow-injection biosensing of zinc(II) ions with use of an immobilized alkaline phosphatase reactor. *In* Technical Digest of the Ninth Sensor Symposium (Tokyo, Japan): 197–200. Institute of Electrical Engineers of Japan. Tokyo, Japan.
7. SATOH, I. 1990. Apoenzyme reactivation microassay for zinc(II) ions in combination with flow-through transducers. *In* Proceedings of the Third International Meeting on Chemical Sensors (Cleveland, Ohio): P/106–P/107. The Organizing Committee of the Third International Meeting on Chemical Sensors. Cleveland, Ohio.
8. SATOH, I. & Y. AOKI. 1990. Biosensing of zinc(II) ions using an apoenzyme reactor and an ISFET detector in flow streams. Denki Kagaku **58**(12): 1114–1118.
9. SATOH, I. 1990. Calorimetric biosensing of heavy metal ions with the reactors containing the immobilized apoenzymes. Ann. N.Y. Acad. Sci. **613:** 401–404.
10. SATOH, I. 1991. Flow-injection calorimetry of heavy metal ions using apoenzyme-reactors. Netsusokutei (Calorimetry and Thermal Analysis) (in Japanese) **18**(2): 89–96.
11. SATOH, I. & T. NAMBU. 1991. Flow-injection photometric biosensing of copper(II) ions with use of an immobilized ascorbate oxidase column. *In* Technical Digest of the Tenth Sensor Symposium (Tokyo, Japan): 77–80. Institute of Electrical Engineers of Japan. Tokyo, Japan.
12. SATOH, I. & Y. YAMADA. 1991. Flow-injection biosensing of cobalt(II) ions with use of an immobilized alkaline phosphatase reactor. *In* Digest of Technical Papers of the Sixth International Conference on Solid-State Sensors and Actuators (Transducers '91) (San Francisco, Calif.): 699–702. The Institute of Electrical and Electronics Engineers, Inc. Piscataway, N.J.
13. SATOH, I. 1991. An apoenzyme thermistor microanalysis for zinc(II) ions with use of an immobilized alkaline phosphatase reactor in a flow system. Biosensors Bioelectron. **6**(4): 375–379.
14. SATOH, I., S. ARAKAWA & A. OKAMOTO. 1988. Flow-injection determination of glutathione with amperometric monitoring of the enzymatic reaction. Anal. Chim. Acta **214:** 415–419.
15. DANIELSSON, B. & K. MOSBACH. 1987. Theory and applications of calorimetric biosensors. *In* Biosensors: Fundamentals and Applications. A. P. F. Turner, I. Karube & G. S. Wilson, Eds.: 572–595. Oxford University Press. Oxford, England.

Neural Networks in Enzyme Engineering

PEKKA LINKO AND YI-HONG ZHU

Laboratory of Biotechnology and Food Engineering
Helsinki University of Technology
SF-02150 Espoo, Finland

INTRODUCTION

In the field of enzyme engineering, the modeling, optimization, and control of the enzyme production process is often difficult owing to the uncertainties involved, and to insufficient or noisy data. Further, difficulties in the measurement of critical process variables complicate on-line state prediction.[1] Although conventional model structures such as ARX (autoregressive with exogenous input) may be used in the bioprocess variable prediction through an input/output model, the construction of the model and the estimation of model parameters are difficult owing to the random changes in physiology during the enzyme production processes. Neural network programming offers exciting novel approaches in the overcoming of such problems.[2] Although neural network programming has, during the last few years, become one of the hottest subjects of artificial intelligence,[3,4] neural networks have been little applied to biotechnology. The special characteristics of neural networks, such as learning from past data in order to adapt to varying dynamics, dealing with noisy and approximate data, and executing actions quickly once the network model has been trained with past time-course data of an appropriate bioprocess, have been previously described in detail and shown to offer a great potential in bioprocess variable prediction and advanced control strategies.[2,5]

In the present study, the potential of neural network models in enzyme engineering is demonstrated by applying different neural network architectures with varying inputs and outputs in the real-time estimation of major state variables in the glucoamylase fermentation. The neural network models were trained with various example data files including series of input/output vector pairs from some batch glucoamylase fermentations, and their estimation ability and adaptability were tested with other different fermentation experiments.

THE NEURAL NETWORK

The human brain is made up of tens of billions of highly interconnected neurons. The relative synaptic strengths of the junctions between an agglomeration of neurons are modified when knowledge is being stored in the brain. A synapse can thus be considered as the basic memory unit of the brain. Similarly, an artificial neural network mimics brain with a layered network structure of a number of interconnected artificial neurons as processing units. In this work, feed-forward three-layered neural network models with an input layer, a hidden layer, and an output layer (FIGURE 1) were used. The number of neurons used in the hidden layer in this

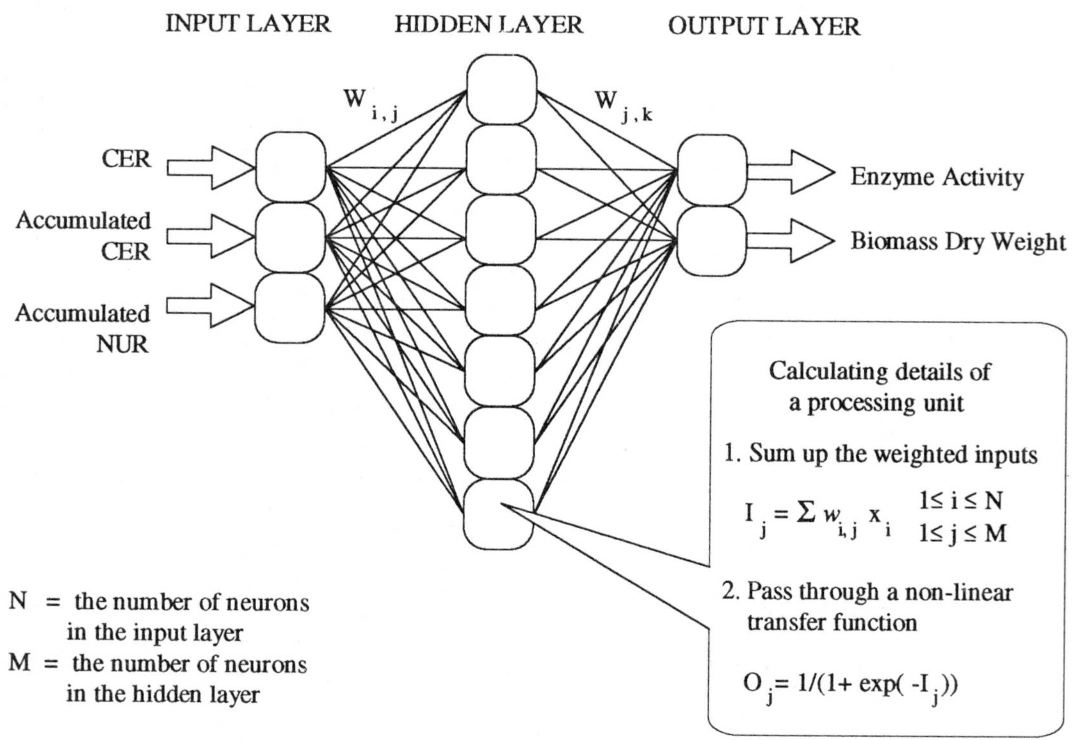

FIGURE 1. The complete architecture of the neural network used for the real-time estimation of glucoamylase activity and biomass.

investigation—seven—was chosen by trial and error on the basis of the best converging performance in training as described in an earlier paper.[2]

The weights of the interconnections between the neurons were described as real values w_{ij} (FIGURE 1). The information in a neural network is stored in the form of weight matrices $W_{i,j}$ and $W_{j,k}$ of the connections between the artificial neurons in adjacent layers. In this way, the relative importance of each "arm" of information received by a neuron could be weighted. After accepting of data in a coded form as an input, the neurons typically sum up the weighted inputs, and compress them through a nonlinear transfer or squashing function before passing the information as the output. In the current work, a sigmoid transfer function of the form $f(\upsilon) = 1/[1 + \exp(-\upsilon)]$, in which υ is equal to the weighted sum of the inputs, was used.

The information fed into the network as the input was coded within a range [0,1]. When an input vector entered the network, an output vector was produced as a response of the propagation by the network. The weight matrices of the neural network thus describe the input to output vector relationships of unknown complexity in the multilayer network.

APPLICATION OF NEURAL NETWORKS

The Working Environment

The training and testing program of the neural network was constructed in Microsoft® Quick C® version 2.5 on the basis of the theory of back-propagation.[6] The program was running in an IBM/PC compatible computer with an 80486/33 MHz processor, 170 MB hard disk, 4 MB memory on the main board, and a Super-VGA display.

Training of the Neural Network

A back-propagation training algorithm was used based on the delta rule or LMS procedure through gradient descent search technique to minimize a cost function equal to the mean square of the output error. In practice, a training data set consists of a series of input/output vector pairs (or pair patterns) relevant to the characteristics of the network presented repeatedly to the network. A back-propagation training algorithm, in which the training rate is increased without leading to oscillation by modifying the back-propagation rule of error by including a momentum term, was used in this investigation.[2,7]

The training of the neural network was started by assigning small random numbers as weights to each of the network's interconnections w_{ij}. The second stage of training involved a presentation of an input to the network, followed by a comparison of the actual output produced by the network through back-propagation with the desired output. The difference, i.e., the output error, between two outputs was then used to increase or decrease the value of each weight w_{ij} in such a way that the response of the network approached the correct, i.e., the desired, output when given the same information repeatedly. Finally, for a well-trained neural network, a set of ideal weight values was obtained after an appropriate number of iterations.

Glucoamylase Fermentations

Glucoamylase fermentations were carried out with an industrial strain of *Aspergillus niger,* using a starch-based medium and Biostat® E fermentors (B. Braun Melsungen, Germany) connected to a mass spectrometer gas analyzer MM8-80 (VG Gas Analysis, United Kingdom) and to the MFCS (Rintekno, Espoo, Finland) data collection and analysis system. The most valuable on-line data in the development of a fuzzy expert control system were previously shown to be the oxygen uptake rate (OUR), the carbon dioxide evolution rate (CER), and the nitrogen (ammonia) utilization rate (NUR).[8] Additional information on the process was provided by varying the pH and the agitation rate. The accumulated value of the carbon dioxide evolution rate (A.CER) and the nitrogen utilization rate (A.NUR) were calculated on-line. The biomass concentration (as dry weight) and the glucoamylase activity were analyzed off-line two to three times a day.

The data of a number of glucoamylase fermentation batches were used for generating an example data file for the training of the neural network model using Microsoft® Excel 3.0 software. The data file for training was formed by designating the time-course data series of the on-line measured CER, A.CER, and A.NUR as an input vector and a series of sampling data of the glucoamylase activity and the biomass dry weight as an output vector using a constant sampling interval. Linear interpolation was used to calculate the missing values. Time-course data of other batches not included in the training were employed in the justifying of the adaptability of the well-trained neural network model in the real-time estimation of the state variables, the enzyme activity, and the biomass dry weight.

Real-Time State Estimation

In the present investigation, four different neural network models of the general form as shown in FIGURE 1 were developed only by varying the input and output vectors and keeping the number of neurons in the hidden layer fixed as in TABLE 1.

FIGURE 2 illustrates the results of the real-time estimation of the enzyme activity and the biomass dry weight using models 1 and 2, respectively. One can clearly see that a state variable as the output in both model 1 and model 2 could be estimated in real time on the basis of the on-line measured data, when the neural network was well trained with a set of data from one successful batch fermentation. The neural network outputs approached well the actual data obtained by the off-line analyses. The accuracy of the estimation was clearly very satisfactory.

TABLE 1

Model	Input Vector	Output Vector	Data for Training
Model 1	A.CER and A.NUR	Enzyme activity	One batch
Model 2	CER and A.CER	Biomass dry weight	One batch
Model 3	CER, A.CER and A.NUR	Enzyme activity and biomass dry weight	One batch
Model 4	CER, A.CER and A.NUR	Enzyme activity and biomass dry weight	Six batches

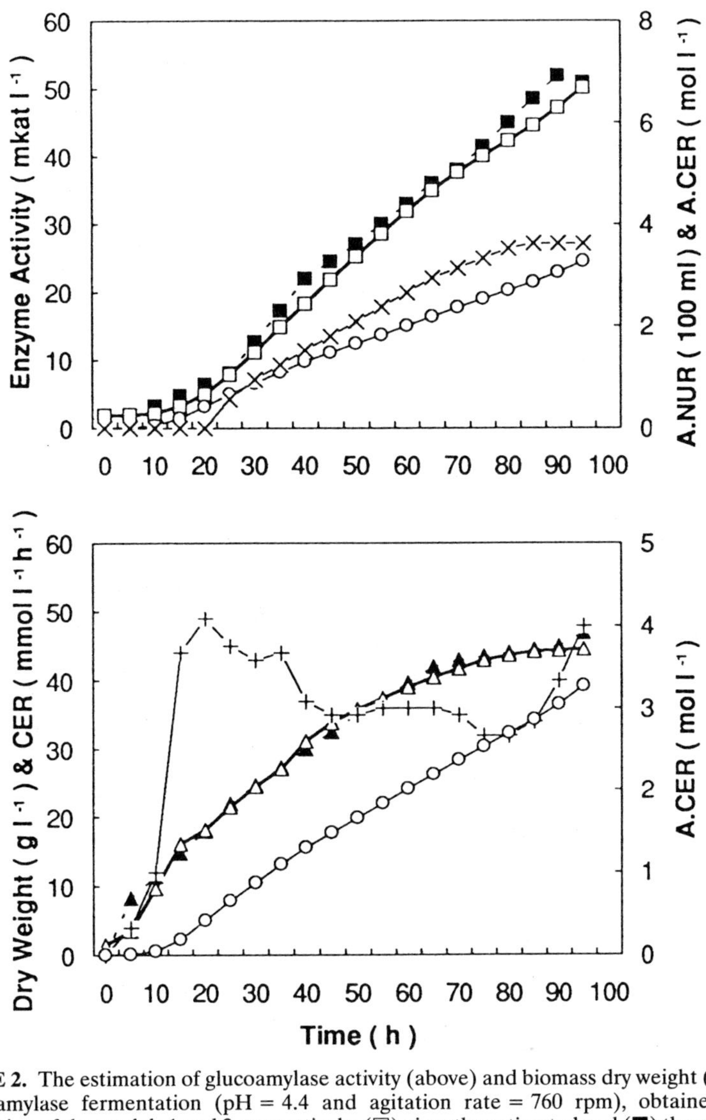

FIGURE 2. The estimation of glucoamylase activity (above) and biomass dry weight (below) in a glucoamylase fermentation (pH = 4.4 and agitation rate = 760 rpm), obtained by the propagation of the models 1 and 2, respectively. (□) gives the estimated and (■) the real values of the enzyme activity for model 1. (△) gives the estimated and (▲) the real values of biomass dry weight for model 2. (×), (○), and (+) give the time courses of the on-line measured A.NUR, A.CER, and CER, respectively.

FIGURE 3 illustrates the case when the complete network model architecture as shown in FIGURE 1 was used in the estimation both of the enzyme activity and biomass dry weight simultaneously in models 3 and 4. In this case the on-line obtained CER, A.CER, and A.NUR were employed as an input vector for the neural

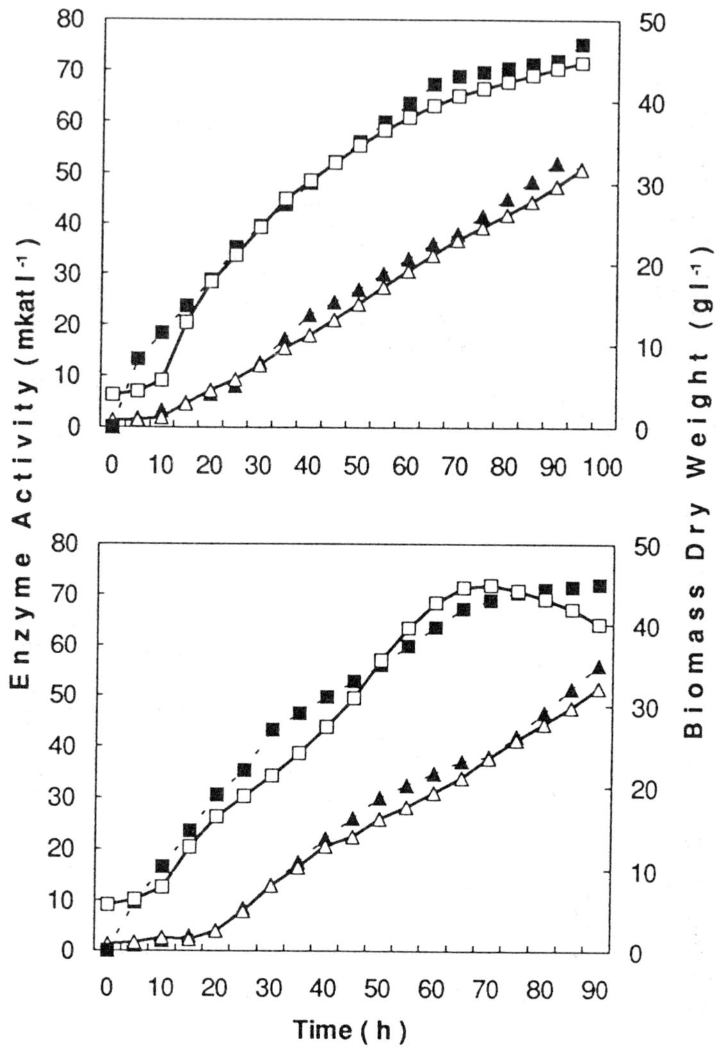

FIGURE 3. The estimation of glucoamylase activity and biomass dry matter by model 3 (above) and model 4 (below) using a complete neural network architecture. (□) gives the estimated and (■) the real values of the enzyme activity. (△) gives the estimated and (▲) the real values of biomass dry weight.

network models used. The major difference between the neural network models 3 and 4 in their training was the use of a different example data file by varying the number of the input/output vector pairs from 18 in the case of only one batch of data for training to 92 in the case of 6 batches of data. In comparing the network model architectures, their estimation ability of the function of the models 1, 2, and 3 depended only on how well trained with an appropriate example data file the neural network was, but not on how complex the network architecture used was. Thus, in this case, the estimating accuracy was not related to the complexity of the network architecture. The accuracy of these models in estimation was influenced entirely by degree of sufficiency of the information include in the example data file, mirroring the relationship between the on-line measured data and the state variables regardless of the size of the example data file.

Further, the neural network model 3 trained with the example data file from one batch fermentation only appeared to be as good in the estimation as the network model 4 trained with data from 6 fermentations. Consequently, the estimation ability of the neural network models generated by the knowledge extracted from the example data file could be applied to situations that are similar, but not identical to, the example data file, thus demonstrating their ability to extrapolate. The accuracy of the estimations of the enzyme activity and the biomass dry weight by on-line measured data in the glucoamylase fermentation was very satisfactory.

EPILOGUE

We have demonstrated above that well-trained neural networks can be successfully used for real-time estimation of critical bioprocess output variables on the basis of on-line measured time-course data sets. In the present work, the enzyme activity and the biomass dry matter in glucoamylase fermentations were successfully estimated in real time from the carbon dioxide evolution rate, the accumulated carbon dioxide evolution, and the ammonia utilization rate. The neural networks developed also had a good adaptability in their estimative capability, and a quick response. Such neural networks may also be incorporated as a part of a process controller, and be used to aid in bioprocess optimization. We are currently working on the construction of an adaptive neural network model and on the integration of neural network programming and object-oriented knowledge-based system for on-line, real-time bioprocess control.

REFERENCES

1. LINKO, P. 1988. Uncertainties, fuzzy reasoning, and expert systems in bioengineering. Ann. N.Y. Acad. Sci. **542:** 83–101.
2. LINKO, P. & Y-H. ZHU. 1991. Neural network programming in bioprocess variable estimation and state prediction. J. Biotechnol. **21:** 253–270.
3. JANSSON, P. A. 1991. Neural networks: an overview. Anal. Chem. **63**(6): 357–362.
4. RICH, E. 1990. Expert system and neural networks can work together. IEEE Expert **5**(5): 5–7.
5. THIBAULT, J., V. VAN BREUSEGEM & A. CHÉRUY. 1990. On-line prediction of fermentation variables using neural networks. Biotechnol. Bioeng. **36:** 1041–1048.
6. LIPPMANN, R. P. 1987. An introduction to computing with neural nets. IEEE ASSP Magazine **4**(2): 4–22.
7. MCCLELLAND, J. L. & D. E. RUMELHART. 1988. Explorations in Parallel Distributed Processing. A Handbook of Models, Programs, and Exercises. The MIT Press. Cambridge, Mass.
8. AARTS, R. J., A. SUVIRANTA, P. RAUMAN-AALTO & P. LINKO. 1990. An expert system for enzyme production control. Food Biotechnol. **4:** 301–315.

Measurement of HIV-1 Reverse Transcriptase by a Nonradioactive Assay System

F. HAMMERSCHMID, E. MAIER, AND B. ROSENWIRTH

Sandoz Forschungsinstitut Wien
Brunner Strasse 59
A-1235 Wien, Austria

HIV-1, the etiological agent of AIDS, codes for an RNA-directed DNA-polymerase (reverse transcriptase) that is essential for proviral DNA synthesis. Therefore, this enzyme is an excellent molecular target for antiviral chemotherapy. Several compounds have been described that interact with reverse transcriptase and block the replication of the virus in cell culture. One of these inhibitors of HIV-1 reverse transcriptase, namely, 2′-azido-2′-deoxythymidine (AZT), proved clinically useful as a treatment for AIDS, though its administration has been limited by severe side effects. To identify substances with increased potency against HIV-1 that might be developed to drugs showing fewer side effects than, e.g., AZT, we wanted to apply a high throughput *in vitro* screening system. Hitherto, all described test methods for HIV-1 reverse transcriptase have used incorporation of radiolabeled nucleotides into acid-precipitable nucleic acid. To minimize expenses for both purchase and waste disposal of radioactive chemicals, we evaluated alternative methods.

Labeling of DNA molecules with dUMP-digoxigenin by means of procaryotic and eucaryotic DNA polymerase reaction has been described[1] and proved useful in detection of, e.g., specific sequences in the human genome on Southern blots, human mRNA on Northern blots, recombinant bacterial colonies, recombinant phage plaques, viral gene sequences in metaphase chromosomes by *in situ* hybridization, mRNA in rat tissue sections by *in situ* hybridization with a labeled oligonucleotide probe, viral DNA in biological fluids by slot blot analysis.

We investigated whether the incorporation of digoxigenin-dUMP into different template primers by HIV reverse transcriptase could be used as a nonradioactive assay for this enzyme.

HIV-1, strain HTLV III B, reverse transcriptase was expressed in *Escherichia coli* SMH50 containing the viral gene on expression vector pkk-233-2. The plasmid and the *E. coli* strain were kindly provided by R. Goody. The enzyme was expressed and purified to homogeneity by a modified version of the procedure described by Mueller *et al.*[2] and was characterized using incorporation of 3H-dTMP into poly A-oligo dT10 template primer as assay.

All chemicals used were of analytical grade purity and RNase free. The template primers, namely, poly A-oligo dT10, and DNA of high molecular weight from calf thymus, DNase I from bovine pancreas grade II, digoxigenin-11-dUTP, and antidigoxigenin peroxidase-conjugated Fab fragments, were purchased from Boehringer Mannheim. The nicked DNA to be used as template primer was prepared as follows.

The content of 3 vials of each 10 mg DNA per 3.8 ml were dialyzed against 2 liters of 50 mM Tris-HCl pH 7.6; 2 mM EDTA; 10 mM $MgCl_2$; 1 mM dithiothreitol (DTT) at 4°C for 16 hours. The sample was adjusted to E 260 = 0.6 by dilution with fresh dialysis buffer. Bovine serum albumin (BSA) was added up to a final concentration of 200 μg/ml, and finally DNase I was added to a final concentration of 100 ng/ml. The sample was incubated at 37°C for 60 minutes, and the reaction was stopped by incubation at 50°C for 20 minutes. The sample was divided in 750-μl aliquots and stored at −20°C.

Conditions for the reverse transcriptase standard assay had been optimized and were used for the digoxigenin assay as well: final concentrations were 100 mM Tris-HCl pH 7.6; 5 mM $MgCl_2$; 100 mM KCl; 4 mM DTT; 0.1 mg/ml BSA; either 10 μM dTTP substrate or 1 μM DIG-dUPT/12 μM dTTP substrate mixture; 0.1 U/ml poly A-oligo dT10 and 257 nM reverse transcriptase. The reaction was started by mixing equal volumes of solutions A and B: both A and B contained equal concentrations of buffer, $MgCl_2$, KCl, and DTT; A contained in addition twice the final concentration of BSA, the template primer, and the reverse transcriptase; B contained in addition twice the final concentration of the respective substrates. When using digoxigenin-dUTP/dTTP as substrate, the reaction was stopped by addition of EDTA to a final concentration of 20 mM. An aliquot was applied onto nitrocellulose membrane, and the amount of incorporated digoxigenin-dUMP was detected by peroxidase-conjugated antidigoxigenin-Fab fragment following standard immunochemical dot-blot procedures. The whole assay was performed in microtiter plates with the aid of a laboratory robotic system. The product of the peroxidase reaction was determined quantitatively by laser densitometry.

For standard assay conditions, 3H dTTP was used instead of the digoxigenin-labeled dUTP and the acid-precipitable (TCA) material was collected on glass fiber filter discs. The bound radioactivity was measured by liquid scintillation counting.

When using nicked DNA as template/primer, in addition to the respective ingredients for the digoxigenin assay or the standard assay, 10 μM of each dATP, dCTP, dGTP, and 1.5 μg/ml nicked DNA were added.

According to the manufacturer (Boehringer-Mannheim), distinct labeling profiles of the synthesized nucleic acids can be obtained depending on spacing of the antigen digoxigenin. In the case of homologeous template primers such as poly A-oligo dT10, the spacing of the antigen will depend on the ratio of digoxigenin-dUTP to dTTP in the reaction mixture. Therefore, we investigated the influence of increasing amounts of dTTP on the incorporation and detection of digoxigenin-dUMP during the reverse transcriptase reaction. The concentration of dTTP in the reaction mixture was increased relative to a constant concentration of digoxigenin-dUTP (1 μM). Under the assumption of linear incorporation of both digoxigenin-dUMP and dTMP into the template primer within a certain range, an increasing displacement of digoxigenin-dUMP by dTMP should be expected. However (FIGURE 1), at low concentrations of dTTP, only marginal incorporation of digoxigenin-dUMP is detectable by means of the immunochemical method; with increasing dTTP concentrations, the amount of incorporated antibody-recognizable dUMP as measured by peroxidase reaction increases considerably. After reaching the optimal relative concentration of digoxigenin-dUTP to dTTP, the expected gradual displacement is observed. This result can be interpreted as follows. If digoxigenin-dUMP

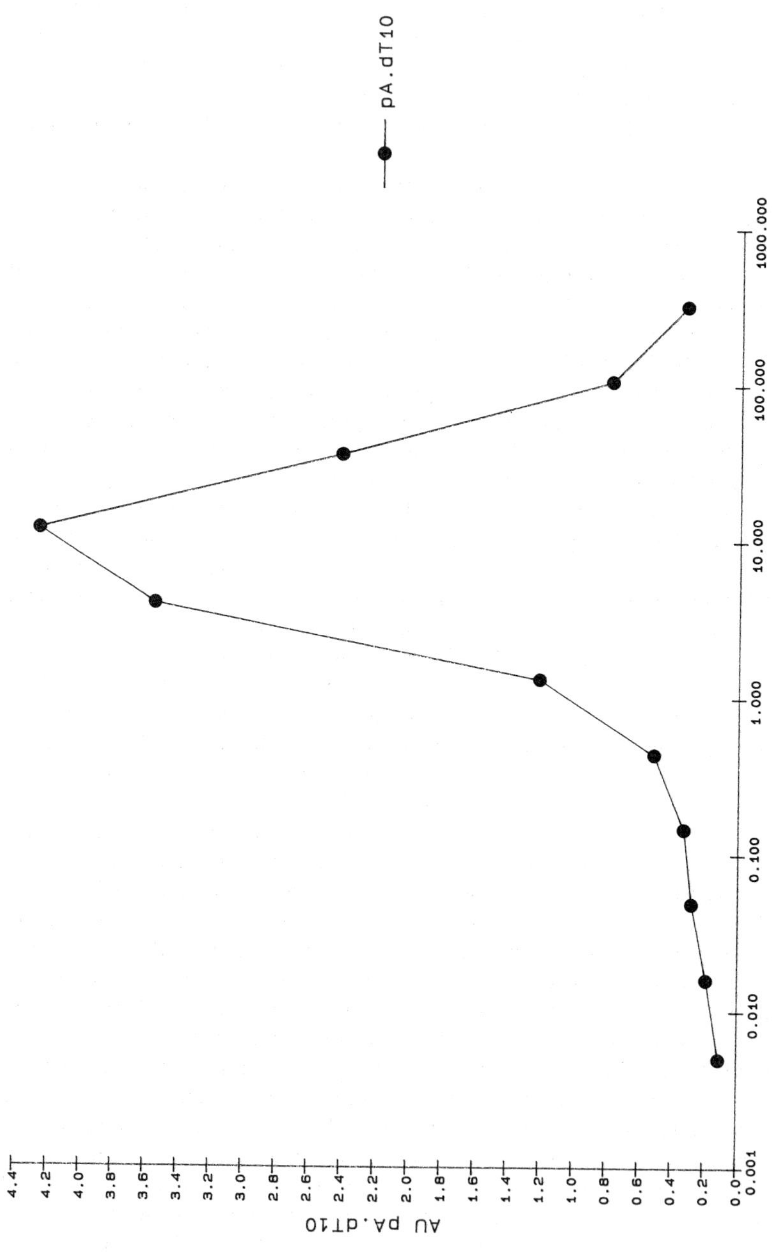

FIGURE 1. Influence of dTTP in DIG Assay on 1-μM Substrate.

incorporation does not lead to chain termination, which obviously is not the case, it is incorporated into the homologous template primer in a space-filling manner in the presence of low concentrations of dTTP. Thus sterical hindrance of the antigen-antibody reaction is assumed to occur. Concentrations higher than 0.8 μM dTTP constantly lead to enhancement of the antibody binding by spacing the antigen to the optimal distance where a detection maximum is reached. Higher concentrations of dTTP show normal displacement kinetics. The optimal concentration of dTTP in the case of poly A-oligo dT10 as template-primer was found to be 12 μM at 1 μM digoxigenin-dUTP concentration.

In order to evaluate the applicability of the digoxigenin assay for testing of potential inhibitors of reverse transcriptase, we compared it with the standard assay by measuring the IC_{50} values of the inhibitors 2′-azido-2′deoxythymidine-triphosphate (AZTTP) and phosphonoformic acid (PFA) (TABLE 1). By using poly A-oligo dT10 as template primer, both the standard assay and the digoxigenin assay gave identical results for PFA and AZTTP.

When using nicked DNA as template primer, a 20-fold higher IC_{50} value for AZTTP as compared to poly A-oligo dT10 was observed in the standard assay. The digoxigenin assay gave a 4-fold increase of the IC_{50} value with this inhibitor. These

TABLE 1. Inhibition Assay

	pA-dT10		Nicked DNA	
	Standard Assay IC_{50} (μM)	DIG Assay IC_{50} (μM)	Standard Assay IC_{50} (μM)	DIG Assay IC_{50} (μM)
PFA	2.79	2.55	NT[a]	NT[a]
AZTTP	0.06	0.08	1.200	0.350

[a]Not tested.

differences may be understood if one takes in consideration the findings of Huebscher *et al.*,[3] who reported that the DNA-dependent DNA-polymerase activity of the HIV-1 reverse transcriptase is confined to the p15 moiety of the p66 molecule and thus different and remote from the active site of the RNA-dependent DNA-polymerase.

CONCLUSION

We conclude that the digoxigenin assay is useful in the evaluation of potential reverse transcriptase inhibitors. Especially for high throughput screening, the immunochemical detection method offers a considerable advantage over the standard measurement of radioactivity, namely, the possibility to avoid use of expensive and environment endangering chemicals.

SUMMARY

Reverse transcriptase activity was measured by incorporation of dUMP linked to digoxigenin into a suitable template-primer molecule. Incorporation was monitored by using peroxidase-conjugated Fab fragments directed against digoxigenin. The standard assay measuring incorporation of radiolabeled nucleotides into acid-precipitable material was compared with this new immunochemical assay with regard to its usefulness for testing inhibitors of reverse transcriptase.

REFERENCES

1. DNA Labeling and Detection: Nonradioactive Application Manual, 1989. Boehringer Mannheim GmbH. Mannheim, Germany.
2. Mueller, B., T. Restle, J. Reinstein & R. S. Goody. 1991. Biochemistry **30**(15): 3709–3715.
3. Hafkemeyer, B., E. Ferrari, J. Brecher & U. Huebscher. 1991. Proc. Natl. Acad. Sci. USA **88:** 5262–5266.

Assay of Urea by Immobilized Urease Coupled to a Differential pH-Meter[a]

PIER GIORGIO PIETTA, PIER LUIGI MAURI, CLAUDIO GARDANA, MARIO PACE, AND DARIO AGNELLINI

Dipartimento di Scienze e Tecnologie Biomediche
Sez. Chimica Organica
Universita' di Milano
Via G. Celoria, 2
20133 Milano, Italy

INTRODUCTION

Urea is the main end product of protein metabolism, and its concentration in blood, urine, and dialyzed fluids is therefore a very important parameter in clinical chemistry for the diagnosis of kidney disorders.

The methods commonly used for the determination of urea are based on the hydrolysis catalyzed by urease (EC 3.5.1.5) according to the reaction:

$$H_2N\text{-}CO\text{-}NH_2 + H_2O \xrightarrow{\text{urease}} 2\,NH_3 + CO_2 \quad (1)$$

Most methods of urea determination rely on the assay of ammonia[1,2] by spectrophotometric or potentiometric procedures, both by kinetic or end-point methods. Methods based on the pH change must be corrected for the following reactions:

$$NH_3 + H_2O \rightleftharpoons NH_4^+ + OH^- \quad (2)$$

$$CO_2 + H_2O \rightleftharpoons HCO_3^- + H^+ \quad (3)$$

With regard to potentiometric methods, a technique based on the differential measurement of pH with urease in homogeneous solution has been already described.[3]

While the assays utilizing soluble enzymes take advantage of the homogeneous media, immobilized enzymes are generally more economic because of the possibility of reusing the same catalyst several times. Thus urease was bound[4] to a number of matrices in order to obtain enzyme reactors suitable for the routine assay of urea in serum or urine.

This work describes a new assay based on the differential measurement of pH following the hydrolysis of urea in serum and hemolyzed blood by means of immobilized urease. The enzyme was conjugated to Ab-antiurease immobilized on nylon-coated magnetic stirrers, and the resulting bioreactors were connected to a differential pH-meter.

[a] This work was supported by the Target Project on Biotechnology and Bioinstrumentation of the Italian National Council of Researches (CNR).

This approach permitted easy resolution of the bioreactor when the enzyme activity was exhausted, by replacing the denatured urease with fresh enzyme.

MATERIALS AND METHODS

Urease activity (IU) is defined as the amount of enzyme that hydrolyzes 1 μmole of urea per minute in the assay conditions of temperature and pH.

Urease (EC 3.5.1.5) with a specific activity of 15 IU/mg and 680 IU/mg was purchased from Sigma Chemical Co. (St. Louis, Mo.). Urea (Carlo Erba, Milan, Italy) was crystallized from ethanol before use, and fresh solutions were made each day. All other reagents were of analytical grade.

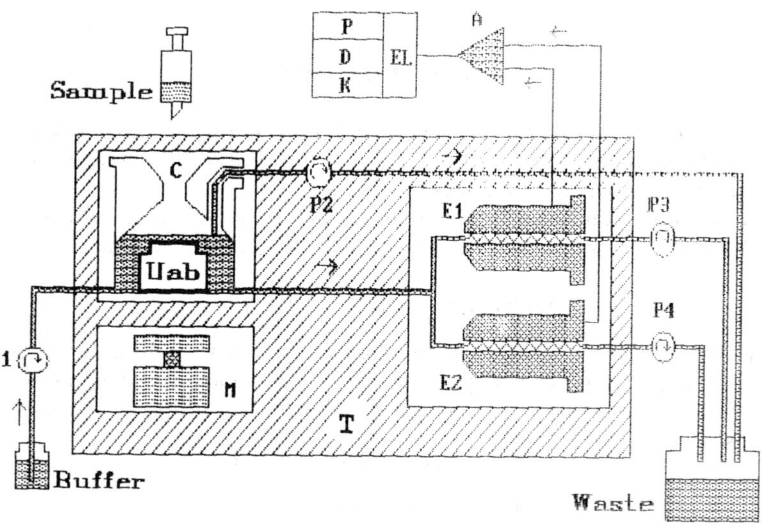

FIGURE 1. Diagram of differential pH-meter. P1, P2, P3, P4 = pumps; T = thermostated block; Uab = bioreactor; E1, E2 = electrode; M = stirrer; A = amplifier; EL = printer, display, keyboard; C = mixing chamber.

The bioreactors used were magnetic stirrers coated with nylon-6 in a cylindric shape (7.5-mm diameter, 6.2-mm height).

A differential pH-meter (Delpas CL; Kontron S.p.a., Milan, Italy), coupled to the urease bioreactor (Uab), was used for the determination of urea in serum and hemolyzed blood (FIGURE 1).

Immobilization of Proteins on Nylon

The stirrers were treated for 15 minutes with 10% triethyloxonium tetrafluoroborate[5] dissolved in dichloromethane and then washed with dichloromethane, methanol-dichloromethane (1:1), and methanol. After reaction with a 10% solution of hexamethylenediamine, the stirrers were treated with 5% glutaraldehyde in borate buffer for 15 minutes. The activated stirrers were washed with 25-mM phosphate buffer pH 7 and reacted overnight at 4°C with a solution of 5 mg/ml of antibodies to urease in 0.1-M phosphate buffer pH 7.0.

Urease (10 mg/ml) was conjugated to the antibody-nylon complex in phosphate buffer pH 7.0 at 4°C for 2 hours, and the unbound enzyme was exhaustively washed with phosphate buffer and water.

Preparation of Polyclonal Antibodies to Urease

Polyclonal antibodies to urease were obtained after four injections in rabbits of 200 μl of enzyme solution (1 mg/ml) emulsified with complete Freund's adjuvant. They were isolated by the Rivanol method[6] and then further purified by affinity chromatography, after immobilization of urease on Affi-Gel 10 (Bio-Rad Laboratories, Richmond, Calif.) according to the procedure suggested by the purchaser. The solution of polyclonal antibodies in 0.1-M phosphate buffer pH 7.0 was loaded onto the affinity column which was washed with 0.1-M phosphate buffer until the eluate showed no absorbance at 280 nm. The antibodies were then eluted with 10% dioxane in 20-mM HCl pH 2.5, dialyzed against water, and stored at 4°C until used. A sample of antibodies was tested by enzyme-linked immunosorbent assay (ELISA).

Enzyme-Linked Immunosorbent Assay

Polyclonal antibodies were linked to alkaline phosphatase by means of glutaraldehyde[7] in order to assay their specificity.

The antibodies (1 μg/ml) were allowed to adsorb for 18 hours at 4°C in a plastic plate and, after washing with phosphate buffer pH 7.5 containing 5% bovine serum albumin and 0.05% Tween-20, 200 μl of antigen (5 μg urease/ml) was added to each well, allowed one hour at 37°C, and then washed with phosphate buffer. After addition of 200 μl of antibodies linked to alkaline phosphatase (in the range 1–7 nM), the plate was incubated one hour at 37°C and 200 μl of a solution of p-nitrophenylphosphate (1 mg/ml) in 10% diethanolamine pH 9.8 containing 1 mM $MgCl_2$ was added to each well. A control was made with wells containing only antibodies without addition of antigen. After washing with phosphate buffer pH 7.5, the plate was incubated 30 minutes at 37°C and the absorbance at 405 nm was measured.

Enzyme Replacement in the Antibody-Urease Bioreactors

After exhaustion, the bioreactors were continuously washed for two hours with a 10% solution of dioxane in 20-mM HCl pH 2.5 in order to remove the inactive enzyme. Fresh enzyme solution in phosphate buffer pH 7.0 was then replaced in each reactor, and allowed to rest overnight at 4°C.

Assay of Urea

In a Delpas differential pH-meter, samples of 1 ml of urea were introduced into a reaction chamber containing 13-mM Tris, 13-mM phosphate buffer pH 7.5, 1% Triton X-100, and a nylon-urease bioreactor. At fixed times, a small amount of sample (200 μl) was drawn into a capillary pH electrode while another pH electrode was filled with a reference solution.

At fixed times, the delta-pH's were recorded and the data plotted in order to obtain the enzyme kinetics. The reaction was carried out at 25°C or 37°C. Calibration curves for urea were obtained with different amounts of urea recording the delta-pH at the end of the reaction.

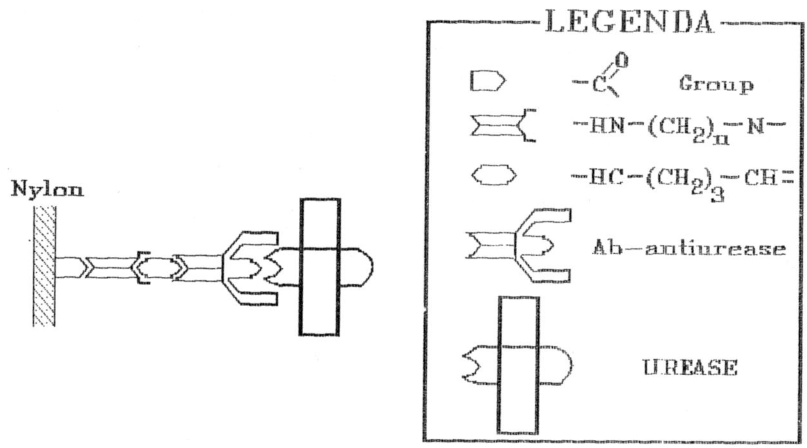

FIGURE 2. Coupling of urease to nylon through antibodies (Ab).

RESULTS AND DISCUSSION

Urea hydrolysis in human serum has been investigated using urease conjugated to polyclonal antiurease antibody covalently bound to nylon-coated stirrers (Uab) (FIGURE 2).

Urease activity immobilized on magnetic stirrers was about 7 IU, and no significant difference in activity was found at 25°C and 37°C.

Kinetic determination of urea in the range of 20–200 mg/dl allowed us to establish that the enzymatic reaction reaches the end point after 1 minute (90% hydrolysis) (FIGURE 3). Thus, this time was chosen for the routine assay.

Urea standard solutions were assayed within the same concentration range, and the results were in good agreement (±3.2%) with those obtained with free urease.

The assays were reproducible and reliable, as was evidenced by comparing calibration curves obtained from standard urea solutions and urea-added human sera (FIGURE 4).

Tests on urea standard solutions gave recoveries of 98 ± 1.5% and, with different

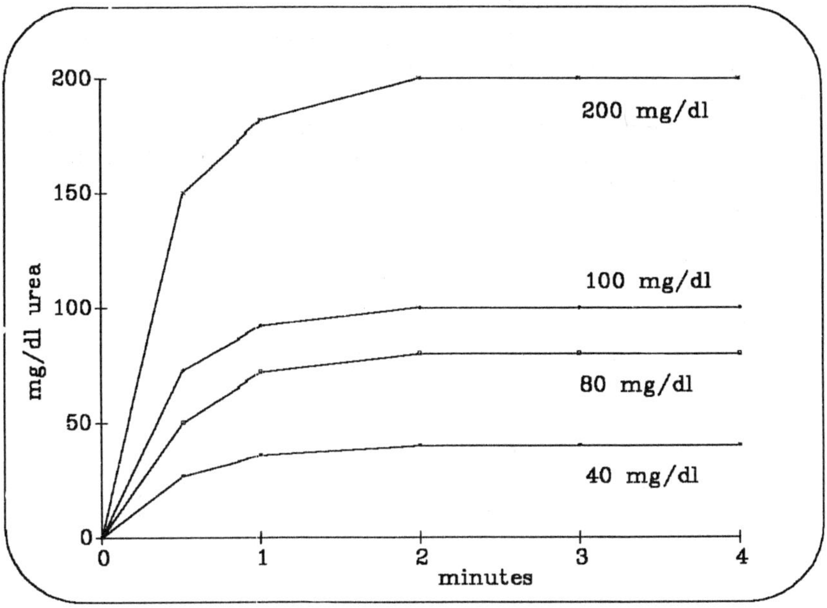

FIGURE 3. Typical kinetics of a bioreactor.

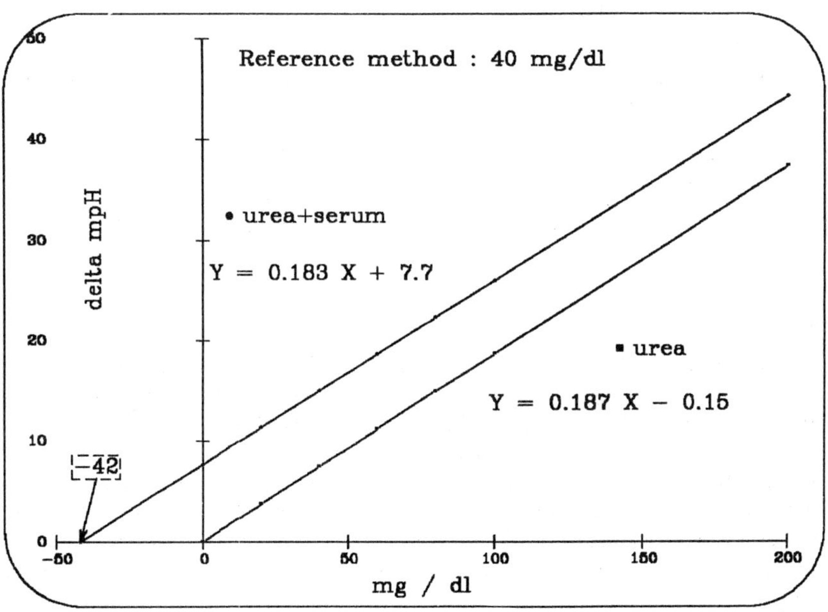

FIGURE 4. Calibration curves obtained from standard urea solutions and urea-added human sera.

amounts of urea (0.2–1 mg/dl) added to serum, 96 ± 3.4%. Intraday relative standard deviation was determined by assaying the samples 12 times within the same day and was 2%. Interday relative standard deviation was obtained by analyzing the samples daily for two weeks, and the result was 3.2%. Most of the urease activity was lost after about 100 assays, and the inactivated enzyme could be removed by washing with a 10% solution of dioxane pH 2.5. Then fresh urease conjugated again. By this procedure, the initial bioreactor activity could be kept almost constant for 5 cycles (500 analyses) (FIGURE 5).

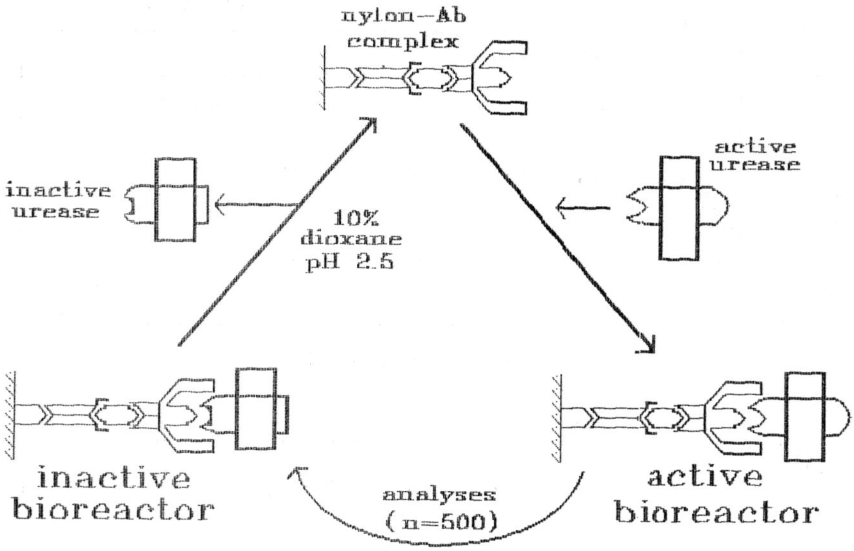

CYCLES	0	1	2	3	4	5
ACTIVITY (%)	100	80	85	95	80	90

FIGURE 5. Regeneration of exhausted bioreactors.

Dithiotreitol (DTT) helps keep the urease activity under storage; nevertheless, a decrease in the replacement of exausted enzyme was noted. This could be due to the reduction of disulfide bonds in the antibody moiety, resulting in a lower affinity to the antigen. Better results were obtained when the bioreactors were stored in 0.1-M phosphate buffer pH 7.2–50% glycerol. Under these conditions, the activity remained approximately constant up to 100 assays. The determination of urea in hemolyzed blood gave reproducible data, but the resulting urea contents were higher (20%) than those arising from assays with free urease. The cause of this difference is still under investigation.

In conclusion, the conjugation of urease through its antibodies bound to nylon represents a valuable alternative for the immobilization of enzymatic activity and an interesting tool to regenerate bioreactors. Further, urease bioreactors connected to a differential pH-meter permit the assay of urea in standard solutions and human sera. However, more work is needed for a reliable analysis of urea in hemolyzed blood.

REFERENCES

1. KERSCHER, L. & J. ZIEGENHORN. 1985. Urea determination. *In* Methods of Enzymatic Analysis. edited by H. U. Bergmeyer, J. Bergmeyer & M. Grassl, Eds. 3rd Edit. **8:** 444–453. VCH Verlagsgesellschaft mbH. Weinheim, Germany.
2. BLAKELEY, R. L., E. C. WEBB & B. ZERNER. 1969. Biochemistry **8:** 1984–1990.
3. BONUCCHI, D., A. CASTELLANI, G. C. ORLANDINI, M. RIPAMONTI, M. LUZZANA & A. MOSCA. 1987. Int. J. Artif. Organs **10:** 352–356.
4. MOYNIHAN, H. J., C. K. LEE, W. CLARK & N. H. L. WANG. 1989. Biotechnol. Bioeng. **34:** 951–963.
5. PIETTA, P. G., D. AGNELLINI & M. PACE. 1984. *In* Proceedings III European Congress on Biotechnology **2:** 591–596. Verlag Chemie GmbH, Weinheim, Germany.
6. HOREJSI, J. & R. SMETANA. 1956. Acta Med. Scand. **155:** 65–70.
7. AVRAMEAS, S., T. TERNYNCK & J. L. GUESDON. 1978. Scand. J. Immunol. **8**(Suppl. 7): 7–23.

Amperometric Assay of Creatinine in Urine by Flow Injection Analysis Based on Conjugated Reactions of Immobilized Enzymes

Simultaneous Compensation of Endogenous Ammonia

CHANG SHENG RUI, HIROAKI I. OGAWA,
KENJI SONOMOTO,[a] AND YASUHIKO KATO[b]

Department of Applied Chemistry
Faculty of Engineering
Kyushu Institute of Technology
Sensui-cho 1-1, Tobata-ku
Kitakyushu 804, Japan

[a]*Department of Biochemical Engineering and Science*
Faculty of Computer Science and Systems Engineering
Kyushu Institute of Technology
Kawazu 680-4
Iizuka 820, Japan

INTRODUCTION

Creatinine is one of the diagnostically important substances indexing human kidney function, and its assay is exclusively performed by a chemical method based on the Jaffe reaction.[1-3] Since the reaction lacks specificity for creatinine and is subject to many interferences, biosensors based on immobilized creatinine deiminase (CRDI) and an ammonia electrode have been developed.[4-6] Though in these sensors endogenous ammonia was removed with immobilized glutamate dehydrogenase (GLDH), the problem of nonspecificity still remained due to potentiometric detection by the ammonia electrode. To get higher selectivity, an amperometric assay is desirable. Thus an amperometric biosensor based on CRDI and nitrifying bacteria was developed.[7,8] However, the reproducibility was poor and the assay time was too long (more than 15 minutes). Moreover, as endogenous ammonia was not removed, the method could not be applied to the assay of creatinine in real samples.

In this paper we report an amperometric flow injection analysis (FIA) biosensor system for the determination of creatinine based on the coupled reactions of immobilized CRDI, GLDH, and glutamate oxidase (GLOD). The principles are shown in FIGURE 1. An oxygen electrode was used as the detection device. An approach based on the principle of split and subsequent confluence of flow was proposed for the simultaneous compensation of endogenous ammonia and glutamate.

[b]Author to whom correspondence should be addressed.

MATERIALS AND METHODS

Apparatus

A Garvanic oxygen electrode equipped with a flow cell was purchased from Able Co. (Tokyo, Japan); the injector (model SVA-7U7H) was from Sanuki Industry Co. (Tokyo); the peristaltic pump was obtained from Ato Co. (Tokyo); the potentio/Garvano stat (model HA-501) was from Hokuto Electronics Ltd. (Japan), and the recorder (model FBR-253A) from Toa Electronics Ltd. (Japan).

Materials

CRDI from microorganisms was obtained from Sigma; GLDH from beef liver and NADH were from Oriental Yeast Co. Ltd. (Osaka, Japan); GLOD from

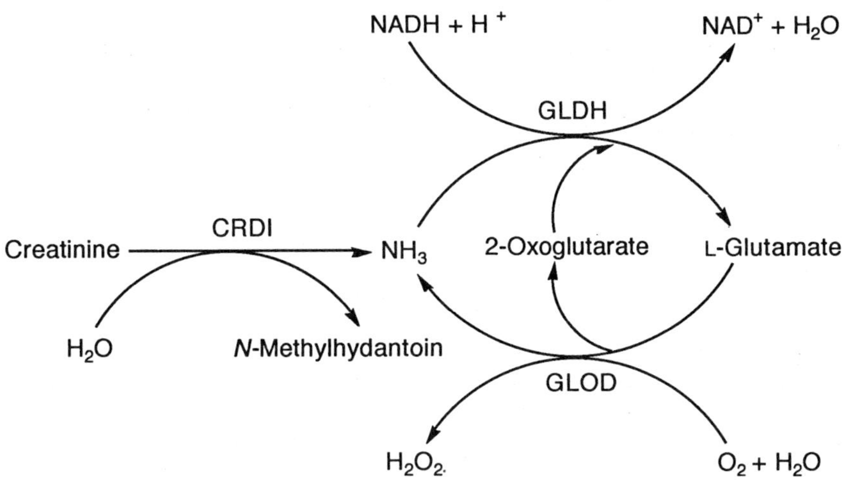

FIGURE 1. Scheme for enzyme reactions of the FIA biosensor system.

Streptomyces sp. was from Yamasa Shoyu Co. Ltd. (Choshi, Chiba, Japan). Controlled porous glass (CPG, 120–200 mesh, 500 Å pore size) and its derivatives (propylamine-CPG and succinate-CPG) were purchased from Pierce Chemical Co. (Rockford, Ill.). All other chemicals were commercially available and of the highest grade.

Preparation of Immobilized Enzymes

CRDI and GLOD were immobilized on propylamine-CPG using glutaraldehyde according to the method of Weetall.[9] GLDH was immobilized on succinate-CPG activated with 1-ethyl-3-(3-dimethylaminopropyl)carbodiimide according to Cho and Swaisgood.[10] Coupling yields based on protein were 100% for CRDI, 78.3% for GLDH, and 97.7% for GLOD.

Enzyme Assay

Assay of native enzymes was performed by the described methods.[11–13] A continuous flow-through method was used for the assay of immobilized enzymes. Reaction mixture was allowed to flow through a column packed with the sample at a rate of about 10 ml/minute. Then the effluent was sampled to determine the concentration of ammonia (for CRDI), of NADH (for GLOD), or of hydrogen peroxide (for GLDH).

Principles of Analysis

The immobilized enzymes were packed into glass tubes (i.d. 3.4 mm); 10 mm of the column contains 36 mg of dry beads. FIGURE 2 illustrates the schematic diagram of the FIA biosensor system for the amperometric assay of creatinine. Split of the flow after sample injection and subsequent confluence before the GLDH column resulted in a two-channel system. By setting a CPG reference column and a delay coil at the channel other than where the CRDI column was located, two peaks were obtained for one injection. The first peak gave the sum response of creatinine, endogenous ammonia, and glutamate, and the second peak that of endogenous ammonia and glutamate. By this method, endogenous ammonia and glutamate were autocompensated.

RESULTS AND DISCUSSION

Enzymatic Studies

TABLE 1 shows the pH optima and K_m s of both native and immobilized enzymes. By immobilization, GLDH and GLOD extended their optimum pH areas to the alkaline side and shared a common value at 8.0 with CRDI. This shift made it possible to use a single working solution to run the FIA system. Immobilized CRDI and GLOD gave similar K_m values with those of native enzymes. In contrast, the K_m of immobilized GLDH for ammonia was much lower than that of native GLDH. It is a great advantage that the K_m of GLDH for ammonia was reduced by immobilization, since low K_m is preferable with regard to the conversion of ammonia to glutamate.

Performance of Enzyme Columns

To get the optimum length of each enzyme column, conversion rates of creatinine to ammonia and of ammonia to glutamate were investigated under different lengths of enzyme columns using the FIA system with a single channel. Under a flow rate of 1.6 ml/minutes and for 35 μl injection of 1.0 mM substrate, 20 mm of CRDI column and 35 mm of GLDH column were found to be necessary for the complete conversion of creatinine to ammonia and of ammonia to glutamate, respectively. For the purposes of substrate conversion, long enzyme columns are preferable; but longer columns will result in a fall of sensitivity and an increase of assay time. Therefore 20 mm and 30 mm were used for CRDI and GLDH columns, respectively.

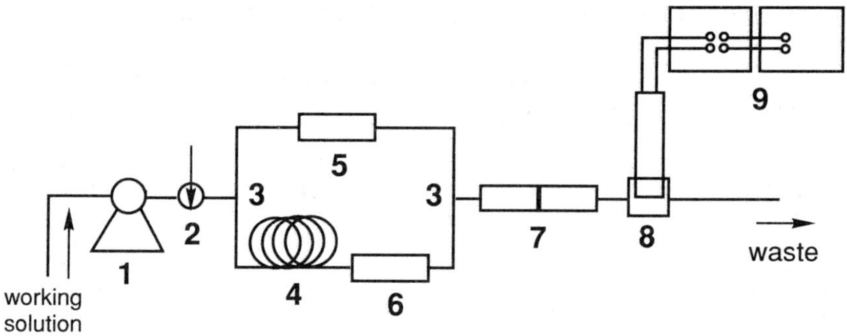

FIGURE 2. Schematic diagram of the FIA system for the amperometric assay of creatinine in urine (two-channel system): (1) pump; (2) injector; (3) T-connector; (4) delay coil, i.d. 0.8 mm, 1.0 m long; (5) CRDI column, 20 mm long; (6) CPG reference column, 18 mm; (7) GLDH/GLOD column, 30/20 mm; (8) oxygen electrode and flow cell; (9) potentiostat and recorder. The inner diameter of all columns was 3.4 mm. When not used, the columns were stored at 4°C in 66.7 mM phosphate buffer (pH 7.5) containing 0.1 mM EDTA. Columns and electrode were kept at 25°C with a water bath. Total flow rate was 1.6 ml/minute. Injection volume was 35 μl. Working solution: 66.7 mM potassium–sodium phosphate buffer (pH 8.0) containing 0.1 mM EDTA, 0.3 mM NADH, and 5.0 mM 2-oxoglutarate. In a single-channel system, the channel involving the CPG reference column was closed.

Oxidation of glutamate was complete when the length of GLOD column was more than 15 mm.

Interferences

The effects of some compounds on the creatinine sensor (single-channel system) are shown in TABLE 2. Of 12 compounds tested, only ascorbate gave meaningful interference. Since ascorbate in aqueous solution is readily air oxidized, it is considered that the positive effect was due to the decrease of background oxygen level and not a change of enzyme activity. In fact, the interference of ascorbate was overcome in the two-channel system (TABLE 3). In biosensors using a hydrogen peroxide electrode as the detector, uric acid and glutathione are great interfering factors since the two compounds are oxidized electrochemically together with hydrogen peroxide. As shown in TABLE 2, however, they scarcely gave effects when an oxygen electrode was used.

TABLE 1. Some Properties of Native and Immobilized Enzymes

	Apparent K_m (mM)		Optimum pH[a]	
Enzyme	Native	Immobilized	Native	Immobilized
CRDI	1.3	3.6	8.0–9.0	8.0–9.0
GLDH	12[b]	4.2[b]	7.5	7.5–8.0
GLOD	0.22	0.35	7.5	7.0–8.5

[a]Area within which relative activity is higher than 95%.
[b]For ammonia, in the absence of ADP.

TABLE 2. Effect of Interfering Compounds on Creatinine Assay (Single-Channel System)

Compound	Concentration in Sample (mM)	Creatinine Found (mM)	Interference (%)
None		2.00	
Allantoin	0.6	2.00	0.0
Ascorbate	5.0	2.49	24.5
Bilirubin	20[a]	2.07	3.5
BSA	200[a]	2.01	0.5
Creatine	2.0	2.04	2.0
Galactose	5.0	1.96	−2.0
Glucose	5.0	1.97	−1.5
Glutathione	1.0	1.97	−1.5
Hippurate	2.0	1.96	−2.0
Pyruvate	5.0	2.07	3.5
Urate	2.0	1.96	−2.0
Urea	200	2.01	0.5

[a]In mg/l.

Flow Distribution and Delay Coil

When other conditions are fixed, the separability of the two peaks and the width of the second peak are dependent on the distribution of working solution over the two-flow path, and the length and inner diameter of the delay coil. Flow distribution was controlled with a reference CPG column set at the second channel. The length of the CPG column was selected so that the flow rate through the second channel was a little greater than that through the first. To get a sharp peak (and therefore high assay speed), the delay coil must be short and thin. For complete separation of the two peaks, 1.2 m of teflon tube with an inner diameter of 0.8 mm was needed.

Typical Response and Calibration

The typical response patterns and the corresponding calibration curves for creatinine and ammonia are shown in FIGURE 3. Response time was about 30 seconds for creatinine and the first peak of ammonia, and 60 seconds for the second peak of ammonia. One run was completed within two minutes. The reproducibility

TABLE 3. Assay of Creatinine in Standard Samples Containing Ammonia, Glutamate, and Ascorbate (Two-Channel System)

Compound Added (mM)	Creatinine Found (mM)
None	1.00
Ammonia 0.5	1.03
1.0	0.98
Glutamate 0.5	0.99
1.0	1.04
Ascorbate 5.0	0.98

was good, and the standard deaviation was less than 3%. The lower detection limit was 0.1 mM, which is 50- to 200-fold lower than creatinine concentration in urine.

Compensation of Endogenous Compounds

To test the potential for ammonia compensation, mixed samples containing ammonia and glutamate were assayed. As showed in TABLE 3, good recovery of creatinine was observed in the case of samples containing not only ammonia, but also glutamate and even the nonsubstrate ascorbate. Note that all results were obtained

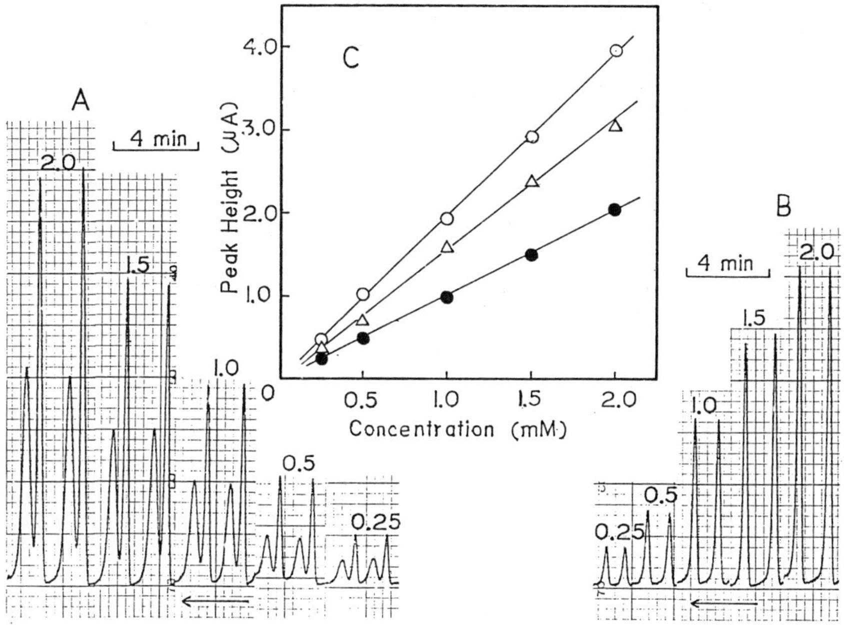

FIGURE 3. Typical response patterns of the two-channel FIA system for ammonia (A) and creatinine (B), and corresponding calibration curves (C) for the first peak (○) and the second peak (●) of ammonia and for creatinine (△). The concentrations in the recordings (A and B) are given in mM.

from calibrations with creatinine and ammonia. In other word, we do not need calibration with glutamate for the compensation of endogenous glutamate. We thus concluded that the interference of all endogenous factors that do not specifically affect the activity of CRDI could be overcome by the present method.

Storage Stability and Operational Stability

Both immobilized CRDI and GLOD retained more than 90% of their original activities over 6 months of storage at 4°C. The activity of GLDH decreased to less than 90% within a month, but GLDH still kept 70% of its original activity after about one year's storage. The response of the FIA system began to decrease after operation for about three weeks (three times a week, approximately 1350 runs). However, linear calibrations could still be obtained over 6 months' occasional operation and the system could be applied to urine assay without exchange of columns for more than 5 months.

Assay of Creatinine in Urine

The FIA system was applied to the assay of creatinine in urine. After dilution (10 to 20-fold), samples were assayed without any pretreatment. For comparison, samples were also assayed by the Jaffe method.[14] As shown in TABLE 4, good correlation was obtained between the two methods.

In conclusion, the FIA system reported here provides a new method for the amperometric assay of creatinine, with the simultaneous compensation of endogenous ammonia and glutamate. Good agreement observed between the results from this system and those from the Jaffe method proved that the system can be used for clinical assay. Since the concentration of glutamate in urine is much lower than that of ammonia, the system can also be considered as a bifunctional biosensor for the simultaneous determination of creatinine and ammonia. This work might not only offer a general approach for overcoming the problem of interfering endogenous factors that do not affect the first enzymatic reaction step in the FIA system based on multienzyme reactions, but may also provide a strategy for multipeak recording of FIA biosensor systems.

TABLE 4. Assay of Creatinine in Urine as Determined by the Two-Channel FIA System and by the Jaffe Method

	Creatinine Found (mM)			Creatinine Found (mM)	
Sample No.	FIA	Jaffe	Sample No.	FIA	Jaffe
1	13.1	15.7	9	8.27	8.08
2	9.49	9.90	10	16.6	17.0
3	10.0	9.51	11	14.8	15.6
4	13.6	13.5	12	5.76	5.75
5	10.9	10.4	13	7.26	7.87
6	17.2	18.4	14	5.14	5.25
7	10.7	11.5	15	10.6	11.0
8	8.27	8.08	16	20.9	22.3

SUMMARY

A flow-injection analysis biosensor system was developed for the amperometric assay of creatinine based on coupled reactions of three immobilized enzymes, using an oxygen electrode as the detection device. The ammonia produced by creatinine deiminase–catalyzed hydrolysis of creatinine was further converted into L-glutamate with two sequentially aligned enzyme reactors: glutamate dehydrogenase and glutamate oxidase. Endogenous ammonia was simultaneously compensated with a double peak recording system, where the flow was split after sample injection and rejoined before the glutamate dehydrogenase reactor. The system gave linear calibration in a range of 0.1–2.0 mM for creatinine and the first peak of ammonia, and 0.1–3.0 mM for the second peak of ammonia. One run was completed within two minutes. The system can be readily applied to the assay of creatinine in urine and showed good correlation with that from the currently used Jaffe method.

ACKNOWLEDGMENTS

The authors express their gratitude to Dr. K. Yoshinaga for many helpful discussions on enzyme immobilization and Dr. T. Okuda for kind assistance on the chemical assay of creatinine in urine.

REFERENCES

1. FOLIN, O. & H. WU. 1919. J. Biol. Chem. **38:** 81–110.
2. FABINY, D. L. & G. ERTINGSHAUSEN. 1971. Clin. Chem. **16:** 696–700.
3. SOLDIN, S. J., L. HENDERSON & J. G. HILL. 1978. Clin. Biochem. **11:** 82–86.
4. MASCINI, M., S. FORTUNATI, D. MOSCONE & G. PALLESCHI. 1985. Anal. Chim. Acta **171:** 175–184.
5. WINQUIST, F. & I. LUNDSTROM. 1986. Anal. Chem. **58:** 145–148.
6. KIHARA, K. & E. YASUKAWA. 1986. Anal. Chim. Acta **183:** 75–80.
7. KUBO, I., I. KARUBE & S. SUZUKI. 1983. Anal. Chim. Acta **151:** 371–376.
8. KUBO, I. & I. KARUBE. 1986. Anal. Chim. Acta **187:** 31–37.
9. WEETALL, H. H. 1975. Immobilized Enzymes, Antibodies and Peptides: 13. Marcel Dekker. New York, N.Y.
10. CHO, I. C. & S. SWAISGOOD. 1974. Biochim. Biophys. Acta **334:** 243–256.
11. UWAJIMA, T. & O. TERADA. 1977. Agric. Biol. Chem. **41:** 339–344.
12. SMITH, E. L., B. M. AUSTEIN, K. M. BLUMENTHAL & J. F. NYC. 1975. The Enzymes. P. D. Boyer, Ed. 3rd edit. **11A:** 293–367. Academic Press. New York, N.Y.
13. BOHMER, A., A. MULLER, M. PASSARGE, P. LIEBS, H. HONECK & H.-G. MULLER. 1989. Eur. J. Biochem. **182:** 327–332.
14. FISTER, H. J. 1950. Manual of Standards for Spectrophotometric Chemistry, Method C: 43. Standard Specific Supply Corp. New York, N.Y.

Ultramicrobiosensors for Monitoring of Neurotransmitters

EIICHI TAMIYA AND ISAO KARUBE

Research Center for Advanced Science and Technology
University of Tokyo
4-6-1 Komaba, Meguro-ku
Tokyo 153, Japan

INTRODUCTION

Neurotransmitters play important roles in the brain, since they are the key link in communication between neurons. Acetylcholine is the transmitter of motor neurons in the spinal cord as well as in all the nerve-sketal junctions in vertebrates. It is located diffusely throughout the brain. The activity of the enzyme choline acetyltransferase in the brain decreases, often significantly, with age. This is the enzyme responsible for synthesis of acetylcholine. Glutamate is the transmitter that is related to expression of long-term memory. Mechanisms of long-term potentiation (LTP) in the hippocampus and long-term depression (LTD) in the cerebellar cortex are unclear because of difficulty in measurement of glutamate release from presynaptic membrane and sensitivity of glutamate receptor in postsynaptic membrane.

In vivo acetylcholine and glutamate sensors are powerful tools to elucidate the sites of action in the brain where they operate. A smaller electrode causes less damage to tissue during insertion into brain or nerve tissue. An extremely small environment can be examined with a microelectrode whose diameter is of the order of a micrometer. The carbon fiber electrode is considered to be one of the most useful transducers for *in vivo* biosensors, because carbon fibers form ultramicroelectrodes with high strength and electrochemically pretreated fibers have much greater sensitivity for dopamine and catecholamine. Usually acetylcholine and glutamate cannot be directly oxidized on an electrode, but the application of carbon fibers using immobilized enzymes has proven successful.

Several acetylcholine sensors have been developed based on enzyme electrodes with potentiometric detectors or amperometric detection of oxygen and hydrogen peroxide. However these sensors are not suitable for *in vivo* analysis. The detection of hydrogen peroxide is an established approach to the construction of biosensors based on electrodes containing immobilized oxidases. In previous work, it was possible to obtain a high sensitivity to dopamine and an excellent reproducibility using microcomputer-controlled potentiostatic pulse polarization techniques.[1]

In this study, electrochemical operations for preelectrolysis and measurement were employed for sensitive determination of hydrogen peroxide. Then carbon fiber electrodes were modified with acetylcholine esterase and choline oxidase for an acetylcholine sensor and glutamate oxidase for a glutamate sensor.

FABRICATION OF MICROBIOSENSORS

Choline oxidase, acetylcholine esterase, and acetylcholine were purchased from Sigma. Glutamate oxidase was donated by Yamasa Company. Poly(vinyl alcohol)-styrylpyridinium (PVA-SbQ, degree of polymerization 1700, degree of saponification 88, SbQ content 1.3 mol%, solid content 11%, pH 7) was kindly donated by Toyo Kagaku Co. Other reagents were of analytical grade.

The material and fabrication of carbon fiber electrodes have been described in a previous paper.[1] The diameter of the electrode was 7 μm. All electrochemical experiments were done with a three-electrode system enclosed in a steel Faraday cage. A microcomputer-controlled potentiostat system was used for precise control of applied potential. This system was newly designed and constructed for determination of H_2O_2. The basic method for the construction of this system was shown in a previous paper.[1]

Our polarization technique was applied to analyze H_2O_2. An activation signal consisting of an anodic-cathodic triangular wave was introduced immediately before every measuring pulse. The parameters for measuring pulses were set as follows. The length of the first pulse (750 mV) was 1000 mseconds. The current was measured every 625 μseconds for 160 mseconds to the end of the first pulse. The average of 256 measured points for 160 mseconds was employed for calculation of current intensity. The length of second pulse (1100 mV) was also 1000 mseconds. The average value was obtained in similar manner as above. The 32 measured points were obtained for 20 mseconds, which is one cycle of the electric line (50 cycles/second). Use of the average improves sensitivity and reproducibility because of noise reduction. The current intensity was defined as difference in currents between first and second pulses. A platinized carbon fiber disk (PCD) electrode was also used as a transducer of the micro–glutamate sensor.

RESPONSE TO HYDROGEN PEROXIDE

Oxidation of H_2O_2 was observed over 850 mV vs. Ag/AgCl based on voltammogram of H_2O_2 in phosphate buffer. When 1100 mV of potential was applied to the carbon electrode, stationary current was obtained within 1 second. Therefore 1 second is suitable for the length of pulse potential. FIGURE 1 shows relationships between the current intensity and H_2O_2 concentration. A single pulse of 750 mV was not available for H_2O_2 determination because current fluctuated randomly against H_2O_2 concentration. Pulse operation noted in the no. 4 gave an excellent calibration for H_2O_2. Activation triangular potential was necessary to obtain a reproducible response. The subsequent response was gradually reduced without activation potential (FIGURE 2). Aoki *et al.* indicated that oxidation of H_2O_2 began at 0.5 V based on cyclic voltamograph and the peak current was related to H_2O_2 concentration higher than 0.1 mM.[3] More sensitive determination was not successful because the oxidation was masked by background current.

Our polarization technique has been established for sensitive detection of dopamine and its derivative.[1,2] This technique is based on use of the activation triangular wave before every measuring pulse. The activation wave included negative potential where oxygen was electrochemically reduced possibly to produce hydrogen peroxide. However, current intensity was not affected by dissolved oxygen concentration, because background current based on oxygen was compensated by our double-pulse method. The detection limit is less than 0.1 μM. The sensitivity of our carbon fiber is better than that of micro–gold electrodes reported previously.[4]

Use of the activation potential was effective for the improvement of sensitivity. A linear relationship was obtained between current and hydrogen peroxide concentration in the range from 0.1 μM to 1 mM; slope, *y*-intercept, and linear correlation coefficient were 0.638 nA/μM, 10.3 nA, and 0.9997, respectively. Reproducibility was tested by repeated 25-μl injections of 50-mM hydrogen peroxide solutions; the relative standard deviation for a set of 10 injections was 0.68%. Current value became saturated above 4 mM. This sensor is stable for at least 1 month. Slope and background currents were different between electrodes probably because of different surface areas and microenvironmental diffusion conditions.

Albumin protein which is rich in body fluid adsorbed nonspecificaly to the surface of the electrode followed by reduction of active surface area. In our electrode system, reproducible determination of H_2O_2 is possible even in samples containing albumin. Our triangular wave produces oxygen and hydrogen gases and etches the surface of the carbon fiber. These phenomena may strip adsorbed protein from the electrode surface. The slope value in albumin solution was reduced to 73% of that in albumin-free solution. Diffusion speed of hydrogen peroxide became small due to high viscosity of 0.5 g/ml albumin solution.

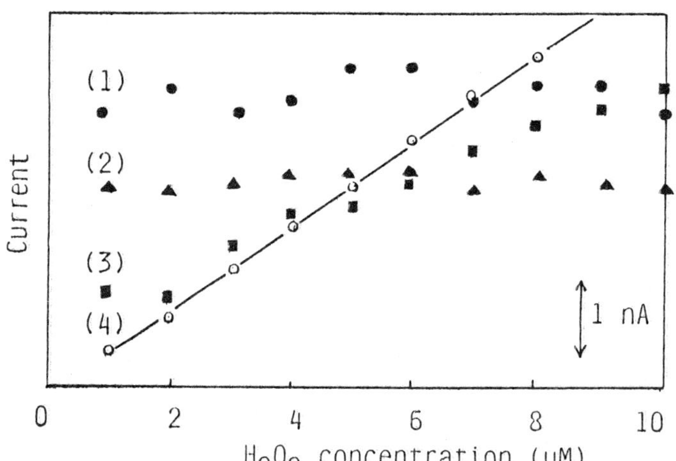

FIGURE 1. Relationship between intensity and hydrogen peroxide concentration under various kinds of pulse operation. Single pulse: (1) P_1, 750 mV; T_1, 0.5 seconds. Double pulse: (2) P_1, 750 mV; P_2, 850 mV; T_1, 0.5 seconds; T_2, 0.5 seconds. (3) P_1, 600 mV; P_2, 950 mV; T_1, 1 second; T_2, 1 second. (4) P_1, 750 mV; P_2, 1100 mV; T_1, 1 second; T_2, 1 second. Currents of 1, 2, 3, and 4 in 1 μM H_2O_2 are 16.5, 1.2, 6.2, and 10.5 nA respectively.

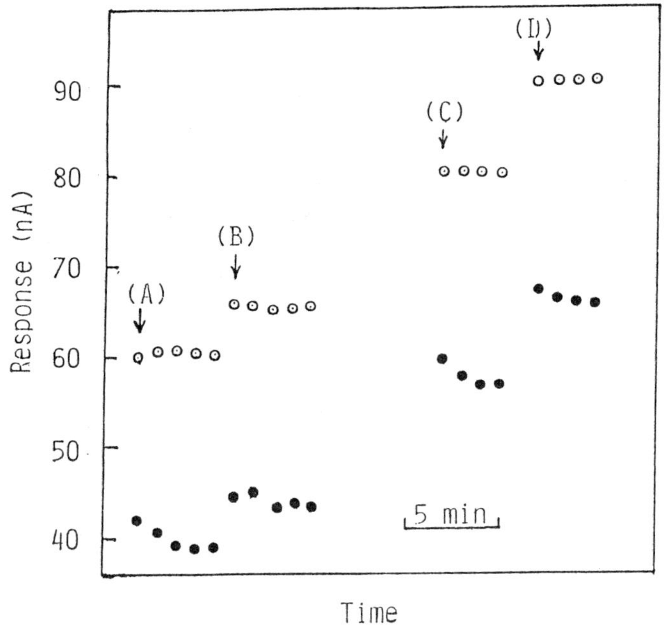

FIGURE 2. Time course of current intensity after addition of hydrogen peroxide. Double pulse no. 4 in FIGURE 1 was used with (⊙) and without (●) activation wave. A, B, C and D indicate addition of hydrogen peroxide of 40 μM, 50 μM, 80 μM, and 100 μM (final concentration), respectively.

CHARACTERISTICS OF THE MICRO–ACETYLCHOLINE SENSOR

Choline oxidase (1 U) and acetylcholine esterase (0.5 U) dissolved in 0.5 ml of water were added to 0.2 mg of 11% aqueous solution of PVA-SbQ (enzyme-PVA mixture).[5] A carbon fiber electrode was dipped into enzyme-PVA mixture and air dried for 5 hours at room temperature. The air-dried carbon fiber was exposed to normal fluorescent room light for 1 hour.

The micro–acetylcholine sensor gave a linear calibration plot for the range 0.1–1.0 mM with a linear correlation coefficient of 0.9842.[6] The sensitivity with PVA-SbQ membrane is better than that obtained by using immobilized enzymes by cross-linking with albumin and glutaraldehyde. The determination time for acetylcholine was 5 seconds. Previous acetylcholine sensors needed at least 1 minute of response time because diffusion of acetylcholine and hydrogen peroxide occurred slowly in thick enzyme-immobilized membrane. The determination limit of our sensor is 0.05 mM (S/N = 2). Sensitivity is strongly related to the amount of enzyme immobilized in the membrane that covers the electrode surface. Scanning electron micrographs indicated that the immobilized enzyme membrane is very thin. The half-life of our micro–acetylcholine sensor was approximately 1 week, which depended on use and storage conditions.

CHARACTERISTICS OF THE MICRO–GLUTAMATE SENSOR

The micro–glutamate sensor consists of a platinized carbon fiber disk (PCD) electrode and an immobilized glutamate oxidase membrane. The PCD electrode with a diameter of 15 μm gave a more stable response to hydrogen peroxide because the background current of the PCD electrode is smaller than that of carbon fiber electrode. Glutamate oxidase was immobilized using a photo-cross-linking method with PVA-SbQ. The micro–glutamate sensor gave a linear calibration for the range from 2 μM to 1.2 mM. This sensor was applied to monitor release of glutamate from the synapse in the cerebellar cortex (FIGURE 3). The amount of glutamate was dependent on calcium concentration in the perfusate. Release of glutamate after electrical stimulation was also monitored by our sensor.

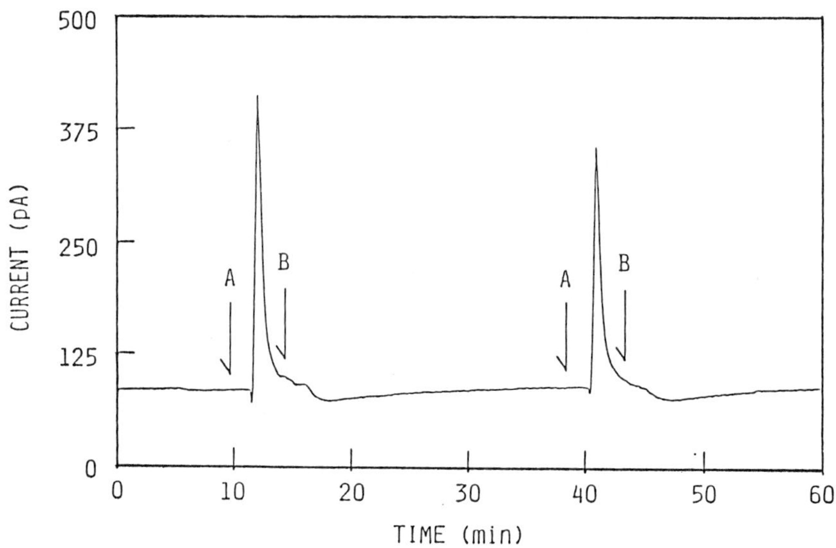

FIGURE 3. Response of the sensor for applying high potassium stimulation to a cerebellar slice. At A, perfusate was changed from normal Ringer solution (NR) to high potassium containing Ringer solution (high K^+); and at B, that was changed from high K^+ to NR.

SUMMARY

Carbon fiber electrodes are used to construct ultramicrobiosensors with 7–15 μm diameter. Electrochemical operations for preelectrolysis and measuring were examined for sensitive determination of hydrogen peroxide. Determination limit was 0.1 μM of hydrogen peroxide. Reproducible determination of hydrogen peroxide is possible even in samples containing albumin protein. A micro–acetylcholine sensor was fabricated by immobilizing acetylcholine esterase and choline oxidase on the carbon fiber by entrapment with PVA-SbQ. This sensor gave a linear calibration plot for the range from 0.1 to 1.0 mM with a linear correlation coefficient of 0.9842. A micro–glutamate sensor consisted of a platinized carbon fiber disk electrode modified with immobilized glutamate oxidase membrane. This sensor gave a linear calibration for the range 2 μM to 1.2 mM. Release of glutamate in the cerebellar cortex was detected after potassium stimulation.

REFERENCES

1. AKIYAMA, A., T. KATO & E. YASUDA. 1985. Anal. Chem. **57:** 1518.
2. NAKAZATO, T. & A. AKIYAMA. 1988. J. Neurochem. **51:** 1007.
3. AOKI, K., M. ISHIDA, K. TOKUDA & K. HASEBE. 1988. J. Electroanal. Chem. **251:** 63.
4. TAMIYA, E. & I. KARUBE. 1988. Sensors Actuators **15:** 199.
5. ICHIMURA, K. 1987. Makromol. Chem. **188:** 763.
6. TAMIYA, E., *et al.* 1991. Anal. Chim. Acta **251:** 129.

Effect of Enzyme Hydration on the Catalytic Activity of Chymotrypsin in Nearly Anhydrous Organic Suspensions[a]

M. E. PAULAITIS, M. J. SOWA, AND J. H. McMINN

Department of Chemical Engineering
University of Delaware
Newark, Delaware 19716

INTRODUCTION

The advantages of enzyme catalysis in organic solvent suspensions include increased substrate solubility, enhanced enzyme thermostability, efficient product recovery, and the ability to catalyze reactions that are suppressed in aqueous solutions.[1] Before such advantages can be fully exploited, however, a fundamental understanding of the effect of solvent medium on enzyme activity is required. One important factor is the amount of water sorbed by the enzyme, which has been observed to have direct impact on both the kinetics and the selectivity of enzymatic reactions. Previous work with lysozyme[2] and urease[3] powders hydrated by exposure to water vapor suggests that a threshold level of enzyme hydration is required before catalytic activity is obtained. For enzyme suspensions in organic solvents, the concept of "essential water" has been used to suggest that a small amount of water is needed in the solvent to establish this threshold hydration level for the onset of enzyme activity.[4]

In this study, the effect of enzyme hydration on the catalytic activity of chymotrypsin powders suspended in 1-M *n*-propanol/toluene has been investigated. The work is unique in the sense that extremely low levels of enzyme hydration have been investigated, i.e., less than 10 wt% water. The following experiments are described: (1) measurements of chymotrypsin activity for the transesterification of *N*-acetyl-L-phenylalanine ethyl ester (APEE) to the corresponding propyl ester, (2) a novel two-step active site titration for the determination of the number of active enzymes in the organic suspension, and (3) enzyme-inhibition studies using the transition state analogue inhibitor, (*R*)-1-acetamido-2-phenylethane boronic acid. All measurements were made as a function of enzyme hydration.

[a]The authors gratefully acknowledge financial support provided by the National Science Foundation (CPE 8351228) and the following sponsors: Air Products and Chemicals Company, E. I. du Pont de Nemours and Company, Dow Chemical Company Foundation, Exxon Education Foundation, Imperial Chemical Industries, Merck and Company, and Procter and Gamble Company.

EXPERIMENTAL SECTION

Prior to this work, a detailed study of water sorption by enzyme powders suspended in organic solvents at low water activities was conducted in our laboratory.[5] The water sorption isotherm at 27°C was measured for the suspension of chymotrypsin in 1 M *n*-propanol/benzene, and these results are presented in FIGURE 1. This sorption isotherm is used here to determine the amount of water sorbed by chymotrypsin as a function of the measured amount of water in the 1 M *n*-propanol/toluene solvent mixture.

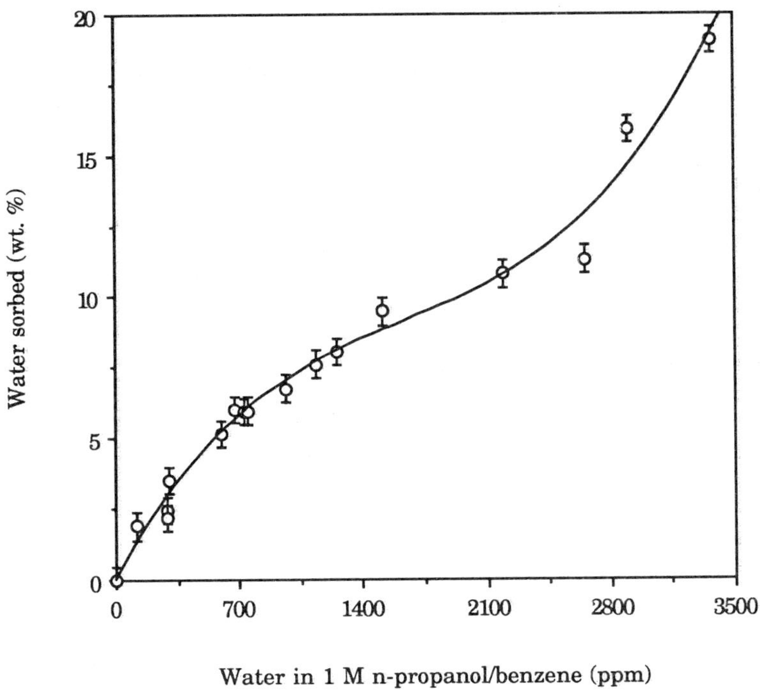

FIGURE 1. Measured water sorption isotherm for chymotrypsin in 1 M *n*-propanol/benzene at 27°C.

Apparent Enzyme Activity

Details of the experimental technique are given elsewhere.[6] Chymotrypsin powders were prepared by lyophilizing aqueous solutions containing the enzyme and a 40-fold molar excess of *N*-acetyl-L-phenylalanine at a pH of 7.8. The APEE conversion in the presence of excess *n*-propanol at 27°C was measured to give the initial reaction velocity. Michaelis-Menten kinetics were assumed to hold and the apparent enzyme activity, k_c/K_m, was obtained from a conventional double reciprocal

enzyme hydration was observed for the onset of catalytic activity, as has been reported for enzyme powders hydrated from the vapor phase. Approximately 30% of the chymotrypsin in the organic solvent suspension was found to be catalytically active over this range of enzyme hydration. Since the number of active sites is independent of enzyme hydration, it follows that the increase in apparent enzyme activity with enzyme hydration results from an increase in the intrinsic catalytic activity of the enzyme, rather than an increase in the population of active sites. Finally, the inhibition measurements suggest that one role of water in activating nearly anhydrous enzyme suspensions is to enhance the intrinsic catalytic activity of chymotrypsin by stabilizing the transition state for the transesterification reaction.

SUMMARY

The relationship between enzyme hydration and the catalytic activity of chymotrypsin powders suspended in nearly anhydrous 1 M *n*-propanol/toluene solutions has been studied. The enzyme hydration range examined extends from approximately 2 to 10 wt% water sorbed. Measurements of apparent enzyme activity and the number of active sites as a function of enzyme hydration were used to calculate intrinsic enzyme activity as a function of enzyme hydration. It was found that the apparent enzyme activity increased linearly over this hydration range and that approximately 30% of the enzyme was catalytically active independent of the hydration level. Thus, the observed increase in apparent catalytic activity with increasing enzyme hydration results from an increase in the catalytic efficiency of a fixed number of active sites, rather than an increase in the population of active sites. Inhibitor binding in the organic suspension was also measured using the transition state analogue, (*R*)-1-acetamido-2-phenylethane boronic acid, which binds reversibly to the active site of the enzyme. The inhibition results suggest that one role of water in activating nearly dry enzyme suspensions is to enhance the intrinsic catalytic activity through transition state stabilization.

REFERENCES

1. KLIBANOV, A. M. 1986. Enzymes work in organic solvents. *CHEMTECH* **16:** 354–359.
2. RUPLEY, J. A., P. H. YANG & G. TOLLIN. 1980. Thermodynamic and related studies of water interacting with proteins. ACS Symp. Ser. No. 127: 111–132.
3. SKUJINS, J. J. & A. D. MCLAREN. 1967. Enzyme reaction rates at limited water activities. Science **158:** 1569–1570.
4. ZAKS, A. & A. M. KLIBANOV. 1988. The effect of water on enzyme action in organic media. J. Biol. Chem. **263:** 8017–8021.
5. MCMINN, J. H., M. J. SOWA, S. B. CHARNICK & M. E. PAULAITIS. The hydration of proteins in nearly anhydrous organic solvent suspensions. Biopolymers. (In press.)
6. SOWA, M. J., J. H. MCMINN & M. E. PAULAITIS. The effect of enzyme hydration on the kinetic and inhibition behaviors of chymotrypsin suspensions in organic solvents. J. Biol. Chem. (Submitted.)
7. MATTESON, D. S., K. M. SADHU & G. E. LIENHARD. 1981. (*R*)-1-Acetamido-2-phenylethane boronic acid. A specific transition state analogue for chymotrypsin. J. Am. Chem. Soc. **103:** 5241–5242.

Solubilization and Activity of Proteins in Supercritical Fluids

GUADALUPE A. AYALA,[a,b] SANJAY KAMAT,[a,b]
CLAIRE KOMIVES,[a] ERIC J. BECKMAN,[a]
AND ALAN J. RUSSELL[a,b,c]

[a]*Department of Chemical Engineering*
[b]*Center for Biotechnology and Bioengineering*
University of Pittsburgh
Pittsburgh, Pennsylvania 15261

INTRODUCTION

The remarkable specificity of enzymes for catalyzing a wide variety of reactions has great potential in many industrial applications. Enzymes offer the possibility of stereo- and regioselectivity for the formation of novel chemicals for drugs, food additives, agricultural chemicals, and other products. For the past 15 years, research in enzymology has included studies in aqueous and organic media. This was motivated principally by the need to extend the applicability of enzymes to catalysis of reactions whose reactants and/or products are not water soluble, as in many industrially important reactions. If the substrate is insoluble or only slightly soluble in water, the maximum activity of the enzyme cannot be achieved in an aqueous solution. Other factors make nonaqueous enzymology attractive. The absence of water in some cases can prevent undesirable side reactions. It has been shown that reduced water content improves the thermostability of enzymes.[1]

Because organic solvents are not the native environment for an enzyme, the activity in most cases is reduced from that in water, often by as much as a 10^6-fold reduction for an enzyme suspended in an anhydrous organic solvent. It has been shown that this activity reduction is not necessarily due to a change in the enzyme mechanism and gross structure, but is principally due to partitioning of substrates to and from the enzyme surface, localized changes in the rigidity of the protein, increased K_ms for most substrates, and mass transfer limitations which are characteristic of all solid-phase catalysts (enzymes are, of course, insoluble in most organic solvents). Various methods to enhance the activity of organic phase enzymatic reactions have been sought. The use of biphasic mixtures of aqueous and organic solvents has been studied for many years. In these systems, the enzyme is contained in the aqueous phase and the substrate is introduced in the organic phase. While this approach enables a higher substrate concentration as well as offering a convenient method for product removal via the organic phase, several disadvantages are associated with it.[2] Because of the low solubility of the substrates in the aqueous phase where the enzyme is located, very high concentrations of the substrate (on the

[c]Author to whom correspondence should be addressed, at 1235 Benedum Hall, Department of Chemical Engineering, University of Pittsburgh, Pittsburgh, Pa. 15261.

order of moles/liter) are necessary to enhance product formation. The use of very low volumes of the water phase can achieve the same result. The use of these systems for preparative chemistry has, however, been described[3] and remains of much interest. Indeed, the innovative work of Arnold and colleagues has shown that it is possible to increase the activity of enzymes in water-cosolvent mixtures by protein engineering.[4]

Another route to achieving some of the advantages of nonaqueous enzymology while maintaining the aqueous environment around the enzyme is the use of reverse micelles, which are spherical aggregates of amphiphilic molecules encapsulating a water pool within bulk organic solvent. The hydrophilic surfactant heads remain at the water/organic interface, while the nonpolar tails extend into the organic phase. Solutions containing reverse micelles are optically transparent, and hence have been described by the term microemulsions. They are thermodynamically stable within a critical range of water, oil, and surfactant concentration. Some surfactants also require an accompanying cosurfactant, such as low to medium chain length alcohols, for enhanced stability. These reverse-micellar solutions can accommodate varying quantities of water, depending on the concentration and properties of the surfactant, the properties of the bulk organic solvent, the overall physical properties of the aqueous phase, and in particular the ionic strength of the aqueous phase. The molar ratio of water to surfactant in the solution is indicative of the size of the water pool, and is commonly described by the parameter W_0. Reverse micelles are dynamic entities, that is, there is a continuous process of collision and exchange of the contents of the water pools. Because of this, reverse micelles are appropriate systems for catalysis of both water-soluble and apolar compounds.

PROTEIN EXTRACTION

Enzymes have been shown to be soluble in reverse micelles. This has important consequences for the applications of enzyme separation and purification as well as biocatalysis. A number of reviews discuss the history and progress of the work.[5–8] Under the appropriate conditions, reverse micelles are capable of extracting proteins out of an adjacent aqueous bulk phase. The type and quantity of protein solubilized are dependent on the surfactant properties including concentration, as well as the pH and ionic strength of the aqueous phase. The type of salt in the aqueous phase is also a contributing factor.[9] The effects of the aqueous solution parameters are principally related to the interactions among the polar surfactant head groups, as well as those between the head groups and the proteins. In the case of anionic surfactants, for example, when the solution pH is below the pI of the protein, the protein will have a net positive charge and the solubilization will be favorable. The effect of pH on the extraction of proteins with reverse micelles of nonionic surfactants has not been studied extensively. It is assumed that pH will not have such a clearly defined role as in ionic surfactant systems.

Recent work has involved a study of protein solubilization in nonionic reverse micelles.[10,11] The choice of a nonionic surfactant results in an important advantage over ionic surfactants, namely, the absence of strong charges at the aqueous/organic interface provides a more favorable environment in which the protein can reside.

The majority of work to date with reverse-micellar enzymology has been with the anionic surfactant, bis(2-ethylhexyl) sodium sulfosuccinate (AOT). Experiments have shown that micelles of AOT can cause rapid loss of the activity of solubilized enzymes.[12–14] The surfactant chosen for this work was polyoxyethylene sorbitan trioleate (Tween-85), which forms stable microemulsions in hexane with isopropanol as a cosurfactant. Micelles formed with this surfactant can accommodate larger water pools than AOT and other detergents previously studied. As predicted, the extraction of cytochrome C was found to be indirectly related to the ionic strength of the aqueous phase. The optimum conditions for extraction of this protein were identified as 14% Tween-85/13% isopropanol/73% hexane (v:v:v) at ionic strength of the aqueous phase at 0.001 μ.

Some useful applications can arise from these techniques. Reverse micelles can be used to transfer proteins from one bulk aqueous phase to another. Some of the practical considerations of this operation have been described by Hatton.[6] Extraction systems of a mixer/settler type have been studied,[15] as well as a hollow-fiber design.[16] Protein separation using reverse micelles is also a viable technique. The efficiency of extraction of a protein depends on its size, hydrophobicity, pI, and other factors. Selection of the optimum conditions for a given protein in a mixture may result in a preferential recovery of the protein of interest. This technique has not yet been extensively studied on a preparative scale.

A notable disadvantage of the system just described is that changes in the physical properties of the aqueous solution are the driving force for the extraction process. Many proteins are inactivated by even small changes in their environment, and one can not extract these proteins into a reverse-micelle system by changing radically their aqueous environment while also retaining activity. Thus, there is a need to develop an extraction system in which the driving force for protein separation comes from changes in the physical properties of the organic phase. This will mean that the protein environment will not change, other than the fact that the protein will be solubilized in a micelle. We have proposed that supercritical fluids could be used as solvents in such a system.

When above their critical pressures and temperatures, gases exhibit radical changes their physical properties. The density of a supercritical fluid lies in between that of gaseous and liquid states, and is a sensitive function of pressure. Compounds such as ethane, ethylene, sulfur hexafluoride, fluoroform, and carbon dioxide have critical temperatures sufficiently low to admit a supercritical state amenable to enzymatic catalysis. In addition, gases such as propane with critical temperatures in excess of 75°C can be used as liquids at high pressure (in the near critical state). FIGURE 1 shows a schematic of the apparatus used to test this hypothesis. Of particular interest is that the high-pressure cells that we have designed can fit directly into a Perkin Elmer spectrophotometer. Given that we have two such cells, we can now perform difference spectrophotometry with materials in supercritical fluids. We routinely work with nontoxic (carbon dioxide and sulfur hexafluoride) and flammable gases (propane, ethane, ethylene). We wish to stress that significant safety issues must be addressed when working with these systems. We have found that in the presence of Tween-85 and isopropanol, near-critical propane (40°C and 3000 psi) forms a stable, optically clear, one-phase mixture with 10% aqueous buffer. This is extremely strong evidence that a microemulsion exists, although we are currently

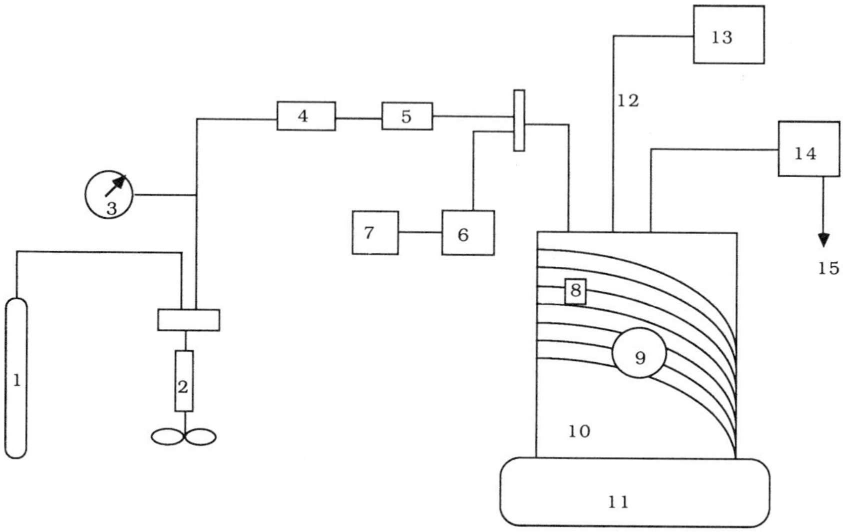

FIGURE 1. Reactor configuration for high-pressure biocatalytic reactions. (1) Gas cylinder; (2) syringe pump; (3) pressure gauge; (4) rupture disc; (5) filter; (6) pressure transducer; (7) pressure read out; (8) heating coils; (9) sapphire windows; (10) stainless steel reactor; (11) magnetic stirrer; (12) thermocouple; (13) temperature readout; (14) sample valve; (15) sample collection port.

confirming this observation with light-scattering techniques. A schematic representing the phase behavior and its dependence on pressure is given in FIGURE 2. Further evidence that microemulsions form is shown in TABLE 1, which demonstrates that cytochrome C can be solubilized in a supercritical fluid in the presence of buffer, Tween-85, and isopropanol. Furthermore, the degree of extraction depends strongly on the pressure of the system. It should be stressed that the protein, fluid, buffer mixture is also optically clear and forms a single, stable, phase. At this time, we are developing the techniques so that we can study the degree of protein solubilization in two-phase systems as a function of pressure. Finally, we are attempting to stabilize and solubilize membrane proteins in these nonaqueous environments so as to develop a general approach to membrane protein purification.

Another important advantage of the use of supercritical fluids as a solvent is that the detergent can be easily separated from the protein. Release of pressure will produce either solid enzyme or enzyme dissolved in aqueous media (depending on the initial amount of water in the system). By utilizing Tween-85, which is insoluble in water, a simple filtration step will separate the enzyme and detergent.

ENZYME ACTIVITY

Although protein extraction into reverse micelles has received much attention in the literature, the most probable area for development of this technology is in the field of biocatalysis in such media. The activity and function of several enzymes in reverse micelles have been analyzed by a number of research groups worldwide. Overviews of the extensive studies carried out in this area can be found in many reviews.[17–20] Once again, the system most often encountered in these discussions concerns reverse micelles of AOT. While this surfactant can promote degradation of the protein, it has also been shown to enhance the activity of some enzymes.[21] The negative charges on the surfactant heads serve to concentrate negatively charged substrates in the protein-containing reverse-micelle interior by repulsive forces. This is applicable only in cases where the substrate is mainly in the aqueous phase such that the enzyme does not rely on partitioning of the substrate through the surfactant layer into the micelles. Nonionic surfactants such as Triton X-100 and Brij varieties have also been used with more success, with respect to increasing enzyme stability, in general. The activity of α-amylase and cellulase enzymes was shown to be enhanced in reverse micelles of Tween-80 in hexane over that in water.[22] Extensive character-

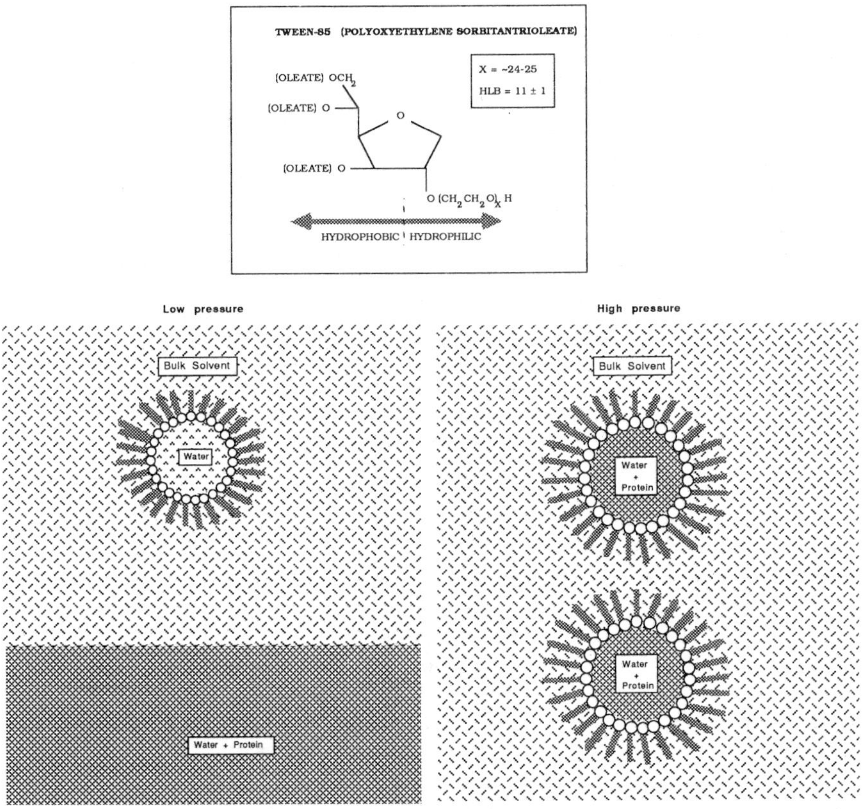

FIGURE 2. Solubilization of proteins in propane. As the density of the fluid increases with increasing pressure, so do the size and number of micelles. With cytochrome C as the protein in the aqueous phase, the propane/buffer mixture eventually becomes a single phase, optically clear and red, indicating the solubilization of protein within reverse micelles.

ization of the physicochemical properties and dynamic functions of reverse micelles for enzymology has been carried out and is reviewed by Luisi *et al.*[23]

Reverse micellar systems have several advantages over the biphasic systems without the presence of detergents. Because of surfactant-protein interactions, high concentrations of solubilized enzyme can be achieved resulting in greater product formation than in simple biphasic systems. In addition, the dependence of protein solubilization on the salt concentration in the aqueous phase of a micellar system can be exploited for the convenient removal of enzyme from the micelles. In most systems, there exists a maximum salt concentration for maintaining the enzyme in the solubilized form. When this concentration is surpassed, the ions compete with the proteins for hydration in the aqueous pools. The proteins are salted out and can be removed from the solution in solid form. Product recovery can also be simplified by altering the proportions of the aqueous/organic volumes. The micelles can be disrupted in this way and the products can be extracted from either the aqueous or organic layers.

TABLE 1. Extraction of Cytochrome C into Propane[a]

Pressure (psi)	Cytochrome C 30 °C	Pressure (psi)	Cytochrome C 45 °C
250	2 ph, o, ppt, sr	225	2 ph o, ppt, sr
520	2 ph, o, ppt, it, sr	560	2 ph, o, ppt, sr
750	2 ph, o, ppt, it, sr	1000	2 ph, o, ppt, sr
1025	2 ph, o, ppt, it, sr	1600	2 ph, t, r, ppt, saq
1400	2 ph, o, ppt, it, sr	2600	2 ph, t, r,ppt, it
1650	2 ph, o, ppt, it, sr	4700	2 ph t, r, ppt, it
3250	1 ph, oc, r	5000	2 ph, t, r, ppt, it

[a]The reactor described in FIGURE 1 was pressurized with propane containing 14% Tween-85, 4 ml cytochrome C in phosphate buffer pH 7.0, 13% (v/v) isopropanol. The appearance of the microemulsion was noted during pressurization: it = increasing transparency relative to lower pressure; o = opaque; oc = optically clear solution; ph = number of phases; ppt = appearance of protein precipitate; r = red color in organic phase; saq = single aqueous phase; sr = overall appearance is red; t = transparent; w = red color in aqueous phase.

Of critical importance for the viability of these systems is the retention of enzyme activity in the micelles. Studies in our group have included ester hydrolysis by subtilisin in micelles of Tween-85 in hexane with isopropanol. With this system, the enzyme showed slightly enhanced stability relative to that in buffer alone (see FIGURE 3). We interpret this increased stability as a decrease in the autolysis rate in the micellar system (as suggested by Luisi). In addition, the degradation of the pesticide paraoxon has been carried out with phosphotriesterase[24] in the same system. The stability of this enzyme in the reverse micelles was comparable to that in buffer.

Given our interest in solubilizing proteins in supercritical fluids, we must show that in such an environment the protein retains its catalytic functions. We have reported enzyme activity of dry enzyme preparations in supercritical ethylene, ethane, sulfur hexafluoride, carbon dioxide, fluoroform, and near-critical propane.[25]

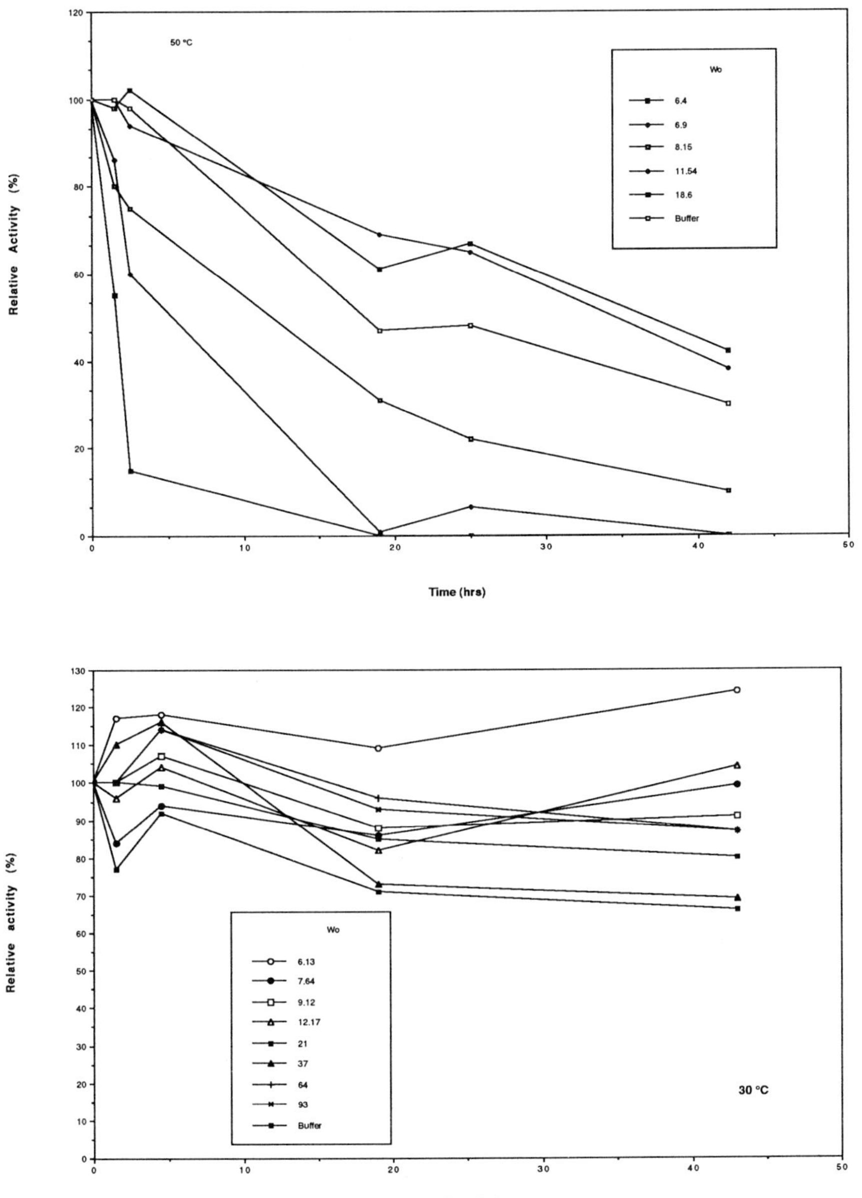

FIGURE 3. Stability of subtilisin solubilized in Tween-85/isopropanol/hexane. Subtilisin (10 μl) in either buffer or microemulsion was assayed as described previously[10] after incubation at either 30 or 50°C.

The advantages of utilizing supercritical fluids as a solvent for enzyme-catalyzed reactions are several. Because the density of supercritical fluids is strongly dependent on pressure, it should be possible to control enzyme activity (and perhaps specificity) as a function of pressure. The pressure can be optimized for the process needs. In addition, reducing the pressure below the critical value results in a return of the fluid to the gaseous state and thus complete separation and recycling of the solvent is possible. A mixture of enzyme and products in concentrated form remains, which results in ease of downstream processing and product recovery, as well as recycling of the fluid for further use. This has attractive implications for environmentally friendly industrial processes.

Using the method we have described above for solubilization of protein in near-critical propane, we have investigated the activity of two proteins in this environment. FIGURE 4 presents an example of the activity of subtilisin in the hydrolysis of *p*-nitrophenyl butyrate. The enzyme exhibits almost all of its activity in water in this system.

The overall structure of cytochrome C can be analyzed by performing difference spectra between the oxidized and reduced forms of the protein. In aqueous solution, redox transition with ascorbate results in characteristic shifts of the absorption peaks and isobestic points for the Soret region and the 500–600 nm region.[4] Our experiments showed that an *in situ* reduction with ascorbate in a near-critical propane–Tween-85 microemulsion did not differ significantly from the same experiment in water (see FIGURE 5), and previous experiments in a phospholipid-toluene microemulsion.[26] These results are in marked contrast to a similar experiment performed by Brochette and colleagues for the AOT/isooctane/water/cytochrome C system,

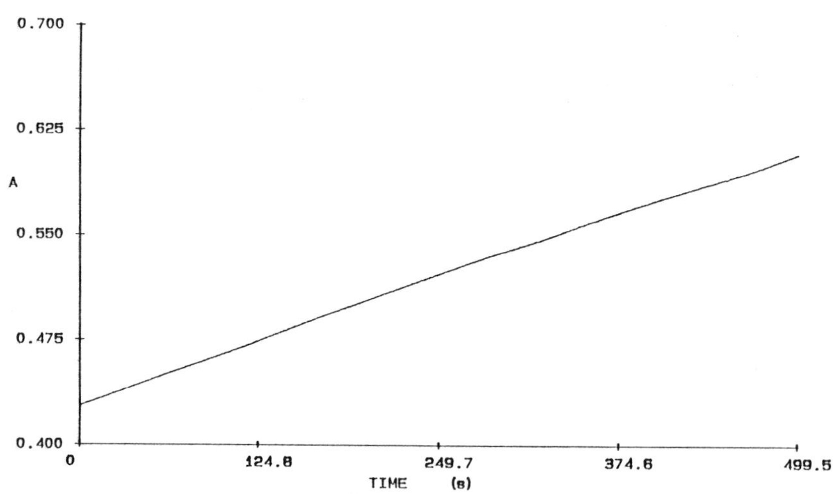

FIGURE 4. Activity of subtilisin solubilized in propane. Subtilisin (14.67 mg/ml) was dissolved in 0.01-M phosphate buffer, pH 7.8, and 20 μl was added to a pressurized propane solution containing 0.01-M buffer, 0.1-mM *p*-nitrophenyl butyrate, 14% Tween-85, and 13% (v/v) isopropanol. The increase in absorbance at 412 nm of the single-phase optically clear solution was followed spectrophotometrically. The figure presents the change in absorbance with time.

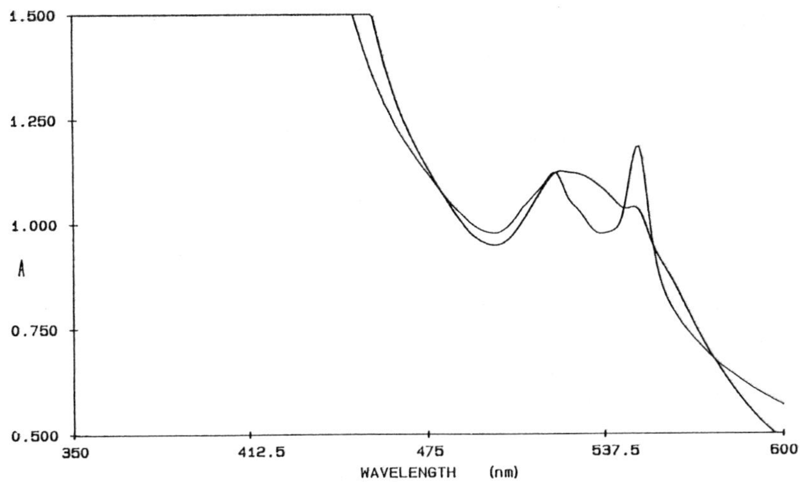

FIGURE 5. Reduction of cytochrome C in propane by ascorbate. Cytochrome C, solubilized as described above [4 ml buffer, 14% Tween-85, 13% (v/v) isopropanol, propane], was reduced *in situ* by the addition of 10 μl of 0.1-M ascorbate. The absorption spectra of the reduced and native cytochrome C are presented. (Reduced cytochrome C has two peaks, whereas native has only one major peak.)

where W_0s greater than 15 produced a microemulsion with significantly reduced ascorbate reduction rates. This reduction in rate was attributed to adsorption of the cytochrome C onto the ionic surfactant head groups at the micellar interface.

Thus, it is reasonable to assume that proteins solubilized in supercritical fluids containing microemulsions of Tween-85 are stable, active, and in their native conformation. Work is now in progress to assess the effect of pressure on the physical properties of the micelles and protein in the system we have described.

CONCLUDING REMARKS

Successful protein purification depends on the availability of multiple approaches. We do not suggest that extraction of proteins into supercritical fluids will become the new standard in protein purification. Nevertheless, there are a number of important features of the system that we describe that offer significant advantages over conventional extraction of proteins into reverse micellar solutions.

Conventional extraction of proteins into predominantly organic media is dependent on changes in the physical properties of the aqueous, protein-containing, solution. As such, the driving force for extraction may damage the protein. The conditions under which the enzyme is most stable, or active, may not mimic the conditions that enable transfer of the protein from the aqueous to the micellar phase. Extraction of proteins into super- and subcritical fluids is dependent not on the aqueous solution properties but on the properties of the organic fluid. As such, extraction should not necessarily affect the protein adversely. This clear difference is particularly important when considering attempts to purify membrane proteins. Membrane proteins are notoriously sensitive to small changes in their environment, and thus developing an extraction system that does not depend on, for instance, changes in pH is of importance.

Finally, extraction of proteins into reverse micelles is of no use if the protein cannot be transferred back to water easily. The presence of organic solvents in conventional approaches makes it difficult to transfer the protein to water without solvent contamination. The use of pressurized gases makes this process relatively straightforward. Release of pressure will result in the rapid precipitation of both protein and detergent. Careful choice of detergent will then enable removal of protein since only the protein is soluble in water. The only problems associated with this method relate to the tendency of the detergent to foam when the pressure is released. We are currently optimizing the extraction process with an adapted supercritical fluid extractor.

REFERENCES

1. Zaks, A. & A. M. Klibanov. 1984. Science **224:** 1249–1251.
2. Cambou, B. & A. M. Klibanov. 1985. Biotechnol. Bioeng. **27:** 1449.
3. Lilly, M. D., S. Harbron & T. J. Narendranathan. 1987. Methods Enzymol. **136:** 138–149.
4. Chen & F. Arnold. Bio/technology. (In press.)
5. Dekker, M., R. Hilhorst & C. Laane. 1989. Anal. Biochem. **178:** 217–226.
6. Hatton, T. A. 1989. *In* Surfactant-Based Separation Processes. J. F. Scamehorn & J. H. Harwell, Eds.: 55–90. Marcel Dekker, Inc. New York, N.Y.
7. Leser, M. E. & P. L. Luisi. 1990. Chimia **44:** 270–282.
8. Luisi, P. L. & L. J. Magid. 1988. CRC Crit. Rev. in Biochem. **20**(4): 409–474.
9. Leser, M. E., G. Wei, P. L. Luisi & M. Maestro. 1986. Biochem. Biophys. Res. Commun. **135**(2): 629–635.
10. Ayala, G. & G. Mendoza-Hernandez. 1990. Biochem. Int. **22:** 717–723.
11. Ayala, G. A., S. J. Kamat, E. J. Beckman & A. J. Russell. 1992. Biotechnol. Bioeng. **40:** 158–166.
12. Samana, J. P. & K. M. Lee. 1987. Eur. J. Biochem. **163:** 609–617.
13. Larsson, H., J. Aldercreuttz & B. Mattiasson. 1987. Eur. J. Biochem. **166:** 157–161.
14. Han, D. & J. S. Rhee. 1986. Biotechnol. Bioeng. **28:** 1250–1255.
15. Dekker, M., K. Van't Riet, S. R. Weijers, J. W. A. Baltussen, C. Laane & B. H. Bijsterbosch. 1986. Chem. Eng. J. **33:** B27.
16. Dahuron, L. & E. L. Cussler. 1988. AICHE J. **34**(1): 130–136.
17. Luisi, P. L. & B. Steinmann-Hofmann. 1987. Methods Enzymol. **136:** 188–216.
18. Shield, J. W., H. D. Ferguson, A. S. Bommarius & T. A. Hatton. 1986. Ind. Eng. Chem. Fundam. **25:** 603–612.
19. Luisi, P. L. & L. J. Magid. 1988. CRC Crit. Rev. Biochem. **20**(4): 409–474.
20. Martinek, K., A. V. Levashov, N. Klyachko, Y. L. Khmelnitski & L. V. Berezin. 1986. Eur. J. Biochem. **155:** 453–468.
21. Karpe, P. & Ruckenstein. 1991. J. Colloid Interface Sci. **141**(2): 534–552.
22. Srivastava, R. C., D. B. Madamwar & V. V. Vyas. 1987. Biotechnol. Bioeng. **29:** 901–902.
23. Luisi, P. L., M. Giomini, M. P. Pileni & B. H. Robinson. 1988. Biochim. Biophys. Acta **947:** 209–246.
24. Dumas, D. P., S. R. Caldwell, J. R. Wild & F. M. Raushel. 1989. J. Biol. Chem. **264**(33): 19659–19665.
25. Kamat, S. J., J. Barrera, E. J. Beckman & A. J. Russell. 1992. Biotechnol. Bioeng. **39:** 806–814.
26. Escamilla, E., G. Ayala, A. Gomez-Puyou, L. Millan & A. Darzon. 1989. Arch. Biochem. Biophys. **272:** 332–343.

Enzyme Reactions under High Pressure and Their Applications

SHIGERU KUNUGI

Department of Polymer Science and Engineering
Kyoto Institute of Technology
Matsugasaki, Sakyo
Kyoto 606, Japan

INTRODUCTION

Pressure is a thermodynamically intensive parameter which has lately gained importance in various fields related to biology and biochemistry. In fundamental studies, pressure has been used as an initiator of change in higher-order structures of biopolymers in order to study their refolding and dissociation processes. In applied fields, pressure is now used for food processing, as well as for sterilization of medicines and other things.

Such uses are essentially based on the structural perturbation of macromolecules, including proteins, caused by increases in pressure. We have studied the pressure dependence of several enzyme reactions[1–9] and found that pressure affects the activities of some enzymes in considerably different ways depending on the nature of the scissile bond of the substrate and the type of substrate. Also, certain enzymes show a considerable apparent pressure activation. These findings encouraged us to explore the use of pressure in various aspects of enzyme engineering.[10–15]

Here some pressure effects in enzyme reactions are reviewed and some examples of the utilization of high-pressure enzyme processes are shown.

PRESSURE DEPENDENCE OF PROTEASE CATALYSIS

The pressure dependencies of chemical equilibrium and reaction velocity are analyzed with Equations 1 and 2, including reaction volume and activation volume, while the temperature dependencies of these parameters are described by the well-known enthalpy of reaction (Equation 3) and activation enthalpy or activation energy (Equations 4 and 5).

$$\partial \ln K/\partial P = \Delta V/\mathrm{RT} \quad (1)$$

$$\partial \ln k/\partial P - \Delta V^{\ddagger}/\mathrm{RT} \quad (2)$$

$$\partial \ln K/\partial T = \Delta H/\mathrm{RT}^2 \quad (3)$$

$$\partial \ln k/\partial T = (\mathrm{RT} + \Delta H^{\ddagger})/\mathrm{RT}^2 \quad (4)$$

$$\ln k = \ln A - E_a/\mathrm{RT} \quad (5)$$

TABLE 1. Volume Changes for the K_m or K_i Parameters of Protease Reactions[a]

Enzyme	Substrate or Inhibitor[b]	$\Delta V_{(1/K)}$ (ml/mol)
Trypsin	BzArgNH$_2$, BzAgrpNA, Thionine	+9–+17
α-Chymotrypsin	Indole, Fua-D-TrpOMe	0–−3
Thermolysin	FuaGlyPheAla, FuaGlyPheNH$_2$ ZPhe, ZGlyLeu, PheLeuNH$_2$ PhePhe	−20–−40
Neutral protease from *B. subtilis*	FuaGlyLeuAla, ZPhe, ZGlyLeu	≃0
Carboxypeptidase A	FuaGlyPhe, FuaPhePhe, GlyTyr ZPhe, Phe, PhePhe	+20–+35
Carboxypeptidase Y	FuaGlyPhe, FuaPhePhe BzGlyOPhe	+15–+30
Carboxypeptidase P	FuaPhePhe, BzGlyOPhe	+35
Carboxypeptidase from wheat	FuaPhePhe, BzGlyOPhe	+25–+35
Leucine aminopeptidase	LeupNA	≃0

[a] ΔV is described for association direction.
[b] pNA denotes *p*-nitroanilide and Fua- is furylacryloyl-.

As for the rate process, it is to be noted that the activation volumes can be either positive or negative, while the activation energy is, for simple reactions, always positive. This means that the reaction velocity can become either greater or lower with increases in pressure.

Fundamental enzyme kinetics tell us that the enzyme catalytic process will be described by K_m and k_{cat} parameters and each process is characterized by the corresponding volumetric parameter, ΔV_{K_m} and $\Delta V^{\ddagger}_{k_{cat}}$.

TABLE 2. Activation Volumes for the Hydrolytic Reactions Catalyzed by Proteases[a]

Enzyme	Substrate	$\Delta V^{\ddagger}$ (ml/mol)
Trypsin	BzArgNH$_2$	+3–+6
	BzArgOEt	−2
Thermolysin	FuaGlyPheAla	−23
	FuaGlyLeuAla	−26
	FuaGlyPheNH$_2$	−35
	FuaGlyLeuNH$_2$	−35
Neutral protease from *B. subtilis*	FuaGlyLeuAla	−29
Carboxypeptidase A	FuaGlyPhe	−4
	FuaPhePhe	−8
Carboxypeptidase Y	FuaGlyPhe	+27
	FuaPhePhe	+10
	BzGlyOPhe	+17
	FuaPheOEt	−1.5
	FuaPheOMe	−3.7
Carboxypeptidase P	FuaPhePhe	−7
	BzGlyOPhe	−8
Carboxypeptidase from wheat	FuaPhePhe	−8
	BzGlyOPhe	−10
Leucine aminopeptidase	LeupNA	−9

[a] Mostly for k_{cat} and some for k_{cat}/K_m.

TABLE 3. Volume Changes for Weak Acid Dissociations

Acid	ΔV (ml/mol)
$H_2PO_4^-$	−23.8
Cacodylate	−13.2
Acetate	−11.2
Tris H+	+1
Taps	+0.4
Hepes	+2.7
Mes	+5.0
Phenol	−11−−17
Imidazole H^+	−2
Amines H^+	+4–+7

TABLES 1 and 2 are the summary of these two parameters which we have measured for several proteolytic enzymes. Both ΔV_{K_m} and $\Delta V^{\ddagger}_{k_{cat}}$ can have a positive or a negative sign. [Hereafter ΔV_{K_m} is described in $-\Delta V_{K_m}$ ($= \Delta V_{1/K_m}$) to show the volumetric change during the progress of the enzyme reaction. Note that the negative activation volume, e.g., means the acceleration of the reaction by an increase in pressure.]

Pressure dependencies of various types of low molecular interactions and reactions have long been studied by a number of pioneers. Some of the interactions important for enzymatic processes respond to pressure perturbations as summarized in TABLES 3, 4, and 5. Roughly speaking the promotion of charged species in aqueous medium is favored by a high-pressure environment, because the electrostriction of water around the charges decreases the molar volume of water. This means that an increase in pressure will weaken the electrostatic or coulombic interactions. On the other hand, hydrogen bonding will be slightly strengthened by an increase in pressure, because of the decrease in the interatomic distance leading to a smaller molecular size. Hydrophobic interaction has much more complicated characteristics. As in TABLE 5, the interactions between aromatic compounds are generally understood to have negative reaction volumes and thus to be strengthened at high pressure. However, the pressure dependence of micell formation studied by several groups[16–18] (FIGURE 1) and also of the dimerization of alkyl carboxylic acid[19] clearly showed that the association between alkyl chains has positive reaction volume, mostly 1 to 2 ml/mol increment per methylene unit.

A complex combination of these fundamental processes results in the volumetric parameters observed for actual enzyme reactions. Since an enzyme is a macromolecular substance and it has higher-order structure, pressure will also influence the

TABLE 4. Volume Changes Accompanying Formation of Hydrogen Bonds

Compounds	Solvent	Bond	ΔV (ml/mol)
Phenol dimer	CCl_4	O..H—O	−2.4
Phenol + Dioxane	Hexane	O..H—O	−3
1-Butanol dimer	CS_2	O..H—O	−2.5
Benzoic Acid dimer	Heptane	O..H—O	−1.3
ε-Caprolactam	CCl_4	O..H—N	−0.65

TABLE 5. Volume Changes Accompanying Association of Aromatic Compounds

Process	V (ml/mol)
9-Methylpurine self-association	−4
1-Methyl-3-carbomethoxypyridinium + 8-chlorotheopherine	−4
Rhodamine self-association	−10
Methylene blue self-association	−11

structure of larger dimensions. In the observed parameters such as given above, only those influences that affect rate or equilibrium process can be detected, but these influences on macromolecular structure will not be ignored even for very simple enzymes composed of monomeric proteins. One typical illustration of such an aspect was given by Low and Somero.[20] When ionic groups are exposed during the reaction process, at the state of transition, the corresponding activation volumes will be in the negative sign since the electrostriction of water will be promoted.

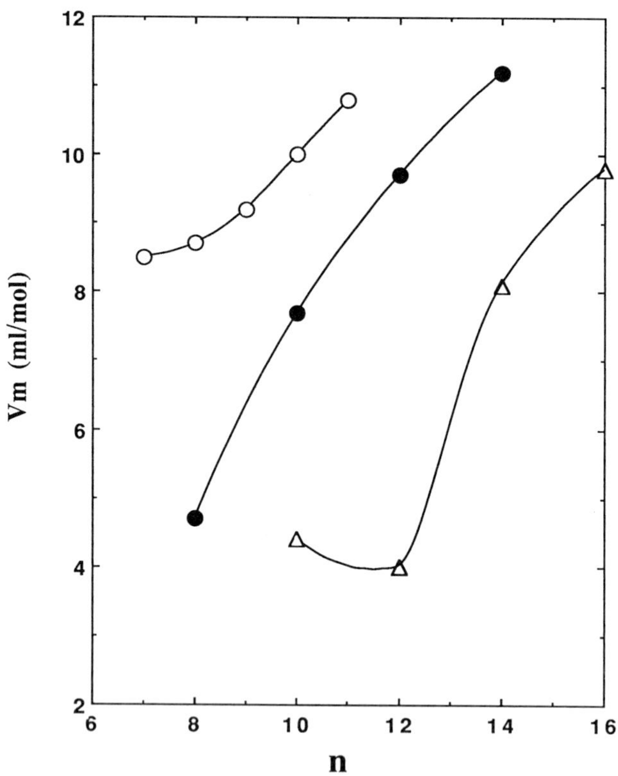

FIGURE 1. Dependence of volume change for micell formation on alkyl chain length in various ionic surfactants. (Taken from References 16–18.) ●, sodium alkylsulfate; ○, sodium alkylcarboxylate; △, alkyltrimethylammonium bromide.

As an example, we can cite the result with thermolysin. On increasing pressure, this metal-containing neutral protease shows characteristic activation of the second-order rate coefficient as illustrated in FIGURE 2a. At 100 MP_a, it reached apparent activity around 15 times higher than that of the atmospheric pressure. This is in high contrast to the result with another metal-containing peptidase, carboxypeptidase A (FIGURE 2b), in which a 100 MP_a increase in pressure depressed the reaction rate to almost one-third of the control. This prominent acceleration—a large and negative apparent activation volume—comes from both K_m and k_{cat} processes (TABLES 1 and 2). These two prominent negative volume parameters cannot be explained without consideration of the change in the hydration state of total protein during the catalysis.

PEPTIDE FORMATION UNDER HIGH PRESSURE

The high catalytic activity of thermolysin could be used for some productive processes. The first target was the condensation of amino acids to form peptides. Thermolysin is actually well known as a catalyst to produce the precursor of a synthetic sweetener. However the peptide condensation from the N-protected amino acid and the C-protected amino acid is mostly a thermodynamically controlled process. As given in TABLE 6, the condensation equilibrium declined to the decomposed side at the elevated pressure, because the ionization of $-COOH$ and $-NH_2$ groups (mainly the former is the key issue) are favored to promote ionic hydration and to decrease molar volume. Therefore, only when the formation has a kinetically controlled character does the thermolysin-catalyzed peptide formation under high pressure have some meaning. This was the case in the formation of the aspartame precursor, because Cbz-Asp is a relatively poor acyl component for this enzyme. (Detailed results were given at the previous enzyme engineering conference.)[12]

Other kinetically controlled peptide formations, especially those containing the aminolytic process of an acylated enzyme intermediate, have some advantages in being processed under high pressure.

One example is that with carboxypeptidase Y. This enzyme was isolated from yeast and classified as a serine-containing carboxypeptidase. It shows esterase (Reaction 7) and amidase (Reaction 8) activities besides the intrinsic carboxypeptidase activity (Equation 6) and hence can be used for peptide synthesis by way of acyl-transfer (Reaction 9) as well as the simple amide condensation (reverse of Equation 1).

$$\text{———}AA_1\text{—}AA_2 \rightarrow \text{———}AA_1 + AA_2 \qquad (6)$$

$$\text{———}AA_1\text{—}OR \rightarrow \text{———}AA_1 + ROH \qquad (7)$$

$$\text{———}AA_1\text{—}NH_2 \rightarrow \text{———}AA_1 + NH_3 \qquad (8)$$

$$\text{———}AA_1\text{—}XX + AA_2\text{———} \rightarrow \text{———}AA_1\text{—}AA_2\text{———} + XX \qquad (9)$$

(AA_i, R, and XX denote amino acid residue, alkyl group, and leaving group.)

Though CPase Y-catalyzed hydrolysis is not highly accelerated by pressure, each of these reactions showed characteristic dependencies on pressure. Reaction 6 is

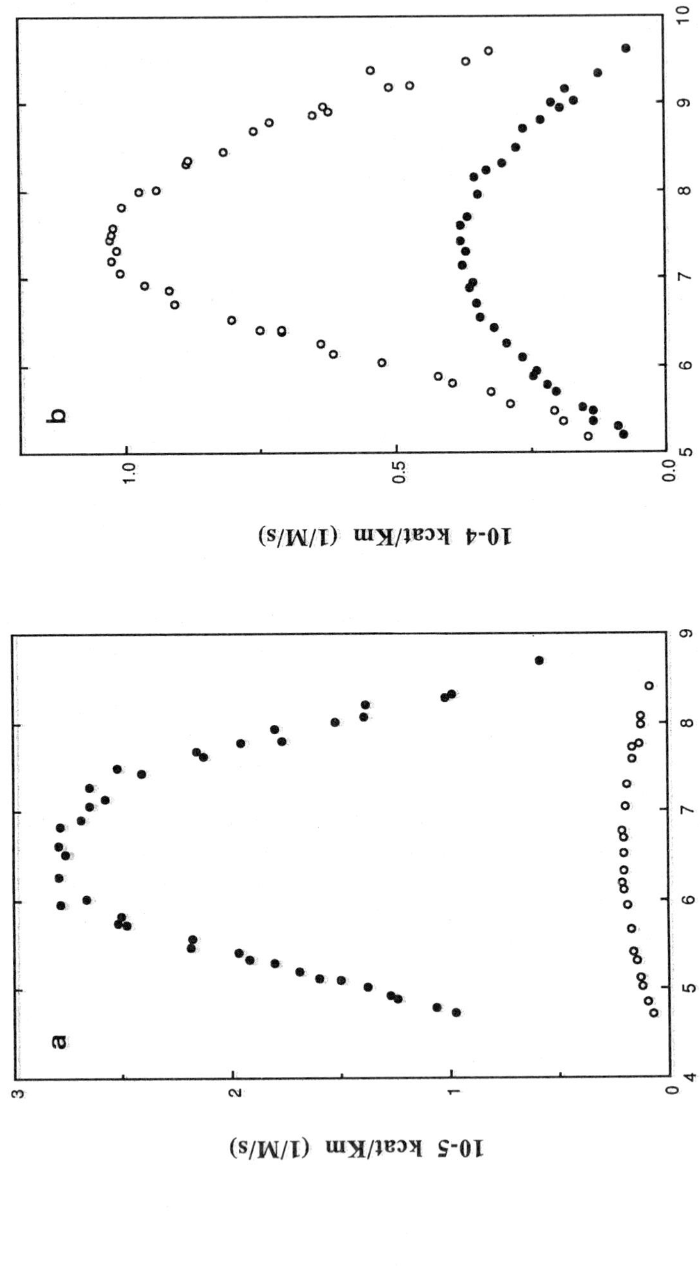

FIGURE 2. Effect of pressure on oH dependence of k_{cat}/K_m of hydrolytic reactions of thermolysin (a) and carboxypeptidase A (b). ○, at atmospheric pressure; ●, at 100 MP_a.

TABLE 6. Effect of Pressure on Thermolysin-Catalyzed Peptide Condensation Reaction from FuaGly and $LeuNH_2$[a]

pH	Yield (%)	
	0.1 MP_a	200 MP_a
5	1	0
6	47	0.9
7	85	53
8	77	64
9	4.2	0.8

[a][FuaGly] = 1mM; [$LeuNH_2$] = 50mM; [TLN] = 1.1 μM. Reaction time = 6 hours; 37°C.

highly depressed by increasing the pressure, while Reaction 7 or 8 was less influenced by pressure. This means that the product distribution of the peptide synthesis with CPase Y can be controlled by pressure. (Detailed results of this process were also reported in the previous conference.)[12] The total peptide yield was increased approximately sixfold by increasing pressure to 200 MP_a for the reaction of *N*-furylacryolylphenylalanine ethyl ester + phenylalanine amide or glycine amide. This result is explained by considering the effect of pressure on the hydrolysis of acyl-dipeptide amides by this enzyme; increase in pressure retards the carboxypeptidase activity more rigorously than the amidase or the carboxamidopeptidase activity and, as a result, a dipeptide with a carboxylate end was obtained in higher yield. Peptide yield was also improved in the case of *N*-acyldipeptide substrate + amino acid amide to form an *N*-acyldipeptide amide or in the case of the reaction between *N*-acylamino acid ester and amino acid ester.

PROTEOLYSIS UNDER HIGH PRESSURE

Another and traditional application of proteases is the cleavage or digestion of proteins. Hayashi *et al.* for the first time applied the thermolysin reaction accelerated by increasing pressure on nonspecific digestion of proteins in various origins and structures (TABLE 7).[13] We found that one protein in a set of proteins (like β-lactoglobulin in whey) can be selectively digested by increasing pressure. The later studies revealed that such a result is brought about not solely by the increased

TABLE 7. Thermolysin Digestion of Various Proteins at 0.1 and 200 MP_a

Proteins	Percent Digestion[a]	
	0.1 MP_a	200 MP_a
Soy protein	20.0	30.0
Casein	68.0	61.0
α-Lactalbumin	13.0	10.3
β-Lactoglobulin	19.4	25.0
Alcohol dehydrogenase	18.1	45.3
Hemoglobin	13.0	34.6
Myoglobin	12.0	15.0

[a]Determined by TCA precipitation (5% except for β-LG).

TABLE 8. Effect of Pressure on Protease-Catalyzed Protein Digestion[a]

Protein	Protease	Digestion at				
		0.1	100	200	300	400
Casein	Thermolysin	1.05	1.09	0.72	0.85	0.51
	Subtilisin	0.60	0.88	0.73	0.65	0.48
BSA	Thermolysin	0.19	0.20	0.23	0.16	0.14
	Subtilisin	0.16	0.24	0.55	0.69	0.44

[a]Digestion is described in the absorbance at 280 nm of TCA-supernatant after treatment with 0.1 mg enzyme/10 mg protein at pH 6.5, 30°C for 1 hour (Casein) or 3 hours (BSA). Pressure is in MP_a.

activity of the protease at high pressure, but by the deformed or denatured structure of the target substrate proteins at high pressure. TABLE 8 shows one of the results. At high pressure, promotion of the digestion was more evidently observed for a structured protein (bovine serum albumin, BSA) and this is not restricted to thermolysin but also is observed for more general proteases.

This result affords us the use of pressure as a sort of denaturation-inducing agent. We would like to call it a "reagentless denaturant." Actually, this "denaturant" has great advantage over conventional chemical denaturing reagents such as urea in that there is no necessity of separation of the "denaturant" after the process: one just opens the screw stop-valve. This principle can be applied in many fields of biochemistry, and some limited aspects are being studied by our group now. As for the relation with protease, limited proteolyses of proteins under pressure-denatured conditions[14] are one example and a chemical modification under pressure-denatured conditions[15] is another.

OTHER HYDROLYTIC ENZYMES

TABLE 9 contains the results of pressure dependence of other hydrolytic enzymes.[a] The number and also the intensity of studies are relatively limited for these enzymes.

As for the hydrolysis of phosphate esters, there are two major factors to be considered. Phosphate always carries negative charge(s), and in most of the cases the counterpart enzymes have positive groups at the binding site and electrostatic force primarily drives the substrate binding. But as mentioned above, ionic binding is not favored at elevated pressure. Secondly, phosphate hydrolysis contains the transition state of penta-coordinated configuration in which charge separation or polarization is increased. This kind of transition state is more stabilized by increased hydration at high pressure, and hence the reaction velocity is accelerated. As long as k_{cat} and K_m parameters are separately evaluated, these two principles seem to be valid. In the apparent activation volumes derived from the second-order rate measurement, mostly the former contribution is dominant and deceleration by pressure is observed.

[a]For the sake of saving space, individual references are not given here, except for the name of main authors and the year of publication.

In the glycolytic enzymes, negative activation volumes are also commonly observed, which might indicate the significance of charge separation at the state of activated complex. Though the substrates listed here for saccharide hydrolases have no charges, the reported reaction volumes for substrate binding are positive, which may be related to dehydration from the substrate upon binding to the enzyme, and these general characters are not drastically changed even when the enzymes are immobilized on solid support.

As for the restriction endonuclease, we have gained fairly different results from McGregor.[21] He reported that EcoRI shows very large rate depression on increasing pressure and hence very large and positive activation volume. FIGURE 3 shows our results on the effect of pressure on EcoRI (a) and BamHI (b) enzymes hydrolyzing λDNA. Though no kinetic analysis is made on these results, it is clear that EcoRI is highly pressure resistant and BamHI is more pressure sensitive. λDNA has multiple cleavage sites for these two enzymes, but fundamentally similar results were obtained for the cleavage of linearized (PstI-treated) pBR322 DNA, which contains only one cleavage point for both enzymes.

For these restriction nucleases, much more interesting aspects will be specificity on the base sequence recognition. Characteristic effects of pressure on sequence specificity were seen in the restoration of the so-called staractivity. These enzymes show relatively loosened sequence specificity under, e.g., lower ionic strength or higher organic solvent content. For example, EcoRI and BamHI do not cleave X174 DNA under normal conditions, but at lower salt content they cleave this DNA at 9 and 25 positions, respectively. These relaxed specificities were tightened again by increasing pressure up to 200 MP_a. The lowering ionic strength and increasing pressure have counteractions. This kind of observation is very important both for the understanding of the mechanisms of these enzyme reactions and for practical application of high-pressure enzyme technology in various specific fields.

TABLE 9. Reported Volumetric Parameters on Nucleic Acid and Polysaccharide Hydrolytic Enzymes

Enzyme	Substrate	$\Delta V_{K_m^{-1}}$	$\Delta V_{k_{cat}}$	Reference
Ribonuclease	cCMP	+6	−20	Williams & Shen (1972)
5′-Nucleotidase	ATP		−18	Penniston (1971)
Endonuclease P1	ApUp		+25[a]	Kunugi *et al.* (1989)
Endonuclease potato	ApUp		+10[a]	
-cyclodextrine[b]	cAMP	+5	−15	Kunugi *et al.* (1991)
Endonuclease (EcoRI)	pBR322		+80[a]	McGregor, Jr. (1990)
Amylase	Starch		−28	Laidler (1951)
Invertase	Sucrose		−8	
	Sucrose	+8	−4	Ludwig & Greulich (1978)
(immobilized)	Sucrose	+20	−29	Sato *et al.* (1987)
Dextranase	Dextran	+13	−4	
-galactosidase	Lactose		−8[a]	Kawade *et al.* (1991)
(immobilized)	Lactose		−8[a]	

[a]For apparent rate parameters.
[b]Enzyme-mimetic catalysis.

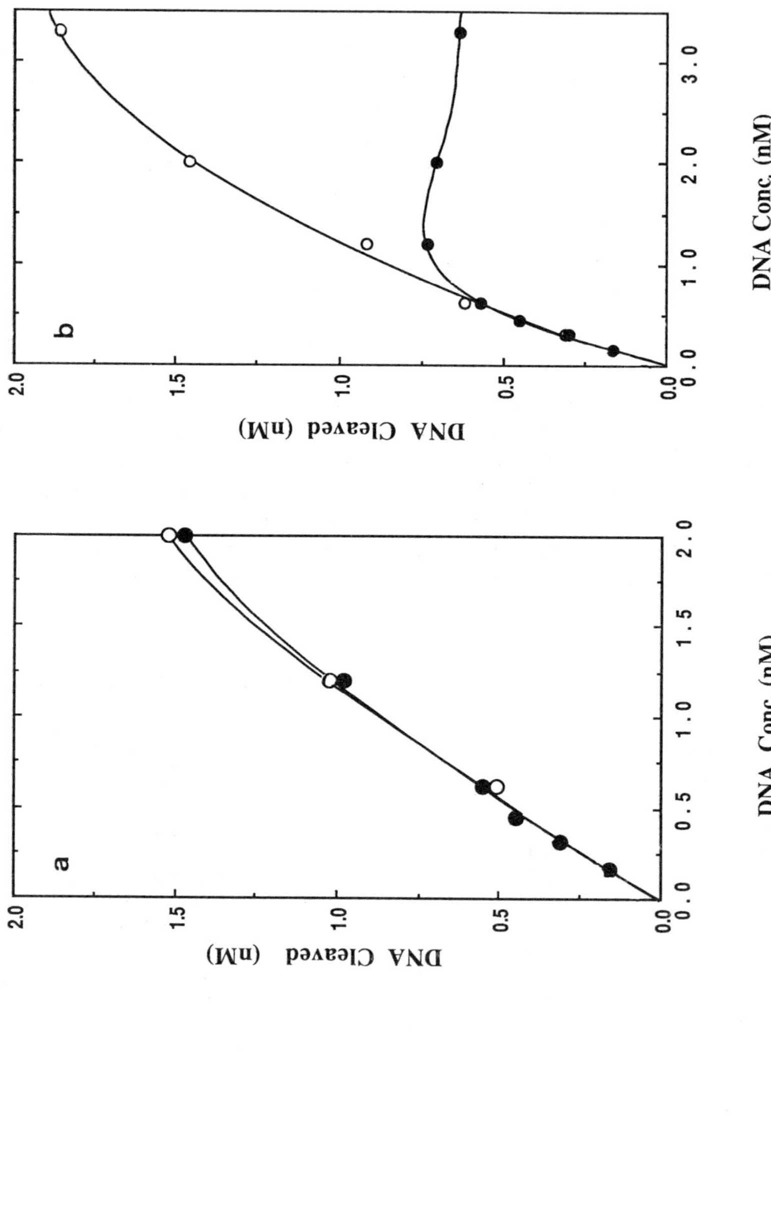

FIGURE 3. Effect of pressure on EcoRI (a) and BamHI (b) catalyzed cleavage of λDNA. ○, at atmospheric pressure; ●, at 200 MP_a.

CONCLUSION

Pressure has been used to influence and improve various organic (or inorganic) syntheses for many years. The present examples of high-pressure enzyme engineering, however, are unique in that the effect of pressure is exerted through molecular recognition of enzyme-substrate combinations and amplified through their catalytic processes, which enable us to control the processes at relatively lower pressure than the ordinary small-molecule chemistry. Thus this technique can be applied in many fields with relatively lower costs and higher flexibility.

REFERENCES

1. KUNUGI, S., M. FUKUDA & N. ISE. 1980. Pressure dependence of trypsin-catalyzed hydrolysis of specific substrates. Biochim. Biophys. Acta **704:** 107–113.
2. FUKUDA, M., S. KUNUGI & N. ISE. 1983. Pressure dependence of carboxypeptidase A action. Bull. Chem. Soc. Jpn. **56:** 3308–3313.
3. FUKUDA, M. & S. KUNUGI. 1984. Pressure dependence of thermolysin catalysis. Eur. J. Biochem. **142:** 565–570.
4. FUKUDA, M., H. SHIMA & S. KUNUGI. 1985. Pressure dependence of L-leucine-*p*-nitroanilide hydrolysis by leucine aminopeptidase, Bull. Chem. Soc. Jpn. **58:** 1349–1350.
5. FUKUDA, M. & S. KUNUGI. 1985. Mechanism of carboxypeptidase Y–catalyzed reaction deduced from a pressure dependence study. Eur. J. Biochem. **149:** 657–662.
6. FUKUDA, M., H. SHIMA & S. KUNUGI. 1985. Kinetic study of carboxypeptidase P–catalyzed reaction. Pressure and temperature dependence of kinetic parameter. J. Biochem. Tokyo **98:** 517–525.
7. FUKUDA, M. & S. KUNUGI. 1987. Kinetic studies of wheat carboxypeptidase-catalyzed reactions. Difference in pressure and temperature dependences of peptidase and esterase actions. J. Biochem. Tokyo **101:** 233–240.
8. KUNUGI, S., K. IWASAKI & A. NOMURA. 1989. Effect of pressure on plant endonuclease reactions. Nucleic Acids Symp. Ser. **21:** 133–134.
9. KAWADE, T., M. SAKAKIBARA, K. SUZUKI, A. NOMURA & S. KUNUGI. 1991. Effect of pressure on free and immobilized β-galactosidase, Res. Rep. Facul. Eng. Fukui Univ. **39:** 45–54.
10. KUNUGI, S., K. TANABE, M. FUKUDA, S. MAKIMOTO & Y. TANIGUCHI. 1987. The effect of pressure on protease-catalyzed peptide formation. J. Chem. Soc. Chem. Commun.: 1335–1336.
11. KUNUGI, S., K. TANABE, K. YAMASHITA, Y. MORIKAWA, T. ITO, T. KONDOH, K. HIRATA & A. NOMURA. 1989. Protease-catalyzed peptide formation under high pressure. Bull. Chem. Soc. Jpn. **62:** 514–518.
12. KUNUGI, S. & A. NOMURA. 1990. Pressure as a control factor of enzymatic synthetic reactions. Ann. N.Y. Acad. Sci. **613:** 501–505.
13. HAYASHI, R., Y. KAWAMURA & S. KUNUGI. 1987. Introduction of high pressure to food processing: preferential proteolysis of β-lactoglobulin in milk whey. J. Food Sci. **52:** 1107–1108.
14. NOMURA, A., K. OKADA & S. KUNUGI. 1990. Pressure as a reagentless denaturant. Limited digestion of BSA by protease. Res. Rep. Facul. Eng. Fukui Univ. **38:** 179–184.
15. KUNUGI, S. 1991. Preparation of proteins carrying electron mediators under pressure-denatured conditions. *In* Preprints of 2nd Pacific Polymer Conference, Otsu, Japan: 273–274.
16. KANESHINA, S., M. TANAKA, T. TOMITA & R. MATSUURA. 1974. Micell formation of sodium alkylsulfate under high pressure. J. Colloid Interface Sci. **48:** 450–460.
17. VIKINGSTAD, E., A. SKAUGE & H. HØILAND. 1978. Partial molar volumes and compressibilities of the homologous series of sodium alkylcarbonates, R_6COONa-$R_{13}COONa$, in aqueous solution. J. Colloid Interface Sci. **66:** 240–246.

18. KANESHINA, M. 1984. Characteristics of surface active substances under high pressures and their application. Hyomen (Surface, Japanese) **22:** 248–261.
19. SUZUKI, K., Y. TANIGUCHI & T. WATANABE. 1973. Effect of pressure on the dimerization of carboxylic acids in aqueous solution, J. Phys. Chem. **77:** 1918–1922.
20. LOW, P. S. & G. N. SOMERO. 1975. Activation volumes in enzymatic catalysis: their sources and modification by low-molecular-weight solutes. Proc. Natl. Acad. Sci. USA **72:** 3014–3018.
21. MCGREGOR, R. B. JR. 1990. Reversible inhibition of EcoRI with elevated pressure. Biochem. Biophys. Res. Commun. **170:** 775–778.

^1H and ^{13}C NMR Studies of Cutinase-Catalyzed Synthesis of Ester in Liquid-Solid Media

Crucial Role of Water[a]

C. SARAZIN,[b] G. GOETHALS,[b] J. P. SEGUIN,[b]
M. D. LEGOY,[c] AND J. N. BARBOTIN[b,c]

[b]*Laboratoire de Génie Cellulaire*
Faculté des Sciences
Université de Picardie
33 rue Saint-Leu
80039 Amiens Cedex, France

[c]*Laboratoire de Technologie Enzymatique*
Université de Technologie de Compiègne
BP 649
60206 Compiègne, France

INTRODUCTION

During the last few years, numerous enzymatic transformations in nonaqueous media have attracted considerable interest.[1] Reactions catalyzed by lipase such as enantiomeric esterifications and transesterifications have been successfully utilized for the synthesis of organic compounds in biphasic aqueous systems or in organic solvents. It has been recognized that water is absolutely essential for enzymatic catalysis, and its role has been discussed in terms of water activity of the enzyme preparation, hydration of the enzyme and solvent, water concentration, and polarity of the medium.[2] Especially in solid/liquid biphasic media, the water produced by esterification reactions was able to activate lipase by continuously changing the hydration state of the enzyme preparation. In order to better understand the crucial role played by water, the lipase-catalyzed esterifications of caprylic acid and butanol were studied through the use of proton and carbon NMR spectroscopy.

MATERIALS AND METHODS

The reaction was studied with a thermostable enzyme, a fungal *Fusarium solani* cutinase,[3,4] expressed in *Escherichia coli* (generous gift from PGS-CORVAS); 6 mg of enzyme in a powder form (previously equilibrated at a given a_w) were transferred to a 5 mm NMR tube, and 300 μl of equimolar mixture of caprylic acid and butanol were then added. The water activity (a_w) of the enzyme preparation and the

[a]This work was supported by the CEE through contract BRIDGE: BIOT-CT91-0274.

substrates was measured at 30°C using a Novasina hygrometric sensor. NMR spectra were recorded by using a Bruker AM-WB, at 30°C. The proton spectrum is recorded at 300 MHz with a spectral width of 6000 Hz over 16 K data points and the carbon spectrum at 75 MHz with a spectral width of 18000 Hz over 32 K data points. The carbon spectra immediately followed by the proton spectra are recorded automatically with 15 minutes delay time between each experiment.

RESULTS

¹H NMR Studies

FIGURE 1 shows a time course of ¹H NMR spectra obtained when cutinase (a_w = 0.37) was incubated with an equimolar mixture of caprylic acid and butanol (a_w = 0.28). The triplet signal at 3.7 ppm is from the α methylene protons of the butanol. As the reaction occurs, the triplet signal area at 4 ppm of the methylene protons near the oxygen of ester grows. A single signal is observed for acidic hydroxylic hydrogen because a fast exchange occurs between the hydroxylic hydrogen of acid, alcohol, and water. The chemical shift varies from 8 to 6.5 ppm as a function of substrate consumption and water production. A signal at 4.7 ppm is then observed at 10 hours characterizing no miscible water.

The ratio between the area of the ester methylene at 4 ppm and the initial area of the methylene protons of butanol at 3.7 ppm has been calculated from these spectra to determine the ester production. The time course of butyl caprylate synthesis is reported in FIGURE 2, showing that the reaction occurs after a time lag of 5 hours and that the ester production reaches 80% in 16 hours.

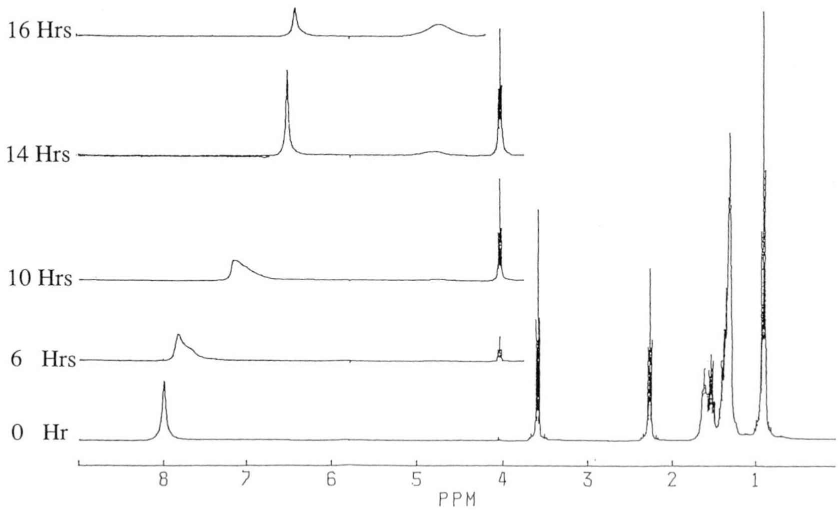

FIGURE 1. Time course of ¹H spectra of the enzymatic esterification reaction of caprylic acid and butanol. Cutinase a_w = 0.37; Substrate a_w = 0.28.

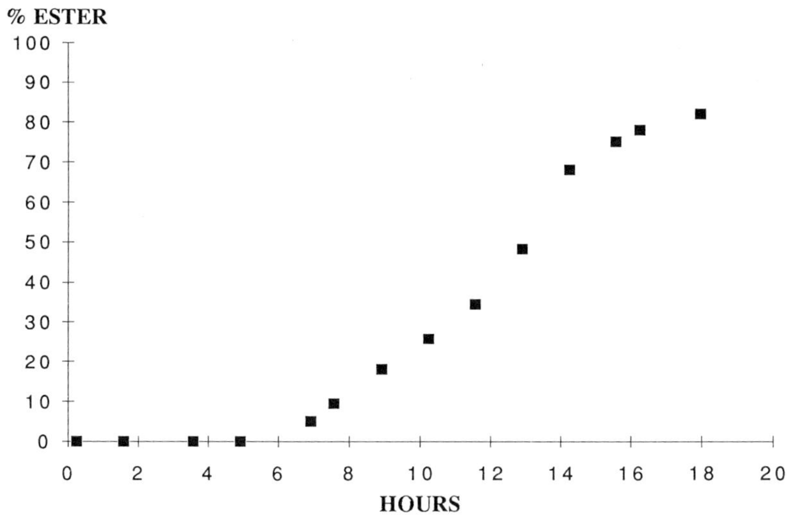

FIGURE 2. Kinetic of the reaction. Sigmoidal curve showing an exaltation of the reaction rate with the *in situ* water formation.

It is well known[1,2,5–9] that the time lag and the slope of such a curve depend on the initial amount of water present whether in the organic phase or in the solid phase. Especially when a_w of each phase is increased, the time lag decreases and the initial rate increases. In a previous paper, the water activity of substrates and the hydroxylic chemical shift were correlated in order to estimate the initial hydration state of the reaction media at the initial point and to investigate water exchange between the two phases.[10]

In the present case where cutinase and substrate equimolar mixture have been previously equilibrated at $a_w = 0.37$ and 0.28 respectively, the hydroxylic chemical shift is constant during the time lag, since a_w values of the two phases are nearly the same so that water exchange does not occur (FIGURE 3, a). When the a_w of the cutinase is greater than the a_w of substrates, a shielding effect on the hydroxylic chemical shift can be observed, suggesting that the enzyme gives water to the substrates. On the other hand, a deshielding effect is observed when the a_w is greater for the substrates. Then a shielding effect of the hydroxylic chemical shift is obtained due to the disappearance of acid and alcohol and to the water production (FIGURE 3, b). When 25% of ester is produced, water appears as a distinct phase (FIGURE 3, c) characterized by the signal at 4.7 ppm of pure water. This phase constitutes some droplets on the glass wall of the tube in the organic phase and gives the wet appearance of the enzyme.

^{13}C NMR Studies

^{13}C NMR spectra have been recorded in the same conditions as described above, and FIGURE 4 shows the typical time course for the carbonyl signals. The peak

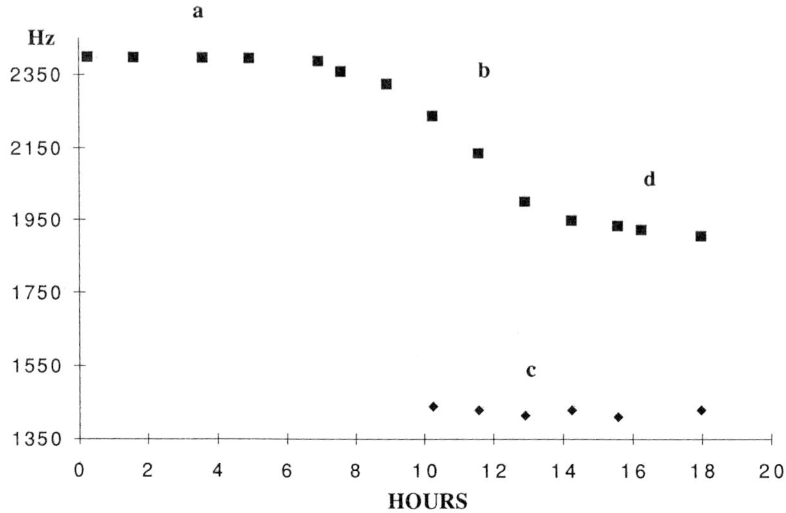

FIGURE 3. Hydroxylic signal chemical shifts evolution according to time. a: OH chemical shift constancy during time lag. b: Consumption of acid-alcohol and production of water induce a shielding effect of the OH chemical shift. c: OH signal of no miscible water. d: End of the reaction.

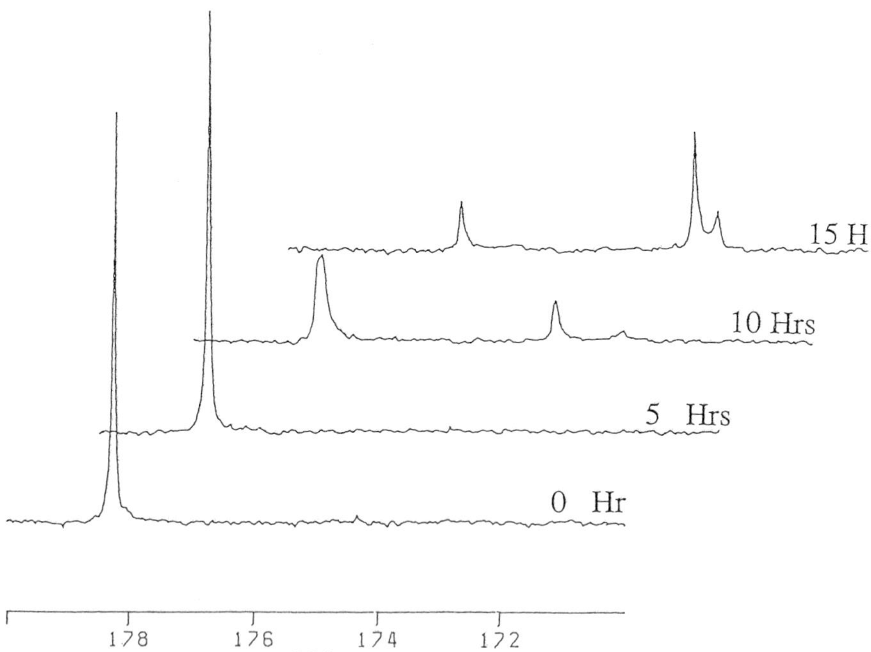

FIGURE 4. Carbonyl part of the ^{13}C spectra. The intensity of the acid carbonyl peak decreases while the ester peak grows. Dating from 10 hours, a second ester peak is detected with a constant chemical shift.

intensity and the chemical shift corresponding to the caprylic acid carbonyl carbon at about 177 ppm decrease during the reaction, while the ester carbonyl carbon signal area increases.

The chemical shift evolution of carbonyl groups according to the amount of synthetized ester is shown in FIGURE 5. A decrease of these chemical shifts is observed as a function of ester production, suggesting a diminution of the acid-alcohol complex in favor of the formation of an acid-ester complex. At the beginning of the reaction, all the synthesized ester is associated with the acid, inducing a deshielding effect on the ester carbonyl carbon (neat ester, 172.5 ppm). As the reaction goes on, the amount of acid decreases, involving a diminution of the ratio between ester complex and free ester and then inducing a shielding effect on the ester carbonyl carbon. During the reaction, the acid carbonyl carbon peak shifts to

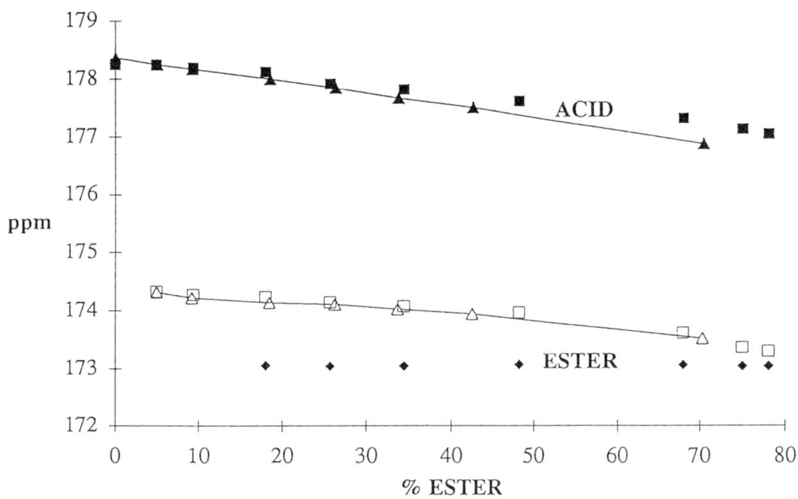

FIGURE 5. Chemical shift evolution of carbonyl groups. Reaction media: ■ acid carbonyl carbon; □ ester carbonyl carbon. Standard mixtures (acid-alcohol-ester-water without enzyme): ▲ acid carbonyl carbon; △ ester carbonyl carbon. Complete agreement between the results obtained with reaction media and standard mixtures.

high fields relative to the increasing ratio of acid complexed with ester.[11] These results are in complete agreement with those obtained with standard mixtures of acid-alcohol-ester and water in the absence of enzyme[10] (FIGURE 5).

Furthermore, a second peak at 173 ppm is observed dating from 10 hours of reaction. Such a time corresponds to the production of 25% of ester but also to the appearance of a distinct water signal recorded on proton spectra (FIGURES 1 and 2). This peak splitting is also present in the methylene region at 64 ppm. The chemical shifts of these new signals, at 173 and 64 ppm, remain unchanged during the reaction and can be attributed to ester molecules in a different environment. These ester peaks, not observed in the standard mixtures, seem to be induced by the presence of the cutinase. Indeed, in experiment A, glass beads have been put above the cutinase powder, in order to keep the enzyme under the detection coil region (FIGURE 6); in

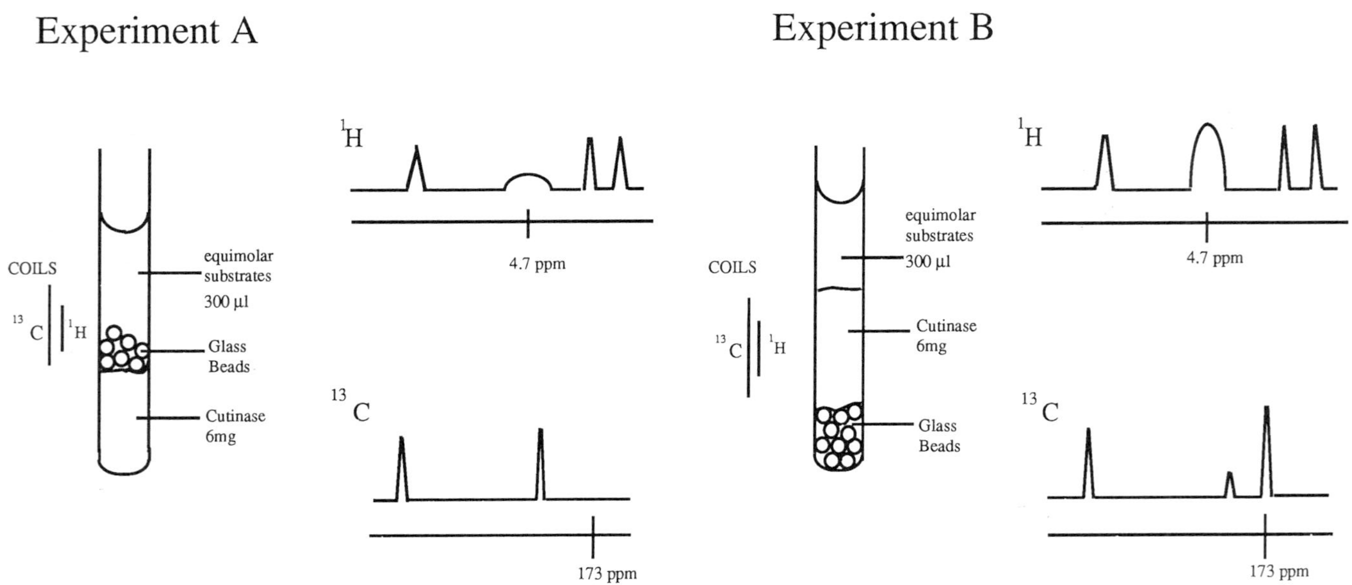

FIGURE 6. Variation of ^{1}H and ^{13}C spectra by changing the detection zone. Experiment A: the cutinase is kept under the detection coils region: a weak water signal is recorded on proton spectra, and a single peak is observed for the ester carbonyl carbon. Experiment B: when the cutinase is in the detection zone, the second ester carbonyl form, at 173 ppm, is observed and the no-miscible-water signal is important.

the experiment B (FIGURE 6), the cutinase is kept in the volume detected by the receiver coil by putting glass beads at the bottom of the NMR tube. In the first case, the water signal intensity is weak and the peak at 173 ppm is not observed. From experiment B, it can be observed that the area of the no-miscible-water signal is very important and that the peak at 173 ppm is always present. Moreover, such a signal is more important that the ester signal in the organic phase.

DISCUSSION

Various factors affecting the catalytic activity of lipases, i.e., additional water, produced water, hydration state of both enzyme and subtrates, have been already studied by numerous authors.[1,2,5–9,12–17] The present study demonstrates that ^{1}H and ^{13}C NMR spectroscopy of cutinase-catalyzed synthesis of ester provides important information on the enzyme reaction mechanism. Whatever the initial hydration state of enzyme and substrates for the synthesis of butyl caprylate, three observations should be analyzed:

a. *A deformation of the hydroxylic hydrogen signal.* At the beginning of the ester synthesis reaction between 0 and 20%, the hydroxylic hydrogen peak is narrow, then it becomes asymmetric dating from 5 hours. The asymmetry is due to the presence of cutinase which contributes to slow hydrogen exchanges, since a study from model media (data not shown) without cutinase shows only a broadening of the peak when the amount of ester and water is increased to 40%.
b. *A distinct aqueous phase.* As the organic phase becomes water saturated, produced water leads to the formation of a discrete water phase. The corresponding signal at 4.7 ppm is very large and may be characterized by a doublet structure signal characterizing the structured water.[18] From experiments with glass beads, it has been seen that water is largely localized within the cutinase phase. Furthermore, as already described by numerous authors,[14,16,17] a change in the appearance of the reaction medium occurs, i.e., the cutinase forms aggregate on the glass walls of the tube. As for other substrates (oleic acid/heptanol; caprylic acid/heptanol), the more the substrates are at high initial a_w, the more the enzyme forms aggregates.
c. From ^{13}C spectra, *the ester appears in two states* corresponding for the first one to the ester in the substrate phase (by comparison with carbonyl chemical shift in model mixtures), and for the second state to the ester in the environment of the enzyme, characterized by a constant chemical shift at 173 ppm. It can be noted that the ester characterized by the constant chemical shift (173 ppm) in the enzyme zone is not accessible by the substrates because the chemical shift is not modified by the substrate concentration decreasing.

It might be thought that the achievement, with high yield, of esterification is due to the partition[19] of water and ester between the liquid and solid phase, allowing the continuous diffusion of substrates. Esterification, in this heterogeneous system, appears to take place by transfer of ester into the cavity formed by the cutinase, driving the reaction to completion, unlike the case of a biphasic system described by Monot,[17] who concluded there was a transfer of ester into the organic phase.

In view of a relation between water and ester in the solid phase, it is quite reasonable to think that the reorganization of the enzyme structure seems to be helped by adsorbed water.[16]

CONCLUSION

The NMR method presented is a simple and direct method for determining ester and water production. The hydroxylic chemical shift can constitute a suitable probe to estimate the initial hydration state of the reaction medium and to investigate water exchange between cutinase and substrates during the time lag.

It has been suggested that at first, an acid-ester complex is formed according to the carbonyl chemical shift evolution and then, a part of the ester and water are localized in the cutinase phase.

The splitting of the ester signals and the formation of a distinct aqueous phase are both observed for such other substrates as oleic acid/heptanol; oleic acid/butanol; caprylic acid/heptanol at different initial a_w.

REFERENCES

1. KLIBANOV, A. M. 1989. Enzymatic catalysis in anhydrous organic solvents. TIBS **14:** 141–144.
2. GOLDBERG, M., D. THOMAS & M. D. LEGOY. 1990. The control of lipase-catalysed transesterification and esterification reaction rate. Effects of substrate polarity, water activity and water molecules on enzyme activity. Eur. J. Biochem. **190:** 603–609.
3. MARTINEZ, C., C. ALBERGEL, C. CAMBILLAU, P. DE GEUS & M. LAUWEREYS. 1990. Crystallographic study of a recombinant cutinase from *Fusarium solani pisi.* GBF Monogr. **16:** 67–70.
4. LAUWEREYS, M., P. DE GEUS, J. DE MEUTTER, P. STANSSENS & G. MATTYSSENS. 1990. Cloning, expression and characterization of cutinase, a fungal lipolytic enzyme. GBF Monogr. **16:** 243–251.
5. GOLDBERG, M., F. PARVARESH, D. THOMAS & M. D. LEGOY. 1988. Enzymatic ester synthesis with continuous measurement of water activity. Biochim. Biophys. Acta. **957:** 359–362.
6. GOLDBERG, M., D. THOMAS & M. D. LEGOY. 1990. Water activity as a key parameter of synthesis reactions: the example of lipase in biphasic (liquid/solid) media. Enzyme Microb. Technol. **12:** 976–981.
7. ZAKS, A. & A. M. KLIBANOV. 1988. The effect of water on enzyme action in organic media. J. Biol. Chem. **263:** 8017–8021.
8. KITAGUCHI, H., I. ITOH & M. ONO. 1990. Effects of water and water-mimicking solvents on the lipase-catalyzed esterification in an apolar solvent. Chem. Lett. 1203–1206.
9. KANERVA, L. T., J. VIHANTO, M. H. HALME, J. M. LOPONEN & E. K. EURANTO. 1990. Solvent effects in lipase-catalysed transesterification reactions. Acta Chem. Scand. **44:** 1032–1035.
10. SARAZIN, C., G. GOETHALS, J. P. SEGUIN, M. D. LEGOY & J. N. BARBOTIN. 1992. Etude par spectroscopie RMN ^1H et ^{13}C de la synthèse d'esters en milieu organique catalysée par une lipase: influence de l'eau. J. Chim. Phys. **89:** 541–547.
11. BREITMAIER, E. & W. VOELTER, Eds. 1987. Carbon-13 NMR Spectroscopy. 3rd edit.: 226–227. VCH. Weinheim, Germany.
12. HIRATA, H., K. HIGUCHI & T. YAMASHINA. 1990. Lipase-catalyzed transesterification in organic solvent: effects of water and solvent, thermal stability and some applications. J. Biotechnol. **14:** 157–167.

13. HEISLER, A., C. RABILLER & L. HUBLIN. 1991. Lipase catalysed isomerisation of 1,2-(2,3)-diglyceride into 1,3-diglyceride. The crucial role of water. Biotechnol. Lett. **13:** 327–332.
14. BOYER, J. L., B. GILOT & R. GUIRAUD. 1990. Heterogeneous enzymatic esterification: analysis of the effect of water. Appl. Microbiol. Biotechnol. **33:** 372–376.
15. CHULALAKSANANUKUL W., J. S. CONDORET, P. DELORME & R. M. WILLEMOT. 1990. Kinetic study of esterification by immobilized lipase in *n*-hexane. FEBS Lett. **276:** 181–184.
16. GATFIELD, I. L. 1984. The enzymatic synthesis of esters in nonaqueous systems. Ann. N.Y. Acad. Sci. **434:** 569–572.
17. MONOT, F., F. BORZEIX, M. BARDIN & J. P. VANDECASTEELE. 1991. Enzymatic esterification in organic media: role of water and organic solvent in kinetics and yield of butyl butyrate synthesis. Appl. Microbiol. Biotechnol. **35:** 759–765.
18. LENK, R., M. BONZON & H. GREPPIN. 1980. Dynamically oriented biological water as studied by NMR. Chem. Phys. lett. **76:** 175–177.
19. MARTINEK, K., A. N. SEMENOV & I. V. BEREZIN. 1981. Enzymatic synthesis in biphasic aqueous-organic systems. I. Chemical equilibrium shift. Biochim. Biophys. Acta **658:** 76–89.

Effects of the Support Material and the Thermodynamic Water Activity on Enzymatic Synthesis in Organic Media[a]

PATRICK ADLERCREUTZ

Department of Biotechnology
Chemical Center
Lund University
Post Office Box 124
S-221 00 Lund, Sweden

INTRODUCTION

In this investigation, enzymes deposited on porous supports were used as catalysts in organic media. It has earlier been shown that the nature of the support profoundly influences the catalytic properties of the enzymes.[1] As in other systems containing enzymes in predominantly organic media, water plays a crucial role. Water partitions between the enzyme, the support, and the solvent. The amount of water bound to the enzyme governs its catalytic activity to a large extent. The amount of water that is dissolved in the solvent or adsorbed on the support can therefore indirectly influence the enzymatic activity. It is not known to what extent these indirect effects have been the cause of the differences in reaction rates obtained with different supports. A good measure of the degree of hydration of a biocatalytic reaction system is the thermodynamic water activity.[2] In this study, the thermodynamic water activity of the enzyme preparations and the reaction media was adjusted prior to the reactions. In this way, rates obtained with the enzyme at the same hydration state but on different supports were compared, and it was possible to study the direct influence of the support on the catalytic properties of the enzyme.

EXPERIMENTAL

Derivatives of controlled pore glass (CPG) were prepared as previously described.[1] Celite was obtained from BDH, and Accurel PA6 from Akzo. Aqueous buffer solutions containing enzyme (and NAD^+ in the case of horse liver alcohol dehydrogenase) were mixed with the support, and water was removed at reduced pressure. Adjustment of the thermodynamic water activity of enzyme preparations and reaction media was achieved by vapor phase equilibration with saturated salt solutions.[3] Enzymatic conversions (FIGURE 1) were carried out in 2 ml scale at 25°C

[a]This project was supported by the Swedish Natural Science Research Council (NFR) and the Biotechnology Research Foundation (SBF).

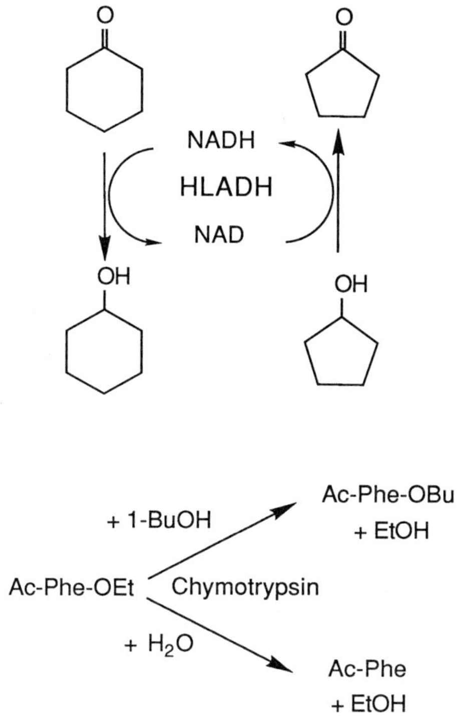

FIGURE 1. Schemes for the reactions catalyzed by horse liver alcohol dehydrogenase (HLADH) and α-chymotrypsin.

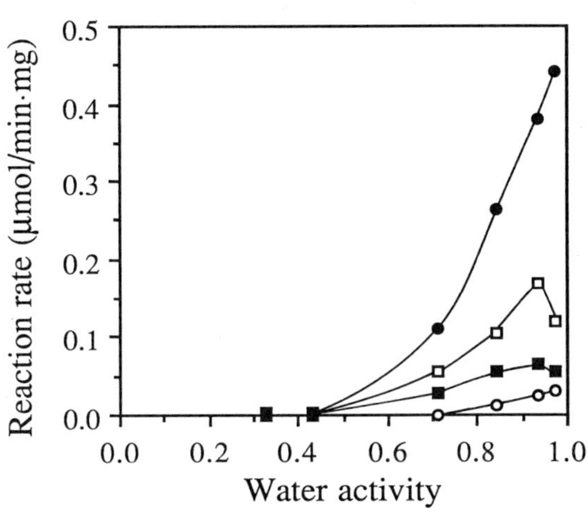

FIGURE 2. Initial catalytic activity of horse liver alcohol dehydrogenase (HLADH) as a function of the water activity. Enzyme (7.3 mg/g) and NAD (2.3 mg/g) were immobilized on different supports: Celite (●); Accurel PA6 (■); glucose-CPG (□) and hexyl-CPG (○). The reaction rate was measured in diisopropyl ether containing 100 mM cyclohexanone and 100 mM cyclopentanol (FIGURE 1).

with reciprocal shaking. The reactions were monitored by high-performance liquid chromatography or gas-liquid chromatography.[4]

RESULTS AND DISCUSSION

Horse liver alcohol dehydrogenase was used to reduce cyclohexanone to cyclohexanol with the concomitant oxidation of cyclopentanol to cyclopentanone (FIGURE 1). No significant reaction was detected at water activities below 0.45, and the reaction rate increased with increasing water activity (FIGURE 2).[4] The highest catalytic activity was obtained with Celite as support. The reaction rate obtained with glucose-CPG as support was much higher than that obtained with hexyl-CPG. Apparently this enzyme is quite sensitive to the surface properties of the support and

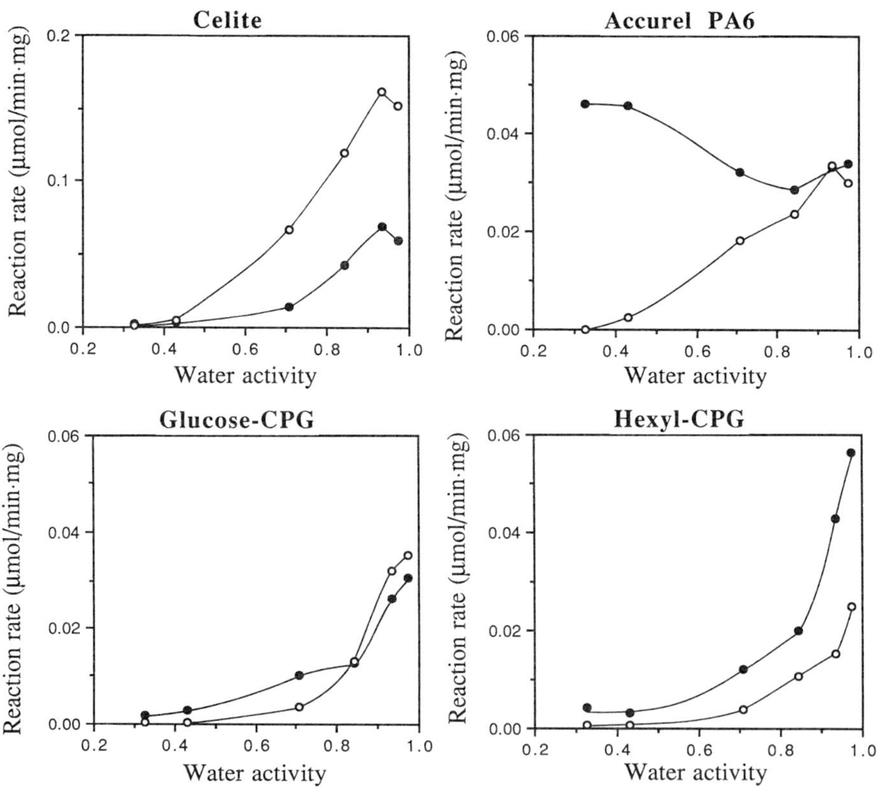

FIGURE 3. Initial catalytic activity of α-chymotrypsin as a function of the water activity. The enzyme was immobilized on different supports: Celite, Accurel PA6, glucose-CPG, and hexyl-CPG. The reaction rate was measured in diisopropyl ether containing 10 mM Ac-Phe-OEt and 100 mM 1-butanol. According to FIGURE 1, Ac-Phe-OEt reacted in an alcoholysis reaction (●) or in a hydrolysis reaction (○).

the hydrophobic hexyl groups produced an unfavorable environment for the enzyme. α-Chymotrypsin was used to catalyze the alcoholysis of *N*-acetyl-L-phenylalanine ethyl ester with 1-butanol (FIGURE 1). Hydrolysis was a competing reaction. In the alcoholysis reaction, the highest rates were observed with Celite as support and a water activity of 0.94 (FIGURE 3).[4] However, with this support only about one-third of the ethyl ester was converted to butyl ester; the rest was hydrolyzed. With the polyamide support, Accurel PA6, alcoholysis was the dominating reaction, and by using a low water activity (0.33), hydrolysis was completely suppressed while still maintaining a high alcoholysis activity. Glucose-CPG favored hydrolysis and hexyl-CPG favored alcoholysis at water activities exceeding 0.8.

CONCLUSIONS

In systems containing enzymes immobilized on support materials in organic media, water plays a crucial role. In conversions carried out at fixed water activity (fixed enzyme hydration), different rates are obtained depending on the support used. The support influences both the total enzymatic activity and the relative rates of the different enzyme-catalyzed reactions. By chosing a suitable support, Accurel PA6, and a low water activity (0.33), it was possible to completely suppress hydrolysis while still maintaining a high alcoholysis activity.

ACKNOWLEDGMENTS

Thanks are due to Ulrika Thornberg and Åsa Lundström for skillful technical assistance.

REFERENCES

1. RESLOW, M., P. ADLERCREUTZ & B. MATTIASSON. 1988. On the importance of the support material for bioorganic synthesis. Influence of water partition between solvent, enzyme and solid support in water-poor reaction media. Eur. J. Biochem. **172:** 573–578.
2. HALLING, P. J. 1987. Water activity in biphasic reaction systems. *In* Biocatalysis in Organic Media, C. Laane, J. Tramper & M. D. Lilly, Eds.: 125–132. Elsevier. Amsterdam, the Netherlands.
3. GREENSPAN, L. 1977. Humidity fixed points of binary saturated aqueous solutions. J. Res. Natl. Bur. Stand. A Phys. Chem. **81A:** 89–96.
4. ADLERCREUTZ, P. 1991. On the importance of the support material for enzymatic synthesis in organic media. Support effects at controlled water activity. Eur. J. Biochem. **199:** 609–614.

Enzymatic Hydrolysis of Cellulose in Water-Immiscible Solvent Systems

M. CANTARELLA, L. CANTARELLA,[a]
A. GALLIFUOCO, AND F. ALFANI

Department of Chemistry, Chemical Engineering and Materials
University of L'Aquila
67040 Monteluco di Roio, Italy

[a]*Department of Industrial Engineering*
University of Cassino
03043 Cassino, Italy

INTRODUCTION

Many studies reported in the literature show that enzymes can efficiently work in organic solvents.[1–3] However, hydrolytic reactions in organic media have received little attention for many years because the addition of organic solvents into aqueous environments was considered a possible source of severe damage to the enzymatic structure.

A previous study showed that stability of hydrolases in two-phase media could be improved if positive interactions, which retard enzyme inactivation, really occurred.[4] In addition, a higher activity could be attained because of more favorable biochemical conditions that involve the environment as well. The results supported the hypothesis that even the properties of enzymes involved in hydrolytic reactions can be controlled by modifications of the reaction media rather than the enzyme itself. On the basis of these findings, it was decided to extend our investigation to the hydrolysis of cellulose, a key step in biochemical routes for biomass utilization.

Several parameters may affect optimization of the enzymatic hydrolysis of cellulose. Some of them are strictly linked to the biomass: pretreatment, chemical composition, weight percent in the liquid phase, particle-size distribution. Other ones are specific to the enzymatic reaction: enzyme-to-biomass ratio, stability and activity of each cellulase component, inhibition of enzymes, reaction temperature, reaction time, pH, stirring speed. Finally, some parameters refer to the two-phase system used as reaction medium, solvent polarity, chemical properties of the solvent, weight composition of the reaction medium, chemical interactions with other biomass components (mainly lignin).

Previous studies also elucidated stability and activity of commercial cellulases in biphasic systems prepared with water and a water-immiscible organic solvent.[5,6] Enough water should be present in these reaction media in order to both preserve the water layers needed to assure the active structure of the protein and provide enough reactant to the system (hydrolytic reaction). This work aimed at investigating phenomena that could be relevant to hydrolysis of lignocellulosic biomass. Experiments were programmed to work at different biomass loading in the reactor and with various acetates as organic phase.

The rate of hydrolysis of an abundant native cellulose (wheat straw) was studied. The yield of glucose production from cellobiose was also investigated. Experiments were also carried out with this model lignocellulosic substrate, either in the native form or variously pretreated. Composites of cellulase complexes from two different strains were used.[7] The data show clearly that biomass conversion is largely improved working with organic solvents. The enhancement of reaction rate depends on medium composition, weight percent of biomass.

WHEAT STRAW HYDROLYSIS

The enzymatic reaction was carried out in stirred batch reactors at 45°C. The concentration of reducing groups[5] reached at the end of 48 hours of biosaccharification with 0.03 g/L (referred to the aqueous phase) of enzyme composite was monitored. The ratio of *Trichoderma viride* (Miles) to *Aspergillus niger* (SIGMA) was 3:1. Wheat straw was ground and sieved. The particle fraction between 32 and 48 mesh was then pretreated at 90°C for 6 hours with 1% w/v H_2SO_4 and at 80°C for 3 hours with 0.8-N NaOH. Samples were tested for enzymatic hydrolysis in pure sodium acetate buffer (50 mM, pH 4.8) and in biphasic media prepared with buffer and organic chemicals (75% v/v). The data of FIGURE 1 refer to experiments with propyl acetate ($\log P = 1.2$), butyl acetate ($\log P = 1.7$), and *i*-amyl acetate ($\log P = 2.2$). The amount of reducing groups depends on the biomass loading and environment composition. The highest production is attained at 25–30 g/L of wheat straw in the presence of either butyl or *i*-amyl acetate. Moreover, in the lower range (7–15 g/L), the advantage of using an organic solvent is indeed nonexistent while it is significant in the upper range (30–50 g/L). These findings mainly suggest that high quantities of biomass in the reactor vary the rheology of the system and limit the transfer of enzyme, thus determining a maximum in the overall kinetics. The same behavior can also be considered a result of denaturation by mechanical stress of the enzyme into contact with the biomass. In fact, according to both these hypotheses, catalytic action of the enzymes is facilitated by biomass availability but is also depressed by the biomass itself, and therefore a maximum should be expected at some intermediate biomass loading. On the other hand, the different yields observed between hydrolysis in pure buffer and in biphasic media, when a large amount of biomass is present, necessitate also the hypothesis of variations in the kinetic mechanism. Both surface interactions between enzyme and cellulose and product inhibition can change depending on the environment.

The hypothesis that the solvent can modify chemical and physical properties of the native cellulose during the reaction is unsound since previous studies showed that the rate of hydrolysis of wheat straw pretreated with butyl acetate (50 hours, 80°C, biomass to solvent ratio 0.1 w/v) is the same or worse than that of the untreated biomass.

In order to check the correctness of the previous explanations of the observed phenomenon, new experiments were set up. The hydrolysis of chemically pretreated wheat straw and of the major intermediate product, cellobiose, was carried out at

45°C with 75% v/v butyl acetate in the reaction medium and 0.03 g/L of cellulase composite (*T. viride* to *A. niger* equal to 3:1). The concentration of glucose is reported in FIGURE 2 as a function of process time. Initial substrate concentration was 10 mM. The rate of cellobiose hydrolysis in Na-acetate buffer approaches almost zero after 40–45 hours of reaction, and the highest cellobiose conversion is 38.9%. If the reaction is carried out in the presence of solvent, the hydrolysis rate is still significantly different from zero when the same process time is elapsed, and cellobiose conversion reaches 75.4%. Therefore, it is evident that β-glucosidase is more active in biphasic media than in pure buffer as result of a better operational stability and/or a reduced inhibition by glucose.

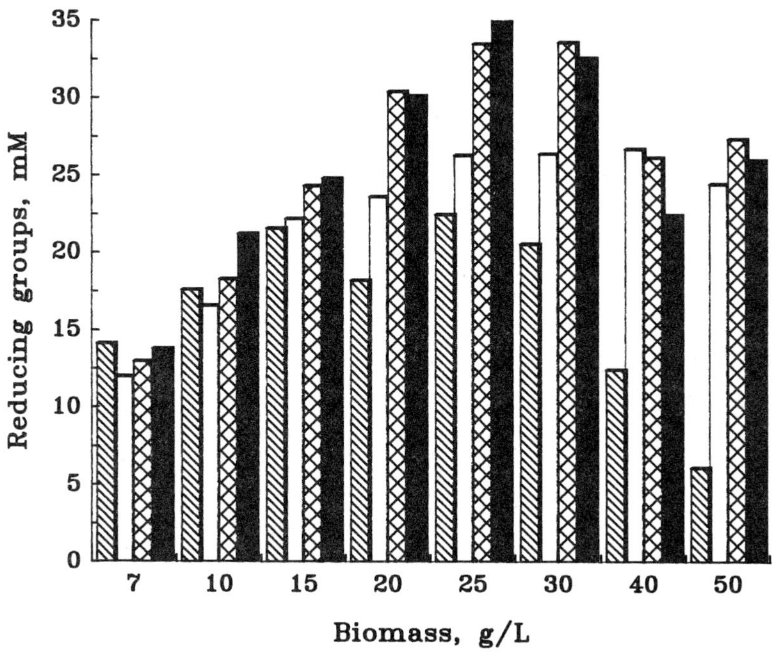

FIGURE 1. Reducing group formation as function of biomass loading in different reaction media. Series sequence: Na-acetate buffer, propyl acetate, butyl acetate, *i*-amyl acetate.

In a different experiment, the amount of reducing group formed during hydrolysis of wheat straw was monitored. Initial loading of the biomass was 25 g/L. Biosaccharification was carried out in stirred batch reactors. Aliquots of the reaction volume were drawn at various time intervals and centrifuged to remove traces of suspended solids. The surnatant was then assayed to determine the concentration of reducing sugars and of total proteins. The results are reported in FIGURE 3. The data on product formation (left-hand axis) confirm that hydrolysis rate and yield of biomass saccharification are much better in the presence of butyl acetate (75% v/v) than in pure buffer. Meanwhile, the concentration of total proteins (right-hand axis) that are free in solution is much higher in butyl acetate in comparison with that in

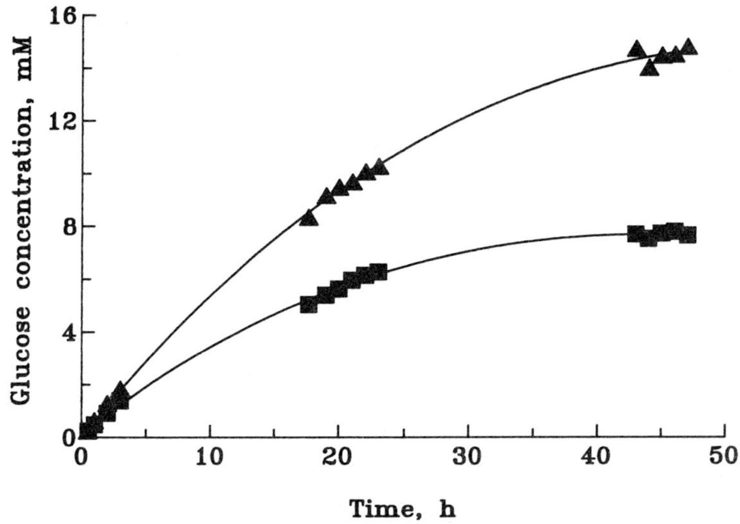

FIGURE 2. Cellobiose hydrolysis: glucose production vs. process time. Reaction medium: ■, Na-acetate buffer; ▲, 75% v/v butyl acetate.

pure buffer. The concentration of cellulase abruptly (10 minutes) drops from 30 μg/mL to 8.17 μg/mL in the biphasic system and to 5.58 μg/mL in the buffered system. Then this continues to decrease quite rapidly during the next three hours. Finally, this remains almost unchanged in biphasic media while it tends to zero in

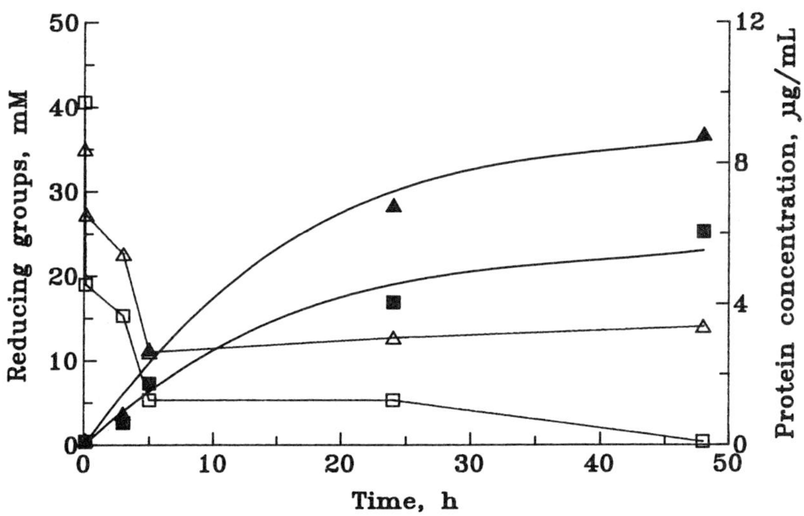

FIGURE 3. Wheat straw hydrolysis: reducing groups (filled symbols) and protein concentration in the surnatant (open symbols) vs. process time. Reaction medium: ■/□, Na-acetate buffer; ▲/△, 75% v/v butyl acetate.

buffer. The presence of butyl acetate apparently modifies surface interactions between cellulose and enzymes in that desorption of the protein from the exhausted biomass appears to be facilitated and a higher concentration of enzyme is available for a subsequent catalytic event.

This behavior is consistent with the observed disappearance of hydrolysis rate after 40–50 hours of operation in buffer and with the reduced but still present rate of hydrolysis in biphasic systems.

CONCLUSIONS

The results of this investigation confirm that enzymatic hydrolysis of lignocellulosic biomass can be conveniently carried out in the presence of organic chemicals (acetate). This finding further supports the evidence that stability and activity of hydrolases can be controlled by modifications induced in two-phase media.

As far as the hydrolysis of wheat straw is concerned, several hypotheses can be formulated to explain the observed higher rate of saccharification. Some of them were checked experimentally in the course of the present study. Activity and stability of the β-glucosidase component in the composite of *T. viride* and *A. niger* are improved by the presence of butyl acetate. Glucose inhibition seems to play a less important role in cellobiose hydrolysis. In addition, since a feedback inhibition is known to affect the overall mechanism of cellulose hydrolysis, if a higher conversion of the intermediate products (mainly cellobiose) is observed, a higher yield of biomass saccharification should also be expected.

Rheological properties of the reaction medium are important but cannot alone explain the differences between hydrolysis in the two media.

In conclusion, the yield of biomass hydrolysis in butyl acetate is up to 49.4% greater than in pure buffer because of more favorable kinetic and fluidodynamic conditions.

REFERENCES

1. Tramper, J., H. C. Van der Plas & P. Linko, Eds. 1985. Biocatalysts in Organic Synthesis. Elsevier. Amsterdam & New York.
2. Dordick, J. S. 1989. Enzymatic catalysis in monophasic organic solvents. Enzyme Microb. Technol. **11:** 194–211.
3. Zaks, A. & A. Klibanov. 1985. Enzyme-catalyzed processes in organic solvents. Proc. Natl. Acad. Sci. USA **82:** 3192–3196.
4. Cantarella, M., L. Cantarella & F. Alfani. 1991. Hydrolytic reactions in two-phase systems. Effect of water immiscible organic solvents on stability and activity of acid phosphatase, β-glucosidase and β-fructofuranosidase. Enzyme Microb. Technol. **13:** 547–553.
5. Cantarella, M., L. Cantarella, A. Gallifuoco & F. Alfani. 1990. Bioconversion of lignocellulosic biomass in organic solvents. *In* Biomass for Energy and Industry. G. Grassi, G. Gosse & G. dos Santos, Eds. Elsevier Appl. Sci. **2:** 23–30.
6. Cantarella, M., L. Cantarella, A. Gallifuoco & F. Alfani. 1992. Water immiscible organic solvent two-phase systems for enzymatic cellulose hydrolysis. *In* Biomass for Energy, Industry and Environment. G. Grassi, A. Collins & H. Zibette, Eds. Elsevier Appl. Sci.: 485–490.
7. Cantarella, M., A. Di Trana, L. Pezzullo & F. Alfani. 1986. Increased saccharification yield by cellulases coupling. Int. J. Solar Energy **4:** 81–95.

Carboxypeptidase Y Reactions in Organic Solvents

SHIGERU KUNUGI,[a] NORIKO SUZUKI,[b] MASAHITO YOKOYAMA,[b] AND AKIHIKO NOMURA[b]

[a]*Department of Polymer Science and Engineering*
Kyoto Institute of Technology
Sakyo
Kyoto 606, Japan

[b]*Department of Applied Chemistry and Biotechnology*
Fukui University
Fukui 910, Japan

INTRODUCTION

Carboxypeptidase Y (E.C.3.4.16.1) was isolated from yeast and classified as a serine-containing carboxypeptidase.[1] It shows esterase and amidase activities besides the intrinsic carboxypeptidase activity[2,3] and hence has been used in peptide synthesis.[4–6] In many cases, peptide synthesis from two component amino acids or peptides is much facilitated in a system containing less water, mainly because of two factors: a low fraction of water is thermodynamically favorable for a condensation reaction, and the coexistence of nonaqueous solvent is very favorable for less polar substrates. Thus some protease-catalyzed peptide syntheses have been successfully carried out in systems containing organic solvents.[7] Furthermore it is becoming clearer that some new enzymatic activities can be exerted in nonnatural conditions such as in organic solvent,[8,9] and under high pressure.[10,11]

Considering this background and the versatility of the carboxypeptidase Y reaction, our knowledge on the reaction of this enzyme in organic solvents is insufficient, mainly because this enzyme has been considered labile to the existence of organic cosolvent.[2,3] During our physicochemical studies on the enzymes of this group,[12–16] we came to consider it necessary to study systematically the effect of organic solvent on this enzyme.

We have already studied the effect of organic solvent for the kinetics of carboxypeptidase Y–catalyzed hydrolysis of amino acid ester and peptide synthesis from, e.g., *N*-acyl amino acid ester and amino acid amide.[17] The result of the ester hydrolysis was analyzed in terms of the K_m value (which increased with increase in the solvent content, with an order of dioxane > MeCN = DMF > DMSO) and the k_{cat} (which showed a slightly increasing tendency in DMF and DMSO). The effect in the peptide formation reaction from Fua-Phe-OEt and Gly-NH_2 was also accounted for in terms of the parameters k_n and K_n, nucleophile reactivity and nucleophile binding, respectively.

In this paper the effects of solvent on the apparent substrate specificity of both hydrolysis and peptide formation are described. These are to be compared with the effects of other factors controlling enzyme structure/function.

MATERIALS AND METHODS

Materials

Carboxypeptidase Y was kindly donated by Oriental Yeast Co. (Osaka, Japan. Lot 2100361). Enzyme was first dissolved in deionized water and then diluted with proper buffer solution. Its concentration was determined from the absorbance at 280 nm using $A^{1\%}_{1cm}$ = 15.0 and M_r = 61000, the stock solutions being stored at 4°C. Glycinamide (Gly-NH_2) was obtained from Tokyo Kasei Co (Tokyo, Japan). *N*-[2-Hydroxyethyl]piperazine-*N*-2-ethanesulfonic acid (HEPES) was purchased from Nacalai Tesque Co. (Kyoto, Japan). *N*-[3-(2-Furyl)acryloyl]-L-phenylalanine ethyl ester (Fu*a*-Phe-OEt) was as used before. Its concentration was determined using ϵ_{305} = 25500 ($M^{-1}cm^{-1}$). *N,N*-Dimethylformamide (DMF), acetonitrile (MeCN), 1,4-dioxane, and dimethyl sulfoxide (DMSO) were analytical grade (UVASOL; E. Merk Co.; obtained through Kanto Kagaku Co., Tokyo, Japan). Other reagents are commercially available.

Methods

The hydrolytic reaction was carried out normally at 25°C in 0.1-M HEPES buffer mixed with a proper organic cosolvent of the indicated amount. The pH of the aqueous buffer was adjusted as 8.0, around which carboxypeptidase Y is known to show highest esterase activity and almost flat pH profiles for both k_{cat}/K_m and k_{cat}. This insures only a small perturbation through pH change caused by the existence of cosolvent.

Hydrolysis was recorded by observing the absorbance decrease at 320–340 nm by a UV spectrophotometer (UVIDEC-610B; JASCO, Japan) with thermostated-water circulation. The difference spectrum between the ester and the product acid in the presence of the cosolvent was determined initially, and a correction was made in the analysis of the kinetic data.

The dipeptide synthesis was carried out at 30°C in the presence of 0.2 M sodium chloride in a small sample tube held in a wet incubator with shaking. The pH was 8.0. After incubation for the indicated time, a portion was taken out, quenched by mixing with a 19-fold volume of MeCN, filtered (Millipore filter, HV-0.45 μm), and then analyzed on reversed-phase chromatography (Shimadzu, LC-6A with Cosmosil 5C18P), detected at 305 nm, and integrated (Shimadzu Chromatopack CR6A). The eluent is a 1:1 mixture of MeCN and water containing 50-mM triethylamine and 5% phosphoric acid.

RESULTS AND DISCUSSION

Effect on Ester Hydrolysis

TABLE 1 shows the effect of organic cosolvent on carboxypeptidase Y–catalyzed hydrolysis for Fua-amino acid methyl esters. It is seen that k_{cat}/K_m values strongly depended on the solvent content and generally decreased with increasing solvent content. Dioxane had the greatest effect and DMSO the least.

In a previous report, the k_{cat} parameter of Fua-PheOEt was measured to show increasing tendency in DMF and DMSO with increasing cosolvent content in a concentration range comparable to the present one. As in that example, the effect on the Michaelis constant is dominant for the second-order rate constant.

Effects of organic solvent on the kinetic constants were studied in several proteases including α-chymotrypsin,[18] in which a considerable increase in K_m was reported upon increase in the solvent content. In the amidase reaction, a slight increase in k_{cat} was observed, while k_{cat} of ester hydrolysis showed a considerable decrease.

As a result, the apparent substrate specificity among these four esters changed slightly by adding organic solvent. Though Trp ester showed the highest rate constant under more aqueous conditions, Phe ester showed a somewhat higher rate parameter under more organic conditions. Furthermore the superiority of Tyr ester over Leu ester somehow depended on the solvent species and the content.

TABLE 1. Apparent Second-Order Reaction Rate Constant of Ester Hydrolysis by Carboxypeptidase Y in Organic-Solvent-Containing Systems[a]

	Solvent and Content (%)											
	MeCN			Dioxane			DMF			DMSO		
Substrate	2.5	10	20	2.5	10	20	2.5	10	20	2.5	10	20
FuaPheOMe	11.7	5.8	1.0	16.8	5.0	0.9	21.7	9.2	2.6	18.3	17.6	7.6
FuaTyrOMe	4.5	1.9	0.3	5.6	1.6	0.2	9.6	3.3	1.3	8.9	6.0	2.6
FuaTrpOMe	12.2	5.3	0.7	17.2	3.6	0.5	19.8	6.2	1.1	20.3	12.4	6.0
FuaLeuOMe	3.0	0.7	0.2	5.0	1.5	0.4	6.4	1.8	0.7	6.4	4.0	1.5

[a] $[E]_o$ = 60 nM; [S] = 40 μM; pH 8.0, 25°C.

Effect on Peptide Synthesis

TABLE 2 contains the variation of peptide yield with substrate ester, nucleophilic amino acid amide, and species and content of organic solvent.

In the combination of FuaPheOMe + $GlyNH_2$, we found that dioxane and MeCN showed high product yield at content smaller than ca. 15%, but the yield showed a steep decrease at a higher fraction and the peptide synthetic activity diminishes at solvent content higher than 20%. DMSO gave rather low peptide yield and selectivity even at a lower fraction, and it decreases less steeply. DMF showed an intermediate tendency.

These characteristics seen in FuaPheOMe + $GlyNH_2$ are generally observed in the present examples. However, when $GlyNH_2$ is used as the nucleophile, the order of the yield of the four better esters (Phe, Trp, Tyr, Leu) did not correspond to the specificity in the hydrolytic reaction. Trp becomes the poorest of them. With respect to this, $AlaNH_2$ is very similar, and of all the results compiled here under the present reaction conditions, the highest peptide yield was recorded with this nucleophile and FuaPheOMe, at MeCN content of 10%.

The variation of the yield with a different amino acid amide became less prominent in DMF and DMSO. For example, $AlaNH_2$ was a three- to sixfold better nucleophile than $IleNH_2$ or $PheNH_2$ in 10 or 20% MeCN, while $AlaNH_2$ gave a slightly smaller yield in 30% DMSO. β-Branched amino acid esters are very poor acylating substrates of this enzyme, and the corresponding amides are also inefficient nucleophiles for peptide formation. Indeed, no detectable yield was observed in $ValNH_2$ with FuaPHeOMe, FuaLeuOMe, and FuaTrpOMe, and hence they were deleted from the table. These are combinations of a better ester substrate and a poor nucleophile, and almost of all the ester was simply hydrolyzed before any trace of

TABLE 2. Peptide Synthesis Yield from Fua-Xxx-OMe and $YyyNH_2$ by Carboxypepetidase Y in an Aqueous-Organic System[a]

		Solvent and Content (%)											
		MeCN			Dioxane			DMF			DMSO		
Xxx	Yyy	10	20	30	10	20	30	10	20	30	10	20	30
Phe	Ala	89.7	20.3	0	73.8	42.3	6.1	53.9	40.6	10.0	43.2	37.0	30.0
	Gly	62.7	5.1	0	61.4	22.7	1.1	39.0	18.7	0	34.9	27.3	19.3
	Leu	34.9	1.5	0	44.7	35.3	2.8	33.2	43.9	15.8	28.7	31.8	31.2
	Phe	19.1	6.7	17.0	32.8	23.5	2.1	45.8	36.7	9.8	34.9	33.5	35.1
	Ile	25.1	3.1	0	40.5	21.4	2.1	32.3	11.2	0	34.4	32.4	36.0
Tyr	Ala	6.5	1.9	0	12.1	7.0	2.4	32.4	20.4	8.5	29.8	33.3	23.6
	Val	0.5	0	0	0	0	0	6.6	10.9	11.1	4.9	6.8	13.2
	Gly	32.6	0.8	0	25.2	6.2	0.5	24.7	14.4	9.4	19.3	14.2	7.3
	Leu	4.2	2.7	2.6	4.6	3.3	2.8	9.9	14.7	7.0	6.4	10.3	12.1
	Phe	14.5	0	0	17.9	6.3	0.4	20.9	11.8	2.1	25.8	25.8	12.5
	Ile	0	0	3.6	26.7	7.9	0	36.1	20.4	3.2	10.1	14.8	6.9
Leu	Ala	17.2	0.7	0	37.6	9.9	0	39.2	12.9	3.0	49.3	26.2	13.0
	Gly	27.9	0	0	37.7	11.2	0.8	30.1	11.0	3.1	32.3	18.8	10.2
	Leu	23.9	0	0	5.9	12.7	0	25.2	17.1	2.7	5.6	4.4	2.6
	Phe	17.5	1.0	0	35.7	16.2	1.9	39.1	23.7	4.5	25.0	31.2	21.6
	Ile	6.2	0	0	11.9	3.8	0	20.7	22.8	6.4	9.6	8.3	6.3
Trp	Ala	18.3	16.8	7.6	13.3	5.3	0	29.1	11.9	1.0	17.8	23.1	23.4
	Gly	2.4	1.8	0	21.1	17.4	2.8	24.0	13.7	2.6	16.2	14.8	15.7
	Leu	5.3	1.5	0	29.2	13.1	0	6.4	18.3	4.3	0	0.2	0.5
	Phe	9.5	1.1	0	2.6	11.7	0	28.8	4.2	28.0	0.6	2.4	5.3
	Ile	1.1	0.9	0	0.3	0.8	0.6	5.4	4.3	1.5	0.4	1.4	5.6
Ile	Ala	3.4	0	0	2.0	0.4	0	2.2	1.2	0.5	2.1	1.0	0.4
	Val	3.8	0	0	4.7	0.9	0	5.4	1.6	0	6.2	3.6	1.6
	Gly	2.4	0	0	1.1	0	0	2.9	1.4	1.2	0.9	0.6	0.4
	Leu	0	0	0	0	0	0	0.5	0	0	0	0	0
	Phe	2.0	0	0	2.8	0.6	0	4.6	2.1	0	2.3	2.3	1.7
	Ile	0.5	0	0	0.8	0	0	4.9	1.4	0	1.8	1.0	0.6
Val	Ala	1.4	0	0	1.7	0.5	0.0	2.9	2.0	0	2.0	0.8	0.6
	Val	1.3	0	0	2.7	0.8	0	3.1	1.2	0.7	4.0	1.9	1.0
	Gly	0.6	0	0	0.5	0	0	2.4	3.0	0	5.2	0.5	0
	Leu	1.3	0	0	1.5	0.5	0	1.5	0.9	0	1.0	1.1	1.0
	Phe	0	0	0	0	0	0	0.6	1.5	0	0	0	0
	Ile	0	0	0	0	0	0	0	0	0	1.1	0.8	0

[a][Fua-xxx-OMe] = 1 mM; [$yyyNH_2$] = 0.1 M; $[E]_o$ = 4 nM; 30°C. Reaction time, 5 hours, pH 8.

TABLE 3. Conversion in the Peptide Synthesis from Fua-Xxx-OMe and $YyyNH_2$ Carboxypeptidase Y in an Aqueous-Organic System[a]

		Solvent and Content (%)											
		MeCN			Dioxane			DMF			DMSO		
Xxx	Yyy	10	20	30	10	20	30	10	20	30	10	20	30
Phe	Ala	100.	22.8	0.5	98.8	60.1	9.4	99.6	76.6	19.9	100.	100.	91.1
	Gly	74.8	8.4	0.5	95.3	35.7	3.2	100.	28.5	1.3	100.	98.8	83.1
	Leu	38.0	3.3	0.7	56.0	44.0	5.1	46.1	62.1	25.3	39.0	45.9	52.3
	Phe	23.6	9.0	19.6	42.4	29.8	3.6	55.1	50.8	16.4	50.7	50.2	56.6
	Ile	29.9	5.0	0.4	46.5	28.2	3.8	41.0	15.0	0.6	42.3	46.2	54.2
Tyr	Ala	44.0	3.8	0	47.9	16.4	3.7	78.9	37.5	12.0	76.2	59.1	32.5
	Val	38.4	3.1	0.6	41.7	14.4	1.1	65.9	39.7	19.2	59.0	51.7	24.8
	Gly	47.1	1.8	3.5	46.7	12.8	1.6	76.8	63.2	18.1	62.2	57.9	32.9
	Leu	30.1	3.2	2.6	23.1	10.1	2.8	31.2	30.3	8.2	18.1	22.5	20.8
	Phe	16.1	0.4	2.7	20.8	7.4	1.6	26.5	16.9	5.3	30.6	35.1	20.3
	Ile	0	12.5	8.4	29.9	9.9	0.7	42.0	26.9	5.5	13.0	26.3	16.4
Leu	Ala	18.9	0.7	0	43.4	12.4	0	47.2	16.6	4.0	56.5	33.0	18.6
	Gly	32.6	0.6	0	57.0	17.0	1.0	63.4	21.7	5.8	71.9	48.5	27.3
	Leu	25.1	0	0	7.3	16.1	1.2	28.7	23.4	2.7	8.0	5.8	3.9
	Phe	20.4	1.8	0	39.6	19.3	3.6	43.0	26.7	6.7	29.1	36.1	25.5
	Ile	7.6	1.1	0	14.3	4.6	0.4	23.2	24.9	7.0	11.1	9.2	7.5
Trp	Ala	23.0	26.7	14.7	17.8	7.7	0	40.9	18.2	1.7	27.9	41.9	48.6
	Gly	3.1	2.9	0	41.3	36.0	6.4	59.4	36.2	6.5	37.0	41.6	45.4
	Leu	6.2	3.5	0.6	32.0	14.9	0	8.6	21.3	6.5	0.2	0.6	1.1
	Phe	10.9	2.7	0	3.2	15.9	1.0	38.8	8.4	36.7	0.8	3.1	7.7
	Ile	16.1	2.0	0	2.0	4.6	0.8	30.7	23.6	4.8	1.6	7.4	23.0
Ile	Ala	4.2	0	0	6.1	1.7	0	8.3	3.2	0.5	8.5	5.4	2.7
	Val	3.8	0	0	6.3	0.9	0	8.3	2.9	0	8.0	5.4	2.6
	Gly	5.5	0	0	8.4	2.5	0	13.9	5.3	2.2	13.4	7.9	4.0
	Leu	0	0	0	0	0	0	0.5	0	0	0.4	0	0
	Phe	2.0	0	0	3.8	0.6	0	5.2	2.1	0	2.3	2.8	1.7
	Ile	0.5	0	0	0.8	0	0	6.0	1.4	0	1.8	1.0	0.6
Val	Ala	2.0	0	0	4.8	1.3	0.0	6.2	3.6	0.5	5.5	3.5	1.9
	Val	1.3	0	0	4.0	1.4	0	4.6	2.0	1.2	6.5	3.3	2.1
	Gly	1.7	0	0	3.2	1.0	0	7.8	4.4	0	6.5	3.8	1.4
	Leu	1.3	0	0	2.7	0.5	0	2.1	1.3	0	1.0	1.6	1.9
	Phe	0	0	0	0	0	0	1.3	1.5	0	0.6	0	0.5
	Ile	0	0	0	0.5	0	0	1.0	0	0	1.9	1.3	0

[a]As in TABLE 2.

peptide formation was recorded. Such a situation can be understood by looking at the conversion of the reaction (percentage of the substrate used for the reaction; hydrolysis + peptide formation) in TABLE 3.

With a combination of poor ester and poor nucleophile, some trace of peptide formation can be detected, though the conversion was also very low. In these cases, however, inactivation of the enzyme becomes a significant problem at higher solvent content. DMSO or DMF showed an effect of relaxation of the substrate or nucleophile recognition (rejection). This seems a more general phenomenon, since some of the highly specific enzyme reactions such as aminoacyl-tRNA synthetases or restriction endonucleases have been reported to show less specific or erroneous recognition with high content of DMSO and others.

REFERENCES

1. HAYASHI, R. 1976. Carboxypeptidase Y. Methods Enzymol. **45:** 568–587.
2. HAYASHI, R., Y. BAI & T. HATA. 1975. Kinetic studies of carboxypeptidase Y. I. Kinetic parameter for hydrolysis of synthetic substrates. J. Biochem. Tokyo **77:** 69–79.
3. BAI, Y., R. HAYASHI & T. HATA. 1975. Kinetic studies of carboxypeptidase Y. II. Effects of substrates and product analogs on peptidase and esterase activities. J. Biochem. Tokyo **77:** 81–88.
4. BREDDAM, K., J. T. JOHANSEN & M. OTESEN. 1984. Carboxypeptidase Y catalyzed transpeptidation and condensation reactions. Carlsberg Res. Commun. **49:** 457–462.
5. BREDDAM, K., F. WIDMER & J. T. JOHANSEN. 1980. Carboxypeptidase Y catalyzed transpeptidations and enzymatic peptide synthesis, Carlsberg Res. Commun. **45:** 237–247.
6. SHIMA, H., M. FUKUDA, K. TANABE, T. ITO & S. KUNUGI. 1987. Wheat carboxypeptidase-catalyzed peptide synthesis by aminolysis of *N*-acyl amino acid ester. Indication of the acyl-enzyme mechanism. Bull. Chem. Soc. Jpn. **60:** 1403–1406.
7. FRUTON, J. S. 1982. Proteinase-catalyzed synthesis of peptide bonds. Adv. Enzymol. **53:** 239–230.
8. KLIBANOV, A. M. 1986. Enzymes that work in organic solvent. Chemtech: 354–359.
9. GRUNWALD, J., B. WIRRZ, M. P. SCOLLAR & A. M. KLIBANOV. 1986. Asymmetric oxidoreductions catalyzed by alcohol dehydrogenase in organic solvents. J. Am. Chem. Soc. **108:** 6732–6734.
10. KUNUGI, S., K. TANABE, M. FUKUDA, S. MAKIMOTO & Y. TANIGUCHI. 1987. The effect of pressure on protease-catalyzed peptide formation. J. Chem. Soc. Chem. Commun.: 1335–1336.
11. KUNUGI, S., K. TANABE, K. YAMASHITA, Y. MORIKAWA, T. ITO, T. KONDOH, K. HIRATA & A. NOMURA. 1989. Protease-catalyzed peptide formation under high pressure. Bull. Chem. Soc. Jpn. **62:** 514–518.
12. FUKUDA, M. & S. KUNUGI. 1985. Mechanism of carboxypeptidase Y–catalyzed reaction deduced from a pressure dependence study. Eur. J. Biochem. **149:** 657–662.
13. FUKUDA, M., H. SHIMA & S. KUNUGI. 1985. Kinetic study of carboxypeptidase P–catalyzed reaction. Pressure and temperature dependence of kinetic parameter. J. Biochem. Tokyo **98:** 517–525.
14. KUNUGI, S., M. FUKUDA & R. HAYASHI. 1985. Action of serine carboxypeptidases on endopeptidase substrates, peptide-4-methyl-coumaryl-7-amides. Eur. J. Biochem. **153:** 37–40.
15. FUKUDA, M. & S. KUNUGI. 1987. Kinetic studies of wheat carboxypeptidase-catalyzed reactions. Difference in pressure and temperature dependences of peptidase and esterase actions. J. Biochem. Tokyo **101:** 233–240.
16. KUNUGI, S., K. TANABE, K. YAMASHITA & M. FUKUDA. 1987. Action of serine carboxypeptidases on endopeptidase substrates, *N*-acyl-dipeptideamides. Bull. Chem. Soc. Jpn. **60:** 1399–1402.
17. KUNUGI, S., T. KONDOH, H. HAYASHIDA & A. NOMURA. 1990. Carboxypeptidase Y–catalyzed reaction in systems containing organic solvent. Biocatalysis **4:** 141–151.
18. CLEMENT, G. E. & M. L. BENDER. 1963. The effect of aprotic dipolar organic solvents on the kinetics of α-chymotrypsin-catalyzed hydrolysis. Biochemistry **2:** 836–843.

Orientation of Enzyme Catalysis and Specificity by Water-Soluble Additives

PATRICIA HERTMANNI, CHRISTINE POURPLANCHE, AND VERONIQUE LARRETA-GARDE[a]

U.R.A. 1442 du CNRS
Laboratoire de Technologie Enzymatique
Université de Technologie de Compiègne
B.P. 649
60206 Compiègne Cedex, France

INTRODUCTION

Molecular interactions between an enzyme and its substrate result from a complex and still unclear interrelation between various interactions whose equilibrium determines specificity.[1] Both in the cell and under industrial conditions, the biocatalysts react in heterogeneous media; the relative importance of the diverse forces regulating enzyme behavior is thus modulated, resulting in modifications on activity,[2] stability,[3] nature of the reaction catalyzed[4,5] and also on specificity of the concerned enzymes.[6,7]

To mimic biological media[8] and try to understand the related enzymatic mechanisms, aqueous media with restricted water content have been used. Little has been done about the influence of water on enzyme specificity when the biocatalyst reacts in an aqueous medium, though it has been shown that bound water molecules play a determining role on affinity and specificity.[9] Water activity was decreased by addition of polyols at high concentrations to the reaction media. In such water-limited media, the usual reaction scheme "one substrate, one enzyme, one product" is no longer true and an orientation of the enzyme selectivity may be observed. The interaction between the nature of the reaction medium and the enzyme catalysis has been studied with two different types of enzymes, allowing the recovery of various products from a single substrate, depending on the expressed enzyme selectivity. First, proteases were used, as the action of the enzyme on a known peptide or protein may give rise to various products depending on the cleavage priority expressed by the biocatalyst. Then, soybean lipoxygenase was chosen, as the enzyme may oxidize linoleic acid either on carbon 9 or on carbon 13 allowing the recovery of different hydroperoxides and derivatives.

MATERIALS AND METHODS

Thermolysin and insulin were obtained from Sigma; pepsin from Merck. Insulin and gliadins were purified[7]; their hydrolysis were performed at 35°C in a 0.1-M

[a] Author to whom correspondence should be addressed.

diethanolamine-HCl pH 8 buffer by thermolysin and at 20°C in a 0.1-M acetic acid–HCl pH 2 buffer by pepsin, respectively. Glycerol, up to 10 M, was added to these media. Various substrate/enzyme ratios (S/E) were used. A reverse-phase high-performance liquid chromatography (HPLC) method using a C_{18} column and a water/acetonitrile gradient was used for peptidic product detection.

Soybean lipoxygenase (Type 1-S) and linoleic acid were purchased from Sigma. The reaction was carried out at 25°C in an air-saturated NaOH solution diluted to pH 9. Polarographic and spectroscopic (234 nm absorbance) methods were applied to measure the oxygen uptake and the conjugated dienoic fatty acid formation respectively.[10] An HPLC method was used to separate 9- and 13-hydroperoxides.

PROTEASES

The hydrolysis of peptides and proteins by thermolysin and pepsin has been studied in glycerol-containing media.

First, the hydrolysis of the oxidized B chain of insulin by thermolysin was followed in the presence of increasing glycerol concentrations. For glycerol concentrations up to 2 M (S/E = 4000), the initial activity was only slightly decreased and 95% hydrolysis was attained in each case within 10 minutes. For higher additive concentrations, an inactivation was observed, due to limiting diffusional constraints. In limited proteolysis (S/E = 10^5), when comparing the elution profiles after a 30-minute proteolysis, it even appeared that the activity of thermolysin was increased for polyol concentrations up to 1 M (FIGURE 1).

Not only the hydrolysis degree, but the kinetics of the reaction were modified by the addition of glycerol. The hydrolysis products have been separated according to their hydrophobicity. Sixteen different peptides were observed (S/E = 4000), which have been divided into two classes depending on their evolution versus time:

- Class A includes the 7 hydrophilic peptides that continuously appeared.
- Class B is the 9 hydrophobic and large peptides that appeared at the beginning of the hydrolysis and were then hydrolyzed to give rise to peptides identical to those of class A.

The evolution as a function of time of these two classes was followed in the absence and in the presence of glycerol (FIGURE 2). This second hydrolysis was more rapid in buffer, and small products were obtained with a greater ratio. While at a 5-M glycerol concentration, the more hydrophobic peptides were hardly cleaved. Similar global kinetics (from 0- to 2-M glycerol) were thus observed though the hydrolysis products were different as the cleavage order was modified. This example well illustrates the possibility of modulating the selectivity of a protease, acting on a peptide, by a change in the reaction medium.

The experiment was extended to proteins. The kinetics of gliadin hydrolysis by pepsin were followed in buffer and in glycerol-containing media. As in the previous example, the activity of the protease was first increased by glycerol (by up to 60%) and then inhibited. In glycerol-containing media, gliadin hydrolysis followed an order different from that observed in buffer (TABLE 1): fraction D, the most hydrophobic, hydrolyzed first in buffer, did not react while fraction A, the most hydrophilic, was more hydrolyzed than in the reference medium. The increase in retention time of gliadin peaks indicated an increase in hydrophobicity. Of the 9 hydrolysis products, only the more hydrophobic peptides show different behavior in the presence of glycerol: the peak corresponding to the site hydrolyzed first in buffer revealed an inhibition of this cleavage by glycerol; new peaks appeared depending on the polyol concentration and the hydrolysis duration; modifications in hydrophobicity of larger peptides were observed.

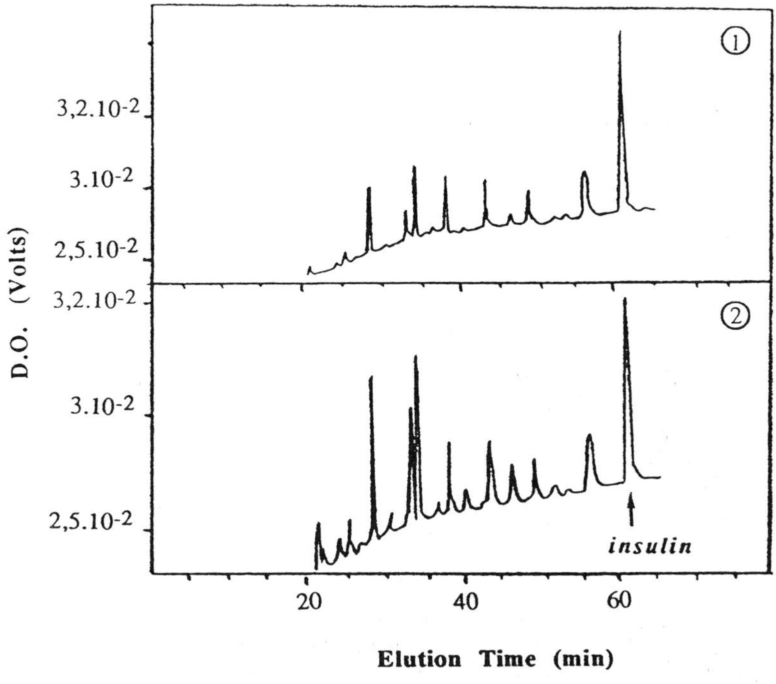

FIGURE 1. Hydrolysis products observed in limited proteolysis of insulin by thermolysin in buffer (1) and in a reaction medium containing 1-M glycerol (2).

From the present results, it appears that the hydrolysis order was affected by water-structuring molecules in the enzyme microenvironement and that new peptides may appear whatever the size of the substrate.

LIPOXYGENASE

Lipoxygenases catalyze the oxidation by molecular oxygen of fatty acids owning unconjugated Z-Z pentadienic double bonds. The initial product of the reaction is a hydroperoxide with conjugated Z-E double bonds.

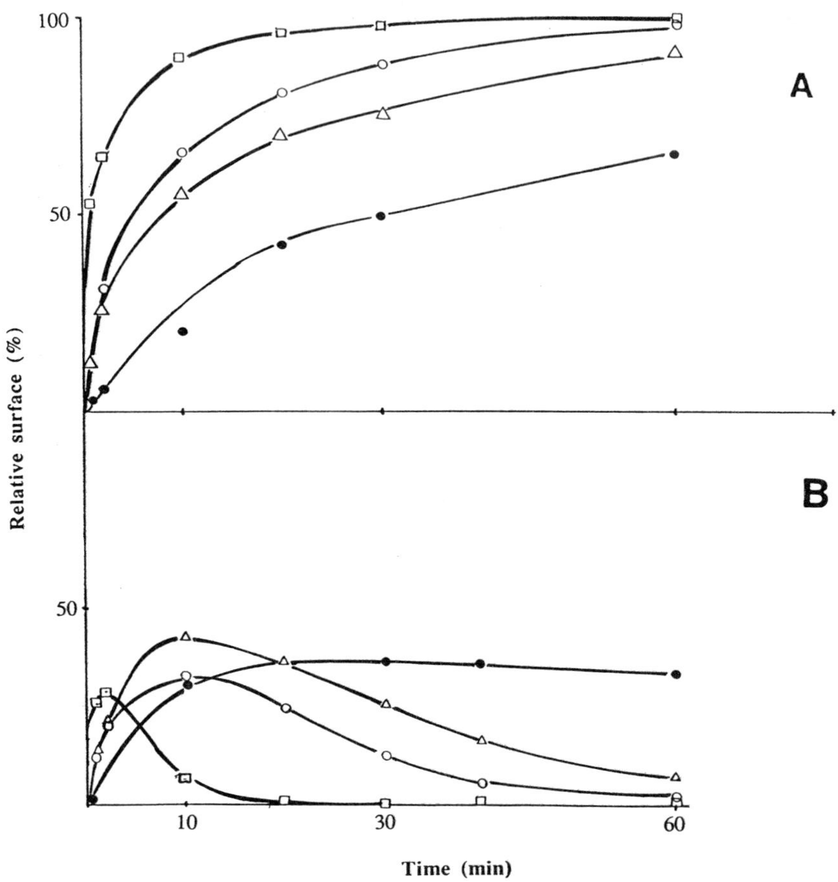

FIGURE 2. Evolution as a function of time of hydrophilic (A) and hydrophobic (B) peptides observed in proteolysis of insulin by thermolysin in buffer (□) and in reaction media containing 1-M (○), 2-M (△), and 5-M (●) glycerol; 100% corresponds to the total hydrolysis of insulin.

TABLE 1. Residual Gliadins after 3 Minutes Hydrolysis by Pepsin as a Function of Glycerol Concentration in the Reaction Medium

Glycerol (M)	Percent of Total Gliadin Surface				
	0	0.1	0.5	1	10
Peak A	32	25	26	12	12
Peak B	16	20	17	32	16
Peak C	23	24	24	7	12
Peak D	18	21	26	45	42

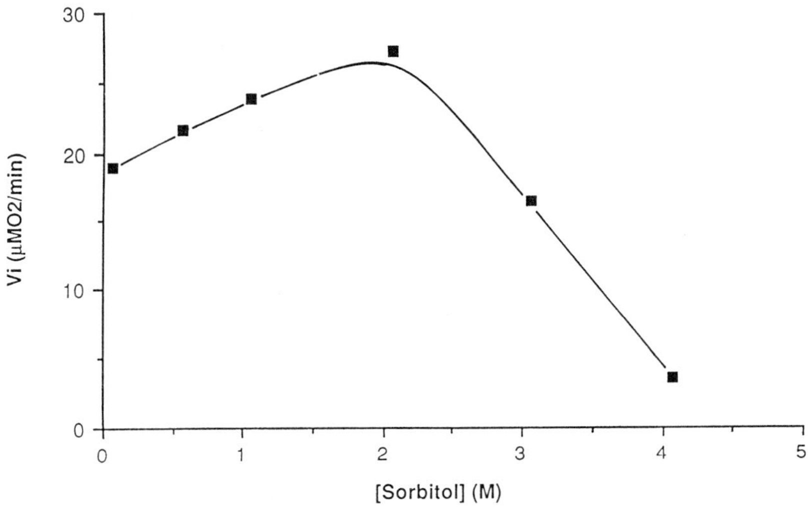

FIGURE 3. Soybean lipoxygenase activity measured with a polarographic method as a function of sorbitol concentration in the reaction medium.

The soybean 1-lipoxygenase–catalyzed oxidation of linoleic acid was measured in the presence of sorbitol. The enzyme activity was measured using both a polarographic and a spectrophotometric method. When dealing with the enzyme activity (FIGURE 3), an activation was observed when increasing the sorbitol concentration

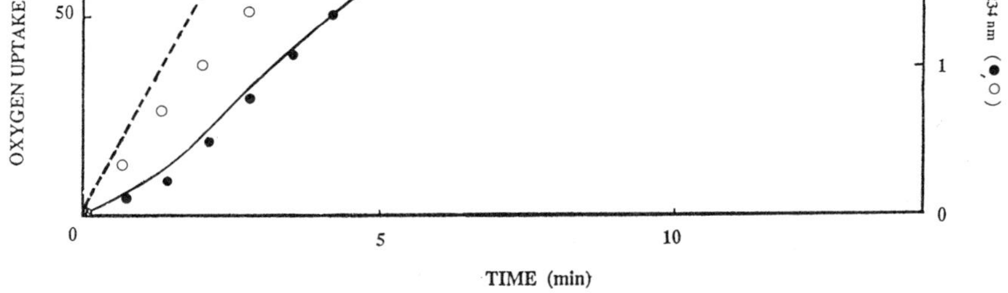

FIGURE 4. Soybean lipoxygenase activity measured either through the oxygen uptake followed by a polarographic method (—, ---) or through the hydroperoxide concentration followed spectrophotometrically (●, ○), in buffer (—, ●) and in a reaction medium containing 2-M sorbitol (---, ○).

up to 2 M, though at that polyol concentration, the reaction medium presented non-Newtonian characteristics and a low oxygen concentration.[10] For sorbitol concentrations higher than 2 M, the diffusion constraints made the activity decrease.

While in buffer, the correlation between polarographic and spectroscopic assays was good; in a medium containing 2-M sorbitol, a significative difference was observed (FIGURE 4). The oxygen uptake was found higher than the hydroperoxide production, indicating that sorbitol favored secondary reactions. This hypothesis was verified spectrophotometrically by a kinetical measurement of degradation product

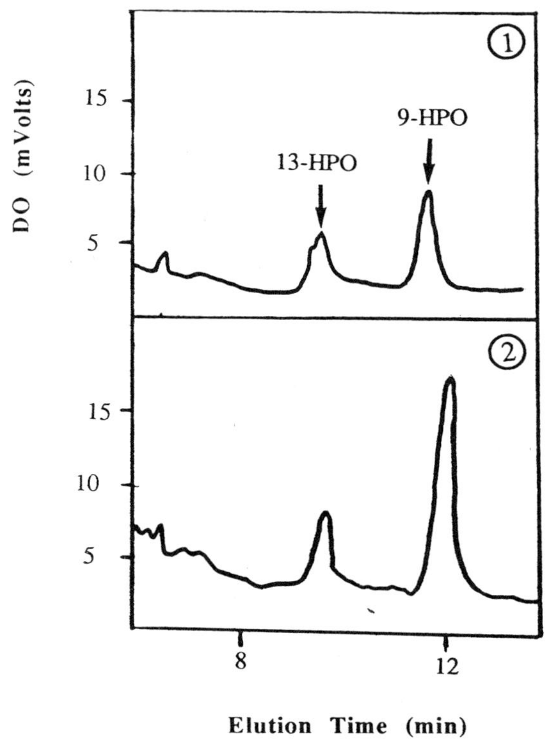

FIGURE 5. Hydroperoxides obtained through the oxidation of linoleic acid by soybean lipoxygenase in buffer (1) and in a reaction medium containing 2-M sorbitol (2).

appearance. The hydroperoxides were found less stable in sorbitol than in buffer during the first minutes of reaction, but they were then less degraded.

Mechanistically, the decision to form 13- or 9-hydroperoxide from linoleic acid depends upon which of the two hydrogens attached to the asymmetric carbon-11 is removed. In this work, the enzyme specificity is defined by the positioning of the hydroperoxide resulting from the lipoxygenase-catalyzed insertion of oxygen.

The synthesis of hydroperoxides in the presence and absence of sorbitol was measured using an HPLC method allowing the separation of the 9- and 13-isomers

(FIGURE 5). While in standard conditions the two isomers were formed with a similar ratio (9/13: 55%/45%), in the presence of 2-M sorbitol, the reaction was orientated towards the synthesis of the 9-hydroperoxide (9/13: 80%/20%).

CONCLUSIONS

The presence of water-soluble additives was demonstrated to have great influence on enzyme activity. The physicochemical properties of the solvent, particularly viscosity, were modified by the presence of polyols, but surprisingly, with non-Newtonian reaction media, important activations have been observed. This phenomenon may be attributed to slight enzyme and/or substrate conformational changes induced by the increased reaction medium hydrophobicity, which have been demonstrated with another enzyme.[11]

The medium modification also affected the selectivity of proteases. With thermolysin hydrolyzing insulin, the kinetics of proteolysis were modulated, giving rise to various peptide ratios in the reaction medium. With pepsin hydrolyzing gliadins, the cleavage order was modified by glycerol and new products appeared. The enzyme microenvironment was thus demonstrated to partly control the selectivity of proteases. This result is of importance as it contributes to the hypothesis that *in vivo* protease specificity could be phenotypic rather than genotypic.

Soybean lipoxygenase specificity was known to be sensitive to medium conditions such as pH, temperature, and oxygen concentration. We have demonstrated that hydroperoxide synthesis was also dependent on the state of water in the enzyme microenvironment. An orientation of the insertion of oxygen on the substrate was obtained when polyols were added to the reaction medium, which were also shown to directly react on the product. As differences were observed in hydroperoxide degradation, this phenomenon seems to affect the oxygen radical reactions.

REFERENCES

1. FERSHT, A. 1985. *In* Enzyme Structure and Mechanism: 293–369. Freeman. San Francisco, Calif.
2. LARRETA-GARDE, V., Z. XU, L. LAMY, M. MATHLOUTHI & D. THOMAS. 1988. Biochem. Biophys. Res. Commun. **155**(2): 816–822.
3. ANDERSSON, E. & B. HAHN-HAGERDAL. 1987. Biochim. Biophys. Acta **912:** 317–324.
4. DRAPRON, R. 1985. *In* Properties of Water in Foods. D. Simatos & J-L. Multon, Eds.: 171–190. Martinus Nijhoff Publish. Dordrecht, the Netherlands.
5. ZAKS, A. & A. KLIBANOV. 1988. J. Biol. Chem. **263**(7): 3194–3201.
6. XU, Z. F., V. LARRETA-GARDE & D. THOMAS. 1989. Biochim. Biophys. Acta **998:** 251–257.
7. HERTMANNI, P., E. PICQUE, D. THOMAS & V. LARRETA-GARDE. 1991. FEBS Lett. **279**(1): 123–131.
8. WATTERSON, J. 1987. Biochem. J. **248:** 615–617.
9. QUIOCHO, F., D. WILSON & N. VYAS. 1989. Nature **340:** 404–407.
10. POURPLANCHE, C., V. LARRETA-GARDE & D. THOMAS. 1991. Anal. Biochem. **198:** 160–164.
11. XU, Z. F., D. THOMAS & V. LARRETA-GARDE. 1990. Ann. N.Y. Acad. Sci. **613:** 506–510.

A Comparison of Enzymatic Reactions in Aqueous, Organic, and Supercritical Phases[a]

U. BORNSCHEUER, A. CAPEWELL, T. SCHEPER,
H. H. MEYER,[b] AND F. KOLISIS[c]

Institut für Technische Chemie
Callinstrasse 3
3000 Hannover 1, Germany

[b]*Institut für Organische Chemie*
Schneiderberg 1B
3000 Hannover 1, Germany

[c]*Department of Chemical Engineering*
The National Technical University of Athens
9, Heroon Polytechniou Str.
Zografou Campus
15773 Athens, Greece

INTRODUCTION

There has been an impressive increase in the number of papers dealing with enzymatic reactions in unconventional phases like organic solvents or supercritical CO_2 over the last few years.[1–8] However, most applications are still empirical, and fundamentals of technical engineering aspects for these reactions are frequently missing. In order to gain more knowledge of these new reaction systems, model enzyme reactions in aqueous, organic solvents, and in supercritical CO_2 were studied in detail. In our investigations we tested three possible ways for the enantioselective, lipase-catalyzed reaction of β-hydroxyesters, as shown in FIGURE 1. The comparison of hydrolysis (a), esterification (b), and transesterification (c + d) is intended to help in finding the best method for enantiomerically pure substances taking into account the advantages of the solvent system used. Cases c and d were conducted in organic solvents and supercritical carbon dioxide ($SCCO_2$). In all tests, the reaction rate and enantioselectivity of these reactions were of interest. The effects of the solvent system and the reaction process on these parameters were studied.

MATERIALS AND METHODS

The lipase from *Pseudomonas cepacia* (PCL) (LPS AP02520, Amano) was used as received. Racemic mixtures of 3-hydroxyhexanoic acid methylester or 3-hydroxyoctanoic acid methylester were synthesized according to Santaniello and Manzocchi.[9]

[a]The authors thank the European Community for supporting this project (Bridge Program).

The determination of conversion and enantiomeric excess was performed by gas chromatography using a chiral stationary phase (Lipodex E™, Macherey & Nagel). The samples were prepared for gas chromatography by derivation with trifluoroacetic acid anhydride (TFAA) as follows: 200 μl of *n*-hexane, 50 μl TFAA, and 5 μl of the reaction solution were mixed. After volatilizing of all solvent, 100 μl of *n*-hexane were added and the solution was analyzed. The conversion was calculated in two ways: first by the method described by Chen and Sih,[10] and second by using a standard. Therefore the product ester was synthesized as follows to calibrate the chromatographic procedure: 150 mg of substrates (1 mmol) and 100 mg acetic acid chloride were mixed in 200 μl pyridine and 500 μl diethyl ether for 24 hours. Five milliliters of saturated $(NH_4)_2CO_3$ were added and extracted three times with 10 ml diethyl ether. The collected organic phases were dried with sodium sulfate. The product ester was isolated by flash chromatography (ether:petrol ether 1:1). The

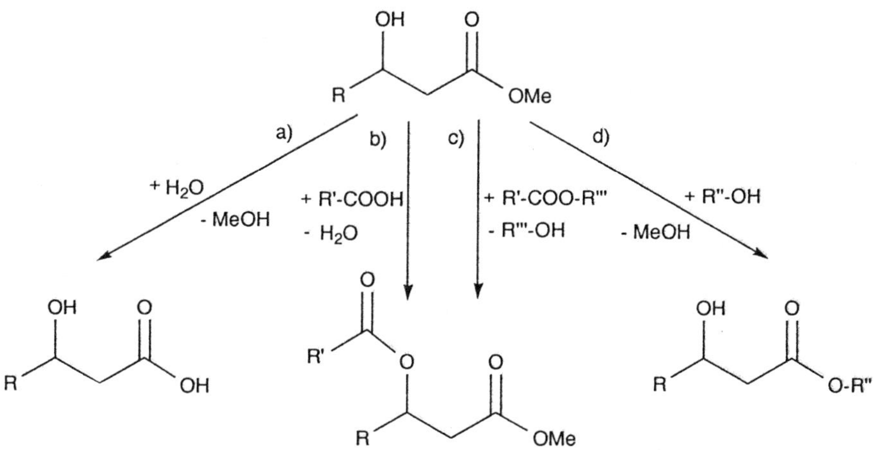

FIGURE 1. Possibilities for lipase-catalyzed reactions: **(a)** hydrolysis, **(b)** esterification, **(c)** and **(d)** transesterification.

reactions in supercritical carbon dioxide were carried out in the apparatus shown in FIGURE 2. The reactor (J&W High Pressure Soxhlet Extractor) was filled with 500 μl cyclohexyl acetate and 200 mg of PCL; the other container (Whitey) was filled with 200 μl of 3-hydroxyhexanoic acid methylester. The liquid carbon dioxide was pumped into the reactor and into the separate chamber with a high-pressure liquid pump (Milton-Roy, Mini-Pump Duplex 24/483 ml/hour) until the desired pressures were reached and all components were dissolved (determination by UV spectrometer and recorder). Samples of the reaction mixtures were taken via the 4-way valve (4 in FIGURE 2). The stainless steel sampling tube was flushed with *n*-hexane. In order to stop the reaction, the reactor was cooled with liquid nitrogen until the pressure dropped. The reactor was opened and the remaining solution was analyzed. The reactions in the organic solvents were conducted in 10-ml round-bottom flasks. The samples taken were centrifugated at 14,000 rpm for 5 minutes to separate the enzyme.

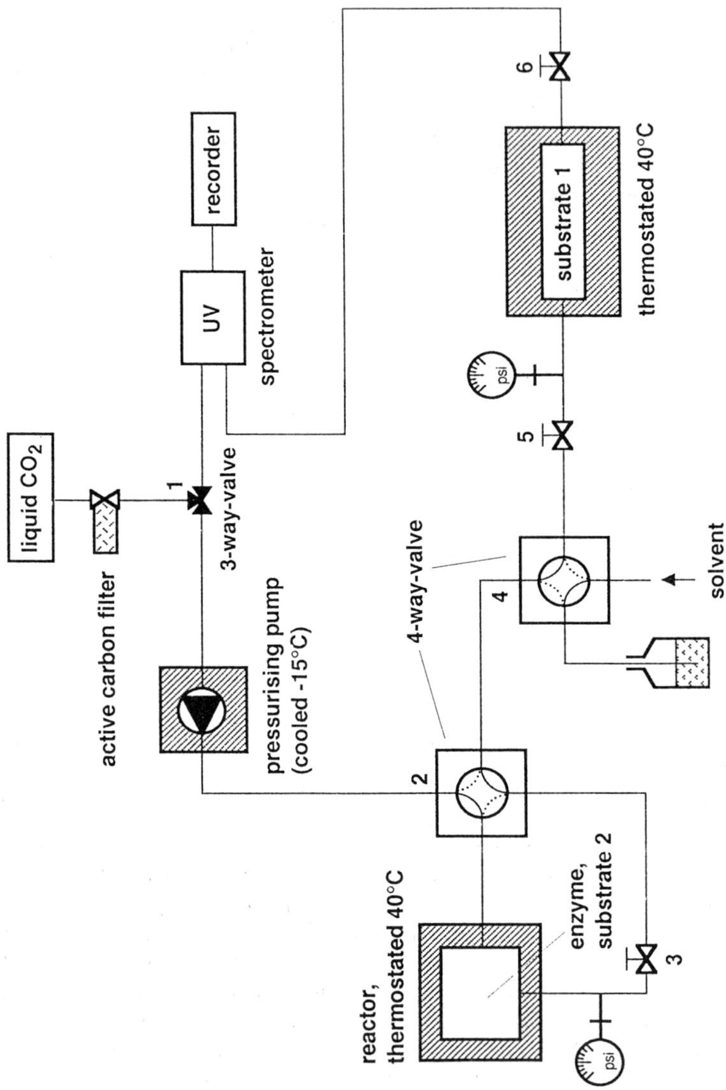

FIGURE 2. Scheme of the experimental apparatus.

RESULTS AND DISCUSSION

Hydrolysis

The hydrolyses were carried out with 3-hydroxyhexanoic acid methylester or 3-hydroxyoctanoic acid methylester suspended without solvent in 4 ml 1.0-M sodium phosphate buffer at pH 7 with 6 mg of lipase from *Pseudomonas cepacia* at room temperature. The results are shown in TABLE 1. The *R* (−) ester was hydrolyzed by preference, so that the remaining ester had an *S* (+) configuration.

Esterification

An esterification reaction was carried out at 40°C with 500 μl acetic acid, 100 μl 3-hydroxyhexanoic acid methylester, and 100 mg PCL in 2 ml *n*-hexane. The analysis of the reaction showed a stereochemically unspecific reaction. After 200 hours, 52% conversion was obtained, but the enantiomeric excesses of substrate and product were zero.

TABLE 1. Comparison of Hydrolysis (H), Esterification (E), and Transesterification [in Organic Solvents (T-O); in $SCCO_2$ (T-CO_2)][a]

Reaction	Compound	Enantiomeric Excess (%eeS)	(%eeP)	Conversion (%)	Time (hours)
H	1	64	43	60	125
E	1	0	0	52	200
T-O	1	42	23	63	340
T-CO_2	2	20	96	17	400
H	2	96	70	58	19
T-O	2	78	82	49	238

[a]Compound 1: 3-hydroxyhexanoic acid methylester; compound 2: 3-hydroxyoctanoic acid methylester; eeS, eeP: enantiomeric excess of substrate and product.

Transesterification

For transesterification reactions, organic solvents and supercritical CO_2 were chosen as solvents. The reactions were carried out at 40°C with different acyl donors, PCL, and racemic substrates. The results of these tests are shown in FIGURES 3 and 4. It is obvious that the transesterification with 3-hydroxyoctanoic acid methylester and cyclohexyl acetate (FIGURE 3) led to a product and a remaining substrate with higher enantiomeric purity than the reaction with 3-hydroxyhexanoic acid methylester and cyclohexyl acetate (FIGURE 4). In the second case, the conversion was much higher (more than 60%). At low conversions, it was very difficult to determine the enantiomeric excesses of the product, so that the conversion was only calculated from the standard peak. During the reaction the enantiomeric excess of the substrate increased, while the conversion plateaued (FIGURES 3 and 4). Normally, one would expect that at the equilibrium, the enantiomeric excess of the substrate would stagnate or decrease. Probably the use of crude lipase was responsible for these observations.

In all cases described above, the β-hydroxyester was acylated by another ester. Furthermore, it was of interest whether a transesterification of the ester function would be possible. Therefore, 3-hydroxyhexanoic acid methylester was converted in *n*-hexane using ethanol and PCL. After 435 hours at 51% conversion, 24%ee for the substrate and 25%ee for the product were obtained, as shown in FIGURE 5. In order to find out more about the equilibrium, the reaction was performed with 3-hydroxyhexanoic acid ethylester and methanol under the same conditions. After 400 hours, no conversion could be ascertained. In supercritical carbon dioxide at 1200 to 1400

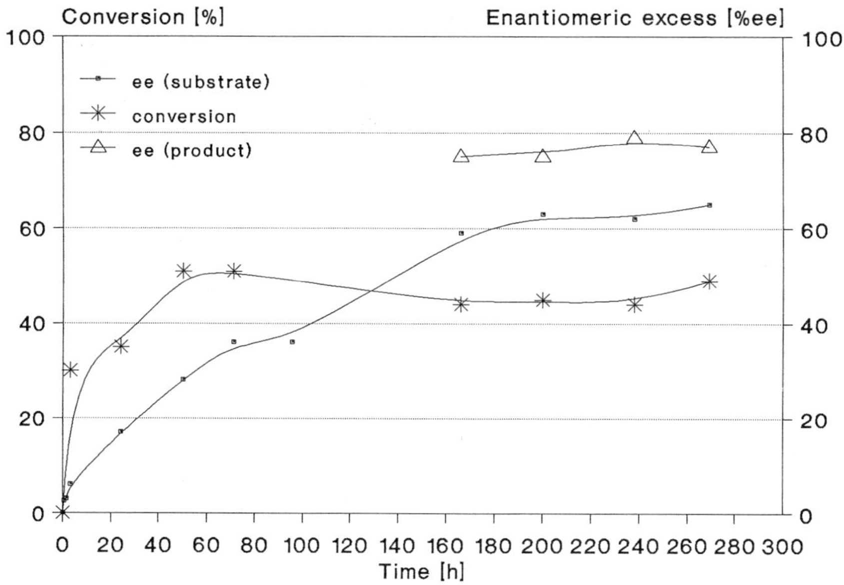

FIGURE 3. Conversion and enantiomeric excesses for the transesterification reactions with 3-hydroxyoctanoic acid methylester, cyclohexyl acetate, and PCL.

psi, only low conversions could be detected up to now, although several parameters have been altered, such as temperature, pressure, humidity, and amounts of compounds and lipase.

COMPARISON AND CONCLUSION

TABLE 1 shows the comparison of all reactions carried out in the different solvent systems. It is obvious that the thermodynamically favored hydrolyses led to higher enantiomeric excesses with less enzyme in a shorter time. The transesterification

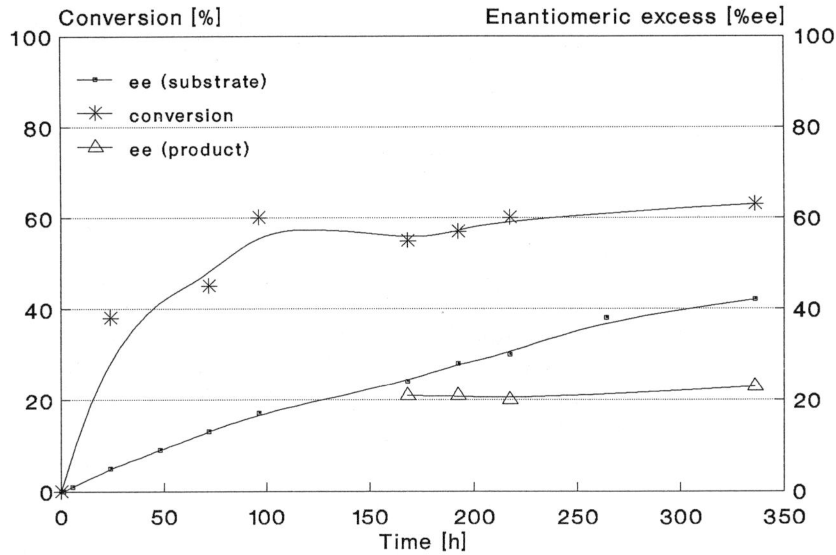

FIGURE 4. Conversion and enantiomeric excesses for the transesterification reactions with 3-hydroxyhexanoic acid methylester, cyclohexyl acetate, and PCL.

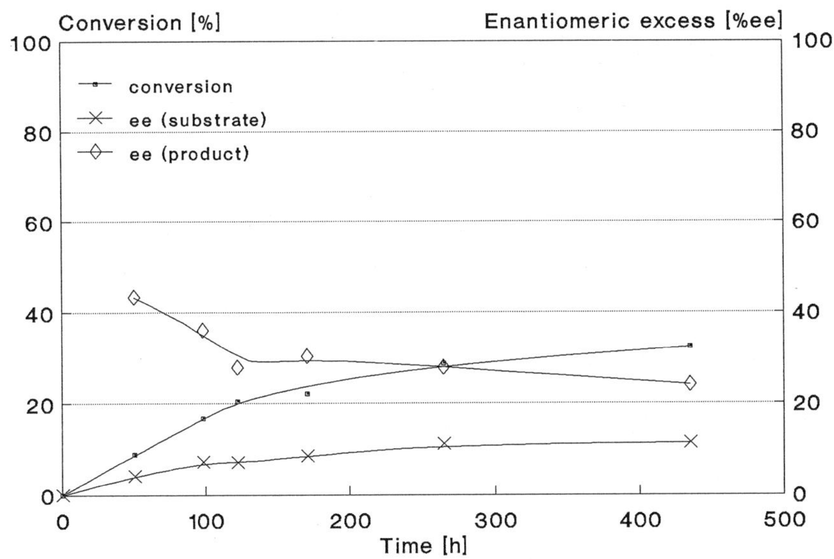

FIGURE 5. Conversion and enantiomeric excesses for the transesterification with 3-hydroxyhexanoic acid methylester, ethanol and lipase from *Pseudomonas cepacia* in *n*-hexane.

reactions in organic solvents were able to compete with the hydrolysis under consideration of further technical aspects, such as easier product recovery or the reuse of the enzyme.

SUMMARY

Reactions of different racemic β-hydroxyesters were performed in aqueous and organic phases with lipase from *Pseudomonas cepacia* (PCL). The hydrolyses led to a remaining substrate with higher enantiomeric purity in shorter reaction time with less enzyme compared to the transesterification reactions in organic solvents. The transesterification of 3-hydroxyhexanoic acid methylester into the corresponding ethylester in *n*-hexane was also catalyzed by PCL, but no reaction of the ethylester was observed employing methanol. Furthermore, the application of supercritical carbon dioxide for lipase-catalyzed reactions was tested.

REFERENCES

1. STEYTLER, D. C., P. S. MOULSON & J. REYNOLDS. 1991. Enzyme Microb. Technol. **13:** 221–226.
2. MARTY, A., W. CHULALAKSANANUKUL, J. S. CONDORET, R. M. WILLEMOT & G. DURAND. 1991. Biotechnol. Lett. **12**(1): 11–16.
3. KLIBANOV, A. M. 1986. Chem. Tech.: 354–359.
4. HALLING, P. J. 1988. DECHEMA Biotechnol. Conf. **1:** 35–40.
5. SCHAPÖHLER, S. 1990. Ph.D. Thesis. University of Hannover. Hannover, Germany.
6. XIE, Z. F., H. SUEMUNE & K. SAKAI. 1990. Tetrahedron Asymmetry **1**(2): 395–402.
7. BORNSCHEUER, U., S. SCHAPÖHLER, T. SCHEPER & K. SCHÜGERL. 1991. Tetrahedron Asymmetry **2**(10): 1011–1014.
8. MILLER, D. A., H. W. BLANCH & J. M. PRAUSNITZ. 1991. Ind. Eng. Chem. Res. **30:** 939–946.
9. SANTANIELLO, E. & A. MANZOCCHI. 1977. Synthesis **10:** 698–699.
10. CHEN, CH.-S. & CH. J. SIH. 1989. Angew. Chem. **101:** 711–724.

Developing Better Enzymes for Organic Synthesis[a]

CHI-HUEY WONG, KEVIN K.-C. LIU, TETSUYA KAJIMOTO, LIHREN CHEN, ZIYANG ZHONG, YOSHITAKA ICHIKAWA, AND GWO-JENN SHEN

Department of Chemistry
The Scripps Research Institute
10666 North Torrey Pines Road
La Jolla, California 92037

Glycoconjugates that exist on cell surfaces are involved in many types of biochemical recognitions.[1] These molecules have been difficult to isolate, characterize, and synthesize, and are the least explored of the major classes of biomolecules. Carbohydrate-related structures therefore offer opportunities for the study of molecular recognition and for the development of therapeutic agents. The recently identified tetrasaccharide sialyl Lewis x, as the ligand of the endothelial leukocyte adhesion molecule,[2] stimulates new interests in the development of practical methods for the synthesis of carbohydrates and mimetics as ligands or antagonists of carbohydrate-binding proteins, or as inhibitors of specific glycoenzymes such as glycosidases and glycosyltransferases. Our interest in this area is to develop enzyme-based technology for the synthesis of carbohydrates and related substances, to make these materials readily available to study their structure-function relation, and to evaluate their therapeutic potential.

SYNTHESIS OF UNCOMMON AND AZASUGARS BASED ON ALDOLASES

More than 20 aldolases are known. These enzymes hold great potential for the synthesis of common and uncommon monosaccharides.[3] Of particular important application of aldolases is the synthesis of deoxyazasugars—a class of molecules useful as glycosidase inhibitors. We have examined five aldolases [i.e., fructose-1,6-diphosphate aldolase from rabbit muscle or *Escherichia coli,*[4] rhamnulose (Rham) and fuculose (Fuc) 1-phosphate aldolases,[5,6] 2-deoxyribose-5-phosphate aldolase (DERA),[5] and sialic acid aldolase[7]] for utility in asymmetric aldol condensation. All these aldolases possess two common features. First, they are highly specific for the donor substrate but flexible for the acceptor component. Second, the stereoselectivity is controlled by the enzyme, not by the substrate, with some exceptions observed in sialic acid aldolase reactions. A general strategy was therefore developed for the synthesis of deoxyazasugars based on aldolase-catalyzed condensation of azidoalde-

[a]This work was kindly supported by the National Institutes of Health, the National Science Foundation, and the Office of Naval Research.

hydes followed by Pd-mediated reductive amination. Both thermodynamic and kinetic approaches have been explored for the aldolase reactions. FIGURES 1–6 illustrate representative syntheses of azasugars based on this combined chemical-enzymatic approach. With this rapid stereocontrolled process, a number of azasugars have been synthesized and evaluated as glycosidase inhibitors, and the research has led to a new direction of inhibitor design.[4–6] The structure-inhibition study indicates that the five-membered-ring aza sugars are good mimics of the transition state of glycosidic bond cleavage, due to their charged half-chair conformation.[6] We believe these aza sugars will be useful building blocks for the synthesis of many sequence-specific glycosidase and glycosyltransferase inhibitors, and the synthetic strategy illustrated in these processes should be applicable to the synthesis of other types of aza sugars based on different aldolases.

REGENERATION OF SUGAR NUCLEOTIDES FOR OLIGOSACCHARIDE SYNTHESIS

Glycosidases and glycosyltransferases have been used in oligosaccharide synthesis.[3,8–11] The sugar-nucleotide-dependent glycosyltransferases seem to be more suitable for the synthesis of complex oligosaccharides as the enzymatic reactions are regioselective for various complex acceptor structures. The major problems are that glycosyltransferases are not readily available and that sugar nucleotides are too expensive to be used as stoichiometric reagents. Furthermore, the reactions often exhibit product inhibition caused by the released nucleoside phosphates. A simple

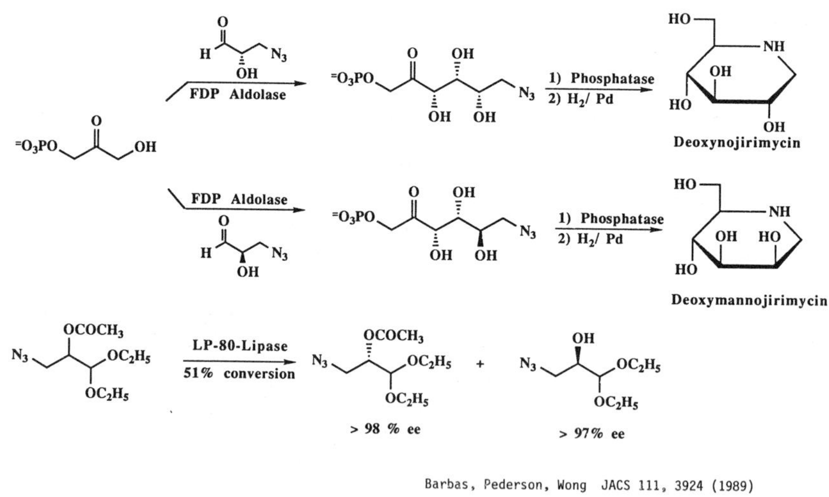

FIGURE 1. Synthesis of deoxynojirimycin and deoxymannojirimycin based on fructose-1,6-diphosphate aldolase from rabbit muscle or *E. coli.* The enantiomerically pure aldehydes were prepared via lipase-catalyzed resolution.

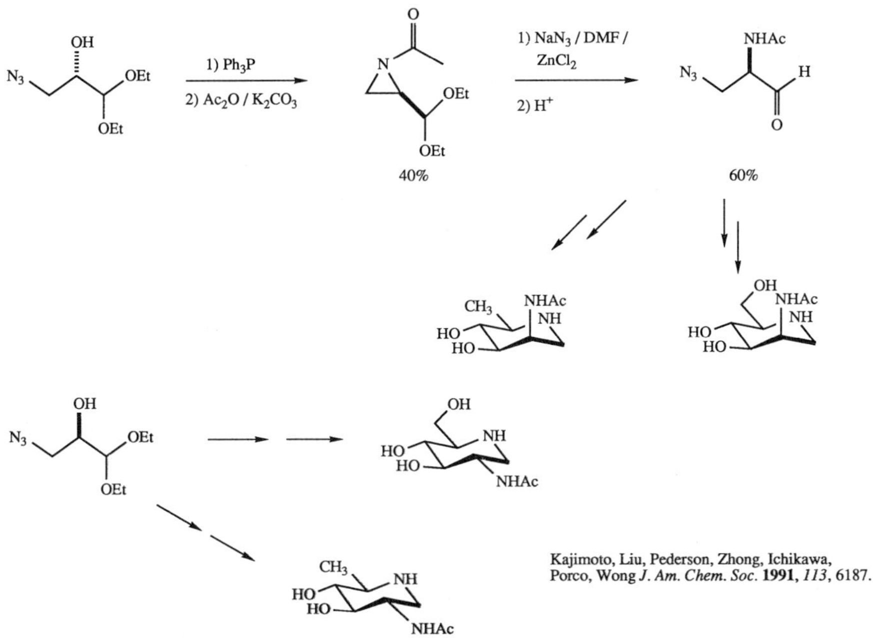

FIGURE 2. Synthesis of deoxy aza sugars structurally related to *N*-acetylmannosamine and *N*-acetylglucosamine, and their 6-deoxy derivatives based on fructose-1,6-diphosphate aldolase reactions as indicated in FIGURE 1. Reductive amination of the enzyme products without removal of the phosphate moiety gave the 6-deoxy aza sugars.

solution to these problems is to regenerate sugar nucleotide during the enzymatic reaction. It has been demonstrated previously that UDP-glucose and UDP-galactose can be regenerated *in situ* in a 50-g synthesis of disaccharide.[12] We have recently developed a new procedure for the regeneration of CMP-sialic acid in a sialyltransferase-catalyzed synthesis of sialyl oligosaccharide (FIGURE 7).[13] We have also demonstrated that two glycosyltransferases can be used in one pot, coupled with regeneration of more than two sugar nucleotides, for the formation of two glycosidic bonds starting from three unactivated monosaccharides, and no product inhibition has been observed.[14] The enzymes can be immobilized in a polymer support and recovered for reuse. We believe the strategies and principles illustrated in these syntheses can be utilized in the synthesis of many oligosaccharides. The remaining issues will be to make glycosyltransferases available. Although several cloning systems have been developed, there is still a need for the development of practical expression systems other than those based on CHO cells.[15] Cloning and expression of the soluble active domains of glycosyltransferases remain an interesting subject for investigation. We are currently applying the phage vector system used for the overproduction of CMP-sialic acid synthetase[13] to express the active domains of several glycosyltransferases in *E. coli.*

E: FDP-Aldolase

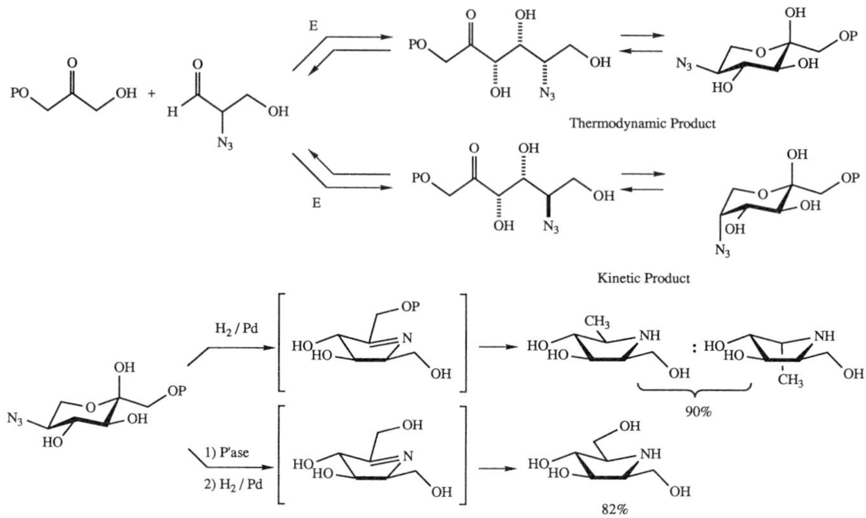

Kajimoto, Chen, Liu, Wong *J. Am. Chem. Soc.* **1991**, *113*, 6678.
Liu, Kajimoto, Zhong, Ichikawa, Wong *J. Org. Chem.* **1991**, *56*, 6280.

FIGURE 3. Thermodynamic and kinetic approaches to the synthesis of azido ketoses for use in the preparation of five-membered-ring aza sugars.

E, fucolose-1-phosphate aldolase

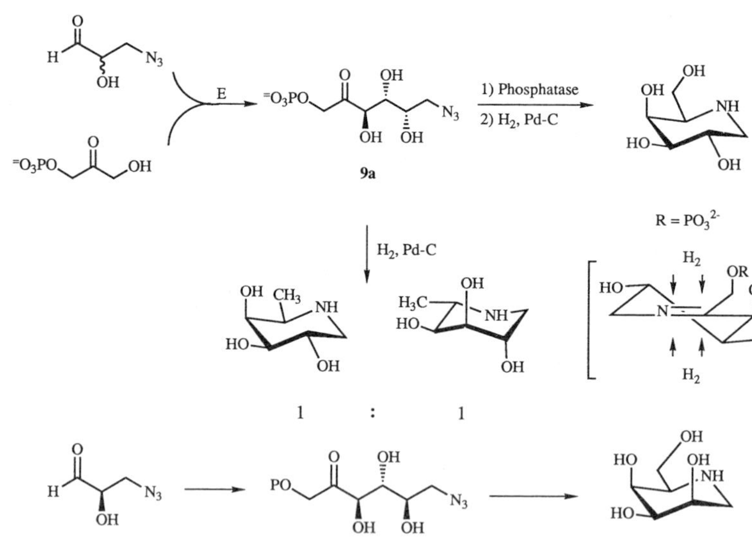

Liu, Kajimoto, Chen, Zhong, Ichikawa, Wong *J. Org. Chem.* **1991**, *56*, 6280.

FIGURE 4. Fuculose-1-phosphate aldolase catalyzed synthesis. Racemic or enantiomerically pure 2-hydroxy-3-azido-propanal was used as substrate.

DEVELOPMENT OF STABLE AND ACTIVE ENZYMES IN POLAR ORGANIC SOLVENTS

In our work on aldolase reactions, O-acylated monossacharides were used as acceptors for sialic acid aldolase–catalyzed reactions. The acylated sugars were prepared via subtilisin-catalyzed transesterification in anhydrous dimethylformamide using enol esters as acylating reagents.[16] We also conducted condensation of peptide fragments in DMF or a mixture of DMF and water via subtilisin-catalyzed aminolysis of N-protected peptide esters. This latter process is aimed to develop new methods for the enzymatic coupling of glycosylpeptides. A major problem encountered in this work is the instability of the enzyme in polar organic solvents which are

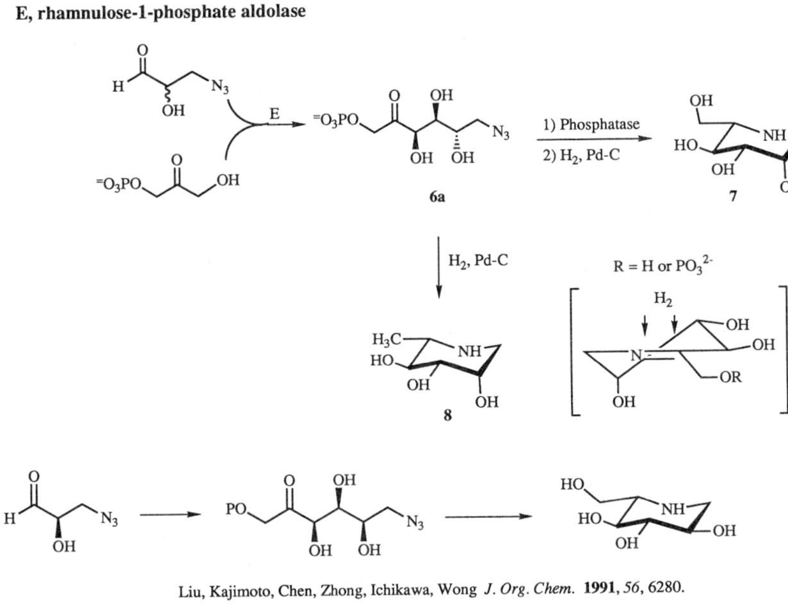

FIGURE 5. Rhamnulose-1-phosphate aldolase catalyzed synthesis.

often required for the reactions. The half-life of subtilisin BPN′ in dry DMF, for example, is about 20 minutes, which is not enough for large-scale processes. A first improvement of the enzyme stability has been accomplished with the use of site-directed mutagenesis to increase the half-life to about one day at room temperature.[17] A second variant of subtilisin BPN′ with a half-life of about 14 days in dry DMF has recently been prepared.[18] TABLE 1 lists the relative stabilities of several subtilisin variants we have prepared. This work has provided a substantial amount of experimental information useful for the design of enzymes to be used in polar organic solvents. Based on the subtilisin work, we feel that several modifications must be carried out simultaneously to make enzymes stable and active in polar organic solvents. These include the minimization of surface charges, the enhance-

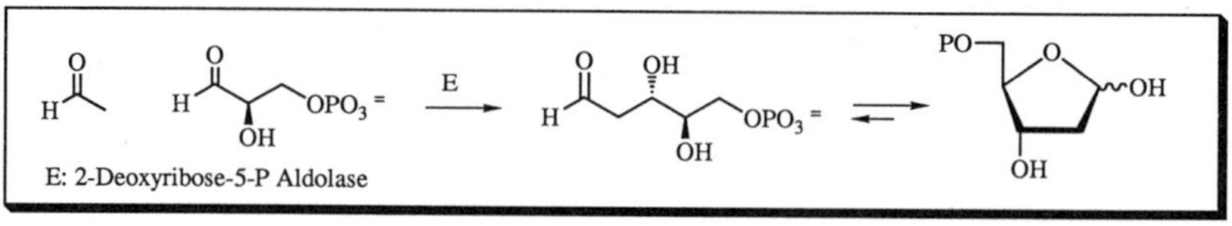

Barbas, Wang, Wong, *J. Am. Chem. Soc.* **1990**, *112*, 2013.
Chen, Dumas, Wong, *J. Am. Chem. Soc.* **1991.**

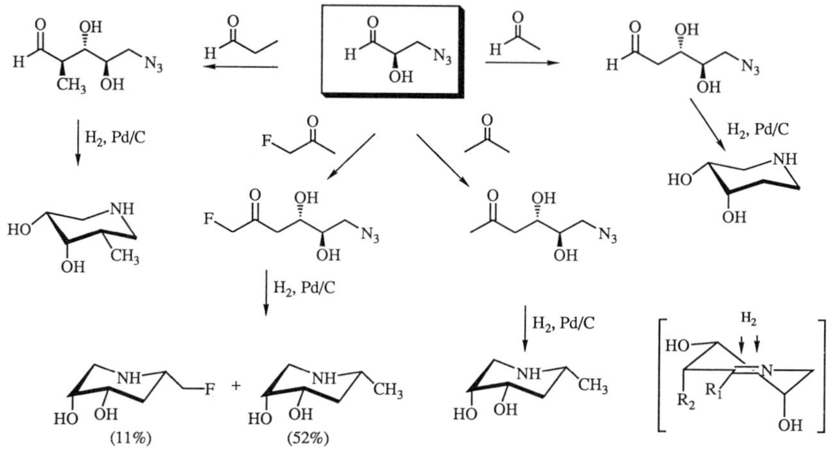

FIGURE 6. 2-Deoxyribose-5-phosphate aldolase catalyzed synthesis using different donors.

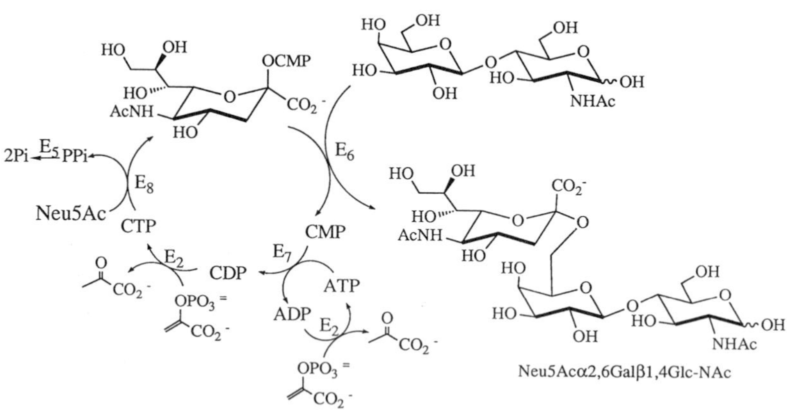

E_5: pyrophosphatase E_6: α2,6-sialyltransferase E_7: Nucleoside monophosphate kinase
E_8: CMP-sialic acid synthetase

FIGURE 7. Regeneration of CMP-sialic acid for use in the synthesis of sialyl oligosaccharides.

TABLE 1. Half-lives of Subtilisin BPN′ and Variants at 25°C in a Phosphate Buffer (0.1 M, pH 8.4) and in Anhydrous Dimethylformamide[a]

Enzyme	$t_{1/2}$ (hours)	
	H_2O (pH 8.4)	DMF
BPN′	15	0.3
8397	1600	350
8350	1500	22
8399	1000	43

[a]The stability was measured by taking aliquots of the enzyme at different periods of time for activity assay in aqueous solution (0.1 M phosphate, pH 8.4). The half-life was thus determined based on the irreversible inactivation.

8399: Met50Phe, Asn76Asp, Gln206Cys, Asn218Ser
8350: Met50Phe, Gly169Ala, Asn76Asp, Gln206Cys, Tyr217Lys, Asn218Ser
8397: Met50Phe, Gly169Ala, Asn76Asp, Gln206Cys, Asn218Ser

ment of internal polar interactions (e.g., H-bonding, ionic, and metal-coordinating forces), the optimization of internal hydrophobic interactions, and the enhancement of conformational restrictions. In the case of subtilisin, surface charge minimization (e.g., Lys217 → Tyr) and conformational restriction (e.g., Gly169 → Ala) are proven to be the most effective. Each of the changes improves the enzyme stability in DMF

TABLE 2. Ratios of Aminolysis to Hydrolysis of Modified Serine Proteases[a]

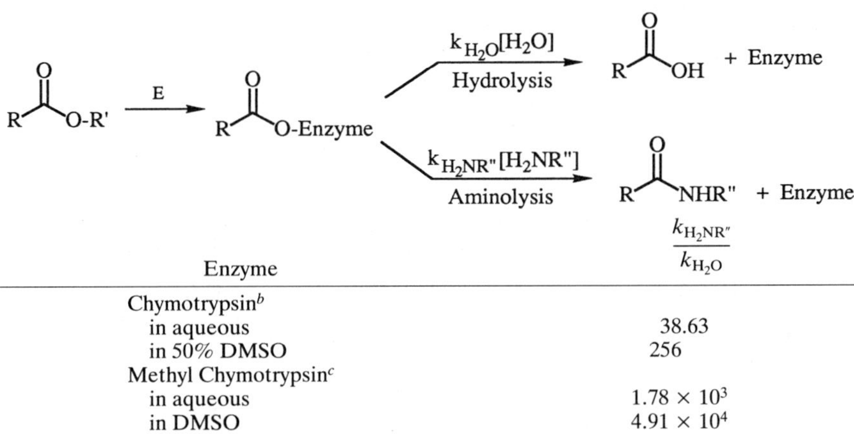

Enzyme	$\frac{k_{H_2NR''}}{k_{H_2O}}$
Chymotrypsin[b]	
in aqueous	38.63
in 50% DMSO	256
Methyl Chymotrypsin[c]	
in aqueous	1.78×10^3
in DMSO	4.91×10^4
Subtilisin BPN′[b]	
in aqueous	55.66
in DMSO	88.5
Thiosbtilisin[b]	
in aqueous	1.7×10^4
in DMSO	1.01×10^5

[a]Activities were determined at 25°C in aqueous solution (0.1 M phosphate, pH 8.4) in the presence of dimethylsulfoxide (50% v/v). Methylchymotrypsin: the active-site His was methylated at the ε2-nitrogen group. Thiosubtilisin: the active-site Ser was converted to Cys.

[b]Cinnamoyl enzyme was prepared and used as acyl donor.

[c]Suc-Ala-Ala-Pro-Phe-SBzl was used as acyl donor. Gly-NH_2 was used as nucleophile.

corresponding to approximately 1 kcal/mol as measured with its irreversible inactivation. Surface charge minimization was also proven to be effective for the stabilization of α-lytic protease[19] and subtilisin E[20] in aqueous DMF.

CONCLUSION

We have demonstrated the use of aldolases, glycosyltransferases, and engineered subtilisin in the synthesis of carbohydrates and related substances. This enzyme-based technology is complementary to nonenzymatic technology, and is well suited for the synthesis of polyhydroxylated structures. Perhaps large-scale synthesis of various monosaccharides, oligosaccharides, and their related substances and analogues will become a routine process in the near future when the enzymes are readily available.

REFERENCES

1. 1989. Carbohydrate Recognition in Cellular Function. Ciba Found. Symp. **145.** John Wiley & Sons. Chichester, England.
2. Phillips, M. L., E. Nudleman, F. C. A. Gaeta, M. Perez, A. L. Singhal, S.-I. Hakomori & J. C. Paulson. 1990. ELAM-1 mediates cell adhesion by recognition of a carbohydrate ligand, sialyl-Le^x. Science **250:** 1130–1132.
3. Toone, E. J., E. S. Simon, M. D. Bednarski & G. M. Whitesides. 1989. Enzyme-catalyzed synthesis of carbohydrates. Tetrahedron **45:** 5365–5421.
4. Kajimoto, T., K. K.-C. Liu, R. L. Pederson, Z. Zhong, Y. Ichikawa, J. A. Porco & C.-H. Wong. 1991. Enzyme-catalyzed aldol condensation for asymmetric synthesis of azasugars: synthesis, evaluation, and modeling of glycosidase inhibitors. J. Am. Chem. Soc. **113:** 6187–6196.
5. Kajimoto, T., L. Chen, K. K.-C. Liu & C.-H. Wong. 1991. Palladium mediated stereocontrolled reductive amination of azido sugars prepared from enzymatic aldol condensation: a general approach to the synthesis of deoxy aza sugars. J. Am. Chem. Soc. **113:** 6678–6680.
6. Liu, K. K.-C., T. Kajimoto, L. Chen, Z. Zhong, Y. Ichikawa & C.-H. Wong. 1991. Use of dihydroxyacetone phosphate dependent aldolases in the synthesis of deoxyazasugars. J. Org. Chem. **56:** 6280–6289.
7. Gautheron-Le Narvor, C., Y. Ichikawa & C.-H. Wong. 1991. A complete change of stereoselectivity in sialic acid aldolase reactions: a novel synthesis route to the KDO type of nine-carbon L-sugars. J. Am. Chem. Soc. **113:** 7816–7818.
8. Nilsson, K. G. 1988. Enzymatic synthesis of oligosaccharides. TIBTECH **6:** 256–264.
9. Hindsgaul, O., K. J. Kaur, U. B. Gokhale, G. Srivastava, G. Alton & M. M. Palcic. 1991. Use of glycosyltransferases in synthesis of unnatural oligosaccharide analogs. Eds., ACS Symp. Ser. **466:** 38–50.
10. Stangier, P. & J. Thiem. 1991. Glycosylation by use of glycohydrolases and glycosyltransferases in preparative scale. ACS Symp. Ser. **466:** 63–78.
11. Drueckhammer, D. G., W. J. Hennen, R. L. Pederson, C. F. Barbas, C. M. Gautheron, T. Krach & C.-H. Wong. 1991. Enzyme catalysis in synthetic carbohydrate chemistry. Synthesis **7:** 499–526.
12. Wong, C.-H., S. L. Haynie & G. M. Whitesides. 1982. Enzyme-catalyzed synthesis of *N*-acetyllactosamine with *in situ* regeneration of uridine 5′-diphosphate glucose and uridine 5′-diphosphate galactose. J. Org. Chem. **47:** 5416–5418.
13. Ichikawa, Y., G.-J. Shen & C.-H. Wong. 1991. Enzyme-catalyzed synthesis of sialyl oligosaccharide with *in situ* regeneration of CMP-sialic acid. J. Am. Chem. Soc. **113:** 4698–4700.

14. ICHIKAWA, Y., J. L.-C. LIU, G.-J. SHEN & C.-H. WONG. 1991. A highly efficient multienzyme system for the one-step synthesis of a sialyl trisaccharide: *in situ* generation of sialic acid and *N*-acetyllactosamine coupled with regeneration of UDP-glucose, UDP-galactose and CMP-sialic acid. J. Am. Chem. Soc. **113:** 6300–6302.

15. PAULSON, J., K. J. COLLEY. 1989. Glycosyltransferases. J. Biol. Chem. **264:** 17615–17618.

16. KIM, M.-J., W. J. HENNEN, H. M. SWEERS & C.-H. WONG. 1988. Enzymes in carbohydrate synthesis: *N*-acetylneuraminic acid aldolase catalyzed reactions and preparation of *N*-acetyl-2-deoxy-D-neuraminic acid derivatives. J. Am. Chem. Soc. **110:** 6481–6486.

17. WONG, C.-H., S.-T. CHEN, W. J. HENNEN, J. A. BIBBS, Y.-F. WANG, J. L.-C. LIU, M. W. PANTOLIANO, M. WHITLOW & P. N. BRYAN. 1990. Enzymes in organic synthesis: use of subtilisin and a highly stable mutant derived from multiple site-specific mutations. J. Am. Chem. Soc. **112:** 945–953.

18. ZHONG, Z., J. L.-C. LIU, L. M. DINTERMAN, M. A. J. FINKELMAN, W. T. MUELLER, M. L. ROLLENCE, M. WHITLOW & C.-H. WONG. 1991. Engineering subtilisin for reaction in dimethylformamide. J. Am. Chem. Soc. **113:** 683–684.

19. MARTINEZ, P. & F. H. ARNOLD. 1991. Surface charge substitutions increase the stability of α-lytic protease in organic solvents. J. Am. Chem. Soc. **113:** 6336–6337.

20. CHEN, K., A. C. ROBINSON, M. E. VAN PAM, P. MARTINEZ, C. ECONOMOU & F. H. ARNOLD. 1991. Enzyme engineering for nonaqueous solvents II. Additive effects of mutation on the stability and activity of subtilisin E in polar organic media. Biotechnol. Prog. **7:** 125–129.

Enzymatic and Chemoenzymatic Approaches to Polymer Synthesis and Modification[a]

JONATHAN S. DORDICK

Department of Chemical and Biochemical Engineering
University of Iowa
Iowa City, Iowa 52242

INTRODUCTION

The application of enzymes in organic solvents is now well established.[1–3] The exquisite selectivities and fast catalytic rates of enzymes under ambient conditions provide commercially competitive alternatives to traditional organic synthesis to be developed in many diverse areas including oxidoreductions and condensations, among others.[4] Nowhere is this more evident than in the use of nonaqueous biocatalysis for the synthesis and modification of polymers.

It is difficult to believe that enzymes can compete with chemical catalysts for the production of polymers. In most respects, enzymes appear poorly suited to function as effective catalysts in this area due to the high productivities required for economical production of polymers. Indeed, most polymers have been and will continue to be produced via simple chemical catalysis because little selectivity is required. For example, the simple stepwise polycondensation of diols with diacids is efficiently catalyzed by base.[5] Polyfunctional monomers and precursors, however, provide challenging problems in the use of traditional catalysis, as reaction stoichiometry—and hence polymer molecular weight and physicochemical properties—is controlled by catalyst selectivity. An increasing number of novel materials are synthesized from polyfunctional precursors. These materials include chiral polymers, photooptical materials (e.g., to provide control of refractive index), and biodegradable/biocompatible polymers. Highly selective (e.g., regio-, stereo-, and chemo-) syntheses are often required for suitable polymeric properties. Specific examples of this requirement are the synthesis of water-absorbent, biocompatible/biodegradable polymers from sugars and the synthesis of modified lignins with free phenolics to produce reactive phenolic resins. In both instances, the biological reactant is multifunctional and well suited for enzymic modification. This paper provides a review of our research efforts with the use of enzymes in organic solvents for the synthesis and modification of polymers from renewable resources.

[a]This work has been funded in part by grants from the Sugar Association, National Science Foundation, U.S. Department of Agriculture, State of Iowa, and the Mead Corporation. This paper describes research previously published (References 6, 8, and 9).

MATERIALS AND METHODS

Poly(sucrose adipate) was synthesized in the following manner:[6] 0.85 g sucrose (0.1 M) was dissolved in 25 mL pyridine containing 0.1-M di-(2,2,2-trifluoroethyl) adipate (synthesized according to Reference 7). The reaction was commenced upon addition of 15 mg/mL alkaline protease (Proleather). The suspension was magnetically stirred under nitrogen at 150 rpm and 45°C. The reaction was terminated by filtering off the enzyme and evaporating the pyridine. The residual di-(2,2,2-trifluoroethyl) adipate was removed by washing with 100 mL diethyl ether leaving 0.97 g (80% yield) of white powder. One hundred milligrams of the sucrose polyester was dissolved in water and analyzed by gel permeation chromatography (GPC). The crude sucrose polyester was fractionated on a Sephadex G-10 molecular weight column (MW cutoff = 700) with water as eluent. The void fraction was freeze-dried, and 78 mg of a white powder was obtained. In addition to di-(2,2,2-trifluoroethyl) adipate, commercially available vinyl adipate can be used to prepare poly(sucrose adipate). Once again, equimolar ratios of sucrose to the adipic acid derivative are employed.

Poly(sucrose adipamide) was synthesized chemoenzymatically following an improvement over our published findings:[8] 0.86 g (0.1 M) sucrose was dissolved in 25 mL pyridine containing 3.1 g (0.4 M) di-(2,2,2-trifluoroethyl) adipate. Excess di-(2,2,2-trifluoroethyl) adipate was used to improve the yield of diester (relative to monoester). Higher sucrose concentrations could not be used due to the limited solubility of sucrose in pyridine. The reaction was initiated by the addition of 100 mg/mL Proleather, and the suspension was magnetically stirred under nitrogen at 150 rpm for 24 hours at 45°C. The reaction was terminated by filtering off the enzyme and evaporating the pyridine and unreacted adipate derivative. The sucrose 6,1′-di(trifluoroethyl) adipate was purified using silica gel chromatography with an eluent consisting of ethyl acetate:methanol:water (18:1.25:1). The sucrose diester was obtained in 20% yield. No triester was formed. Polymerization was carried out by adding 15 mg (0.125 M) ethylenediamine to 0.19 g (0.125 M) of diester in 2 ml of *N*-methylpyrrolidone (NMP). The solution was stirred at 35°C for 24 hours.

Poly(sucrose acrylate) was prepared by dissolving 3.42 g (0.1 M) sucrose in 100 mL pyridine containing 5.88 g (0.6 M) vinyl acrylate.[8] Hydroquinone (0.5% w/v) was added to inhibit polymerization of the vinyl acrylate during sucrose acrylate synthesis. The reaction was initiated by addition of 100 mg/mL alcalase 2-T (an alkaline protease from Novo which was found to be superior to proleather for the monoacylation of sucrose in pyridine), and the suspension was magnetically stirred under nitrogen at 150 rpm for 24 hours at 45°C. The reactions were terminated by filtering off the enzyme, evaporating the pyridine and unreacted vinyl acrylate, and the product was purified and separated by silica gel chromatography with an eluent consisting of ethyl acetate:methanol:water (18:1.25:1). The sucrose 1′-acrylate (as identified by ^{13}c-NMR) was obtained in 30% yield, 1.2 g. The ester was an amorphous solid, mp = 78°C; $[\alpha]_D^{25}$ = 50.4 (cl, H_2O). Poly(sucrose acrylate) synthesis was carried out by dissolving 1.0 g (0.25 M) of the sucrose monoester in 10 ml H_2O, and the solution was sparged with N_2 for 10 minutes. Potassium persulfate (0.15%) and 0.2% hydrogen peroxide were added, and the solution was stirred at 25°C for 24 hours. The resulting poly(sucrose 1′-acrylate) was recovered by precipitation with acetone, filtered, and dried under vacuum at 45°C.

Lignin modification was carried out as described in our recent paper.[9] The reaction vessel contained 1 mg horseradish peroxidase (HRP) in a volume of 0.5 ml. A high initial concentration of HRP was used because of the subsequent dilution. Solutions (in 50% dioxane) of H_2O_2 (4.4 ml of 1.7 mM) and *p*-cresol (4.4 ml of a 2.25-mM solution containing a total of 0.5 μCi) were pumped into the reaction vessel at approximately 0.05 ml/minute with constant stirring. The lignin (2 mg, 10 μmol assuming $M_n = 198$ per phenylpropane unit) was dissolved in the H_2O_2 solution. The reaction mixture was stirred for at least an additional 30 minutes after addition of the solutions was complete. Dimethylformamide (DMF, 2 ml) was then added and the dioxane:water removed by vacuum evaporation. The sample was further concentrated to about 0.5 ml and particulates removed by centrifugation. The sample was loaded onto a column (1 × 23 cm) of Sephadex LH-60 equilibrated in DMF containing 0.1 M LiCl. The same solvent was used for elution, and 0.75-ml fractions were collected in scintillation vials containing 8 ml of Poly-fluor scintillation cocktail (Packard Instrument Co., Downer's Grove, Ill.). The fractions were counted in a Packard Tricarb Model 5430 scintillation counter. All results were verified in at least one repeat of the experiment.

RESULTS AND DISCUSSION

Sugar-Based Polymers

The incorporation of sugars into polymers such as poly(esters, amides, or acrylates) is a difficult task in traditional chemistry due to the large number of equally reactive hydroxyl groups on the sugar molecule. For example, sucrose contains 8 hydroxyl groups: 3 primary and 5 secondary. Regioselective acylation of sucrose via chemical means (e.g., through the use of acid chlorides and anhydrides) is somewhat selective for primary groups, and can be shifted to secondary groups if the primary hydroxyls are blocked with bulky reagents (e.g., trityl, tosyl, or sulfuryl chlorides).[10–12] However, little or no differentiation between the three primary hydroxyls is observed. This cross-reactivity poses dire consequences for polycondensation reactions. For example, the traditional synthesis of polyesters involves the condensation of a diol with a diacid (or diacid derivative) in an equimolar ratio. The reactivity of additional hydroxyl functionalities generates cross-linked polyesters that are typically low molecular weight and brittle in nature, yet with little practical value.[13] This can be prevented through the use of bulky blocking reagents to limit polycondensation to sterically unhindered hydroxyl moieties. However, the overall process of blocking, polymerization, and subsequent deblocking is tedious, time-consuming, and expensive. Alternatives are needed.

As opposed to chemical catalysts, the use of enzymes in organic solvents provided us with an opportunity to develop a synthetic strategy for sugar-based polyester synthesis. Specifically, enzyme-catalyzed regioselective acylation of sugars offers an alternative to the poor selectivity of chemical synthesis. Our first synthetic strategy was to incorporate sucrose (as a diol) into a polyester backbone by condensing the sugar with a diacid derivative.[6] A screen of 45 commercially available hydrolases (e.g., lipases, esterases, proteases, and carbohydrases) resulted in the identification of proleather (an alkaline protease from *Bacillus sp.,* sold by Amano) as a suitable enzyme for this synthesis. The activity of this enzyme in pyridine was found to be highly dependent upon the pH from which the enzyme was lyophilized from with an optimal pH of 9.5.

The proleather-catalyzed polyester synthesis was performed by suspending 15 mg/mL of the pH-adjusted enzyme in 25 mL pyridine containing 0.1-M sucrose and

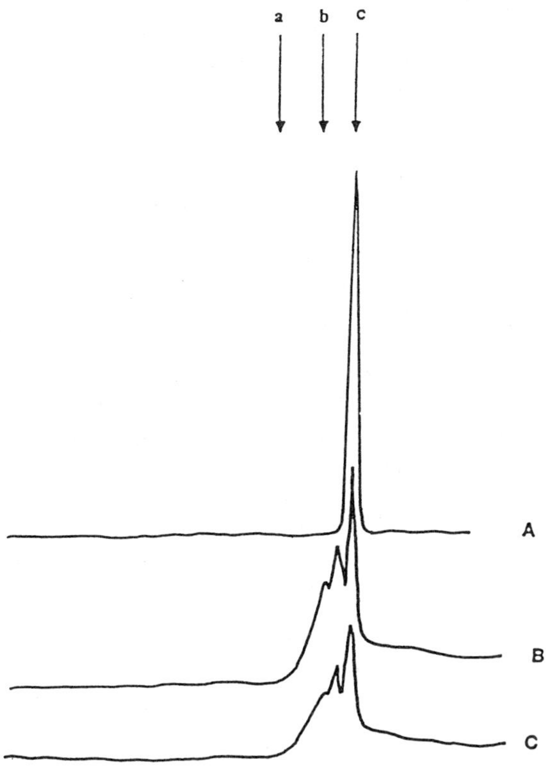

FIGURE 1. Gel permeation chromatographs [120Å and 250Å Ultrahydrogel (Waters)] of proleather-catalyzed polymerization of sucrose with di-(2,2,2-trifluoroethyl) adipate. Arrows indicate molecular weight markers for polyethylene glycol (12,500 and 1100) and sucrose (342). The mobile phase consisted of 0.1-M $NaNO_3$ at a flow rate of 1.0 ml/minute. The time course of the reaction is shown with A, B, and C representing time zero, 7 days, and 25 days, respectively.

0.1-M di-(2,2,2-trifluoroethyl) adipate. This adipate derivative provided for a highly reactive leaving group (trifluoroethanol) which is known to improve the rate of transesterification.[14] The reaction time course and molecular weight profile are shown in FIGURE 1. After 25 days, the polymeric material had an M_W of 2100 and an M_n of 1600 for a polydispersity of 1.3. Attempts to increase the reaction rate by addition of more enzyme failed, as above 15 mg/mL, the proleather swelled to give a thick and unstirrable mixture. This problem was eliminated when vinyl adipate was used as the diacid derivative. Using vinyl adipate, it was possible to increase the enzyme loading to 100 mg/mL and the reaction time shortened to 7 days.

The enzymatically produced poly(sucrose adipate) from vinyl adipate had an $M_n = 11{,}150$ and an $M_w = 14{,}500$ with some polymers as large as 40,000 daltons. Analysis of the polymer by ^{13}C-NMR revealed the highly selective nature of proleather catalysis, as only the 6 and 1′ hydroxyls were acylated with the adipate derivative to produce the polyester with the structure shown in FIGURE 2. The polymer was highly soluble in water, yet could be cross-linked with hexamethylene diisocyanate to give water-insoluble gels that absorbed nearly three times their weight in water. In addition to sucrose, raffinose could also be polymerized by this approach. FIGURE 3 shows the structure of the poly(raffinose adipate) with ester linkages at the 6 and 1′ positions.

The poly(sucrose adipate) product was also biodegradable. To test this, 10 mg/mL polyester [prepared from the di-(trifluoroethyl) adipate reaction] was dissolved in 0.1-M phosphate buffer, pH 7, and 2.0 mg/mL Proleather was added and the reaction shaken at 100 rpm at 25°C. Aliquots of the reaction were periodically removed and analyzed by GPC. Within 48 hours, 2.65 mg/mL free sucrose was liberated from the polymer, representing a biodegradative yield of 31%. The formation of free sucrose indicates that the proleather can hydrolyze *both* the 6 and 1′ ester linkages during biodegradation. Furthermore, the resulting biodegraded material had an $M_w = 1300$ as compared to 2100 for the fractionated sucrose polyester starting material. Continued biodegradation occurred with values of M_w ca. 900 after 9 days. Therefore, significant depolymerization of the sucrose polyester takes place enzymatically in aqueous solution, a prerequisite for successful biodegradability. Interestingly, the same enzyme that catalyzed the synthesis of the polymer in pyridine, catalyzed its breakdown in aqueous buffer.

The power of selective enzymic synthesis of sugar-based polymers can be enhanced by taking advantage of chemoenzymatic synthesis. In this manner, the high regioselectivity of enzymatic catalysis can be coupled with the high reaction productivity and synthetic versatility of chemical catalyst. In this regard, we have synthesized two sucrose-containing polymers.[8] The first is a poly(sucrose adipamide) with sucrose remaining as a component of the polymer backbone; and the second is a

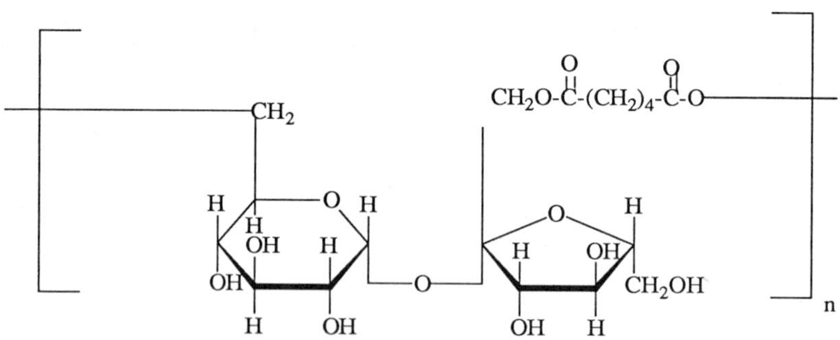

FIGURE 2. Molecular structure of poly(sucrose adipate).

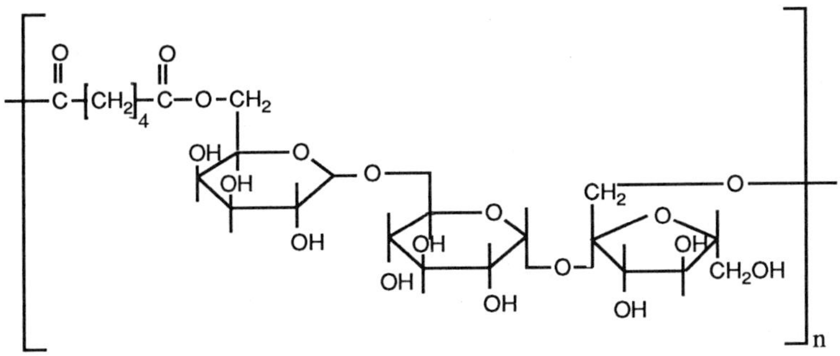

FIGURE 3. Molecular structure of poly(raffinose adipate).

poly(sucrose acrylate) wherein the sucrose exists as a polymer side chain. In both instances, the high degree of regioselectivity afforded by enzymatic catalysis is used to modify the sucrose before chemical polymerization is performed.

The synthesis of each polymer involved an initial enzymatic reaction. For the poly(sucrose adipamide) proleather was employed to catalyze the formation of sucrose 6,1′-di-(trifluoroethyl) adipate followed by polycondensation with ethylenediamine. FIGURE 4 depicts the synthetic strategy employed in this synthesis, and TABLE 1 outlines the resulting polymer properties. The enzymatic portion of this synthesis is complete (using 100 mg/mL enzyme) in 24 hours, while the chemical polymerization is complete in an additional 24 hours.

The synthesis of poly(sucrose acrylate) was also performed by an initial acylation of sucrose using proleather. In this case, vinyl acrylate (0.3 M) was dissolved along with 0.1 M sucrose, and 15 mg/mL proleather was added. The suspension was stirred at 45°C for 24 hours, and the sucrose 1′-acrylate product isolated by silica gel chromatography. The sucrose monoacrylate was then polymerized in water via hydrogen peroxide and potassium persulfate. FIGURE 5 depicts the reaction scheme, and the properties of the polymer are described in TABLE 1. Clearly, the molecular weights obtained specifically with the poly(sucrose acrylate) are substantially higher than those obtained by the purely enzymic synthetic route.

Monosaccharide derivatives could be used for polyacrylate synthesis in addition to sucrose. The range of acrylate products include galactose, glucose, and mannose *O*-glycosides. Acylation is consistently, and solely, at the 6-position. Polymerization with the initiator AIBN yields polymers with molecular weights > 100,000. Chemoenzymatic synthesis represents a significant advantage over either purely chemical approaches (enhanced regioselectivity) or purely enzymatic routes [greater speed and, in the case of poly(sucrose acrylate), substantially higher molecular weights].

Lignin Modification

Phenolic resins have applications as adhesives, laminates, foams, conditioning agents, etc.[15] The most common phenolic resins are phenol-formaldehyde polymers, which, however, pose health and environmental hazards due to the toxicity associated with formaldehyde.[16] Also, the polymerization reactions can be difficult to control both in terms of the reactivities of different phenolic monomers and the molecular weights attained. Peroxidase-catalyzed phenoxy radical polymerization is an alternative method of phenolic resin production and has been used to prepare phenolic resins from phenolic starting materials.[17,18]

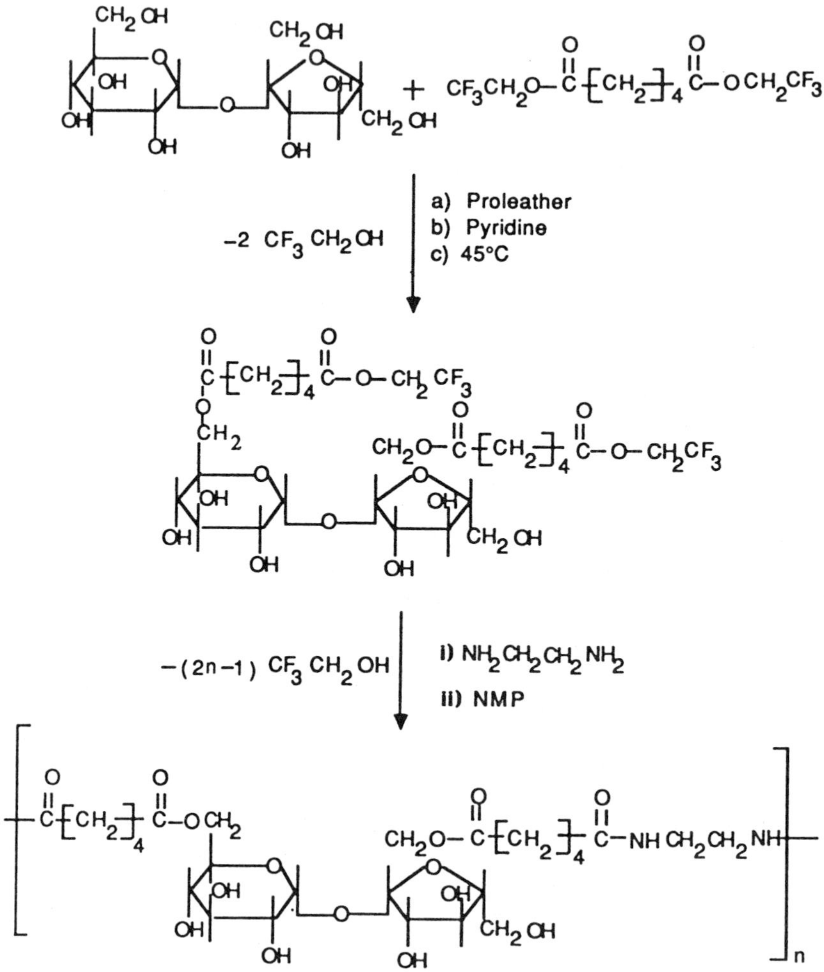

FIGURE 4. Chemoenzymatic synthesis of poly(sucrose adipamide).

Lignin, the second most abundant biopolymer after cellulose, is a natural phenolic resin. This material is a polydisperse, irregular polymer of methoxyl- and hydroxyl-substituted phenylpropane units.[19] Several methods have been developed for the production of lignin-containing phenolic resins,[20,21] many of which employ

TABLE 1. Properties of Polymers Prepared via Chemoenzymatic Methods.[a]

Poly(Sucrose Adipamide)	Poly(Sucrose Acrylate)
$M_w = 8100$	$M_w = 91{,}000$
$M_n = 4800$	$M_n = 57{,}000$
$M_w/M_n = 1.7$	$M_w/M_n = 1.6$
Water insoluble	Highly water soluble
Soluble in polar organics	Soluble in polar organics

[a]See text for details.

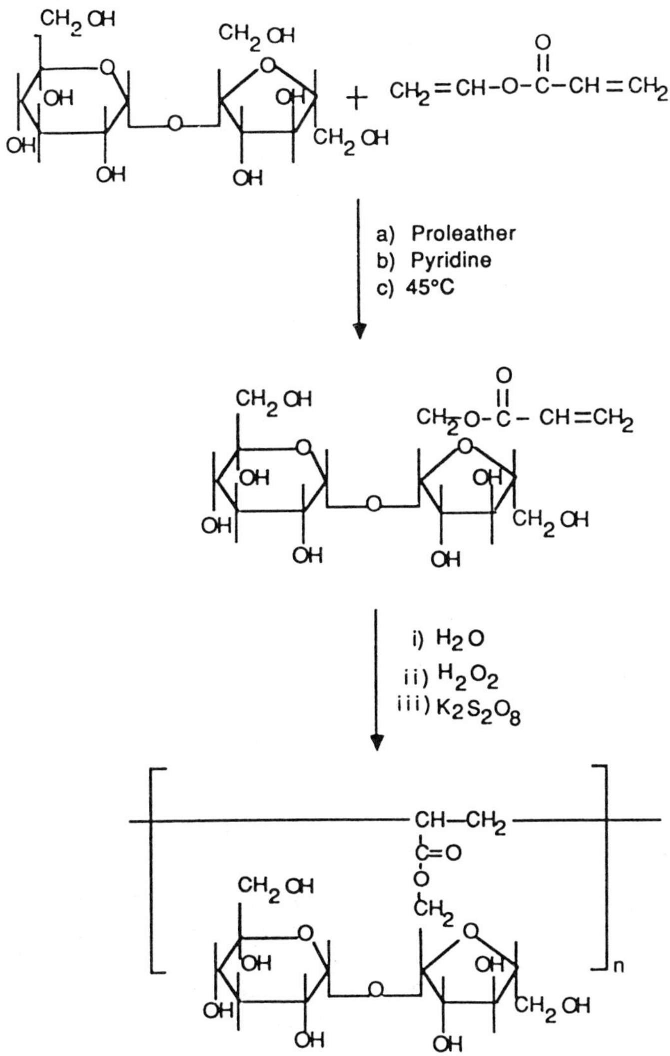

FIGURE 5. Chemoenzymatic synthesis of poly(sucrose acrylate).

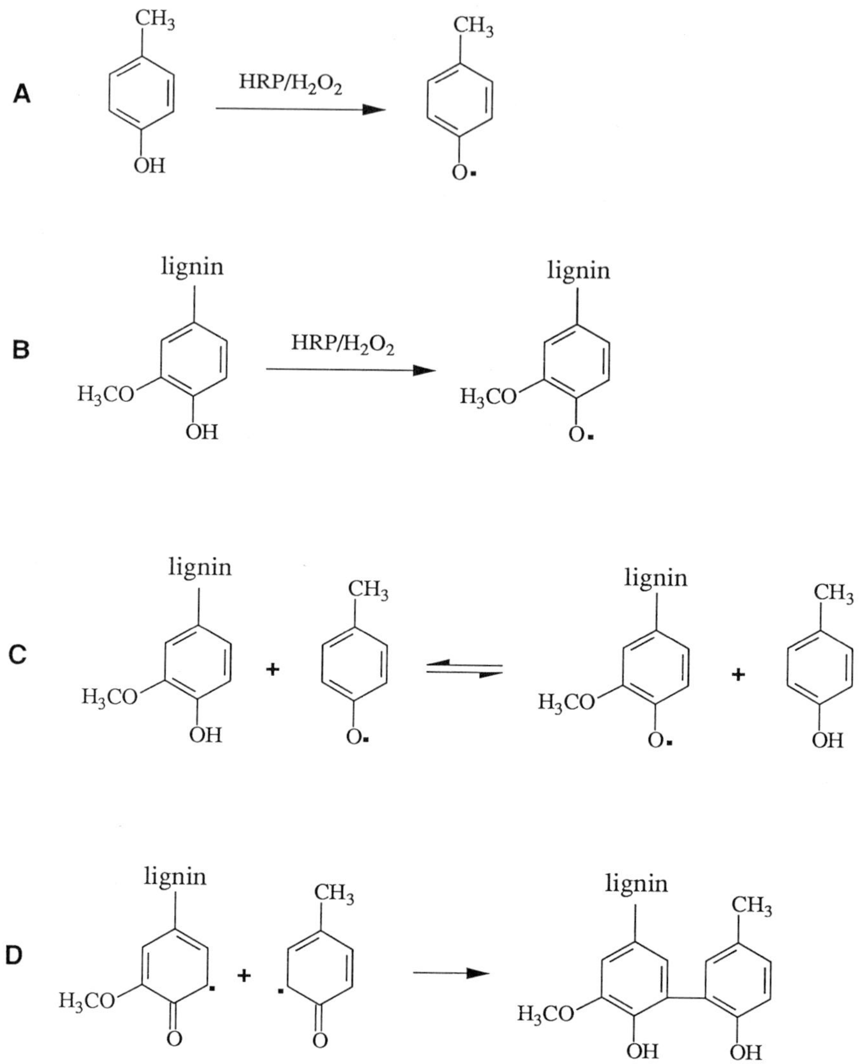

FIGURE 6. Proposed reaction mechanism of peroxidase-catalyzed grafting of phenols onto lignin.

formaldehyde to methylolate the lignin and/or phenol in order to cross-link the polymers. Unfortunately, the same problems of formaldehyde toxicity noted above apply to these lignin-phenol resins. Instead of formaldehyde as a cross-linking agent, we have used horseradish peroxidase to catalyze copolymerization of phenols and lignin. This so-called grafting of phenols onto lignins should result in the modification of lignin's properties, at least partly due to the bound phenols. Hence, modified lignins with defined properties can be prepared as potential replacements of conventional phenolic resins. The mechanism shown in FIGURE 6 describes the proposed grafting reaction.

The results of peroxidase-catalyzed copolymerization (grafting) of ^{14}C-labeled *p*-cresol onto birch (hardwood) milled wood lignin in 50% aqueous dioxane are shown in FIGURE 7. Also shown is a homopolymerization of *p*-cresol used as a control to indicate the size of homopolymers of *p*-cresol formed during a copolymerization reaction. Clearly, the copolymerization resulted in the incorporation of radioactivity in the highest molecular weight fractions of the LH-60 Sephadex column. Based on radioactive incorporation into the high-molecular-weight milled-wood lignin fraction, ca. 20% of the phenylpropane units reacted with *p*-cresol. Blocking of the lignin's free phenolic groups via acetylation with acetic anhydride resulted in no grafting of *p*-cresol onto the lignin. Moreover, the reaction was sensitive to the solvent. In 85% dioxane, only minimal grafting was observed, perhaps due to the poor activity of peroxidase in this solvent.[22] In addition to birch milled-wood lignin, several other lignins could be modified by this approach including milled-wood lignin from Loblolly pine (softwood), lignin from the kraft pulping process, and lignin from brown-rot treatment of lignocellulose.

In conclusion, the high specificity of enzymatic catalysis has now been extended to the synthesis and modification of polymers. Polyfunctional monomers or polymers can now be used to prepare novel materials via wholly enzymatic or combined

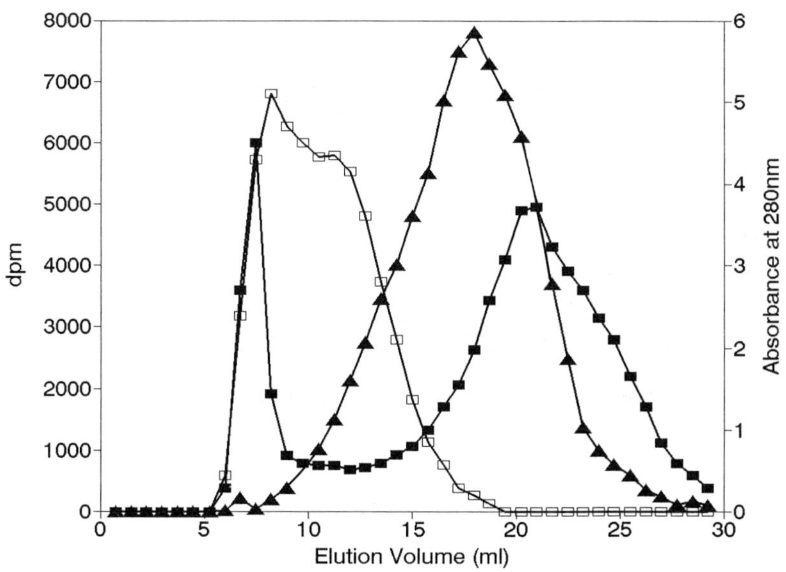

FIGURE 7. Gel permeation chromatography on Sephadex LH-60 of birch MWL detected by A_{280} (□), and products of a homopolymerization of [U-^{14}C]-*p*-cresol (▲) and a copolymerization reaction of birch MWL and [U-^{14}C]-*p*-cresol (■) detected by determination of radioactivity (dpm) in the fractions. All the lignin was added at the start of the reaction. Arrows indicate peak elution of molecular weight standards, from left to right: blue dextran, 2 × 10^6; polystyrenes, 9000, 4000, 2150; and veratraldehyde, 166.

chemoenzymatic syntheses. These materials have a variety of potential uses ranging from biodegradable plastics to drug-delivery matrices to phenolic resins.

ACKNOWLEDGMENTS

The author acknowledges Prof. David G. Rethwisch, Dr. D. R. Patil, and Gao Yi at the University of Iowa for collaboration on sucrose polymerization, and Dr. T. Kent Kirk (USDA Forest Products Laboratory, Madison, Wis.) and Dr. Janet L. Popp (now at Sterling Research Group) for collaboration on lignin modification.

REFERENCES

1. Dordick, J. S. 1991. Principles and selected applications of nonaquoeus enzymology. *In* Applied Biocatalysis. H. W. Blanch & D. S. Clark, Eds.: 1–51. Marcel Dekker. New York, N.Y.
2. Klibanov, A. M. 1986. Enzymes that work in organic solvents. Chemtech **16:** 354.
3. Zaks, A. & A. J. Russel. 1988. J. Biotechnol. **8:** 259.
4. Dordick, J. S. 1989. Enzyme Microb. Technol. **11:** 194.
5. Klibanov, A. M. 1990. Acc. Chem. Res. **23:** 114.
6. Patil, D. R., D. G. Rethwisch & J. S. Dordick. 1991. Biotechnol. Bioeng. **37:** 639.
7. Carrea, G., S. Riva, F. Secundo & B. Danielli. 1989. J. Chem. Soc. Perkins Trans.: 1057.
8. Patil, D. R., J. S. Dordick & D. G. Rethwisch. 1991. Macromolecules. **24:** 3462.
9. Popp, J. L., T. K. Kirk & J. S. Dordick. 1991. Enzyme Microb. Technol. **13:** 964.
10. Otake, T. 1972. Bull. Chem. Soc. Jpn. **43:** 3199.
11. Ballard, J. M., L. Hough & A. C. Richardson. 1980. Carbohydr. Res. **83:** 138.
12. Hough, L., S. D. Phadnis & E. Tarelli. 1975. Carbohydr. Res. **44:** 37.
13. G. Odier. 1981. Principles of Polymerization. 2nd Edit. Wiley Interscience. New York, N.Y.
14. Koskikallio, J. 1969. *In* The Chemistry of Carboxylic Acids and Esters. S. Patai, Ed. (Chapter 3). Interscience. London, England.
15. Whitehouse, A. A. K., E. G. K. Pritchett & G. Barnett. 1967. Phenolic Resins: 6–50. American Elsevier. New York, N.Y.
16. Clary, J. J., J. E. Gibson & R. S. Waritz, eds. 1983. Formaldehyde: Toxicology, Epidemiology, Mechanisms. Marcel Dekker. New York, N.Y.
17. Ryu, K. & J. S. Dordick. 1990. Resources, Conserv. Recycl. **3:** 177.
18. Pokora, A. R. & W. L. Cyrus. 1987. U.S. Patent 4,647,952.
19. Sarkanen, K. V. & C. H. Ludwig, Eds. 1971. Lignins. Wiley. New York, N.Y.
20. Tock, R. W., R. S. J. Chen & C. R. Richardson. 1987. Chem. Eng. Commun. **56:** 229.
21. Obst, J. R. & T. K. Kirk. 1988. Methods Enzymol. **161:** 3.
22. Ryu, K. & J. S. Dordick. 1989. J. Am. Chem. Soc. **111:** 8026.

Factors Affecting Performance of a Proteinase-Catalyzed Bioreactor for Oligopeptide Syntheses

RYUICHI MATSUNO, YUKITAKA KIMURA,
SHUJI ADACHI, AND KAZUHIRO NAKANISHI[a]

Department of Food Science and Technology
Faculty of Agriculture
Kyoto University
Sakyo-ku
Kyoto 606-01, Japan

[a]*Department of Biotechnology*
Faculty of Engineering
Okayama University
Tsushima-naka
Okayama 700, Japan

INTRODUCTION

Oligopeptides are physiologically, nutritionally, and physicochemically different from free amino acids and proteins. Many of both endogeneous and exogeneous oligopeptides with physiological activity have been discovered so far, but they are confined to a relatively short length, consisting of less than 5 amino acid residues. Among the endogeneous oligopeptides, those of food origin are very interesting. Physiologically active oligopeptides exist in the primary structure of food protein and show their activity concomitant with the digestion by poteinases.[1] There is also the possibility of direct absorption of such oligopeptides at the intestinal tract.[2] The high assimilation rate at the intestinal tract and the lower osmotic pressure of oligopeptides than those of their constituent amino acids encourage us to construct nutritionally functional oligopeptides. The sweetener Aspartame, a very famous organoleptic oligopeptide,[3] and a salty oligopeptide, ornithyltaurine,[4] also have been discovered. The antioxidant property of oligopeptides consisting of histidine is attractive,[5] and a surfactant property might be useful for fabrication of a variety of foods.

TABLE 1 shows the list of work done in our laboratory on oligopeptide syntheses. In these instances, each synthesis reaction had unique characteristics. The uniqueness resulted from differences in reaction type, reaction system, kind of enzyme, state of the enzyme, properties of substrate and product, and strategy of synthesis, etc. This means that we learned something different from each experience, which was either predictable or unpredictable. In this article various factors affecting the performance of proteinase-catalyzed oligopeptide syntheses are described, based upon our previous experience.

TABLE 1. Work on Oligopeptide Syntheses

Objective	Enzyme	Reaction	Remark	Reference
Precursor of Aspartame[a]	Thermolysin	Condensation	H_2O-AcOEt, I.E. long term continuous	6,7,8
Precursor of des Tyr[1]-Leu-enkephalin[b]	Thermolysin	Condensation	H_2O-AcOEt, free	11
Precursor of Leu-enkephalin[c]	Papain	Condensation	H_2O-AcOEt, I.E.	12
	α-Chymotrypsin			
	Thermolysin			
Z-GlyPheLeuNH_2	Thermolysin	Condensation	H_2O-AcOEt, I.E.	13
	α-Chymotrypsin	Transesterification	Two series column	
Z-Lys-LysNH_2	Trypsin	Transesterification	H_2O-Acetonitrile	
Oligopeptides consisting of essential amino acids + His	Thermolysin	Condensation	H_2O-AcOEt, I.E.	15
	α-Chymotrypsin	Transesterification		
	Papain			
Oligopeptide group of high Fischer ratio[d]	Thermolysin	Hydrolysis + adsorptive separation	Free	17
	Papain			

[a] Z-AspPheOMe.
[b] Z-GlyGlyPheLeuOEt.
[c] Z-TyrGlyGlyPheLeuOEt.
[d] (Val + Leu + Ile)/(Tyr + Phe), therapy of cirrhosis.

FACTORS AFFECTING DIPEPTIDE SYNTHESIS THROUGH CONDENSATION: THE SYNTHESIS OF ASPARTAME PRECURSOR

Z-AspPheOMe, the precursor of Aspartame, was synthesized from acid substrate Z-Asp and amine substrate PheOMe through condensation catalyzed by thermolysin in an H_2O-AcOEt system. The method for estimating the product yield and the knowledge necessary for designing the practical, continuous production process were from previous experiments.[6–8]

From extensive batchwise experiments in an aqueous-organic solvent biphasic system with various conditions, it was concluded that in this system, a yield was predictable based on kinetic constant, partition coefficients of reactants, volume ratio of organic to aqueous phases, and pH as suggested by Klibanov *et al.*[9] and Martinek and Semenov.[10]

The predicted dependence on pH of the equilibrium yield of synthesis of Z-AspPheOMe by thermolysin in a biphasic buffer-AcOEt system at 40°C was compared with an experimental one. The predicted result indicated that the optimum pH of synthesis in the biphasic system was lower than that in an aqueous phasic system. Increasing the volume ratio of organic solvent to aqueous phase, α, and increasing the amine substrate, PheOMe, increased the yield. The prediction coincided well with the experimental yield. The effect of organic solvent was also predictable, and it was proven theoretically and experimentally that ethyl acetate was superior to 1,2-dichloroethane and chloroform both in equilibrium yield and initial rate.

Information on initial velocity, final yield, and stability of an enzyme are necessary to construct a long-term continuous production process. The information obtained with the enzyme thermolysin immobilized onto mechanically strong Amberlite XAD 7 indicated that the range of pH should be kept between 5 and 6 during the continuous reaction. Since the partition coefficients of substrate and product between resin (aqueous) and bulk (organic solvent) phases are different, the concentrations of charged reactants in the resin phase, where the enzymes are, change with the progress of the reaction in the case of batch reaction. Or these concentrations change along the column in the continuous reaction with the column reactor. Consequently the pH in the resin phase changes. The pH in the resin phase was calculated at 0 and 90% conversions with $\alpha = 15$. In the calculation, Z-Asp concentration was fixed at 80 mM while that of PheOMe was changed. With the initial concentration of PheOMe at 200 mM or more, pH was kept between 5 and 6 during the conversion from 0 to 90%. Since optimal conditions were determined, the continuous reaction was carried out in a plug flow type reactor (PFR) with substrate concentrations of 80 mM for Z-Asp and 200 mM for PheOMe in water-saturated AcOEt with space velocity 0.95/hour.

Unfortunately, although 95% yield was attained at the initial stage, yield decreased to about 30% after 80 hours. We had not anticipated this low operational stability. The reasons for this were sought, and we came to the conclusion that the problem was related to enzyme properties as shown in TABLE 2.

Calcium ion is essential for stability of the enzyme, thermolysin, and an acid substrate, Z-Asp, removes Ca^{2+} from thermolysin by the chelating action of two carboxylic acid groups. This resulted in inactivation of the enzyme. The reactor was

PFR, so that the high concentration of Z-Asp at the inlet of the reactor resulted in the deactivation of the enzyme. This deactivation might spread toward the outlet during continuous operation. Two attempts were made to resolve this difficulty. The first was the addition of Ca^{2+} ion to the substrate solution, and the second was the use of a continuous stirred tank reactor (CSTR). Ca^{2+} addition was effective, but not greatly compared with CSTR used without addition of Ca^{2+}.

However, PFR is superior to CSTR with respect to efficiency of synthesis and ease of operation. Therefore, three attempts were made to improve operational stability in PFR. They were decrease in operational temperature, optimization of water concentration in AcOEt for highest reaction rate, and decrease in Z-Asp concentration. These attempts were effective.

However, we encountered another difficulty as shown in TABLE 2. The problem was related to a substrate property. An amine substrate, PheOMe, was hydrolyzed

TABLE 2. Difficulty Encountered in the Continuous Synthesis of the Precursor of Aspartame

A. Difficulty related to enzyme.
- Ca^{2+} Essential for stability.
- Z-Asp Removal of Ca^{2+} by chelating action, resulted in inactivation of enzyme.

Resolution
1. Ca^{2+} addition to substrate solution.
2. Use of CSTR.
3. Decrease in operational temperature.
4. Optimum water concentration for highest reaction rate.
5. Decrease in Z-Asp concentration.

B. Difficulty related to substrate.
1. Nonenzymatic hydrolysis of PheOMe.
2. Low solubility of Phe into AcOEt → precipitate → channeling.

Resolution
1. Avoid high pH.
2. Increase in enzyme activity (short reaction time).
   Increase in solubility of thermolysin by 5-M NaBr at immobilization of enzyme.
3. Washing of the column to remove precipitated Phe.

nonenzymatically to form Phe, which was less soluble in AcOEt. This formed a precipitate, which resulted in channeling of the reactor.

Use of an organic solvent system alone is adequate. High pH should also be avoided in order to reduce the rate of hydrolysis. Our first effort to overcome hydrolysis was to increase the activity of immobilized enzyme, since a short reaction time reduces the opportunity for hydrolysis. Solubility of thermolysin was not high. Therefore, it was increased with addition of 5-M NaBr at immobilization of the enzyme so as to obtain immobilized enzyme of high activity. This was also necessary to overcome a disadvantage of the organic solvent system itself. In a biphasic system, high product yield through the condensation reaction is attained only in conjunction with a decreasing rate of synthesis, because of low partition of substrates in the aqueous phase where the enzymes are. However, this attempt to increase the specific rate of reaction of enzyme was not sufficient for the long-term reaction, because a small amount of precipitated Phe gradually accumulated on the surface of immobilized enzyme during long-term continuous operation. Therefore, the column was washed every 24 hours first with buffer-saturated AcOEt and next with AcOEt-saturated buffer each for 15 minutes.

Finally, Z-AspPheOMe was continuously synthesized in a PFR at the substrate concentrations 40 mM for Z-Asp and 100 mM for PheOMe in AcOEt containing 2.5% Mes buffer and 0.5-mM $CaCl_2$ with intermittent washing as described. The space velocity was as high as 1.85/hour, and operational temperature was 25°C. The results were successful and the yield was maintained at more than 90% for at least 500 hours. In these operating conditions, the productivity for Z-AspPheOMe synthesis on the basis of the amount of the immobilized enzyme loaded was evaluated as 0.62 g/g wet immobilized enzyme per day.

FACTORS AFFECTING THE SYNTHESIS OF OLIGOPEPTIDES HAVING MULTIPLE PEPTIDE BONDS SUSCEPTIBLE TO ENZYME ATTACK: SYNTHESIS OF DES-TYR¹-LEU-ENKEPHALIN PRECURSOR

In the previous example, the effect of organic solvent was dramatic in increasing product yield. Through synthesizing the precursor of des-Tyr^1-Leu-enkephalin, another advantage to the use of organic solvent was discovered, that is, that organic solvent increases not only yield but also the purity of the product.[11] The desired product was synthesized from Z-GlyGly and PheLeuOEt through a condensation reaction catalyzed by thermolysin in H_2O-AcOEt system. In this reaction system, the desired product was cleaved by thermolysin between Phe and Leu residues and the resultant Z-GlyGlyPhe condensed with PheLeuOEt to form an undesirable by-product. Through the use of organic solvent, by-product formation was reduced to obtain high product purity.

In the buffer system, many kinds of by-products—initiated by the cleavage of the product's peptide bond between Phe and Leu—were formed, and consequently, equilibrium product yield and purity were almost negligible. The yield and purity were improved dramatically in the biphasic system with a volume ratio of organic solvent phase to aqueous phase α of 1, and the substrate concentrations at 20 mM Z-GlyGly and 20 mM PheLeuOEt, because of extraction of desired product by AcOEt. With an increase in α to 15 and an increase in concentration of amine substrate PheLeuOEt three times as high as that of acid substrate, the yield and purity further increased to 90% and 97%, respectively.

EFFECT OF SUPPORT USED FOR IMMOBILIZATION OF ENZYME

Through these two experiments, the use of a support for immobilization of enzyme was shown to have a clear effect on the performance of peptide synthesis. Stabilization of enzyme in the organic solvent is also a very important factor. Other than providing stability, the support influences partition of reactants. If this effect is appropriately used, the product yield is enhanced dramatically. The third effect is mass transfer, which is a disadvantage to be avoided.

TABLE 3. Effect of Support for Immobilization of Enzyme on Reaction Performance in the Synthesis of Leu-Enkephalin Precursor[a]

	Free Enzyme	Immobilized Enzyme	
	Biphasic System (C_w/C_{org})	Buffer System (C_{in}/C_w)	Organic Solvent System (C_{in}/C_{org})
Z-TyrGlyGly	1.2	35	2.1
PheLeuOEt	0.57	16	7.5
Z-TyrGlyGlyPheLeuOEt	1.2×10^{-3}	$>8.1 \times 10^3$	1.0
Product yield	0.3–3.4%	40%[b]	20–60%

[a] C_w, C_{org}, and C_{in}: concentrations in aqueous, organic solvent, and solid phases.
[b] Formed in solid phase.

Partition Effect: Synthesis of the Leu-Enkephalin Precursor

The precursor of Leu-enkephalin, Z-TyrGlyGlyPheLeuOEt, was synthesized from Z-TyrGlyGly and PheLeuOEt through condensation catalyzed by thermolysin in an H_2O-AcOEt system.[12] A difficulty in this reaction system was that an acid substrate, Z-TyrGlyGly, was hydrolyzed by thermolysin to form Z-Tyr. Synthesis using free enzyme even in a biphasic system was unsuccessful. However, use of immobilized thermolysin dramatically increased the product yield. The reason is clearly shown in TABLE 3 as the effect of the support on the partition of reactants. The support for immobilization of enzyme was a rather hydrophobic resin, Amberlite XAD-7. As shown in the bottom row, the product yield is very low for free enzyme even in the biphasic system. In contrast, the yield improved significantly for immobilized enzyme. Product was not found in the buffer phase but in the solid phase for the buffer system. The most striking feature in the table is the high partition of PheLeuOEt in the solid phase where the buffer and enzyme are. High partition of PheLeuOEt in the solid phase caused an enhancement of condensation reaction rate while the rate of Z-TyrGlyGly hydrolysis was lowered due to inhibition by a high concentration of PheLeuOEt.

Mass Transfer Effect: Synthesis of Z-GlyPheLeuNH$_2$

Z-GlyPheLeuNH$_2$ was synthesized from Z-GlyPheOMe and LeuNH$_2$ through transesterification catalyzed by immobilized α-chymotrypsin onto Dowex MSC-1 in an H_2O-AcOEt system.[13]

FIGURE 1 shows the dependencies of the concentration of water in the resin and initial activity of enzyme on the concentration of water in the solution phase. The equilibrium relations of water for resin without enzyme are the same as those for IMC and freeze-dried IMC. Saturation water concentration in AcOEt was 32.4 μl/ml. Just above the saturation water content, water concentration in resin increases rapidly. An increase in enzyme activity with an increase in water concentration in the low water concentration range is reasonable. The sharp decreases in activity just above the saturation water concentration cannot be ascribable to the increase of water concentration as a nucleophile, but can be ascribable to mass transfer resistance because of the weak dependence of enzyme activity on temperature, the recovery of enzyme activity by grinding of immobilized enzyme, and the low partition of acid substrate, Z-GlyPheOMe, to the aqueous phase where the enzymes are.

FIGURE 2 shows the explanation of mass transfer resistance. The left figure shows the scheme in the pore of the support at low water concentrations, while right is at high water concentrations. As shown in the figure, partition of Z-GlyPheOMe was in favor of the organic solvent phase. At low water concentration, the organic solvent phase might penetrate into the pore and be continuous, while at high water concentrations, the organic solvent phase might decrease in the pore and be discontinuous. Because of very low partition of Z-GlyPheOMe in the water phase, mass transfer of Z-GlyPheOMe from bulk phase to enzyme might be difficult in high water concentrations.

FACTORS AFFECTING THE SYNTHESIS OF TRIPEPTIDE PRECURSOR THROUGH SUCCESSIVE TRANSESTERIFICATION AND CONDENSATION REACTIONS: SYNTHESIS OF Z-GLYPHELEUNH$_2$

The tripeptide precursor Z-GlyPheLeuNH$_2$ was synthesized through condensation with immobilized thermolysin followed by transesterification with immobilized

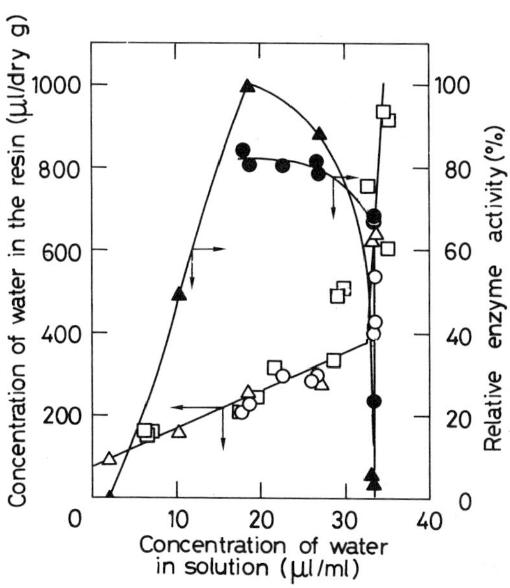

FIGURE 1. Effect of concentration of water in reaction solution on enzyme activity for the synthesis of Z-GlyPheLeuNH$_2$ by immobilized α-chymotrypsin (IMC) and concentration of water in the resin (Dowex MSC-1): ○, △, and □ show the relationship between the concentrations of the water in solution and in the resin; ● and ▲ show the relative enzyme activity at various concentration of water. (○, ●) IMC, 1.54 μmol Z-GlyPheLeuNH$_2$/g wet IMC × minutes; (△, ▲) freeze-dried IMC; (□) resin without enzyme.

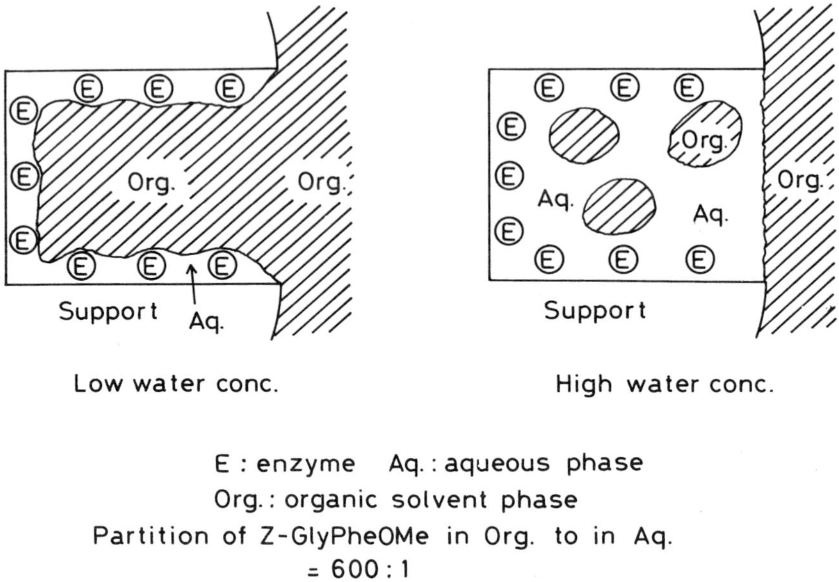

FIGURE 2. Illustration of mass transfer effect in the synthesis of Z-GlyPheLeuNH_2 by α-chymotrypsin immobilized onto Dowex MSC-1.

α-chymotrypsin in an H_2O-AcOEt system.[13] The first virtue of this system is the high reaction rate of transesterification. The rate of transesterification is 10 to 100 times greater than that of condensation. The second virtue is that the product of the first reaction can be used directly for the second reaction without removing methoxy group and without a separation step. The scheme of the continuous and successive reaction system is shown in FIGURE 3. Thermolysin immobilized onto Amberlite XAD-7 was used for the first column, and α-chymotrypsin immobilized onto Dowex MSC-1 for the second one. The columns were made of glass with diameter 11 mm and length 150 mm. The first column loaded 5 g of immobilized thermolysin, and the second, 2.5 g of immobilized α-chymotrypsin. Ethyl acetate saturated with 50-mM Mes-NaOH buffer at pH 7 containing 5-mM $CaCl_2$ and substrate Z-Gly and PheOMe was used as feed solvent. AcOEt containing LeuNH_2 and 1.6% water and the effluent from the first column were fed into the second column.

Difficulties in this reaction system are the formation of Z-GlyPhePheOMe in the second reaction from Z-GlyPheOMe and PheOMe, which is the unreacted substrate in the first reaction, and the removal of the amide group of the final product. The first problem was overcome by adopting an excess concentration of LeuNH_2 in the second reactor. The second problem might be resolved by constructing an amidase reactor.[14] Continuous synthesis was carried out under following conditions.

First column: Concentrations of substrates in water-saturated AcOEt, Z-Gly 40 mM, and PheOMe 80 mM; immobilized thermolysin, 5.0 g; and retention time, 6.8 hours.

Second column: Concentrations of substrates in AcOEt with water 18 μl/ml,

Z-GlyPhyOMe ca. 5 mM, and $LeuNH_2$ 25 mM; immobilized α-chymotrypsin 2.5g; and retention time, 19 minutes.

Byproduct Z-GlyPhePheOMe formation was negligible, and the yield was maintained at more than 80% for at least 200 hours.

FACTORS AFFECTING THE SYNTHESIS OF PEPTIDE GROUP HAVING DEFINITE CHARACTERISTICS: SYNTHESIS OF OLIGOPEPTIDES CONSISTING OF ESSENTIAL AMINO ACIDS CONTAINING HISTIDINE WITH A FEW TYPES OF ENZYMES

Finally, we would like to point out a rather peculiar factor. The objective of synthesis might be important. For instance, for the synthesis of physiologically active oligopeptides, the number, kind, and sequence of amino acids must be strictly specified. However, for the synthesis of oligopeptides for food use as a nutritional substance, the specification is not strict.

In this context, oligopeptides consisting of essential amino acids containing histidine were synthesized with a few types of enzymes.[15] These syntheses were challenged because of the superiority of oligopeptides to the constituent amino acids, that is, high assimilation rate at the brush border membrane of epithelial cells in intestinal tract[16] and low osmotic pressure.

For design of the syntheses of these oligopeptides, restriction conditions might be as follows: the first is estimates of amino acid requirements of adults by FAO/WHO/UNU in 1985; the second, taste; and the third, solubility. In this investigation, only the first restriction was taken into consideration. Through experience, difficulty was found in the formation of peptide bonds between branched chain amino acid residues. Taking into account these recommendations and enzyme specificity, we first aimed at syntheses of three oligopeptide precursors: Z-ThrTrpOEt, Z-LysIleVal-NH_2, and Z-MetLeuPheOMe. However, because of difficulty in making peptide

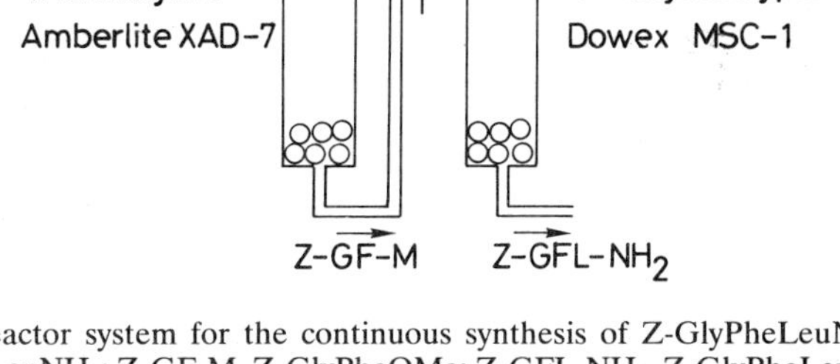

FIGURE 3. Reactor system for the continuous synthesis of Z-GlyPheLeuNH_2 from Z-Gly, PheOMe, and $LeuNH_2$: Z-GF-M, Z-GlyPheOMe; Z-GFL-NH_2, Z-GlyPheLeuNH_2.

TABLE 4. Maximum Yields of Syntheses of Peptides Consisting of Essential Amino Acids Attained So Far with a Free Proteinase in a Biphasic System or with an Immobilized Proteinase in an Organic Solvent[a]

Reaction Product	Yield (%)		Enzyme
	Free E[b]	IME[c]	
1. Z-Lys(Z)IleOMe	≃100	≃100	Thermolysin
Z-ThrTrpOEt	73	43	Papain
2. Z-ThrTrpValNH$_2$	≃100	≃100	α-Chymotrypsin
Z-LeuPheOMe	≃100	≃100	Thermolysin
3. Z-MetLeuPheOMe	≃100	≃100	Thermolysin
4. Boc-His(Tos)ValOEt	82	91	Papain

[a]The Ministry of Health and Welfare, Japan, has not yet recommended L-histidine as an essential amino acid for adults.
[b]Enzyme.
[c]Immobilized enzyme.

bonds between branched side chain amino acids, that is, between Ile and Val, these syntheses were abandoned. Then the syntheses of the second most desirable set of oligopeptides were aimed at. The results are summarized in TABLE 4.

The results of synthesis of a peptide containing histidine are also shown in the table, since a lot of evidence was accumulated for histidine to be included as an essential amino acid. As shown in TABLE 4, although we need further optimization, results are successful for both free and immobilized enzymes. This system of producing oligopeptides consisting of essential amino acids by a few types of enzyme reactors might become an example of the so-called multiproduct process with low volume, which is the recent trend in biochemical processes.

In this article, we describe the various factors affecting performance of bioreactor for oligopeptide syntheses, which were gleaned from unique previous experiences. Some were predictable factors, and others, unpredictable. In future, further experience will be necessary to uncover more unpredictable factors.

REFERENCES

1. YOSHIKAWA, M. & H. CHIBA. 1990. Physiologically active peptides of food origin (in Japanese). Shokuhin Kogyo (Food Industry) **33:** 20–25.
2. SUETSUNA, K. & K. OSAJIMA. 1989. Blood pressure reduction and vasodilatory effects in vivo of peptides originating from sardine muscles. J. Jpn Soc. Nutr. Food Sci. **42:** 47–54.
3. OYAMA, K., S. NISHIMURA, Y. NONAKA, K. KIHARA & T. HASHIMOTO. 1981. Synthesis of an aspartame precursor by immobilized thermolysin in an organic solvent. J. Org. Chem. **46:** 5242–5244.
4. TADA, M., I. SHINODA & H. OKAI. 1984. L-Ornithyltaurine, a new salty peptide. J. Agric. Food Chem. **32:** 992–996.
5. KOHEN, R., Y. YAMAMOTO, K. C. CUNDY & B. N. AMES. 1988. Antioxidant activity of carnosine, homocarnosine, and anserine present in muscle and brain. Proc. Natl. Acad. Sci. USA **85:** 3175–3179.
6. NAKANISHI, K., T. KAMIKUBO & R. MATSUNO. 1985. Continuous synthesis of *N*-(benzyloxycarbonyl)-L-asparatyl-L-phenylalanine methyl ester with immobilized thermolysin in an organic solvent. Bio/Technology **3:** 459–464.
7. NAKANISHI, K., Y. KIMURA & R. MATSUNO. 1986. Kinetics and equilibrium of enzymatic synthesis of peptides in aqueous/organic biphasic system. Thermolysin-catalyzed synthe-

sis of *N*-(benzyloxycarbonyl)-L-asparatyl-L-phenylalanine methyl ester. Eur. J. Biochem. **161:** 541–549.

8. NAKANISHI, K., A. TAKEUCHI & R. MATSUNO. 1990. Long-term continuous synthesis of aspartame precursor in a column reactor with immobilized thermolysin. Appl. Microbiol. Biotechnol. **32:** 633–636.
9. KLIBANOV, A. M., G. P. SAMOKIN, K. MARTINEK & I. V. BEREZIN. 1977. A new approach to preparative enzymatic synthesis. Biotechnol. Bioeng. **19:** 1351–1361.
10. MARTINEK, K. & A. N. SEMENOV. 1981. Enzymatic synthesis in biphasic aqueous organic systems. II. Shift of ionic equilibria. Biochim. Biophys. Acta **658:** 90–101.
11. NAKANISHI, K., Y. KIMURA & R. MATSUNO. 1986. Design of proteinase-catalyzed synthesis of oligopeptides in an aqueous-organic biphasic system. Bio/Technology **4:** 452–454.
12. KIMURA, Y., K. NAKANISHI & R. MATSUNO. 1990. Enzymatic synthesis of the precursor of Leu-enkephalin in water-immiscible organic solvent systems. Enzyme Microb. Technol. **12:** 272–280.
13. KIMURA, Y., T. YOSHIDA, K. MURAYA, K. NAKANISHI & R. MATSUNO. 1990. Continuous synthesis of a tripeptide by successive condensation and transesterification catalyzed by two immobilized proteinases in organic solvent. Agric. Biol. Chem. **54:** 1433–1440.
14. STEINKE, D. & M-R. KULA. 1990. Selective deamination of peptide amides. Angew. Chem. Int. Ed. Engl. **29:** 1139–1140.
15. KIMURA, Y., K. MURAYA, Y. ARAKI, H. MATSUOKA, K. NAKANISHI & R. MATSUNO. 1990. Synthesis of peptides consisting of essential amino acids by a reactor system using three proteinases and an organic solvent. Agric. Biol. Chem. **54:** 3331–3333.
16. GANAPATHY, V., G. BURCKHARDT & F. H. LEIBACH. 1984. Characteristics of glycylsarcosine transport in rabbit intestinal brush-border membrane vesicles. J. Biol. Chem. **259:** 8954–8959.
17. ADACHI, S., Y. KIMURA, K. MURAKAMI, R. MATSUNO & H. YOKOGOSHI. 1991. Separation of peptide groups with definite characteristics from enzymatic protein hydrolysis. Agric. Biol. Chem. **55:** 925–932.

Synthesis of Chiral Intermediates for D-Pantothenate Production by Microbial Enzymes

HIDEAKI YAMADA AND SAKAYU SHIMIZU

Department of Agricultural Chemistry
Kyoto University
Kitashirakawa, Sakyo-ku
Kyoto 606, Japan

INTRODUCTION

The commercial production of D-pantothenate depends exclusively on chemical synthesis. The conventional chemical process involves reactions yielding racemic pantoyl lactone from isobutyraldehyde, formaldehyde, and cyanide; optical resolution of the racemic pantoyl lactone to D-(−)-pantoyl lactone with expensive alkaloids, such as quinine, quinidine, cinchonidine, brucine, and so on; and condensation of D-(−)-pantoyl lactone with β-alanine. This is followed by isolation as the calcium salt and drying to obtain the final product. A problem of this chemical process apart from the use of poisonous cyanide is the troublesome resolution of the racemic pantoyl lactone and the racemization of the remaining L-(+)-isomer. Therefore, most of the recent studies in this area have concentrated on the development of an efficient method to obtain D-(−)-pantoyl lactone.

Here we report the results of our recent work on the enzymatic production of D-pantothenate, which includes the evaluation of useful enzymatic reactions for the synthesis, screening of potent microorganisms as catalysts, characterization of the responsible enzymes for the conversions, and determination of reaction conditions for practical preparation.[1–3]

SCREENING OF USEFUL MICROBIAL REACTIONS

Using several possible starting substrates (i.e., ketopantoyl lactone and its derivatives, and DL-pantoyl lactone) and a variety of microorganisms as catalysts, we searched for microbial conversions to useful chiral intermediates for D-pantothenate synthesis, and we found several useful reactions for skipping the above-mentioned resolution-racemization step. They are summarized in FIGURE 1.

ASYMMETRIC REDUCTION OF PROCHIRAL KETONES BY CARBONYL REDUCTASES

Recently we developed an efficient one-pot synthesis method of ketopantoyl lactone, in which it is synthesized from isobutyraldehyde, sodium methoxide, diethyl oxalate, and formalin (FIGURE 2). The reaction is performed in one step at room temperature with a yield of 81.0%.[4] Ketopantoyl lactone is a very promising starting material for the synthesis of D-(−)-pantoyl lactone, because it might permit several microbiological approaches leading to D-(−)-pantoyl lactone or D-(+)-pantothenate, as shown in FIGURE 1. Thus, we assayed a variety of microorganisms as to their reducing ability using several prochiral carbonyl compounds, such as ketopantoyl lactone, ketopantoic acid, ethyl 2′-ketopantothenate, and 2′-ketopantothenonitrile.

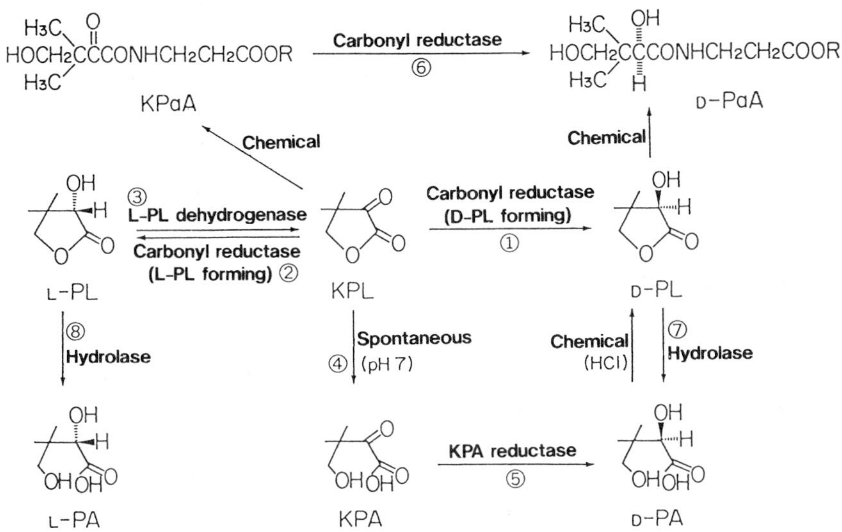

FIGURE 1. Reactions involved in the enzymatic transformation to D-(−)-pantoyl lactone or D-(+)-pantothenate. D-PL, D-(−)-pantoyl lactone; L-PL, L-(+)-pantoyl lactone; KPL, ketopantoyl lactone; KPA, ketopantoic acid; D-PA, D-(−)-pantoic acid; L-PA, L-(+)-pantoic acid; KPaA, 2′-ketopantothenate; D-PaA, D-(+)-pantothenate.

Conversion of Ketopantoyl Lactone to D-(−)-Pantoyl Lactone

This conversion (FIGURE 1, reaction 1) was assayed at pH 4 to 6 by incubating ketopantoyl lactone (10 mg/ml) in the culture broth, which had been grown with each test microorganism for 2 days at 28°C. Many microorganisms including bacteria, actinomycetes, yeasts, molds, and basidiomycetes converted the added ketopantoyl lactone to pantoyl lactone (FIGURE 3a). However, the ratios of D- and L-isomers of formed pantoyl lactone were randomly distributed among the strains tested and the stereospecificity shown by the tested strains almost completely showed no relation to the genera or sources. For example, *Mucor racemosus* produced almost specifically the L-isomer with more than 85% molar yield. On the other hand, *Mucor javanicus* gave a racemic mixture with 59% yield. Among 9 strains of *Rhodotorula glutinis,* which produced pantoyl lactone with greater than 90% molar yields, 5 gave racemic mixtures, 2 gave the L-isomer predominantly, and the remaining 2 gave the D-isomer with more than 70% ee.[5]

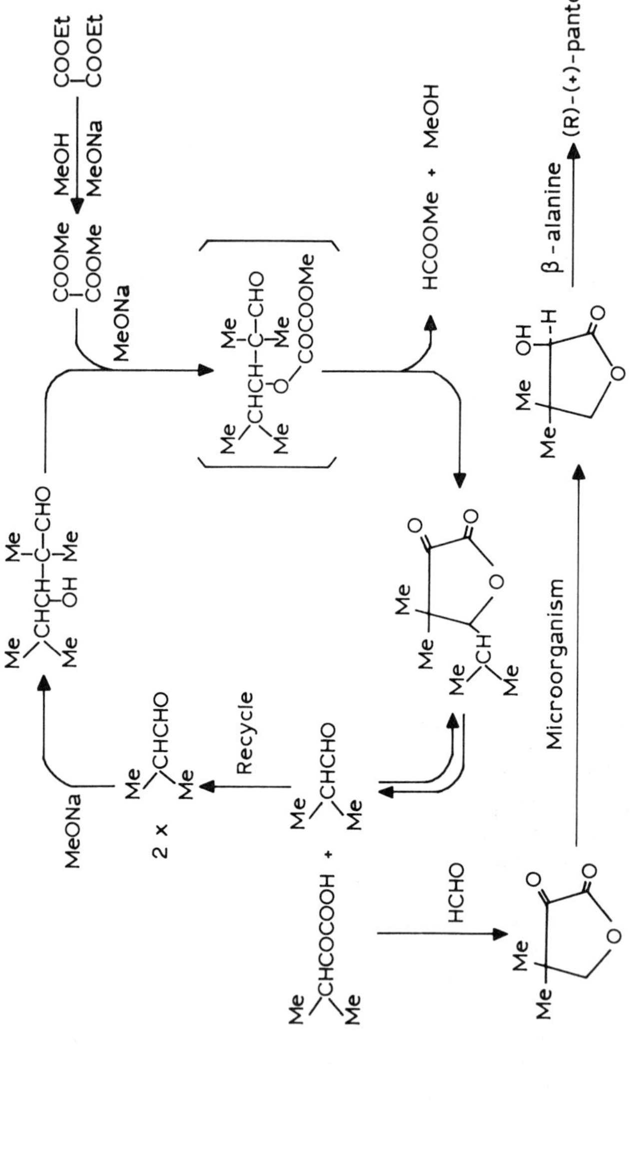

FIGURE 2. The proposed reaction pathway for the one-pot synthesis of ketopantoyl lactone from isobutyraldehyde, diethyl oxalate, and formalin.

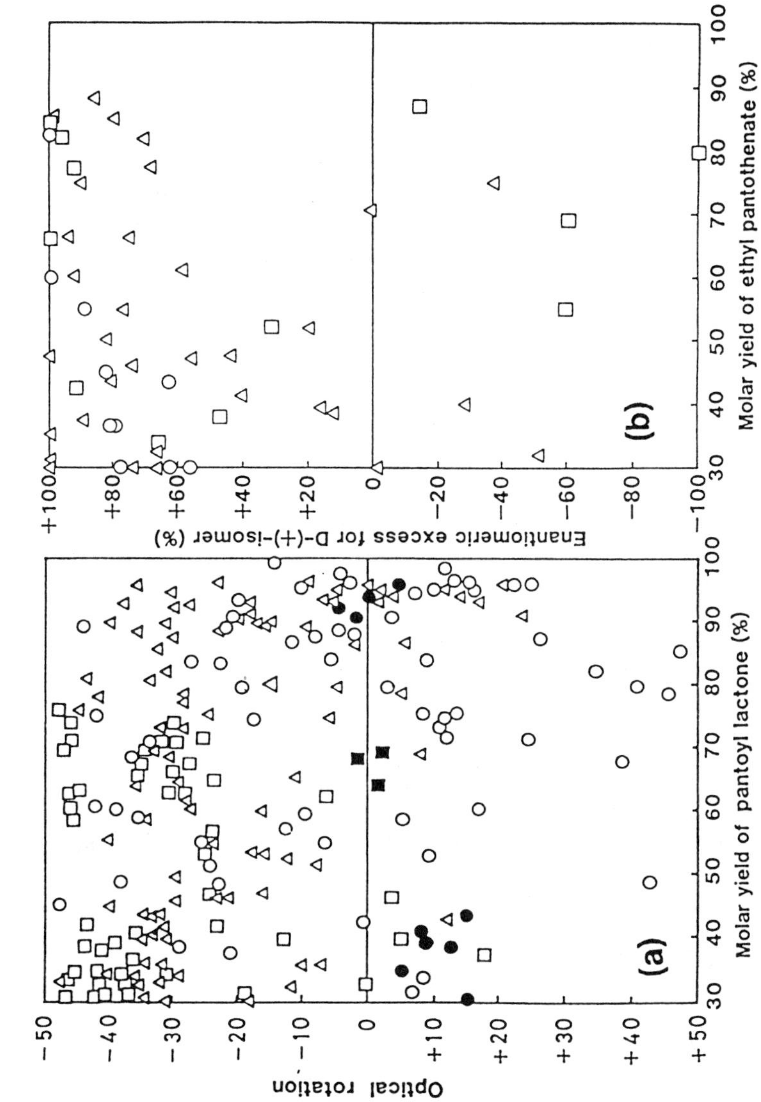

FIGURE 3. Diversity of microbial reduction of ketopantoyl lactone (a) and ethyl 2′-ketopantothenate (b). △, yeasts; ○, molds; □, bacteria; ■, actinomycetes; ●, basidiomycetes.

Practical stereospecific reduction of ketopantoyl lactone to D-(−)-pantoyl lactone was carried out with washed cells of *Candida parapsilosis* or *Rhodotorula minuta* as a catalyst and glucose as energy for the reduction. About 90 or 50 g/liter of D-(−)-pantoyl lactone (94 or 98% ee, respectively) was produced with a molar yield of nearly 100% by *Candida parapsilosis* or *Rhodotorula minuta,* respectively (FIGURE 4).[4,6]

The enzyme catalyzing the asymmetric reduction of ketopantoyl lactone was isolated in a crystalline form from the cells of *Candida parapsilosis* and characterized in some detail[7] (also see TABLE 1). It is a novel NADPH-dependent carbonyl reductase with a molecular mass of about 40,000. In addition to ketopantoyl lactone, the enzyme catalyzed reductions of a variety of cyclic diketones, including derivatives of ketopantoyl lactone, isatin, camphorquinone, and so on, and gave the corresponding *R*-alcohols (FIGURE 5).[7,8] We named the enzyme "conjugated polyketone reductase," since the enzyme catalyzes only the reduction of conjugated polyketones as follows,

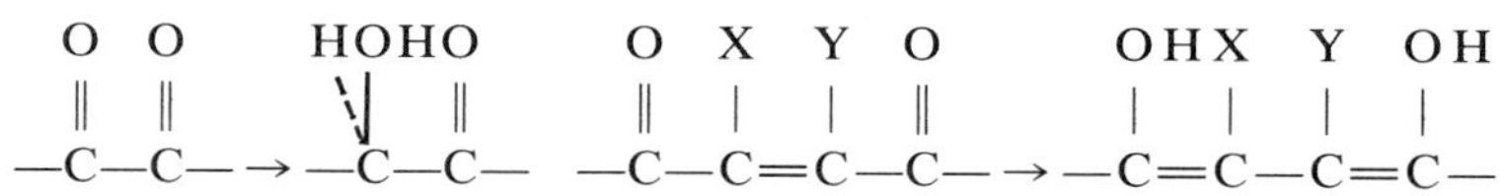

X,Y=H or alkyl

The enzyme yielding the antipode, L-(+)-pantoyl lactone, from ketopantoyl lactone (FIGURE 1, reaction 2), was also isolated from *Mucor ambiguus* cells.[9] It is also a kind of "conjugated polyketone reductase" and consists of two polypeptide chains with an identical molecular mass of about 27,000 (see TABLE 1). The occurrence of two kinds of enzymes that show similar substrate specificity but are different from each other in their stereospecificity may be one of the possible reasons why the reduction of unnatural ketopantoyl lactone resulted in the formation of the D- and L-isomers in varying ratios as shown in FIGURE 3a.

Since the above-mentioned enzymes require NADPH for the reduction of ketopantoyl lactone, there must be regenerating reaction(s) for NADPH coupled with the reduction reaction in the cells. Comparative enzymatic studies using cell-free extracts of *Candida parapsilosis, Mucor ambiguus,* and baker's yeast (*Saccharomyces cerevisiae*) demonstrated that hexokinase, glucose 6-phosphate dehydrogenase, and 6-phosphogluconate dehydrogenase were involved in the regeneration of NADPH. Cellular levels of these enzymes and "conjugated polyketone reductase" in the former two microorganisms were considerably higher than those in baker's yeast. FIGURE 6 outlines the mechanism for this reduction.[10]

Conversion of Ketopantoic Acid to D-(−)-Pantoic Acid

The stereospecific reduction of ketopantoic acid to D-(−)-pantoic acid by ketopantoic acid reductase (EC 1.1.1.169) (FIGURE 1, reaction 5), which is involved in the pantothenate biosynthesis pathway, is also a promising reaction for the same purpose, because the ring opening of ketopantoyl lactone to yield ketopantoic acid is easy and ketopantoic acid reductase shows absolute stereospecificity for D-(−)-pantoic acid.[11] When the same screening as described above was performed at pH 7

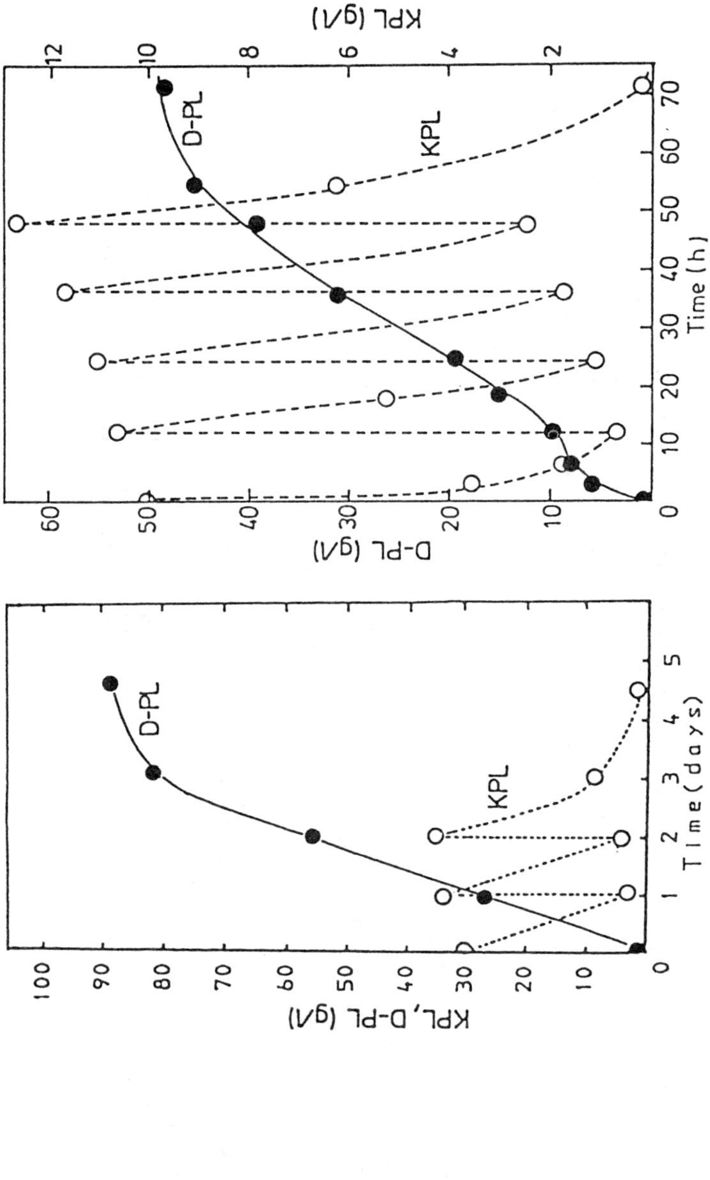

FIGURE 4. Conversion of ketopantoyl lactone (KPL) to D-(−)-pantoyl lactone (D-PL) by *Candida parapsilosis* (left) and *Rhodotorula minuta* (right).

TABLE 1. Properties of the Carbonyl Reductases and Alcohol Dehydrogenase Purified from Various Microorganisms[a]

	Polyketone Reductase (*C. parapsilosis*)	Polyketone Reductase (*M. ambiguus*)	KPA Reductase (*P. maltophilia*)	K-PaOEt Reductase (*C. macedoniens*)	L-PL Dehydrogenase (*N. asteroides*)
Native M_r	37,000	56,000	116,000	45,000	600,000
Subunit M_r	41,600	27,000	30,500	42,000	42,000
$s_{20,w}$ (S)	4.8	—	7.75	—	—
pI	6.3	6.4	3.5	5.5	4.3
Absorption maximum (nm)	278	276	276	278	276, 370, 456
E (1%)	8.3	—	20.0	—	—
COOH amino acid	—	—	Phe	Thr	—
K_m (mM)	0.33 (KPL)	0.71 (KPL)	0.40 (KPL)	2.50 (K-PaOEt)	26.8 (L-PL)
V_{max} (μmol/min/mg)	481 (KPL)	541 (KPL)	1,310 (KPL)	120 (K-PaOEt)	—
Cofactor	NADPH	NADPH	NADPH	NADPH	FMN
Optimum pH	7.0	6.0	6.0	6.5	9.0–9.5
Optimum temperature (°C)	40	40	37	40	40
pH stability	6.0–7.5	5.5–7.0	6.0–10	4.5–10.5	9.0–10
Thermal stability	42% (40°C, 10 min)	75% (45°C, 10 min)	90% (60°C, 10 min)	100% (55°C, min)	60% (45°C, 10 min)
Inhibitor	Quercetin	Quercetin	—	—	—
Reaction mechanism	—	—	Ordered Bi-Bi	—	—
Enzyme formation	Constitutive	Constitutive	Constitutive	Constitutive	Inducible

[a]KPA, ketopantoic acid; KPL, ketopantoyl lactone; K-PaOEt, ethyl 2′-ketopantothenate; L-PL, L-(+)-pantoyl lactone.

to 8, under which the added ketopantoyl lactone underwent rapid and spontaneous hydrolysis to ketopantoic acid (FIGURE 1, reaction 4), we observed a quite different distribution profile of the reducing activity. As expected, most of the microorganisms that showed high reducing activity produced only the D-isomer. Through this screening, we found that many bacteria belonging to the genus *Agrobacterium* almost specifically produce the D-isomer (>96% ee) in high yields. On incubation with a soil isolate, *Agrobacterium* sp. S-246, and fructose as an energy source for the reduction, the yield of D-(−)-pantoic acid reached 119 g/liter (molar yield, 90%; optical purity, >98% ee).[12]

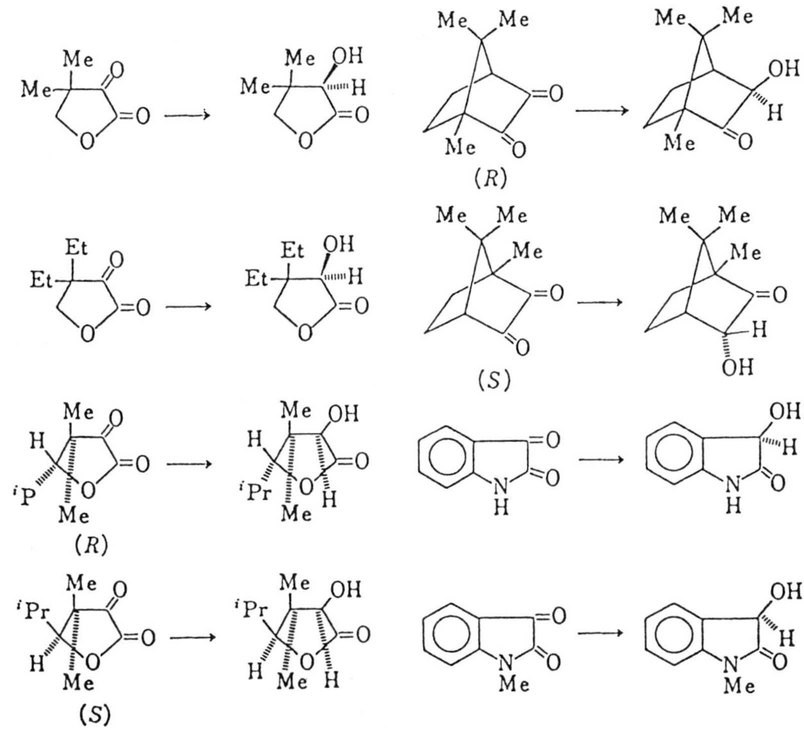

FIGURE 5. Reduction of various conjugated diketones by *Candida parapsilosis* carbonyl reductase.

Ketopantoic acid reductase was isolated in a crystalline form from one of the potent D-(−)-pantoic acid producers, *Pseudomonas maltophilia,* and characterized in some detail[11] (see also TABLE 1). It is an NADPH-dependent enzyme and is almost strictly specific to ketopantoic acid. The observation that mutants lacking this enzyme require either D-(−)-pantoic acid or pantothenic acid for growth and the revertants regain this activity indicates that it is involved in the pantothenate biosynthesis.

Reduction of 2′-Ketopantothenate Derivatives

We recently found that the rate of condensation of ketopantoyl lactone or D-(−)-pantoyl lactone with ethyl β-alanine, which yields ethyl 2′-ketopantothenate or ethyl D-(+)-pantothenate, respectively, is quite fast in comparison with that of the condensation of ketopantoyl lactone or D-(−)-pantoyl lactone with β-alanine, and that it proceeds more stoichiometrically.[13] Since the enzymatic hydrolysis of ethyl D-(+)-pantothenate has already been established,[14] if the stereoselective reduction of ethyl 2′-ketopantothenate to ethyl D-(+)-pantothenate is possible, both the troublesome resolution and the incomplete condensation might be avoided. Thus, we assayed the reducing ability toward ethyl 2′-ketopantothenate (FIGURE 1, reaction 6) of a variety of microorganisms (see FIGURE 3b). Several yeast strains, such as *Pichia aganobii, Hansenula miso,* and *Candida macedoniensis,* exhibited high ability to reduce the 2′-ketopantothenate ester. Under optimal conditions, *Candida macedoniensis* converted ethyl 2′-ketopantothenate (80 g/liter) almost specifically to ethyl

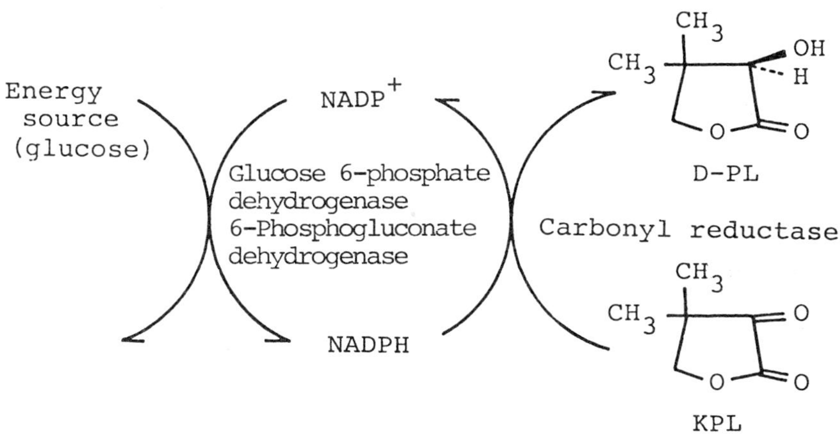

FIGURE 6. The proposed mechanism for the reduction of ketopantoyl lactone (KPL) to D-(−)-pantoyl lactone (D-PL) coupled with the regeneration of NADPH.

D-(+)-pantothenate (>98% ee), with a molar yield of 97.2%.[15] In a similar manner, 2′-ketopantothenonitrile (50 g/liter) was converted to D-(+)-pantothenonitrile (93.6% ee), with a molar yield of 95.6%, on incubation with *Sporidiobolus salmonicolor* cells as a catalyst.[16]

The enzyme catalyzing these conversions was isolated from *Candida macedoniensis* and characterized in some detail (see TABLE 1). The enzyme shows broad substrate specificity; not only conjugated polyketones, but also aromatic aldehydes and 4-haloacetoacetic acid esters are reduced. The enzyme gives only the D-isomer on reduction of ethyl 2′-ketopantothenate, whereas it gives a mixture of the D- and L-isomers in a ratio of 4:1 on reduction of ketopantoyl lactone. The reduction product from ethyl 4-chloroacetoacetate is ethyl (*S*)-4-chloro-3-hydroxybutanoate (95% ee).[17]

TABLE 2. Properties of Aldonolactonase of *Fusarium oxysporum*

Native M_r		K_m & V_m	
TSK G-3000SW	125,000	Hydrolysis (pH 7.0)	
Asahipak GS-520	120,000	D-Pantoyl lactone	120 mM, 653 units/mg
Subunit M_r (SDS-PAGE)	60,000	Lactonization (pH 6.0)	
Number of subunits	2	D-Pantoic acid	9.1 mM, 193 units/mg
E (1 mg/ml, 280 nm, 1 cm light path)	0.784	Optimum pH	Hydrolysis 7-7.5 Lactonization 6.0
pI	4.7		
N-Terminal amino acid	Ala	Optimum temperature	50°C
Metal content (Ca)	2.40 mol/mol	pH stability (30°C, 30 min)	2–11.5
Sugar content (as glucose)	15.4% (w/w)	Thermal stability (pH 7.0, 20 min)	50°C

These enzymatic methods are simple and require no racemization step, which is necessary for the conventional chemical resolution.

CONVERSION FROM RACEMIC PANTOYL LACTONE BY STEREOSPECIFIC OXIDATION AND REDUCTION ENZYMES

Racemic pantoyl lactone can also be used as a starting substrate.[18,19] We found that *Nocardia asteroides* cells specifically oxidize the L-(+)-isomer in a racemic mixture of pantoyl lactone to ketopantoyl lactone, which is then converted to D-(−)-pantoyl lactone by the reduction with *Candida parapsilosis* cells as described above. In these two enzymatic steps, the coexisting D-(−)-isomer remains without any modification (FIGURE1, reactions 3 and 1). Under suitable conditions, 72 g/liter of D-(−)-pantoyl lactone was obtained from 80 g/liter of DL-pantoyl lactone.[18] *Agrobacterium radiobacter* can also be used in place of *Candida parapsilosis* (yield from 80 g/liter DL-pantoyl lactone, 71 g/liter)[19] (FIGURE 1, reactions 3 and 5). Similar specific oxidation and reduction reactions can also be carried out with a single microorganism as catalyst.[20] On incubation with washed cells of *Rhodococcus erythropolis*, D-(−)-pantoyl lactone in the reaction mixture reached 18.2 g/liter with a molar yield of 90.5% (optical purity, 94.4% ee). This unique conversion proceeds through the successive reactions as follows: (1) the enzymatic oxidation of L-(+)-

TABLE 3. Substrate Specificity of *Fusarium* Aldonolactonase

Substrate	K_m (mM)	V_m (units/mg)	V_m/K_m
D-Pantoyl lactone	120	653	5.44
D-Galactono lactone	3.6	1,440	400
L-Mannono lactone	23	1,300	56.5
D-Gulono lactone	29	1,120	38.6
D-Ribono lactone	2.5	307	123
α-D-Glucoheptanoic γ-lactone	118	1,400	11.9
D-Glycero-L-manno heptanoic γ-lactone	167	2,200	13.2

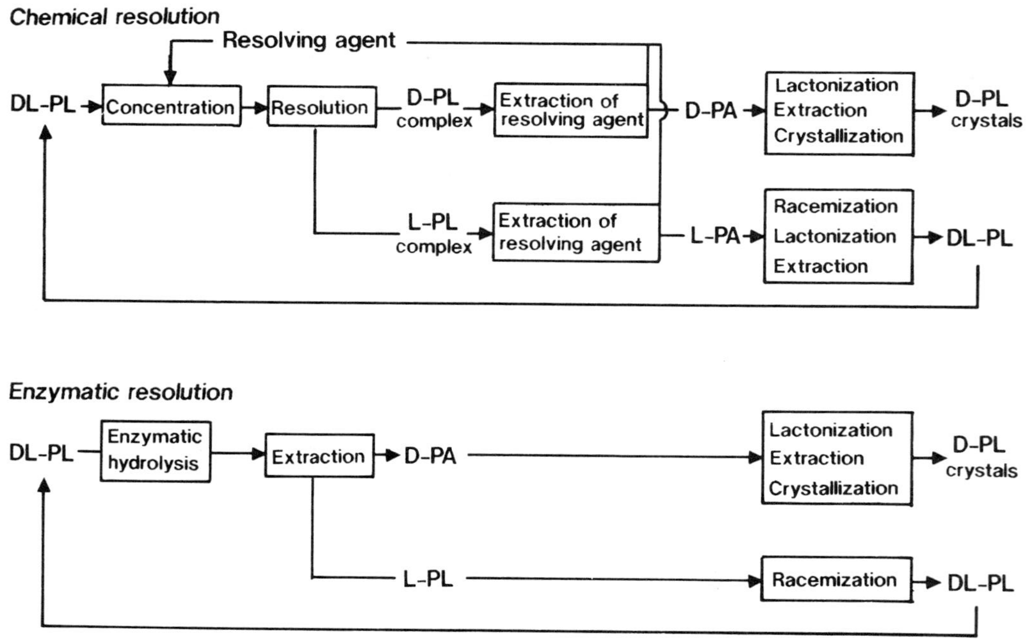

FIGURE 7. Comparison of enzymatic and conventional chemical resolution processes for DL-pantoyl lactone. DL-PL, DL-pantoyl lactone; D-PL, D-(−)-pantoyl lactone; L-PL, L-(+)-pantoyl lactone; D-PA, D-(−)-pantoic acid; L-PA, L-(+)-pantoic acid.

pantoyl lactone to ketopantoyl lactone, the same enzyme as that in *Nocardia asteroides,* has been suggested to be the responsible enzyme for this oxidation; (2) the rapid and spontaneous hydrolysis of ketopantoyl lactone to ketopantoic acid; and (3) the enzymatic reduction of the ketopantoic acid to D-(−)-pantoic acid (see FIGURE 1, reactions 3, 4, and 5). The enzyme catalyzing this reduction is assumed to be ketopantoic acid reductase, because *Rhodococcus erythropolis* cells could not utilize ketopantoyl lactone as substrate, unlike *Candida parapsilosis.*

The enzyme catalyzing oxidation of L-(+)-pantoyl lactone to ketopantoyl lactone was purified from *Nocardia asteroides* cells. It is a novel membrane-bound flavoprotein with a molecular mass of about 600,000. Noncovalently bound FMN was identified as the responsible cofactor for the oxidation[21] (see TABLE 1).

OPTICAL RESOLUTION OF RACEMIC PANTOYL LACTONE BY A SPECIFIC HYDROLASE

Kinetic resolution of unmodified DL-pantoyl lactone can be carried out by specific fungal hydrolases (FIGURE 1, reaction 7 or 8). We found that many mold strains belonging to the genera *Fusarium, Giberella,* and *Cylindrocarpon* specifically hydrolyze D-(−)-pantoyl lactone to D-pantoic acid. On the other hand, several yeast strains hydrolyzed only the L-isomer. For practical purposes, the former reaction (i.e., reaction 7 in FIGURE 1) is more advantageous than the latter (i.e., reaction 8 in FIGURE 1), because, in the latter case, optical purity of the remaining D-(−)-pantoyl lactone is low unless the hydrolysis of L-(+)-pantoyl lactone is complete.

When *Fusarium oxysporum* cells were incubated in 70% (w/v) aqueous solution of DL-pantoyl lactone for 24 hours at 30°C with automatic pH control (pH 6.8–7.2), about 90% of the D-isomer was hydrolyzed. The resultant D-(−)-pantoic acid in the reaction mixture showed a high optical purity (96% ee), and the coexisting L-isomer remained without any modification.[3]

The enzyme responsible for this hydrolysis was isolated from *Fusarium oxysporum* cells and crystallized. It is a kind of aldonolactonase with a molecular mass of 125,000 and contains about 2 mol of Ca^{2+} per mol of enzyme (see TABLE 2). It catalyzes reversible hydrolysis of several sugar lactones, such as D-galactonolactone, L-mannonolactone, and D-gulonolactone. All the substrate lactones have a downward hydroxyl group at the 2-position, when the lactone rings are drawn according to the Haworth system (TABLE 3), and the corresponding enantiomers are competitive inhibitors.[22]

A comparison of the enzymatic process proposed by the present study and the conventional chemical process for the resolution of DL-pantoyl lactone is shown in FIGURE 7.

REFERENCES

1. YAMADA, H. & S. SHIMIZU. 1988. Microbial and enzymatic processes for the production of biologically and chemically important compounds. Angew. Chem. Int. Ed. Engl. **27:** 622–642.
2. SHIMIZU, S. & H. YAMADA. 1989. Pantothenic acid (vitamin B_5), coenzyme A and related compounds. *In* Biotechnology of Vitamins, Pigments, and Growth Factors. E. J. Vandamme, Ed.: 199–219. Elsevier Applied Science. London, England.
3. SHIMIZU, S. & H. YAMADA. 1990. Stereoselective synthesis of biologically and pharmacologically important chemicals with microbial enzymes. *In* Chemical Aspects of Enzyme

Biotechnology: Fundamentals. T. O. Baldwin, F. M. Raushel & A. I. Scott, Eds.: 151–163. Plenum Press. New York, N.Y.

4. SHIMIZU, S., H. YAMADA, H. HATA, T. MORISHITA, S. AKUTSU & M. KAWAMURA. 1987. Novel chemoenzymatic synthesis of D-(−)-pantoyl lactone. Agric. Biol. Chem. **51:** 289–290.
5. SHIMIZU, S., H. HATA & H. YAMADA. 1984. Reduction of ketopantoyl lactone to D-(−)-pantoyl lactone by microorganisms. Agric. Biol. Chem. **48:** 2285–2291.
6. HATA, H., S. SHIMIZU & H. YAMADA. 1987. Enzymatic production of D-(−)-pantoyl lactone from ketopantoyl lactone. Agric. Biol. Chem. **51:** 3011–3016.
7. HATA, H., S. SHIMIZU, S. HATTORI & H. YAMADA. 1989. Ketopantoyl-lactone reductase from *Candida parapsilosis:* purification and characterization. Biochim. Biophys. Acta **990:** 175–181.
8. HATA, H., S. SHIMIZU, S. HATTORI & H. YAMADA. 1990. Stereoselective reduction of diketones by a novel carbonyl reductase from *Candida parapsilosis.* J. Org. Chem. **55:** 4377–4380.
9. SHIMIZU, S., S. HATTORI, H. HATA & H. YAMADA. 1988. A novel fungal enzyme, NADPH-dependent carbonyl reductase, showing high specificity to conjugated polyketones: purification and characterization. Eur. J. Biochem. **174:** 37–44.
10. KATAOKA, M., Y. NOMURA, S. SHIMIZU & H. YAMADA. 1992. Enzymes involved in the NADPH regeneration system coupled with asymmetric reduction of carbonyl compounds in microorganisms. Biosci. Biotechnol. Biochem. **56:** 820–821.
11. SHIMIZU, S., M. KATAOKA, M. C. M. CHUNG & H. YAMADA. 1988. Ketopantoic acid reductase of *Pseudomonas maltophilia* 845: purification, characterization, and role in pantothenate biosynthesis. J. Biol. Chem. **263:** 12077–12084.
12. KATAOKA, M., S. SHIMIZU & H. YAMADA. 1990. Novel enzymatic production of D-(−)-pantoyl lactone through the stereospecific reduction of ketopantoic acid. Agric. Biol. Chem. **54:** 177–182.
13. SAKAMOTO, K., S. KITA, T. MORIKAWA, S. SHIMIZU & H. YAMADA. 1989. Preparation of ethyl 2′-ketopantothenate and ethyl pantothenate. Japanese patent application H1-45407.
14. SHIMIZU, S., K. SAKAMOTO & H. YAMADA. 1987. Enzymatic hydrolysis of pantothenate esters (in Japanese). Nippon Nogeikagaku Kaishi **62:** 283.
15. KATAOKA, M., S. SHIMIZU, Y. DOI & H. YAMADA. 1990. Stereospecific reduction of ethyl 2′-ketopantothenate to ethyl D-(+)-pantothenate with microbial cells as a catalyst. Appl. Environ. Microbiol. **56:** 3595–3597.
16. KATAOKA, M., S. SHIMIZU, Y. DOI, K. SAKAMOTO & H. YAMADA. 1990. Microbial production of chiral pantothenonitrile through stereospecific reduction of 2′-ketopantothenonitrile. Biotechnol. Lett. **12:** 357–360.
17. KATAOKA, M., Y. DOI, T-S. SIM, S. SHIMIZU & H. YAMADA. 1992. A novel NADPH-dependent carbonyl reductase of *Candida macedoniensis:* purification and characterization. Arch. Biochem. Biophys. **294:** 469–474.
18. SHIMIZU, S., S. HATTORI, H. HATA & H. YAMADA. 1987. Stereoselective enzymatic oxidation and reduction system for the production of D-(−)-pantoyl lactone from a racemic mixture of pantoyl lactone. Enzyme Microb. Technol. **9:** 411–416.
19. KATAOKA, M., S. SHIMIZU & H. YAMADA. 1991. Stereospecific conversion of a racemic pantoyl lactone to D-(−)-pantoyl lactone through microbial oxidation and reduction reactions. Recl. Trav. Chim. Pays-Bas Netherlands **110:** 155–157.
20. SHIMIZU, S., S. HATTORI, H. HATA & H. YAMADA. 1987. One-step microbial conversion of a racemic mixture of pantoyl lactone to optically active D-(−)-pantoyl lactone. Appl. Environ. Microbiol. **53:** 519–522.
21. KATAOKA, M., S. SHIMIZU & H. YAMADA. 1992. Purification and characterization of a novel FMN-dependent enzyme, membrane-bound L-(+)-pantoyl lactone dehydrogenase, from *Nocardia asteroides.* Eur. J. Biochem. **204:** 799–806.
22. SHIMIZU, S., M. KATAOKA, K. SHIMIZU, M. HIRAKATA, K. SAKAMOTO & H. YAMADA. A novel lactonohydrolase, catalyzing the hydrolysis of aldonate lactones and aromatic lactones, from *Fusarium oxysporum:* purification and characterization. Eur. J. Biochem. (In press.)

Enzymatic Peptide Synthesis

Systematic Studies of Proteases[a]

ROBERT G. WHITTAKER, KATHERINE J. BRYANT,[b,c]
ELIZABETH A. HAMILTON, LOUISE T. McVITTIE,[b,d]
AND PAUL A. SCHOBER[b]

CSIRO Division of Biomolecular Engineering
Post Office Box 184
North Ryde, New South Wales 2113, Australia

[b]*Peptide Technology Limited*
4-10 Inman Road
Dee Why, New South Wales 2099, Australia

INTRODUCTION

The earliest suggestions that proteases could be used to synthesize peptide bonds were made by van't Hoff in 1898,[1] and in the 1930s, Bergmann and associates synthesized dipeptide products using the proteases papain and chymotrypsin.[2,3] In fact for many years proteases were thought to be the enzymes responsible for the *in vivo* biosynthesis of proteins and a great deal of research effort was directed towards elucidation of a possible mechanism. These ideas survived into the 1950s until the roles of m-RNA, t-RNA, and ribosomes were recognized.

A resurgence of interest in enzymatic synthesis occurred in the 1970s as a possible preparatory method for the increasing numbers of identified biologically active peptides, and since then enzymes have been used for the full or partial synthesis of a number of peptides.[4–9] Peptides as large as 29 amino acid residues have been produced by a combination of enzymatic and traditional chemical methods (Widmer *et al.* in the synthesis of a fragment of epidermal growth factor),[10,11] and this methodology is used commercially for the kilogram production of valuable peptides such as gonadotropin-releasing hormone.

Enzymatic peptide bond formation can occur by a condensation mechanism, the simple reversal of the proteolytic reaction as performed by Bergmann and Fraenkal-Conrat in 1937,[2] where the reaction is driven in the unfavored direction by removal of the synthesis product via precipitation or extraction into organic solvents. Bonds can also be formed by transpeptidation reactions, where a peptide or protein is cleaved by an enzyme in the presence of a large excess of another amino acid or peptide. For example porcine insulin can be converted to human insulin in this manner by the replacement of a single amino acid on the C-terminal of the B chain.[12,13]

[a] Grants from Peptide Technology Limited have supported some of these studies.

[c] Current affiliation: MAF Technology, Ruakara Agriculture Center, Private Bag, Hamilton, New Zealand.

[d] Current affiliation: Royal Alexandra Hospital for Children, Pyrmount Bridge Road, Camperdown, NSW 2050, Australia.

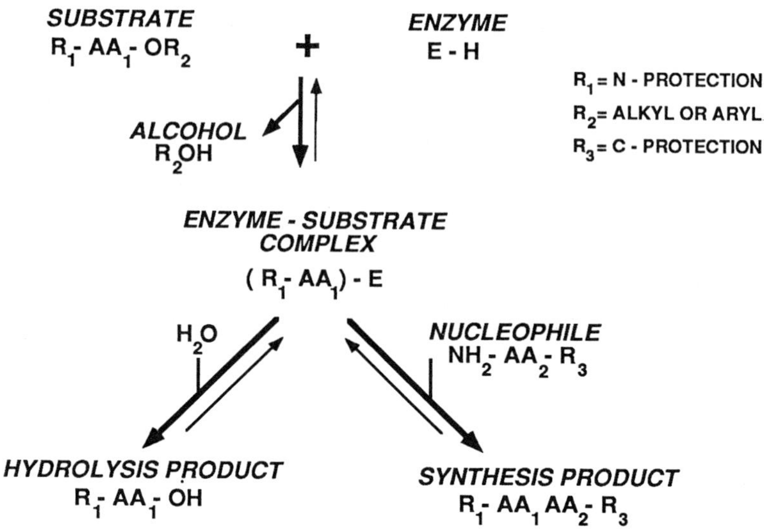

FIGURE 1. "Kinetic" enzymatic synthesis reaction scheme.

In our hands the most useful procedure of enzymatic peptide synthesis is the kinetic method, which takes advantage of the esterase activity of the serine and thiol proteases. As shown in FIGURE 1, the carboxyl group of an N-terminally protected amino acid or peptide is modified by esterification. This allows for rapid formation of a transient acyl-enzyme complex by a transesterification reaction. This complex is then cleaved, either by a water molecule to give an undesired hydrolysis product, or by nucleophilic attack by the amino group of a second amino acid or peptide to form a new peptide bond. The efficiency of the reaction is determined by the relative affinity of the nucleophile for the enzyme-substrate complex. Proteolytic degradation of the synthesis product is much slower than the synthetic step, and by stopping the reaction to take advantage of the kinetics, high synthetic yields are possible.

Kinetic synthesis has advantages over enzymatic condensation in that the reactions proceed much faster and therefore require less enzyme; the reactions do not require precipitation or extraction to drive them; and, because of the speed of the kinetic synthesis, it is often possible to join complex peptide substrates that contain sites normally susceptible to the enzyme on prolonged exposure.

The major limitations of enzymatic synthesis are the limited specificities of the available serine and thiol proteases and, although a number of studies have been carried out (e.g., References 14 and 15), the relatively sparse data on the use of these enzymes in the synthetic mode. Proteolytic preference is unfortunately only a guide to synthetic preference, necessitating an in-depth examination of the synthetic capability of each enzyme. Commencing in 1984, our research has concentrated on a systematic study of commonly available serine and thiol proteases (trypsin, chymotrypsin, papain, chymopapain, bromelain, elastase, subtilisin, protease K, endoproteinase Glu C, alpha lytic protease) to optimize their condition of use and to more accurately define the amino acids preferred on either side of the synthetic bond. As the volume of our accumulated studies is large, this paper concentrates on chymotrypsin as an example and generalizations are made for the other enzymes. Many of our the findings for the remaining enzymes have been presented at conferences[16–19] and will be published in the near future.

METHODS

The enzymes are studied under standard reaction conditions of 100 mM ester substrate and 200 mM nucleophile at pH 8.5, in 50% dimethylformamide (DMF) and 25°C unless otherwise specified; chymotrypsin 0.5 to 1.0 μM (note these are not optimal conditions). For each enzyme, an ester substrate and nucleophile are selected that give a reaction of approximately 60 to 80% under the standard conditions. Reactions are maintained on a pH stat (Radiometer) and reactions followed by high-performance liquid chromatography (HPLC, Waters) using standard C18 columns using pH 3 triethylamine phosphate:acetonitrile gradient systems.

Reactions are monitored for efficiency (the amount of synthesis product produced relative to the amount of ester substrate utilized) and reaction rate. Efficiencies can be determined throughout the course of a reaction, while rates are determined at the point of half ester substrate utilization under the standard reaction conditions; efficiencies quoted are as determined at the 50% point. FIGURE 2 shows a typical profile for the synthesis of benzoyl-tyrosylserine-ethyl ester (Bz-Tyr-Ser-OEt) by chymotrypsin from Bz-Tyr-OEt and Ser-OEt. As we are interested in overall reaction kinetics and characteristics, the rates are calculated at the reaction

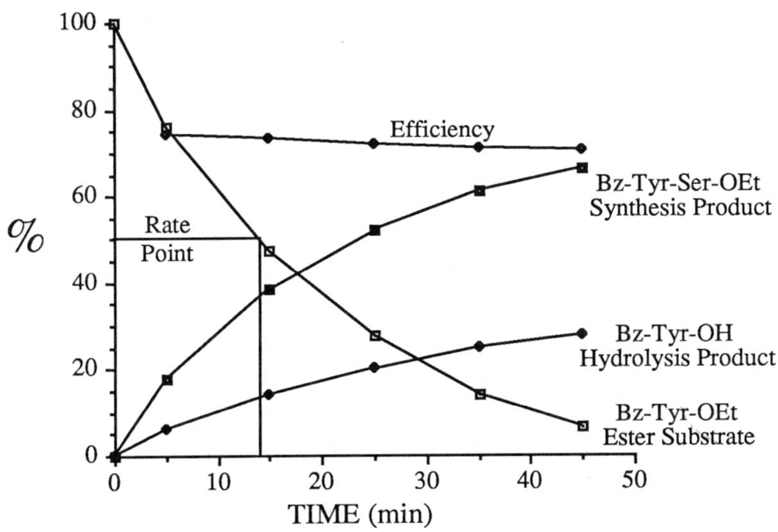

FIGURE 2. "Kinetic synthesis" reaction profile for the chymotrypsin-catalyzed reaction between benzoyl-tyrosine ethyl ester and serine ethyl ester under standard conditions. Efficiency is the ratio of synthesis product to the amount of ester substrate used expressed as a percentage.

midpoint rather than initially; this also helps to overcome inaccuracies in HPLC integrators.

Substrates and enzymes are purchased from a variety of sources but predominantly the Sigma Chemical Co., Boehringer Mannheim, Aldrich, and Novachem. When necessary, simple esters are prepared by the traditional acidic alcohol procedure.[20] Enzymes are used as purchased without further purification. Chymotrypsin was from the Sigma Chemical Company, Product No. C4129.

RESULTS

The effects of pH, temperature, solvent (DMF and ethanol), and nucleophile on synthesis efficiency and rate were studied by varying the individual parameter in the standard reaction.

As shown in FIGURE 3a, the pH optimum of chymotrypsin for synthesis is 8.5; efficiency remains high above 8.5, but the rate declines due to a combination of factors including substrate and enzyme instability or structural alteration at the higher pHs. Below pH 8.5, both efficiency and reaction rate are lower, presumably as a function of the declining concentration of the uncharged form of the amino group.

Increasing temperature (FIGURE 3b) results in slight but reproducible decline in reaction efficiency with chymotrypsin. The reaction rate does not increase as expected, with the low rates being due to enzyme instability at higher temperatures under the standard conditions.

DMF and ethanol have opposite effects on reaction efficiency (FIGURE 3c and d). The expected theoretical[21] result is found in DMF, that is, a slight increase in efficiency with increasing solvent due to reduction in the water concentration. The decline observed with ethanol (FIGURE 3d) is most likely due to solvent-induced deleterious structural alteration. Both solvents markedly reduce reaction rates with the practicable maximum solvent level for single enzyme addition being approximately 55%. Reactions can be carried out at higher concentrations if desired by using multiple enzyme addition.

Nucleophile concentration (FIGURE 3e) alters efficiency as expected from the reaction mechanism and as predicted by theory.[21] In practical terms, this means that only one test reaction for a given synthetic step needs to be performed before calculations to determine the optimal nucleophile concentration (involving such factors as nucleophile availability, recycling costs, etc.) are carried out.

The affinity of the nucleophilic component for the enzyme substrate complex (FIGURE 1) is the major determining factor on reaction efficiency. The efficiencies obtained with various amino acid derivatives (amides and methyl, ethyl, and benzyl esters) are shown in FIGURE 4. It is important to note that these figures were obtained under standardized conditions for comparison and that much higher efficiencies would be observed upon optimization, predominantly by alteration of the substrate and nucleophile concentrations. No simple pattern emerges for chymotrypsin. As a generalization, amino acid amides are preferred to esters, with ethyl and benzyl esters being preferred to methyl esters. Comparison of the four amino acids where every derivative is examined (Ala, Gly, Leu, Ser) shows that the influence of the C-terminal modification on reaction efficiency is variable; it can be of

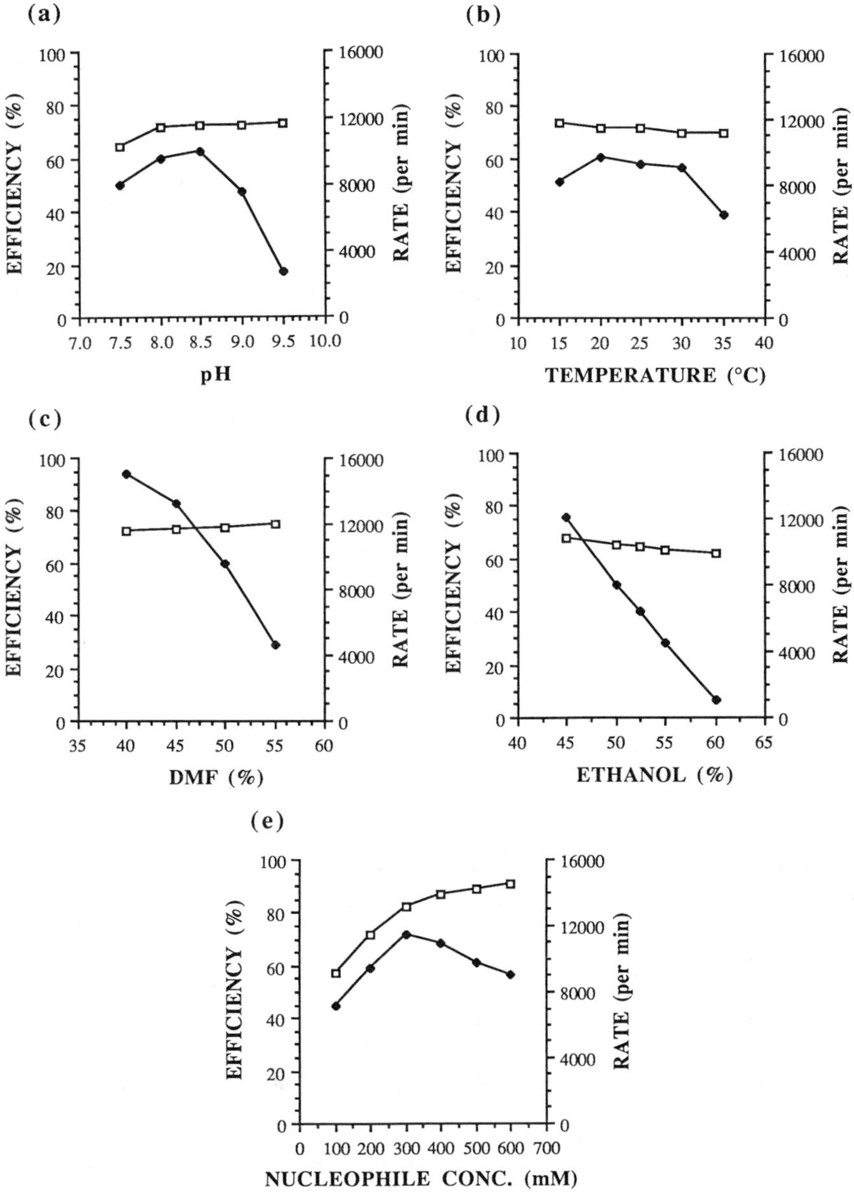

FIGURE 3. Chymotrypsin-catalyzed synthesis of Bz-Tyr-Ser-OEt; variation of the standard reaction parameters: **(a)** pH; **(b)** temperature; **(c)** dimethylformamide concentration; **(d)** ethanol concentration; and **(e)** nucleophile concentration. □, efficiency; ■, rate.

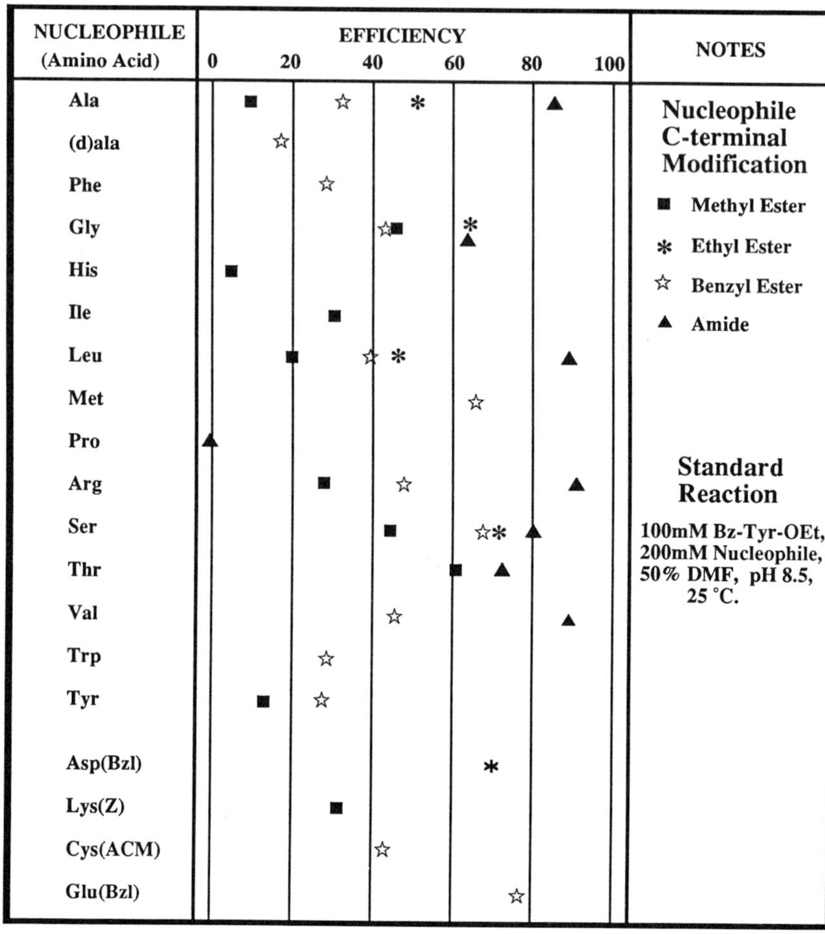

FIGURE 4. Nucleophile variations. Efficiency of chymotrypsin-catalyzed synthesis reactions with benzoyl-tyrosine ethyl ester as substrate; reactions under standard reaction conditions.

marked (Ala, Leu) or of little effect (Gly). As judged by D-alanine benzyl ester, D amino acids can be accepted as nucleophiles. This broad acceptance of amino acid derivatives at the S'_1 site[22] makes chymotrypsin particularly useful as a synthetic enzyme.

From the limited number of ester substrates studied (TABLE 1), chymotrypsin specificity at the S_1 site more or less follows its proteolytic preferences, i.e., hydrolysis after aromatic residues (Tyr, Phe, Trp) and to a lesser degree bulky hydrophobic residues. Both Tyr and Phe (not shown) substrates are acceptable, giving good synthetic efficiencies at high rates. Tryptophan substrates are less acceptable, giving lower yields and slower reaction rates. Leucine and aspartic acid modified by side chain benzylation also give high efficiencies but at very slow rates; hydrolysis at other sites in complex peptide substrates and nucleophiles could be expected at these rates. As shown by a comparison of ester substrates with Tyr at the P_1 position, changes upstream mainly affect reaction rates; reaction efficiencies are only marginally altered (compare Bz-Tyr-OEt, Ac-Tyr-OEt, Z-Val-Tyr-OEt and Z-Ser-Tyr-OEt).

TABLE 1 demonstrates the major limitation in the use of chymotrypsin. The enzyme's S_1 preference for hydrophobic residues and the need to block the N-terminal of the ester substrate (to prevent polymerization) result in substrates of limited solubility under the solvent conditions tolerated. This results in high nucleophile-to-substrate ratios to obtain satisfactory yields, necessitating nucleophile recycling for economical production. One way to overcome this problem would be to genetically engineer chymotrypsin to improve its solvent tolerance.

DISCUSSION

The results obtained with chymotrypsin are typical of most of the proteases studied to date (papain, chymopapain, bromelain, elastase, trypsin, subtilisin, and proteinase K; endoproteinase Glu C and alpha lytic protease in progress).

pH optima are generally around 8.5, and, while slightly higher efficiencies can be obtained at higher pH's, problems are often encountered due to reduced rates caused by enzyme instability and ester substrate instability.

The slight decline in efficiency with increasing temperature is also a common observation, and elevated temperatures should only be used when substrate solubility is a problem. Rate generally increases with temperature until the point where the enzyme's stability is affected.

The behavior of the enzymes with solvent is variable, with each enzyme having its own level of tolerance; chymotrypsin is one of the least stable and levels up to 70% are tolerated by some enzymes. The effect on efficiency is solvent and enzyme dependent. With chymotrypsin, DMF increases efficiency while ethanol causes a decrease; the reverse occurs with papain and chymopapain. Subtilisin, proteinase K,

TABLE 1. Ester Substrate Variation

Substrate	Yield (%)	Rate (min^{-1})
Ac-Trp-OMe (100mM)	41	1800
Ac-Tyr-OEt (100 mM)	81	2600
Bz-Tyr-OEt (100 mM)	73	12600
Z-Ser-Tyr-OEt (100 mM)	760	390
Bz-Ala-OMe (50 mM)	80	~2
Z-Gly-Asp (Bzl)-OBzl (10 mM)[b]	58	~7
Z-Leu-Arg-Trp-OMe (20mM)	30	245
Z-Gly-Trp-OMe (20 mM)[c]	46	750
Z-Leu-OMe (20 mM)	70	~3
Z-Val-Tyr-OEt (25 mM)[c]	75	4800

[a]Chymotrypsin synthesis reactions varying the ester substrates. Other reaction conditions as per standard with serine ethyl ester as nucleophile.

[b]Product unstable.

[c]Product precipitates.

and α-lytic protease show higher than expected increases in reaction efficiency with increasing DMF, presumably due to solvent-induced structural changes that increase the relative affinity of the nucleophile for the enzyme-substrate complex.

The effect of increasing nucleophile concentration is typical for all the enzymes studied so far, including the decline in rate at higher concentrations. This is presumably due to substrate (nucleophile) inhibition.

Nucleophile preference is, as expected, enzyme specific. Enzymes with broad specificity at the S'_1 position include papain, chymopapain, bromelain, trypsin, and chymotrypsin. Elastase and α-lytic protease also exhibit broad specificity but give lower efficiencies under the standard reaction conditions (α-lytic protease is superior to elastase). Subtilisin and proteinase K, despite their broad proteolytic activity, in synthesis show a strong preference for glycine-based nucleophiles and, apart from catalyzing the formation of X-glycine bonds, they are of little use in enzymatic peptide synthesis. Endoproteinase Glu C (still under study) requires more complex nucleophiles.

As with chymotrypsin, the enzyme's preference at the S_1 site generally follows that determined from proteolytic studies. The most useful general enzymes therefore are the papain family, where specificity is determined by a bulky hydrophobic group at the S_2 position. In synthesis, this group can be occupied by commonly used protecting chemical groups such as carbobenzoxy and butyloxycarbonyl.[4] The papain family tolerate many amino acids at the S_1 position, although smaller residues are more suitable since larger residues give problems with the generation of elongation products.

In summary, a subjective assessment of the enzymes' general suitability for synthesis based on our experience with the synthesis of large peptides and considering such factors such as substrate and nucleophile preference, rate, and general handling properties under final reaction conditions is (in order of best to worst): {chymopapain, papain, chymotrypsin, trypsin}; {α-lytic protease, elastase}; {subtilisin and proteinase K}. The latter two are more useful if nucleophiles with N-terminal glycine are used. Endoproteinase Glu C is still under study but is useful for fragment coupling. Bromelain has similar specificity properties to papain and chymopapain but, as commercial grades give relatively poor reaction rates, it is considered unsuitable.

ACKNOWLEDGMENTS

The authors would like to acknowledge[a] Trish Hayes, Ellen Antinino, Ken Moon, and Prof. E. O. P. Thompson who were involved in studies on enzymes other than chymotrypsin.

REFERENCES

1. VAN'T HOFF, J. H. 1898. Uber die zunehmende Bedeutung der anorganaschen Chemie. Z. Anorg. Chem. **18:** 1–13.
2. BERGMANN, M. & H. FRAENKAL-CONRAT. 1937. The role of specificity in the enzymatic synthesis of peptides. J. Biol. Chem. **119:** 707–720.

3. BERGMANN, M. & J. S. FRUTON. 1938. Some synthetic and hydrolytic experiments with chymotrypsin. J. Biol. Chem. **124:** 321–329.
4. FRUTON, J. S. 1982. Proteinase-cataysed synthesis of peptide bonds. Adv. Enzymol. **53:** 239–306.
5. JAKUBKE, H.-D., P. KUHL & A. KONNECKE. 1985. Basic principles of protease catalysed peptide bond formation. Angew. Chem. Int. Ed. Engl. **24:** 85–93.
6. KULLMANN, W. 1986. Enzymatic Peptide Synthesis. CRC Press Inc. Boca Raton, Fla.
7. MORIHARA, K. 1987. Using proteases in peptide synthesis. Trends Biotechnol. **5:** 164–170.
8. JAKUBKE, H.-D. 1987. Enzymatic peptide synthesis. *In* The Peptides. J. Meienhofer & S. Udenfriend, Eds. **9:** 103–165. Academic Press. New York, N.Y.
9. WHITTAKER, R. G. & P. A. SCHOBER. 1990. Synthesizing peptides with enzymes. Today's Life Sci. **2:** 60–64.
10. WIDMER, F., S. BAYNE, G. HOUEN, B. A. MOSS, R. D. RIGBY, R. G. WHITTAKER & J. T. JOHANSEN. 1985. Use of proteolytic enzymes for the synthesis of fragments of mouse epidermal growth factor. *In* Peptides 1984. U. Ragnarsson, Ed.: 193–200. Almquist and Wiksell. Stockholm, Sweden.
11. WIDMER, F., S. BAYNE, G. HOUEN, B. A. MOSS, R. D. RIGBY, R. G. WHITTAKER & J. T. JOHANSEN. 1985. Use of proteolytic enzymes for the synthesis of fragments of mouse epidermal growth factor, part 2. *In* Forum Peptides Le Cap d'Agde, Proc. 1st Int. Forum Peptides. J. B. Castro & J. Martinez, Eds.: 439–442. Centre de Pharmacologie-Endocrinology. Montpellier, France.
12. BREDDAM, K. & J. JOHANSEN. 1984. Semisynthesis of human insulin utilizing chemically modified carboxypeptidase Y. Carlsberg Res. Commun. **49:** 463–472.
13. GATTNER, H. G., W. DANHO, R. KNORR, V. K. NAITHANI & H. ZAHN. 1981. Trypsin catalysed peptide synthesis: modification of the B-chain C-terminal region of insulin. *In* Peptides 1980. Proceedings of the 16th European Peptide Symposium. K. Brumfeldt, Ed.: 372–377. Scriptor. Copenhagen, Denmark.
14. PETKOV, D. D. & I. B. STOINEVA. 1984. Nucleophile specificity in chymotrypsin peptide synthesis. Biochem. Biophys. Res. Commun. **118:** 317–323.
15. MITIN, YU. V., N. P. ZAPEVALOVA & E. YU. GORBUNOVA. 1984. Peptide synthesis catalysed by papain at alkaline pH values. Int. J. Peptide Protein Res. **23:** 528–534.
16. WHITTAKER, R. G., L. T. MCVITTIE, K. E. MOON, E. O. P. THOMPSON & B. A. MOSS. 1987. Papain catalyzed peptide synthesis. Proceedings of the 12th Annual Conference on Protein Structure and Function, Lorne, Australia.
17. WHITTAKER, R. G., L. T. MCVITTIE, K. J. BRYANT & E. A. HAMILTON. 1988. Trypsin catalyzed peptide synthesis. Proceedings of the 13th Annual Conference on Protein Structure and Function, Lorne, Australia.
18. WHITTAKER, R. G., E. A. HAMILTON, K. J. BRYANT, L. T. MCVITTIE & P. A. SCHOBER. 1988. Chymotrypsin catalyzed peptide bond synthesis. European Peptide Symposium, Tubingen, Germany.
19. SCHOBER, P. A., K. J. BRYANT, L. T. MCVITTIE, M. E. ANTONINO, F. WIDMER & R. G. WHITTAKER. 1991. Elastase catalyzed peptide synthesis. Proceedings of the 16th Annual Conference on Protein Structure and Function, Lorne, Australia.
20. BODANSZKY, M. & A. BODANSZKY. 1984. *In* The Practice of Peptide Synthesis. K. Hafner, J.-M. Lehn, C. W. Rees, P. von Rague Schleyer, B. M. Trost & R. Zahradnik, Eds: 33–35. Springer-Verlag. Berlin, Germany.
21. PETKOV, D. D. & I. B. STOINEVA. 1984. Enzyme peptide synthesis by an iterative procedure in a nucleophile pool. Tetrahedron Lett. **25:** 3751–3754.
22. SCHECHTER, I. & A. BERGER. 1967. On the active site in proteases. I. Papain. BBRC **27:** 157–162.

Enzymes for Epimerization of Isopenicillin N, Ring Expansion of Penicillin N, and 3′-Hydroxylation of Deacetoxycephalosporin C

Function, Evolution, Refolding, and Enzyme Engineering

W. K. YEH, S. K. GHAG, AND S. W. QUEENER

Lilly Research Laboratories
A Division of Eli Lilly and Company
Lilly Corporate Center
Indianapolis, Indiana 46285

Progress toward industrial use of cephalosporin biosynthetic enzymes was recently summarized.[1] We report here extensions of that work: elucidation of the nature of the active site involved in the ring expansion of penicillin N and 3′-hydroxylation of deacetoxycephalosporin C (DAOC); expressions of hybrid isopenicillin N (IPN) epimerase and penicillin N expandase genes in *Penicillium chrysogenum* and production of DAOC in the resulting recombinant organism; and finally, further progress in refolding of an insoluble form of fungal expandase/hydroxylase. Evolutionary implications of the concept of a single active site for ring expansion and 3′-hydroxylation are presented. Practical uses for the DAOC-producing *P. chrysogenum* recombinant strain and for the refolding technology are discussed. Reviews on molecular biology of β-lactam biosynthetic genes[2–5] are recommended for background reading.

RELATIVE BIFUNCTIONALITY OF FUNGAL EXPANDASE/HYDROXYLASE, BACTERIAL EXPANDASE, AND BACTERIAL HYDROXYLASE

Previously, enzymology[6] and gene cloning[7] have established that a *cef*EF gene in *Cephalosporium acremonium* encodes a bifunctional enzyme, penicillin N expandase/DAOC hydroxylase. The enzyme catalyzes ring expansion of penicillin N and 3′-hydroxylation of DAOC at nearly equal efficiencies (FIGURE 1). In *Streptomyces clavuligerus* separate *cef*E and *cef*F genes encode penicillin N expandase and DAOC hydroxylase,[8–10] respectively. Recently, direct comparisons of these three enzymes have shown that each catalyzes ring expansion of penicillin N and 3′-hydroxylation of DAOC although the ratios of the two activities vary markedly among the three enzymes:[11] the fungal enzyme exhibits efficient expandase and hydroxylase activities; the bacterial expandase has a minor hydroxylase activity; and the bacterial hydroxylase has a minor expandase activity (FIGURE 1).

The bacterial expandase was competitively inhibited by 3-exomethylenecephalosporin C (EMCC), and the hydroxylase was competitively inhibited by penicillin N

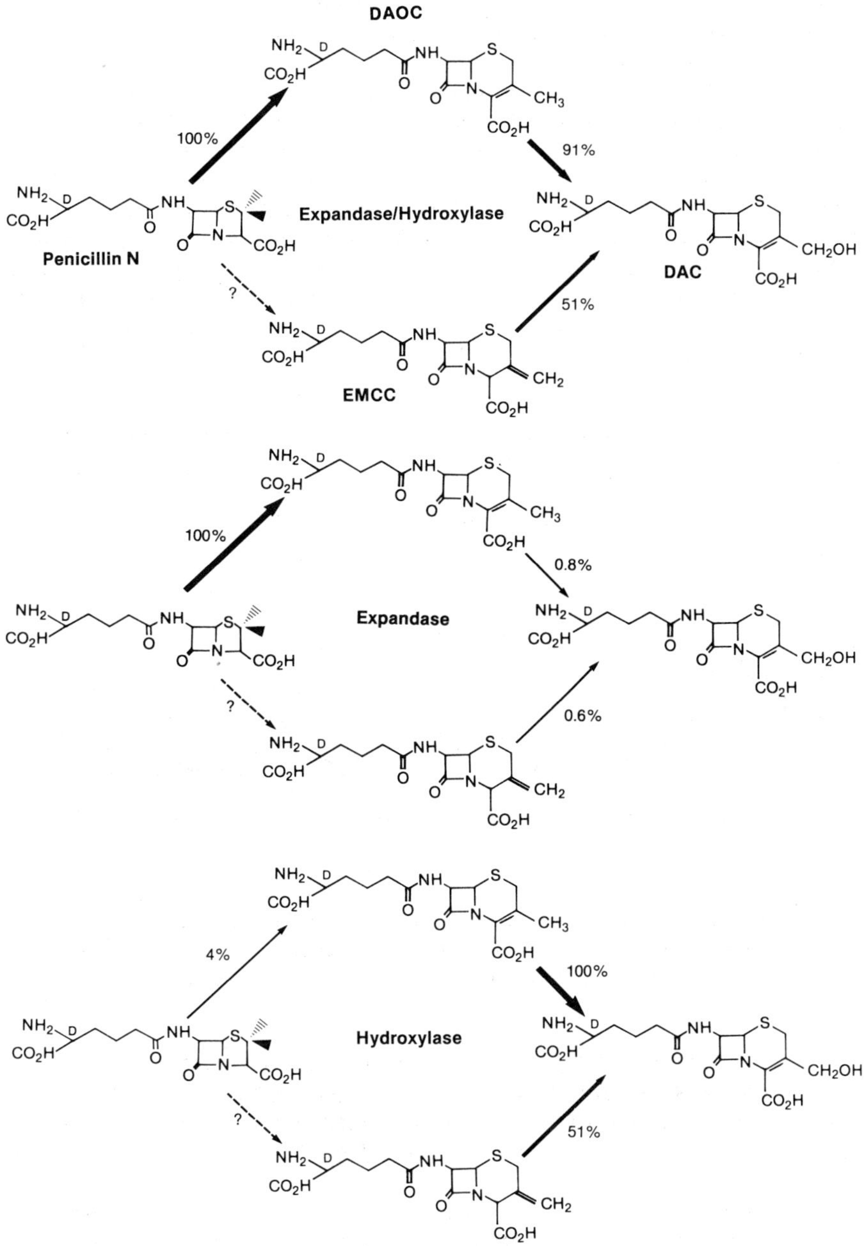

FIGURE 1. Bifunctionality of *C. acremonium* expandase/hydroxylase and *S. clavuligerus* expandase and hydroxylase.

(FIGURES 1 and 2).[9,10] The observation of the two competitive inhibitions is consistent with a single active site for both ring expansion and 3′-hydroxylation in the fungal enzyme and in the two bacterial enzymes. In addition, the three enzymes were similarly susceptible to inhibition by many analogues of penicillin N and DAOC.[11] The bifunctionality of the three enzymes and the associated kinetic inhibition data strongly argue for a single active site for ring expansion and 3′-hydroxylation. The concept of a single active site for ring expansion of penicillin N and 3′-hydroxylation of DAOC is consistent with the observations that the two enzymatic reactions require the same cosubstrates, α-ketoglutarate and molecular oxygen, and the same cofactor, ferrous iron. In both reactions, α-ketoglutarate is converted to succinate.[12]

Earlier the idea of fundamentally different active sites for ring expansion and 3′-hydroxylation was suggested, just after the cloning and characterization of the *C. acremonium* expandase/hydroxylase gene.[7] The idea of two different active sites, in turn, influenced hypotheses on the evolution of these genes, e.g., the hypothesis of

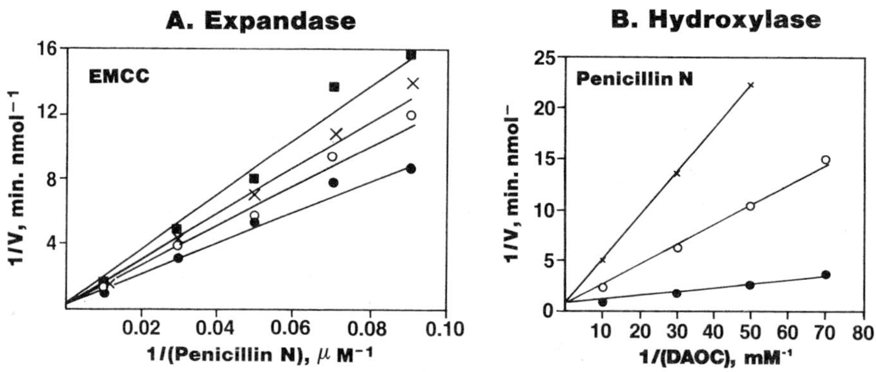

FIGURE 2. Inhibition of expandase by EMCC (A) and inhibition of hydroxylase by penicillin N (B).

fusion of an expandase gene with a hydroxylase gene to form the bifunctional fungal expandase/hydroxylase was suggested.[4,7] The direct evidence for a single active site refutes this hypothesis and requires another enzyme evolution model.

A NEW MODEL FOR EVOLUTION OF FUNGAL EXPANDASE/HYDROXYLASE, BACTERIAL EXPANDASE, AND HYDROXYLASE; INCLUDING RELATIONSHIPS OF *cef*EF, *cef*E, *cef*F GENES TO *pcb*C AND *cef*D GENES

There are very high sequence similarities among *C. acremonium* expandase/hydroxylase and *S. clavuligerus* expandase and hydroxylase. Comparisions of deduced amino acid sequences for the fungal expandase/hydroxylase[7] and the bacterial expandase[13] and hydroxylase[14] reveal 55–60% similarities among the enzymes.[10] Regions of high similarity are spread evenly along the sequences (FIGURE 3).[11] Such

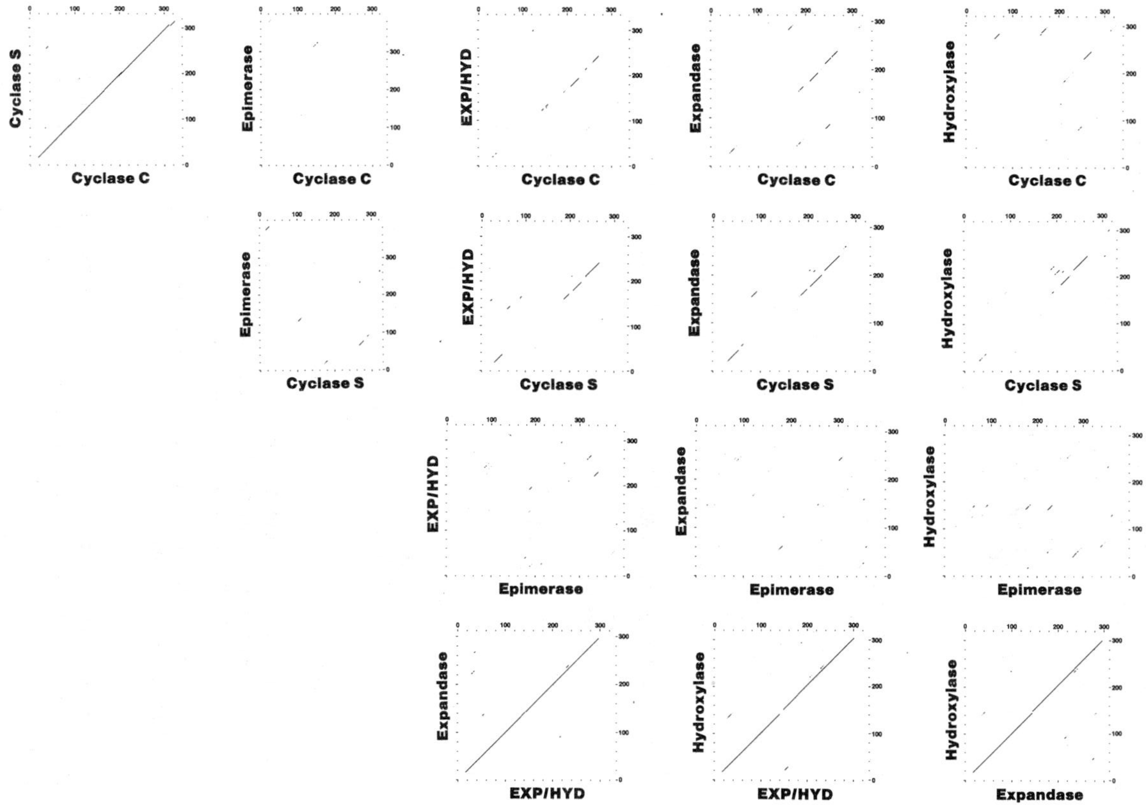

FIGURE 3. DOTPLOT for sequence relatedness among β-lactam cyclases, epimerase, expandase/hydroxylase, expandase, and hydroxylase. Cyclase C, from *C. acremonium;* Cyclase S, from *S. clavuligerus;* and EXP/HYD = expandase/hydroxylase.

a high degree of sequence similarity for two antibiotic enzymes that catalyze sequential metabolic reactions is rare; we are not aware of a comparable example.

The *cef*EF gene encodes *C. acremonium* expandase/hydroxylase. The *cef*E and *cef*F genes encode *Streptomyces clavuligerus* expandase and hydroxylase, respectively. The very high sequence similarities of the *cef*EF, and *cef*E, and *cef*F genes strongly suggest that they all were derived from a common ancestral gene.[2,10,14] Each of these genes encode an α-ketoglutarate-dependent oxygenase that carries out—with vary-

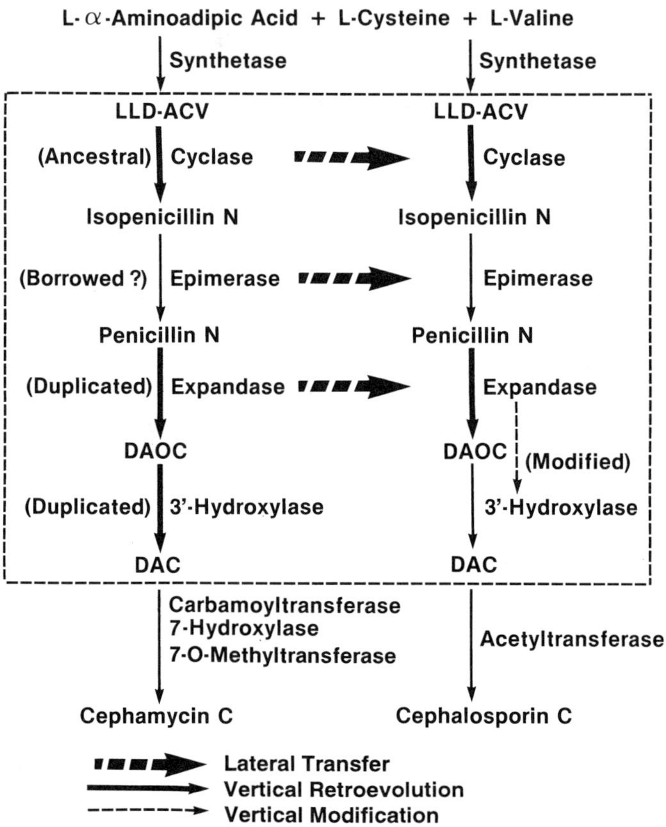

FIGURE 4. Enzyme-based metabolic evolution for β-lactam cyclase, epimerase, expandase, hydroxylase, and expandase/hydroxylase.

ing efficiencies at one active site (discussed above)—ring expansion of penicillin N and 3′-hydroxylation of DAOC. Thus, a common ancestral gene probably exhibited these properties.

We offer a new model for divergence from this ancestral gene to *C. acremonium* *cef*EF and *S. clavuligerus* *cef*E and *cef*F. The model has five basic features (FIGURE 4). (1) An ancestral bacterial expandase with a minor hydroxylase activity evolved first; the modern bacterial expandase is directly descended from that ancestral enzyme. (2) After gene duplication of a primitive *cef*E gene, modification of the duplicated gene in bacteria improved the hydroxylase activity at the expense of some of the expandase activity (i.e., retroevolution). (3) The fungal expandase/hydroxylase with balanced bifunctionality arose from a primitive bacterial gene by lateral (horizontal) transfer. This occurred in one of two ways. (a) After lateral gene transfer from bacteria to fungi, an ancient *cef*E gene encoding expandase with a minor hydroxylase activity might have evolved via mutations at the single active site that increased hydroxylase activity without substantially lowering expandase activity. (b) Alternatively, a bacterial gene encoding an enzyme with balanced bifunctionality (nonexistent today) evolved as mutations increased hydroxylase activity and decreased expandase activity (relative to those activities as encoded by the common ancestral gene). In fungi that received this gene, the efficiency of the ring expansion and 3′-hydroxylation activities improved in a balanced fashion. (4) The evolution of increasing hydroxylase activity continued in the ancestral bifunctional bacterial expandase/hydroxylase gene until hydroxylase activity was dominant. (5) Some unknown selective pressure caused bacteria with one enzyme responsible, predominantly, for ring expansion of penicillin N and one enzyme responsible, primarily, for 3′-hydroxylation of DAOC to become more prevalent than bacteria containing a gene encoding an expandase/hydroxylase with balanced activities.

If one assumes that deacetylcephalosporin C (DAC) provided cells with greater selective advantage than DAOC, then mechanism 3b provides the evolutionary pressure for balanced improvement of both expandase and hydroxylase activities in fungi. A mutation that dramatically improved expandase activity at the expense of hydroxylase activity would decrease overall production of the end product. Likewise a mutation that significantly increased hydroxylase activity at the expense of expandase activity would reduce the net flow to DAC or a derivative of DAC. This restriction would not occur in bacteria containing an expandase gene plus an expandase/hydroxylase gene; the latter gene would be free to evolve to a better hydroxylase activity. If 3a correctly predicts a frequent evolutionary route, then separable fungal expandase and hydroxylase enzymes may be expected to be discovered in the future. If 3b predicts the predominant mode of evolution, fungi with separate expandase and hydroxylase enzymes may be hard to find.

C. acremonium expandase/hydroxylase and *S. clavuligerus* expandase and hydroxylase share moderate amino acid sequence similarities with the fungal and bacterial δ-(L-α-aminoadipyl)-L-cysteinyl-D-valine (ACV) cyclases but not to the bacterial IPN epimerase—the enzyme that converts IPN to penicillin N (FIGURE 3).[11] Likewise, the *cef*EF, *cef*E, and *cef*F genes share modest nucleotide sequence similarities with the *pcb*C gene, which encodes ACV cyclase but not the cefD gene, which encodes IPN epimerase.[2,15]

The three bifunctional α-ketoglutarate-dependent oxygenases encoded by *cef*EF, *cef*E, and *cef*F also are related to the ACV cyclases in size and catalytic requirements: each enzyme has a molecular weight within the range of 30,000–40,000; each enzyme uses molecular oxygen as substrate and ferrous iron as cofactor. On the other hand, IPN epimerase has a molecular weight of 50,000 and requires pyridoxal-5′-phosphate as cofactor.[16]

The ancestral gene that gave rise to *cef*E, *cef*EF, and *cef*F probably arose from the *pcb*C gene (FIGURE 4) or an earlier gene that evolved to the *pcb*C gene. It is likely that this early oxygenase gene obtained a binding site by recombination with a primitive gene that encoded α-ketoglutarate-dependent oxygenase unrelated to β-lactam biosynthesis. The bacterial IPN epimerase gene, *cef*D, most likely arose separately from a fundamentally different ancestral gene (FIGURE 4). That gene might have encoded an enzyme involved in primitive primary amino acid metabolism. One cannot rule out the possibility that this ancestral gene for IPN epimerase recombined with a primitive β-lactam biosynthetic gene to pick up a motif that enhanced the binding of IPN and penicillin N. However, one must then assume that genetic drift subsequently minimized or destroyed the sequence similarity to that motif during the final evolution of the *cef*D gene.

METABOLIC ENGINEERING VIA HYBRID EPIMERASE AND EXPANDASE GENES

We have discussed above scientific implications of new biochemical data on β-lactam biosynthetic enzymes. Since our previous overview,[1] practical applications of β-lactam enzymes/genes have also been extended. Previously, an extra copy of the fungal expandase/hydroxylase gene was cloned back to a production strain of *C. acremonium.*[17] The gene cloning led to a significant yield improvement of cephalosporin C. Regulatory approvals for using the improved fungal strain have been secured, and tests at industrial scale have begun. More recently, two cloned genes have been used toward development of a simple, aqueous enzymatic process for making 7-aminodeacetoxycephalosporanic acid (7-ADCA) (FIGURE 5). Modified forms of the *cef*D and *cef*E[18] genes from *S. clavuligerus* were transformed into and expressed in *P. chrysogenum.* The hybrid genes were constructed by replacing the promoters of the bacterial epimerase and expandase genes with promoters from *P. chrysogenum.* In addition, termination sequences from *P. chrysogenum* were placed adjacent to the termination codon of the open reading frames of the *cef*E and *cef*D genes to improve expression of those genes in *P. chrysogenum.* The resulting recombinant *P. chrysogenum* produced a moderate level of DAOC and a trace of DAC (TABLE 1). Natural *P. chrysogenum* lacks penicillin N expandase activity and does not produce DAOC. No positive correlation for the epimerase and expandase activities from crude extracts to DAOC levels of the two transformants was observed. The low epimerase activities, which may result from poor *cef*D expression or enzyme instability, were likely a contributory factor to the moderate DAOC levels. To increase DAOC biosynthesis steps must also be taken to block the function of the *pen*DE gene which encodes acylCoA:isopenicillin N in *P. chrysogenum* (FIGURE 5), an enzyme that competes for the substrate isopenicillin in the recombinant strains.

An enzyme system that forms 7-ADCA from DAOC by removing the D-α-aminodipyl side chain has been improved (J. Vicenzi, Lilly, personal communication).

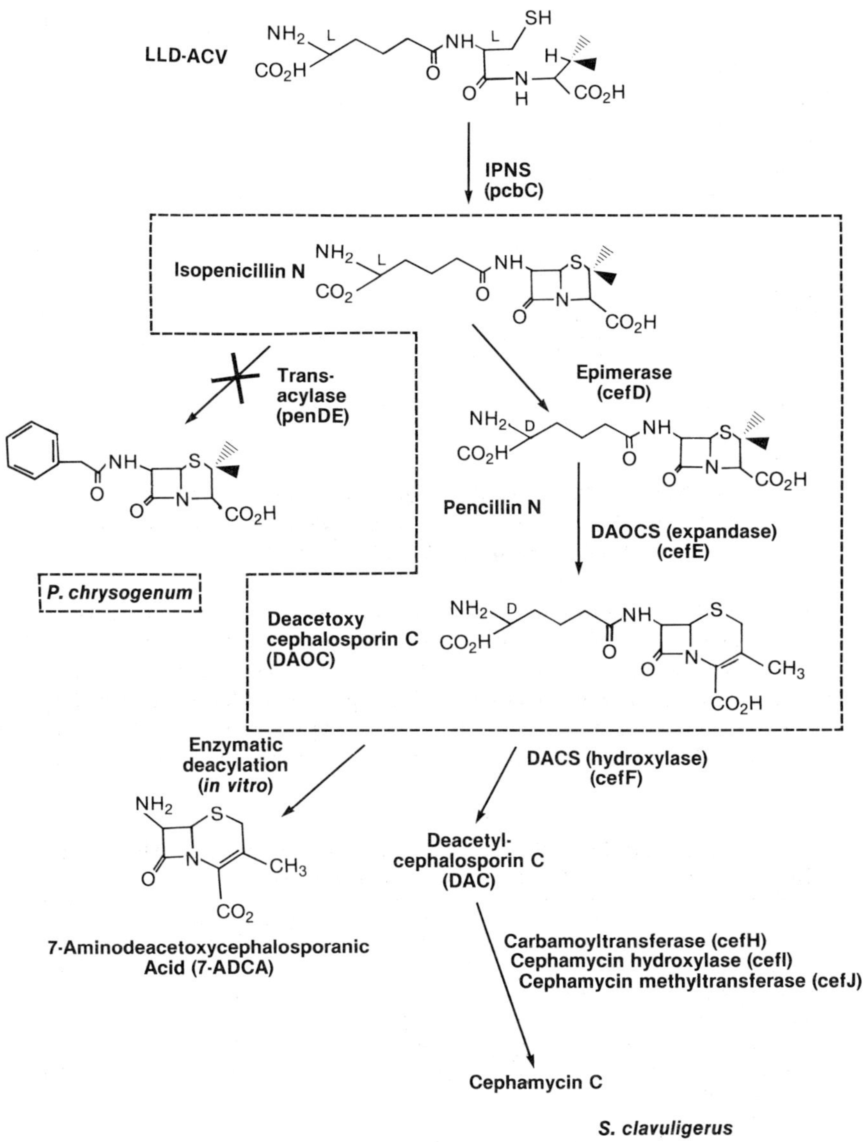

FIGURE 5. Metabolic engineering for DAOC biosynthesis in *P. chrysogenum.*

TABLE 1. Correlation of Epimerase and Expandase Activities to DAOC Levels of *P. chrysogenum* Transformants

β-Lactam Producer	Specific Activity (mU/mg)						DAOC[c] (mg/ml)
	Epimerase[a]			Expandase[b]			
	48h	72h	96h	48h	72h	96h	168h
P. chrysogenum							
Untransformed	0.7	0.5	0.4	0	0	0	0
Transformed							
6MY-1-D1	1.1	0.9	1.5	39	39	13	0.55
6MY-21-6-Dc14	0.6	0.7	1.2	6	6	3	2.54
C. acremonium							
394-4		Not done		35	122	87	N/A
LU-4-79-6		Not done		55	182	136	N/A
S. clavuligerus							
ATCC27064	5.1	7.8	8.7	5	5	3	N/A

[a]S. Usui and C. A. Yu, personal communication.
[b]J. E. Dotzlaf and W. K. Yeh, unpublished results.
[c]P. A. Whiteman and E. P. Abraham, personal communication.

POTENTIAL APPLICATIONS FOR ENZYME ENGINEERING OF FUNGAL EXPANDASE/HYDROXYLASE AND BACTERIAL EXPANDASE AND HYDROXYLASE

Site-directed mutagenesis may be able to minimize the hydroxylase activity in the fungal expandase/hydroxylase to allow DAOC biosynthesis in *C. acremonium.* Also, site-directed mutagenesis may allow modification of bacterial expandase to a form able to accept penicillin V as substrate for ring expansion to deacetoxycephalosporin V.[1] DAOV could then be deacylated enzymatically to form 7-ADCA. An even greater potential for industrial application may come from alteration of the bacterial

TABLE 2. Purification of Partially Folded Recombinant Expandase/Hydroxylase from Insoluble Granules of Recombinant *Escherichia coli*

Step	Protein (mg)	Activity (mU)	Specific Activity (mU/mg)	Yield (%)
Granular extract				
Solubilized[a]	25	2450	98	—
pH jumped[b]	25	2625	105	—
Folded[c]	25	4900	196	—
Filtered[d]	25	4700	188	100
Mono Q eluate				
Peak 1	5.1	1652	324	35
Peak 2	6.1	1694	278	36

[a]At 5 M urea.
[b]From pH 9 to 115.
[c]In 5 mM Tris-HCI, pH 9.5, in the presence of 0.5 mM DTT and 5 M urea.
[d]Through 0.2 μ filter paper.

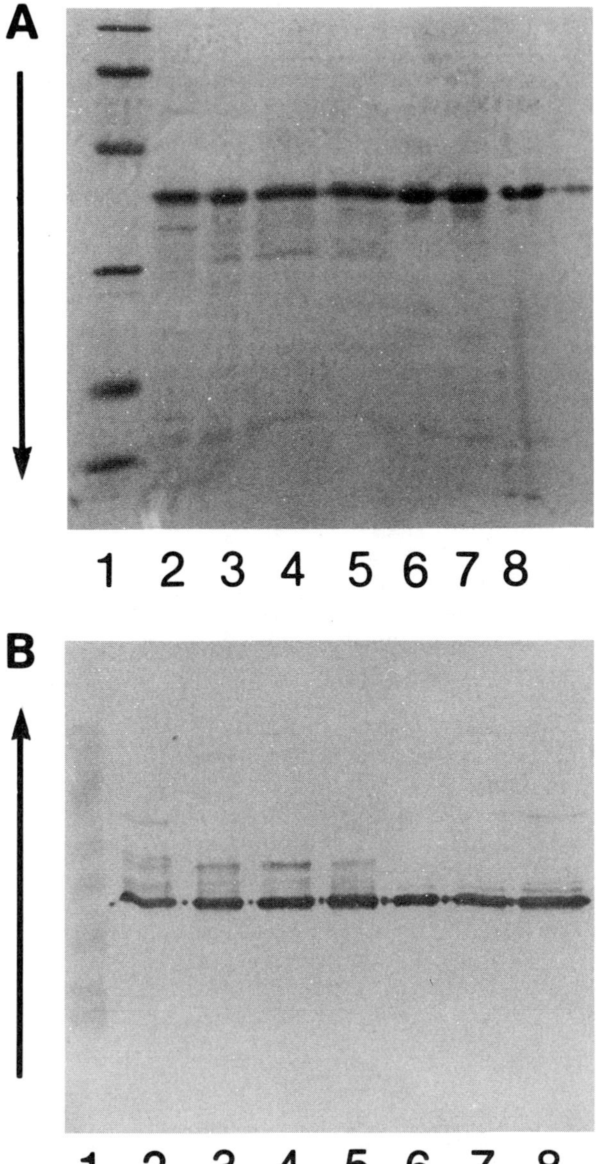

FIGURE 6. Sodium dodecyl sulfate-polyacrylamide gel electrophoresis (A) and Western blotting (B) of refolded recombinant expandase/hydroxylase. Lanes 1–8 in both A and B are protein standards (1), urea-solubilized granular extract (2), pH-jumped extract (3), partially folded extract (4), filtered extract (5), Mono Q eluate (6), DTT-dialyzed eluate (7), and GSH-dialyzed eluate (8).

expandase or hydroxylase for conversion of penicillin N to EMCC (FIGURE 1), a key intermediate in the manufacture of cefaclor, a clinically important oral antibiotic.

Ready access to pure enzyme by a single-step purification procedure would greatly aid enzyme engineering of fungal expandase/hydroxylase or bacterial expandase. This ready access would facilitate analysis of the enzyme active site which could be chemically modified with affinity labels. Purifying and crystalizing several recombinant enzymes for x-ray structural determinations clearly require a simple procedure for obtaining large amounts of pure enzymes.

REFOLDING OF FUNGAL EXPANDASE/HYDROXYLASE

The *C. acremonium cef*EF gene has been overexpressed in recombinant *Escherichia coli,*[7] but nearly all of the expandase/hydroxylase produced in *E. coli* was in the form of insoluble granules (inclusion bodies). Efficient refolding of the fungal expandase/hydroxylase from the granules and one-step purification of the refolded enzyme would dramatically enhance ability to analyze genetically engineered forms of the enzyme as outlined above.

Previously, several major variables for enzyme refolding solutions were identified and evaluated statistically to maximize the expandase activity recovery.[19] The enzyme could be refolded in initial granular extracts with an overall expandase activity recovery of 40–60% (calculations assuming equal specific activity for native fungal expandase/hydroxylase and the properly refolded enzyme produced from recombinant *E. coli*). The partially folded enzyme was purified 90–95% in the presence of dithiothreitol (DTT) or reduced glutathione (GSH) by a single-step Mono Q FPLC (TABLE 2, FIGURE 6). Final folding of the purified enzyme by a stepwise dialysis for urea removal with DTT only led to moderately active and stable enzyme (TABLE 3). Recently, evaluation for further improvement of the enzyme refolding has been conducted by a structure approach using circular dichroism, fluorescence, and gel filtration high-performance liquid chromatography (HPLC). As an example, a monomer/dimer transition of the recombinant enzyme occurred within 4 to 2 M urea (FIGURE 7) and the resulting dimer became either aggregated or degraded from 3 to 1 M urea (not shown). A major change in maximal fluorescence wavelength was

TABLE 3. Final Refolding of Purified Recombinant Expandase/Hydroxylase with DTT or GSH[a]

[Urea] (M)	Specific Activity (mU/mg)			
	DTT	GSH	GSH	DTT
Prior to dialysis (for purified enzyme)				
5	234	—	78	—
Postdialysis (to DTT or to GSH)				
3	230	233	54	153
1	259	143	39	290
0.5	264	129	45	274
0	323	149	87	303

[a]A stepwise dialysis of 5→3→1→0.5→0 urea was conducted in 5-mM Tris-HCI, pH 9.5, in the presence of 0.5-mM DTT or 1-mM GSH.

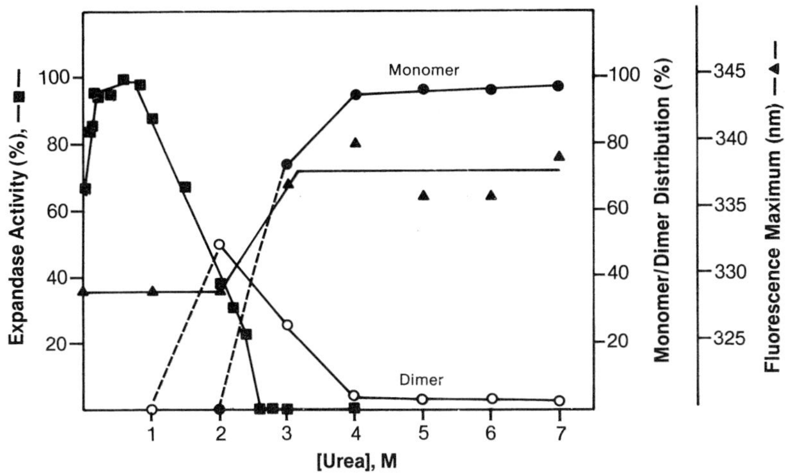

FIGURE 7. Urea-dependent structure/activity relationships of purified recombinant expandase/hydroxylase. The monomer/dimer distribution (percent of total protein) was analyzed by gel filtration HPLC.

observed within 4 to 1 M urea (FIGURE 7). The enzyme activity was determined under enzymatic reaction conditions that were different from those conditions for the two structural determinations. It appeared first around 2.5 M urea, reached maximum around 0.5 M urea, and decreased about 30% in the absence of urea (FIGURE 7). Since the native enzyme from *C. acremonium* is monomeric, the monomer of the recombinant enzyme is likely more active than its dimer. Thus, refolding of the recombinant enzyme may improve significantly by keeping it in monomeric form (i.e., preventing the monomer/dimer conversion). We expect that scientific insights from those and other structure/activity relationships will contribute to effective refolding of this and related enzymes. Effective refolding of granular expandase/hydroxylase will allow a large quantity of highly active, pure, and stable enzyme to be prepared by a single chromatographic step.

ACKNOWLEDGMENTS

We thank S. Usui and C. A. Yu for the epimerase activity analysis and J. E. Dotzlaf for the expandase activity analysis. We are grateful to C. A. Cantwell for use of *cef*E and *cef*F cloning information and to P. A. Whiteman and E. P. Abraham for use of DAOC/DAC biosynthesis data prior to publication.

REFERENCES

1. YEH, W.-K. & S. W. QUEENER. 1990. Potential industrial use of cephalosporin biosynthetic enzymes and genes: an overview. Ann. N.Y. Acad. Sci. **613:** 128–141.
2. SKATRUD, P. L. 1991. Molecular biology of the beta-lactam producing fungi. *In* More Gene

Manipulations in Fungi. J. W. Bennet & L. Lasure, Eds.: 364–395. Academic Press. New York, N.Y.

3. QUEENER, S. W. 1990. Molecular biology of penicillin and cephalosporin biosynthesis. Antimicrob. Agents Chemother. **34**(6): 943–948.
4. INGOLIA, D. T. & S. W. QUEENER. 1989. Beta-lactam biosynthetic genes. Med. Res. Rev. **9**(2): 245–264.
5. MILLER, J. R. & T. D. INGOLIA. 1989. Cloning β-lactam genes from *Streptomyces spp.* and fungi. *In* Genetics and Molecular Biology of Industrial Microorganisms. C. L. Hershberger & S. W. Queener, Eds.: 246–255. American Society for Microbiology. Washington, D.C.
6. DOTZLAF, J. E. & W.-K. YEH. 1987. Copurification and characterization of deacetoxycephalosporin C synthetase/hydroxylase from *Cephalosporium acremonium.* J. Bacteriol. **169**(4): 1611–1618.
7. SAMSON, S. M., J. E. DOTZLAF, M. L. SLITZ, G. W. BECKER, R. M. VAN FRANK, L. E. VEAL, W.-K. YEH, J. R. MILLER, S. W. QUEENER & T. D. INGOLIA. 1987. Cloning and expression of the fungal expandase/hydroxylase gene involved in cephalosporin biosynthesis. Biotechnology **5:** 1207–1214.
8. JENSEN, S. E., D. W. S. WESTLAKE & S. WOLFE. 1985. Deacetoxycephalosporin C synthetase and deacetoxycephalosporin C hydroxylase are two separate enzymes in *Streptomyces clavuligerus.* J. Antibiot. Tokyo **38**(2): 263–265.
9. DOTZLAF, J. E. & W.-K. YEH. 1989. Purification and properties of deacetoxycephalosporin C synthase from recombinant *Escherichia coli* and its comparison with the native enzyme purified from *Streptomyces clavuligerus.* J. Biol. Chem. **264**(17): 10219–10227.
10. BAKER, B. J., J. E. DOTZLAF & W.-K. YEH. 1991. Deacetoxycephalosporin C hydroxylase of *Streptomyces clavuligerus:* purification, characterization, bifunctionality and evolutionary implication. J. Biol. Chem. **266**(8): 5087–5093.
11. YEH, W.-K., J. E. DOTZLAF & G. W. HUFFMAN. 1992. Biochemical characterization and evolutionary implication of β-lactam expandase/hydroxylase, expandase and hydroxylase. *In* 50 Years of Penicillin Application (Symposium in Honour of Sir Edward P. Abraham). H. Kleinkauf & H. von Dohren, Eds.: 180–197.
12. BALDWIN, J. E., R. M. ADLINGTON, C. J. SCHOFIELD, W. J. SOBEY & M. E. WOOD. 1989. The role of α-ketoglutarate in cephalosporin biosynthesis. J. Chem. Soc. Chem. Commun.: 1012–1215.
13. KOVACEVIC, S., B. J. WEIGEL, M. B. TOBIN, T. D. INGOLIA & J. R. MILLER. 1989. Cloning, characterization, and expression in *Escherichia coli* of the *Streptomyces clavuligerus* gene encoding deacetoxycephalosporin C synthetase. J. Bacteriol. **171**(2): 754–760.
14. KOVACEVIC, S. & J. R. MILLER. 1991. Cloning and sequencing of the β-lactam hydroxylase gene (cefF) from *Streptomyces clavuligerus:* gene duplication may have led to separate hydroxylase and expandase activities in the Actinomyces. J. Bacteriol. **173**(1): 398–400.
15. WEIGEL, B. J., S. G. BURGETT, V. J. CHEN, P. L. SKATRUD, C. A. FROLIK, S. W. QUEENER & T. D. INGOLIA. 1988. Cloning and expression in *Escherichia coli* of isopenicillin N synthetase gene from *Streptomyces lipmanii* and *Aspergillus nidulans.* J. Bacteriol. **170**(9): 3817–3826.
16. USUI, S. & C.-A. YU. 1989. Purification and properties of isopenicillin N epimerase from *Streptomyces clavuligerus.* Biochem. Biophys. Acta **999:** 78–85.
17. SKATRUD, P. L., A. J. TIETZ, T. D. INGOLIA, C. A. CANTWELL, D. F. FISHER, J. L. CHAPMAN & S. W. QUEENER. 1989. Use of recombinant DNA to improve production of cephalosporin C by *Cephalosporium acremonium.* Biotechnology **7:** 477–485.
18. CANTWELL, C. A., R. J. BECKMANN, J. E. DOTZLAF, D. F. FISHER, P. L. SKATRUD, W.-K. YEH & S. W. QUEENER. 1990. Cloning and expression of a hybrid *Streptomyces clavuligerus* cefE gene in *Penicillium chrysogenum.* Curr. Genet. **17:** 213–221.
19. GHAG, S. K. & W.-K. YEH. 1990. Folding of the bifunctional β-lactam expandase/hydroxylase from granules of recombinant *E. coli.* AAAS Annual Meeting on Protein Folding, New Orleans, Louisiana.

α-Chymotrypsin in Plastein Synthesis

Influence of Water Activity[a]

DIDIER COMBES AND PEDRO LOZANO

Departement de Genie Biochimique et Alimentaire
UA-CNRS 544
Institut National des Sciences Appliquées
Complexe Scientifique de Rangueil
F-31077 Toulouse. France

INTRODUCTION

The plastein reaction i.e., the incubation of a highly concentrated solution of protein hydrolyzate with proteases, which produces water-insoluble and gel-forming products, could be included in the list of the unusual reactions catalyzed by proteases.[1] This process is of great interest for the food industry, such as in the debittering of hydrolyzed protein from unconventional sources,[2] modifying the functional properties of proteins incorporated to different food products,[3] and offering a potential means of supplementing food proteins with the desired amino acids.[4] Three conditions are required for the plastein reaction to take place: a low molecular weight for the peptidic substrate, a substrate concentration in the reaction medium between 20 and 40% (w/v), and an optimum pH value for the "synthetic" activity of the protease. These conditions suggest that a large number of factors may be involved in plastein formation simultaneously to the enzyme action, such as hydrophobic and ionic bonding. The role of proteolytic enzymes in the plastein reaction, and the kinetic mechanism have been very controversial.[1,2] Thus, several authors have concluded that a mechanism of direct condensation of the peptides is responsible for the plastein synthesis,[5,6] while other authors have proposed a transpeptidation mechanism, where the intermediate covalent acyl-enzyme is deacylated by a kinetically controlled process of peptide synthesis, together with an important participation of the hydrophobic and ionic interaction in the plastein production.[7–9]

In the plastein reaction, the substrate concentration is the most important limiting factor to the catalytic efficiency of the protease. This parameter determines simultaneously the total free amino group concentration, as nucleophilic acyl acceptors of the enzymic complex, and the water activity (A_w) in the reaction medium.[1,2] In a previous work, where the influence of the substrate concentration on the plastein reaction catalyzed by α-chymotrypsin was studied, we have observed that both transpeptidation and condensation mechanisms could be simultaneously involved.[10] In order to ascertain the role of A_w on the enzyme action in the plastein reaction, we have studied the effect of the substrate concentration as a function of this parameter. Additionally, the presence of different hydroxylated additives, which are known to be enzyme-stabilizing and water-activity-reducing agents,[11] has also been analyzed.

[a] This work was partially supported by the B/BIOT-900024 BRIDGE CEE grant.

MATERIALS AND METHODS

Substrate Preparation

A 10% (w/v) bovine serum albumin (Sigma Chemical Co.) solution (pH = 1.6) was treated by pepsin (Sigma Chemical Co.) (1% w/w) with magnetic stirring for 48 hours at 40°C. The hydrolyzate solution was ultrafiltered through polysulphone membranes (10,000-Da cutoff) and then concentrated in a Rotavapor to 55% (w/v). This substrate showed a free amino group concentration of 1.494 μmol NH_3^+/mg albumin.[10]

Plastein Reaction

Into an Eppendorf tube of 1 ml total volume, amounts of 100, 200, or 350 mg of the concentrated hydrolyzate of albumin were placed, and 3.5 mg of α-chymotrypsin (Sigma Chem. Co.) were added. The reaction volume was adjusted to 1 ml by addition of water or aqueous polyol solution, and the reaction mixture incubated without stirring at 40°C. Aliquots of 50 μl were extracted from the reaction mixture at different times, and the plastein products were determined by precipitation in a 10% (w/v) trichloroacetic acid solution and quantified spectrophotometrically at 280 nm, by redissolving in 50% (v/v) acetic acid solution.[10]

One unit of plastein activity was defined as the amount of enzyme that produced 1 mg of plastein per minute under optimum assay conditions.

Determination of Water Activity

Water activity was determined using a humidity and temperature digital indicator HUMIDAT—IC II (Novasina), with a humidity sensor model BS-3(4)/PP (Novasina) at a temperature of 20°C. The humidity sensor was checked and periodically recalibrated at three points, with control saturated salt solutions [LiCl, A_w = 11.3%; $Mg(NO_3)_2$, A_w = 54.4%; and $BaCl_2$, A_w = 90.5%] for the overall measuring range.

RESULTS AND DISCUSSION

In most cases, enzymatic hydrolytic reactions are equilibria reactions, with the equilibrium shifted in the direction of hydrolysis, that consume one or more water molecules. It is thus quite reasonable to assume that a medium with a decreased water activity, produced by high substrate concentration or water activity reducing agents, will shift the equilibrium to the synthetic reaction.[11,12]

The effect of substrate concentration in the range 2.5–52.5% (w/v) on the plastein activity of α-chymotrypsin was analyzed as a function of water activity (A_w). FIGURE 1 shows the dependence of the plastein yield on the A_w induced by the substrate concentration [$A_w = 100 - 3.83 \cdot 10^{-3} \cdot [S]^2 - 0.125 \cdot [S]$, where [S] is measured in percent (w/v)], determined at two different times (4 and 24 hours). As can be seen, after 4 hours of reaction time, the plastein yield profile showed two different stages. First, a "negative" stage of yield ($A_w > 97\%$), which implied a hydrolytic activity for the enzyme. Second, there was the stage in which plastein synthesis was higher than the hydrolytic activity. In this case, the decrease in A_w and the concomitant increase in the free amino group concentration allowed clearly a "positive" synthetic activity, which could be related to a transpeptidation mechanism.[10] Additionally, when plastein yield was measured after 24 hours reaction, two phenomena were observed. First, at A_w higher than 97%, a "positive" plastein yield was observed, as a consequence of the initial hydrolytic activity which increased the concentration of free amino groups, which acted as nucleophile acceptors to the acyl

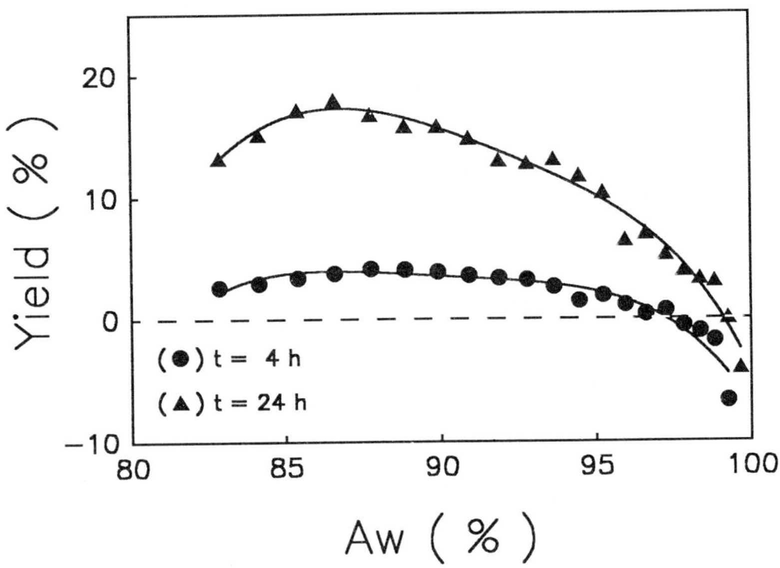

FIGURE 1. Effect of water activity (A_w), induced by substrate concentration in the reaction media, on the plastein yield produced by α-chymotrypsin at two reaction times (4 and 24 hours), using a peptic albumin hydrolyzate as substrate under optimal conditions of assay (40°C, pH 7.0).

groups transferred by the enzyme. Second, the plastein yield was maximum at 86% A_w. These facts showed that both free amino groups and A_w are key parameters to the plastein synthesis control.

In this order, if the plastein reaction occurred exclusively by a transpeptidation mechanism, the free amino group concentration should be constant during the reaction, while a condensation pathway would imply a decrease in this parameter. TABLE 1 shows the analysis of the free amino group concentration in the plastein reaction after 24 hours, at four different substrate concentrations (10, 20, 35, and 42.5% w/v). As can be seen, an increase in this parameter is observed in all cases except for the highest substrate concentration, which can be related to a decrease in the hydrolytic and an increase in the transpeptidation actions of the enzyme. However, the decrease in this parameter for the 42.5% w/v substrate concentration is explained only if a condensation pathway was present, in agreement with the results obtained by Yamashita *et al.* when a 50% w/v substrate concentration was assayed.[6] Additionally, in all cases where a depressing water activity agent was present in the reaction media, the free amino group concentration was reduced with respect to the control reaction without additive, producing a similar effect to the increase in substrate concentration, and therefore showing an increase in the condensation pathway occurrence.

In order to discriminate the actual effect of substrate concentration on the enzyme action, we have studied the effect of several hydroxylated additives (methanol, ethylene glycol, glycerol, erythritol, xylitol, and sorbitol) at different concentrations (0–4 M) on the plastein synthesis. FIGURE 2 shows the evolution of the plastein activity of α-chymotrypsin, at three different substrate concentrations (10, 20, 35% w/v), as a function of the A_w in the reaction media. Initially, for the lowest substrate concentration, a clear activation effect on the plastein activity is observed when the A_w is reduced by the presence of the additives. However, the increase in the mass substrate concentration into the reaction media involves a decrease in this activation power of the additives. Moreover, in the case of the most concentrated substrate solution, this effect is finally inversed. This unexpected change in the behavior of the reaction could be related to the observed raise of the viscosity in the media, producing a limitation in the microenvironmental conditions to the diffusion of the substrate and products. In TABLE 2, the effect of the different additives is analyzed as a function of the number of hydroxylated carbons per molecule. At low substrate concentration, this table shows how the increase in the molecular weight of the additive produces an increase in the plastein activity; and, also, the increase in the substrate concentration reduces this activation effect. Both methanol and sorbitol showed an exceptional behavior. In the case of methanol, a deactivation effect was observed, which was reduced by the increase in the substrate concentration. Sorbitol showed an exceptional activation effect when the highest substrate concentration was assayed. However, the increase in the plastein synthesis produced by the presence of additives was less remarkable than that involved by the increase in the substrate concentration. Thus, the plastein reaction should be regarded as being pushed by high concentration of reactants, rather than being pulled by low water activity.[12]

TABLE 1. Analysis of the Free Amino Group Concentration in the Plastein Reaction Catalyzed by α-Chymotrypsin after 24 Hours, at Four Different Substrate Concentrations (10, 20, 35, and 42.5% w/v), with or without the Presence of 3 M Xylitol in the Reaction Media

[S] (% w/v)	[Amino Groups] (μmol/mg) t = 0 hours	t = 24 hours, Without Xylitol	t = 24 hours, 3 M Xylitol
10	1.494	1.655	1.598
20	1.494	1.605	1.521
35	1.494	1.591	1.166
42.5	1.494	1.317	0.933

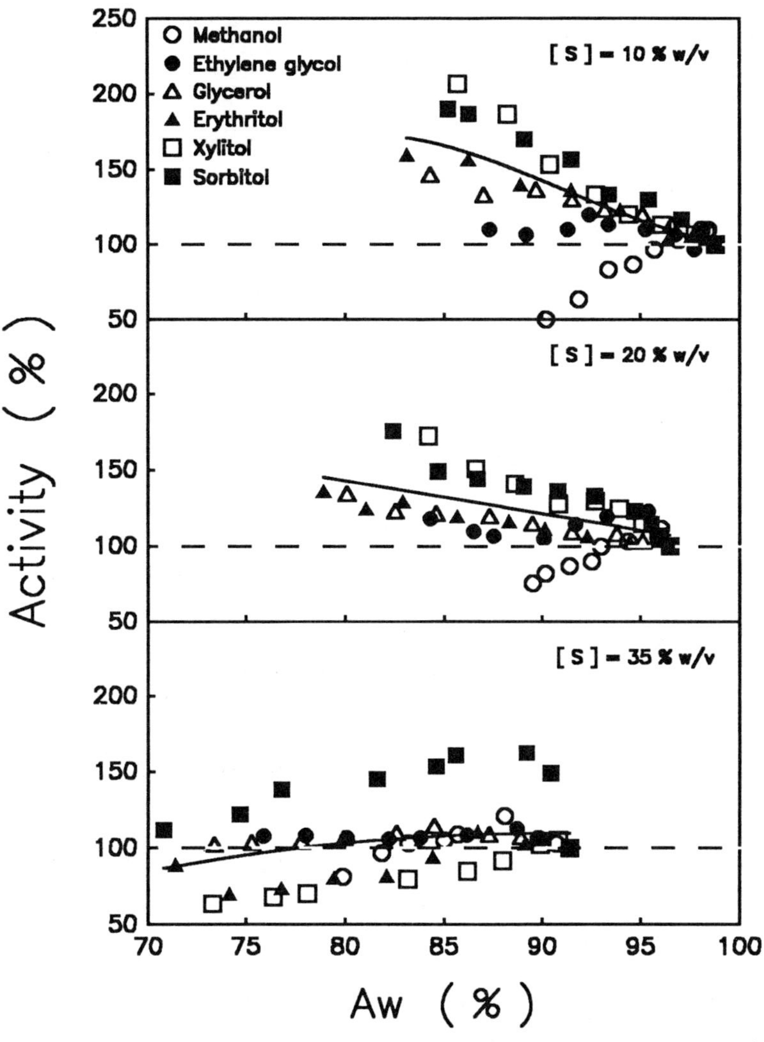

FIGURE 2. Effect of water activity (A_w), induced by the presence of different hydroxylated additives at several concentrations (0–4 M) in the reaction media, on the plastein reaction catalyzed by α-chymotrypsin using a peptic albumin hydrolyzate as substrate, at three different substrate concentrations (10, 20, and 35% w/v) under optimal conditions of assay (40°C, pH 7.0).

CONCLUSIONS

This study allows us to conclude that all possible biocatalytic pathways, i.e., hydrolysis, transpeptidation, and condensation, can occur simultaneously. The increase in substrate concentration, as acyl-nucleophilic acceptors, shifts the enzyme action from hydrolysis to plastein synthesis. A_w reduction by both high substrate

TABLE 2. Influence of Hydroxylated Additives on the Enzymatic Activity of the α-Chymotrypsin in the Plastein Synthesis at Three Different Substrate Concentrations (10, 20, and 35% w/v) under Optimal Conditions of Assay

[S] (% w/v)	Activity (mU/mg Enzyme)						
	Without Additive	3 M Hydroxylated Additive[a]					
		C1	C2	C3	C4	C5	C6
10	0.75	0.63	0.80	1.03	1.05	1.16	1.28
20	2.95	2.56	3.14	3.62	3.81	4.15	4.25
35	11.79	12.16	12.58	12.32	8.71	8.28	16.30

[a]C1, methanol; C2, ethylene glycol; C3, glycerol; C4, erythritol; C5, xylitol; C6, sorbitol.

concentration and presence of additives enhances the condensation/transpeptidation ratio of plastein product. However, A_w should be regarded as one "motor" of the reaction, but substrate concentration is the most important factor in the control of the plastein synthesis catalyzed by the proteases.

REFERENCES

1. ARAI, S., M. YAMASHITA & M. FUJIMAKI. 1981. Protease action on protein at low water concentration. *In* Water activity: Influence on Food Quality. L. B. Rockland & G. F. Stewart, Eds.: 489–510. Academic Press. New York, N.Y.
2. FUJIMAKI, M., S. ARAI & M. YAMASHITA. 1977. Enzymatic protein degradation and resynthesis for protein improvement. *In* Food Proteins: Improvement through Chemical and Enzymatic Modification. R. E. Feeney & J. R. Whitaker, Eds: 156–183. American Chemical Society 160. Washington, D.C.
3. CHOBERT, J. M., C. BERTRAND-HARB & M. G. NICOLAS. 1988. Solubility and emulsifying properties of caseins and whey proteins modified enzymatically by trypsin J. Agric. Food Chem. **36:** 883–892.
4. HAJÓS, G. Y., I. ELIÁS & A. HALÁSZ. 1988. Methionine enrichment of milk protein by enzymatic peptide modification. J. Food Sci. **53:** 739–742.
5. DETERMAN, H., S. EGGENSCHWILER & W. MICHEL. 1965. Plastein reaction. VII. Molecular weight distribution of the product of enzymic condensation. Ann. Chem. **690:** 182–188.
6. YAMASHITA, M., S. ARAI, S. Y. TANIMOTO & M. FUJIMAKI. 1973. Condensation reaction occurring during plastein formation by α-chymotrypsin. Agric. Biol. Chem. **37:** 953–954.
7. SUKAN, G. & A. T. ANDREWS. 1982. Application of the plastein reaction to caseins and to skim-milk powder. II. Chemical and physical properties of the plastein and the mechanism of plastein formation. J. Diary Res. **49:** 279–293.
8. EDWARDS, J. H. & W. F. SHIPE. 1978. Characterization of plastein reaction product formed by pepsin, α-chymotrypsin and papain treatment of egg albumin hydrolysates. J. Food Sci. **43:** 1215–1218.
9. VON HOFSTEN, B. & G. LALASIDIS. 1976. Protease-catalyzed formation of plastein products and some of their properties. J. Agric. Food Chem. **24:** 460–465.
10. LOZANO, P. & D. COMBES. 1991. α-Chymotrypsin in plastein synthesis: influence of substrate concentration on enzyme activity. Biotechnol. Appl. Biochem. **14:** 212–221.
11. MONSAN, P. & D. COMBES. 1984. Effect of water activity on enzyme action and stability. Ann. N.Y. Acad. Sci. **434:** 48–60.
12. HAHN-HÄGERDAL, B. 1986. Water activity: a possible external regulator in biotechnical processes. Enzyme Microb. Technol. **8:** 322–327.

Stereoselective Enzymatic Esterification of 3-Benzoylthio-2-methylpropanoic Acid

RAMESH N. PATEL, JEFFREY M. HOWELL, AMIT BANERJEE, KENNETH F. FORTNEY, AND LASZLO J. SZARKA

Department of Microbial Technology
Bristol-Myers Squibb
Pharmaceutical Research Institute
New Brunswick, New Jersey 08903

INTRODUCTION

Captopril is designated chemically as 1-[(2*S*)-3-mercapto-2-methylpropionyl]-L-proline *2*. It is used as an antihypertensive agent through suppression of the renin-angiotensin-aldosterone system.[1,2] Captopril prevents the conversion of angiotensin I to angiotensin II by inhibition of ACE (angiotensin-converting enzyme).[3] The potency of captopril *2* as an inhibitor of ACE depends critically on the configuration of the mercaptoalkanoyl moiety; the compound with the *S*-configuration is about 100 times more active than its corresponding *R*-enantiomer.[4]

The required 3-mercapto-(2*S*)-methylpropionic acid moiety has been prepared from the microbially derived chiral 3-hydroxy-(2*R*)-methylpropionic acid, which is obtained by the hydroxylation of isobutyric acid.[5–7]

The use of extracellular lipases of microbial origin to catalyze the enantioselective hydrolysis of 3-acylthio-2-methylpropanoic acid ester in an aqueous system has been demonstrated to produce optically active 3-acylthio-2-methyl propanoic acid.[8–10]

In this report, we describe the stereoselective esterification of racemic 3-benzoylthio-2-methylpropanoic acid in an organic solvent system to yield *R*-(+) methyl ester and unreacted acid enriched in the desired *S*-(−) enantiomer. *S*-(−) 3-benzoylthio-2-methylpropanoic acid is a key intermediate for the synthesis of captopril[11] or zofenopril[3,12] both are antihypertensive drugs (SCHEME 1).

MATERIALS AND METHODS

Enzyme Source

Crude lipases AK, P-30, CES-30, and CE from *Pseudomonas* sp., lipase N from *Rhizopus* sp., lipase MAP from *Mucor* sp., lipase AY from *Candida* sp., lipase CE from *Humicola langinosa,* lipase G from *Penicillium* sp., and lipase APF-12 from *Aspergillus niger* were obtained from the Amano International Enzyme Co., Inc., Troy, Virginia. Lipase B from *Pseudomonas fluorescens* and lipase F9 from *Aspergillus* sp. were obtained from Enzymatics Ltd., Cambridge, United Kingdom.

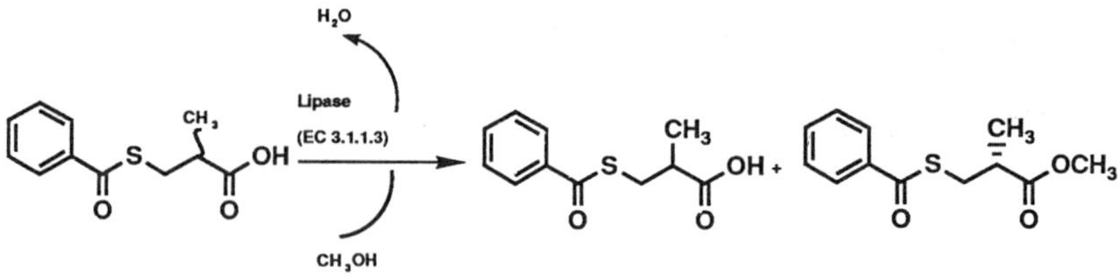

Racemic 3-benzoylthio-2-methylpropanoic acid (Racemic 1)

S-(-)-3-benzylthio-2-methylpropanoic acid (S-(-) 1)

R-(+)-3-benzoyl-thio-2-methyl-propanoic acid methyl ester

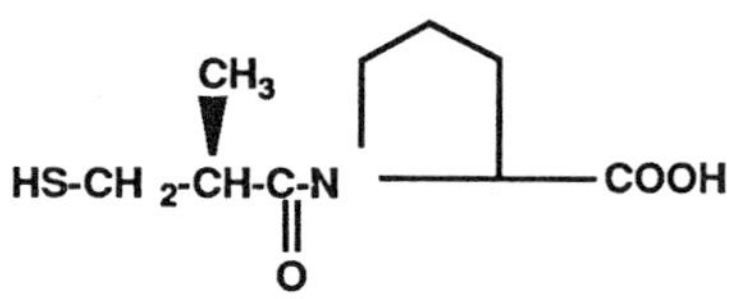

1-[(2 S)-3-mercapto-2-methyl 1-oxopropyl]-L-proline

2 Captopril

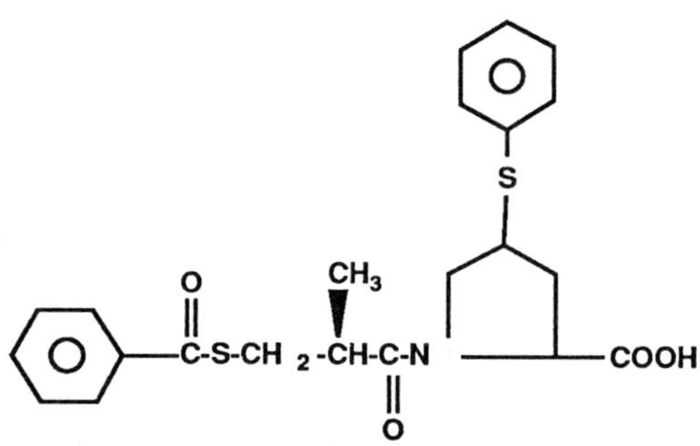

[1(R [1]),2 α,4 α]-1-[3-(benzoylthio)-2-methyl-1-oxopropyl]-4-(phenylthio)-L-proline

3 Zofenopril

SCHEME 1

Lipase Activity Assay

The lipase activity was determined by titrimetric assay as recommended by Sigma Chemical Co. One unit will hydrolyze 1.0 microequivalent of fatty acids from a triglyceride in one hour at pH 7.7 and 37°C.

Immobilization of Lipase

Three different resins, XAD-7 (Amberlite nonionic polymeric adsorbent, 20–60 mesh, polyacrylate), XAD-2 (Amberlite nonionic polymeric adsorbent, 20–60 mesh, polystyrene), and Accurel polypropylene (PP) (nonionic polymeric adsorbant, 200–400 mesh polypropylene), were used for the immobilization of lipase. Crude Amano P-30 lipase from *Pseudomonas* sp. (30 g) was dissolved in 75 ml of distilled water and centrifuged at 15,000 × *g* for 15 minutes at 4°C to obtain a clear supernatant. Twenty five milliliters of the supernatant was added to 1.0 g each of resin previously washed three times with 25 ml of methanol. Resins and enzyme solutions were gently agitated on a gyrotory shaker at room temperature for 24 hours. Adsorption of enzyme to each resin was checked periodically by determining the lipase activity and protein content remaining in the supernatant obtained after centrifugation of the enzyme-resin mixture. After 24 hours, the slurry was filtered and the cake was washed with 400 ml of distilled water. The resins containing immobilized enzyme were dried under vacuum for 24 hours.

Crude Amano P-30 *Pseudomonas* sp. lipase was also immobilized on Accurel PP on a preparative scale. The enzyme (500 g) was dissolved in 3.0 L of distilled water and centrifuged at 15,000 × *g* for 15 minutes at 40°C to obtain a clear supernatant. Accurel PP (50 g) previously washed with methanol was added to the enzyme solution, and the mixture was agitated gently on a gyrotory shaker at room temperature for 24 hours. Periodically, samples were taken to check lipase activity and protein in the supernatant obtained after centrifugation. After 24 hours, the enzyme and resin slurry was filtered and the cake was washed with 15 L of distilled water. The resin containing immobilized enzyme was dried under vacuum at 25°C for 72 hours.

Analytical Methods

The esterification of 3-benzoylthio-2-methylpropanoic acid was monitored by gas chromatography (flame ionization detector) of reaction mixture samples using a Hewlett Packard fused silica capillary column (cross-linked methyl silicone, 15 m long, 1.0 μm film thickness, 0.31 mm i.d., HP 190912-215) at 215°C oven tempeature, 250°C injection temperature, and 250°C detection temperature. The retention time for 3-benzoylthio-2-methylpropanoic acid was 3.8 minutes and for the corresponding methyl ester was 3.1 minutes.

The optical purity of 3-benzoylthio-2-methylpropanoic acid was determined by treating the samples with thionyl chloride and reacting the acid chloride formed with *d*-2-octanol to yield the diastereomers.[13] In the procedure used, 1 ml of reaction mixture filtrate (3 mg of 3-benzoylthio-2-methylpropanoic acid) was evaporated to remove solvent. One ml of 10% thionyl chloride solution in *n*-hexane containing 2.5 μl of dimethylformamide was added to the vial, mixed thoroughly on a vortex mixer, and incubated at room temperature for at least 30 minutes. The reagent was subsequently evaporated under a gentle stream of nitrogen at 50°C. To the dried residue, 0.3 ml of *d*-2-octanol was added, mixed thoroughly on a vortex mixer, and incubated at 60°C for 30 minutes in a heating block. Octanol was evaporated under a stream of nitrogen at 50°C, and the residue was dissolved in 0.5 ml of methylene chloride. Diastereomers were analyzed by gas chromatography (flame ionization detector) using a Hewlett Packard fused silica capillary column (HP No. 190912-215) at 215°C oven temperature, 250°C injection temperature, and 250°C detector temperature. The retention times for the esters of the L- and D-enantiomers were 14.8 minutes and 16.0 minutes, respectively.

Enzymatic Reaction

The enzymatic reaction mixture for the esterification contained 0.1-M (22.4 mg/ml) 3-benzoylthio-2-methylpropanoic acid, 0.4-M methanol, 0.1% water, and 1 g of crude lipase in 10 ml of toluene. The reaction was conducted at 40°C and 280. Periodically, samples were taken and analyzed by GC to determine the yield and optical purity of the product.

Evaluation of Immobilized Amano P-30 Lipase

Immobilized crude Amano P-30 lipase on Accurel PP was evaluated in a 3-ml and 50-ml reactor volume. The reaction mixture contained 0.1-M 3-benzoylthio-2-methylpropanoic acid (22.4 mg/ml), 0.4-M methanol, 0.1% water and 0.3 g and 6 g of immobilized enzyme in 3 ml or 50 ml of toluene, respectively. The reaction was conducted at 28°C and 280 RPM.

Lipase Fermentation

Pseudomonas sp. ATCC 21808 was used in the fermentation. One frozen vial of culture was used to inoculate 100 ml of F7 medium (1% w/v malt extract, 1% w/v yeast extract, 0.3% w/v peptone, and 2% w/v glucose) in a 500-ml Erlenmeyer flask. The flask was incubated at 28°C, 280 RPM for 72 hours and the entire contents used to inoculate 1.5 L of F7 medium in a 4-L flask. The flask was incubated at 28°C, 180 for 24 hours before being used to inoculate the germinator to grow seed culture in a 380-L tank. The germinator contained 150 L of medium L (3% w/v cornsteep liquor, 1.1% w/v cerelose hydrate, 0.6% w/v urea, 0.2% w/v potassium dihydrogen phosphate, 0.05% w/v potassium chloride, 0.05% w/v magnesium sulfate, 1% w/v soybean oil, and 0.1% w/v SAG antifoam). The culture was grown at 280 agitation, 150 LPM aeration, 10 psig pressure, 26°C temperature. Germinator-grown culture (12.5 L) was used to inoculate 250 L of medium L contained in a 380-L fermentor. The fermentation conditions were 250 agitation, 250 LPM aeration, 10 psig pressure, and 26°C.

Enzyme Recovery

At the end of the fermentation, the culture broth was centrifuged through a high-speed Sharples centrifuge to remove cells. The filtrate was concentrated under vacuum at 50°C to one-tenth volume in a climbing film evaporator. Cold ethanol (250 L, 5–10°C) was added to 25 L of stirred concentrate, and the mixture was kept for 12 hours at 0–5°C without agitation. The supernatant was decanted, and the slurry was filtered. The filtered precipitate was washed with cold ethanol, and the cake was vacuum-dried at 40°C for 48 hours to obtain the crude lipase. This was immobilized on Accurel PP as described earlier.

RESULTS AND DISCUSSION

Screening of Lipases

Commercially available lipases were screened for the stereoselective esterification of racemic *1* to yield the acid R-(+) methyl ester and unreacted acid enriched in the desired *S*-(−) **1**. Lipases (P-30, AK, CES-30, B1) from various *Pseudomonas* sp. catalyzed the stereoselective esterification (TABLE 1). *S*(−) **1** was obtained with a reaction yield of 37% (based on racemic substrate initially present) and an optical purity of 97.2% using Amano Lipase P-30. Poor selectivity was observed with lipases from *Candida cylindracea, Rhizopus* sp., *Mucor* sp., *Penicillium* sp., and *Humicola langinosa.* Further research was conducted using lipase P-30.

TABLE 1. Evaluation of Lipases for the Stereoselective Estrification of 3-Benzylthio-2-methyl Propanoic Acid

Lipase Source	Reaction Time (hours)	Conversion (%)	Yield (%)	Optical Purity (%)
Candida cylindraceae	1.25	64	36	38.7
Rhizopus sp. (Amano N)	26.5	58	42	56
Mucor sp. (Amano MAP)	0.5	78	22	58.4
Penicillin sp. (Amano P)	99	61	39	42
Pseudomonas sp. (Biacatalyst Co.)	27	68	32	95.9
Pseudomonas sp. (Enzymatics Co.)	27	58	42	91.7
Aspergillus niger (Amano APF)	99	86	14	68.6
Pseudomonas sp. (Amano AK)	27	56	44	86.4
Pseudomonas sp. (Amano CES)	27	59	41	82.6
Humicola langinosa (Amano CE)	27	78	22	56.1
Pseudomonas sp. (Amano P-30)	27	62	38	97.0

[a]Reaction mixture in 5 ml of toluene contained 0.025-M 3-benzylthio-2-methyl propanoic acid, 0.1-M methanol, 0.1% water, and 2 grams of crude lipase. Reaction was conducted at 40°C. The reaction yield and optical purities were determined by GC.

Screening of Alcohols

Six different alcohols were screened in the esterification reaction using lipase P-30 (TABLE 2). Methanol exhibited the highest stereoselectivity. Benzyl alcohol and 1-octanol also exhibited comparable stereoselectivity. Trifluoroethanol and 2-amino-1-butanol exhibited poor selectivity.

Immobilization and Evaluation of Lipase P-30

Crude lipase P-30 was immobilized on three different resins, XAD-7, XAD-2, and Accurel PP. The absorption efficiencies were about 68%, 71%, and 98.5%, respectively. These immobilized lipases were evaluated for the ability to stereoselectively esterify racemic **1.** As shown in TABLE 3, enzyme immobilized on Accurel PP

TABLE 2. Screening of Alcohols for the Estrification Reaction[a]

Alcohol	Reaction Time (hours)	Conversion (%)	Yield (%)	Optical Purity (%)
	112	50.3	49.7	91.7
Methanol	136	55.1	44.9	95.4
	162	63.0	37.0	97.3
2-Amino-1-ethanol	16	68.3	31.7	50.0
	40	93.3	6.7	49.9
Trifluoroethanol	162	28.8	71.2	61.8
	328	38.4	61.6	63.0
Isopropylidene	112	33.0	67.0	68.6
glycerol	280	58.3	41.7	86.7
Benzyl alcohol	112	39.1	60.9	71.9
	280	66.9	33.1	96.0
1-Octanol	112	34.0	66.0	68.0
	280	68.0	32.0	95.1

[a]Reaction mixture in 25 ml of toluene contained 0.1-M (22.4 mg/ml) of 3-benzylthio-2-methyl propanoic acid, 0.4-M alcohol as indicated, 0.1% water, and 1 gram of Amano lipase P-30. Reaction was conducted at 28°C, 280 RPM on a rotary shaker.

catalyzed efficient esterification, giving 36–45% reaction yield and 97.7% optical purity of $S(-)$ **1**. The enzyme under identical conditions gave similar optical purity and yield of product in two additional reaction cycles.

Immobilized enzyme was evaluated for reusability in 3-ml and preparative 50-ml reactor volumes. Enzyme was reused over 23 cycles without loss of activity in a 3-ml reactor and over 15 cycles in a 50-ml reactor. The optical purity of 97.7% and reaction yield of 40% was obtained for S-$(-)$ **1.**

Evaluation of Solvents for Esterification

Various solvents such as heptane, cyclohexane, n-octane, and toluene were used in the esterification reaction. Toluene was the best based on the yield and optical

TABLE 3. Evaluation of Immobilized Lipase P-30 on Various Supports for the Esterification Reaction[a]

Cycle No.	Immobilized Lipase on Support	Reaction Time (hours)	Conversion (%)	Yield (%)	Optical Purity (%)
1	XAD-7	17.5	70.6	29.4	96.7
	XAD-2	17.5	28.0	72.0	—
		25.0	35.5	64.5	—
		41.5	50.0	50.0	82.5
	Accurel PP	17.5	64.1	35.9	97.5
		25.0	66.1	33.1	97.7
2	XAD-7	17.0	61.2	33.1	91.8
	Accurel PP	17.0	49.1	50.9	—
		22.5	54.2	45.8	97.7
3	XAD-7	16.0	62.8	37.2	90.2
		16.0	55.6	46.4	96.7

[a]Reaction mixture in 3 ml of toluene contained 0.1-M substrate, 0.4-M methanol, 0.1% (v/v) water, and 0.3 gram of immobillzed Amano P-30 lipase. Reaction was conducted at 28°C.

purity of the product. Octane and cyclohexane gave products with high optical purities but in lower yields (TABLE 4).

Effect of Temperature on Esterification

The esterification was conducted at 28°C, 40°C, and 60°C. On increasing the reaction temperature from 28°C to 40°C and to 60°C, the reaction time decreased from 18 hours to 4 hours and to 2 hours, respectively. The optical purity (ca. 82%) obtained at 40°C and 60°C was inferior; however, a higher yield (46%) was obtained. The reaction yield of 37% and optical purity of 97.7% were obtained at 28°C.

Effect of Substrate Concentration on Esterification

The racemic acid was evaluated at 0.25-M, 0.5-M, and 1.0-M substrate concentrations with 1-M, 2-M, and 4-M methanol, respectively, in the esterification reaction using 0.3 g of lipase P-30 immobilized on Accurel PP. The yield (38–47%) was similar; however, the optical purity decreased with increasing substrate concentration (TABLE 5).

TABLE 4. Effect of Organic Solvent on the Esterification Reaction[a]

Solvent	Reaction Time (hours)	Conversion (%)	Yield (%)	Optical Purity (%)
Heptane	16.5	82.9	17.1	92.2
Cyclohexane	16.5	90.6	9.4	97.8
N-Octane	16.5	91.4	8.6	97.3
Toluene	16.5	63.0	37.0	97.2

[a]Reaction mixture in 4 ml of solvent contained 0.1-M substrate, 0.4-M methanol, 0.1% (v/v) water, and 0.48 gram of immobilized Amano P-30 lipase. Reactions were conducted at 40°C, 280 RPM.

TABLE 5. Effect of Substrate Concentration on the Esterification Reaction[a]

Substrate Concentration (M)	Methanol Concentration (M)	Reaction Time (hours)	Conversion (%)	Yield (%)	Optical Purity (%)
0.1	0.4	6.0	55.7	44.3	97.1
0.25	1.0	17.0	61.7	38.3	96.5
0.5	2.0	23.5	53.8	46.2	92.0
1.0	4.0	48.0	52.6	47.4	90.0

[a]Reaction mixture in 10 ml of toluene contained substrate and methanol as indicated, 0.1% (v/v) water, and 1.2 grams of Amano P-30 lipase immobilized on Accurel PP. Reaction was conducted at 40°C.

Effect of Methanol Concentration

Various methanol concentrations (0.5 M, 1.0 M, 2.0 M, and 4.0 M) were evaluated at 0.5-M substrate concentration. The rate of esterification decreased as the methanol-to-substrate ratio was increased from 1:1 to 4:1 (TABLE 6). Higher methanol concentrations may inhibit the esterification reaction by stripping the essential water from the enzyme. Note the marked decrease in optical purity at >3 M methanol, indicating a marked change in the enzyme.

Lipase Fermentations

The highest lipase activity achieved in batch fermentation using *Pseudomonas* sp. SC13809 was 56 units/ml. A significant increase in lipase production was achieved using a fed-batch culture in which soybean oil was added to the fermentor at a level of 1% w/v. After 24 hours of batch culture, additional soybean oil was supplied by adding 250 ml of soybean oil every hour from 24 to 48 hours.

As seen in FIGURE 1, very little lipase was produced during the growth phase (up to 37 hours). Most of the glucose was consumed during the first 37 hours, which corresponded to the period of rapid cell growth. Although lipase activity was low during the first 45 hours of fermentation, secretion of lipase quickly accelerated between 45 and 71 hours of fermentation. The concentration of lipase, 128 units/ml, was increasing steeply at harvest and would undoubtedly have been greater had the

TABLE 6. Effect of Methanol Concentration on the Esterification Reaction[a]

Methanol Concentration (M)	Reaction Time (hours)	Conversion (%)	Yield (%)	Optical Purity (%)
0.5	5.5	55.0	45.0	93.6
1.0	8.0	61.2	38.8	96.0
1.5	21.0	60.0	40.0	96.5
2.0	30.0	62.0	38.0	95.5
3.0	64.0	13.0	87.0	56.0
4.0	64.0	6.0	94.0	56.0

[a]Reaction mixture in 10 ml of toluene contained 0.5-M substrate, 0.1% (v/v) water, and 1 gram of Amano P-30 lipase immobilized on Accurel PP. Reaction was conducted at 40°C.

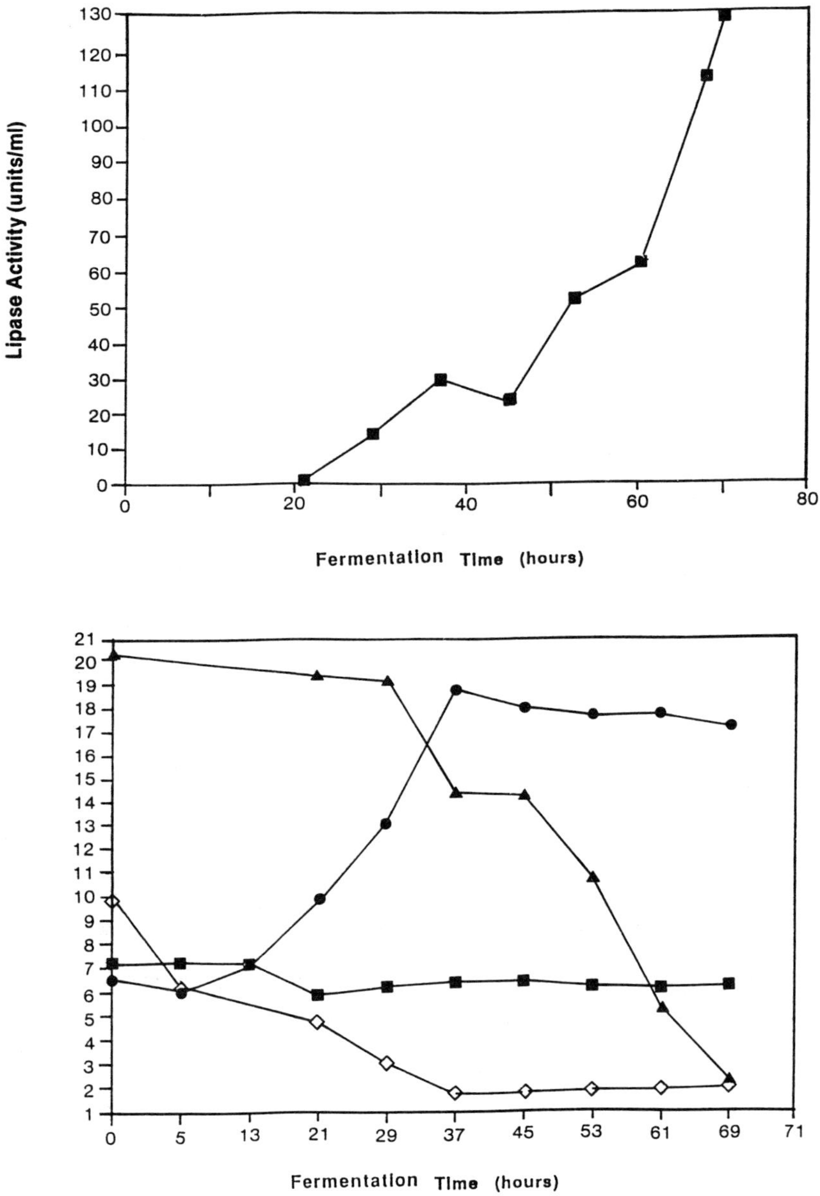
Lipase Activity (units/ml)
130
120
110
100
90
80
70
60
50
40
30
20
10
0
0
20
40
60
80
Fermentation Time (hours)
21
20
19
18
17
16
15
14
13
12
11
10
9
8
7
6
5
4
3
2
1
0
5
13
21
29
37
45
53
61
69
71
Fermentation Time (hours)

FIGURE 1

fermentation been extended. The lipase activity achieved represents an increase of 129% over the highest we have previously achieved in a batch fermentation (56 units/ml).

Crude lipase (Bristol-Myers Squibb lipase, or BMS) was recovered from the fermentation broth by ethanol precipitation and was also immobilized on Accurel PP. Crude lipase from three different fermentation batches (NWM0589, 0506, and 0889) and immobilized lipase from batch NWM0589 were evaluated in the esterification reaction. A yield of 35% and the optical purity of 96.5% were obtained using crude lipase batch NWM-0889), while a yield of 48% and optical purity of 87.7% were obtained with immobilized lipase (batch NWM0589).

REFERENCES

1. ONDETTI, M. A., B. RUBIN & D. W. CUSHMAN. 1977. Design of specific inhibitors of angiotensin-converting enzyme: new class of orally active antihypertensive agents. Science **196:** 441–444.
2. ONDETTI, M. A. & D. W. CUSHMAN. 1981. Inhibition of renin-angiotensin system: a new approach to the theory of hypertension. J. Med. Chem. **24:** 355–361.
3. CUSHMAN, D. W. & M. A. ONDETTI. 1980. Inhibitors of angiotensin-converting enzyme for treatment of hypertension. Biochem. Pharmacol. **29:** 1871–1877.
4. CUSHMAN, D. W., M. S. CHEUNG, E. F. SABO & M. A. ONDETTI. 1977. Design of potent competitive inhibitors of angiotensin-converting enzyme. Carboxyalkanoyl and mercaptoalkanoyl amino acids. Biochemistry **16:** 5484–5491.
5. GOODHUE, C. T. & J. R. SCHAEFFER. 1971. Preparation of L(+) β-hydroxyisobutyric acid by bacterial oxidation of isobutyric acid. Biotechnol. Bioeng. **13:** 203–214.
6. SCHIMAZAKI, M., J. HASEGAWA, K. KAN, K. NEMURA, Y. NOSE, H. KONDO, T. OHASHI & K. WATANABE. 1982. Synthesis of captopril from an optically active β-hydroxy acid. Chem. Pharmacol. Bull. **30:** 3139–3146.
7. HASEGAWA, J., M. OGURA, H. KANEMA, N. NODA, H. KAWAHARADA & K. WATANABE. 1982. Production of D-β-hydroxyisobutyric acid from isobutyric acid by *Candida rugosa* and its mutant. J. Ferment. Technol. **60:** 501–508.
8. SIH, C. T. 1987. Process for preparing optically active 3-acylthio-2-methyl propionic acid derivatives. European Patent 87264125.
9. GU, Q. M., D. R. REDDY & C. J. SIH. 1986. Bifunctional chiral synthons via biochemical methods. VIII. Optically active 3-aroylthio-2-methylpropionic acids. Tetrahedron Lett. **27:** 5203–5206.
10. SAKIMAS, A., K. YURI, N. RYOZO & O. HISAO. 1986. Process for preparing optically active carboxylic acids and antipode esters thereof. U.S. Patent 4629701.
11. MONIOT, J. L. 1988. Preparation of *N*-[2-(mercaptomethyl)propionyl]-L-prolines. U.S. Patent Application CN 88-100862.
12. ONDETTI, M. A. & J. KRAPCHO. 1982. Mercaptoacyl derivatives of substituted prolines. U.S. Patent 431906.
13. JEMAL, M. & A. T. COHEN. 1987. Determination of enantiomeric purity of Z-oxylysine by capillary gas chromatography. J. Chromatography **394:**388–394.

Dextran and Oligosaccharide Production with Glucosyltransferases from Different Strains of *Leuconostoc mesenteroides*

E. CASTILLO,[a] F. ITURBE,[a]
A. LOPEZ-MUNGUIA,[a,c] V. PELENC,[b]
F. PAUL,[b] AND P. MONSAN[b]

[a] *Instituto de Biotecnología UNAM*
Apartado Postal 510-3
Cuernavaca, Mor.
62271, Mexico

[b] *BioEurope S.A.*
4 impasse Didier Daurat, ZI Montaudran
31400, Toulouse, France

INTRODUCTION

The glucosyltransferases are one of the most studied group of enzymes, and there is an increasing interest in their application for the synthesis of novel sugars and oligosaccharides.[1] Among these enzymes, dextransucrase (E.C. 2.4.1.5) produced by *Leuconostoc* species is capable of efficiently synthesizing oligosaccharides from sucrose and low molecular weight sugar acceptors and has been extensively studied for the industrial production of dextran.[2] However, as dextran structure depends on the glucosyltransferase-producing strain, it is interesting to consider strains other than the industrial for potential applications in the food and pharmaceutical industries.

In this work a screening of the main properties of 12 *Leuconostoc mesenteroides* strains is reported. The strains belong to the original collection of Jeanes (1954)[3] and were kindly provided by the NRRL in Peoria, Ill. They were characterized in terms of glucosyltransferase activity, the localization of the activity (soluble or associated to the insoluble products of the culture), the solubility of the polysaccharides synthesized from sucrose, and the type of oligosaccharides produced when acceptor sugars were used in the reaction. In all cases, reference is made to the behavior of dextransucrase from the industrial strain (*Leuconostoc mesenteroides* B-512 F).

[c] Author to whom correspondence should be addressed.

MATERIALS AND METHODS

Culture Conditions

All strains were grown at 27–29°C, initial pH of 6.9, with the following medium: sucrose 2%, yeast extract 2%, K2HPO4 2%, MgSO4 0.02%. CaCl2 0.005%, and FeSO4, NaCl, MnSO4 at 0.001%, as described by Monsan *et al.*[4]

Fermentations were carried out in erlenmeyer flasks and in 2-liter fermenters. When stated, the pH was regulated either with 2 N NaOH or with an alkaline sucrose solution (400 g/l sucrose in 2 N NaOH) at pH = 6.7. Agitation was 550 to 600 rpm, and aeration rate 1 vvm. When stated, maltose was added to the fermentation medium at a sucrose/maltose ratio of 2 (w/w).

Enzyme Activity

Total glucosyltransferase activity was measured by following the initial release of reducing sugars by the DNS methods.[5] One activity unit (UI) is defined as the amount of enzyme that produces one μmol of fructose per minute at 10% sucrose concentration, pH = 5.2 and 30°C.

Sugar and Polymer Analysis

Polymers obtained by enzymatic synthesis were recovered by ethanol precipitation and dried for quantification. Insoluble polymers were directly recovered by centrifugation.

Acceptor reaction products were analyzed by high-performance liquid chromatography (HPLC) in a Beckman instrument with a refraction index detector and a Spherogel C18 column with water at 0.5 ml/minute as the elution solvent.

RESULTS AND DISCUSSION

The main results concerning the first analysis of 12 *Leuconostoc mesenteroides* strains is presented in TABLE 1. In erlenmeyer experiments all the strains showed a similar behavior, with growth in around 8 hours and with final optical densities of 5–8 units. The exceptions were strains 1297 and 1120 which needed more than 24 hours. It is shown that almost all the strains produce enzymes both in the supernatant and associated to cells and insoluble polymers. The enzymes in the soluble part of the culture were used for the synthesis of polymers in the presence of sucrose. As may be observed several enzymes produce insoluble polymers, some in considerable quantities, and in some cases they also produce gels. It is interesting to observe the high glucosyltransferase activities of strains 1499 and 1501. This activity is even higher than the one obtained with the industrial strain under similar production conditions.

Production System for Glucosyltransferase from L. mesenteroides *Strains*

We have already shown that fed-batch cultures with pH regulation with alkaline solution is the best production system for *L. mesenteroides* NRRL B-512F and B-1355.[2,6] In FIGURE 1, the production of glucosyltransferase from the strain IB-101 is shown under various conditions. It is shown that enzyme production increases more than twofold when fed batch is used instead of batch cultures. This conclusion applies to almost all strains and is illustrated in TABLE 2.

Synthesis of Oligosaccharides in the Presence of Maltose

Synthesis reactions in the presence of maltose were also carried out. In all cases, a ratio sucrose/maltose concentration of 2 (w/w) was used. It is interesting to point out that, besides the usual oligosaccharides obtained with dextransucrase from the B-512F strains, the enzymes from the strains B-523 and IB-101 also form insoluble products. This may be due either to low transfer capacity of the enzyme to maltose or, eventually, insoluble products of the acceptor reaction.

TABLE 1. Preliminary Evaluation of Different Strains of *Leuconostoc mesenteroides*

Strain NRRL B-	Final pH	Growth after 12 Hours (o.d. = 650 nm)	Total GTF Activity: Soluble (UI/ml)	Total GTF Activity: Insoluble (UI/mg)[d]	Percent Insoluble Polymer[c]: Sucrose 10%	Percent Insoluble Polymer[c]: Sucrose 10% Maltose 5%
523	4.71	7.50	0.650	[b]	90	—
1438	4.79	6.71	0.185	0.446	12	16
1433	4.94	1.99	0.787	0.282	17	18
1149	6.04	5.52	0.335	0.623	11	20
1399	4.99	8.01	1.330	0.226	6	19
1254	6.13	5.26	0.028	0.024	11	18
1120	6.70	0.10[a]	[b]	[b]	9	—
1297	6.00	2.29[a]	0.281	0.321	9	—
1498	4.48	15.20	6.700	[b]	0	0
1501	4.45	15.20	5.100	[b]	0	0
1299	4.70	17.64	0.370	0.400	0	0
1B-101	5.17	7.23	0.430	0.310	23	39

[a]After 24 hours.
[b]Not determined because of low enzymatic activity.
[c]Units/mg of insolubles.
[d]Reaction conditions: 1 UI/ml, 30°C, pH = 5.2.

The chromatograms of the soluble products of the acceptor reactions of some strains show that they all produce the usual products of dextransucrase from *L. mesenteroides* B-512F (Panose, tetrasaccharide, and pentasaccharide). Only the strains IB-101, B-1299 and B-1355 produce dextransucrases able to synthesize oligosaccharides different from those of the B-512 F strain in significant amounts.

Products of the strain B-1299 have been identified as glucooligosaccharides containing α(1-2) oligosaccharides.[7] This activity complements that of alternansu-

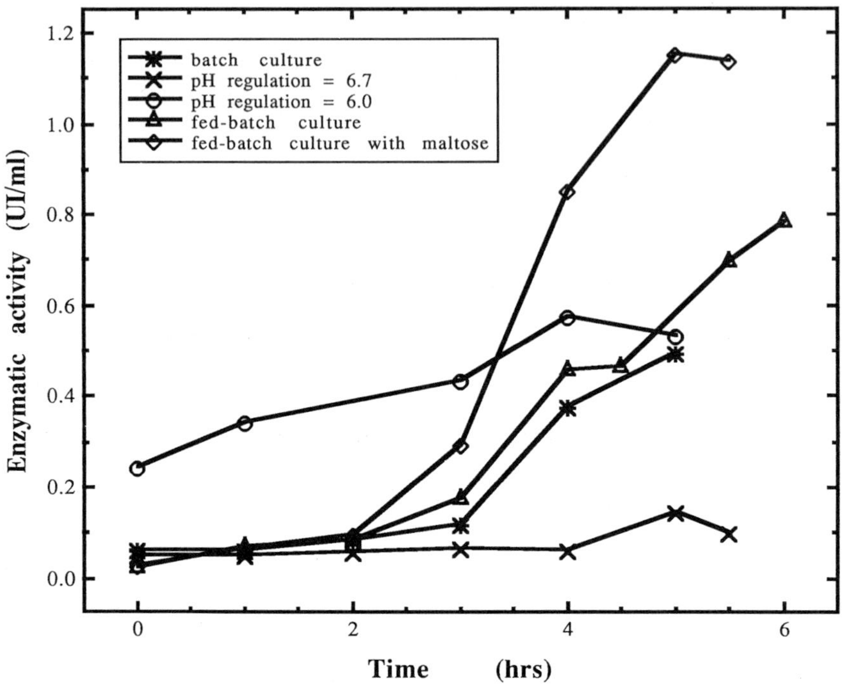

FIGURE 1. Effect of the type of culture on the glucosyltransferase production from *L. mesenteroides* IB-101.

crase from the B-1355 strain[8] for the production of glucose oligosaccharides from sucrose, containing unusual linkages, as illustrated in FIGURE 2.

CONCLUSIONS

Glucosyltransferases from *Leuconostoc mesenteroides* are efficiently produced by fed-batch culture with pH regulation with alkaline sucrose. It is shown that the

TABLE 2. Effect of Fed-batch Culture in the Production of Glucosyltransferase Activity in Various *L. mesenteroides* Strains

Strain NRRL B-	Enzymatic Activity (UI/ml)	
	Batch Culture	Fed-Batch Culture
512	2.0	8.0
523	0.6	2.2
1299	0.3	0.5
1355	0.2	0.8
IB-101	0.5	1.2

activity may be found both in the supernatant and associated to the insoluble products of the culture (cells and polymer). There are enzymes like glucosyltransferase from *L. mesenteroides* B-523 that produce mainly insoluble polymers. Considering the type of linkages synthesized with some glucosyltransferases in the presence of acceptors, these enzymes may be an efficient tool for the controlled transfer of glucosyl groups to various acceptors.

SUMMARY

Twelve *Leuconostoc mesenteroides* strains from the collection of the NRRL (Peoria, Ill.) have been compared with respect to enzyme production. It is shown that

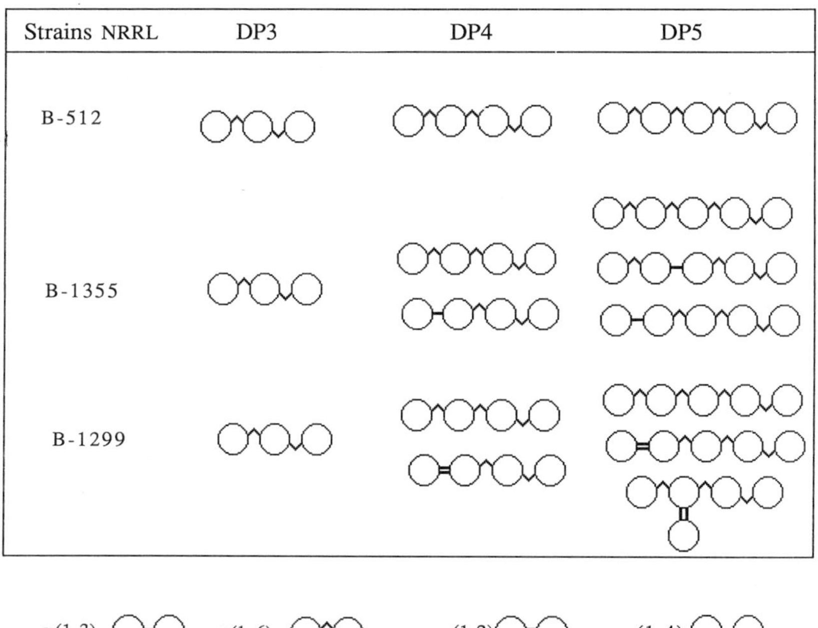

FIGURE 2. Main products of acceptor reactions with maltose using various *L. mesenteroides* dextransucrases.

the fed-batch fermentation is the best production method for glucosyltransferases from *L. mesenteroides* Three strains exhibited interesting properties: the production of highly insoluble glucan containing a considerable amount of branching linkages and glucose oligosaccharides different from those obtained with *L. mesenteroides* NRRL B-512F.

REFERENCES

1. MONSAN, P., F. PAUL, M. REMAUD & A. LOPEZ-MUNGUIA. 1989. Novel enzymatic synthesis of oligosaccharides. Food Biotechnol. **3:** 111–130.
2. MONSAN, P. & A. LOPEZ-MUNGUIA. 1981. On the production of dextran by free and immobilized dextransucrase. Biotechnol Bioeng. **23:** 2027–2037.
3. JEANES, A. R., W. C. HAYNES, C. A. WILHAM, J. C. RANKIN, E. H. MELVIN, M. J. AUSTIN, J. E. CLUSKEY, B. E. FISHER, H. M. TSUCHIYA & C. E. RIST. 1954. Characterization and classification of dextrans from ninety-six strains of bacteria. J. Am. Chem. Soc. **76:** 5041–5052.
4. MONSAN, P., F. PAUL, D. AURIOL & A. LOPEZ-MUNGUIA. 1987. Dextran synthesis using immobilized dextransucrase. Methods Enzymol. **136:** 239–253.
5. SUMNER, J. B. & S. F. HOWELL. 1935. A method for the determination of invertase activity. J. Biol. Chem. **108:** 51–54.
6. LOPEZ-MUNGUIA, A., V. PELENC, M. REMAUD, F. PAUL, P. MONSAN, J. BITON, J. M. MICHEL & C. LANG. 1990. Production and purification of *Leuconostoc mesenteroides* NRRL B-1355 Alternansucrase. Ann. N.Y. Acad. Sci. **613:** 717–722.
7. PAUL, F., A. LOPEZ-MUNGUIA, M. REMAUD, V. PELENC & P. MONSAN. 1988. French patent 2 626,583.
8. PELENC, V., A. LOPEZ-MUNGUIA, M. REMAUD, J. BITON, J. M. MICHEL, F. PAUL & P. MONSAN. 1991. Enzymatic synthesis of oligoalternans. Sci. Aliments **11:** 465–476.

Bioconversion of Nonnatural Organic Compounds

Esterification and Dehydrogenation of Organosilicon Compounds

ATSUO TANAKA, TOSHIAKI FUKUI, AKINORI UEJIMA, MIN-HUA ZONG, AND TAKUO KAWAMOTO

Laboratory of Industrial Biochemistry
Department of Industrial Chemistry
Faculty of Engineering
Kyoto University
Yoshida, Sakyo-ku
Kyoto 606, Japan

INTRODUCTION

Recently, the use of nonnatural organic compounds as enzyme substrates has been becoming more and more important in the field of biotechnology, thereby increasing the potential and expanding the use of enzymes. For example, bioconversion of organosilicon compounds is of great interest both in the fundamental study of enzymology and in the field of industrial biochemistry because organosilicon compounds play important roles in organic chemistry.[1] The production of useful organosilicon compounds is also very important. We have already reported the use of organosilicon compounds as nonnatural substrates of hydrolases and showed that trimethylsilylmethanol was an efficient nonnatural substrate for stereoselective esterification of 2-(4-chlorophenoxy)propanoic acid with yeast lipase in organic media.[2]

However, optical resolution of organosilicon compounds themselves is also of interest and importance in synthetic organic chemistry. Therefore, optical resolution of organosilicon compounds by hydrolases was studied. As model nonnatural organic compounds, three kinds of propanol derivatives containing the silicon atom—that is, 1-trimethylsilyl-2-propanol, 1-trimethylsilyl-1-propanol, and 2-trimethylsilyl-1-propanol—have been selected, and they were examined as the substrates in the stereoselective esterification with 5-phenylvaleric acid by hydrolases.

Furthermore, bioconversion of organosilicon compounds by alcohol dehydrogenase is very attractive because of their importance in asymmetric alcohol/carbonyl oxidoreduction in synthetic organic chemistry.[3] Then the enzymatic activity of horse liver alcohol dehydrogenase (HLADH) for the dehydrogenation of organosilicon compounds, $Me_3Si(CH_2)_nOH$ (n = 1–3), was examined. Furthermore, the effect of the silicon atom in these compounds on the HLADH-catalyzed reaction was discussed on the basis of the physical properties of the silicon atom, comparing it with the corresponding carbon compounds, $Me_3C(CH_2)_nOH$ (n = 1–2).

METHODS

For stereoselective esterification of organosilicon compounds, hydrolases adsorbed on Celite No. 535 (hydrolase, 100 mg; deionized water, 0.1 ml; Celite No. 535, 250 mg) were used in water-saturated isooctane containing 100 mM substrates. In the case of dehydrogenation of organosilicon compounds, 0.408 mg of HLADH was used in 3.4 ml/tetrahydrofuran, 50 mM Tris-HCl buffer (pH 8.8) containing 5% 1.2 mM NAD^+, and substrate.

RESULTS AND DISCUSSION

For the optical resolution of three organosilicon compounds, that is, 1-trimethylsilyl-2-propanol, 1-trimethylsilyl-1-propanol, and 2-trimethylsilyl-1-propanol, the stereoselective esterification of these compounds with 5-phenylvaleric acid was examined. We used the following hydrolases: lipase OF 360 from *Candida cylindracea,* lipase Saiken 100 from *Rhizopus japonicus,* Steapsin from hog pancreas, lipoprotein lipase type A from *Pseudomonas* sp., and cholesterol esterase type A from *Pseudomonas* sp.

TABLE 1. Esterification of Racemic 1-Trimethylsilyl-2-propanol by Hydrolase

Hydrolase	Reaction Rate (μmol $\times$ hour^{-1} $\times$ mg powder^{-1})	Percent ee[a]
Lipase OF 360	0.21	8
Lipase Saiken 100	Nil	—
Steapsin	0.01	—
Lipoprotein lipase type A	0.37	91
Cholesterol esterase type A	0.08	93

[a]Percent ee of remaining 1-trimethylsilyl-2-propanol at about 50% conversion.

Among these hydrolases, lipase OF 360, lipoprotein lipase type A, and cholesterol esterase type A could esterify 1-trimethylsilyl-2-propanol, which easily undergoes β-elimination in the case of esterification with carboxylic acid by chemical catalysts (TABLE 1). Lipase OF 360 and cholesterol esterase type A showed low stereoselectivity and low activity, respectively, while the optical resolution could be successfully carried out by lipoprotein lipase type A.

Stereoselective esterification of 1-trimethylsilyl-1-propanol could be also achieved by lipase OF 360 and lipoprotein lipase type A (TABLE 2). In the case of 1-trimethylsilyl-1-propanol, reaction rate was smaller than that on 1-trimethylsilyl-2-propanol but stereoselectivity was higher, probably because of its steric hindrance.

Natural primary alcohols are in general esterified very well but not so stereoselectively.[4,5] However, 2-trimethylsilyl-1-propanol, which is a nonnatural organic compound, was esterified with high stereoselectivity by lipase Saiken 100 (TABLE 3).

These results not only represented the first demonstration of useful stereoselective esterification of organosilicon compounds by hydrolases but also indicated that

TABLE 2. Esterification of Racemic 1-Trimethylsilyl-1-propanol by Hydrolase

Hydrolase	Reaction Rate (μmol × $hour^{-1}$ × mg $powder^{-1}$)	Percent ee[a]
Lipase OF 360	0.17	89
Lipase Saiken 100	0.02	—
Steapsin	0.05	—
Lipoprotein lipase type A	0.10	96
Cholesterol esterase type A	0.04	—

[a]Percent ee of remaining 1-trimethylsilyl-1-propanol at about 50% conversion.

enzymatic conversion of organosilicon compounds would be useful for synthetic organosilicon chemistry.

We also examined the dehydrogenation of organosilicon compounds by alcohol dehydrogenases. HLADH and yeast alcohol dehydrogenase (YADH) were tested for the dehydrogenation of $Me_3Si(CH_2)_nOH$ (n = 1–3). HLADH showed enzymatic activities on two of them, 2-trimethylsilylethanol (n = 2) and 3-trimethylsilyl-1-propanol (n = 3) (TABLE 4), while YADH did not act on them under the conditions employed. Thus, HLADH was found to be capable of catalyzing the dehydrogenation of organosilicon compounds.

HLADH activity on $Me_3Si(CH_2)_nOH$ (n = 1–3) is shown in TABLE 4 in comparison with the carbon analogues. Among the organosilicon compounds investigated, 2-trimethylsilylethanol (n = 2) and 3-trimethylsilyl-1-propanol (n = 3) were better substrates for HLADH than ethanol was. The influence of the silicon atom on substrate reactivity was quite different depending on the number of methylene groups between the trimethylsilyl group and the hydroxyl group. 2-Trimethylsilylethanol was more reactive than its carbon counterpart 3,3-dimethyl-1-butanol, whereas trimethylsilylmethanol was inert in contrast to its carbon analogue 2,2-dimethyl-1-propanol.

The influence of the silicon atom on the kinetic parameters K_m and V_{max} of HLADH-catalyzed dehydrogenation of $Me_3El(CH_2)_nOH$ (E1 = Si or C) was also studied. The results obtained are shown in TABLE 4, which demonstrates that the value of V_{max} increased with the decrease in K_m, and that silicon substitution for the carbon atom in 3,3-dimethyl-1-butanol lowered the value of K_m and raised the value

TABLE 3. Esterification of Racemic 2-Trimethylsilyl-1-propanol by Hydrolase

Hydrolase	Reaction Rate (μmol × $hour^{-1}$ × mg $powder^{-1}$)	Percent ee[a]
Lipase OF 360	0.17	17
Lipase Saiken 100	0.18	95
Steapsin	0.09	61
Lipoprotein lipase type A	0.34	0
Cholesterol esterase type A	0.10	48

[a]Percent ee of remaining 2-trimethylsilyl-1-propanol at about 50% conversion.

TABLE 4. HLADH Activity and Kinetic Parameters of HLADH-Catalyzed Dehydrogenation Reaction

Substrate	n^a	RA[b] (%)	K_m (mM)	V_{max} (μmol × minute^{-1})
Ethanol	—	100	1.55	0.194
Trimethylsilylmethanol	1	0	—	—
2,2-Dimethyl-1-propanol	1	75	3.40	0.148
2-Trimethylsilylethanol	2	209	0.520	0.360
3,3-Dimethyl-1-butanol	2	146	0.773	0.283
3-Trimethylsilyl-1-propanol	3	145	0.881	0.268

[a]Number of methylene group(s) between the trimethylsilyl/*tert*-butyl group and the hydroxyl group.

[b]Relative activity. The activity on ethanol was expressed as 100%.

of V_{max}. That is, such a replacement caused an enhancement in substrate reactivity. According to the value of K_m, 2-trimethylsilylethanol had the strongest affinity toward HLADH, giving the highest HLADH activity. The structure of HLADH has been determined crystallographically, and it has been revealed that HLADH has a hydrophobic active center when NAD^+ is bound.[6] The higher affinity of HLADH toward 2-trimethylsilylethanol is probably due to the greater hydrophobicity of the trimethylsilyl group[7] and also to the appropriate distance between the trimethylsilyl group and the hydroxyl group caused by the bigger radius of the silicon atom.

For more comprehensive study on the effect of the silicon atom on the dehydrogenation reaction of organosilicon compounds by HLADH, the activation energy (E_A) and the frequency factor (A) of HLADH-catalyzed dehydrogenation reaction were measured. The dehydrogenation of 2-trimethylsilylethanol by HLADH had a lower E_A and A than those of 3,3-dimethyl-1-butanol (TABLE 5). These results indicate that the silicon substitution for the carbon atom in 3,3-dimethyl-1-butanol decreases both the E_A and A. As E_A influences the reaction rate exponentially, it plays a much more important role in the reactivity of substrate than A does in this case. Accordingly, the enzymatic activity of HLADH on 2-trimethylsilylethanol became higher than that on 3,3-dimethyl-1-butanol, even though the A of 3,3-dimethyl-1-butanol was higher.

It is well known that the silicon atom has small electronegativity and can stabilize negative charge at the α-position and positive charge at the β-position of the carbon atom from the silicon atom.[1] Owing to the smaller electronegativity of silicon atom, the α-carbon atom of trimethylsilylmethanol ($n = 1$) is more electronegative than that of 2,2-dimethyl-1-propanol. This makes it more difficult for H^- anion to be

TABLE 5. Activation Energy (E_A) and Frequency Factor (A) of HLADH-Catalyzed Dehydrogenation Reaction

Substrate	n^a	E_A (J × mol^{-1} × K^{-1})	A
2-Trimethylsilylethanol	2	2.07×10^4	1.11
3,3-Dimethyl-1-butanol	2	2.55×10^4	6.07

[a]Number of methylene groups between the trimethylsilyl/*tert*-butyl group and the hydroxyl group.

removed from the α-carbon atom, which is the first step of dehydrogenation of alcohol catalyzed by alcohol dehydrogenase.[3] As a result, the E_A of dehydrogenation of trimethylsilylmethanol by HLADH probably becomes extraordinarily higher than that of 2,2-dimethyl-1-propanol. And thus, trimethylsilylmethanol is not a substrate, in spite of the fact that its carbon analogue is a good one. Conversely, as the silicon atom stabilizes the positive charge of the β-carbon, H^- anion can be removed from the β-carbon atom of 2-trimethylsilylethanol ($n = 2$) more easily than from that of 3,3-dimethyl-1-butanol. This was supported by an observed smaller E_A of the dehydrogenation of 2-trimethylsilylethanol by HLADH. In the case of 3-trimethylsilyl-1-propanol ($n = 3$), the effect of the silicon atom will become weaker because of the long distance between the silicon atom and the carbon atom binding hydroxyl group. The trimethylsilyl group is more bulky than the *tert*-butyl group due to the bigger radius of the silicon atom, so A of 2-trimethylsilylethanol may become smaller.

In conclusion, this study represents the first demonstration of useful stereoselective esterification of organosilicon compounds with hydrolases and the dehydrogenation of organosilicon compounds with HLADH. The silicon atom in substrates has great effect on enzymatic reaction, and the effect of the silicon atom could be explained in terms of its specific characteristics. This work not only shows the utility of bioconversion of organosilicon compounds in synthetic organosilicon chemistry, but also gives a clue to further research on recognition and reaction of enzymes.

REFERENCES

1. Colvin, E. 1981. Silicon in Organic Synthesis. Butterworth. London, England.
2. Kawamoto, T., K. Sonomoto & A. Tanaka. 1991. Efficient optical resolution of 2-(4-chlorophenoxy)propanoic acid with lipase by the use of organosilicon compounds as substrate: the role of silicon atom in enzymatic recognition. J. Biotechnol. **18:** 85–91.
3. Bentley, R. 1970. Molecular Asymmetry in Biology. Vol. **2.** Academic Press. New York, N.Y.
4. Kawamoto, T., K. Sonomoto & A. Tanaka. 1987. Esterification in organic solvents: selection of hydrolases and effects of reaction conditions. Biocatalysis **1:** 137–145.
5. Pan, S.-H., T. Kawamoto, T. Fukui, K. Sonomoto & A. Tanaka. 1990. Stereoselective esterification of halogen-containing carboxylic acids by lipase in organic solvent: effects of alcohol chain length. Appl. Microbiol. Biotechnol. **34:** 47–51.
6. Eklund, H., J. P. Samana, L. Wallën & C. I. Brändën. 1981. Structure of a triclinic ternary complex of horse liver alcohol dehydrogenase at 2.9 Å resolution. J. Mol. Biol. **146:** 561–587.
7. Fessenden, R. J. & J. S. Fessenden. 1980. Trends in organosilicon biological research. Adv. Organomet. Chem. **16:** 275–299.

Enzymatic Preparation of Acylated Sucroses[a]

M. A. CRUCES,[b] C. OTERO,[b] M. BERNABE,[c]
M. MARTIN-LOMAS,[c] AND A. BALLESTEROS[b,d]

[b]*Instituto de Catálisis, CSIC*
Universidad Autónoma
Cantoblanco
28049 Madrid, Spain

[c]*Instituto de Química Orgánica, CSIC*
Juan de la Cierva 3
28006 Madrid, Spain

INTRODUCTION

Serine hydrolases are enzymes very useful in industry and in preparative organic chemistry. Their substrate specificity is not very stringent; therefore they have been employed to carry out many hydrolytic and synthetic conversions in a regio- or stereoselective way.[1]

Sugars, once considered just inert molecules, are recognized today as very important components of living processes. Owing to the presence of several primary and/or secondary hydroxylic groups, they constitute interesting starting points for the regioselective action of enzymes.[2]

In this work several biocatalysts have been utilized: subtilisin and lipases, either in the native state or insolubilized on suitable supports (Celite, resins). Two reactions have been followed: (A) selective hydrolysis of peracylated sucrose, (B) acylation of sucrose.

MATERIALS AND METHODS

Subtilisin and octa-*O*-acetylsucrose were purchased from Sigma; the lipases from *Candida cylindracea,* pig pancreas, and wheat germ were also from Sigma. Merck supplied the sucrose. *Mucor* sp. (LIP F7 and LIP F8) and *Rhizopus* sp. (LIP F3 and LIP F4) lipases were products of Enzymatix Ltd. (Babraham, Cambridge CB2 4AT, England). *Mucor miehei* lipase immobilized on Duolite 568, a phenol-formaldehyde macroporous weakly basic anion-exchange resin (Lipozyme), and *Candida antarctica* lipase immobilized on a macroporous acrylic resin (SP 382) were kind gifts from Novo Industri A/S (Copenhagen, Denmark). Celite for gas-liquid chromatography (30–80 mesh) was from BDH (Poledorset, England).

[a]This research was supported by the EEC BRIDGE Program (No. BIOT-CT90-0176) and by the Spanish CICYT (No. BIO88-0241).

[d]Author to whom correspondence should be addressed.

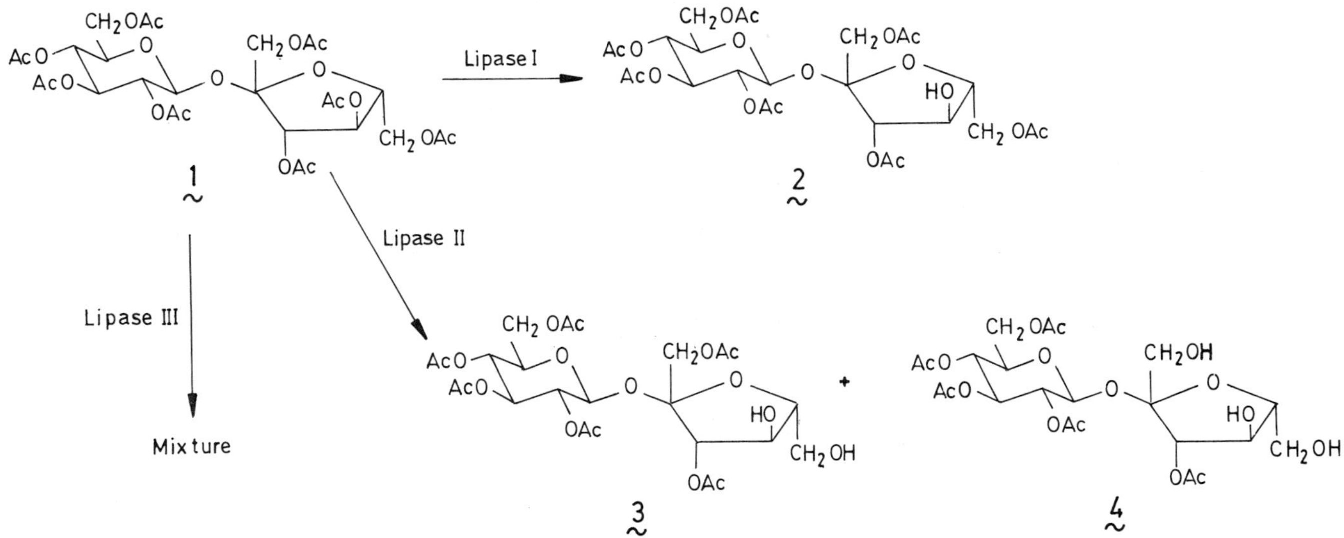

FIGURE 1. Lipase I means *C. cylindracea*; Lipase II, wheat germ; and Lipase III, pig pancreatic lipase.

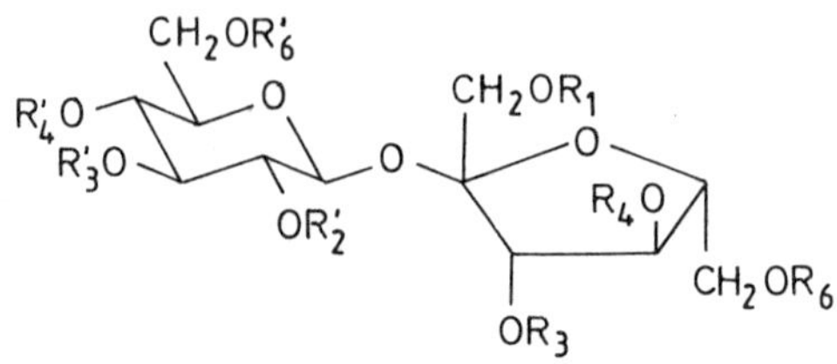

$$R_1 = R_3 = R_4 = R_6 = R'_2 = R'_3 = R'_4 = R'_6 \equiv CH_3CO \qquad (1)$$

$$R_1 = R_3 = R_4 = R_6 = R'_2 = R'_3 = R'_4 = R'_6 \equiv CH_3(CH_2)_2CO \qquad (5)$$

$$R_1 = R_3 = R_4 = R_6 = R'_2 = R'_3 = R'_4 = R'_6 \equiv H \qquad (6)$$

FIGURE 2. Regioselective hydrolysis of octaacetylsucrose and octabutyrylsucrose in two immiscible phases.

Octa-*O*-butyrylsucrose was synthesized in our laboratory following the procedure described by Höfle *et al.*[3] Covalent immobilization of the enzyme (lipase or subtilisin) through its amino groups on Sepharose CL-6B (from Pharmacia, Sweden) activated with 2,3-epoxy-1-propanol was carried out as previously described.[4]

For adsorption of the enzyme on Celite, 200 mg of lipase was dissolved in 10 mL of 0.1-M Na phosphate buffer, pH 7.2, 0.1-M NaCl. After stirring for 1 hour at room temperature, 600 mg of the support was added, and the stirring continued during 2 hours. Then, 10 mL of cold (−15°C) acetone was slowly added, with stirring. The immobilized enzyme was filtered, washed with 2 × 10 mL of acetone, dried 30 minutes *in vacuo,* and stored in a closed vial at 4°C. In the case of subtilisin, 500 mg of

TABLE 1. Screening for Regioselective Hydrolysis

Enzymes	Results
C. cylindracea lipase *Mucor sp.* lipase *C. cylindracea* lipase/celite Subtilisin Subtilisin/Celite	Hydrolysis observed in steps. Intermediate products can be isolated.
Pig pancreas lipase Pig pancreas lipase/Celite	Mixture of products.
Lipozyme SP 382 Wheat germ lipase *C. cylindracea*/Sepharose *Rhizopus sp.* lipase Subtilisin/Sepharose	No reaction.

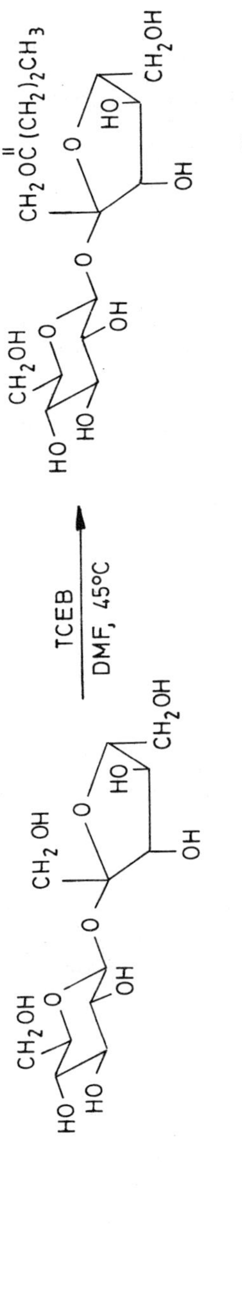

6 7

FIGURE 3. Formation of 1-*O*-butyrylsucrose from sucrose by subtilisin/Celite.

enzyme was dissolved at 0°C in 10 mL 10-mM Na phosphate buffer, pH 7.8, 1.5 g Celite was then added and the mixture stirred for 2 hours. The immobilized enzyme was stored at −18°C.

RESULTS AND DISCUSSION

Selective Hydrolysis of Peracylated Sucrose

The hydrolysis of octaacetyl- and octabutyrylsucrose has been studied at 37°C in several reaction media.

In Water

Treatment in water of octa-*O*-acetylsucrose (**1**) with lipase from *Candida cylindracea* resulted (see FIGURE 1) in the formation of **2** (hydrolysis at R_4 hydroxyl); this

TABLE 2. Acylation of Sucrose to 1-*O*-butyrylsucrose by Native Subtilisin and by Subtilisin/Celite[a]

	Percent Conversion	
Reaction Time (hours)	With Subtilisin	With Subtilisin/Celite
1	2.4	22.6
2	3.4	33.3
4	6.0	48.2
8	7.0	63.6
24	8.5	85.9
48	11.6	93.0

[a]The conversion was determined by gas-liquid chromatography. The reaction conditions were: (1) Subtilisin (0.3 mkat, measured in the hydrolysis of *N*-acetyltyrosine ethyl ester); TCEB (0.6 M); sucrose (0.15 M); DMF 1 ml. (2) Subtilisin/Celite (0.3 mkat); TCEB (0.6 M); sucrose (0.15 M); DMF 1 ml.

coincides with the results of Kloosterman *et al.*[5] with the same lipase but in water-toluene (90:10).[5] Treatment of **1** with wheat germ lipase afforded a minor amount of **3**, with **4** as the major product, whereas pig pancreatic lipase converted **1** into a mixture of products.

In Two Phases

Several enzymatic preparations were screened for regioselective hydrolysis of octaacetylsucrose (**1**) and octabutyrylsucrose (**5**) in two phases (water-toluene for **1** and water-hexane for **5**), cf. FIGURE 2. TABLE 1 classifies the biocatalysts tested in three categories, of which only the first one is useful regarding the sequential hydrolysis of the esters.

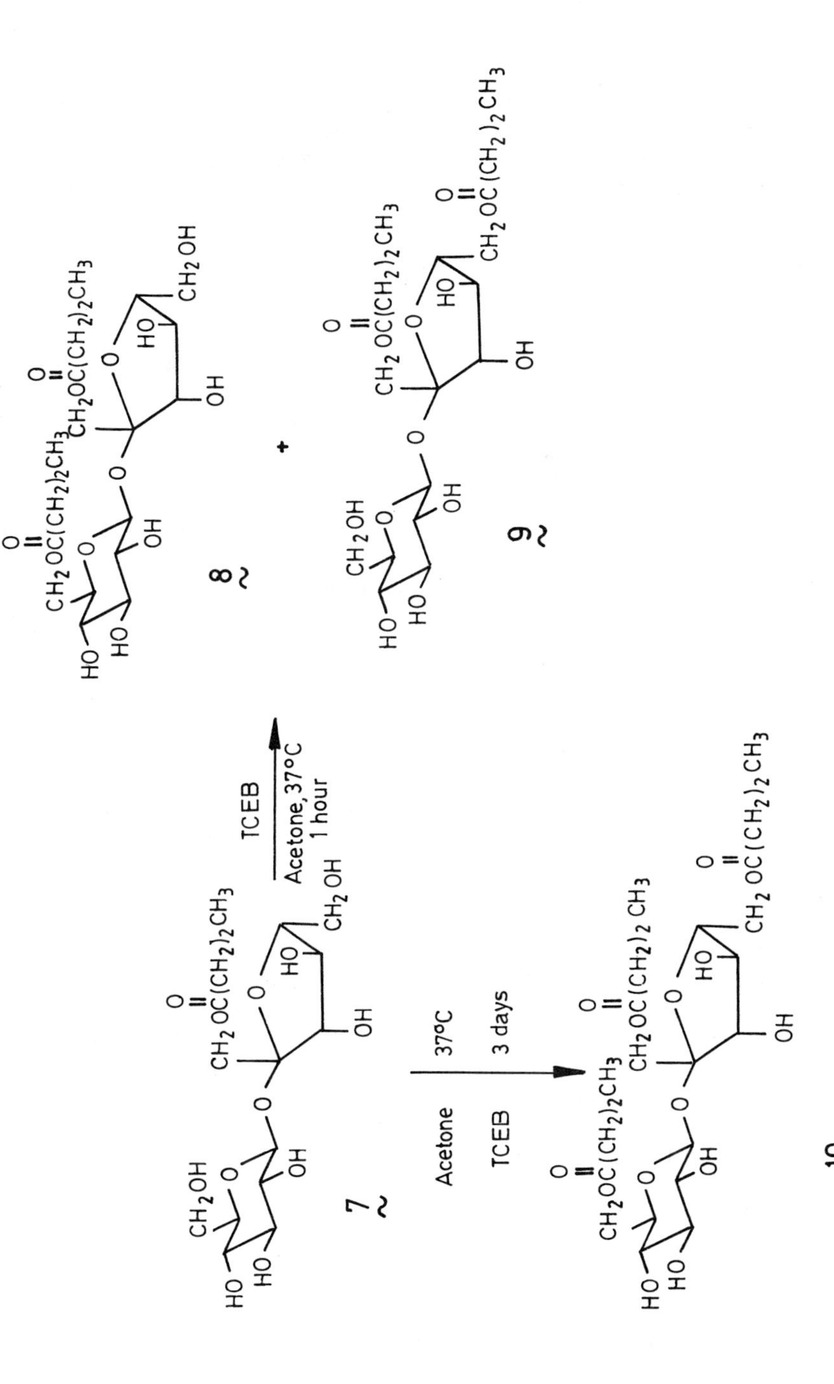

FIGURE 4. Acylation of 1-*O*-butyrylsucrose by SP382 lipase.

In the sequential hydrolysis of **1**, the reactivity order of the ester bonds was always:

$$4 > 6 \geq 6' > 1 \gg 4' > 2' > 3' > 3$$

Treatment of **1** with *C. cylindracea* lipase in water-toluene resulted in the predominant formation of **2** (see above). We investigated whether the addition of different amounts of toluene could change the product yield and the enzyme selectivity. We found that the optimal amount of toluene was 5%. This solvent did not influence enzyme selectivity.

In the hydrolysis of **5**, the reactivity order of the ester bonds was:

$$4 = 6' > 6 > 1 >>> 4', 2', 3', 3$$

With all the enzymatic preparations used, no total hydrolysis of **5** to sucrose (**6**) was observed, even after reaction and times of 20 days.

In Toluene/AOT Microemulsions

Hydrolysis of **1** in water-in-oil microemulsions of toluene/sodium ethylhexylsulfosuccinate (AOT) was probed, using *C. cylindracea* lipase as catalyst. The reaction rate decreased substantially, without any increase in desired selectivity.

Acylation of Sucrose (6)

The synthesis of monobutyryl derivatives of **6** has been attempted using trichloroethylbutyrate (TCEB) as acylating agent, in dimethylformamide (DMF); DMF/toluene; acetone; acetonitrile; and in water-in-oil microemulsions.

Several enzymatic preparations were tested: (1) Lipase from *C. cylindracea* (native and immobilized on Celite), Lipozyme, and SP 382 were not catalytically active in these conditions. (2) In acetone or acetonitrile, subtilisin and subtilisin/Celite converted **6** into a mixture of products without any product prevailing. But it has been reported[6] that subtilisin promoted the esterification of the 1-OH in DMF. In the present work, we have found that the rate and yield of this transesterification (FIGURE 3) can be increased more than one order of magnitude by immobilization on Celite, as shown in TABLE 2.

In water-in-oil microemulsions of toluene/AOT, transesterification only took place with subtilisin as catalyst. However, the use of microemulsions offers no advantage, since the reaction rate decreases considerably.

1-*O*-butyrylsucrose (**7**) has been prepared in DMF on a gram scale with subtilisin/Celite after a short period of time (6 hours). The obtained ester (**7**) was further acylated with lipases in dry acetone.[7] After a preliminary screening, the most useful catalyst was SP 382 (FIGURE 4).

REFERENCES

1. WHITESIDES, G. M. & C. H. WONG. 1985. Enzymes as catalysts in synthetic organic chemistry. Angew. Chem. Int. Ed. **24:** 617–718.

2. RIVA, S. & F. SECUNDO. 1990. Selective enzymatic acylations and deacylations of carbohydrates and related compounds. Chimicaoggi **6:** 9–16.
3. HÖFLE, G., W. STEGLICH & H. VORBRÜGGEN. 1978. 4-Dialkylaminopyridines as highly acylation catalysts. Angew. Chem. Int. Ed. **17:** 569–583.
4. OTERO, C., A. BALLESTEROS & J. M. GUISAN. 1988. Immobilization/stabilization of lipase from *Candida rugosa*. Appl. Biochem. Biotechnol. **19:** 163–175.
5. KLOOSTERMAN, M., J. G. J. WEIJNEN, N. K. DE VRIES, J. MENTECH, I. CARON, G. DESCOTES, H. E. SCHOEMAKER & E. M. MEIJER. 1989. Octa-*O*-acetyl-sucrose: regioselective deacylations by lipolytic enzymes. J. Carbohydrate Chem. **8:** 693–704.
6. RIVA, S., J. CHOPINEAU, A. P. G. KIEBOOM & A. M. KLIBANOV. 1988. Protease-catalyzed regioselective esterification of sugars and related compounds in anhydrous dimethylformamide. J. Am. Chem. Soc. **110:** 584–589.
7. CARREA, G., S. RIVA, F. SECUNDO & B. DANIELLI. 1989. Enzymatic synthesis of various 1-*O*-sucrose and 1-*O*-fructose esters. J. Chem. Soc. Perkin Trans. I: 1057–1061.

Preparative Enzymatic Synthesis of Activated Neuraminic Acid by Using a Microbial Enzyme

M. KITTELMANN,[a] T. KLEIN,[b] U. KRAGL,[b]
C. WANDREY,[b] AND O. GHISALBA[a]

[a]*CIBA-GEIGY Limited*
Pharmaceutical Division
Biotechnology, R-1060.108
CH-4002 Basel, Switzerland

[b]*Research Center Jülich*
Institute of Biotechnology 2
Post Office Box 1913
D-5170 Jülich, Germany

INTRODUCTION

Sialic acids are important building blocks of mammalian glycoproteins. The most important natural sialic acid is *N*-acetylneuraminic acid (Neu5Ac). The biosynthesis and therefore also the *in vitro* enzymatic synthesis of sialylated glycoconjugates via sialyltransferase requires CMP-activated sialic acid (CMP-Neu5Ac). To promote new developments in glycobiology and glycotechnology, Neu5Ac and CMP-Neu5Ac must become available in large quantities at reasonable costs.

Neu5Ac can be synthesized in large scale from *N*-acetylglucosamine (GlcNAc) and pyruvate in a two-step reaction with GlcNAc-epimerase and Neu5Ac-aldolase (continuous production in enzyme membrane reactor).[1] CMP-Neu5Ac-synthetase (see FIGURE 1) for the preparation of CMP-Neu5Ac from CTP and Neu5Ac has so far mainly been isolated from animal tissues, which contain only small amounts of the enzyme and are difficult to obtain in large quantities with constant quality. Therefore we screened for a suitable microbial enzyme and chose *Escherichia coli* K-235 ATCC 13027 as a production strain.

RESULTS

Fermentation of E. coli *K-235*

By developing a sorbitol/yeast extract medium based on the recipe given by Uchida *et al.*[2] for the production of colominic acid with *E. coli* K-235, we could increase the specific activity in the crude extract up to 15 mU/mg (see Assay conditions below), which is about three times the value reported for *E. coli* K1[3] and for calf brain.[6] The composition of the final medium is sorbitol, 6 g/l; yeast extract, 2.5 g/l; $(NH_4)_2SO_4$, 2.5 g/l; $MgSO_4 \times 7\ H_2O$, 1 g/l; K_2HPO_4, 11 g/l; pH value, 6.5.

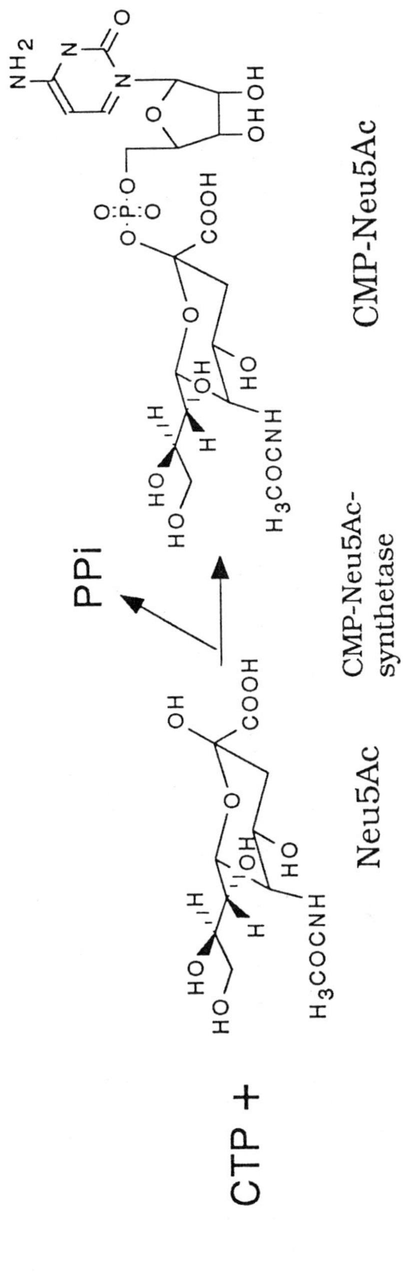

FIGURE 1. Activation of Neu5Ac with CMP-Neu5Ac-synthetase.

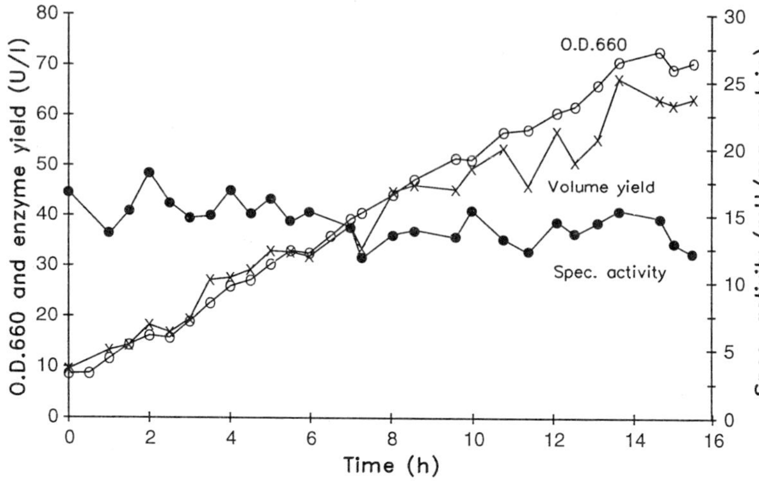

FIGURE 2. Production of CMP-Neu5Ac-synthetase from *E. coli* K-235 by fed-batch fermentation in 10 l scale.

In order to obtain high cell densities in a fed-batch fermentation, we added 9 times the initial amount of sorbitol and yeast extract and once 1 g/l $MgSO_4 \times 7\,H_2O$. In 10 l scale (FIGURE 2), the results were cell density, 195 g wet weight/l = 72.8 O.D.660; enzyme yield, 64 U/l (see Assay conditions below); enzyme concentration, 320–540 U/kg wet weight (calf brain 30–70 U/kg,[4,5] in our hands not reproducible); specific activity, 13.5 mU/mg (calf brain 4.2–6.7 mU/mg,[6] *E. coli* K1 5 mU/mg[3]).

In a 200 l fermentation, we produced 9200 U of CMP-Neu5Ac-synthetase, which is sufficient for the preparation of CMP-Neu5Ac in the kilogram range.

Assay conditions: 150 mM Tris/HCl, 1 mM dithiothreitol, 40 mM $MgCl_2$, 10 mM Neu5Ac and CTP, 1–3 hour, 30°C. Analysis of CMP-Neu5Ac was performed by high-performance liquid chromatography (HPLC) using amino substituted silica gel as the stationary phase, and for elution, a gradient with increasing concentration of potassium phosphate from 3 to 250 mM and decreasing concentration of acetonitrile from 50 to 0% v/v was applied. Definition of one unit (U) of CMP-Neu5Ac-synthetase activity: formation of 1 μMol CMP-Neu5Ac per minute under reaction conditions. The specific activity is expressed in mU/mg protein determined with the Bio-Rad Protein Assay (Biorad, Glattbrugg, Switzerland).

Partial Enrichment of the Synthetase

After cell disruption by ultrasonification, cell debris and nucleic acids precipitated with polyethyleneimine (final concentration of 0.4% w/w) were removed by a single centrifugation step.

In order to achieve high recovery yields of the synthetase as a catalyst for chemical synthesis we concentrated on precipitation and batch adsorption as methods for enzyme enrichment (see TABLE 1).

Precipitation with polyethyleneglycol and batch adsorption to hydroxylapatite provided synthetase preparations almost free of CTP-dephosphorylating activity.

Synthesis of CMP-Neu5Ac

Synthesis of CMP-Neu5Ac from CTP and Neu5Ac

Using increasing equimolar concentrations of CTP and Neu5Ac (FIGURE 3), the maximum yield of CMP-Neu5Ac is 10 g/l or 16.3 mM (reaction conditions: 50 mM Tris/HCl, pH 9, 40 mM $MgCl_2$). The molar yield decreases significantly with increasing concentrations of the educts. Since the CTP (Fluka, Buchs, Switzerland) used in our assays contains about 12% of CDP and 1–2% of CMP, inhibitory effects of these nucleotides besides product and substrate inhibition have to be taken into account. With CTP of higher purity, the molar yield may be shifted to higher concentrations.

In Situ *Generation of Neu5Ac*

Since Neu5Ac-aldolase and GlcNAc-epimerase are not active at alkaline pH values, the reactions including *in situ* generation of Neu5Ac were performed at pH 7.6. Although CMP-Neu5Ac-synthetase is inactive at this pH in the presence of Mg^{2+}, it still retains 100% activity in the presence of 10 mM Mn^{2+} (FIGURE 4).

The CMP-Neu5Ac-synthetase obtained by precipitation with polyethyleneglycol was successfully applied for the synthesis of the target compound from *N*-acetylmannosamine (ManNAc), pyruvate, and CTP in a two-step reaction involving Neu5Ac-aldolase (Toyobo, Osaka, Japan). The molar yield based on ManNAc was 37%

TABLE 1. Enrichment of CMP-Neu5Ac-Synthetase from Crude Extract[a]

Method	Specific Activity (mU/mg)	Purification Factor	Yield (%)
None = crude extract	9–18	1.0	100
Precipitation with			
• Polyethyleneglycol 1000, fraction 0–30% w/v	42	4.7	93
• Polyethyleneglycol 6000, fraction 0–16% w/v	14	2.9	100
" 0–18% w/v	21	4.3	83
Batch adsorption to			
• Butyl-Fractogel TSK (Merck, Darmstadt, FRG), fraction 15–0% $(NH_4)_2SO_4$	11–28	max. 3.2	11–26
• QAE-Sephadex (Pharmacia), fraction 0.17–0.3 M NaCl	21	2.4	33
• Hydroxylapatite (Biorad, Glattbrugg, Switzerland) fraction 0.1–0.2 M Kpi, pH 7.5	68	3.0	48–69

[a]*Experimental conditions:* 50 mM Tris/HCl buffer, 1 mM dithiothreitol, pH 7.5, 4°C.

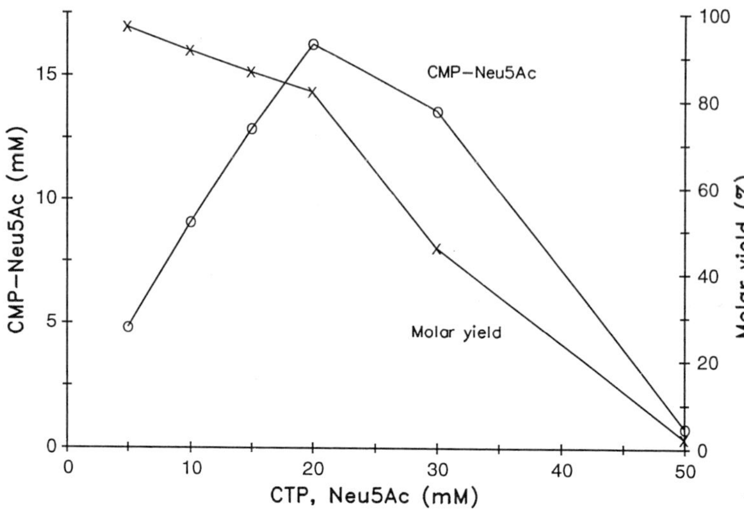

FIGURE 3. Batch synthesis of CMP-Neu5Ac: variation of CTP and Neu5Ac concentration.

(TABLE 2). Purification of CMP-Neu5Ac can be performed by anion exchange chromatography as described by Simon *et al.*[7]

When the synthesis was started from GlcNAc by including GlcNAc-epimerase, CMP-Neu5Ac was obtained in 10.6% yield based on CTP (GlcNAc 200 mM, pyruvate 100 mM, CTP 50 mM).

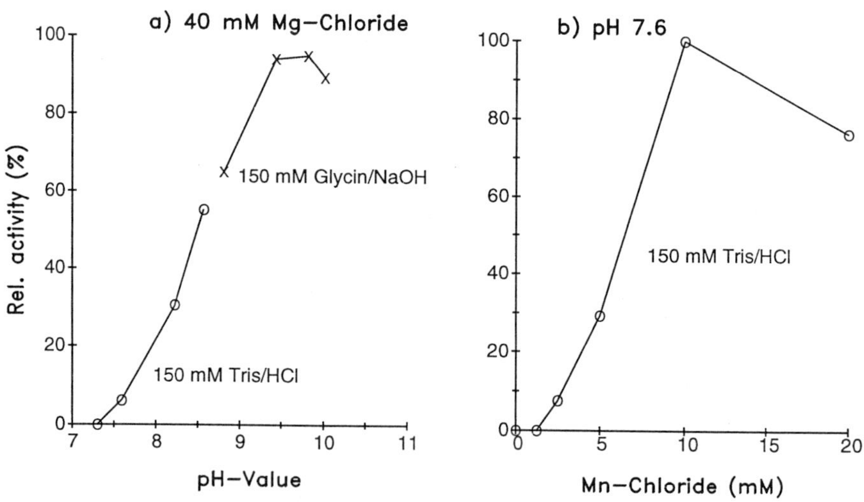

FIGURE 4. Activity of CMP-Neu5Ac-synthetase (a) depending on pH value in the presence of $MgCl_2$, (b) depending on $MnCl_2$ concentration at pH 7.6.

TABLE 2. Synthesis of CMP-Neu5Ac Including *In situ* Generation of Neu5Ac Catalyzed by Neu5Ac-Aldolase and CMP-Neu5Ac-Synthetase[a]

ManNac/CTP (mM)	Pyruvate (mM)	$MnCl_2$ (mM)	CMP-Neu5Ac (mM)	Molar Yield[b] (%)
5	10	10	0.52	10
10	20	15	2.3	23
15	30	20	5.06	34
20	40	25	6.62	30
30	60	35	11.1	37
50	100	55	18.4	37

[a]*Reaction conditions:* Tris/HCl 150 mM, pH 9; $MgCl_2$ 40 mM, pH 7.5; $MnCl_2$ 15 mM, Dithiothreitol 1 mM; enzyme concentrations: 4–5 mU Neu5Ac aldolase and CMP-Neu5Ac synthetase per 1 μmol of CTP to be converted; 20 hours, 30°C.

[b]Molar yields based on ManNAc and CTP.

In situ *Generation of CTP*

Using synthetase preparations obtained by batch adsorption to different chromatographic matrices, CMP-Neu5Ac could directly be synthesized from ATP, CMP, and Neu5Ac or from PEP, CMP, and Neu5Ac without the addition of any other enzyme (40 and 78% yield, respectively, TABLE 3).

OUTLOOK

Our future work will concentrate on kinetic studies with CMP-Neu5Ac synthetase and on the development of a continuous process for the production of CMP-Neu5Ac employing enzyme membrane reactor technology. The optimized process should work with high space time yields and low enzyme consumption.

TABLE 3. Synthesis of CMP-Neu5Ac Including *In Situ* Generation of CTP[a]

Enzyme Preparation	Enzyme Concentration (mU/ml)	pH Value	% CMP-Neu5Ac from Neu5Ac with PEP[b] CMP	ATP[c] CMP	CTP[d]
Crude extract	35	7.5	42	ND[e]	67
"	35	9	39	ND	69
Precipitation with PEG 1000	22	7.5	23	ND	98
"	22	9	0	5	97
Batch adsorption to					
• Butyl-Fraktogel	28	7.5	52	ND	97
"	28	9	78	32	99
• QAE-Sephadex	20	7.5	0	ND	ND
	20	9	0	40	78

[a]*Reaction conditions:* see TABLE 2, Neu5Ac 10 mM.

[b]PEP = phosphoenolpyruvate 25 mM, CMP 10 mM.

[c]ATP, CMP 10 mM.

[d]CTP 10 mM.

[e]ND = not determined.

SUMMARY

Using *Escherichia coli* K-235 as a production strain in a fed-batch fermentation process with an optimized sorbitol/yeast extract medium, we were able to produce 640 U of CMP-Neu5Ac synthetase in 10 l scale (64 U/l) and 9200 U (total enzyme) in 200 l scale (390 U/kg wet weight). By simple one-step purification procedures, enzyme preparations were obtained that could be used efficiently for the synthesis of CMP-Neu5Ac from CTP and Neu5Ac with over 90% yield, from Neu5Ac, CMP, and ATP or phosphoenolpyruvate by *in situ* generation of CTP, and from CTP, pyruvate, and ManNAc or GlcNAc by *in situ* generation of Neu5Ac.

REFERENCES

1. Kragl, U., D. Gygax, O. Ghisalba & C. Wandrey. 1991. Angew. Chem. Int. Ed. Engl. **30**(7): 827–828. European Pat. Appl. 90121478.3/German Pat. Appl. DE 3937891.
2. Uchida, Y., Y. Tsukada & T. Sugimori. 1973. Agric. Biol. Chem. **37**(9): 2105–2110.
3. Vann, W., R. P. Silver, C. Abeijon, C. Chang, W. Aaronson, A. Sutton, C. W. Finn, W. Linder & M. Koasatos. 1987. J. Biol. Chem. **262**(36): 17556–17562.
4. Thiem, J. & P. Stangier. 1990. Liebigs Ann. Chem. **1990:** 1101–1105.
5. Auge, C., R. Fenandez-Fernandez & C. Gautheron. 1990. Carbohydrate Res. **200:** 257–268.
6. Van den Eijnden, D. H. & W. van Dijk. 1972. Hoppe Seyler's Z. Physiol. Chem. **353:** 1817–1820.
7. Simon, E. S., M. D. Bednarski & G. M. Whitesides. 1988. J. Am. Chem. Soc. **110:** 7159–7163.

Studies on the Reaction Catalyzed by Protease for the Acylation of Flavonoids in Organic Solvents

G. KODELIA AND F. N. KOLISIS[a]

Biosystems Technology Laboratory
Department of Chemical Engineering
Division of Process and Product Development (IV)
National Technical University of Athens
Zographou Campus
15 700 Athens, Greece

INTRODUCTION

The development of enzymatic bioconversions in nonaqueous media is one of the main recent achievements of biotechnology. Many enzyme-catalyzed reactions practically impossible in water are readily carried out in organic solvents. This is due not only to a solubility factor and an equilibrium shift, but also to novel catalytic properties and markedly enhanced operational stability displayed by enzymes in low water environments.[1,2] For instance, lipases, the most popular group of enzymes to be used at catalysts in anhydrous organic solvents, have been utilized for a wide range of stereoselective and regioselective transformations (including esterification, tranesterification, aminolysis, thiotransesterification) whereas in water these processes are almost suppressed by hydrolysis.[1] Another class of enzymes that illustrate the synthetic potential of this phenomenon are proteolytic enzymes and specifically serine proteases, which have been shown to catalyze the acylation of carbohydrates in anhydrous dimethylformamide and pyridine.[3] Carbohydrate molecules represent a particularly challenging target for regioselective modifications due to their multiple hydroxyl groups.[4]

In the present work, we have performed a transesterification reaction of the flavonoid glycoside rutin with trichlorethylbutyrate (TCEB) catalyzed by subtilisin in anhydrous pyridine solution. Flavonoid glucosides have been reported to present various interesting biological effects, such as antitumor, hypotensive, synergetic role with ascorbic acid, etc.[5] The addition of a hydrophobic chain on rutin is expected to change the hydrophilic/hydrophobic balance of the molecule giving new properties to this compound.[6] An increasing number of acyl derivatives of flavonoid glycosides has been isolated, invariably esterified at one of the sugar hydroxyls with carboxylic acids such as acetic, malonic, benzoic acids, etc. However these nonspecific acyl glycosides can be obtained only through complex multistep synthetic procedures and not by direct chemical esterification.[7] Specific acylation of flavonoid monoglycosides by proteases has been reported by other research workers.[8] In our investigation, we have also studied in detail several parameters affecting this reaction, such as, the

[a] Author to whom correspondence should be addressed.

kinetic constants, the role of the pH of the aqueous solution from which the enzyme was lyophilized, the nature of the solvents (hydrophobicity combined with their ability to solubilize the substrates), as well as the stability of the enzyme as a function of the temperature during the reaction course.

MATERIALS AND METHODS

Materials

Protease from *Bacillus subtilis* (subtilisin ED3.4.21.14) was purchased from Sigma as lyophilized powder. Commercially obtained pyridine was purified and dried. All other chemicals used in this work were obtained commercially and were of analytical grade. All solvents were of the highest purity commercially available and were used without further purification. When needed, solvents were dried by gentle shaking with 3A molecular sieves (Linde).

Methods

Preparation of Subtilisin

Prior to use, the enzyme was dissolved in 0.1-M phosphate buffer pH 7.8, and then it was lyophilized. This "pH adjustment" markedly increases catalytic activity of subtilisin in organic solvents.

Preparation of Trichlorethylbutyrate

Trichlorethylbutyrate (TCEB) was synthesized from butyryl chloride and 2,2,2-trichlorethanol following a general methodology.[14]

Initial Rate Assays

The enzymatic transesterification reaction was followed by measuring the formation of the new ester in high-performance liquid chromatography (HPLC). A μBondapak C18 125A, 10 μm (3.9× 300 mm) Waters column with a Lambda Max model 481 LC spectrophotometer were used. A Metrohm Karl-Fischer titrator was used for water content measurements of the system.

In a typical procedure, 0.3 mmole of rutin and 2.7 mmoles of TCEB were placed in 1 ml of dry pyridine solution, while, 100 mg of subtilisin were added and sonicated in order to reduce the size of the enzyme suspended molecules, since subtilisin is not dissolved in organic solvents. Then the suspension was placed in 45°C under vigorous stirring. Aliquots were withdrawn periodically, the enzyme was removed, and the sample was subjected to thin-layer chromatography (TLC). The various samples were developed with the system: AcOEt:AcOH:H_2O, 250:1:2. The product was extracted from the plate, dissolved in methanol, filtered, and passed through the

HPLC. A 15-minute linear gradient from 10% to 60% acetonitrile in water was employed. The flow rate was 1 ml/minute, and the detection was done at 254 nm. The chromatograms show that retention time of rutin was 12.2 minutes while that of the product was 15.1 minutes.

The product formation was also identified by IR and NMR.

RESULTS AND DISCUSSION

Rutin Acylation by Subtilisin

The physiological role of *Bacillus subtilis* protease is the hydrolysis of water-soluble proteins.[9] Recently proteases have also been used for catalysis in nonaqueous solvents, replacing hydrolytic with synthetic activity, and a number of important reactions such as peptide synthesis and esterification of sugars and related compounds have been reported.[3,10]

In our studies we used *Bacillus subtilis* protease (subtilisin) to introduce a butyryl moeity to rutin, a flavonoid diglucoside. Subtilisin is a serine protease with a nucleophile group in the active center mechanistically essential for a transesterification reaction. The enzyme is stable and active in numerous anhydrous organic solvents. Another reason for the utilization of this enzyme was that it has been successfully used for acylation of disaccharides and compounds containing both sugars and aglycon moieties.[3] In this type of synthetic reaction, the rather bulky molecule of the substrate does not seem to present any significant difficulty for approaching the enzyme complex with the fatty acid derivative usually used as the acylating agent.

Rutin was used as the nucleophile in the transesterification reaction, and the second substrate of the reaction, butyric acid, was used in its active trichlorethyl ester form. We dissolved 0.3 mmole of rutin and 2.7 mmoles TCEB in 1 ml of anhydrous pyridine, then 100 mg of dry subtilisin were added and the suspension was placed in 45°C under vigorous stirring. The water content of the system was calculated by the Karl-Fischer method and found to be 0.3% (v/v). After about 8 hours of reaction, almost 90% of rutin was converted to 3-*O*-butyrylrutin. No reaction was observed in the absence of the enzyme or if the enzyme was used inactivated. The esterification has been identified to occur exclusively on the sugar moiety of rutin. After completion of the reaction, the insoluble enzyme can be recovered by filtration or centrifugation and used again.

Kinetic studies of the system were undertaken under special conditions in order to minimize the diffusional limitations of the substrates, which is often the case in heterogeneous catalysis. The external mass transfer effects were reduced by extensive shaking, while the internal were dismissed by sonication of the enzyme suspension. Enzymatic rate constants were determined by varying substrate concentrations and measuring initial rates. High concentrations for both substrates were used. As indicated by our measurement, the reaction behaved as a homogeneous one. The enzyme activity was studied at various ester concentrations in the presence of 0.3- and 0.15-M rutin. When the reciprocal initial rates were plotted against the reciprocal ester concentrations at the two different rutin concentrations, the lines were

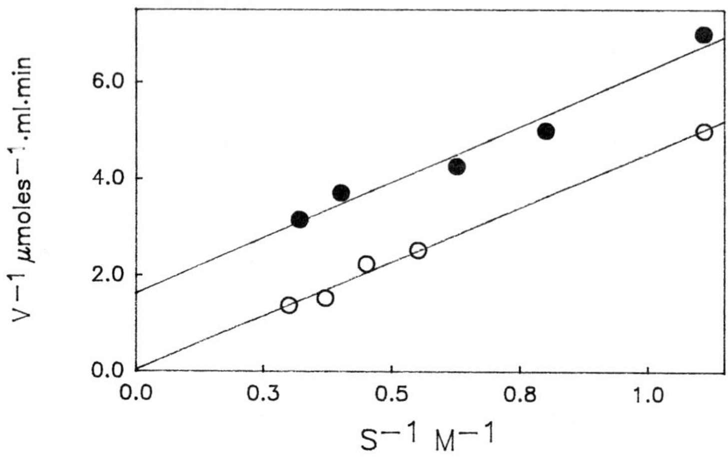

FIGURE 1. Initial rate kinetics of the acylation of rutin with TCEB by substilisin in anhydrous pyridin. Reciprocal rates (V) vs. reciprocal TCEB concentrations at 0.3 M Rutin (○), at 0.15 M rutin (●) concentrations. The initial rates were determined on the basis of four independent measurements. Details in Material and Methods.

parallel and the ratio V/K_m was calculated to 0.22 min)$^{-1}$ (FIGURE 1). These results suggest that the reaction mechanism obeys "Ping-Pong" kinetics as is believed to be the case for protease catalyzed[10] esterification of alcohols, glycerides, even peptide synthesis. Our synthetic reaction, we propose, involves formation of a noncovalent enzyme-TCEB complex, which then transforms to an acyl-subtilisin intermediate with the concomitant release of trichlorethanol. The acyl-subtilisin then interacts with rutin to form another binary complex, which then yields the rutin-ester and the free enzyme.

In the kinetic studies, a substrate inhibition was observed in TCEB concentrations higher than 2.7 M, giving 60% subtilisin inhibition at a concentration of 3.6 M.

Effect of pH on Enzyme Activity

Previous investigations have demonstrated that the activity of subtilisin in organic solvents was greatly enhanced when the enzyme was recovered from an aqueous solution at the pH optimum for the protease activity. This effect is due to the fact that ionogenic groups of the enzyme acquire a certain ionization state in the aqueous solution of a given pH.[1,11] This ionization state (and the enzymatic activity corresponding to it) is retained in the solid state and in organic solvents. This requirement stems from the insolubility of enzymes in nearly all organic solvents.[12] In our studies, subtilisin was dissolved in 0.1-M phosphate buffer, at various pH values, lyophilized, and added to the reaction mixture. The results obtained are depicted in FIGURE 2. It is obvious that the enzymatic acylation of rutin in pyridine very much depends on the pH of the last aqueous solution to which the enzyme was exposed. The optimum pH was found to be 7.8.

Effect of the Nature of the Solvent on Enzymatic Activity

The properties of enzymes can be favorably modified in pure organic solvents. Increased solubility of hydrophobic substrates, reversal of hydrolytic reactions, changes of stereoselectivity, and higher thermostability are among the possible advantages afforded by nonpolar solvents. The choice of solvent can be critical, however, and guidelines for optimizing enzymatic activity in organic solvents have been proposed. As outlined in the literature, biocatalysis in organic solvents having a log $P < 2$ are not suitable in biocatalytic systems since they distort the essential water-biocatalyst reaction, thereby inactivating or denaturating the biocatalyst.[13] Among the various organic solvents, the ones having the ability to solubilize the substrates, rutin and TECB, were chosen to be used as media of the reaction catalyzed by subtilisin. FIGURE 3 shows the activity of subtilisin as a function of DMSO, methanol, DMF, and pyridine. The enzyme appears to be more active in pyridine. In DMF, although the enzyme was almost as active as in pyridine (80%), after one hour incubation at 45°C, it lost about 50% of its activity. Very low or no activity was detected in methanol and dimethylsulfoxide. In the case of dimethylformamide, it seems that the solvent strips the essential water from the enzyme molecule, diminishing its activity, due to the hydrophilic character of DMF (log $P = -1.0$).

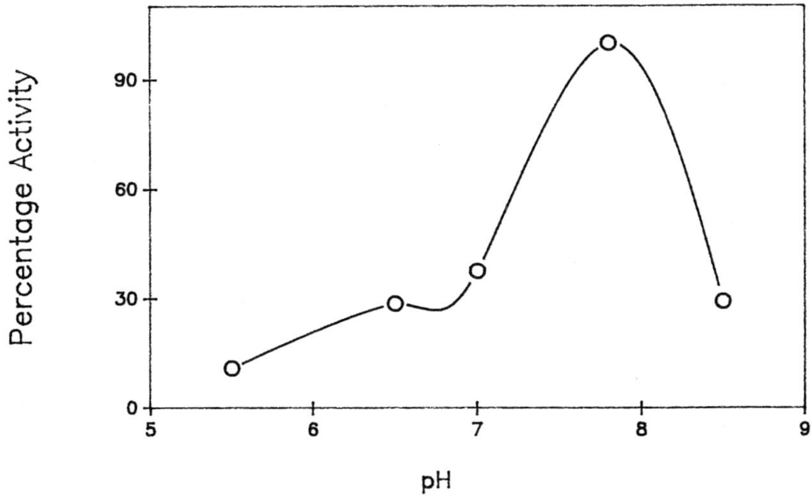

FIGURE 2. Effect of pH on the reaction catalyzed by subtilisin for the acylation of rutin in anhydrous pyridine. Subtilisin was dissolved in 0.1 M phosphate buffer, at various pH values, lyophilized, and added to the reaction mixture. The activity was calculated from the slopes of the initial rates of the product formation as a function of time.

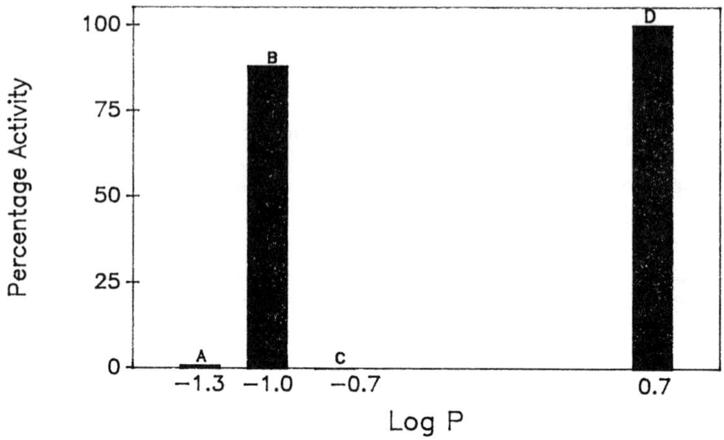

FIGURE 3. The activity of subtilisin as a function of the nature of the solvents. The dry enzyme was placed in the anhydrous organic solvents: (A) DMSO, (B) DMF, (C) methanol, and (D) pyridine. The activity was calculated from the slopes of the initial rates of the product formation as a function of time. The various solvents were chosen for their ability to solubilize the substrates, rutine and TECB. Details in Material and Methods.

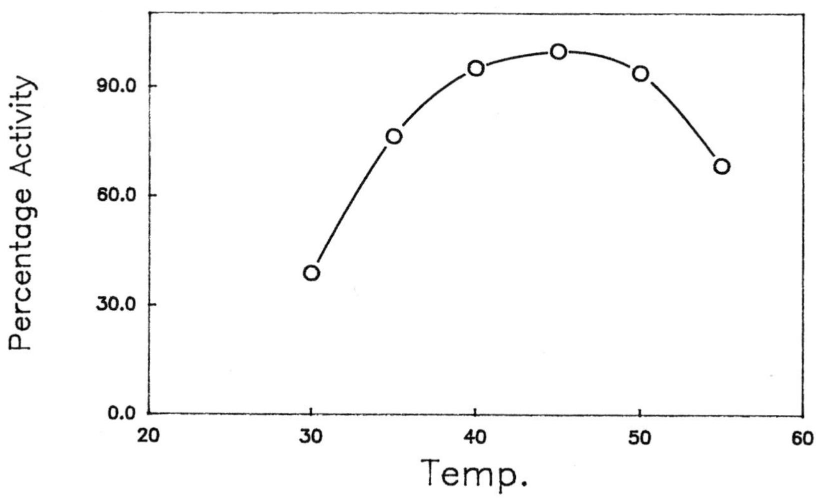

FIGURE 4. The stability of subtilisin in anhydrous pyridine at various temperatures. The dry enzyme after lyophilization at optimum pH was added at the reaction mixture containing dry pyridine, and the activity was studied at different temperatures. The activity was calculated from the slopes of the initial rates of the product formation as a function of time.

Effect of Temperature on Enzymatic Activity

All reactions causing irreversible thermoinactivation of enzymes involve free water. This process can be hindered by dehydration. Detailed data in the recent literature reveal that many enzymes like porcine pancreatic lipase, chymotrypsin, and subtilisin possess a great degree of thermoresistance in anhydrous solvents.[10,15] This thermoresistance is a function of the nature of the solvent and the pH of the aqueous solution from which the enzyme is lyphilized. In the case of porcine pancreatic lipase, a half-life of more than 12 hours at a temperature as high as 100°C has been reported, when the enzyme powder was placed in dried organic solvent in a reaction system containing 0.015% of water. In addition the enzyme activity at 100°C exceeds that at 20°C in the same system by a factor of 5. In our experiments, we tried to identify the optimum temperature at which subtilisin exhibits the higher activity over a time period at least as long as the reaction time. The dry enzyme after lyophilization at optimum pH was added at the reaction mixture containing dry pyridine, and the activity was studied at different temperatures. The results are presented in FIGURE 4; as one can see, the highest activity was found at 45°C. At temperatures higher than 45°C, a decline of the protease activity was observed.

The above results indicate that subtilisin can acylate rutin with high efficiency in anhydrous pyridine solution, in a Ping-Pong bi bi system mechanism. The enzyme was shown to be stable at 45°C with an optimum pH at 7.8.

REFERENCES

1. KLIBANOV, A. M. 1989. Trends Biochem. Sci. **14:** 141–144.
2. KLIBANOV, A. M. 1986. CHEMTECH (June): 354–359.
3. RIVA, S., J. CHOPINEAU, A. P. G. KIEBOOM & A. M. KLIBANOV. 1988. J. Am. Chem. Soc. **110:** 584–589.
4. HAINES, A. H. 1976. Adv. Carbohydr. Chem. Biochem. **39:** 13–70.
5. HARBORNE, J. B. & T. J. MABRY. 1975. The Flavonoids. Chapman and Hall, Eds., London, England.
6. KODELIA, G., K. ATHANASIOU, N. SIDERIS & F. N. KOLISIS. Biochem. Biophys News Lett. (In press).
7. VERMES, B., V. M. CHARI & H. WAGNER. 1981. Helv. Chim. Acta **64:** 1964–1968.
8. DANIELI, B., P. DE BELLIS, G. CARREA & S. RIVA. 1989. Heterocycles **29**(11): 2061–2064.
9. OTTESEN, M. & I. SVENDSEN. 1970. Methods Enzymol. **19:** 199–215.
10. ZAKS, A. & A. M. KLIBANOV. 1988. J. Biol. Chem. **263**(7): 3194–3201.
11. ZAKS, A. & A. M. KLIBANOV. 1985. Proc. Natl. Acad. Sci. USA **82:** 3192–3196.
12. SINGER, S. J. 1962. Adv. Protein Chem. **17:** 1–68.
13. LAANE, C., S. BOEREN, K. VOS & C. VEEGER. 1987. Biotechnol. Bioeng. **30:** 80–87.
14. STEGLICH, W. & G. HOFLE. 1969. Angew Chem. Int. Ed. Engl. **8:** 981.
15. CHINSKY, N., A. L. MARGOLIN & A. M. KLIBANOV. 1989. J. Am. Chem. Soc. **111:** 386–388.

Enzymatic Synthesis of Peptides with Hydrophilic Amino Acid in Water-Miscible Organic Solvent

YUKITAKA KIMURA, YOSHIKO TARI, SHUJI ADACHI,[a] AND RYUICHI MATSUNO

Department of Food Science and Technology
Faculty of Agriculture
Kyoto University
Sakyo-ku
Kyoto 606-01, Japan

INTRODUCTION

An enzymatic synthesis of peptides catalyzed by proteinases has been extensively investigated in some solvent systems. We have successfully synthesized some peptides containing hydrophobic amino acids in high yields using an aqueous-organic biphasic system.[1,2] It is, however, not easy to synthesize peptides consisting of hydrophilic amino acids without protective groups of their side chains. Hydrophilic peptides are excellent in solubility to water and favorably improve the taste of food.[3] In this context, we have attempted to synthesize peptides including a hydrophilic and basic amino acid lysine, which is one of the essential amino acids, through transesterification catalyzed by trypsin. The overall reaction scheme is as follows:

$$\text{Z-K-OHex} + \text{AA-NH}_2 \rightarrow \text{Z-K-AA-NH}_2 + \text{HexOH} \qquad \textbf{(1)}$$

where Z = *N*-benzyloxycarbonyl, K = L-lysine, AA = amino acid, Hex = *n*-hexyl, and NH_2 = amide. A biphasic system may not be adequate to realize a high yield of the product in this reaction because of a low partition of the hydrophilic product to an organic phase. Therefore, the synthesis was performed in water-miscible organic solvents in this study.

MATERIALS AND METHODS

Materials

Trypsin from porcine pancreas was purchased from Wako Pure Chemicals (Japan), and used without further purification. Z-K-OHex was prepared by esterification of α-Z-ε-Boc-K (Kokusan Chemicals, Japan; Boc = *t*-butyloxycarbonyl) with *n*-hexanol[4] and by deblocking with formic acid.[5] Amino acid amides were commercially available products. Solvents used were of analytical grade.

[a] Author to whom correspondence should be directed.

Reaction in Water-Miscible Solvent

An ester substrate was fixed to be 50 mM Z-K-OHex in this study. As an amide substrate, L-NH_2·HCl (L = L-leucine) was first used to find optimum conditions for trypsin-catalyzed reaction in water-miscible organic solvents. W-NH_2·HCl (W = L-tryptophan), V-NH_2·HCl (V = L-valine), and K-NH_2· 2HCl were also used. Their concentrations were 100 mM. Triethylamine was added to the reaction mixture to neutralize hydrochloric acid of the substrates. The reactions were carried out at 30°C with trypsin at 1.0 mg/mL. The content of water-miscible organic solvent was varied from 0 to 100 vol%.

Stability of Trypsin in Acetonitrile-Water Mixture

Trypsin (0.1 mg/mL) was incubated in an acetonitrile-water mixture over 0 to 95 vol% acetonitrile at 30°C. The remaining activity was determined at appropriate intervals from the initial rate of hydrolysis of benzyl-L-arginine ethyl ester (Aldrich) at 10% acetonitrile in Tris-HCl buffer, pH 8.0, at 30°C.

TABLE 1. Relative Initial Reaction Rate and Equilibrium Yield of the Synthesis of Z-KL-NH_2 in Some Water-Miscible Organic Solvents (95 Vol%) at 30°C

Organic Solvent	Relative Initial Reaction Rate (%)	Equilibrium Yield (%)
Acetonitrile	100	72
N,N-Dimethylformamide	13	5
Ethanol	0	0
Butanol	0	0

Analysis

The amounts of substrates and products were measured by high-performance liquid chromatography (HPLC) with an octadecyl silica (ODS) column under a linear gradient elution of acetonitrile. The pH of eluent was adjusted to 2.5. The elution of the components was monitored by a UV detector (254 nm) and recorded.

RESULTS AND DISCUSSION

The initial formation rate and equilibrium yield of Z-KL-NH_2 were examined in some water-miscible organic solvents at 95 vol% (TABLE 1). Among the solvents tested, acetonitrile was the most adequate for both the rate and yield.

The initial formation rates of Z-KL-NH_2 and the initial hydrolytic rates of Z-K-OHex were measured at various contents of acetonitrile. As the content increased, the hydrolytic activity of trypsin decreased gradually and the synthetic activity increased, especially beyond 50% of acetonitrile.

FIGURE 1 shows the equilibrium yield of Z-KL-NH_2 at different contents of acetonitrile. The higher yields were obtained at higher contents of acetonitrile. This may be due to the lower hydrolytic activity at the higher acetonitrile contents.

The syntheses of Z-KW-NH_2, Z-KV-NH_2, and Z-KK-NH_2 were carried out at 95 vol% acetonitrile. Their equilibrium yields were in a range of 25 to 37%. In the syntheses of Z-KW-NH_2 and Z-KV-NH_2, the reaction mixtures were uniphasic. On the other hand, a precipitate appeared in the synthesis of Z-KK-NH_2.

An enzymatic synthesis of Z-KK-NH_2, which is a precursor of our objective KK, was investigated in detail because the peptide is composed of only a hydrophilic amino acid and the method to synthesize it effectively has not been established. The equilibrium yields are plotted against the acetonitrile contents in FIGURE 1. At acetonitrile contents lower than 68 vol%, the reaction mixture was uniphasic and the yield rose with an increase of the content. The reaction medium became biphasic between 68 and 95% acetonitrile content. The appearance of the biphasic system may be ascribed to the hydrophilic properties of the substrate K-NH_2 and the product Z-KK-NH_2. When the content was at 95% or more, a precipitate appeared in the reaction mixture. The yield decreased steeply as the content came near 100%. The decrease seems to be due to the fact that trypsin cannot exhibit its catalytic ability at such high contents of acetonitrile although the enzyme is not inactivated.

TABLE 2 lists the half-lives of trypsin at different acetonitrile contents. A steep decrease in the activity occurred at 25%, while the half-life elongated as the content increased. The half-life at 95% acetonitrile was longer than that in aqueous solution (at 0%).

2-Pyrrolidone has been used as a water-miscible solvent for the lipase-catalyzed syntheses of carbohydrate esters.[6] The feasibility of the solvent for the synthesis of

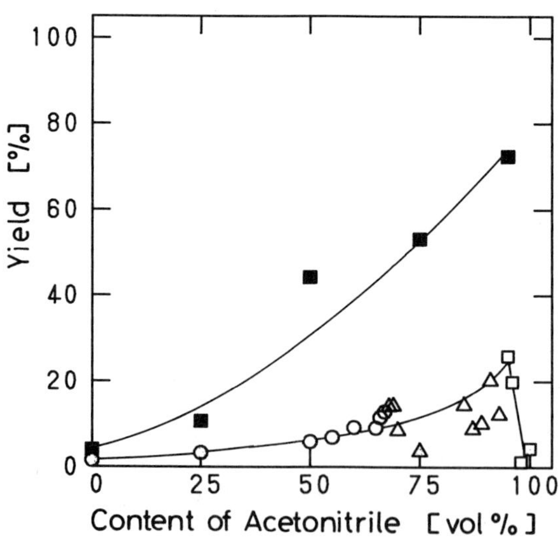

FIGURE 1. Effects of content of acetonitrile on the yields of Z-KL-NH_2 (■) and Z-KK-NH_2. Reaction medium was uniphasic (○), biphasic (△), or included precipitate (□) in the synthesis of Z-KK-NH_2.

TABLE 2. Half-Lives of Trypsin in Acetonitrile-Water Mixture

Content of Acetonitrile (Vol %)	Half-life (hours)
0	32.9
25	5.5
50	13.3
75	21.5
95	39.6

Z-KK-NH_2 by trypsin was examined. The equilibrium yields of Z-KK-NH_2 were observed at different contents of 2-pyrrolidone. In this solvent system, neither biphase nor precipitate appeared over 0 to 100 vol%. The maximum yield (*ca.* 25%), which was almost equal to that in acetonitrile-water system, was obtained at 70% 2-pyrrolidone. The 2-pyrrolidone–water system may provide an operational facility because it makes uniphase at any content. Further studies such as on the stability of the enzyme in the solvent system will be done to find the optimum operating conditions.

REFERENCES

1. NAKANISHI, K., Y. KIMURA & R. MATSUNO. 1986. Kinetics and equilibrium of enzymatic synthesis of peptides in aqueous/organic biphasic systems. Thermolysin-catalyzed synthesis of *N*-benzyloxycarbonyl)-L-aspartyl-L-phenylalanine methyl ester. Eur. J. Biochem. **161:** 541–579.
2. KIMURA, Y., K. MURAYA, Y. ARAKI, H. MATSUOKA, K. NAKANISHI & R. MATSUNO. 1990. Synthesis of peptides consisting of essential amino acids by a reactor system using three proteinases and an organic solvent. Agric. Biol. Chem. **54:** 3331–3333.
3. NOGUCHI, M., M. YAMASHITA, S. ARAI & M. FUJIMAKI. 1975. On the bitter-masking activity of a glutamic acid–rich oligopeptide fraction. J. Food Sci. **40:** 367–369.
4. BOCCHI, V., G. CASNATI, A. DOSSENA & R. MARCHELLI. 1979. Esterification of amino acids and dipeptides under mild conditions. II. Via Sodium salts synthesis. Synthesis **11:** 961–962.
5. TERADA, S., K. SATO, T. KATO & N. IZUMIYA. 1980. Studies on the synthesis of proteinase inhibitors. II. Synthesis of cyclic nonapeptide fragments and analogs related to the reactive sites on soybean Bowman-Birk inhibitor. Int. J. Pept. Protein Res. **15:** 441–454.
6. JANSSEN, A. E. M., C. KLABBERS, M. C. R. FRANSSEN & K. VAN'T RIET. 1991. Enzymatic synthesis of carbohydrate esters in 2-pyrrolidone. Enzyme Microb. Technol. **13:** 565–572.

Kinetic Resolution of Racemic Glycerol Derivatives with Lipase and Quinohemoprotein Alcohol Dehydrogenase[a]

J. B. A. VAN TOL, A. GEERLOF, J. A. JONGEJAN, AND J. A. DUINE[b]

Kluyverlaboratory for Biotechnology
Delft University of Technology
Julianalaan 67
2628 BC Delft, the Netherlands

INTRODUCTION

1,2-*O*-Isopropylidene glycerol, solketal, and 2,3-epoxy-1-propanol, glycidol, are widely applicable synthons for the preparation of biologically active fine chemicals including phospholipids, β-blockers (e.g., propranolol and atenolol), as well as other chiral compounds of industrial interest.[1] To meet the demands for high enantiomeric purity of the final products (typically $ee > 0.99$), (bio-)catalytic kinetic resolution of the racemic starting materials is considered to be an attractive strategy. The economic feasibility of such processes will, however, depend strongly on the degree of enantioselectivity of the catalyst. Kinetic resolution of racemic glycidyl butyrate using porcine pancreatic lipase, PPL, has been described.[2,3] A value of ≈ 7 for the enantiomeric ratio, E, was inferred from the enantiomeric excess value of residual ester after 60% conversion, using the relation developed by Chen and coworkers.[4,5] Application of this relation, however, appeared unjustified since the enantiomeric excess value that was reached at $\xi = 0.6$ remained practically constant as ξ was increased further (FIGURE 1). In order to account for this deviation, we investigated the accuracy and applicability of current methods for the determination of E, possible heterogeneity of commercial PPL, the kinetic mechanism of hydrolysis, and equilibrium effects. A preliminary investigation[6] suggested enantioselective product inhibition inherent to the Ping-Pong kinetics of PPL as a major cause. As similar kinetic schemes can be expected for the oxidation of racemic solketal and glycidol by quinohemoprotein alcohol dehydrogenases, QHADHs, the enantioselective properties of these enzymes were also investigated.

THE ENANTIOMERIC RATIO, E

Chen and coworkers[4,5] showed that for kinetic resolutions of a racemic substrate with enzymes catalyzing reactions according to Michaelis-Menten-type kinetics, the

[a] This work was supported by Andeno BV, Venlo, the Netherlands.
[b] Author to whom correspondence should be addressed.

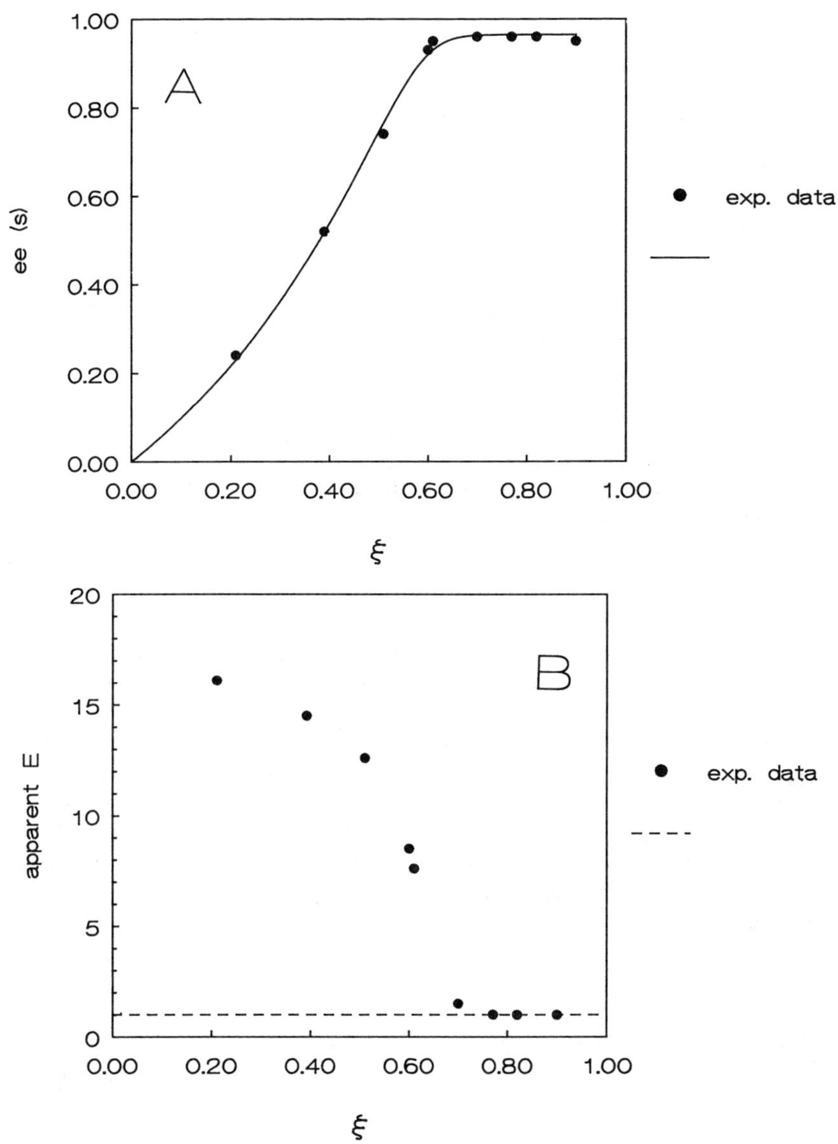

FIGURE 1. Panel A: Enantiomeric excess values of remaining substrate (*ees*) versus the degree of conversion (ξ) in the hydrolysis of glycidyl butyrate at T = 298 K and pH = 7.8. **Panel B:** Actual apparent E values $\{E = (v_S/v_R) \cdot ([R]/[S])\}$ versus ξ in the hydrolysis of glycidyl butyrate at T = 298 K and pH = 7.8.

enantiomeric ratio, E, the enantiomeric excess value of residual substrate, *ees*, and the degree of conversion, ξ, are related by Equation 1:[4,5]

$$E = \frac{\ln [(1 - ees) \cdot (1 - \xi)]}{\ln [(1 + ees) \cdot (1 - \xi)]} \tag{1}$$

E is independent of ξ and equals the ratio of specificity constants for the two enantiomers, Equation 2:

$$E = (k_{cat}/K_M)_S/(k_{cat}/K_M)_R \tag{2}$$

We have shown that for these enzymes the initial reaction rates for mixtures of enantiomers at fixed total-substrate concentration under otherwise identical conditions are related by Equation 3:

$$E = \frac{1 - x}{x} \cdot \frac{v_S\,(v_x - v_R)}{v_R\,(v_S - v_x)} \tag{3}$$

where v_S, v_R, and v_x represent initial reaction rates of homochiral S-enantiomer, homochiral R-enantiomer, and mixtures with (molar) fraction x of S-enantiomer, respectively.

KINETIC RESOLUTION OF GLYCIDYL BUTYRATE USING PORCINE PANCREATIC LIPASE

Partial hydrolysis of racemic glycidyl butyrate with commercial PPL (Sigma Chem. Co.) leads to the enantiomeric enrichment of R-glycidyl butyrate (SCHEME 1).

As is shown in FIGURE 1, a good fit of Equation 1 to the experimental results requires the apparent E values $\{E = (v_S/v_R) \cdot ([R]/[S])\}$ to decrease rapidly for $\xi > 0.5$. Although the accuracy of E value determinations using Equation 1 has been shown to decrease sharply for increasing values of E,[8] the observed deviations are well outside this range.

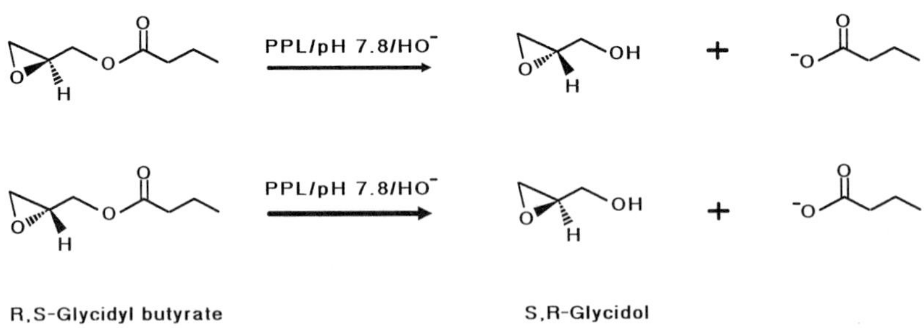

SCHEME 1. Kinetic resolution of racemic glycidyl butyrate by porcine pancreatic lipase, PPL.

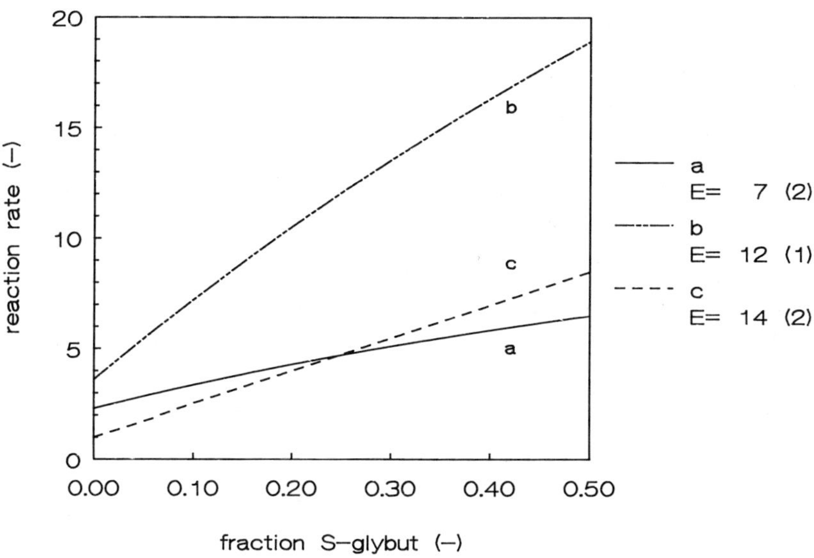

FIGURE 2. Estimation of the apparent E value from initial hydrolysis rates measured at fixed total glycidyl butyrate concentration, according to Equation 3. Standard deviation is given in parentheses. **a:** 5 ml/l glycidyl butyrate (single phase): $E = 7$ (2). **b:** 40 ml/l glycidyl butyrate (two phases): $E = 12(1)$. **c:** 5 ml/l glycidyl butyrate + 25 ml/l CH_2Cl_2 (two phases): $E = 14(2)$. x = fraction S-glycidyl butyrate; v_S, v_R = rate on homochiral enantiomers; $T = 298$ K; pH = 7.8.

INITIAL RATE MEASUREMENTS

Due to the heterogenous reaction conditions, determination of E from the kinetic constants according to Equation 2 is not feasible for the hydrolysis of glycidyl butyrate by PPL. Application of Equation 3, on the other hand, shows that the apparent E value is consistently lower when a single phase is present (FIGURE 2). Yet, E values determined by initial kinetics according to Equation 3 for substrate concentrations that occur for $\xi > 0.5$ are distinctly larger than those observed in kinetic resolution experiments (FIGURE 1).

HETEROGENEITY OF COMMERCIAL PPL

Analysis of commercial PPL (Sigma Chem. Co.) by chromatographic[3] and electrophoretic techniques revealed that several species with hydrolytic activity for glycidyl butyrate are present. Fractions containing significant activities, however, appeared to have similar E values in the hydrolysis of racemic glycidyl butyrate, ranging from $E = 15$ (± 3) for lipase A and B to $E = 11$ (± 2) for an activity that has been tentatively identified as carboxypeptidase B. Time-dependent inactivation of individual fractions had no influence on the overall E value. The hydrolytic activity of total protein that was recovered from the reaction mixture at various time intervals was found to retain the initial enantioselectivity.

EQUILIBRIUM EFFECTS

The overall equilibrium for the hydrolysis of (racemic-)glycidyl butyrate at pH 7.8 is at the product side ($\xi_{eq} > 0.99$). Introduction of K_{eq} as a process parameter according to Chen and coworkers[5] did not improve the fit of the data given in FIGURE 1 for constant E.[7] Since glycidyl esters are poorly soluble in water, a separate organic phase is formed as soon as the concentration exceeds their solubility (approx. 25 g of glycidyl butyrate per liter water at room temperature and atmospheric pressure). As a result of partitioning of the enantiomeric esters between the two phases, a pseudoequilibrium is set up for both ester enantiomers. This equilibrium adapts continuously to the volume of the organic phase which diminishes as the conversion proceeds. As shown in FIGURE 3, this effect contributes to a decrease of the apparent

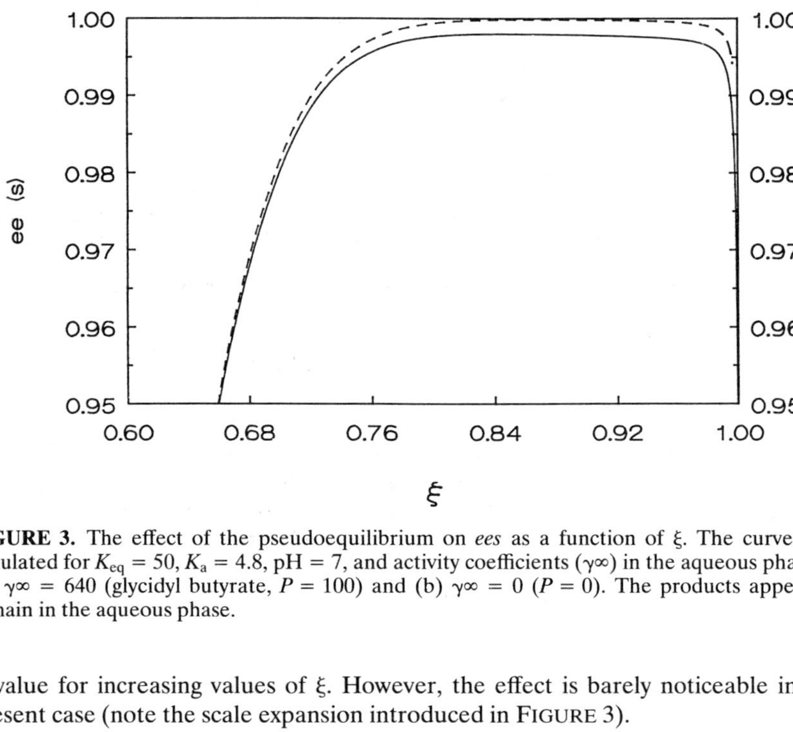

FIGURE 3. The effect of the pseudoequilibrium on *ees* as a function of ξ. The curves are simulated for $K_{eq} = 50$, $K_a = 4.8$, pH = 7, and activity coefficients ($\gamma\infty$) in the aqueous phase of (a) $\gamma\infty = 640$ (glycidyl butyrate, $P = 100$) and (b) $\gamma\infty = 0$ ($P = 0$). The products appear to remain in the aqueous phase.

E value for increasing values of ξ. However, the effect is barely noticeable in the present case (note the scale expansion introduced in FIGURE 3).

KINETICS OF THE PPL-CATALYZED HYDROLYTIC REACTION

We investigated three alternative kinetic schemes: (1) reversible Michaelis-Menten kinetics, (2) Rabin-type "memory" mechanisms,[9] and (3) enantioselective product inhibition resulting from partitioning of acyl-enzyme intermediate between water and enantiomeric glycidols. It appeared that the decrease of the apparent E value (FIGURE 1) could be accounted for on the basis of realistic microscopic

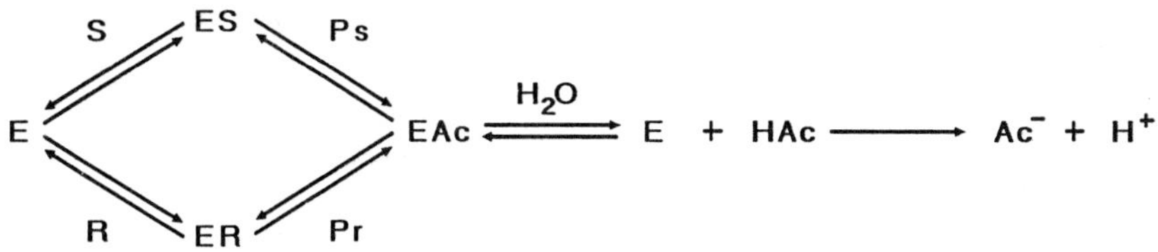

SCHEME 2. Kinetic scheme (pseudo-Ping-Pong BiBi) for the competition of products (P_S and P_R the enantiomeric glycidols) and H_2O for the acyl-enzyme intermediate.

reaction constants in the case of enantioselective product inhibition only. The kinetic scheme is given (SCHEME 2).

Determinations of the microscopic constants for the reaction of *S*- and *R*-glycidol with acyl-enzyme, by independent measurements (e.g., by evaluating *E* from initial rate measurements according to Equation 3 in the presence of varying amounts of *S*- *viz.* *R*-glycidol), are now being conducted. It should be stressed that under the conditions used (pH 7.8), a contribution of the net back-reaction (i.e., enzyme + acid → acyl-enzyme) does not occur to a significant extent for $\xi < 0.95$ due to the low concentration of protonated acid.

OXIDATION OF SOLKETAL AND GLYCIDOL BY QUINOHEMOPROTEIN ALCOHOL DEHYDROGENASES

Extensive screening of bacteria for the presence of enzymes that require the cofactor pyrroloquinoline quinone, PQQ, for activity, resulted in the characterization of a heme c–containing protein from *Comamonas testosteroni* (formerly known as *Pseudomonas testosteroni*) that showed appreciable dye-linked dehydrogenase activity for (primary-)alcohols upon addition of PQQ[10] and PQQ analogues[11] (FIGURE 4).

Subsequent investigations of the enantioselective properties of these enzymes showed that high values of *E* could be observed in the kinetic resolution of racemic solketal and related glycerol alkanonides.[12] Similar enzymatic activities present in

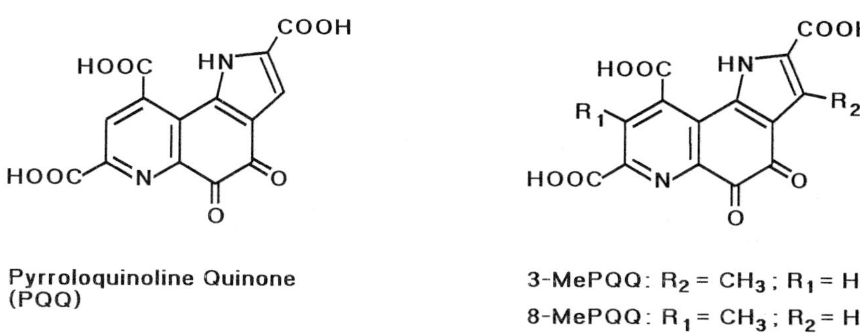

FIGURE 4. The structure of 4,5-dioxo-1*H*-pyrrolo[2,3-*f*]quinoline-2,7,9-tricarboxylic acid, pyrroloquinoline quinone, PQQ, and some analogues.

TABLE 1. Selective Oxidation of Glycidol and Solketal by Quino(hemo)protein Alcohol Dehydrogenases

Enzyme	Microorganism	Glycidol	Solketal
MDH	*Hyphomicrobium*	1.6 (R)	—[a]
MDH	*Pseudomonas BB1*	1	—
ADH	*Pseudomonas aeruginosa (Ps. putida)*	1	—
EDH	*Comamonas testosteroni*	1	> 100 (R)
ADH	*Acetobacter aceti*	10–20 (S)	—
ADH	*Acetobacter peroxydans*	19 (S)	—
ADH	*Gluconobacter oxydans*	~ 30 (S)	12 (R)
ADH	*Rhodopseudomonas acidophila*	1.6 (R)	—
PEG-DH	*Flavobacterium*	1	18 (S)

[a]—Substrate not oxidized.

acetic acid bacteria were found to catalyze the dehydrogenation of glycidol with appreciably high *E* values (TABLE 1).

Since these enzymes catalyze the overall four-electron oxidation of the alcohol to the corresponding acid (SCHEME 3), mechanistic effects on the value of *E* were anticipated. Preliminary results show that such effects do indeed occur. However, contrary to our findings for the kinetic resolution of glycidyl butyrate with PPL, enantioselective product inhibition by the product acids does not exert a major effect. Also, it appears that the eventual substrate inhibition, probably resulting from increased concentrations of intermediate aldehydes, has insignificant effects on the overall *E* value.

ENGINEERING THE ENANTIOSELECTIVITY OF QHADHs BY COFACTOR MODIFICATION

The use of cofactor analogues and derivatives has been advocated for a number of applications, e.g., to act as ligands for affinity chromatography; as (inactive)

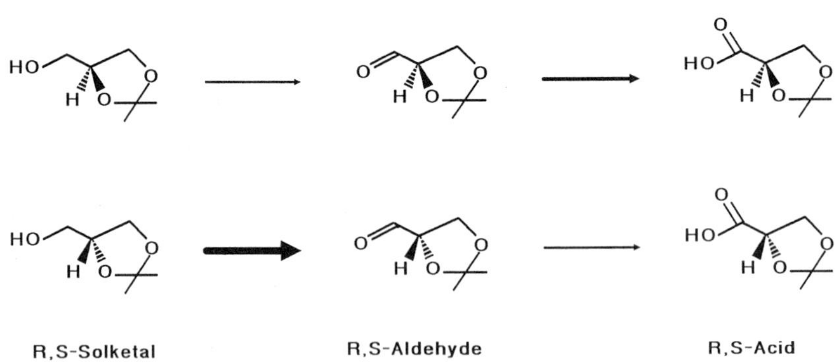

SCHEME 3. Kinetic resolution of racemic solketal by quinohemoprotein alcohol dehydrogenase from *C. testosteroni*. Intermediate aldehydes occur free in solution.

cofactor mimicks for target enzymes in drug design; and to improve solubility or regeneration of the coenzyme of NAD(P)-dependent dehydrogenases. Since enantiomeric substrates differ in geometry only, they represent subtle probes for (small) changes in the topology of the enzyme active site. So far, few examples have been described of the effect of cofactor modification on the enantioselectivity of enzymes.[13] We found that the effect of replacing PQQ by several PQQ analogues in the active site of (apo-)quinohemoprotein alcohol dehydrogenase of *C. testosteroni* resulted in a marked decrease of the activity as well as the enantioselectivity towards solketal as substrate (TABLE 2).[14]

A possible explanation for these results might be that the active site reconstituted with PQQ as a cofactor is highly optimized for the enantioselective oxidation of solketal, in which case even minor changes of the cofactor lead to diminished selectivity (and activity). Preliminary investigations using substrates on which the E value of the QHADH is less pronounced showed that similar, and in a single event even higher, E values could be observed (Geerlof, Jongejan, and Duine, unpublished).

TABLE 2. Effect of Replacement of PQQ by PQQ Analogues on the Enantioselectivity of Quinohemoprotein Alcohol Dehydrogenase from *C. testosteroni* in the Oxidation of Racemic Solketal[a]

PQQ/PQQ Analogue	Activity (%)	E value
PQQ	100	43
3-CH_3-PQQ	12	11
8-CH_3-PQQ	23	7

[a]Activity (%) represents V_{max} initial rate on the preferred enantiomer, *R*-solketal.

CONCLUDING REMARKS

Investigations on the kinetic resolution of racemic glycerol derivatives with lipase and quinohemoprotein alcohol dehydrogenases have shown that the occurrence of *non*-Michaelis-Menten-type kinetics can have an important effect on the enantioselectivity of the catalyst. In order to quantify these effects, we found the accuracy involved in the determination of the enantiomeric ratio to be highly important. In addition, it was found that the determination of E by analysis of progress curves, according to Chen and coworkers,[4,5] may not be conclusive. Instead, alternative measurements, e.g., by analysis of initial-rate data of mixtures of enantiomers at constant concentration,[7] should be considered. The effect of cofactor modification on the enantioselectivity of quinohemoprotein dehydrogenases suggests that practical applications for this type of cofactor engineering may eventually be found.

SUMMARY

Enantiomerically pure glycerol derivatives of industrial interest have been prepared by enzyme-catalyzed kinetic resolution of the racemic compounds. Analysis of the kinetic resolution of racemic glycidyl butyrate by commercial porcine pancreatic lipase showed that the Ping-Pong kinetics of the hydrolytic reaction are a major cause for the decrease of enantioselectivity observed at high degrees of conversion. Oxidation of solketal by quinohemoprotein alcohol dehydrogenase from *Comamonas testosteroni* occurs with $E > 50$ using isolated enzyme as well as cell suspensions. High values of E have been measured for the oxidative kinetic resolution of glycidol using *Acetobacter* sp. The effect of cofactor modification (cofactor engineering) on the enantioselective properties of quinohemoprotein alcohol dehydrogenase enzyme from *C. testosteroni* has been studied by replacing the cofactor PQQ with suitable analogues.

ACKNOWLEDGMENTS

We wish to thank Ms. Diana Kraaijveld for her contribution to the characterization of the lipase preparations.

REFERENCES

1. Tweel, W. J. J. van den., Th. J. G. M. van Dooren, M. Kloosterman, V. H. M. Elferink, L. A. Hulshof, A. Geerlof, J. A. Jongejan, B. W. Groen & J. A. Duine. 1990. *In* 5th European Congress on Biotechnology. C. Christiansen, L. Munck & J. Villadsen, Eds.: 283. Munksgaard International Publisher. Copenhagen, Denmark.
2. Ladner, W. T. & G. M. Whitesides. 1984. J. Am. Chem. Soc. **106:** 7250–7251.
3. Philippi, M. Chr., J. A. Jongejan & J. A. Duine. 1987. *In* Biocatalysis in Organic Media. C. Laane, J. Tramper & M. D. Lilly, Eds.: 279. Elsevier Science Publishers. Amsterdam, the Netherlands.
4. Chen, C.-S., Y. Fujimoto, G. Girdaukas & C. J. Sih. 1982. J. Am. Chem. Soc. **104:** 7194–7299.
5. Chen, C.-S., S.-H Wu, G. Girdaukas & C. J. Sih. 1987. J. Am. Chem. Soc. **109:** 2812–2817.
6. Tol, J. B. A. van, J. A. Jongejan & J. A. Duine. 1990. *In* 5th European Congress on Biotechnology. C. Christiansen, L. Munck & J. Villadsen, Eds.: 267. Munksgaard International Publisher. Copenhagen, Denmark.
7. Jongejan, J. A., J. B. A. van Tol, A. Geerlof & J. A. Duine. 1991. Recl. Trav. Chim. Pays-Bas **110:** 247–254.
8. Tol, J. B. A. van, J. A. Jongejan, A. Geerlof & J. A. Duine. 1991. Recl. Trav. Chim. Pays-Bas **110:** 255–262.
9. Newsholme, E. A. & C. Start. 1973. Regulation in Metabolism: 64–66. John Wiley & Sons. Chichester, England.
10. Groen, B. W., M. A. G. van Kleef & J. A. Duine. 1986. Biochem. J. **234:** 611–615.
11. Jongejan, J. A., B. W. Groen & J. A. Duine. 1989. *In* PQQ and Quinoproteins, Proceedings of the 1st International Symposium on PQQ and Quinoproteins. J. A. Jongejan & J. A. Duine, Eds.: 205–216. Kluwer Academic Publishers. Dordrecht, the Netherlands.
12. Duine, J. A., B. W. Groen & A. Geerlof. 1990. Eur. Patent Appl. EP 90201651.
13. Tate, S. S. & A. Meister. 1969. Biochemistry **8:** 1056–1065.
14. Duine, J. A. 1991. Trends Biochem. Sci. **9:** 343–346.

Some Characteristics of a Serine Proteinase Isolated from an Extreme Thermophile for Use in Kinetically Controlled Peptide Bond Synthesis[a]

KEITH PEEK,[b] SHELLEY-ANN WILSON,[c]
MARK PRESCOTT,[c] AND ROY M. DANIEL[c]

[b]*Pacific Enzymes Limited*
Post Office Box 1496
Hamilton, New Zealand

[c]*Thermophile Research Unit*
University of Waikato
Hamilton, New Zealand

INTRODUCTION

The use of proteinases for the synthesis of peptides and esters has been studied for over 50 years and has increased substantially over recent years. In particular, the use of proteinases in anhydrous organic solvents ($<0.01\%$ water), where they exhibit novel properties, is gaining momentum.[1] However, most of the work on peptide synthesis has been on the use of proteinases in water-miscible solvents. There are a number of advantages in employing water-miscible solvents.[2] For example, some amino acid derivatives are poorly soluble in nonpolar solvents and limitations due to the diffusion of substrates in biphasic systems or enzymes suspended in hydrophobic solvents are avoided. The main disadvantage of employing water-miscible solvents is that at high concentrations, many enzymes are relatively inactive or denatured due to the stripping of enzyme-bound water.[3] The use of proteinases isolated from extreme thermophiles, with their inherent stability to extremes of pH and temperature and to organic solvents, would seem to be one of the obvious ways of solving these problems. In addition, the reactions could be carried out at higher temperatures, which would result in benefits from increased substrate solubilities, accelerated rates of diffusion, and decreased viscosity.

We have isolated a thermostable serine proteinase from an extreme thermophile of the genus *Thermus*. Here we present some characteristics and preliminary results for the use of this enzyme in peptide synthesis.

[a]This work was funded by Pacific Enzymes Limited.

METHODS

Purification and Characterization of Rt41A Proteinase

Thermus sp. strain Rt41A was grown in a 600 I fermenter and the proteinase concentrated from the cell free supernatant as described previously.[4] The enzyme was purified by cation exchange chromatography with S sepharose fast flow (Pharmacia) on a Mono S 10/10 column. The proteinase was pure as determined by native polyacrylamide gel electrophoresis (PAGE) and isoelectric focusing.

The enzyme is a thermostable alkaline serine proteinase, has a molecular weight of 32.5 kDa [sodium dodecyl sulfate (SDS) PAGE] and an isoelectric point >10.5. The proteinase is stable, in the presence of 5 mM $CaCl_2$, at 70°C and has half-lives of 13.5 hours and 20 minutes at 80°C and 90°C respectively. No loss of activity was observed after 4 hours at 70°C in the pH range 5–10.

The hydrolysis of the oxidized insulin B chain by Rt41A proteinase (FIGURE 1) is similar to that for other microbial serine proteinases.[5] Although rather nonspecific, the cleavage pattern indicates a preference for aromatic or aliphatic amino acids at the P'_1 site. The esterase activity also indicates a similar preference at the P_1 site, with the highest activity against CBZ-blocked tyrosine and alanine *p*-nitrophenol esters.

A number of substrates were screened for the production of dipeptides, and the synthesis of Ac-Phe-Tyr-NH_2, Ac-Tyr-Trp-NH_2, Ac-Tyr-Tyr-NH_2, Bz-Ala-Trp-NH_2, Bz-Ala-Tyr-NH_2, and Bz-Ala-Phe-NH_2 was demonstrated. The optimization of Bz-Ala-Tyr-NH_2 synthesis was studied with regard to pH, solvent concentration, ionic strength, and nucleophile concentration.

Reaction Scheme and High-Performance Liquid Chromatography

The reaction under study was as shown in SCHEME 1. The concentration of the hydrolytic (Bz-Ala-OH) and synthetic (Bz-Ala-Tyr-NH_2) products was determined by reverse-phase (C_{18}) high-performance liquid chromatography (HPLC) using a linear gradient 10–100% v/v acetonitrile containing 0.1% v/v trifluoroacetic acid (TFA) over 15 minutes at 1 ml/minute. Reactants and products were detected at 280 nm, and peak areas were used to determine the amount of products formed. Peptide synthesis was confirmed by the amino acid analysis of collected peaks.

Optimization of Bz-Ala-Tyr-NH_2 Synthesis

Bz-Ala-OMe was incubated with Tyr-NH_2 in 50 mM Hepes or Caps buffer containing 0.01% v/v Triton X-100 and 5 mM $CaCl_2$ at 40°C. The pH and concentrations of dimethylformamide (DMF), NaCl, and reactants were varied according to the parameter under study (see FIGURE 3). The reaction was initiated by the addition of 40 μg of Rt41A proteinase to 1 ml of reaction media. Duplicate samples (50 μl) were collected at various time intervals, for HPLC analysis, and stopped by the addition of 50 μl of 0.1% v/v TFA in DMF.

The efficiency of acyl transfer to Tyr-NH_2 was expressed in terms of the partition

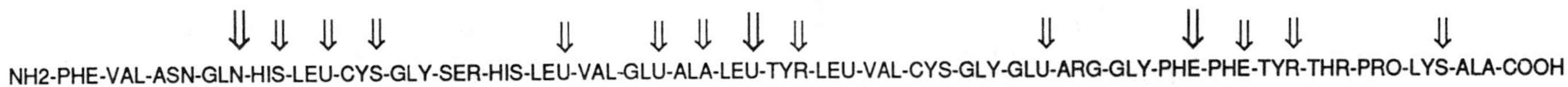

FIGURE 1. Cleavage of the oxidized bovine insulin B chain by *Thermus* strain Rt41A proteinase. Primary cleavage sites are indicated by large arrowheads, and minor sites by small arrowheads.

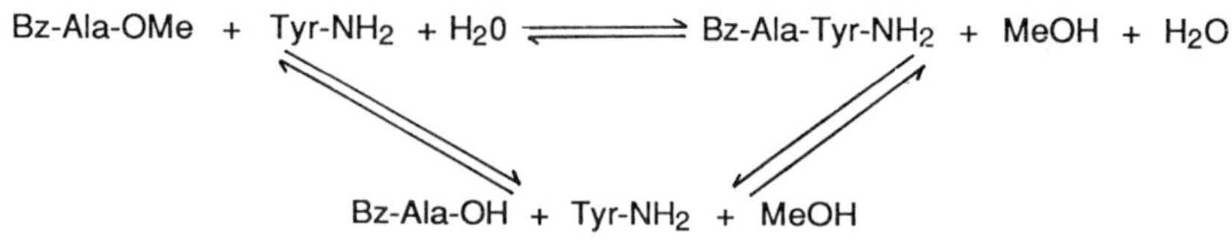

SCHEME 1

constant[6] p as given in the equation:

$$p = \frac{V_H[N]}{V_A} = \frac{[P_2][N]}{[P_2]}$$

and is equal to the nucleophile concentration ($[N]$) at which the initial rates of hydrolysis (V_H) and aminolysis (V_A) are equal. $[P_2]$ and $[P_3]$ are the concentrations of the hydrolytic and aminolytic products respectively. The initial rates were determined from progress curves such as that depicted in FIGURE 2. During the optimization studies, the yield of peptide was always less than 7% and the nucleophile concentration was considered constant during the initial rate determination.

RESULTS AND DISCUSSION

The effect of pH on the partition constant is shown in FIGURE 3. The results show that the rate of aminolysis increases with increasing pH and the lowest p value is

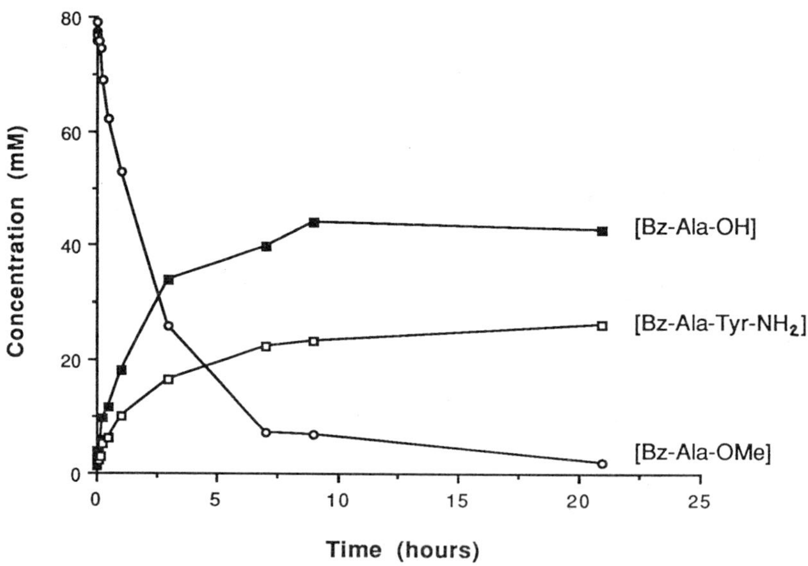

FIGURE 2. Time course for Bz-Ala-Tyr-NH_2 synthesis. Reaction containing 50-mM Caps pH 10, 90% v/v DMF, 100-mM Bz-Ala-OMe, 800-mM Tyr-NH_2, and 40 μg of Rt41A proteinase at 40°C.

obtained at pH 10. The increased rate is the result of deprotonation of the tyrosine amide ($pk_b = 7.7$). At 2 pH units above the pK_b, this will be approximately 98% deprotonated and coincides with the observed pH optimum for synthesis. The increase in p above pH 10 is probably due to increased rate of deacylation by hydroxyl

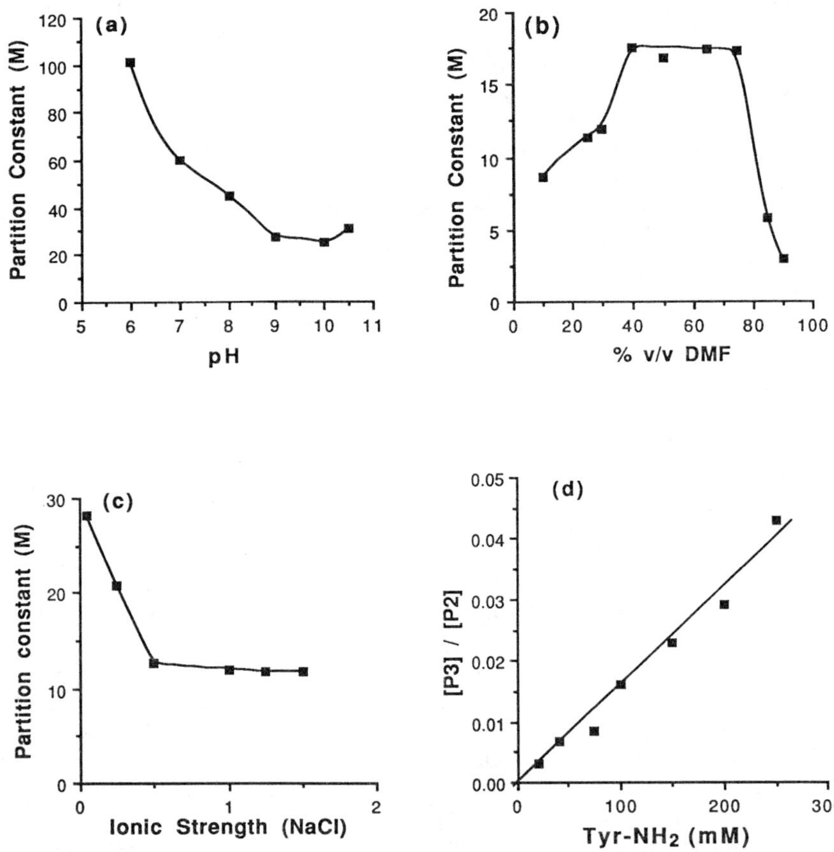

FIGURE 3. Optimization of Bz-Ala-Tyr-NH_2 synthesis. The effect of (a) pH. Reaction containing 50-mM Hepes pH 6–8 or 50-mM Caps pH 8.5–10.5, 50% v/v DMF, 25-mM Bz-Ala-OMe, 100-mM Tyr-NH_2, and 40 μg of Rt41A proteinase at 40°C; (b) dimethylformamide concentration. Reaction conditions as in (a) with 50-mM Caps pH 10 and DMF varied; (c) ionic strength. Reaction conditions as in (a) with 50-mM Caps pH 10 and the appropriate amount of NaCl added; and (d) nucleophile concentration. Reaction condition as in (a) with 50-mM Caps pH 10 except 10-mM Bz-Ala-Tyr-OMe and Tyr-NH_2 varied.

ions. All further optimization studies were performed at pH 10 in 50 mM Caps buffer.

Increasing the solvent concentration shows a number of interesting effects. The initial rates of hydrolysis and aminolysis are reduced with increasing solvent. Decreased activity is a phenomenon that is generally observed with enzymes in hydrophilic solvents due to the stripping of enzyme-bound water.[3] This decreased activity should only affect the time to complete the reaction in the absence of other factors. FIGURE 3b shows that by increasing the solvent concentration, the partitioning of hydrolysis and aminolysis is also affected. Increasing the solvent concentration to 40% results in an increase in p, suggesting that the rate of aminolysis is decreasing faster than hydrolysis, which could be due to solvent interaction with the S'_1 subsite and nucleophile. From 40–75% v/v DMF, the value of p remains constant.

At 75% v/v DMF there is a dramatic drop in the partition constant, which suggests an increased affinity of the S'_1 for tyrosine amide. This could be due to a number of factors, including a change in the conformation of the S'_1 subsite resulting in increased affinity for the nucleophile as a result of water stripping or the binding of DMF molecules to the enzyme. Several studies have shown that bound water molecules coupled to localized changes in conformation can determine substrate specificity.[7] Recently Kasche *et al.* have shown with buffers saturated with water-immiscible solvents that the P'_1–P'_2 specificity is altered by structural changes in the S' subsites.[8] Alternatively, the observed effect could simply be due to substrate partitioning from the bulk solution into the S'_1 binding pocket, which has been used to explain changes in substrate specificity in organic solvents.[1]

The effect of ionic strength on the partition constant is shown in FIGURE 3c. The value of p decreases with increasing ionic strength up to $I = 0.5$ and then remains constant up to $I = 1.5$. A decrease in p is often observed when nucleophiles containing side chains of equal charge to those found on the enzyme are used, due to charge suppression effects.[9] In the case of uncharged nucleophiles such as tyrosine amide, a decrease in p has been shown to be due to salting-out/salting-in effects.[9] The decrease in p observed here occurs at relatively low ionic strength compared to those observed by Schellenberger *et al.*[9] Although salting out could account for the observed results, salt-dependent changes in the S'_1 subsite resulting in increased affinity for the nucleophile may also explain this phenomenon.

The effect of increasing the nucleophile concentration on the ratio of aminolytic and hydrolytic products is shown in FIGURE 3d. The ratio is linearly dependant on the nucleophile concentration as shown by other workers for trypsin and chymotrypsin. The result shows that p is independent of $[N]$ and there is no hydrolysis of the acyl-enzyme/nucleophile complex.[6] The increase in $[N]$ shifts the reaction in favor of synthesis by the law of mass action.

A time course for the synthesis of Bz-Ala-Tyr-NH_2 using a combination of optimized parameters is shown in FIGURE 2 and resulted in an increase in the yield of peptide from 0.75% to 26%.

CONCLUSIONS

Studies on Rt41A proteinase have demonstrated that the enzyme can be used for kinetically controlled peptide synthesis and that many of its properties afford advantages over proteases isolated from mesophilic organisms. The enzyme is stable to extremes of temperatures, pH and organic solvents allowing synthesis under conditions not possible with mesophilic enzymes. Optimization of Bz-Ala-Tyr-NH_2 synthesis has improved the initial yield from 0.75% to 26%. Although the yields of peptide and partition constants reported here do not compare favorably to many of the mesophilic proteases used for peptide synthesis, recent studies in which the effect of different solvents and *c*-terminal blocking of the nucleophile are being investigated show that quantitative yield of peptide ($>95\%$) can be obtained.

REFERENCES

1. KLIBANOV, A. M. 1989. Enzymatic catalysis in anhydrous organic solvents. TIBS **14:** 141–144.
2. KISE, H., A. HAYAKAWA & H. NORITOMI. 1990. Protease catalysed synthetic reactions and immobilization-activation of the enzymes in hydrophilic organic solvents. J. Biotechnol. **14:** 239–254.
3. ZAKS, A. & A. M. KLIBANOV. 1988. The effect of water on enzyme action in organic media. J. Biol. Chem. **263:** 8017–8021.
4. PEEK, K., P. H. JANSSEN, H. W. MORGAN & R. M. DANIEL. 1990. Enhancement of *in vitro* stability and recovery of a proteinase produced by an extreme thermophile. *In* Fermentation Technologies: Industrial Application. Pak-Lam Yu, Ed.: 97–102. Elsevier Applied Science. London, England.
5. MATSUZAWA, H., K. TOKUGAWA, M. HAMAOKI, M. MIZOGUCHI, H. TAGUCHI, I. TERADA, S. T. KWON & T. OHTA. 1988. Purification and characterization of aqualysin 1. Eur. J. Biochem. **171:** 441–447.
6. SCHELLENBERGER, V., H. JAKUBKE, N. P. ZAPEVALOVA & Y. V. MITIN. 1991. Protease-catalyzed peptide synthesis using inverse substrates: the influence of reaction conditions on the acyl transfer efficiency. Biotechnol. Bioeng. **38:** 104–108.
7. QUIOCHO, F. A., D. K. WILSON & N. K. VYAS. 1989. Substrate specificity and affinity of a protein modulated by bound water molecules. Nature **340:** 404–407.
8. KASCHE, V., G. MICHAELIS & B. GALUSKY. 1991. Binding of organic molecules influences the P'_1-P'_2 stereo and sequence specificity of α-chymotrypsin in kinetically controlled peptide synthesis. Biotechnol. Lett. **13:** 75–80.
9. SCHELLENBERGER, V., H. JAKUBKE & V. KASCHE. 1991. Electrostatic effects in the α-chymotrypsin catalysed acyl transfer. II. Efficiency of nucleophiles bearing charged groups in various locations. Biochim. Biophys. Acta **1078:** 8–11.

A New Racemase for 5-Monosubstituted Hydantoins

MARKUS PIETZSCH, CHRISTOPH SYLDATK,
AND FRITZ WAGNER[a]

Institute of Biochemistry and Biotechnology
Technical University of Braunschweig
Konstantin-Uhde-Strasse 5
3300 Braunschweig, Germany

INTRODUCTION

The great advantage of producing L-amino acids from D,L-5-monosubstituted hydantoins is not only the wide substrate specificity of the enzymes involved in this process, but also the fact that racemic substrates can be totally converted to the optically pure products within a short time.[1-3]

Using resting cells of *Arthrobacter* sp. DSM 3747, a productivity of 1.6 g (Trp) × g $(CDM)^{-1}$ × $hours^{-1}$ was determined.

Three enzymes are shown to be involved in this process in *Arthrobacter* sp. DSM 3747 after growth in a medium containing the chemically synthesized D,L-3-*N*-methyl-3′-methylene-indolyl-5-hydantoin as the most effective inducer. Besides an L-hydantoinase and an *N*-carbamoyl-L-amino acid-amidohydrolase, a hydantoin-racemase was found and could be separated from the other two enzymes.

In this report, we present results concerning the isolation and characterization of the racemase as well as studies to clear up the enzymatic reaction mechanism.

MATERIAL AND METHODS

Microorganism

The microorganism *Arthrobacter* sp. DSM 3747 was cultivated in a 70 l open frame bioreactor (type 44401/1, Braun-Diessel Biotech GmbH, Melsungen, Germany) under aerobic conditions at 30°C and pH 6.8 in a minimal medium containing D,L-3-*N*-methyl-3′-methylene-indolyl-5-hydantoin as inductor.[4]

Chemicals

All chemicals used in these experiments were commercially available products of analytical grade. The hydantoin derivatives were prepared by K. Krohn and his coworkers (Institute of Organic Chemistry, University of Braunschweig, Germany) and by Degussa (Degussa AG, Hanau, Germany).

[a] Author to whom correspondence should be addressed.

Isolation of the Enzyme

After harvesting the biomass by centrifugation at 20,000 rpm (Padberg centrifuge type 41, Carl Padberg GmbH, Germany), the cells were desintegrated in a glass bead mill (Netsch LME 4, Netsch-Feinmahltechnik GmbH, Selb, Germany). Ammonium sulfate and a solution of Protamin sulfate in water (pH 7.0) were added to a final concentration of 1.0 M and 1.0 g/l respectively. After centrifugation at 13,000 rpm (4°C, Suprafuge 22, Heraeus, Osterode, Germany), further purification was carried out using hydrophobic interaction chromatography on HiLoad Phenyl-Sepharose 16/10, gel exclusion chromatography on Sephacryl S-300, and ion exchange chromatography on MonoQ HR 5/5.

Measurement of Catalytic Activities

For purification, assays were carried out with 8.724×10^{-4} M of D-3′-methylene-indolyl-5-hydantoin for 5 minutes at 37°C in an 800-μl-volume glycine buffer solution (0.05 M, pH 8.5). The assays were started by addition of 25 μl enzyme solution and stopped by addition of 575 μl methanol. Quantitative determination of substrates and products were performed at room temperature by high-performance liquid chromatography (HPLC) using a Spectra-Physics-pump (SP 8800, Spectra-Physics, Darmstadt, Germany), a chiral L-Pro-Cu column (Si 100 Polyol, 5 μm, 250 mm, Serva, Heidelberg, Germany), filtered and degased $CuSO_4$ solution (0.01 M) as eluent at a flow rate of 1 ml/minute and UV detection at 280 nm.

Substrate specificity was investigated using the following method: from a 60-ml reaction vessel, the reaction solution of 50 ml substrate (8.724×10^{-4} M) and 5 ml enzyme solution were pumped through a YM 10 polysulfon ultrafiltration membrane (Amicon, Witten, Germany). While the filtrate passes a polarimeter (Dr. Kernchen GmbH, Seelze, Germany) first, retentate is pumped directly back into the vessel. The reactions were followed on-line by recording the change of optical rotation.

Method of Studying Enzymatic Reaction Mechanism

After H/D-exchange of buffers and enzyme solution, the reaction of 0.2 g/l D-5-IMH in 1000 ml deuterated phosphate buffer, pD 8.9, and 30 ml enzyme after buffer exchange was carried out for 45 minutes at 37°C in a stirred reactor. The products were isolated carefully by freeze drying, centrifugation and washing with D_2O. For ^1H- and ^{13}C-NMR studies, the lyophilisate was solved in DMSO-d_6.

RESULTS AND DISCUSSION

Purification of the Hydantoin-Racemase

Intensive studies on the stability of the hydantoin-racemase showed that only phosphate buffers at neutral pH are useful for the purification. Stability decreases with increasing purity. Higher salt concentrations have been found to stabilize

Substrate 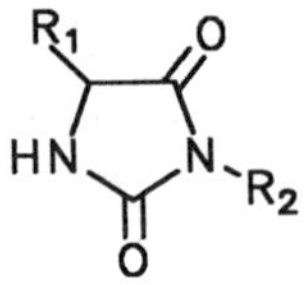 R$_1$	R$_2$	relative Activity [%]
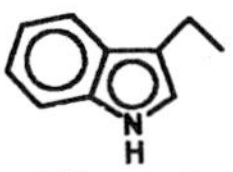	-H	100.0
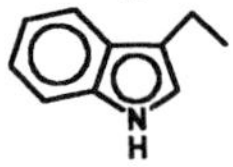	$-CH_3$	20.2
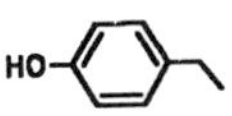	-H	76.7
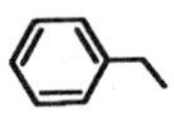	-H	62.7
$-CH-CH_2<^{CH_3}_{CH_3}$	-H	9.8
$-CH_2-CH_2-S-CH_3$	-H	20.4
$-CH_2-CH_2-S-CH_3$	$-CH_3$	0
$-CH<^{CH_3}_{CH_3}$	-H	0
$-CH_3$	-H	0
$-CH_2OH$	-H	0
$-CH_2-CH_2-COOH$	-H	0

FIGURE 1. Substrate specificity of the hydantoin-racemase of *Arthrobacter* sp. DSM 3747.

enzyme activity. Thus, the first purification step was a combined precipitation with protamine sulfate and ammonium sulfate. To minimize the time spent on the purification procedure and to use high salt concentration, hydrophobic interaction chromatography was used; thus, desalting and concentration steps are not necessary. Gel exclusion chromatography was used for purification and desalting. Lastly, anion exchange chromatography was applied. The hydantoin-racemase of *Arthrobacter* sp. DSM 3747 was purified to homogeneity with an enrichment factor of 70.

TABLE 1. Characteristics of the Hydantoin-Racemase of *Arthrobacter* sp. DSM 3747

Isoelectric point	pH 4.5
Molecular weight	84,000 Da
Subunits	4 × 21,000
Optimal pH	8.5
Optimal temperature	37°C

Characterization and Substrate Specificity of the Hydantoin-Racemase

The results presented in FIGURE 1 were prepared using the L-hydantoinase-free enzyme solution after hydrophobic interaction chromatography.

Besides the compounds mentioned in FIGURE 1, several chiral amino acids, *N*-acetyl-amino acids, and dipeptides as well as D-α-amino-ε-caprolactame have been tested as possible substrates but none of them is accepted by the racemase.

In TABLE 1, the characteristics of the purified enzyme are shown.

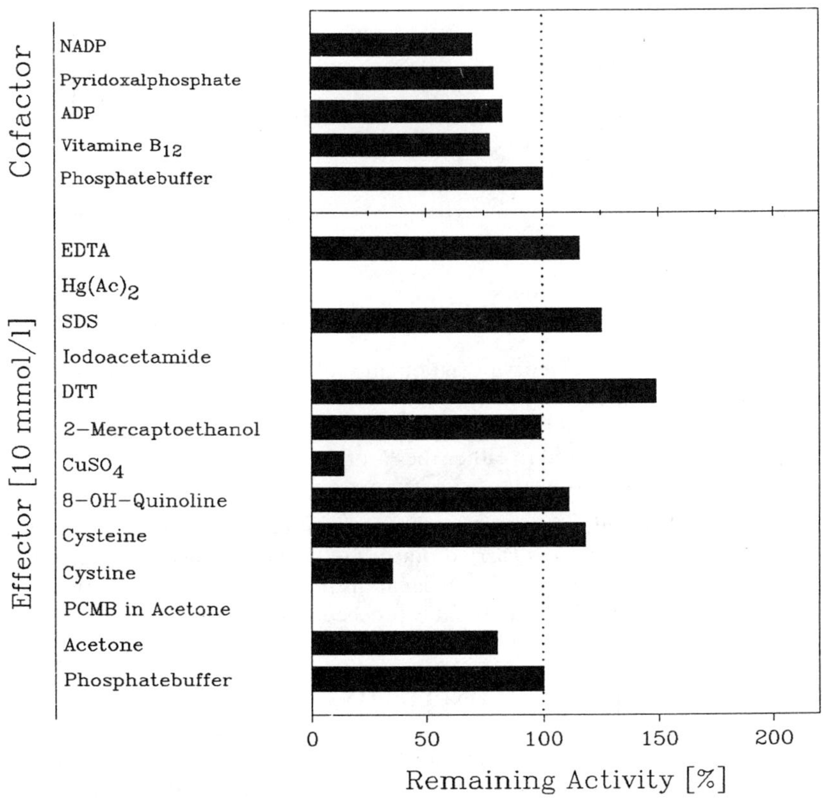

FIGURE 2. Influence of possible effectors and cofactors on the activity of the hydantoin-racemase from *Arthrobacter* sp. DSM 3747 [preincubation of the enzyme (30 minutes) with the effector, final concentration: 10 mM].

Several possible effectors and coenzymes have been tested for their influence on the enzymatic activity. FIGURE 2 shows the results of incubating the enzyme with various effectors in a final concentration of 10 mM.

As can be seen from FIGURE 2, several SH blocking reagents lead to an inactivated enzyme, whereas the addition of, e.g., pyridoxalphosphate, which acts as a cofactor in several amino acid racemases, does not exhibit any positive effect.

Mechanism

Only 5-aryl-hydantoins racemize chemically more or less quickly under reaction conditions pH 8.5 and 37°C. Keto-enol-tautomerism (FIGURE 3) is considered to be responsible for the racemization. For example 5-(4-hydroxy-phenyl)-hydantoin racemizes spontaneously in about 1 hour, because of an important mesomeric stabilization of the enolate border structure.

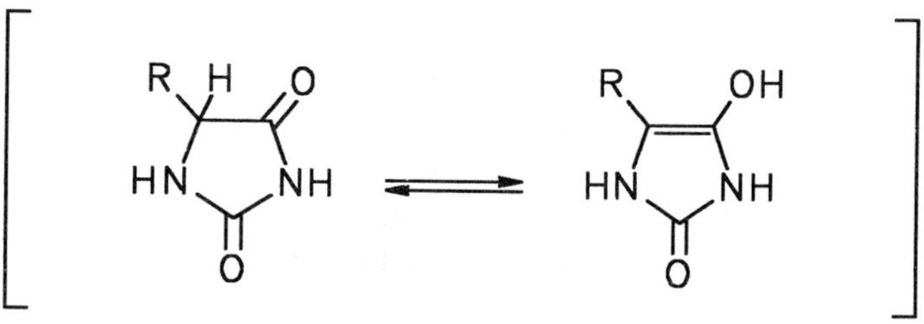

FIGURE 3. Keto-enol tautomerism of 5-monosubstituted hydantoins.

Two enzyme-catalyzed racemization pathways can be discussed:

I. Enolate is formed by enzyme action and released into the medium—solvent takes part in the reaction.
II. Carbanion is formed by enzyme action in 5-position and the enzyme-bound proton is retransferred to either the *R*- or *S*-side of the molecule—no solvent effect should be observed.

For investigations into whether there is a solvent-participation in the racemization reaction, the reaction was carried out in D_2O. The product isolated from the reaction mixture showed no further H signal in the 5-position in H^1-NMR studies. For this reason, the racemization probably proceeds on mechanism I.

CONCLUSIONS

For the racemase to function as substrate, a CH_2 spacer group between the hydantoin ring and the substituent in 5-position is needed. Aromatic substituents are converted about 5 times faster than apolar aliphatic, whereas polar aliphatic substituents are not racemized. As shown by H/D-exchange experiments, the solvent takes part in the racemization of IMH. Application of a process using the three enzymes immobilized on Q-Sepharose to products of industrial processes, e.g., aromatic nonproteinogenic amino acids, is possible and the object of present investigations.

ACKNOWLEDGMENTS

We thank Professor Karsten Krohn and his coworkers from the Institute of Organic Chemistry, TU Braunschweig, Germany, for the synthesis of most of the hydantoin derivatives used in these experiments.

REFERENCES

1. SYLDATK, CH., G. DOMBACH, C. GROSS, R. MÜLLER & F. WAGNER. 1988. Ann. N.Y. Acad. Sci. **542:** 323–329.
2. SYLDATK, CH. & F. WAGNER. 1990. J. Food Biotechnol. **4**(1): 87–95.
3. SYLDATK, CH., M. PIETZSCH, R. MÜLLER & F. WAGNER. Hanser Publishers. (In press.)
4. SYLDATK, C., V. MACKOWIAK, H. HÖKE, C. GROSS, G. DOMBACH & F. WAGNER. 1990. J. Biotechnol. **14:** 345–362.

Synthesis of Uncommon Wax Esters with Immobilized Lipases[a]

FRITZ WAGNER, FRANK KLEPPE,
WOLFRAM LOKOTSCH, ANNETTE ZIEMANN,
AND SIEGMUND LANG

Institute of Biochemistry and Biotechnology
Technical University of Braunschweig
Konstantin-Uhde-Strasse 5
D-3300 Braunschweig, Germany

INTRODUCTION

Lipases from different sources catalyze hydrolysis reactions as well as reactions of synthesis and interesterification of various classes of esters. Whereas the hydrolysis has to be carried out in aqueous media, organic solvents are essential for realization of esterification reactions. The resistance of lipases against the denaturation caused by organic solvents can be increased by immobilization on different supporting materials.[1]

Long-chained esters functionalized with hydroxy groups in the alkyl chain of the fatty acid or alcohol moiety are called "uncommon wax esters" with regard to their special structure. These compounds show interesting surface-active properties.[2] For the synthesis of new wax esters, various diols, triols, and hydroxylated fatty acids were used as substrates.

In this report, the objects of our studies were the testing of two lipases as biocatalyst: (1) *Mucor miehei* lipase, (2) *Ustilago maydis* lipase.

Using the *Mucor miehei* lipase, the reaction conditions were optimized concerning both the choice of suitable solvents for hydroxylated substrates and the prevention of multiple acylation or lactonization. Subsequently the products were investigated in view of a correlation between molecular structure and surface properties. The *Ustilago maydis* lipase was tested with regard to a wax formation after immobilization on different hydrophobic supports.[3,4]

MATERIAL AND METHODS

Enzymes

Commercial Lipozyme™ is an immobilized lipase of *Mucor miehei* (Novo, Bagsvaerd, Denmark) having the activity of 23.2 batch interesterification units (BIU/g).

Lipase of *Ustilago maydis* ATCC 14826 was obtained from the culture supernatant during the cultivation on coconut oil as carbon source. After the separation from

[a]This work was financially supported by the German Government of Research and Technology (BMFT).

the cell mass, the lipase can be immobilized by adding carrier material to the supernatant. (Activity test: hydrolysis of *p*-nitrophenylpalmitate.)

Immobilization

For the immobilization of the *Ustilago maydis* lipase, 3 g of carrier material were added to 100 ml of the culture supernatant containing the enzyme. After 2 hours, the active carrier was filtered off.

Bioconversions

Unless stated otherwise, reactions were carried out in 50 ml stoppered Erlenmeyer flasks at 37°C or 60°C. The reaction mixtures for ester synthesis were composed of 20 ml organic solvent, 0.1–0.9 mmol acid, and 0.1–0.9 mmol alcohol. The bioconversions were started by adding the immobilized lipases to the reaction solutions.

Analysis

The extent of conversion was determined by quantitative thin-layer chromatography (Densitometer CD 60, Desaga, Heidelberg, Germany). The elucidation of the molecular structures was carried out by ^{1}H-NMR, ^{13}C-NMR, and gas chromatography/mass spectrometry.

Physicochemical Methods

The measurements of surface tension of the formed esters were performed in aqueous solutions at 25°C by the ring method of de Nouy (Lauda tensiomat, Lauda, Königshofen, Germany).

Determination of surface film pressure was carried out on an aqueous subphase at 25°C by a Langmuir film balance (Lauda FW 2, Lauda, Königshofen, Germany).

RESULTS AND DISCUSSION

Influence of Various Ratios of Substrates and of the Organic Solvents on Ester Formation

FIGURE 1 shows the esterification of stearic acid with 1,12-dodecanediol catalyzed by Lipozyme™ to the hydroxylated waxes.

The products of this bioconversion were the mono- and distearate, but only the formed monostearate of the diol has interesting surface properties on account of the terminal hydroxy group. In order to find the best reaction conditions for minimizing the yield of distearate, the acylation of 1,12-dodecanediol with different ratios of acid and diol was investigated (FIGURE 2).

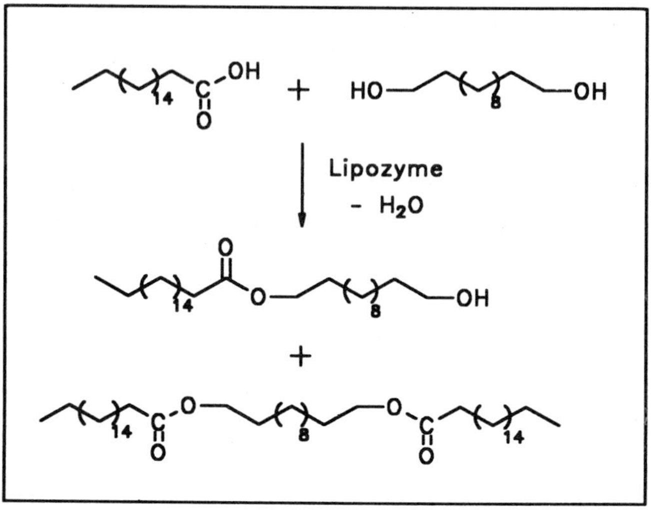

FIGURE 1. Reaction scheme of the acylation of 1,12-dodecanediol with stearic acid.

According to the principle of mass balance, high quantities of α,ω-diols were necessary, forming high yields of 12-hydroxydodecyl-1-stearate.

The regioselectivity of the used enzyme was investigated for the reaction of 1,2,6-hexanetriol and stearic acid. Complete acylation of all three hydroxy groups did not occur according to the 1,3-selectivity of Lipozyme™.[5,6] Monoacylation to 5,6-dihydroxyhexyl-1-stearate as the main product was carried out in polar organic solvents. Reactions in apolar organic solvents like *n*-hexane or isooctane only led to the 2-hydroxyhexyl-1,6-distearate.

No selectivity for one enantiomer of the racemic alcohol could be observed.

Surface Properties of Uncommon Wax Esters

The surface tension of the monostearates was investigated using a tensiomat. TABLE 1 summarizes the results. The dihydroxylated stearates and the stearates containing a short alkyl group in the alcohol moiety showed significantly lower surface tensions on account of their more polar character.

As to the film balance measurement, FIGURE 3 represents three π/A-isotherms of uncommon hydroxylated stearates. A more liquid-expanded behavior of the π/A-isotherms of the wax esters was indicated.

In further experiments, the enzymatic formation of wax esters of the 9,10,16-trihydroxy-palmitic acid (aleuritic acid) and alcohols with different chain length catalyzed by Lipozyme™ was investigated (FIGURE 4). By using tert-butylmethylketone as reaction solvent at a temperature of 60°C and a ratio acid:alcohol of 1:5, it was possible to increase the solubility of the acid in the reaction medium. Also the formation of by-products like lactones was prevented. The yield of the wax esters was approx. 60%.

FIGURE 5 shows three π/A-isotherms of aleuritic acid esters with different chain length.

The 9,10,16-trihydroxy-palmitic acid-decylester and the 9,10,16-trihydroxy-palmitic acid-hexadecylester exhibit gaseouslike behavior at high areas/molecule, analogous with the di-hydroxyhexyl-1-stearates seen above. Both wax esters show a point of inflexion in their respective π/A-isotherms. This can be correlated to a transition state from expanded to condensed behavior. At surface pressure of 60

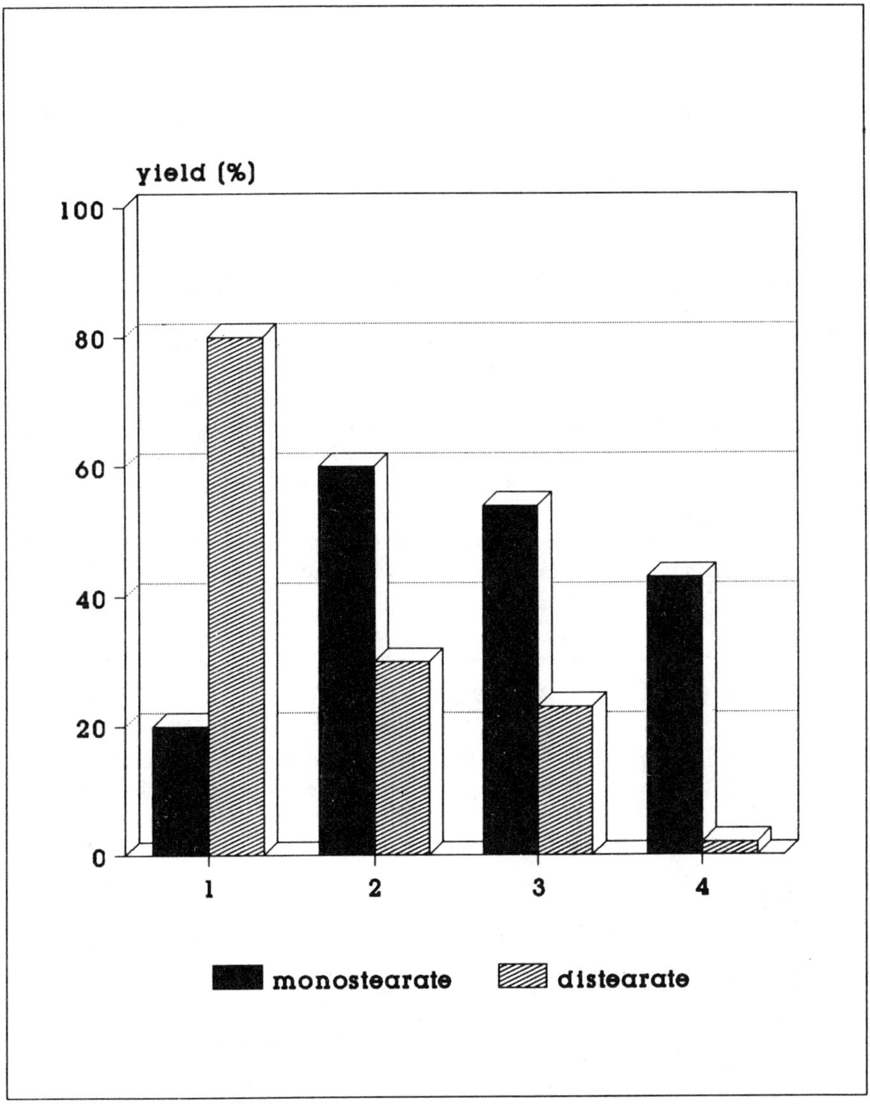

FIGURE 2. Influence of various ratios of 1,12-dodecanediol and stearic acid on the yield of 12-hydroxydodecyl-1-stearate and dodecyl-1,12-distearate. Ratio of acid to diol (time): (1) 1:0.5 (8 hours); (2) 1:2 (8 hours); (3) 1:10 (8 hours); (4) 1:10 (3 hours). Fixed reaction conditions: 0.3 mmol stearic acid, 0.5–3 mmol 1,12-dodecanediol, 100 mg Lipozyme™, 15 ml hexane, $T = 60°C$.

TABLE 1. Surface Tension, σ, and Critical Micelle Concentration (CMC) of Solubilized Wax Esters in Pure Water at 25°C

Wax Ester	CMC (mg/l)	σ (mN/m)
12-Hydroxydodecyl-1-stearate	7	61
4-Hydroxybutyl-1-stearate	7	43.5
3-Hydroxypropyl-1-stearate	8	38
5,6-Dihydroxyhexyl-1-stearate	7	31

mN/m at 20 Å/molecule (smallest possible area of an alkyl-chain) the collapse points of the monolayers were reached.

The 9,10,16-trihydroxy-palmitic acid-butylester was too polar to build up a stable monolayer.

Investigations of the Immobilized Lipase of Ustilago maydis

For the immobilization of the lipase of *Ustilago maydis,* adsorption on different supporting materials was tested (TABLE 2).

The lipolytic activity adsorbs completely on hydrophobic materials like Amberlite XAD-2 or silica gel RP-18. The protein content of the supernatant decreased approx. 50% during the immobilization step. The *Ustilago maydis* lipase cannot be immobilized on a material that forms covalent bondings with the enzyme like the oxiran beads Eupergit C™.

FIGURE 6 shows the wax ester formation with uncommon or common fatty acids and primary alcohols catalyzed by immobilized *U. maydis* lipase.

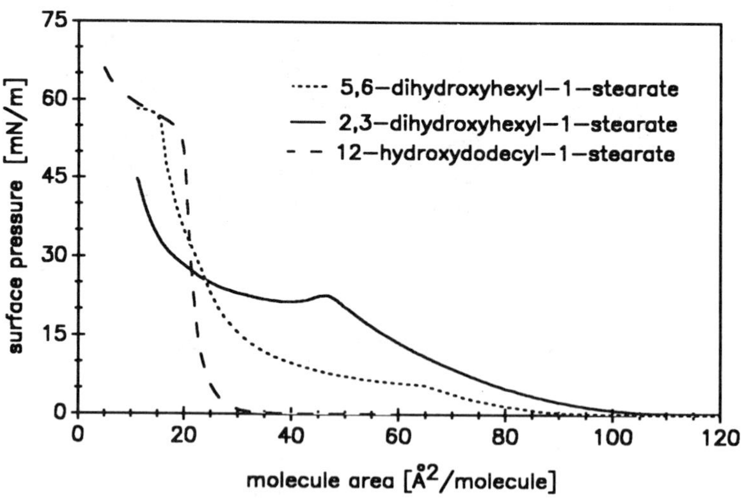

FIGURE 3. π/A-isotherms of uncommon wax esters at 20°C.

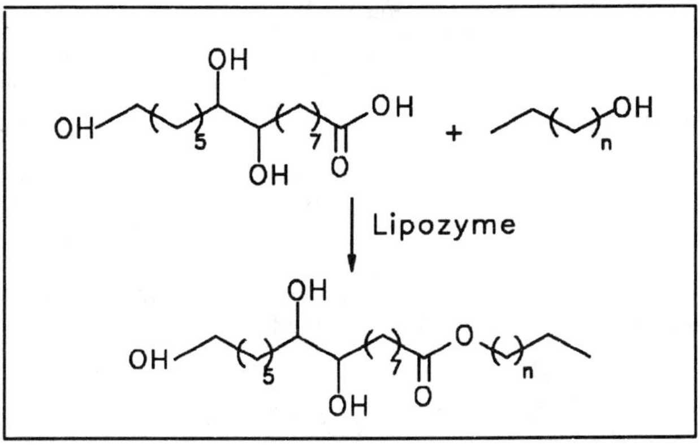

FIGURE 4. Reaction scheme for the esterification of 9,10,16-trihydroxy-palmitic acid (aleuritic acid) and alcohols with different chain length.

The 2-OH-dodecanoic, 2-OH-tetradecanoic, and 2-OH-hexadecanoic acids were esterified with alcohols of different chain length in several organic solvents to form hydroxy wax esters.

The best results for esterification reactions catalyzed by the immobilized *U. maydis* lipase were obtained by using unpolar solvents like *n*-hexane, isooctane, etc., at a temperature of 37°C.

The influence of the hydroxy group near the reactive COOH group has a decreasing effect on the esterification activity compared to nonsubstituted acids. Also the chain length of the used 2-OH-fatty acids shows an effect on the lipase

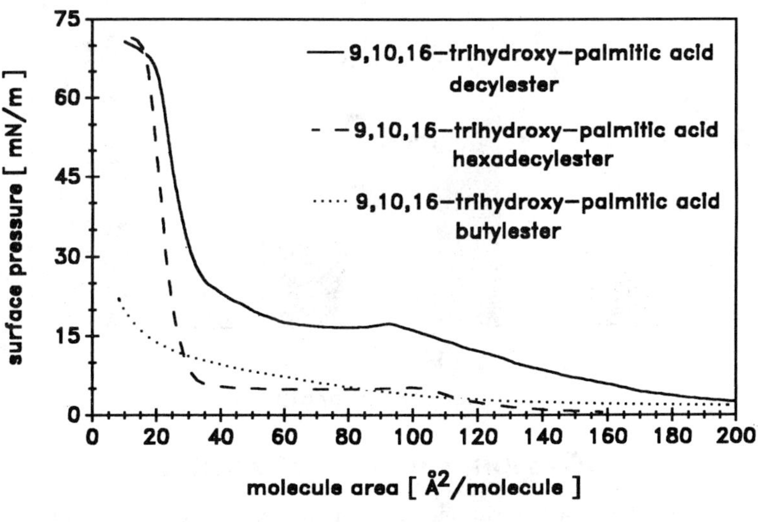

FIGURE 5. π/A-isotherms of different wax esters at 25°C.

TABLE 2. Immobilization of *Ustilago maydis* Lipase

Material	Filtrate Activity (μmol ml^{-1} minute^{-1})	Filtrate Protein (gl^{-1})
Unbounded	0.78	0.265
Amberlite XAD-2	0.0	0.120
Silica gel RP-18	0.0	0.125
Eupergit C(TM)	0.70	0.245

activity concerning ester formation. No enantioselectivity for esterification reactions of racemic hydroxy fatty acids could be observed.

CONCLUSIONS

By using hydroxy derivatives of fatty acids or alcohols, the Lipozyme(TM)-catalyzed wax ester synthesis resulted in more difficult reaction conditions: more polar

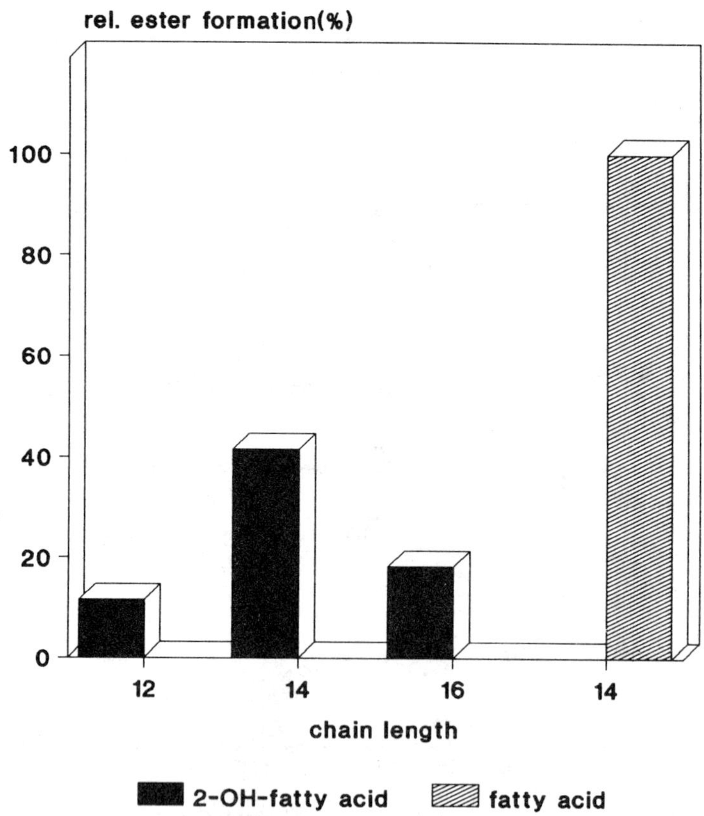

FIGURE 6. Wax ester formation with uncommon or common fatty acids and myristic alcohol in *n*-hexane at 37°C catalyzed by a lipase from *Ustilago maydis.* Reaction conditions: 0.9 mmol acid, 0.3 mmol myristic alcohol, 500 mg immobilized enzyme, 48 hours.

solvents, longer reaction times, higher temperatures, respectively. Side reactions like multiple acylation of diols or triols and lactonization of ω-hydroxylated acids were prevented by using 1:3 or 1:5 ratios of fatty acid and alcohol components. Monolayers of some waxes of polar alcohols (e.g., triols) or fatty acids (e.g., aleuritic acid) showed a transition state from expanded to condensed behavior by compression. This is caused by different hydrophilic behaviors of the polar groups of the wax esters expelled from the interface by increasing pressure.

Ustilago maydis lipase immobilized on Amberlite XAD-2 or silica gel RP-18 is a successful biocatalyst for wax ester formation of uncommon fatty acids.

ACKNOWLEDGMENTS

We thank J. Rürup (GBF Braunschweig) for his help at the film balance.

REFERENCES

1. Tanaka, A., T. Kawamoto, M. Kawase, T. Nanko & K. Sonomoto. 1988. Immobilized lipases in organic solvents. *In* Proceedings-World Conference on Biotechnology for the Fats and Oil Industry. T. H. Applewhite, Ed.: 123–130. American Oil Chemist's Society, Champaign, Ill.
2. Lang, S., R. Multzsch, A. Passeri, A. Schmeichel, B. Steffen, F. Wagner, D. Hamann & H. K. Cammenga. 1991. Acta Biotechnol. **11**(4): 379–386.
3. Hsu, S., S. S. Wu, Y. F. Wang & C. H. Wong. 1990. Lipase-catalyzed irreversible transesterification using enol esters. Tetrahedron Lett. **31**(44): 6403–6406.
4. Carta, G., J. L. Gainer & A. H. Benton. 1991. Enzymatic synthesis of esters using an immobilized lipase. Biotechnol. Bioeng. **37:** 1004–1009.
5. Miller, C., H. Austin, L. Pesorske & J. Gonzalez. 1988. Characteristics of an immobilized lipase for the commercial synthesis of esters. J. Am. Oil Chem. Soc. **65:** 927–931.
6. Monot, F., Y. Benoit, D. Ballerini & J. P. Vandecasteele. 1990. Enzymatic synthesis of neopentylpolyol esters in organic media. Appl. Biochem. Biotechnol. **24/25:** 375–386.

Enzymic Synthesis of Oleic Acid Esters by Various Lipases

YU-YEN LINKO AND HE-CI YU

Laboratory of Biotechnology and Food Engineering
Helsinki University of Technology
SF-02150 Espoo, Finland

INTRODUCTION

There has been great interest in recent years in lipase biocatalysis for the production of fatty acids, modification of fats and oils, synthesis and hydrolysis of esters, optical resolution,[1,2] etc., particularly after the demonstration that lipases (triacylglycerolhydrolase, EC 3.1.1.3) can catalyze reactions in organic solvents.[3] Especially ester biosynthesis from alcohols and fatty acids offers a number of novel applications of lipases. Most prior work on ester biosynthesis has been carried out in organic solvent systems or with immobilized lipases.[4–12] However, for many applications the absence of an organic solvent would be a definite advantage. In the present work, a number of oleic acid esters were synthesized from various primary alcohols without addition of solvent by the *Archromobacter* lipase, which to our knowledge had not been previously reported. Butyl oleate synthesis was then chosen for a more detailed study with 25 lipases. Finally, the effect of water content on ester biosynthesis was investigated by the most suitable lipases.

MATERIALS AND METHODS

Lipases

Commercial lipases from *Achromobacter* sp., *Alcaligenes* sp., *Aspergillus niger* (3 enzymes), *Candida cylindracea* (3 enzymes), *Candida lipolytica, Chromobacterium viscosum, Geotrichum candidum, Humicola lanuginosa, Mucor javanicus, Mucor miehei* (3 enzymes), *Penicillium cyclopium, Penicillium roqueforti, Pseudomonas fluorescens, Rhizopus arrhizus, Rhizopus delemar, Rhizopus japonicus* (2 enzymes), *Rhizopus javanicus, and Rhizopus niveus* were obtained from Sigma Chemical Co. (United States), Meito Sangyo Co., Ltd. (Japan), and Biocatalysts, Ltd. (United Kingdom).

Chemicals

Chemicals used were obtained from Merck (Germany) and were of analytical grade.

Ester Synthesis

Ester syntheses were carried out in 15-cm^3 screw-capped test tubes, containing 0.7 mM (0.2 g) oleic acid and either 7.98 mM of various alcohols, with or without the addition of water, and 0.005 g *Achromobacter* sp. lipase or 3.27 mM butanol and 0.005 g various lipases at 40°C, 200 rpm for different lengths of time.

Determination of the Degree of Ester Synthesis

At a given time the contents were mixed with a blend of diethylether-ethanol (1:1, v/v) and the residual free oleic acid was titrated with 0.1 mM NaOH. The degree (in percent) of ester synthesis was calculated on the basis of oleic acid consumed.

TABLE 1. Ester Synthesis from Oleic Acid and Various Alcohols[a]

Alcohol	Degree of Synthesis (%)
Methanol	3.03
Ethanol	3.51
Propanol	5.27
Butanol	7.39
Pentanol	16.0
Hexanol	26.1
Decanol	17.1

[a] Reaction mixture contained 0.7 mM oleic acid, 7.98 mM alcohol, and 0.005 g *Achromobacter* sp. lipase. The reaction was carried out at 40°C and 200 rpm for 18 hours.

Identification of the Reaction Products

The synthesized esters were identified by thin layer chromatography (TLC), infrared spectroscopy, and NMR. TLC was carried out on silica gel plates (type 60, Merck) with petroleum ether–diethyl ether–acetic acid (70:30:1, v/v) as solvent. The spots were visualized by spraying with 10% phosphomolybdic acid in ethanol.[13]

RESULTS AND DISCUSSION

Ester Synthesis with Oleic Acid and Various Alcohols

Direct ester synthesis for 18 hours with a mixture of 0.7 mM oleic acid, 7.98 mM of various alcohols, and 0.005 g *Achromobacter* sp. lipase without any additional organic solvent or water resulted in different degrees of esterification as shown in TABLE 1. Under the reaction conditions used, the *Achromobacter* sp. lipase catalyzed poorly the esterification of C_1–C_4 primary alcohols as compared with C_5, C_6, and C_{10} alcohols. Omar *et al.* reported similar results with lauric acid and primary alcohols by

Humicola lanuginosa lipase.[5] Nishio *et al.,* on the other hand, reported that oleate synthesis by *Pseudomonas fragi* lipase was not affected by the length of the primary alcohol carbon chain.[13] Ishii *et al.* reported poor esterification of oleic acid with methanol and ethanol by *Rhizopus oligosporus* lipase.[14]

Comparison of Lipases in Butyl Oleate Synthesis

A total of 25 commercial lipases from various sources were compared for their ability to catalyze the synthesis of butyl oleate as a model system in the absence of

TABLE 2. Comparison of Various Lipases for Ester Synthesis from Oleic Acid and Butanol[a]

Lipase	No.	Degree of Synthesis (%)	
		2 Hours	20 Hours
Achromobacter sp.	1	13.8	49.2
Alcaligenes sp.	2	33.1	84.4
Aspergillus niger 1	3	0.86	0
Aspergillus niger 2	4	4.59	64.6
Aspergillus niger 3	5	0.38	9.38
Candida cylindracea 1	6	62.6	85.8
Candida cylindracea 2	7	7.61	48.6
Candida cylindracea 3	8	24.7	78.7
Candida lipolytica	9	7.03	21.5
Chromobacterium viscosum	10	76.5	87.3
Geotrichum candidum	11	0	2.78
Humicola lanuginosa	12	7.63	63.1
Mucor javanicus	13	6.83	62.3
Mucor miehei 1	14	4.35	40.0
Mucor miehei 2	15	11.7	50.8
Mucor miehei 3	16	47.1	84.6
Penicillium cyclopium	17	9.97	67.3
Penicillium roqueforti	18	0	0.79
Pseudomonas fluorescens	19	55.9	87.0
Rhizopus arrhizus	20	2.79	31.8
Rhizopus delemar	21	5.99	61.2
Rhizopus japonicus 1	22	0	1.63
Rhizopus japonicus 2	23	0.22	5.67
Rhizopus javanicus	24	7.14	52.8
Rhizopus niveus	25	2.46	23.7

[a]Reaction mixture contained 0.7 mM oleic acid, 3.27 mM butanol, and 0.005 g lipase. Reaction was carried out at 40°C and 200 rpm for 2 hours and 20 hours.

additional solvent. The reaction mixture contained 0.2 g (0.7 mM) oleic acid and 0.3 cm^3 (3.27 mM) butanol. The degree of ester synthesis was determined after 2 and 20 hours incubation at 40°C. The results are shown in TABLE 2. It can be seen that only 5 of the lipases possessed an excellent ability for ester synthesis under the conditions used. These were *Alcaligenes* sp., *Candida cylidracea* 1, *Chromobacterium viscosum, Mucor miehei* 3, and *Pseudomonas fluorescens* lipases. With these lipases, an about 85% to 87% degree of ester synthesis was obtained in 20 hours. The *Chromobacte-*

TABLE 3. Effect of the Quantity of Enzyme on Ester Synthesis from Oleic Acid and Butanol[a]

Lipase	Degree of Synthesis (%) Lipase 0.005 g	Lipase 0.015 g	Lipase 0.05 g
Aspergillus niger 1	2.91	3.83	5.28
Aspergillus niger 2	66.5	83.5	90.7
Aspergillus niger 3	5.71	6.82	12.6
Candida cylindracea 2	48.0	73.7	83.9
Candida lipolytica	9.70	10.0	8.60
Geotrichum candidum	1.21	1.73	3.94
Humicola lanuginosa	44.9	—	89.4[b]
Penicillium roqueforti	0.7	3.21	4.39
Rhizopus japonicus 2	1.81	6.25	66.9
Rhizopus niveus	5.01	7.76	7.36

[a]Reaction mixture contained 0.7 mM oleic acid, 3.27 mM butanol, and various quantities of lipases. Reaction was carried out at 40°C and 200 rpm for 20 hours.

[b]0.04 g of lipase was used.

rium viscosum lipase yielded 76% of butyl oleate by 2 hours. Eight of the tested lipases gave a conversion of less than 25% even after 20 hours of reaction. In order to clarify whether the low ability for ester synthesis was caused by a low specific lipase activity of the enzyme preparation, further experiments were carried out by increasing the enzyme quantity used in the reaction mixture. It was clearly shown that even at high enzyme levels, some of the lipases synthezied butyl oleate poorly, such as the lipases from *Candida lipolytica, Geotrichum candidum, Penicillium roqueforti, Rhizopus japonicus* and *Aspergillus niger* 1 and 3 (TABLE 3). The lipases from *A. niger* 2, *C. cylidracea* 2, *Humicola lanuginosa,* and *Rhizopus japonicus* gave a satisfactory ester yield at the increased dosage level.

TABLE 4. Effect of Water Content on Ester Synthesis from Oleic Acid and Butanol[a]

Water Added (%)	Degree of Synthesis (%) *Candida cylindracea* 1 (1.82%)[b]	*Alcaligenes* sp. (10.7%)[b]	*Chromobacterium viscosum* (6.35%)[b]	*Pseudomonas fluorescens* (10.1%)[b]	*Mucor miehei* 3 (23.6%)[b]
0	35.3 (0.087)[a]	77.5 (0.119)[c]	90.4 (0.07)[c]	2.19 (0.113)[c]	89.39 (0.260)[c]
2.18	84.8	67.3	87.5	87.1	81.4
6.28	64.8	48.6	65.6	81.2	54.2
18.3	43.7	60.2	77.0	73.2	23.4
40.1	43.0	79.8	83.1	67.5	26.5
69.1	56.0	85.4	78.7	83.7	50.6
81.7	60.5	86.0	71.3	80.2	70.4

[a]Reaction mixture contained 0.7 mM oleic acid, 3.27 mM butanol, 0.005 g lipases, and different quantities of water as indicated in the table. Reaction was carried out at 40°C and 200 rpm for 21 hours.

[b]Enzyme water content.

[c]Quantity of water contributed by the enzyme preparation.

Effect of Water Content on Ester Synthesis

It is well known that water plays a very important role in enzymic reactions. According to the mass action law, a high content of water results in the hydrolysis of the ester, whereas a low water level favors synthesis. The five lipases from various sources with a high ability for ester synthesis were chosen for further investigation on the effect of the water content on ester synthesis. Reactions were carried out in a mixture of oleic acid and butanol at different added water levels (0, 2.18, 6.38, 18.3, 40.1, 69.1, and 81.7%). The initial water content of enzyme preparations was 1.82–23.6% (TABLE 4). In all cases, the quantity of water contributed by the enzyme preparation seemed to be insignificant. It can also be seen from TABLE 4 that water content of the reaction medium strongly affected the ester synthesis, the effect varying with the enzyme. Some lipases showed a high activity without any addition of water, such as *Chromobacterium viscosum* and *Mucor miehei* 3 enzymes, whereas *Pseudomonas fluorescens* and *Candida cylindracea* 1 showed a high activity for ester synthesis only when a small amount (2.18%) of water was added.

REFERENCES

1. LINKO, Y.-Y., U.-M. KOIVISTO & H. KAUTOLA. 1990. Optimization of enzymic production of oleic acid. Ann. N.Y. Acad. Sci. **613:** 691–696.
2. BJÖRKLING, F., S. E. GODTFREDSEN & O. KIRK. 1991. The future impact of industrial lipases. TIBTECH **9:** 360–363.
3. ZAKS, A. & A. M. KLIBANOV. 1984. Enzymatic catalysis in organic media at 100°C. Science **224:** 1249–1251.
4. TAKAHASHI, K., T. YOSHIMOTO, A. AJIMA, Y. TAMAURA & Y. INADA. 1984. Modified lipoprotein lipase catalyzes ester synthesis in benzene. Enzyme **32:** 235–240.
5. OMAR, I. C., N. NISHIO & S. NAGAI. 1987. Fat hydrolysis and esterification by a lipase from *Humicola lanuginosa.* Agric. Biol. Chem. **51:** 2153–2159.
6. GILLIES, B., H. YAMAZAKI & D. W. ARMSTRONG. 1987. Production of flavor esters by immobilized lipase. Biotechnol. Lett. **9:** 709–714.
7. LANGRAND, G., C. TRIANTAPHYLIDES & J. BARATTI. 1988. Lipase catalyzed formation of flavour esters. Biotechnol. Lett. **10:** 549–554.
8. MARTY, A., W. CHULALAKSANANUKUL, J. S. CONDORET, R. M. WILLEMOT & G. DURAND. 1990. Comparison of lipase-catalyzed esterification in supercritical carbon dioxide and in *n*-hexane. Biotechnol. Lett. **12:** 11–16.
9. KNEZ, Z., M. LEITGEB, D. ZAURSNIK & B. LAVRIC. 1990. Synthesis of oleic acid esters with immobilized lipase. Fat Sci. Technol. **92:** 169–172.
10. LANGRAND, G., N. RONDOT, C. TRIANTAPHYLIDES & J. BARATTI. 1990. Short chain flavour ester synthesis by microbial lipases. Biotechnol. Lett. **12:** 581–586.
11. MONOT, F., F. BORZEIX, M. BARDIN & J.-P. VANDECASTEELE. 1991. Enzymatic esterification in organic media: role of water and organic solvent in kinetics and yield of butyl butyrate synthesis. Appl. Microbiol. Biotechnol. **35:** 759–765.
12. CARTA, G., J. L. GAINER & A. H. BENTON. 1991. Enzymatic synthesis of esters using an immobilized lipase. Biotechnol. Bioeng. **37:** 1004–1009.
13. NISHIO, T., T. CHIKANO & M. KAMIMURA. 1988. Ester synthesis by the lipase from *Pseudomonas fragi* 22.39b. Agric. Biol. Chem. **52:** 1203–1208.
14. ISHII, T., T. MORI, J. CHEN, Y. ITOH, S. SHIMURA, K. KIRIMURA & S. USAMI. 1990. Ester synthesis by a crude lipase of *Rhizopus oligosporus* in an aqueous system. J. Ferment. Bioeng. **70:** 188–189.

Evaluation of Process Options for Enzymatic Kinetic Resolution

A. J. J. STRAATHOF, J. L. L. RAKELS, AND J. J. HEIJNEN

Department of Biochemical Engineering
Delft University of Technology
Julianalaan 67
2628 BC Delft, the Netherlands.

Because of their chiral nature, enzymes are useful for converting one of the enantiomers of a racemic compound, leaving the desired substrate enantiomers or forming the desired product enantiomer in an enantiopure form. In many cases, the enzyme is not fully enantioselective. Then, the reaction has to be terminated at the moment that the yield of the desired compound is maximal but its enantiomeric excess is still satisfactory.

In order to predict the practical value of such a kinetic resolution process, a quantitative relation between the conversion and the enantiomeric excess (of substrate or product) is required. This relation may be obtained from the ratios of macroscopic balances of the (*R*) and (*S*)-enantiomers. In this respect the results of Chen *et al.* are well-known.[1,2] They found that a single kinetic parameter, the enantiomeric selectivity E [$= (v_{max}/K_m)^R / (v_{max}/K_m)^S$], suffices to describe the kinetic resolution process (FIGURE 1). Usually research is focused on enhancement of E, by modification of the enzyme, substrate, or reaction medium.

A different approach would be to consider the assumptions on which the results of Chen *et al.*[1,2] are implicitly based. These are:

1. Batch operation of the resolution process.
2. A homogeneous system.
3. Uni-uni reactions obeying Michaelis-Menten kinetics.

When these assumptions are not fulfilled, the resolution process may follow a course that positively or negatively differs from the predictions of Chen *et al.*[1,2] In order to improve existing resolution processes, we are looking systematically at such process options. As shown in the examples below, several cases have been worked out already.

NO BATCH OPERATION

The kinetic resolution may be performed in a continuous flow reactor rather than a batch reactor. Then, the accumulation terms in the macroscopic balances have to be replaced by flow terms. It can be shown[3] that the performance of a plug flow reactor is similar to that of a batch reactor. The length dimension of the tube replaces the time dimension in the batch reactor. A continuous stirred tank reactor (CSTR), however, will show a performance worse than a batch reactor (compare FIGURE 1

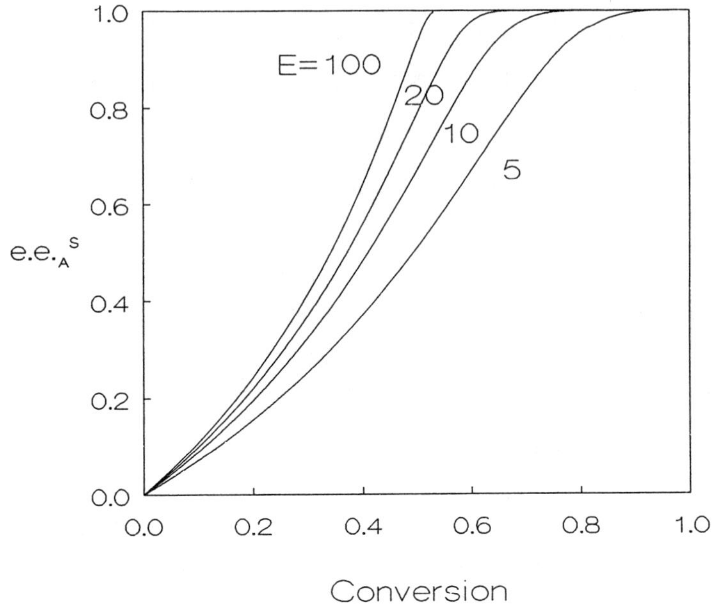

FIGURE 1. Enantiomeric excess of the remaining substrate A^S as a function of the conversion for several values of the enantiomeric selectivity, according to Chen *et al.*[1]

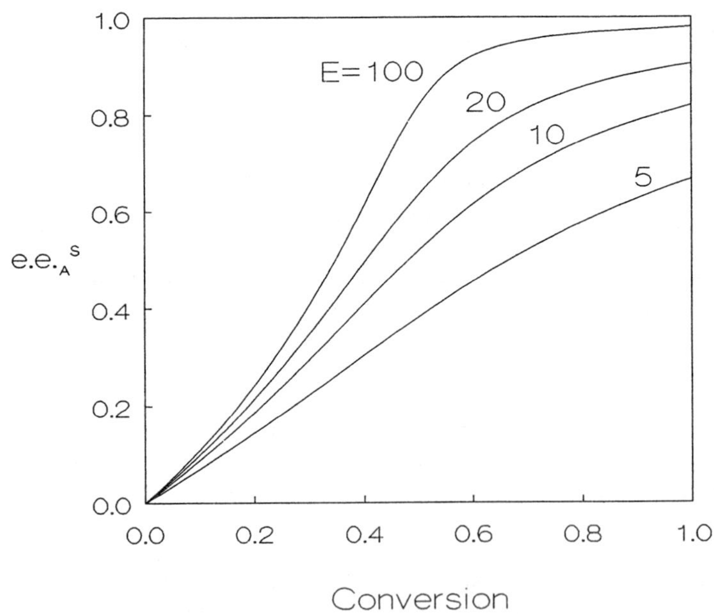

FIGURE 2. Progress of kinetic resolution in a CSTR instead of a batch reactor, for several values of the enantiomeric selectivity.

and FIGURE 2).[3] Since part of the racemic substrate that is fed at the inlet will directly leave the reactor, a high enantiopurity is not to be expected anyway.

NO HOMOGENEOUS SYSTEM

In a heterogeneous system, e.g., when an enzyme is immobilized in a spherical particle or in a membrane, diffusional limitations may occur. Therefore, the concentration of each substrate enantiomer that the enzyme "sees" within the support will

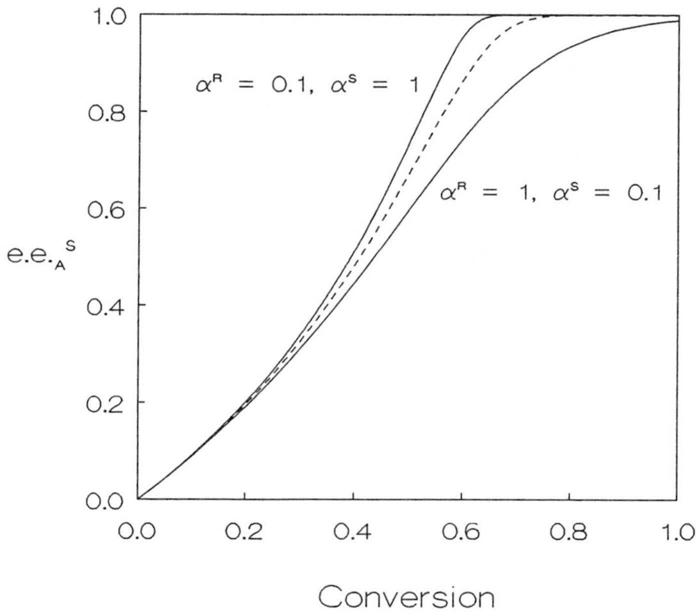

FIGURE 3. Progress of kinetic resolution for irreversible bi-bi A–Q resolution[7] instead of uni-uni. The broken line shows the situation when a large excess of the second substrate is available; the other lines are two situations at an equimolar ratio of substrates. The parameter α is defined by $(V_1/K_{mA})/(V_{-1}/K_{mQ})$, and E was set at 10.

be less than the bulk concentration. This difference with the bulk concentration will be larger for the more reactive enantiomer. The result is that the enantiomeric selectivity of the enzyme will be offset to some degree by the relative availabilities of the two enantiomers to the immobilized enzyme as has been shown recently.[4,5] Thus, diffusional limitations will decrease the efficiency of a kinetic resolution process.

Of course, in a monophasic batch system, temperature or concentration (pH) gradients may occur because of insufficient mixing. This will also have a detrimental effect.

NO UNI-UNI MICHAELIS-MENTEN KINETICS

When multiple substrates or products are present, when more then one reaction is performed simultaneously, or when the assumptions underlying the Michaelis-Menten conditions are not fulfilled, deviations from the Chen curves may be found as indicated below.

Bi-bi reactions (with two substrates and two products) are far more common than uni-uni reactions in resolution processes. Rate equations for bi-bi reactions may be different from those for uni-uni reactions, which has an impact on the performance of a resolution process.[6,7] When a product competes with the second substrate at the

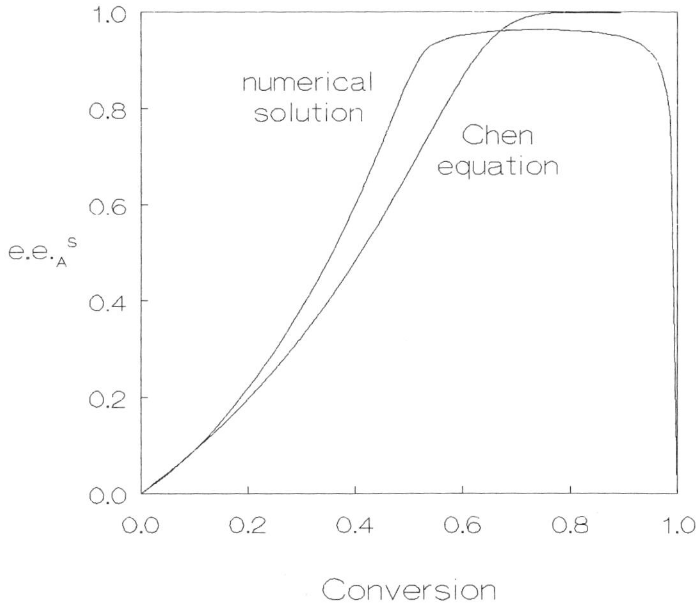

FIGURE 4. Progress of kinetic resolution according to numerical integration of the rate equations. The values of the microscopic rate constants were set at $k_1^R = 200$, $k_{-1}^R = 0.3$, $k_2^R = 0.1$, $k_1^S = 50$, $k_{-1}^S = 1800$, $k_2^S = 200$. The ratio between the initial concentration of racemic substrate and enzyme was 2:1. The Chen equation ($E = 10$) is shown for comparison.

active site, its inhibitory effect on the conversion of the substrate may be different for the (*R*) and (*S*)-enantiomer. Usually a negative effect on the performance of the process is to be expected, but positive deviations from the Chen curves have been found as well[8] (FIGURE 3).

Two simultaneous enzymatic kinetic resolution reactions catalyzed by a single enzyme[9–13] or two different enzymes[14,15] may lead to more efficient processes when performed in the proper manner. When the two reactions are consecutive, the faster formed product should be converted further very selectively. In the parallel mode, the substrate enantiomer that would lead to the unwanted product enantiomer should be removed by a second reaction.

Even when a single substrate is converted to a single product in a single enzymatic reaction, the Michaelis-Menten equation may not be valid. One assumption for the Michaelis-Menten equation is that the initial substrate concentration is very much larger than the concentration of the enzyme. As a rule of thumb, a ratio of 1000 on a molar basis should be taken. For many preparative purposes, this limit is not obeyed, e.g., inside a support particle for immobilized enzyme, the enzyme may be in excess. Below, a hypothetical example will be shown for which deviations from the Chen curve occur because of the invalidity of the Michaelis-Menten equation due to a non–steady state of the enzyme-substrate complex.

Suppose a reaction with the following microscopic rate constants is considered:

$$E + A^R \underset{0.3}{\overset{200}{\rightleftarrows}} EA^R \underset{0}{\overset{0.1}{\rightleftarrows}} E + P^R$$

$$E + A^S \underset{1800}{\overset{50}{\rightleftarrows}} EA^S \underset{0}{\overset{200}{\rightleftarrows}} E + P^S$$

Since this is an irreversible uni-uni reaction, the relations of Chen usually will be valid. The enantiomeric selectivity can be calculated[7] from the microscopic rate constants to be $E = 10$, yielding the Chen curve of FIGURE 4. The course of the reaction may also be calculated by incorporation of the rate equations of A^R, A^S, EA^R, and EA^S in macroscopic balances and numerical integration of the differential equations thus obtained. At a very high enzyme concentration, a large difference between the numerical solution and the Chen curve was found (FIGURE 4).

REFERENCES

1. CHEN, C. S., Y. FUJIMOTO, G. GIRDAUKAS, & C. J. SIH. 1982. J. Am. Chem. Soc. **104:** 7294–7299.
2. CHEN, C. S., S. H. WU, G. GIRDAUKAS & C. J. SIH. 1987. J. Am. Chem. Soc. **109:** 2812–2817.
3. RAKELS, J. L. L., H. T. PAFFEN, A. J. J. STRAATHOF & J. J. HEIJNEN. Manuscript in preparation.
4. MATSON, S. L. & J. L. LOPEZ. 1990. *In* Frontiers in Bioprocessing. S. K. Sikdar, M. Bier & P. Todd, Eds.: 391–403. CRC Press. Boca Raton, Fla.
5. WU, D., G. BELFORT & S. CRAMER. 1990. Ind. Eng. Chem. Res. **29:** 1612–1621.
6. TOL, J. B. A. VAN, A. GEERLOF, J. A. JONGEJAN & J. A. DUINE. Ann. N.Y. Acad. Sci. (This volume.)
7. STRAATHOF, A. J. J., J. L. L. RAKELS & J. J. HEIJNEN. Biocatalysis. (In press.)
8. RAKELS, J. L. L., P. CAILLAT, A. J. J. STRAATHOF & J. J. HEIJNEN. Manuscript in preparation.
9. MACFARLANE, E. L. A., S. M. ROBERTS & N. J. TURNER. 1990. J. Chem. Soc. Chem. Commun.: 569–571.
10. KAZLAUSKAS, R. J. 1989. J. Am. Chem. Soc. **111:** 4953–4959.
11. GUO, Z. W., S. H. WU, C. S. CHEN, G. GIRDAUKAS & C. J. SIH. 1990. J. Am. Chem. Soc. **112:** 4942–4945.
12. ENGEL, K. H., M. BOHNEN & M. DOBE. 1991. Enzyme Microb. Technol. **13:** 655–660.
13. THEIL, F., S. BALLSCHUH, A. KUNATH & H. SCHICK. 1991. Tetrahedron Assym. **2:** 1031–1034.
14. STRAATHOF, A. J. J., J. L. L. RAKELS & J. J. HEIJNEN. 1990. Biocatalysis **4:** 89–102.
15. KAKEYA, H., N. SAKAI, T. SUGAI & H. OHTA. 1991. Tetrahedron Lett. **32:** 1343–1346.

Biocatalytic 7-Aminocephalosporanic Acid Production

W. TISCHER, U. GIESECKE, G. LANG, A. RÖDER,
AND F. WEDEKIND

Boehringer Mannheim GmbH
Nonnenwald 2
Postfach 1152
8122 Penzberg, Germany

INTRODUCTION

7-Aminocephalosporanic acid (7-ACA) is an important intermediate for the production of semisynthetic cephalosporins. Usually it is prepared from cephalosporin C by chemical reactions. The chemical reaction steps need highly reactive chemicals, which can only be handled at low temperatures and require costly safety measures and energy supply, as well as strong environmental controls.

Although for many years attempts have been made to substitute the chemical process by a biocatalytic process, there is still no fully enzymatic 7-ACA process commercialized. This may be caused by lack of an appropriate acylase for a single-step cleavage of cephalosporin C. Furthest in development and on the brink of commercialization is a 3-step/2-enzyme process. This can be performed by enzymatic deamination with D-amino acid oxidase, followed by chemical decarboxylation with hydrogen peroxide, which is produced in the first step. As a third step, the resulting intermediate 7-β-(4-carboxybutanamido)-cephalosporanic acid (glutaryl-7-ACA) is hydrolyzed by a second enzyme, glutaryl-7-ACA acylase (FIGURE 1).

This pathway has been reported in the literature for many years;[1] however, economic competitiveness to the well-established chemical process still awaits proof.

Our experimental results obtained in biocatalyst development let us expect that an economically competitive catalyst system can be provided.[2]

MATERIALS AND METHODS

Reversed-Phase High-Performance Liquid Chromatography Analysis

Eluent: 2% acetonitrile in 20 mM ammonia acetate buffer pH 5.7, 1.3 ml/minute flow rate. *Sample:* 20 μl amount, 1:400 diluted in the eluent buffer. *Column:* LiChrosorb RP 18, 5 μm, 250 mm × 4 mm, 30°C. *Detection:* UV at 250 nm.

Oxygen Electrode Activity Assay for D-Amino Acid Oxidase

Solution: 20 mM cephalosporin C in 100 mM K_iPO_4 buffer pH 8.0, thermostated to 25°C.

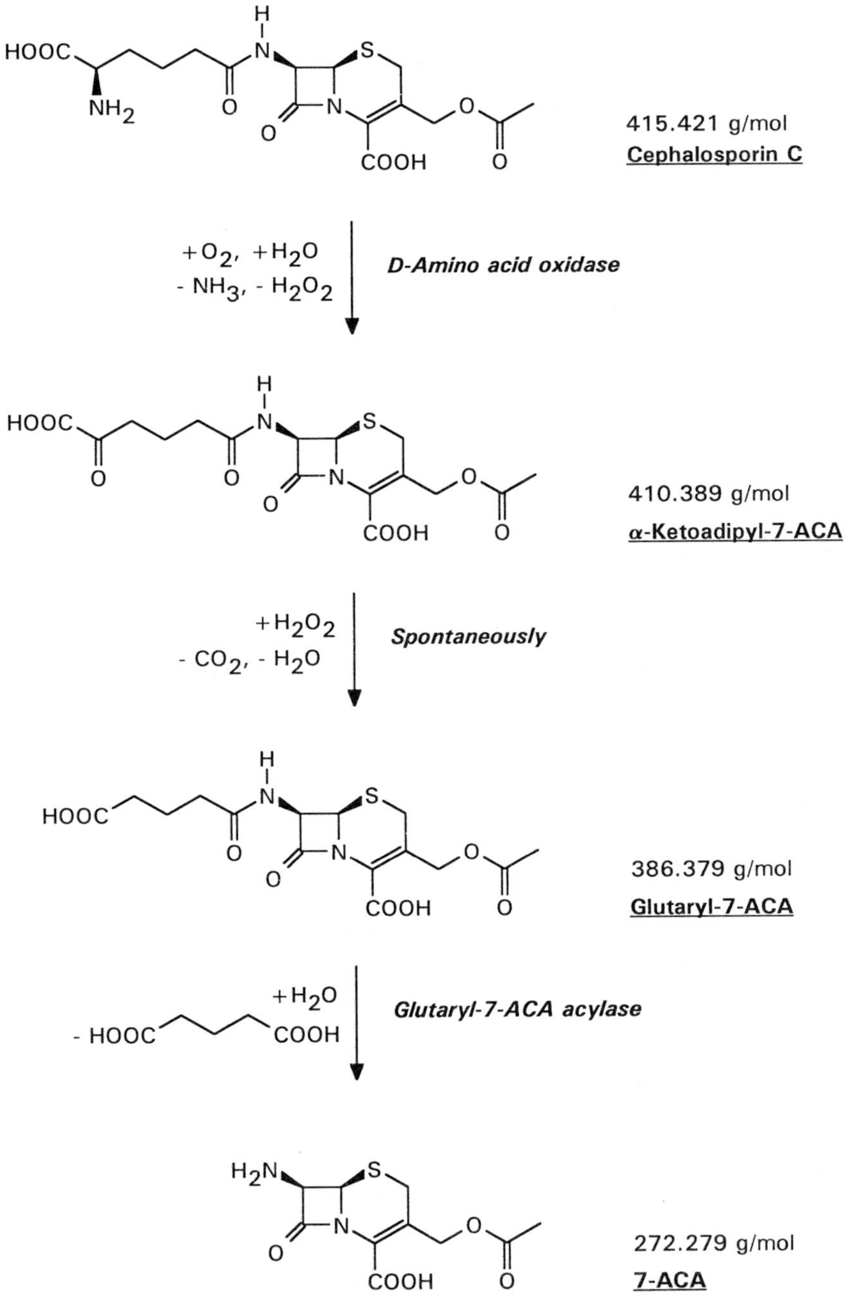

FIGURE 1. Reaction scheme for the conversion of cephalosporin C to 7-aminocephalosporanic acid in 3 steps by D-amino acid oxidase and glutaryl-7-ACA acylase.

Procedure: 50 ml substrate solution is given to a 50-ml thermostated Erlenmeyer flask. The solution is saturated with pure oxygen under magnetic stirring for 10 minutes, and the dissolved oxygen increases to 28–30 mg O_2/I. The oxygen supply and the magnetic stirrer are then stopped. The loss of dissolved oxygen is determined for 5 minutes. The reaction is started by addition of 3–5 U D-amino acid oxidase. The consumption of dissolved oxygen is observed for 10 minutes.

Calculation: 1 U consumes 1 μmol dissolved oxygen per minute.

Autotitrator Activity Assay for Glutaryl-7-ACA Acylase

Solution: 2% glutaryl-7-ACA in 5 mM Na_iPO_4 buffer pH 8.0, thermostated to 37°C, 50 mM NaOH.

Procedure: 50 ml substrate solution is given to a 50-ml thermostated reaction vessel. The pH value is titrated to pH 8.0 with NaOH by an autotitrator. The prereaction is determined for 10 minutes. The reaction is started by addition 10–20 U glutaryl-7-ACA acylase. The consumption of NaOH is observed for 10 minutes.

Calculation: 1 U forms 1 μmol glutaric acid per minute.

Standard Reaction Conditions and Initial Activity Assay for D-Amino Acid Oxidase

Solution: 75 mM cephalosporin C in 100 mM K_iPO_4 buffer pH 8.0, 1-M KOH.

Procedure: 300 ml substrate solution is given to a 500-ml stirred vessel. The vessel contains 2.46 kU/I immobilized D-amino acid oxidase. The conversion is performed at 20°C, 400 rpm and 1 vvm 3.5 bar O_2. The pH value is titrated to pH 8.0 with KOH by an autotitrator. The conversion is controlled by HPLC. When all cephalosporin C is consumed, the reaction solution is filtered off. The remaining catalyst is ready for reuse.

Calculation: 1 U consumes 1 μmol cephalosporin C per minute. This value corresponds well with the oxygen electrode activity assay.

Standard Reaction Conditions for Glutaryl-7-ACA Acylase

Solution: 75 mM glutaryl-7-ACA in 100 mM K_iPO_4 buffer pH 8.0, 3 M NH_3.

Procedure: 600 ml substrate solution is given to a 2400-ml stirred vessel. The vessel contains 5.0 kU/I immobilized glutaryl-7-ACA acylase. The conversion is performed at 28°C and 400 rpm. The pH value is titrated to pH 8.0 with NH_3 by an autotitrator. When the NH_3 consumption decreases to 1% of the initial rate, the reaction solution is filtered off. The remaining catalyst is ready for reuse.

Kinetic Model for D-Amino Acid Oxidase

$$\frac{\mathrm{d}c_{\mathrm{CC}}}{\mathrm{d}t} = -\frac{v_{\max} \times c_{\mathrm{CC}}}{k_{\mathrm{m,CC}} + c_{\mathrm{CC}} \times (1 + c_{\mathrm{CC}}/k_{\mathrm{i}})} \times \frac{c_{\mathrm{O_2}}}{k_{\mathrm{m,O_2}} + c_{\mathrm{O_2}}} - k_1 \times c_{\mathrm{CC}}$$

$$\frac{\mathrm{d}c_{\mathrm{KA}}}{\mathrm{d}t} = \frac{v_{\max} \times c_{\mathrm{CC}}}{k_{\mathrm{m,CC}} + c_{\mathrm{CC}} \times (1 + c_{\mathrm{CC}}/k_{\mathrm{i}})} \times \frac{c_{\mathrm{O_2}}}{k_{\mathrm{m,O_2}} + c_{\mathrm{O_2}}} - k_2 \times c_{\mathrm{H_2O_2}} \times c_{\mathrm{KA}} - k_1 \times c_{\mathrm{KA}}$$

$$\frac{dc_{GA}}{dt} = k_2 \times c_{H_2O_2} \times c_{GA} - k_1 \times c_{GA}$$

$$\frac{dc_{DA}}{dt} = k_1 \times (c_{CC} + c_{KA} + c_{GA})$$

$$\frac{dc_{H_2O_2}}{dt} = \frac{v_{max} \times c_{CC}}{k_{m,CC} + c_{CC} \times (1 + c_{CC}/k_i)} \times \frac{c_{O_2}}{k_{m,O_2} + c_{O_2}} - k_2 \times c_{H_2O_2} \times c_{KA}$$

$$\frac{dc_{O_2}}{dt} = k_La \times (c^*_{O_2} - c_{O_2}) - \frac{v_{max} \times c_{CC}}{k_{m,CC} + c_{CC} \times (1 + c_{CC}/k_i)} \times \frac{c_{O_2}}{k_{m,O_2} + c_{O_2}}$$

Symbols: c_{cc}, cephalosporin C, at 0 minutes, 75.0 mM; c_{DA}, deacetyl side products, at 0 minutes, 1.8 mM; c_{GA}, glutaryl-7-aminocephalosporanic acid, at 0 minutes, 4.4 mM; $c_{H_2}O_2$, hydrogen peroxide, at 0 minutes, 0.0 mM; c_{KA}, α-ketoadipyl-7-aminocephalosporanic acid, at 0 minutes, 2.9 mM; c_{o_2}, dissolved oxygen, at 0 minutes, 1.26 mM; $c^*_{o_2}$, dissolved oxygen, saturated with 3.5 bar oxygen, 4.41 mM; k_1, rate for deacetylation, 13.64 10^{-4}/minute; k_2, rate for decarboxylation, 0.182 mM^{-1} $minutes^{-1}$; k_1, substrate excess inhibition (cephalosporin C), 13.6 mM; k_La, oxygen transfer coefficient, 1.41 $minutes^{-1}$; $k_{m,CC}$, Michaelis-Menten constant (cephalosporin C), 44.5 mM; k_{m,O_2}, Michaelis-Menten constant (dissolved oxygen), 1.70 μM; t, time, minutes; v_{max}, maximum reaction rate, 21.5 mM/min.

Kinetic Model for Glutaryl-7-ACA Acylase

$$\frac{dn_{GA}}{dt} = -\frac{\dfrac{v_{+max} \times n_{GA} \times (V_0 + V_{NH_3})}{k_{m,GA}} - \dfrac{v_{-max} \times n_G \times n_A}{k_{m,G\&A}}}{V_0 + V_{NH_3} + \dfrac{n_{GA}}{k_{m,GA}} + \dfrac{n_G \times n_A}{k_{m,G\&A} \times (V_0 + V_{NH_3})}} - k_{1,GA} \times n_{GA}$$

$$\frac{dn_A}{dt} = \frac{\dfrac{v_{+max} \times n_{GA} \times (V_0 + V_{NH_3})}{k_{m,GA}} - \dfrac{v_{-max} \times n_G \times n_A}{k_{m,G\&A}}}{V_0 + V_{NH_3} + \dfrac{n_{GA}}{k_{m,GA}} + \dfrac{n_G \times n_A}{k_{m,G\&A} \times (V_0 + V_{NH_3})}} - k_{1,A} \times n_A$$

$$\frac{dn_G}{dt} = \frac{\dfrac{v_{+max} \times n_{GA} \times (V_0 + V_{NH_3})}{k_{m,GA}} - \dfrac{v_{-max} \times n_G \times n_A}{k_{m,G\&A}}}{V_0 + V_{NH_3} + \dfrac{n_{GA}}{k_{m,GA}} + \dfrac{n_G \times n_A}{k_{m,G\&A} \times (V_0 + V_{NH_3})}}$$

$$\frac{dV_{NH_3}}{dt} = \frac{\dfrac{\dfrac{v_{+max} \times n_{GA} \times (V_0 + V_{NH_3})}{k_{m,GA}} - \dfrac{v_{-max} \times n_G \times n_A}{k_{m,G\&A}}}{V_0 + V_{NH_3} + \dfrac{n_{GA}}{k_{m,G\&A}} + \dfrac{n_G \times n_A}{k_{m,G\&A} \times (V_0 + V_{NH_3})}} + k_{1,GA} \times n_{GA} + k_{1,A} \times n_A}{c_{NH_3}}$$

Symbols: c_{NH_3}, ammonia solution, 1.86 M; n_A, 7-aminocephalosporanic acid, at 0 minutes, 0.0 mmol; n_G, glutaric acid, at 0 minutes, 0.0 mmol; n_{GA}, glutaryl-7-aminocephalosporanic acid, at 0 minutes, 38.72 mmol; $k_{1,A}$, rate for deacetylation (7-ACA), 5.51 10^{-4}/minute; $k_{1,GA}$, rate for deacetylation (glutaryl-7-ACA), 0.44 10^{-4}/minute; $k_{m,GA}$, Michaelis-Menten constant (glutaryl-7-ACA), 1.97 mM; $k_{m,G\&A}$, Michaelis-Menten constant (glutaric acid & 7-ACA), 221.5 mM^2; t, time, minutes; v_{+max}, maximum forward reaction rate, 8.62 mM/minute; v_{-max}, maximum backward reaction rate, 0.57 mM/minute; V_0, initial volume of enzyme and splitting solution, 587.0 ml; V_{NH_3}, volume of ammonia, at 0 minutes, 0.0 ml.

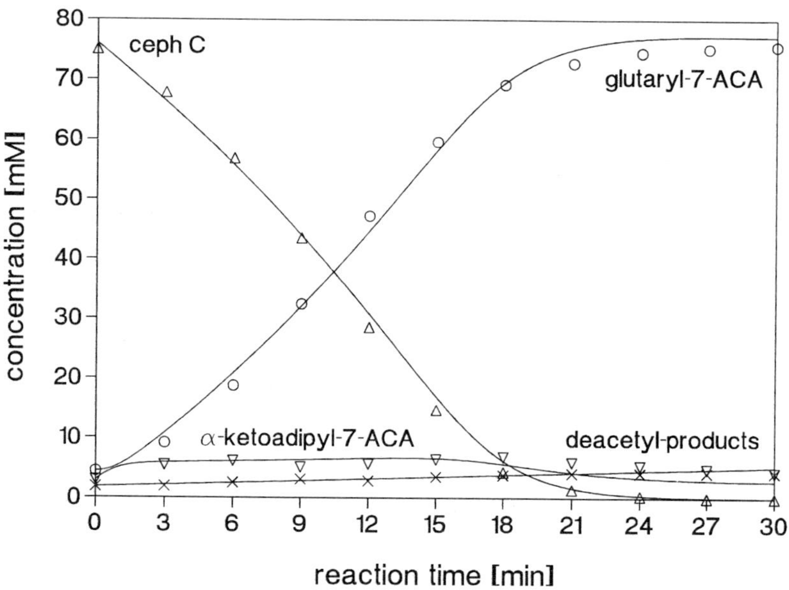

FIGURE 2. Conversion of cephalosporin C to glutaryl-7-ACA by immobilized D-amino acid oxidase under standard reaction conditions.

RESULTS

Conversion of Cephalosporin C to Glutaryl-7-ACA by Immobilized D-Amino Acid Oxidase

Conversion of cephalosporin C to α-ketoadipyl-7-aminocephalosporanic acid as intermediate and glutaryl-7-aminocephalosporanic acid as product is achieved within 30 minutes under the experimental conditions described above. Yield of glutaryl-7-aminocephalosporanic acid is around 94%. At the end of the conversion, the amount of the intermediate α-ketoadipyl-7-aminocephalosporanic acid has increased only 1.7%. No significant loss due to lactam ring cleavage or deacetylation could be detected. Experimentally, data fit well into model calculations as described above

TABLE 1. Specification of Immobilized D-Amino Acid Oxidase and Glutaryl-7-ACA Acylase

Property	D-Amino Acid Oxidase	Glutaryl-7-ACA Acylase
Appearance/Shape	Unsoluble, irregularly formed, wet granulate	Unsoluble, spherically formed, wet granulate
Carrier	Organic polymer	Inorganic polymer
Particle size	0.2–0.4 mm (>70%)	0.1–0.3 mm (>90%)
Immobilization	Covalent cross-linking via functional groups	Covalent cross-linking via functional groups
Enzyme	D-Amino acid oxidase (EC 1.4.3.3) from *Trigonopsis variabilis*	7-β-(4-Carboxybutanamido)-cephalosporanic acid acylase respectively glutaryl-7-ACA acylase (EC 3.5.1.3) from *Escherichia coli* overproducer
Activity	Approximately 25 kU/kg wet weight	Approximately 110 kU/kg wet weight
Side activities (β-lactamase, catalase, and 3-deacetyl esterase)	Under standard reaction conditions <1% side products	Under standard reaction conditions 1% side products
Storage	Aqueous suspension pH 7.5	Aqueous suspension pH 7.0
Stability	>95% during 3 months at 4°C 84% during 6 weeks at 37°C	>95% during 4 months at 4°C >95% during 4 weeks at 37°C

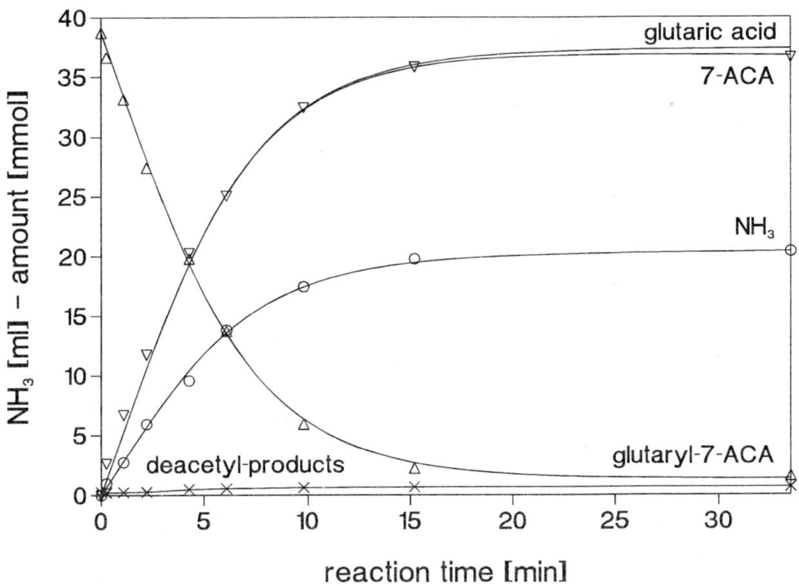

FIGURE 3. Conversion of glutaryl-7-ACA to 7-ACA by immobilized glutaryl-7-ACA acylase under standard reaction conditions.

(FIGURE 2). The specifications of immobilized D-amino acid oxidase are noted in TABLE 1.

Conversion of Glutaryl-7-ACA to 7-ACA by Immobilized Glutaryl-7-ACA Acylase

Conversion of glutaryl-7-aminocephalosporanic to 7-aminocephalosporanic acid is achieved within 30 minutes under the experimental conditions described above. Yield of 7-aminocephalosporanic acid is in the range of 94 to 95%. The final substrate concentration at the end of reaction is approximately 4%. About 1% hydrophilic compounds were formed (here referred to as deacetyl products). This

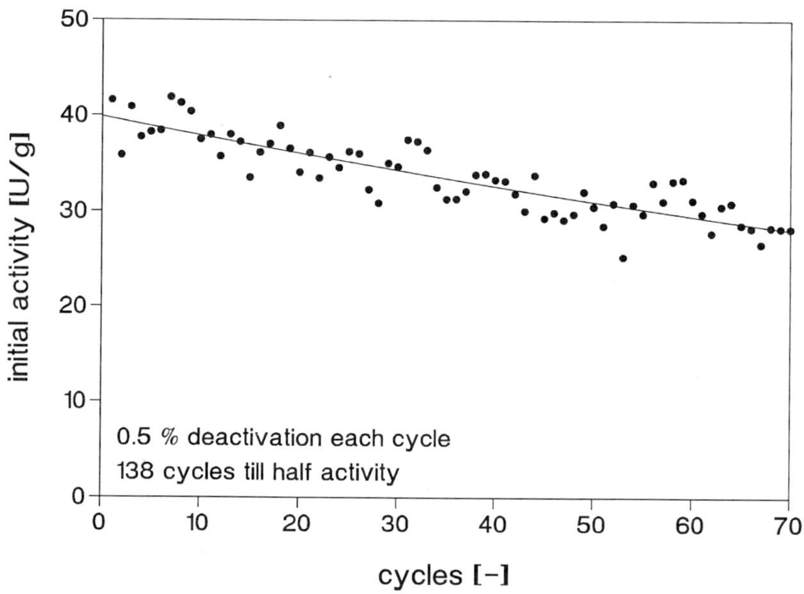

FIGURE 4. Operational stability of immobilized D-amino acid oxidase under standard reaction conditions.

increase is not due to enzymatic activities present in the catalyst, but occurs during incubation of substrate solution under reaction conditions. Experimentally, data fit well into model calculations as described above (FIGURE 3). The specifications of immobilized glutaryl-7-ACA acylase are noted in TABLE 1.

Operational Stability of Immobilized D-Amino Acid Oxidase

Operational stability is analyzed under standard reaction conditions as described above. The catalyst shows 0.5% deactivation per cycle. Therefore half-activity is reached after 138 cycles (FIGURE 4). This corresponds to an enzyme consumption of 1.078 kU/kg 7-ACA.

Operational Stability of Immobilized Glutaryl-7-ACA Acylase

First results show an operational stability of at least 150 cycles until half-activity is reached. This corresponds to an enzyme consumption of 2 kU/kg 7-ACA.

CONCLUSIONS

The experimental data provide evidence for the enzymatic process being an economical alternative to the well-established chemical process with a supposed yield of 7-aminocephalosporanic acid of more than 80% in high purity.

REFERENCES

1. VANDAMME, E. J. 1983. Immobilized enzyme and cell technology to produce peptide antibiotics. *In* Enzyme Technology, 3rd Rotenburg Fermentation Symposium: 237–270. Lafferty, Springer Verlag. New York, N.Y.
2. Product profile (in progress). Boehringer Mannheim GmbH, Biochemical Division NB-PP, Sandhofer Str. 116, D-6800 Mannheim 31, Germany.

New Developments in the Synthesis of Natural and Unnatural Amino Acids

J. KAMPHUIS, E. M. MEIJER,[a] W. H. J. BOESTEN,
T. SONKE, W. J. J. VAN DEN TWEEL,
AND H. E. SCHOEMAKER

DSM Research
Bio-Organic Chemistry Section
Post Office Box 18
6160 MD Geleen, the Netherlands

[a]DSM Fine Chemicals
Mauritslaan 49
6129 EL Urmond, the Netherlands

INTRODUCTION

The synthesis of amino acids has been studied for several years now.[1–4] Several technologies have been commercialized, fermentation being the prime technology to produce natural amino acids on a large scale (L-glutamic acid, L-lysine, L-proline, etc.).

The applications of unnatural amino acids as chiral building blocks for the synthesis of different pharmaceuticals are receiving increasing attention from industrial and scientific chemists.[5,6]

During the last decade, several approaches to the synthesis of various amino acids have been published.[6,7] In this publication, we will restrict ourselves to the developments of DSM in this area.

APPLICATIONS OF AND SYNTHETIC ROUTES TO α-*H*-AMINO ACIDS

The applications of natural amino acids are found in a wide variety of areas, i.e., in the food, feed, agro, and pharmaceutical industries. Mixtures of essential natural L-α-*H*-amino acids are used as physiological salt solution in the so-called injectable market (FIGURE 1). L-Lysine, D,L-methionine, and L-threonine are used as feed additives, whereas L-glutamic acid is used as a seasoning agent.[8] L-Phenylalanine and L-aspartic acid[9] are the chiral building blocks for the artificial sweetener Aspartame.[10]

Nonnatural amino acids have been receiving increasing attention during the last decade.[6,7] Some of them are already used on a commercial scale. D-Phenylglycine and D-*para*hydroxyphenylglycine are the prime examples, reaching several thousands of tons production per year. They are used as the side chains in the semisynthetic broad-spectrum antibiotics Ampicillin and Amoxycillin.[11]

D-Valine is applied in the synthesis of Fluvalinate, a pyrethroid insectide marketed by Zoëcon (FIGURE 2).[12] Other applications are found in the peptide area (FIGURE 2).[13]

In principle, several routes are possible for the synthesis of these amino acids as indicated in FIGURE 1. Starting from prochiral substrates like amino malonic esters or prochiral diesters,[14,15] several types of unnatural amino acids are obtainable. Lately asymmetric transformation variants have been elaborated.[16] At DSM, a diastereomeric resolution of racemic amino acid amides with *in situ* racemization has been developed[17,18] and will be discussed in the sequel. Most of the efforts preparing chiral amino acids have been directed towards the use of (chemo)enzymatic kinetic resolution processes.[19]

New developments in this area are found in the use of biocatalysts whereby a hydrolase activity is combined with some kind of a racemase activity. Asymmetric synthesis approaches have been studied by several research groups around the

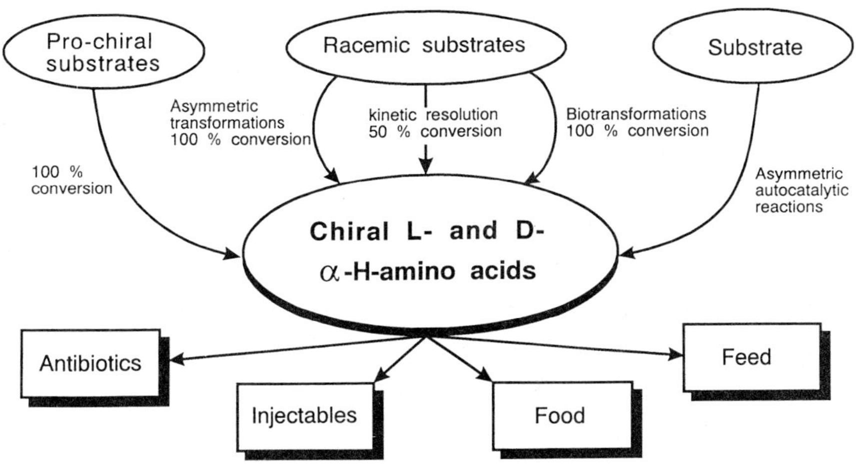

FIGURE 1. General synthetic routes to and applications of α-*H*-amino acids.

world.[16] The asymmetric synthesis approach has the advantage of a stoichiometric conversion, while kinetic resolution results in a maximum yield of 50%. The "unwanted" enantiomer has to be recycled externally or discarded. The ultimate goal will of course be the development of an asymmetric autocatalytic reaction. This has not been achieved yet, but preliminary results in this area are promising.[20–22]

In this publication we will elaborate on three examples of approaches towards the synthesis of chiral L- and D-α-*H*-amino acids studied by DSM.

CHEMOENZYMATIC SYNTHESIS OF CHIRAL L- AND D-α-*H*-AMINO ACIDS

In the midseventies, a research program at DSM was directed to the development of a route to D-phenylglycine, an intermediate in the synthesis of the semisynthetic antibiotic Ampicillin.

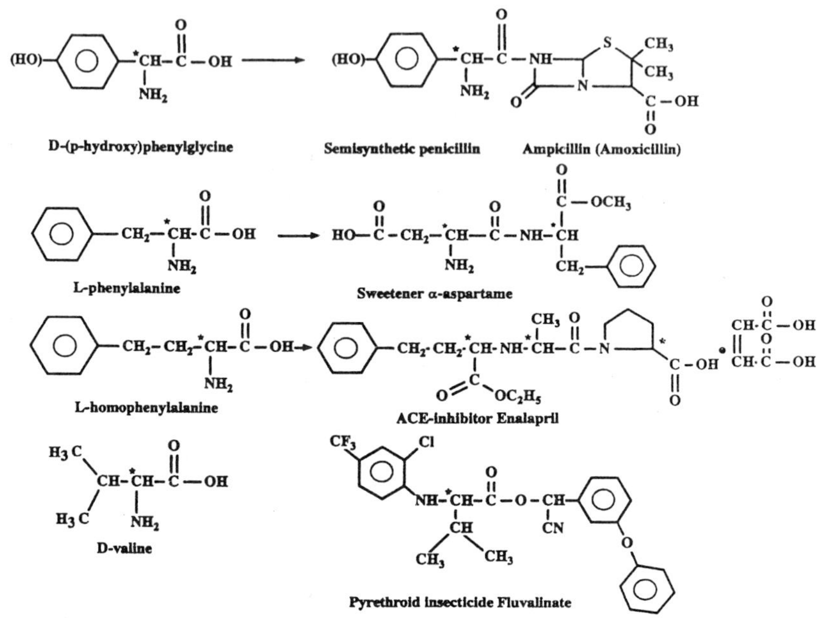

FIGURE 2A. Applications of α-*H*-amino acids.

Due to the idea that in the future other unnatural chiral α-*H*-amino acids and derivatives would find their application, especially in the pharmaceutical industry, the research was targeted at the development of a general approach to the synthesis of structurally different unnatural amino acids. "Simple" organic chemistry would then allow us to prepare a broad range of chiral synthons derived from α-*H*-amino

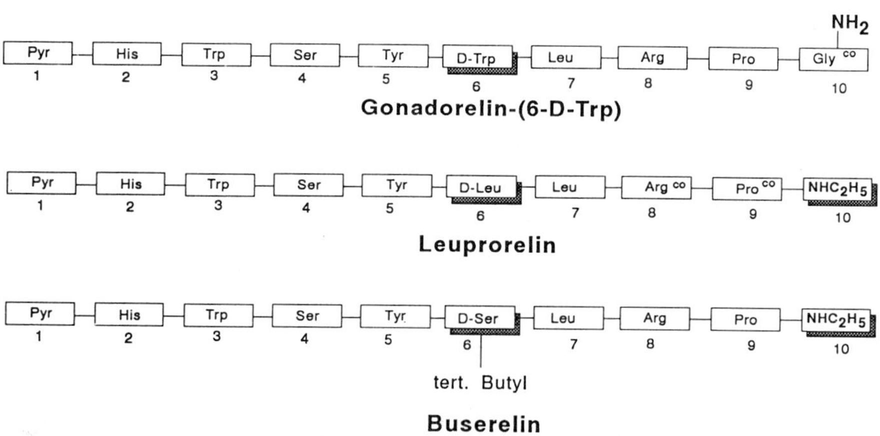

FIGURE 2B. Applications of α-*H*-amino acids in peptides.

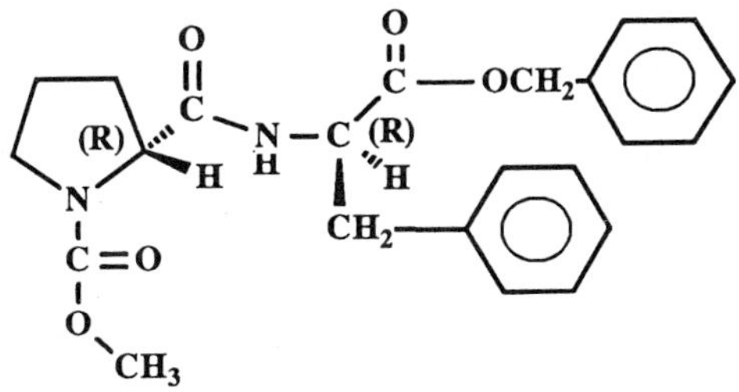

CPF

Dana - Farber Cancer Inst. /Harvard University

Science (july 20th, p287, 1990)

FIGURE 2C. Small peptide blocks HIV infection.

acids, i.e., chiral amino alcohols, diamines, hydroxy acids, diols, imidazolidinones, etc.[23]

Using a biocatalyst, for example, leucine aminopeptidase, for the kinetic resolution of amino acid amides was tried at first. This approach was successful. The application of leucine aminopeptidase, extracted from calf eye lenses, in the hydrolyses of leucine amide gave optically pure L-leucine and D-leucine amide. Subsequently, microorganisms were screened for their activity to stereospecifically hydrolyze amino acid amides, especially the hydrolysis of D,L-phenylglycine amide. A strain from the genus *Pseudomonas* was selected and classified as being *Pseudomonas putida* ATCC 12633. Together with NOVO-NORDISK Industries (Denmark), the fermentation of this microorganism was optimized. The whole-cell enzymatic resolution step combines a substrate specificity towards a broad range of structurally related substrates with an almost 100% enantioselectivity in hydrolyzing the L-amino acid amide (FIGURE 3).[24]

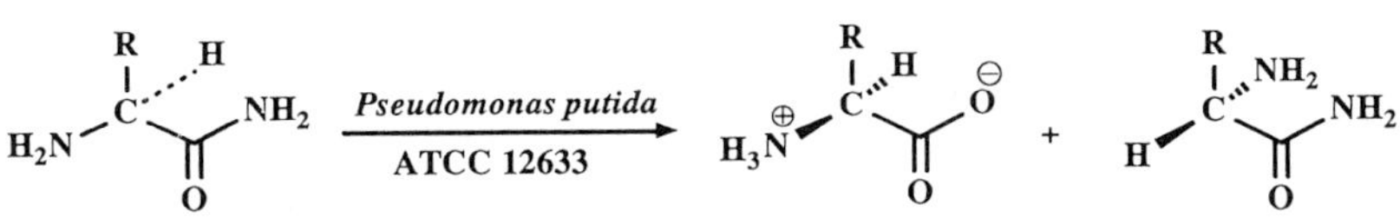

D,L-α-amino acid amide **L-α-amino acid** **D-α-amino acid amide**

W.H.J. Boesten *et al* DSM/Stamicarbon, U.S.Pat. 1.548-032 (1976)

J.Kamphuis, H.F.M.Hermes, J.A.M. van Balken, H.E.Schoemaker, W.H.J.Boesten and E.M.Meijer, In 'Amino Acids : Chemistry, Biology and Medicine ', G.Lubec and G.A.Rosenthal (Eds) (1990) ESCOM Science Publ. , Leiden, p 119

FIGURE 3. Kinetic resolution of D,L-α-*H*-amino acid amides.

Besides this research, a general route to the D,L-α-amino acid amides starting from aldehydes (FIGURE 4) was developed.

By serendipity, a very efficient and generally applicable way to separate the L-amino acid from the D-amino acid amide was found. The addition of one equivalent of benzaldehyde to the enzymatic (with respect to the D-amide) hydrolysate results in the crystallization of the Schiff base of the D-α-*H*-amino acid amide (see FIGURES 5 and 6).[24]

The hydrolysis of the D-amino acid amide into the D-amino acid can be done chemically using excess amounts of acid. This, however, raises a problem in the separation of the amino acid from several equivalents of salts after neutralization.

An elegant alternative to this approach is the use of another amidase from *Rhodococcus erythropolis*[24] to hydrolyze the D-α-*H*-amino acid amide without the formation of salts, and therefore the α-*H*-amino acid can be easily isolated by evaporation of the excessive amount of water present[26] (FIGURE 7).

In principle, the development of a universally applicable process for the synthesis

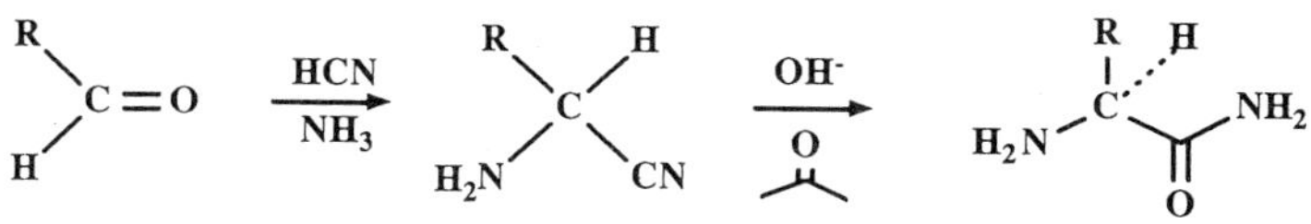

W.H.J. Boesten DSM/Stamicarbon, Brit.Pat. 1.548.032 (1976)
J.Kamphuis, M.Kloosterman, H.E.Schoemaker and E.M.Meijer, Proc. 4th. Eur. Congr. on Biotechn. (1987), 331-348.

FIGURE 4. DSM route to α-*H*-amino acid amides.

of L-α-*H*- and D-α-*H*-amino acids is now complete (racemization of the Schiff base of the D-α-*H*-amino acid in acid and basic media is also possible)[26] (FIGURE 5).

ENZYME PURIFICATION

Here we describe the purification[27] of an L-aminopeptidase from *Pseudomonas putida* and its characterization, especially with respect to classification, substrate specificity, enantioselectivity, and kinetic properties.

The cell envelope extracts were prepared by disrupting the cells (dissolved in a buffer at 4°C) by one passage through a French pressure cell at a cell pressure of 1000 MPa. The cell envelopes were collected by centrifugation (70,000 × g, 45 minutes). The pellet was washed once, centrifuged again, and the membrane proteins were extracted by resuspending the pellet in buffer (supplemented with 100-mM $MgSO_4$). Membranes were removed by centrifugation, yielding a clear supernatant.

The cell envelope extracts were purified by using a PD-10 column [equilibrated with a buffer (50-mM HAc, NH_4OH pH = 5.2, 100-mM $MgSO_4$)]. The proteins

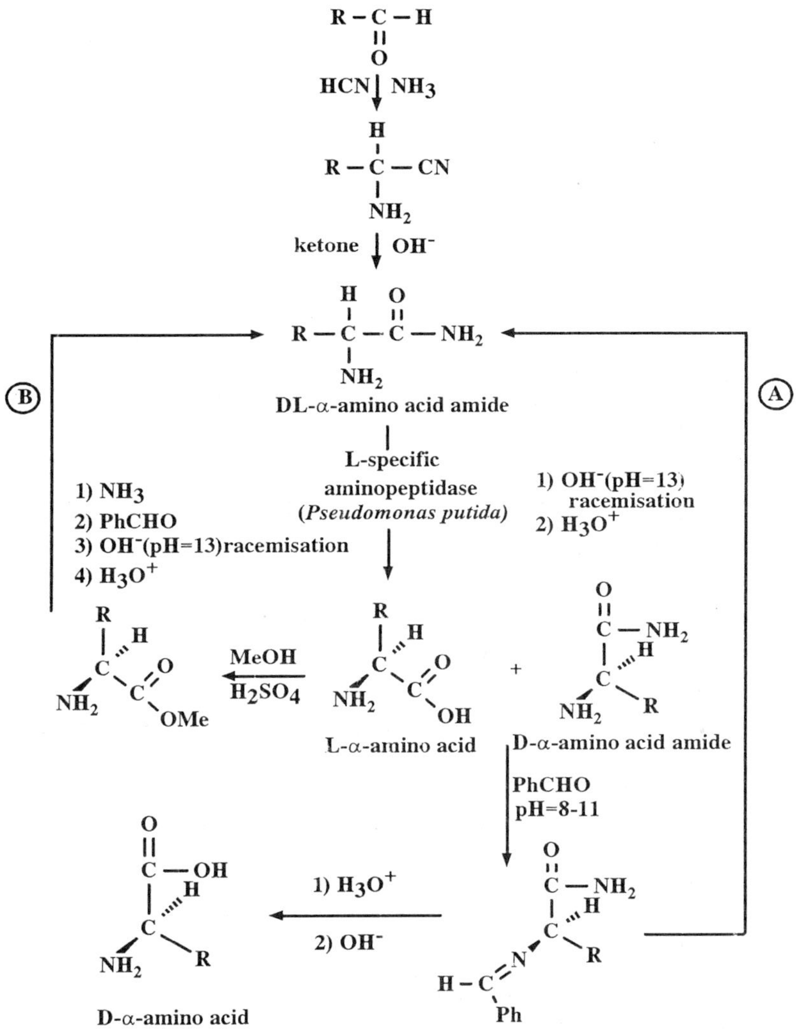

E.Meijer, *et al*; In 'Biocatalysts in Organic Synthesis', J.Tramper, H.C. van der Plas, P.Linko (Eds), Elsevier, A'dam, (1985), p 135
J.Kamphuis *et al*; In 'Advances in Biochemical Engineering/Biotechnology' H.Fiechter (Ed), Springer-Verlag, Berlin (1990), p 134.

FIGURE 5. DSM process for the preparation of L- and D-amino acids.

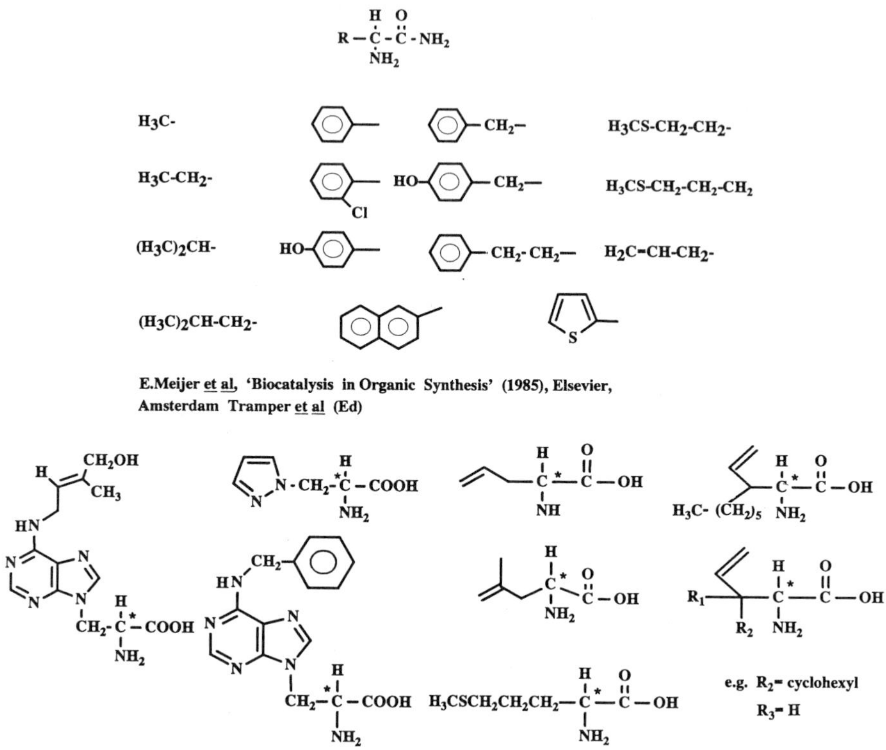

FIGURE 6. Substrate specificity of the whole cells from *Pseudomonas putida.*

eluted were subsequently loaded on a Mono S HR 5/5 cation exchange column (equilibrated with the same buffer as used for the PD-10 column).

In a linear gradient [50–300 mM $(NH_4)_2$ SO_4], the L-aminopeptidase was eluted. Fractions with the highest hydrolysis activity towards L-phenylglycine amide were pooled, desalted on PD-10 columns (buffer), and stored at −80°C. The enzyme

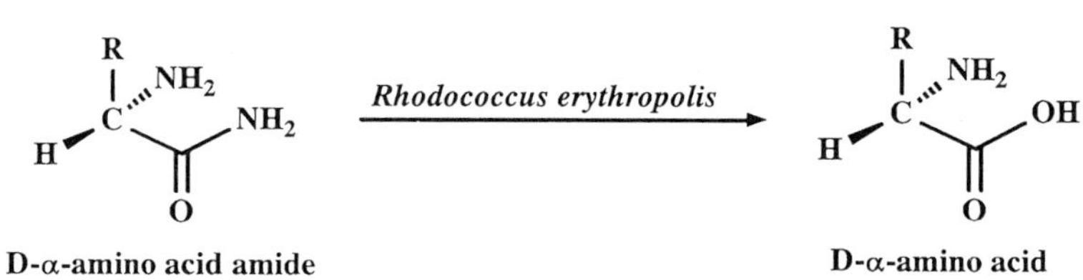

DSM/Stamicarbon, Dutch Pat. 8.403.093 (1984)

FIGURE 7. Enzymatic hydrolysis of D-α-*H*-amino acid amides.

TABLE 1. Purification of L-Aminopeptidase from *Pseudomonas putida* ATCC 12633

Purification Step	Total Activity (μmol × min^{-1})	Total Protein (mg)	Specific Activity (μmol × min^{-1} × mg^{-1})	Recovery (%)	Purification (fold)
Whole cells	4507	657	7	100	1
Cell envelopes	2342	194	12	52	1,7
Cell envelope extract[a]	1453	24	62	32	9,0
Mono S (FPLC)[a]	946	3	338	21	49,0

[a]Variations in activity and protein values were observed in three independent experiments; mean values are presented.

purification is summarized in TABLE 1. Sodium dodecyl sulfate-polyacrylamide gel electrophoresis (SDS-PAGE) analysis of the purified enzyme revealed only a single band corresponding to a molecular weight of approximately 53 kDa. Protein samples were diluted 5-fold (12.5% SDS, 25% β-mercaptoethanol, 0.05% bromophenol blue), and incubuted at 95°C for 10 minutes. Markers in the range of 20–94 kDa were used, and after electrophoresis, the protein bands were visualized by silver staining (see FIGURE 8).

The molecular weight of the native enzyme, estimated by nondenaturing gel

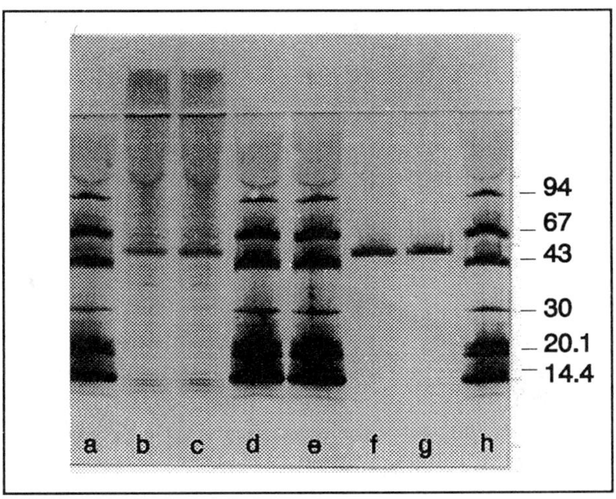

SDS - PAGE (8 - 25 % gradient gel) of the enzyme fractions

Lanes: **a,d,e,h** :mol. weight standards

Lanes: **b and c** :100 ng of protein from the cell envelope extract

Lanes: **f and g** :50 ng protein after mono S column

FIGURE 8. SDS-PAGE (8–25% gradient gel) of the enzyme fractions purified from *Pseudomonas putida* ATC 12633.

filtrations, was 402 kDa. This indicates that the L-amino peptidase most probably consists of eight subunits of equal size.

The isoelectronic point of the purified enzyme was estimated to be 10.5 using IEF 3-9 Phastgel with an expanded pH range.

The effect of the pH on the enzyme activity was studied in the range of 6 to 12 using a universal buffer. Hydrolyzing activity towards L-PG-NH_2 occurs between pH 6.5 and 11.5 with a maximum activity at pH 9.5.

The thermal stability of the purified L-aminopeptidase was determined following preincubation of the enzyme for maximal 60 minutes at temperatures ranging from 30 to 70°C. After a 60-minute incubation period at 30°C, 65% of the initial activity remained, and at 40°C this was 50%. At higher temperatures, the activity loss was rather dramatic; after 15 minutes at 60°C, only 1% of the initial activity remained.

The effect of divalent cations on the enzyme activity showed that preincubation of the enzyme with p.e. 0.02–20 mm Mn^{2+} resulted in a 12-fold increase in activity irrespective of whether sulfate or chloride salts were used, whereas Cu^{2+} and Ca^{2+} inhibit the enzyme by 69% and 39% respectively at 1 mM concentration.

The L-aminopeptidase was sensitive to various protease inhibitors, like 0.1-mM EDTA; *o*-phenanthroline, *p*-chloromercuribenzoate, phenylmethylsulfonyl fluoride, and especially diisopropyl fluorophosphate caused considerable inhibition.

The substrate specificity of the L-aminopeptidase towards various amides is shown in FIGURE 9. The enzyme was only active with α-amino acid amide, and it has the highest activity towards L-Leu-NH_2.

The enzyme displays normal Michaelis-Menten type of kinetics, and k_m and V_{max} values for six L-α-amino acid amides were determined from double-reciprocal plots (TABLE 2).

NEW DEVELOPMENTS

In principle, the biotransformation procedures mentioned above will always result in a 50% conversion of the substrate into the product leaving 50% of the substrate unchanged. A major advantage of this approach is that one obtains both enantiomers in one and the same biotransformation step. Normally, however, the markets for both enantiomers are not in balance and recycling procedures (*vide supra*) have to be developed to obtain a quantitative conversion into one of the enantiomers. Therefore it would be profitable if a (constructed) microorganism was available combining two enzyme activities, a stereoselective hydrolase in combination with a racemase racemizing the unwanted isomer.

In our case, the availability of our L-aminopeptidase or D-aminopeptidase in combination with an amino acid amide racemase would give us the opportunity to convert the racemic amino acid amides quantitatively into the L-amino acid or D-amino acid, respectively (see FIGURE 10).

Until 1989, amino acid amide racemases and D-aminopeptidases were unknown and, as far as we know, not reported in the literature.

A broad and extensive screening program was launched at DSM in 1988 in order to obtain an amino acid amide racemase and a D-aminopeptidase. Unfortunately, no generally applicable amino acid amide racemase activity could be found, while several microorganism showed D-amino peptidase activities.

Substrate	Formula	R^1	R^2	R^3	Relative activity
Acetamide [a]	CH_3-$CONH_2$	-H	-H	-H	<5
Propionamide [a]	CH_3-CH_2-$CONH_2$	-H	-CH_3	-H	<5
Butyramide [a]	CH_3-CH_2-CH_2-$CONH_2$	-H	-CH_2-CH_3	-H	<5
Isobutyramide [a]	$(CH_3)_2CH$-$CONH_2$	-H	-CH_3	-CH_3	<5
Acrylamide [a]	CH_2=CH-$CONH_2$	-H	=CH_2	-	<5
Fluoroacetamide [a]	F-CH_2-$CONH_2$	-H	-H	-F	<5
D,L-Mandelic acid amide [a]	C_6H_5-CH(OH)-$CONH_2$	-H	-Ø	-OH	<5
Pivalamide [a]	$(CH_3)_3C$-$CONH_2$	-CH_3	-CH_3	-CH_3	<5
Methacrylamide [a]	CH_2=C(CH_3)-$CONH_2$	-CH_3	=CH_2	-	<5
Nicotinamide [a]	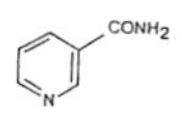				<5
Glycine amide [b]	H-CH(NH_2)-$CONH_2$	-H	-H	-NH_2	<5
L-Alanine amide [b]	CH_3-CH(NH_2)-$CONH_2$	-H	-CH_3	-NH_2	<5
L-α-Aminobutyramide [b]	CH_3-CH_2-CH(NH_2)-$CONH_2$	-H	-CH_2-CH_3	-NH_2	10
L-Valine amide [b]	$(CH_3)_2CH$-CH(NH_2)-$CONH_2$	-H	-CH$(CH_3)_2$	-NH_2	5
L-Leucine amide [b]	$(CH_3)_2CH$-CH_2-CH(NH_2)-$CONH_2$	-H	-CH_2-CH$(CH_3)_2$	-NH_2	215
L-Isoleucine amide [b]	CH_3-CH_2-CH(CH_3)-CH(NH_2)CONH2	-H	-CH(CH_3)-CH_2-CH_3	-NH_2	15
L-Phenyl glycine amide [a,b]	C_6H_5-CH(NH_2)-$CONH_2$	-H	-Ø	-NH_2	100
L-Phenylalanine amide [b]	C_6H_5-CH_2-CH(NH_2)-$CONH_2$	-H	-CH_2-Ø	-NH_2	15
L-Methionine amide [b]	CH_3-S-CH_2-CH_2-CH(NH_2)-$CONH_2$	-H	-CH_2-CH_2-S-CH_3	-NH_2	60
D,L-Proline amide [a]	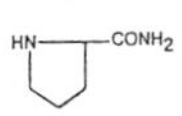	-H	-CH_2-CH_2-CH_2-	-NH-	<5
L-Tryptophan amide [b]	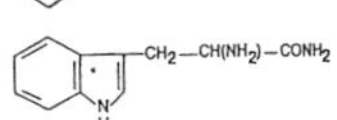	-H	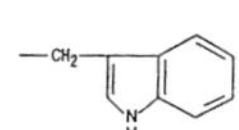	-NH_2	15
L-Serine amide [b]	HO-CH_2-CH(NH_2)-$CONH_2$	-H	-CH_2OH	-NH_2	<5
L-Glutamic acid amide [b]	HOOC-CH_2-CH_2-CH(NH_2)-$CONH_2$	-H	-CH_2-CH_2-COOH	-NH_2	<5
D,L-α-Methyl valine amide [a]	$(CH_3)_2CH$-C(CH_3)(NH_2)-$CONH_2$	-CH_3	-CH$(CH_3)_2$	-NH_2	<5

[a] Reactions were performed in Tris-H_2SO_4 pH 8.0.
[b] Reactions were performed in TAPS-KOH pH 9.0.

FIGURE 9. Substrate specicity of the purified L-Aminopeptidase of *Pseudomonas putida* ATCC 12633. Activity toward L-PG-NH_2 (100%) was 180 (pH 8.0) and 490 (pH 9.0) μmol $min^{-1} \times mg^{-1}$.

TABLE 2. Kinetic Parameters for the Hydrolysis of Different Amino Acid Amides by Purified L-Aminopeptidase of *Pseudomonas putida* ATCC 12633

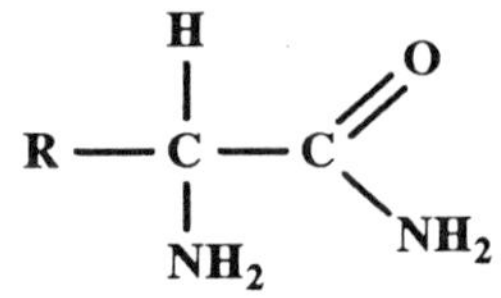

Substrate	-R	K_m[a]	V_{max}[b]	k_{cat}[c]	k_{cat}/K_m[d]
L-PG-NH_2	-C_6H_5	65	1,565	1,380	20,580
L-Phe-NH_2	-CH_2-C_6H_5	15	80	68	4,525
L-Phe-NH_2·HCl	-CH_2-C_6H_5	15	70	65	3,910
L-Val-NH_2·HCl	-$CH(CH_3)_2$	130	110	95	715
L-Leu-NH_2·HCl	-CH_2-$CH(CH_3)_2$	15	1,915	1,690	99,215
L-Ile-NH_2·HCl	-$CH(CH_3)$-CH_2-CH_3	150	270	240	1,604

[a] K_m is expressed in millimolar.
[b] V_{max} is expressed in micromoles of amino acid formed per milligram of protein per minute.
[c] k_{cat} is expressed per seconds.
[d] k_{cat}/K_m is expressed per second per molar.

A different approach, comprising "*in vivo*" protein engineering techniques was investigated for obtaining the so-wanted amino acid amide racemase. Cultivating microorganisms in the laboratory under selective conditions, particularly by using chemostat culture, novel (and unnatural) enzyme activities and metabolic pathways can evolve, using the microbe as an engineer. Possibly this approach used on the *Pseudomonas putida* ATCC 12633 strain would result in the evolution of an amino

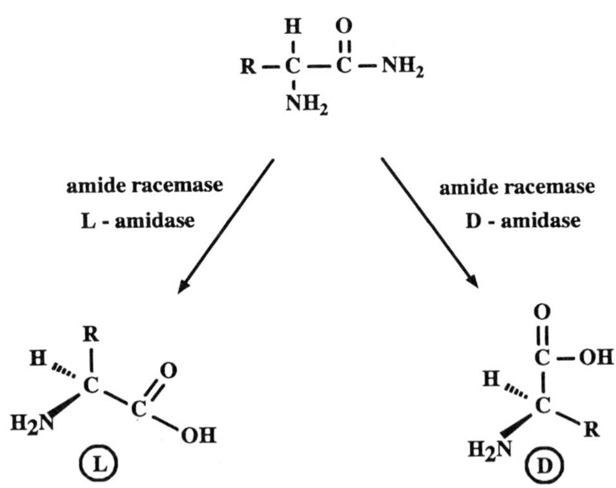

FIGURE 10. Principle of a 100% conversion concept using amide racemase in combination with L- or D-amidase activities.

acid amide racemase and D-amino peptidase.[28] And indeed it worked. Appropriate selective pressure (low nitrogen source) in chemostat culture was found to result in the selection of a mutant strain with, in addition to the original L-specific aminopeptidase activity, novel D-aminopeptidase and amino acid amide racemase activities.

The construction of a biocatalyst with a combination of L-aminopeptidase or D-aminopeptidase with an amino acid amide racemase requires the generation of respectively L- and D-aminopeptidase negative mutants of the obtained *Pseudomonas putida* mutant.

Otherwise all enzyme activities have to be isolated separately and the combined activities have to be cloned, using rDNA techniques, in for example *Escherichia coli.* This research is at the moment in progress at DSM.

ASYMMETRIC TRANSFORMATION

In the foregoing, we have seen that kinetic resolution results in a theoretical one-through yield of 50%. In general the unwanted enantiomer has to be racemized and recycled in order for the process to be economically viable. Preferential crystallization (or kinetic resolution) is a particularly attractive method for industrial synthesis when it is accompanied by spontaneous *in situ* racemization which allows for a theoretical yield of 100%. Such a process is called a crystallization-induced asymmetric transformation and may occur with racemates or diastereomeric mixtures. In the content of industrial synthesis, this is the situation that one is always looking (hoping) for.

In the former case, the process is often referred to as a deracemization. An elegant example is reported by Okada and coworkers.[29]

At DSM, an asymmetric transformation of α-*H*-amino acid amides was successfully developed.[17,18]

Racemization of the Schiff bases of amino acid amides in basic media was already known.[26] Due to the basic nature of the amino acid amides, diastereomer formation would eventually be possible using chiral carboxylic acids. For an asymmetric transformation variant, it is then mandatory that the Schiff bases also racemize under acidic conditions.

This was indeed observed (FIGURE 11), when an optically active amino acid amide was reacted with a carboxylic acid in the presence of a solvent and an aldehyde.[30] Suprisingly a sufficient amount of aromatic or aliphatic aldehyde (0.5–4.0 equivalents relative to amide) and water is needed to obtain a high degree of racemization and high yields. Next to this, the pK_a of the carboxylic acid is of importance and should in general be 3 to 5.

Examples of suitable carboxylic acids are acetic acid, mandelic acid, propionic acid, benzoic acid, and 2-pyrrolidine-5-carboxylic acid. In general, at 85°C complete racemization is observed within 5 hours, (L-valine amide excluded) (TABLE 3) using amino acid amides from leucine, methionine, phenylglycine, and homophenylalanine. This racemization system is the right starting point for further development. Diastereomeric crystallization is needed, to obtain as an overall result an asymmetric transformation.

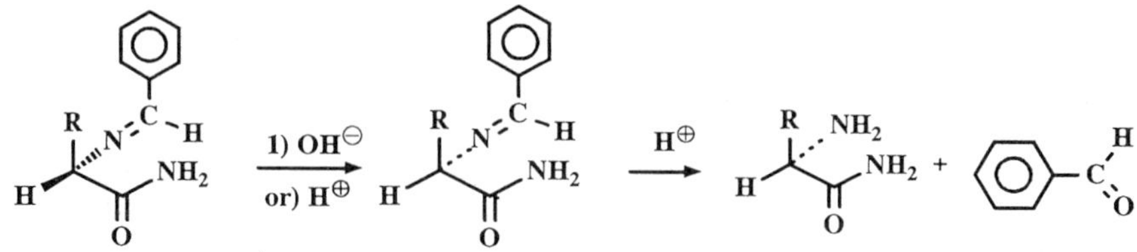

D-Schiff base D,L-Schiff base D,L-α-amino acid amide

W.H.J.Boesten, H.E.Schoemaker, B.H.N.Dassen, Eur. patent 1.994.07, (1987)

FIGURE 11. Racemization of the Schiff bases of amino acid amides.

Several optically active carboxylic acids were screened in order to find the right diastereomeric resolution of a D,L-α-*H* -amino acid amide. Mandelic acid, 2-pyrrolidone-5-carboxylic acid, and L-aspartic acid were selected because efficient diastereomeric resolutions were obtained. The choice of the solvent used is very important[31] (TABLE 4). The invention is now complete.

It is the preparation of optically active amino acid amides, whereby a mixture of the L-amino and D-amino acid amides in the presence of an aldehyde—a suitable solvent—is converted *in whole* by means of an optically active carboxylic acid into the salt of the carboxylic acid (FIGURE 12). Eventually the complete conversion of an L-α-*H*-amino acid amide into D-α-*H*-amino acid amide or *vice versa* can be envisaged.

The slurry concentration of the diastereomeric salt can be as high as 30 wt%. The optically active amino acid amide can be obtained from the separated diastereomeric salt by dissolving the salt in a mixture of water and a virtually equimolar quantity of mineral acid, such as hydrochloric acid, sulfuric acid, or phosphoric acid, and extracting the optically active carboxylic acid by means of an extraction agent.

The optically active amino acid amide obtained can be converted into the corresponding amino acid in a known manner by hydrolysis with an excess of diluted mineral acids.

The scope and limitation of this new concept are still under investigation at DSM.

Remarkably, methionine amide, phenylglycine amide, and *para*hydroxyphenyl glycine amide can be asymmetrically transformed into the pure enantiomers using one and the same chiral carboxylic acid. This is of course of major importance to developing a general route to optically pure α-*H*-amino acids.

AUTOCATALYSIS

The next generation of asymmetric synthesis would of course be autocatalysis.[32] Asymmetric autocatalysis is the process whereby a chiral reaction product is the

TABLE 3. Racemization of the Schiff Bases of Amino Acid Amides[a]

SB	$[\alpha]_{t=0}$	t	$[\alpha]_{t=t}$	Selectivity
D-Leucine amide	−0.525°	30	−0.492°	96.9%
		1210	−0.341°	82.5%
L-Alanine amide	+1.970°	30	+1.450°	86.8%
		1200	+0.066°	51.7%
D-Chlorophenylglycine amide	−0.730°	30	−0.500°	84.2%
		1150	−0.007°	50.5%
D-Homophenylalanine amide	−0.095°	30	−0.063°	83.2%
		1140	−0.050°	76.3%
L-Methionine amide	−0.190°	30	−0.078°	70.5%
		1100	0.000	50.5%
D-Phenylalanine amide	+2.160°	30	+1.790°	91.4%
		1080	+0.327°	57.6%
D-Phenylglycine amide	−1.626°	30	−1.140°	85.1%
		1070	0.000	50.0%

[a]The Schiff base (SB) of different amino acid amides (selectivity at t = 0:100%), as indicated in the table, is racemized as a 2 wt.% solution at a temperature of 25°C in the same manner as in example 1. As solvent is used a mixture of acetic acid and water with an acetic acid/water ratio of 99/1. The angle of rotation $[\alpha]_t$ and the selectivity after 30 and after about 1200 minutes are shown. Also at this low racemization temperature, racemization occurs relatively quickly. *Example 1:* In a flask provided with a stirrer, a thermometer, and a reflux cooler, 23.8 g (0.10 mole) D-*N*-benzylidene phenylglycine amide (selectivity: 99.5%), 15.2 g (0.10 mole) D,L-mandelic acid, 200 ml toluene, 50 ml ethyl acetate, and 2.7 ml (0.15 mole) water are stirred for 4 hours at a temperature of 85°C. After cooling to 25°C, the mandelate of D,L-phenylglycine amide is filtered and washed on the glass filter with 4 × 25 ml toluene. The yield of thin-layer chromatography (TLC) pure mandelate of D,L-phenylglycine amide amounts to 29.9 g, which corresponds to an efficiency of 99.1%. 1.0 g of the mandelate of D,L-phenylglycine amide is added to a mixture of 10 ml water and 10 ml 12 N hydrochloric acid. The D,L-phenylglycine amide·HCl formed is filtered and washed on the glass filter with 4 × 5 ml acetone. The yield of TLC pure D,L-phenylglycine amide·HCl amounts to 0.57 g (efficiency = 92.3%). The specific rotation of the D,L-phenylglycine amide·HCl is: $[\alpha]^{20}_D = -0.6°$ (c = 0.8; water). The selectivity is 50.3% D-enantiomer. *General example:* In a reaction flask provided with a stirrer, a thermometer and a reflux cooler, 15.0 g (0.10 mole) D,L-phenylglycine amide, 15.2 g (0.10 mole) D-mandelic acid, 230 ml toluene, 20 ml benzaldehyde, and 1.8 g (0.10 mole) water are stirred for 2 hours at a temperature of 88°C. After cooling to 20°C (0.5 hour), the resulting diastereoisomeric salt of L-phenylglycine amide and D-mandelic acid (LD salt) is filtered and washed on the glass filter with 4 × 25 ml toluene. The yield of TLC pure LD salt after drying amounts to 29.3 g, which corresponds to an efficiency of 97.0%. The specific rotation of the optically pure LD salt (recrystallization) amounts to: $[\alpha]^{20}_D = +40°$ (c = 1.0; water). 1.0 g of the diastereoisomeric LD salt obtained is suspended in 10 ml water, to which subsequently 10 ml 12 N hydrochloric acid is added with stirring. After filtration and washing with 4 × 10 ml acetone on the glass filter the L-phenylglycine amide·HCl crystal mass is obtained. The specific rotation of the TLC pure L-phenylglycine amide·HCl salt obtained (yield = 0.55 g; efficiency = 88.7%) is: $[\alpha]^{20}_D = +101.7°$ (c = 1.0; water). The LD salt (22.7 g = 0.75 mole) is dissolved at 50°C in a mixture of 40 ml water and 2.7 ml wt.% sulfuric acid (0.05 mole) and, after extraction with 5 × 50 ml methyl-*t*-butyl ether (removal of D-mandelic acid), hydrolyzed to L-phenylglycine at a temperature of 90°C by means of 10 ml 96 wt.% sulfuric acid (0.18 mole) with stirring for 3 hours. After neutralization of the sulfuric L-phenyglycine hydrolizate by means of 28 ml 25 wt.% ammonia to an acidity of pH = 5 and cooling to 30°C, the L-phenylglycine thus obtained is filtered and then washed on the glass filter with 5 × 10 ml water. The yield of L-phenylglycine after drying amounts to 10.5 g, which corresponds to an efficiency of 92.9%. The specific rotation of the TLC pure L-phenylglycine is: $[\alpha]^{20}_D = +157.7°$ (c = 1.6; 2.6 wt% hydrochloric acid). Literature value of specific rotation of L-phenylglycine (Beilstein 14 III, p. 1188); $[\alpha]^{20}_D = +157.7°$ (c = 1.6; 2.6 wt% hydrochloric acid).

TABLE 4. Results of the Asymmetric Transformation Performed under Different Reaction Conditions[a]

Example	Solvent	Aldehyde (eq.)	Water	Slurry Concentration	T	t (hours)	Efficiency	Selectivity[b]
1	Methylisobutyl ketone	1.0 BA	2.0	15%	85°C	3	90.1%	99.8%
2	Cyclohexane	2.0 BA	2.0	11%	77°C	3	96.0%	99.8%
3	Butyl acetate	1.0 BA	1.5	12%	85°C	2	91.0%	99.4%
4	Ethyl acetate/ toluene: 1/3	2.0 BA	1.5	12%	86°C	2	96.0%	99.3%
5	Ethyl acetate/ toluene: 1/3	1.5 AA	1.0	14%	84°C	3.5	96.7%	99.8%
6	Ethyl acetate/ toluene: 1/4	2.0 IBA	1.0	11%	80°C	3	92.4%	99.8%
C	Ethyl acetate/ toluene: 1/3	2.0 BA	0	12%	86°C	3.5	87.7%	99.8%
D	Methylisobutyl ketone	0.1 BA	1.0	15%	70°C	2.5	93.7%	50.5%

[a]Abbreviations: AA = anisaldehyde; BA = benzaldehyde; IBA = isobutyraldehyde
[b] = optical purity of phenylglycine amide·HCl.

Examples 1–6: D,L-Phenylglycine amide is contacted with 1 equivalent optically active D-mandelic acid and the TLC pure L-phenylglycineamide·HCl salt, recovered as in the general example, the quantities of aldehyde and water added being varied. The results are given above and also the temperature (T) in °C and the retention time (t). *Comparative Experiment C:* A procedure analogous to the example is carried out up to the recovery of the TLC pure L-phenylglycine amide·HCl salt, but without addition of water, which results in a lower yield. The results and specific conditions are given above. *Comparative Experiment D:* A procedure analogous to the example is carried out up to the recovery of the TLC pure L-phenylglycine amide·HCl salt, but with only 0.1 equivalent benzaldehyde, which results in a very low selectivity. The results and specific conditions are given above.

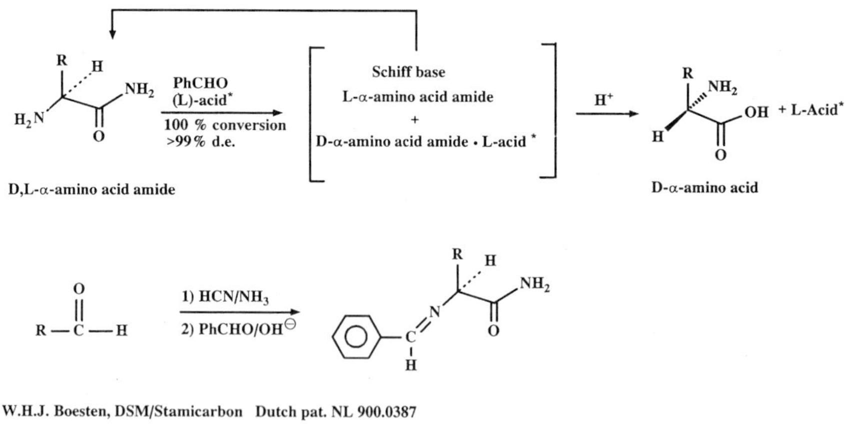

FIGURE 12. Asymmetric transformation (resolution with *in situ* racemization).

catalyst of its own formation from achiral reactants

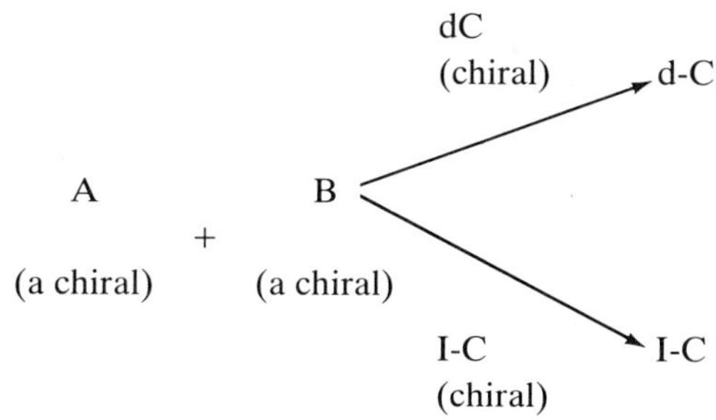

where dC and I-C are the catalysts. The development of such a reaction (self-production of chiral molecules) for example for the synthesis of amino acids or derivatives would of course be the ultimate goal.

Although the implication of asymmetric autocatalysis has been described, only a few examples are known and recently described.[20–22,33–36] We at DSM are currently investigating the possibilities of this interesting approach.

CONCLUSIONS

In conclusion, it can be stated that very efficient procedures have been developed for the synthesis of optically pure unusual amino acids. Both enantiomers can be obtained. The enzyme used was purified to homogeneity, and structure activity relations studied.

Recent new developments like the asymmetric transformation of α-*H*-amino acid amides are very promising (surely from an economic standpoint). Autocatalytic routes are being studied now. The application of chiral amino acids as chiral synthons, chiral ligands in asymmetric synthesis, is currently under investigation.

Therefore, in general, further research in these fascinating and versatile areas is in progress in our laboratories.

SUMMARY

Amino acids play an important role in biochemistry and chemistry. They are the building blocks of proteins and play an essential role in the regulation of the metabolism of living organisms.

In general, it can be stated that microbial processes (fermentation) are the industrial production methods of choice for large-scale production of naturally occuring proteinogenic L-α-*H*-amino acids, while for the production of synthetic D- and/or L-α-*H*-amino acids, several other methods are highly competitive.

At DSM, several routes, i.e., (chemoenzymatic) synthesis, towards L-α-*H* and D-α-*H*-amino acids have been elaborated since the midseventies.

A general process for the synthesis of natural as well as synthetic optically pure amino acids has been developed, using an enzymatic kinetic resolution step on racemic amino acid amides as the key step. In this case, both enantiomers of the α-*H*-amino acids are prepared in one single step. This process has been commercialized since 1988.

More recent developments using L- or D-amino peptidases in combination with amino acid amide racemases and an asymmetric transformation concept are discussed.

REFERENCES

1. AIDA, K., I. CHIBATA, K. NAKAYAMA & K. TAKINAMI, Eds. 1986. Biotechnology of Amino Acid Production: Progress in Industrial Microbiology. **24.** Elsevier. Amsterdam, the Netherlands.
2. WILLIAM, R. M., Ed. 1989. Synthesis of Optically Active Amino Acids, Pergamon. Oxford, England.
3. DALANEY, F. L. 1967. Microbial-production of amino acids, *In* Microbial Technology. H. L. Pepple, Ed.: 308. Reinhold. New York, N.Y.
4. KAMPHUIS, J., W. H. J. BOESTEN, Q. B. BROXTERMAN, H. F. M. HERMES, J. A. M. VAN BALKEN, E. M. MEIJER & H. E. SCHOEMAKER. 1990. New developments in the chemo-enzymatic production of amino acids. Adv. Biochem. Eng. Biotechnol. **42:** 134–182.
5. COPPOLA, G. M. & H. E. SCHUSTER. 1987. Asymmetric Synthesis: Construction of Chiral Molecules Using Amino Acids. Wiley. New York, N.Y.
6. BARRETT, G. C., Ed. 1985. Chemistry and Biochemistry of the Amino Acids. Chapman and Hall. London, England.
7. SCHMIDT-KASTNER, G. & P. EGERER. 1984. *In* 1016A, Biotransformations in Biotechnology. H. J. Rehm & S. B. Reed, Eds.; 387–421. Verlag Chemie. Weinheim, Germany.
8. FUKUMURA, T. 1977. Agric. Biol. Chem. **40:** 1327.
9. CHIBATA, I., T. TOSA & T. SATO. 1986. *In* Aspartic Acid in Biotechnology of Amino Acid Production. K. Aida, I. Chibata, K. Nakayama & H. Yamada, Eds. **24:** 144. Elsevier. Amsterdam, the Netherlands.
10. BOESTEN, W. H. J., B. H. N. DASSEN, J. C. S. KLEINJANS, B. VAN AGEN, S. VAN DER WAL, N. K. DE VRIES, H. E. SCHOEMAKER & E. M. MEIJER. 1991. J. Agric. Food Chem. **39:** 154–158. (And references cited therein.)
11. SHELDON, R. A., L. A. HULSHOF, A. BRUGGINK, F. J. J. LEUSEN, A. D. VAN DER HAEST & H. WIJNBERG. 1991. Chim. Oggi: 23–29.
12. 1986. Farmaceutical Chemical Handbook. 72nd edit. Meister Publishing Co. Willoughby, Ohio.
13. RUTH, H. J. & H. FENNER. 1988. Arzneistoff—Pharmazeutische Chemie III: 202–204. Georg Thieme Verlag. Stuttgart, Germany.
14. BJÖRKLING, F., J. BOUTELJE, S. GATENBECK, K. HULT & T. NORIN. 1985. Tetrahedron Lett. **40:** 4957–4961.
15. VRIESEMA, B. K., W. TEN HOEVE, H. WIJNBERG, R. M. KELLOGG, W. H. J. BOESTEN, E. M. MEIJER & H. E. SCHOEMAKER. 1986. Tetrahedron Lett. **26:** 2045–2049.
16. REIDER, P. J., P. DAVIS, D. L. HUGHES & E. J. GRABOWSKI. 1987. J. Org. Chem. **52:** 957–958.
17. BOESTEN, W. H. J. Dutch Patent App. 900, 0387.
18. BOESTEN, W. H. J. Dutch Patent App. 900, 0386.
19. KAMPHUIS, J., E. M. MEYER, W. H. J. BOESTEN, Q. B. BROXTERMAN, B. KAPTEIN, H. F. M. HERMES & H. E. SCHOEMAKER. 1992. *In* Production of Amino Acids and Derivatives. D. Rozzell & F. Wagner, Eds. Carl Hanser-Verlag Publ. München, Germany.
20. WIJNBERG, H. 1989. J. Macromol. Sci-Chem. **A26**(8): 1033–1041.
21. NOYORI, R. M. & KITAMURA. 1991. Angew. Chem. **103:** 34–55.
22. OGUNI, N., Y. MITSUDA & T. KANEKO. 1988. J. Am. Chem. Soc. **110:** 7877.
23. SHELDON, R. A., H. E. SCHOEMAKER, J. KAMPHUIS, W. H. J. BOESTEN & E. M. MEIJER. 1988. *In* Stereoselectivity of Pesticides. E. J. Ariens, J. J. S. van Rensen & W. Welling, Eds.: 409. Elsevier. Amsterdam, the Netherlands.

24. MEIJER, E. M., W. H. J. BOESTEN, H. E. SCHOEMAKER & J. A. M. VAN BALKEN. 1985. *In* Biocatalysis in Organic Synthesis. J. Tramper, H. C. van der Plas & P. Linko, Eds.: 135–156. Elsevier. Amsterdam, the Netherlands.
25. BOESTEN, W. H. J. Dutch Patent Appl. 8.403.093, 1984.
26. BOESTEN, W. H. J. Eur. Patent Appl. EP 1, 816.75, 1985; Eur. Patent Appl. EP 1.442.585, 1991; US Patent 4.182.846, 1979.
27. HERMES, H. F. M., T. SONKE, R. F. TANDLER, J. A. M. VAN BALKEN, E. M. MEIJER, J. KAMPHUIS & L. DIJKHUIZEN. To be published.
28. DSM/Stamicarbon—NOVO/Nordisk. European Patent Appl. 0.307.023, 1989.
29. OKADA, Y. & T. TAKEBAYASHI. 1988. Chem. Pharm. Bull. **36:** 3787.
30. BOESTEN, W. H. J. Eur. Patent Appl. 0 442 585.
31. BOESTEN, W. H. J. Eur. Patent Appl. 0 442 584.
32. WIJNBERG, H. 1989. Chimia **43:** 150–152.
33. BOLM, C. 1991. Tetrahedron Asymm. **2**(7): 701–704.
34. DANDA, H., H. NISHIKAWA & K. ORAKA. 1991. J. Org. Chem. **56:** 6740–6741.
35. SOAI, K., S. NIWA & H. HORI. 1990. J. Chem. Soc. Chem. Commun.: 982–983.
36. DE MIN, M., G. LEVY & J. C. MICHEAU. 1988. J. Chem. Phys. **85**(5): 603–619.

Enzyme Reaction Engineering for Peptide Synthesis[a]

CHRISTIAN WANDREY,[b] ANDREAS FISCHER,[b,d]
BURKHARD JOKSCH,[b] AND ALEXANDER SCHWARZ[b,c]

[b]*Institute of Biotechnology*
Research Center Jülich
Post Office Box 1913
D-5170 Jülich, Germany

[c]*Weizmann Institute of Science*
Department of Membrane and Molecular Recognition
76100 Rehovot, Israel

INTRODUCTION

Oligopeptides are of interest for nutritional purposes and especially in infusion solutions in order to decrease osmolarity, increase solubility of aromatic amino acids, and enhance the biological uptake. As a case study, the synthesis of Kyotorphin was analysed. Kyotorphin is not used in infusion solutions but shows analgetic properties.[1]

The Kyotorphin example makes one special advantage of enzymatic peptide synthesis quite clear. Both amino acids do not need additional protection in the side chain during enzymatic synthesis.

Enzymatic peptide synthesis is now well established.[2–4] The various advantages and disadvantages of classical peptide synthesis and enzymatic peptide synthesis are summarized in TABLE 1.

In enzymatic peptide synthesis, one distinguishes between the thermodynamically and the kinetically controlled synthesis as shown in FIGURE 1. The best-known example for the thermodynamic approach is the synthesis of Aspartame.[5] Here only the kinetic approach was used.

The enzyme substrate complex, the so-called acylenzyme (ES), is formed from the first activated amino acid (S). After the formation of the acylenzyme, a competitive reaction between aminolysis (second amino acid, nucleophile, N) and hydrolysis (H_2O) takes place. Apart from the product (dipeptide, P), the first amino acid (not activated, S′) is also obtained (FIGURE 2). This phenomenon decreases selectivity, which is defined as the product (P) formed in relation to the substrate used (S): $\sigma_{P,S} = P/(S_0 - S)$. Other important measures closely connected to the selectivity are the conversion (U) and the reaction yield (η) (see also FIGURE 2): $U_S = S_0 - S/S_0$, $\eta_{P,S} = \sigma_{P,S} \times U_S$ (S_0 = initial substrate concentration).

While secondary hydrolysis of the formed peptide can mostly be avoided using

[a]The project was supported financially by a grant from the Bundesministerium für Forschung und Technologie, Germany.

[d]Author to whom correspondence should be addressed.

TABLE 1. Advantages and Disadvantages of Classical Peptide Synthesis and Enzymatic Peptide Synthesis

Chemical		Enzymatic	
Advantages	Disadvantages	Advantages	Disadvantages
• General method of linking • Automation of synthesis • Optimized protecting groups	• Racemization • Side reactions • Maximum protection	• No racemization • No side reactions • Minimum protection • Very fast reaction	• No general method • Problem of selectivity • Excess of nucleophile

alkaline pH, the generated parallel hydrolysis is difficult to avoid completely. Therefore, organic solvents are often used. In such a case, protecting groups have to be used that facilitate solubility in organic solvents. However, organic solvents often have a negative influence on enzyme stability and may cause problems when the product is used for nutritional or therapeutic purposes. By choosing appropriate enzymes or changing the enzyme properties in aqueous solution, the undesired hydrolysis may be discriminated against.[6,7] Here, the use of charged protecting groups that improve the solubility of the substrates in water helps to favor the desired synthesis.

KINETICS

For enzyme reaction engineering, the kinetics of the aminolysis and the hydrolysis were measured, and as expected, increasing substrate concentration does not

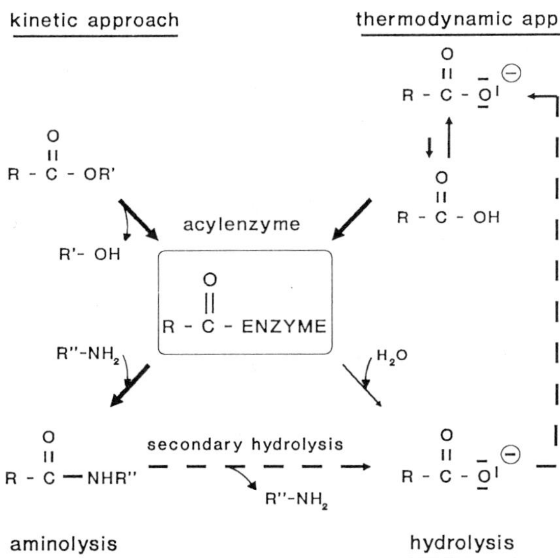

FIGURE 1. Comparison of kinetic and thermodynamic approach.

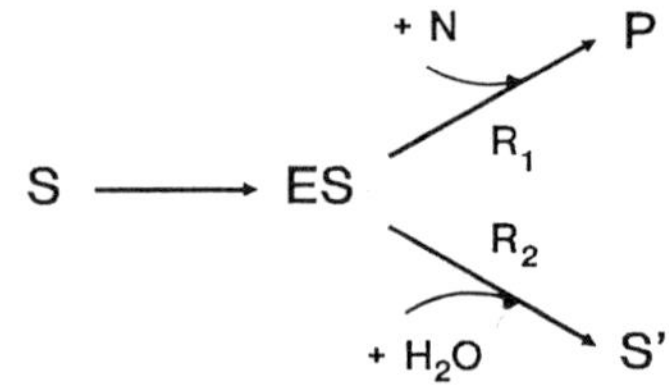

$$\text{conversion} \quad U_S = \frac{S_0 - S}{S_0}$$

$$\text{selectivity} \quad \sigma_{P,S} = \frac{P}{S_0 - S}$$

$$\text{yield} \quad \eta_{P,S} = \frac{P}{S_0}$$

$$\eta_{P,S} = \sigma_{P,S} * U_S$$

FIGURE 2. Definitions of conversion, selectivity, and yield.

influence selectivity (at given nucleophile concentration) but does influence the activity of the enzyme (FIGURE 3). On the other hand, at a given substrate concentration, increasing nucleophile concentration gives increasing activity and selectivity (FIGURE 4).

$$\text{Tyr-OEt} + \text{ArgNH}_2 \xrightarrow{\text{CPD-Y}} \text{TyrArg-NH}_2$$

Variation of Substrate

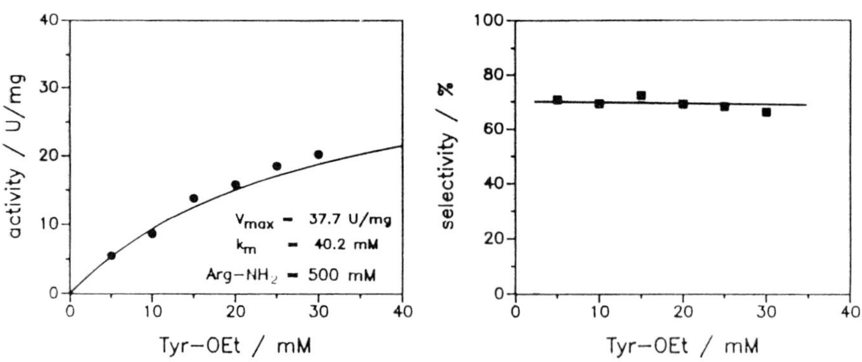

no product inhibition

FIGURE 3. Kinetics of the synthesis reaction Tyr-OEt + Arg-NH_2 (variation of the substrate). Conditions: 500 mM Arg-NH_2, pH 9.0, 0.1 mg/mL CPD-Y, $T = 25°C$.

The nucleophile arginine amide was specifically chosen because an enzyme is known that hydrolyzes the amide group from peptides but not from the nucleophile.[8] The kinetics of the enzymatic deprotection, as well as the kinetics of the secondary hydrolysis which becomes inevitable after deprotection, were measured thoroughly. The kinetic parameters obtained are summarized in FIGURE 5.

An enzyme-membrane reactor, which retains both the dissolved carboxypeptidase Y (CPD-Y) and the peptide amidase from flavedo (PAF), can be used in continuous operation.[9] All mass balances were established and are listed in FIGURE 6. These balances were calculated by solving the nonlinear homogeneous differential equations.[10] The results are shown in FIGURES 7 and 8.

If these enzymes are used in one reactor ("one-pot reaction," FIGURE 7), the yield reaches a maximum, since the simultaneous synthesis and enzymatic deprotection make the consecutive secondary hydrolysis of the deprotected dipeptide possible at high conversions. The initial selectivity is determined from the kinetic parameters. The drop in selectivity is due to the consumption of nucleophile as well as to the subsequent reaction. This situation suggests that the synthesis and the deprotection reaction should be separated in a stirred vessel cascade (FIGURE 8).

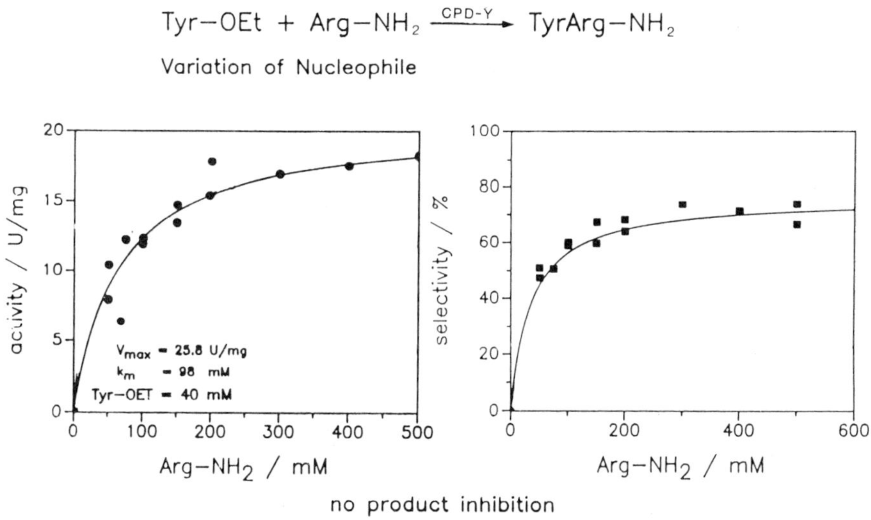

FIGURE 4. Kinetics of the synthesis reaction Tyr-OEt + Arg-NH_2 (variation of the nucleophile). Conditions: 40 mM Tyr-OEt, pH 9.0, 0.1 mg/mL CPD-Y, T = 25°C.

Secondary hydrolysis is now completely avoided, because the carboxypeptidase Y cannot hydrolyze the deprotected dipeptide. Due to the limited stability of CPD-Y, further enzyme was added from time to time during continuous operation, resulting in fluctuations in conversion. Results of the continuous experiments are shown in FIGURES 9 and 10.

Even better results with respect to space-time yield and selectivity were obtained using maleyl-tyrosineethylester (Mal-Tyr-OEt) as the substrate, which can easily be prepared using maleic anhydride,[11] and with arginineethylester (Arg-OEt) as the nucleophile. For this system, the inexpensive enzyme α-chymotrypsin could be used. Once again, all relevant kinetic parameters were measured (FIGURE 11).

The charged protecting group resulted not only in a dramatically increased solubility of the substrate (600 mmol) but also in a remarkably high activity of the enzyme (more than 3,700 U/mg). Solving the appropriate mass balances as described above, it becomes clear that the highest possible concentrations should be used in order to operate at a substrate-nucleophile ratio of 1:1 (FIGURES 12 and 13).The experimental values are in close agreement with the theory (FIGURE 13). These results suggest that a high conversion is possible without a significant drop in selectivity.

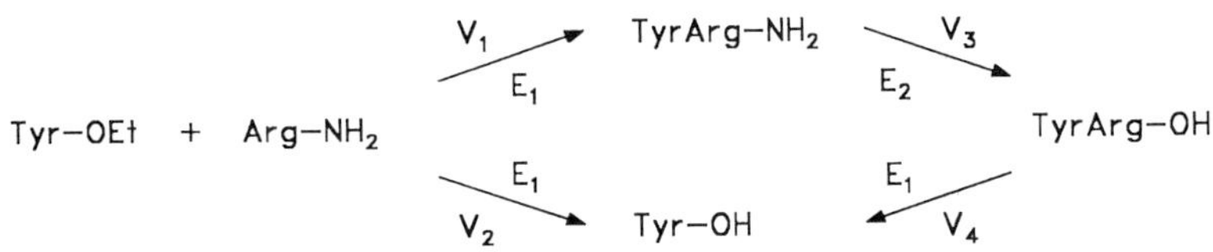

V_1 (E1)	V_{max}	:	37.7 U/mg	V_2 (E1)	$V_{max\,H_2O}$	:	10.2 U/mg
	K_{m_S}	:	40.2 mM		$K_{m\,H_2O}$	:	–
	K_{m_N}	:	98.0 mM				
V_3 (E2)	V_{max}	:	12.2 U/mg	V_4 (E1)	V_{max}	:	0.75 U/mg
	K_m	:	3.2 mM		K_m	:	32.2 mM

FIGURE 5. Kinetic parameters of the enzymatic synthesis reaction (aminolysis, v_1), the parallel reaction (hydrolysis, v_2), the deprotection (v_3) and of the secondary hydrolysis (v_4), refer to FIGURES 7 and 8. Conditions: (v_1, v_2) variation of the substrate: 500 mM Arg-NH$_2$, 5–30 mM Tyr-OEt, pH 9.0, T = 25°C, 0.1 mg/mL CPD-Y. (v_1, v_2) variation of the nucleophile: 40 mM Tyr-OEt, 50–500 mM Arg-NH$_2$, pH 9.0, T = 25°C, 0.1 mg/mL CPD-Y. (v_3) 5–30 mM TyrArg-NH$_2$, pH 9.0, T = 25°C, 0.038 mg/mL PAF. (v_4) 5–30 mM TyrArg, pH 9.0, T = 25°C, 0.1 mg/mL CPD-Y. Abbreviations: S = substrate, N = nucleophile, E = enzyme.

Thus, the reaction conditions in a continuously operated membrane reactor were set for approximately 95% conversion, resulting in a selectivity of nearly 80% (FIGURE 14). A space-time yield of more than 13 kg/(L × d) was achieved. In order to show that this technique can be scaled up, the reaction was also carried out in a 400-L stirred tank reactor. Deprotection is readily achieved using sulfuric acid (pH 0) at 50°C; 12 kg of the final product TyrArg were isolated.[12]

The protected dipeptide Mal-TyrArg-OEt can be used for further synthesis of the tripeptide TyrArgAla. The space-time yield achieved in the continuously oper-

$$\frac{d\,[\text{Tyr-OEt}]}{dt} = \text{conv.} - E_1\,(v_1 + v_2) \qquad \text{conv.} = \frac{c_0 - c_E}{\tau}$$

$$\frac{d\,[\text{Arg-NH}_2]}{dt} = \text{conv.} - E_1 * v_1$$

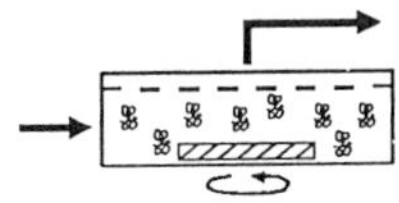

$$\frac{d\,[\text{TyrArg-NH}_2]}{dt} = \text{conv.} + E_1 * v_1 - E_2 * v_3$$

$$\frac{d\,[\text{TyrArg-OH}]}{dt} = \text{conv.} + E_2 * v_3 - E_1 * v_4$$

$$\frac{d\,[\text{Tyr-OH}]}{dt} = \text{conv.} + E_1\,(v_2 + v_4)$$

FIGURE 6. Mass balances of the synthesis reaction and the deprotection for the enzyme-membrane reactor: c_0 = inlet concentration, c_E = outlet concentration, τ = residence time, v_n = reaction rate, E_n = enzyme concentration, conv. = convective term.

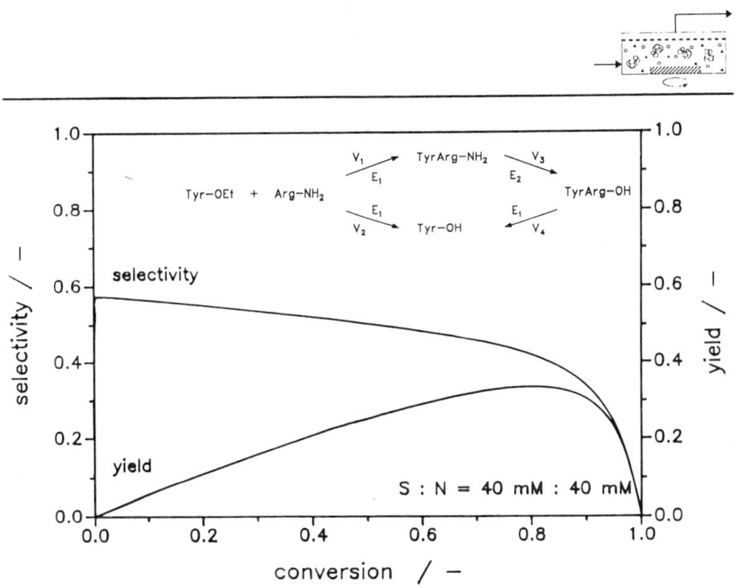

FIGURE 7. Selectivity and yield as function of conversion at continuous operation ("one-pot reaction"). Calculations with 40-mM Tyr-OEt (S) and 40-mM Arg-NH_2 (N). Kinetic parameters used, see FIGURE 5.

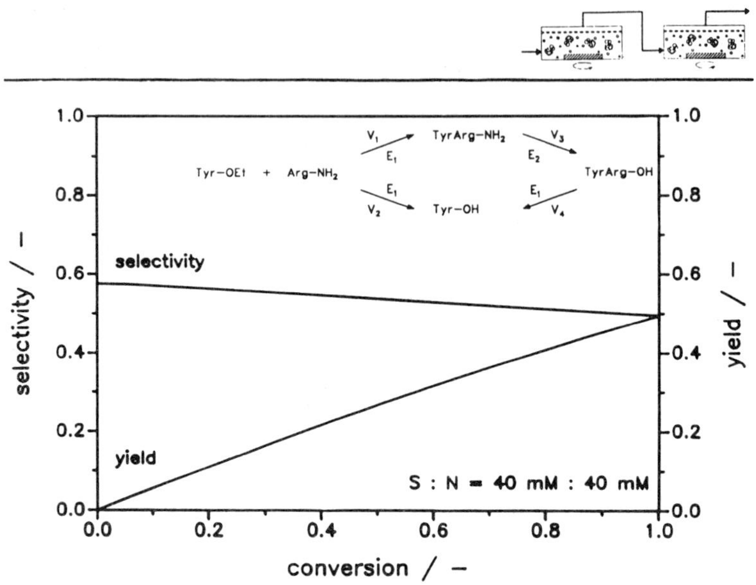

FIGURE 8. Selectivity and yield as function of conversion at continuous operation (stirred vessel cascade; CPD-Y in the first and peptide amidase in the second vessel of the cascade). Calculations, see FIGURE 7. Kinetic parameters used see FIGURE 5.

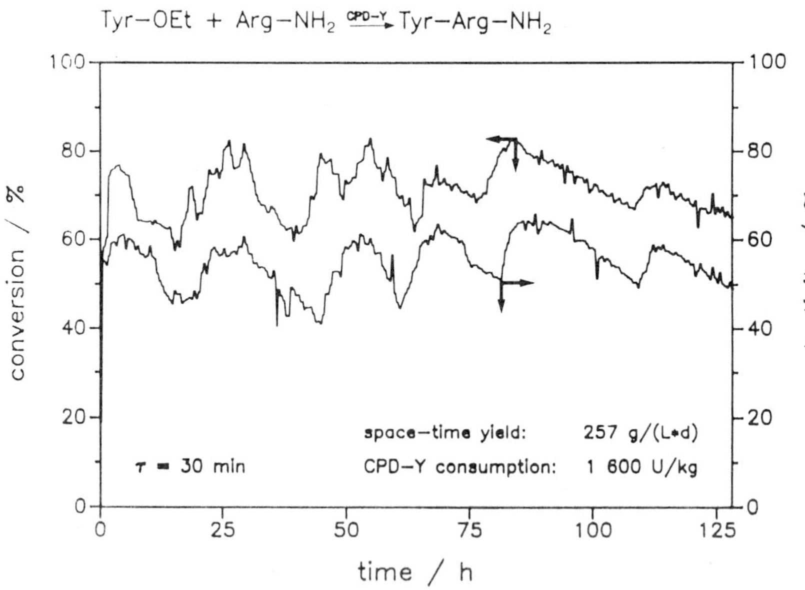

FIGURE 9. Continuous production of Kyotorphin amide in a 2-mL enzyme-membrane reactor. Conditions: 40 mM Tyr-OEt, 120 mM Arg-NH_2, 0.8 mg/mL CPD-Y, pH 9.0, T = 25°C, and residence time = 30 minutes.

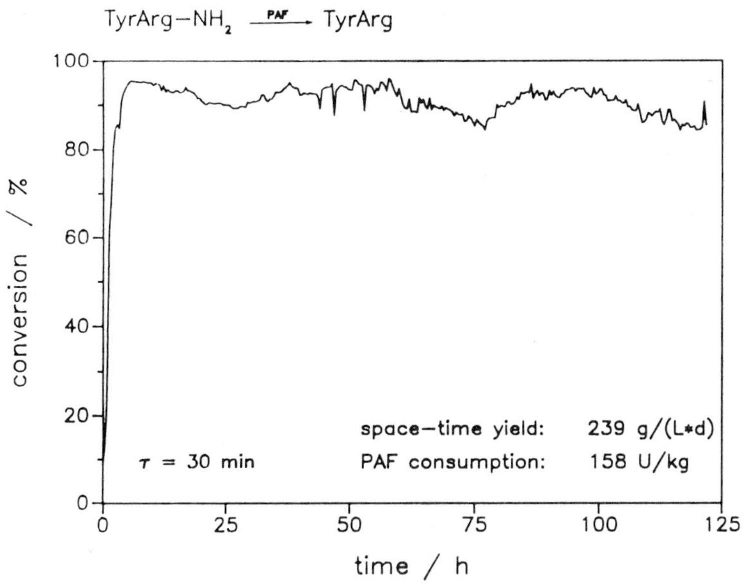

FIGURE 10. Continuous deamidation of Kyotorphin amide in a 2-mL enzyme-membrane reactor. Conditions: Product solution from reactor 1 (FIGURE 9), 0.25 mg/mL PAF, pH 9.0, T = 25°C, and residence time = 30 minutes.

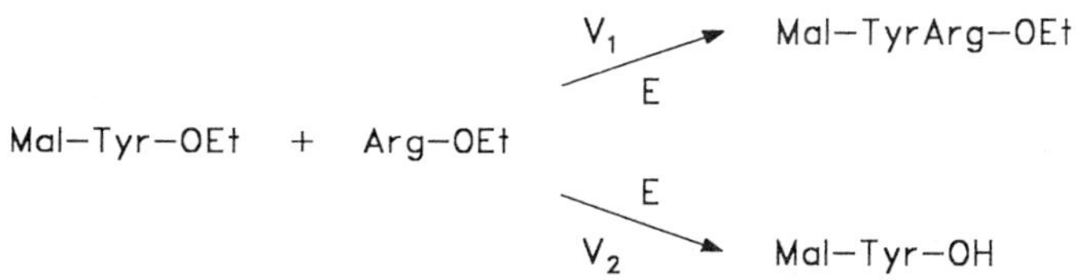

V_1 :	V_{max}	: 3758 U/mg		V_2 :	$V_{max\,H_2O}$	:	537 U/mg
	K_{m_N}	: 251 mM			$K_{m\,H_2O}$	:	–
	K_{m_S}	: 52.7 mM					

FIGURE 11. Kinetic parameters of the synthesis reaction (aminolysis, v_1) and the parallel reaction (hydrolysis, v_2), refer to FIGURES 12 and 13. Conditions: (v_1, v_2) variation of the substrate: 2000 mM Arg-OEt, 100–600 mM Mal-Tyr-OEt, pH 9.5, T = 25°C, 0.0006 mg/mL α-chymotrypsin. (v_1, v_2) variation of the nucleophile: 600 mM Mal-Tyr-OEt, 500–2000 mM Arg-OEt, pH 9.5, T = 25°C, 0.0006 mg/mL α-chymotrypsin. Abbreviations: S = substrate, N = nuclephile, E = enzyme.

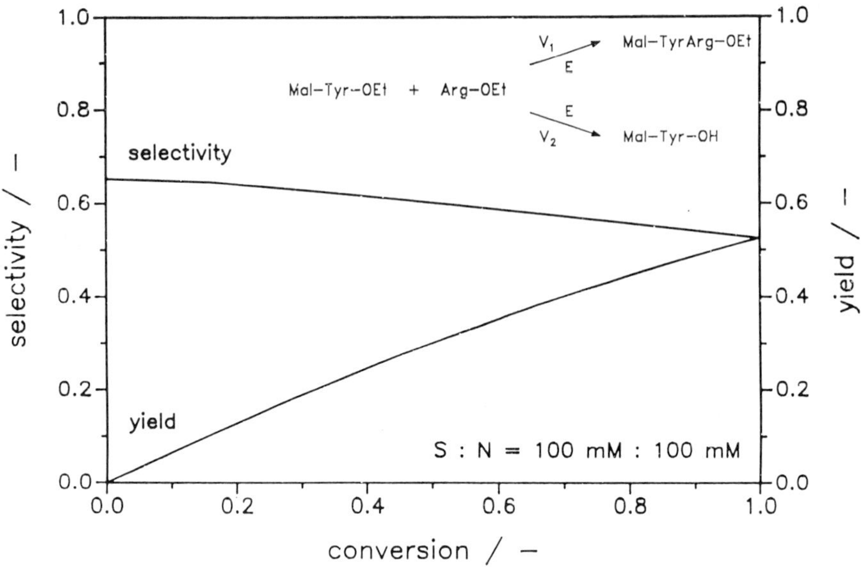

FIGURE 12. Selectivity and yield as function of conversion at continuous operation. Calculations with 100 mM Mal-Tyr-OEt (S) and 100 mM Arg-OEt (N). Kinetic parameters used, see FIGURE 11.

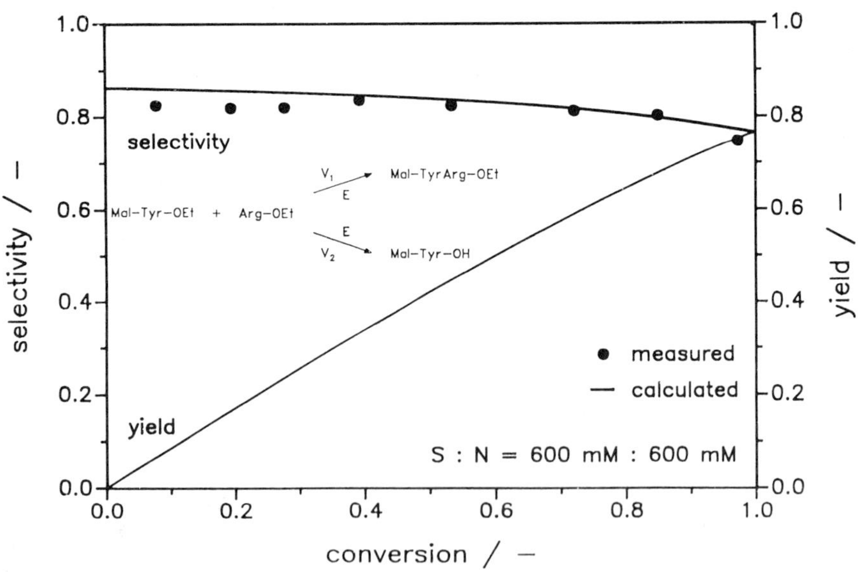

FIGURE 13. Selectivity and yield as function of conversion at continuous operation. Calculations with 600 mM Mal-Tyr-OEt (S) and 600 mM Arg-OEt (N). Kinetic parameters used see FIGURE 11.

ated enzyme-membrane reactor was 0.15 kg/(L × d), at a residence time of 15 minutes.

CONCLUSION

It has been demonstrated that kinetically controlled enzymatic peptide synthesis can be carried out in highly concentrated, homogeneous aqueous solution on a multi-kg scale. There are no technical limitations to further scale-up. Charged protecting groups facilitate solution of the substrates and the activity of some enzymes. Enzymatic deprotection is an elegant way to benefit from the selectivity of enzymes not only in synthesis but also in protection group chemistry. In the future, the field of enzymatic peptide synthesis will certainly gain from screening for sequence-specific peptidases as well as from genetic engineering or chemical modification of enzymes for even greater selectivity. Quantitative modeling of the relevant reactions is possible, as shown in this paper, so that the catalysts can be exploited in an optimal way, by minimizing nucleophile surplus and avoiding organic solvents.

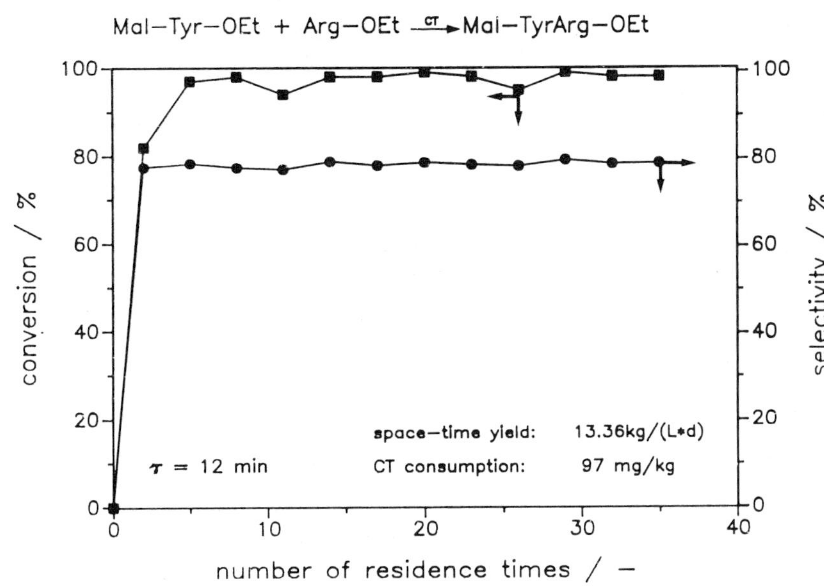

FIGURE 14. Continuous production of Mal-TyrArg-OEt in a 10-mL enzyme-membrane reactor. Conditions: 600 mM Mal-Tyr-OEt, 600 mM Arg-OEt, 0.2 mg/mL α-chymotrypsin, pH 9.5, T = 25°C, and residence time = 12 minutes.

SUMMARY

The kinetically controlled peptide synthesis of TyrArg and TyrArgAla was carried out in aqueous solution. Using the negatively charged N-terminal protecting group maleyl (Mal), solubilities of the substrates were drastically increased. Thus, it became possible to operate with a substrate-nucleophile ratio of 1:1 and in the continuous α-chymotrypsin-catalyzed production of Kyotorphin (TyrArg), a space-time yield of up to 13 kg/(L × d) was achieved. At 95% substrate conversion, the selectivity was 80%. Selectivity and yield as functions of residence time and enzyme concentration can be predicted using thoroughly measured kinetic data.

ACKNOWLEDGMENTS

The vital contribution of our industrial partner Degussa AG, Germany, where the scale-up experiments were carried out, is gratefully acknowledged. The authors also wish to express their special thanks to Prof. Kula and her coworkers for their most valuable support in all the amidase investigations and to Mrs. Burichter for her skillful technical assistance.

REFERENCES

1. TAKAGI, H., H. SHIOMI, H. UEDA & H. AMANO. 1979. Morphine-like analgesia by a new dipeptide, L-tyrosyl-L-arginine (Kyotorphin) and its analogue. Eur. J. of Pharmacol. **55:** 109–111.
2. JAKUBKE, H. D., P. KUHL & A. KÖNNECKE. 1985. Basic principles of protease catalyzed peptide bond formation. Angew. Chem. Int. Ed. Engl. **24:** 85–93.
3. MORIHARA, K. 1987. Using proteases in peptide synthesis. TibTech. **5:** 164.
4. KULLMANN, W. 1987. Enzymatic Peptide Synthesis. CRC Press. Boca Raton, Fla.
5. OYAMA, K. & K. KIHARA. 1984. A new horizon for enzyme technology. ChemTech (Feb.): 100–105.
6. WEST, J. B. & C.-H. WONG. 1987. Use of nonproteases in peptide synthesis. Tetrahedron Lett. **28:** 1629.
7. WEST, J. B., W. J. HENNEN, J. L. LALONDE, J. A. BIBBS, Z. ZHONG, E. F. MEYER, JR. & C.-H. WONG. 1990. Enzymes as synthetic catalysts: mechanistic and active-site considerations of natural and modified chymotrypsin. J. Am. Chem. Soc. **112:** 5313–5320.
8. STEINKE, D. & M. R. KULA. 1990. Selective deamidation of peptide amides. Angew. Chem. Int. Ed. Engl. **29:** 1139–1141.
9. KULA, M. R. & C. WANDREY. 1987. Continuous enzymatic transformation in an enzyme membrane reactor with simultaneous NADH-regeneration. Methods Enzymol. **136:** 9–21.
10. JOKSCH, B. 1991. Personal communication. Research Center Jülich, Germany.
11. BUTLER, P. J. G., J. I. HARRIS, B. S. HARTLEY & R. LEBERMAN. 1969. The use of maleic anhydride for the reverse blocking of amino groups in polypeptide chains. Biochem. J. **112:** 679–689.
12. FISCHER, A., A. SCHWARZ, C. WANDREY, A. S. BOMMARIUS, G. KNAUP & K. DRAUZ. 1991. Preparative-scale enzyme-catalyzed peptide synthesis using solubilizing N-terminal protecting groups. Biomed. Biochim. Acta **10/11:** 169–174.

Membrane-Assisted Synthesis of Chiral Drug Intermediates at the Multikilogram Scale

PAUL VAN EIKEREN, DAVID C. MUCHMORE, DANIEL J. BROSE, AND RAY H. COLTON

Bend Research, Inc.
64550 Research Road
Bend, Oregon 97701-8599

INTRODUCTION

Many biological substances (such as pharmaceuticals, sweeteners, insecticides, herbicides, and pheromones) contain a chiral center and thus exist as two enantiomers. Generally, only one enantiomer in a racemic mixture produces the desired biological activity.[1] Under the best of circumstances, the inactive enantiomer dilutes the biological activity, and the racemic mixture exhibits half-potency. Frequently, however, the unwanted enantiomer inhibits the biological activity of the active enantiomer and/or produces undesirable side effects.[2]

Despite the advantages of enantiomerically pure drugs, most synthetic chiral drugs are marketed as racemic mixtures because enantiomerically pure drugs are difficult and expensive to produce.[3] In light of the pitfalls associated with assessing the pharmacological properties of racemic substances[4,5] and the changing treatment of this problem by regulatory agencies,[6–9] pharmacological evaluations on enantiomerically pure substances have become mandatory. Consequently, a pressing need exists for reliable, efficient, and scalable methods to prepare enantiomerically pure substances.

Chiral alcohols are important synthons, intermediates used in the synthesis of chiral pharmaceuticals and pesticides. Chiral alcohols are produced conventionally by one of two methods: resolution or chiral synthesis. Resolution is accomplished in one of two ways: (1) the alcohols are allowed to react with a chiral derivatizing agent and the resulting diastereomers are separated by physical means, or (2) the alcohols are allowed to undergo a reaction in the presence of an enantioselective catalyst (e.g., kinetic resolution using an enzyme). In chiral synthesis, an asymmetrically substituted ketone is allowed to react with an enantioselective reducing agent that preferentially adds hydride to one face of the prochiral ketone. These methods have inherent drawbacks that affect the yield, chemical purity, and the enantiomeric purity of products when they are used to produce pharmaceutical-quality chiral alcohols (especially at large scale). For example, use of a highly activated substrate in kinetic resolution ensures high yield but lowers the chemical and enantiomeric purity of the product, due to competing reactions.

At Bend Research, we have developed a family of technologies that overcomes the inherent limitations of conventional methods used to make chiral alcohols. The concept behind these technologies, collectively termed membrane-assisted synthesis, is illustrated in FIGURE 1. Membrane-assisted synthesis is based on the premise that the yield and the chemical and enantiomeric purities of products in conventional synthesis reactions can be improved by using an auxiliary process. To achieve synergy between the synthesis reaction and the auxiliary process, the two processes must be chemically coupled—that is, they must share a common intermediate.

As an example, the rates of many enzyme-catalyzed reactions are inhibited by their products and the products are subject to undesired secondary reactions. The yield, productivity, and selectivity of such reactions can be improved through use of an auxiliary process that continuously and irreversibly separates and traps the reaction product (the common intermediate). To ensure that unwanted cross-reactions between the synthesis reaction and the auxiliary process (trapping process) do not occur, the two processes are maintained in separate compartments separated by a membrane. Because the membrane is highly selective, chemical species involved in the coupling process pass through, but all other species that would result in unwanted cross-reactions are rejected.

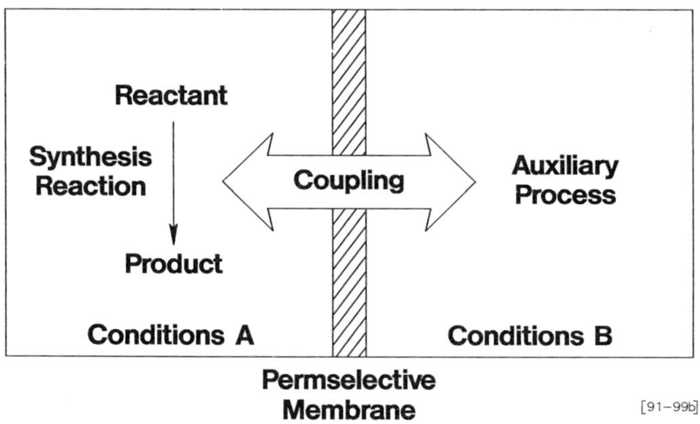

FIGURE 1. Elements of membrane-assisted synthesis technologies.

In this paper, we report results of work directed at producing a chiral alcohol, (*R*)-*endo*-norbornen-2-ol (FIGURE 2), using membrane-assisted synthesis. (*R*)-*endo*-Norbornen-2-ol, subsequently referred to as (*R*)-alcohol, was needed to synthesize a pharmaceutical candidate in clinical trials, so the technology used to produce (*R*)-alcohol had to meet stringent criteria. The process had to be (1) capable of producing (*R*)-alcohol of no less than 98% enantiomeric purity (enantiomeric excess) and no less than 99% chemical purity; (2) capable of producing (*R*)-alcohol at the tens-of-kilograms scale; (3) economical relative to competing methods; and (4) capable of being scaled up to production quantities.

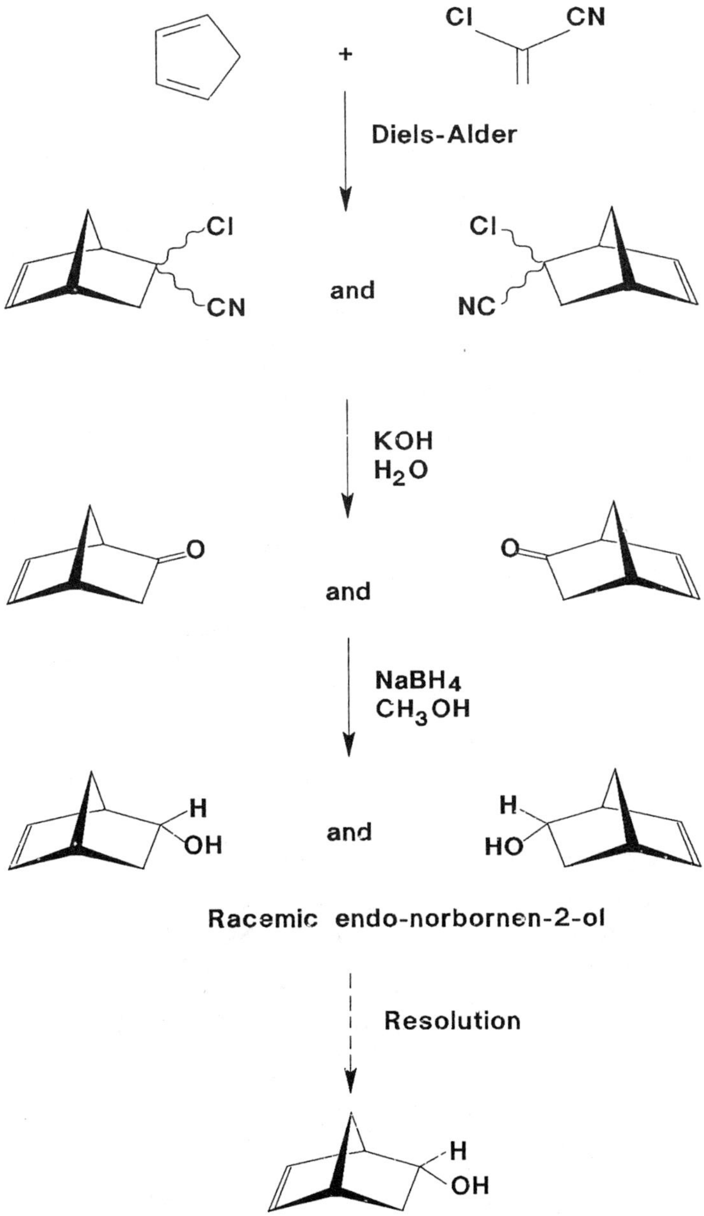

(R)-endo-Norbornen-2-ol [91–99a]

FIGURE 2. Preparation of racemic *endo*-norbornen-2-ol and (*R*)-*endo*-norbornen-2-ol.

TECHNICAL APPROACH

Racemic *endo*-norbornen-2-ol, subsequently referred to as alcohol, was prepared by the route outlined in FIGURE 2. 2-Chloro-2-cyanobicyclo-[2.2.1]-hept-5-ene was prepared via the Diels-Alder reaction of chloroacrylonitrile and cyclopentadiene in 91% yield and subsequently hydrolyzed to norbornen-2-one with potassium hydroxide in 91% yield using a reported procedure.[10] Norbornen-2-one was reduced stereoselectively to alcohol with sodium borohydride in 86% yield using a reported procedure.[11] It is important to recognize that two chiral centers are introduced during the Diels-Alder reaction (at the bridgehead) and that norbornen-2-one is a chiral molecule. Consequently, chiral synthesis of (*R*)-alcohol via the route illustrated in FIGURE 2 requires an enantiospecific Diels-Alder reaction. Since enantiospecific Diels-Alder reactions are difficult to achieve, production of (*R*)-alcohol can be accomplished only via resolution of norbornen-2-one followed by stereoselective reduction or resolution of alcohol. The resolution of norbornen-2-one by dehydrogenase-catalyzed reduction has been reported in the literature[12] but was deemed impractical for this application. Consequently, we chose to prepare (*R*)-alcohol by resolution of a racemic mixture of alcohol.

Kinetic resolution of alcohols is based on the principle that the acylation of one alcohol enantiomer or the hydrolysis of the corresponding ester derivative is more rapid than that of the other enantiomer. Enzymes (e.g., lipases, esterases, proteases) are generally the catalysts of choice for kinetic resolution because they exhibit high enantioselectivities. Enantioselectivity is expressed as the enantiomeric ratio, corresponding to the ratio of V_{max}/K_M for each enantiomer.[13] The enantiomeric purity obtained in kinetic resolutions is expressed as percent enantiomeric excess, which is the difference in weight between the two enantiomers in the mixture divided by the total weight of the two enantiomers.

Three principal considerations dictated the strategy selected for preparing

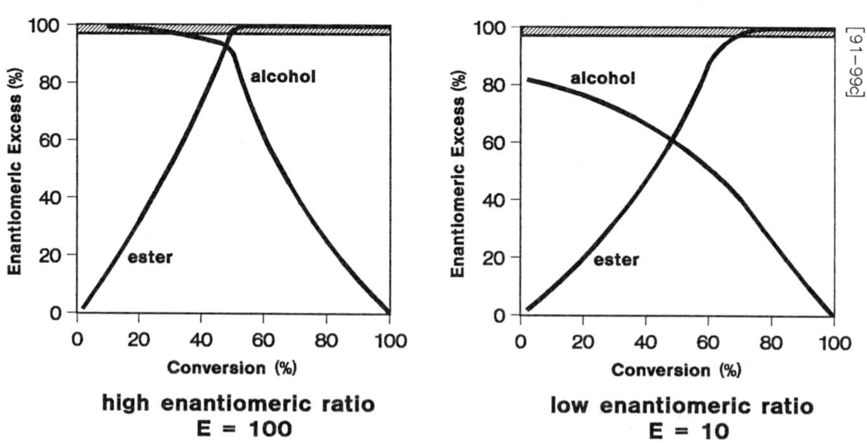

FIGURE 3. Consequences of enantiomeric ratio on the enantiomeric purities of alcohol produced and ester remaining during enzyme-catalyzed hydrolysis of a racemic mixture of alcohol esters.

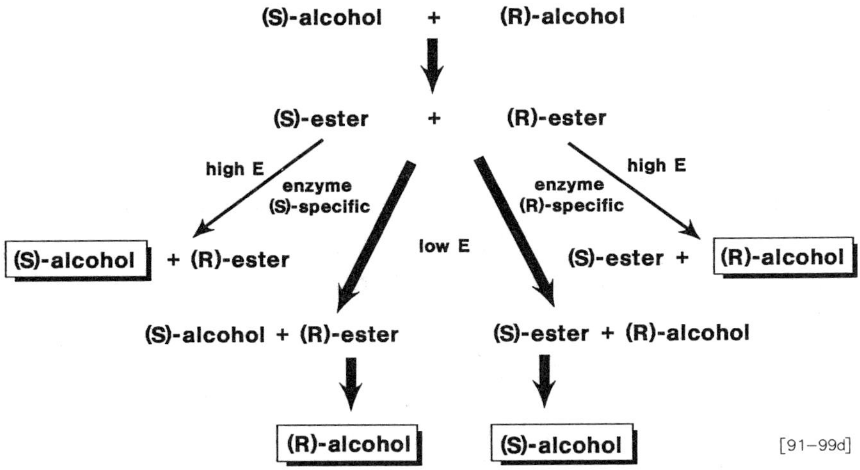

FIGURE 4. Consequences of enzyme specificity and enantiomeric ratio on the selection of routes to preparing (*R*)-alcohol.

(*R*)-alcohol: (1) the magnitude of the enantiomeric ratio in the preferred enzyme catalyst; (2) whether the preferred enzyme exhibits (*R*)- or (*S*)-specificity; and (3) whether the preferred enzyme can catalyze the acylation of alcohols as well as the hydrolysis of the corresponding esters. The consequences of these considerations are further described below.

The consequences of the first consideration (i.e., the magnitude of the enantiomeric ratio in the preferred enzyme catalyst), illustrated for the enzyme-catalyzed hydrolysis of alcohol esters, are shown in FIGURE 3. As this figure shows, by using an enzyme-catalyzed hydrolysis reaction that exhibits a high enantiomeric ratio (E = 100), the alcohol produced in the early stages of reaction ($<30\%$ conversion) is of sufficient enantiomeric purity to satisfy the requirements of pharmaceutical applications (shaded region). However, at higher conversions, the alcohol falls below target enantiomeric purity. By contrast, by using an enzyme-catalyzed hydrolysis reaction that exhibits a low enantiomeric ratio (E = 10), the alcohol produced in the hydrolysis fails to meet requisite criteria for enantiomeric purity at any conversion. It is important to note, however, that in both cases, the remaining ester will achieve the requisite enantiomeric purity by allowing the reaction to reach sufficiently high conversion. This is one of the major practical advantages of kinetic resolution.

The consequences of the second consideration, [i.e., whether the preferred enzyme exhibits (*R*) or (*S*) specificity], are illustrated for the hydrolysis of racemic alcohol esters[a] in FIGURE 4. FIGURE 4 shows that production of (*R*)-alcohol via enzymatic hydrolysis of racemic alcohol esters requires either (1) a high-enantiomeric-ratio (*R*)-specific enzyme operated at low conversion, or (2) a low-enantiomeric-ratio (*S*)-specific enzyme operated at high conversion. As will be described below, despite an extensive search, we were unable to identify a suitable enzyme and thus elected to produce (*R*)-alcohol via the two-stage process described below.

[a] An analogous scheme exists for the acylation of racemic alcohols.

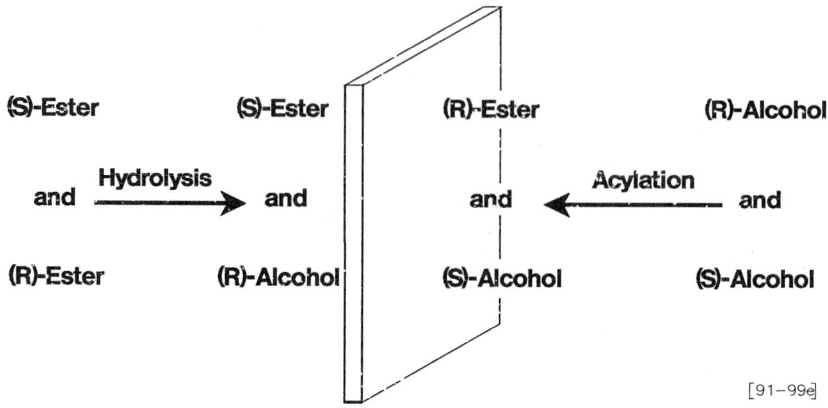

FIGURE 5. Complementary results achieved during hydrolysis and acylation when the same (R)-specific enzyme is used.

The consequences of the third consideration (i.e., whether the preferred enzyme can catalyze the acylation of alcohols as well as the hydrolysis of the corresponding esters) are illustrated in FIGURE 5. As this figure shows, the products of hydrolysis and acylation reactions are mirror images of each other when the same (R)-specific enzyme is used. Thus, both enantiomers can be produced in pure form using a two-stage resolution process involving acylation followed by hydrolysis.

We identified the preferred enzyme by screening our bank of commercially available enzymes (lipases, esterases, and proteases). Screening consisted of measuring the productivity and the enantiomeric ratio during the hydrolysis of racemic endo-norbornen-2-ol butyrate (ester). The lipase from *Candida cylindracea* was specific for the (R)-enantiomer, exhibited excellent productivity, and had an enantiomeric ratio of approximately 20 at 28°C and 40 at 4°C. On the basis of these measurements, cost considerations, and the enzyme's ready availability in large quantities, the lipase from *Candida cylindracea* was selected as the preferred enzyme.

Given the characteristics of the preferred enzyme, the requirement that both alcohol enantiomers were desired [although the (R)-enantiomer was most critical], and the three considerations described above, we chose the scheme outlined in FIGURE 6 as the preferred route to the preparation of (S)- and (R)-alcohol. The preferred route consists of the following sequence: (1) depletion of one enantiomer by enantioselective acylation, and (2) resolution of the resulting enriched ester by enantioselective hydrolysis.

RESULTS

Enzyme-Catalyzed Acylation and Membrane-Based Separation

Enantioselective acylation was achieved using an immobilized lipase in an organic solvent to catalyze acyl transfer from butyric anhydride to (R)-alcohol. Lipase, immobilized onto Celite R649 by adsorption, was suspended in heptane.

Measurements of the enantiomeric purity of (*S*)-alcohol and (*R*)-ester during the course of numerous experiments, illustrated in FIGURE 7, are consistent with an enantiomeric ratio of 40. During large-scale (10- to 15-kg) runs, acylations were carried to 55% conversion, producing "(*R*)-ester" of 78% enantiomeric excess and (*S*)-alcohol of greater than 98% enantiomeric excess. Enantiomeric excess of alcohol was determined by preparing diastereomeric derivatives by reaction between alcohol and (*R*)-methylbenzyl isocyanate and separating the diastereomers by gas chromatography.

Production of (*R*)-alcohol of target enantiomeric purity (no less than 98% enantiomeric excess) by the scheme outlined in FIGURE 6 requires quantitative separation of the (*S*)-alcohol from the "(*R*)-ester" produced in the acylation reaction. Failure to remove all (*S*)-alcohol following acylation contaminates the "(*R*)-ester" with (*S*)-alcohol, which limits the enantiomeric purity of the (*R*)-alcohol produced in the subsequent enzyme-catalyzed hydrolysis. Toward that end, we have devised a novel method[14] to separate the products of the reaction—"(R)-ester," unreacted (*S*)-alcohol, and spent acylating agent—using a membrane that selectively allows the passage of the (*S*)-alcohol and spent acylating agent, but rejects the "(*R*)-ester." The approach, illustrated in FIGURE 8, employs a hybrid membrane system in which the membrane-contactor unit contacts the heptane solution exiting the enzyme reactor with water to extract (*S*)-alcohol, and the reverse-osmosis (RO) trapping unit concentrates the (*S*)-alcohol—that is, traps it in the (*S*)-alcohol

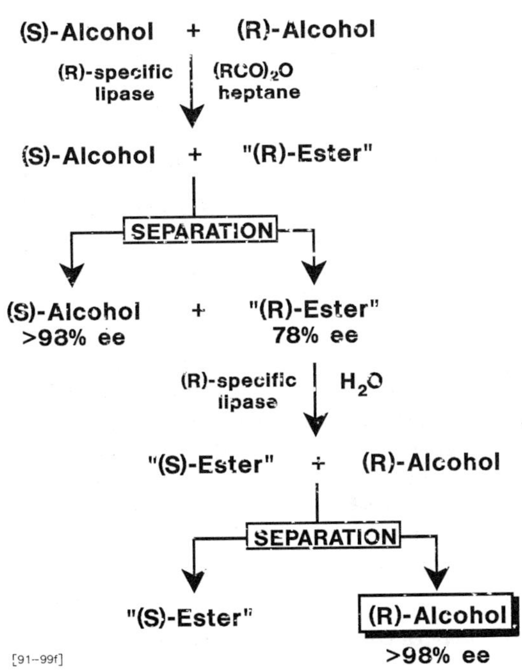

FIGURE 6. Preferred route for the production of (*S*)- and (*R*)-alcohol via an acylation/hydrolysis sequence employing (*R*)-selective lipase.

reservoir—and provides a continuous source of fresh water for further extraction. This unit operation is accomplished by the apparatus shown in FIGURE 9.

A comparison of the performance of the hybrid membrane system and a conventional liquid-liquid extraction scheme illustrates the utility of the hybrid membrane system for the separation of (*S*)-alcohol from (*R*)-ester. For this purpose, consider an enzyme-catalyzed reaction that produces an equimolar mixture of (*S*)-alcohol and (*R*)-ester (enantiomeric ratio of infinity). Given that the distribution coefficients (organic/aqueous) are 100 to 1 for the (*R*)-ester and 1 to 1 for the (*S*)-alcohol, the calculated concentrations of (*R*)-ester and (*S*)-alcohol in the organic phase (heptane) in a series of conventional extraction steps are shown in TABLE 1. TABLE 1 also shows the calculated enantiomeric excess of the (*R*)-alcohol that would be produced by hydrolyzing all the (*R*)-ester present in the organic phase at each extraction stage, taking into account the (*S*)-alcohol that contaminates the

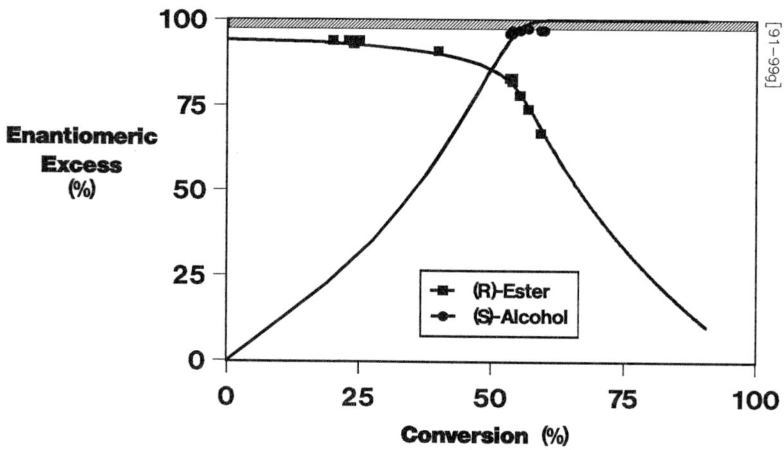

FIGURE 7. Enantiomeric purity of alcohol and ester during lipase-catalyzed acylation using butyric anhydride in heptane. Curves are calculated[13] for a reaction with an enantiomeric ratio of 40.

organic phase. The calculation shows that at least seven extraction steps are required to obtain an (*R*)-alcohol of requisite enantiomeric excess. By contrast, FIGURE 10 shows that by use of the RO trapping unit to trap the (*S*)-alcohol, the (*S*)-alcohol can be quantitatively removed in a single step with minimal added processing time over that required when the unit is not used. FIGURE 10 also shows that if the RO trapping unit is not used, membrane-based extraction (3 volumes of water to 1 volume of heptane) removes approximately 75% of the (*S*)-alcohol, in accordance with predictions based on distribution coefficients.

Enzyme-Catalyzed Hydrolysis and Membrane-Based Separation

Enantioselective hydrolysis of (*R*)-ester, enriched in the (*R*)-enantiomer (78% enantiomeric excess), was achieved using the same (*R*)-specific lipase from *Candida*

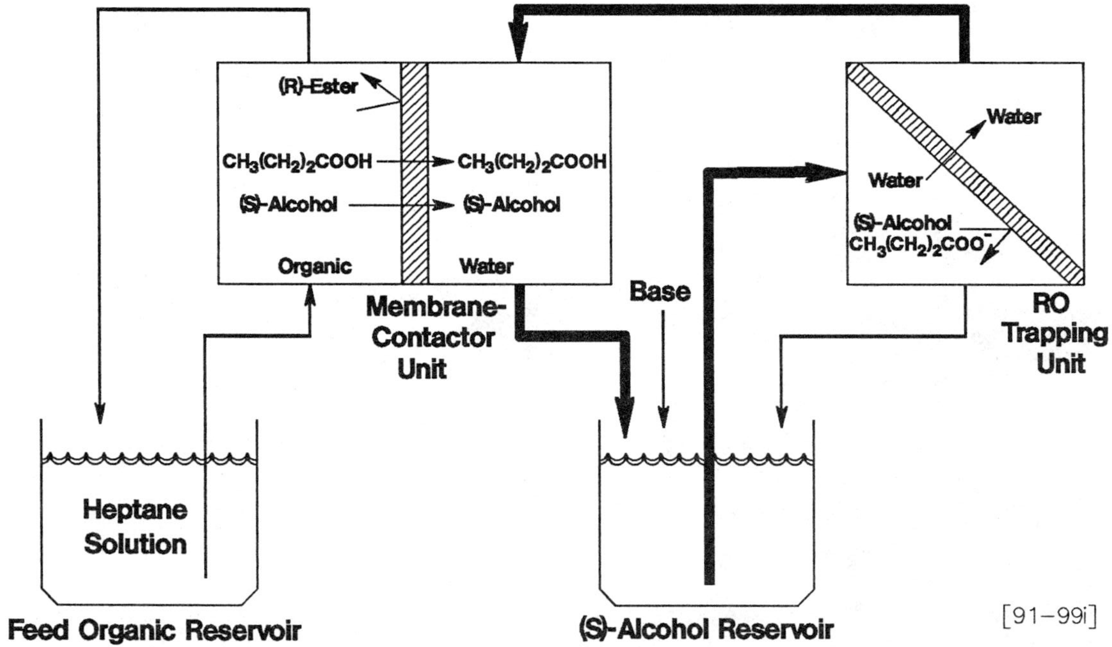

FIGURE 8. Hybrid membrane system for the separation of (*S*)-alcohol from (*R*)-ester.

FIGURE 9. Hybrid membrane system.

cylindracea. However, unlike during the acylation reaction (in which lipase is immobilized), during the hydrolysis reaction lipase is dissolved in water, and the reaction mixture consists of a two-phase (organic-aqueous emulsion) system. FIGURE 11 shows the calculated enantiomeric excess of (*R*)-alcohol and (*S*)-ester during the hydrolysis. (*R*)-Alcohol of target enantiomeric excess was obtained by carrying the hydrolysis to 80% conversion.

Lipase-catalyzed hydrolysis, when carried out in a conventional batch reactor, has a major drawback and carries a significant practical risk. The drawback is that

TABLE 1. Calculated Concentrations of (*R*)-Ester and (*S*)-Alcohol in the Organic Phase in a Series of Conventional Liquid-Liquid Extraction Steps

Extraction Step	Organic Phase (*R*)-Ester (mol)	Organic Phase (*S*)-Alcohol (mol)	Enantiomeric Excess (%)
0	0.500	0.500	0.0
1	0.495	0.250	32.9
2	0.490	0.125	59.4
3	0.485	0.063	77.2
4	0.481	0.031	87.8
5	0.476	0.016	93.6
6	0.471	0.008	96.7
7	0.466	0.004	98.3
8	0.462	0.002	99.2

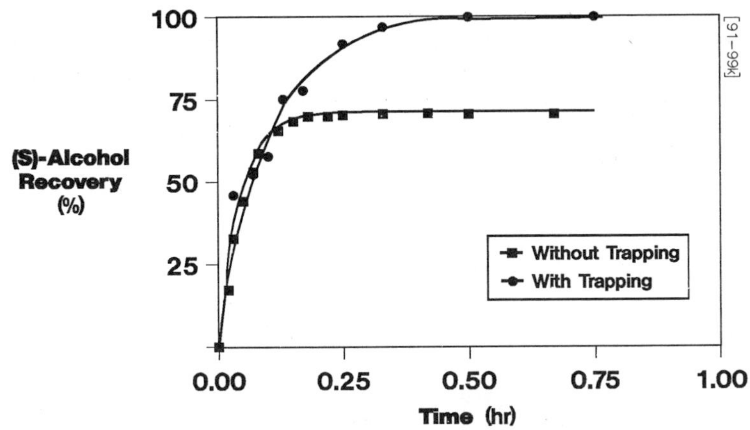

FIGURE 10. Impact of RO trapping unit during the removal of (S)-alcohol.

many lipase-catalyzed hydrolysis reactions are inhibited by the alcohol and carboxylic acid products, which slow down the reaction at higher conversion. Moreover, since it is difficult to measure the progress of the reaction accurately, there is a risk that the hydrolysis reaction will be allowed to go too far (beyond 80% conversion), which, as shown in FIGURE 11, produces (R)-alcohol of insufficient enantiomeric purity. To overcome this drawback and eliminate the risk, the hydrolysis reaction was carried out in the membrane reactor illustrated in FIGURE 12. The membrane reactor employs a permselective membrane that rejects the ester in the emulsion but allows

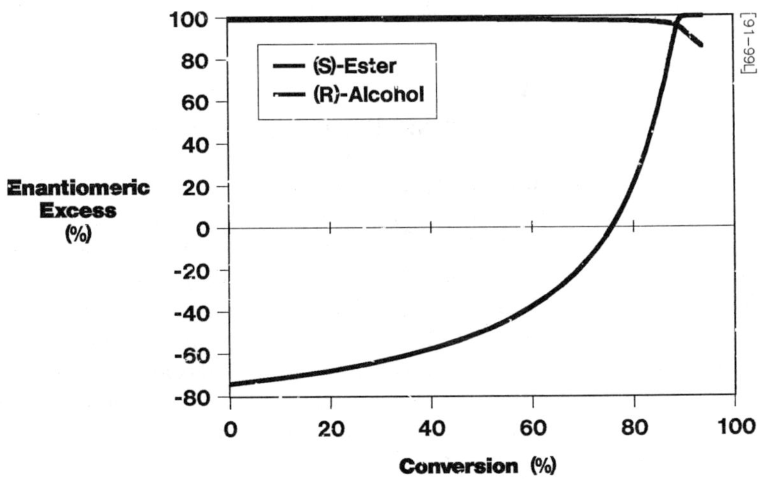

FIGURE 11. Enantiomeric purity of (R)-alcohol and (S)-ester during lipase-catalyzed hydrolysis of ester highly enriched in the (R)-enantiomer. Curves are calculated for a reaction with an enantiomeric ratio of 40.

passage of the (*R*)-alcohol and butyric acid. The (*R*)-alcohol is isolated and concentrated using the membrane-based system shown in FIGURE 9. By operating the hydrolysis reaction in a membrane reactor, we avoided losing enzyme productivity due to product inhibition. Furthermore, since product removal is continuous, we were able to isolate (*R*)-alcohol as a series of fractions. The fractions were then analyzed for enantiomeric purity, and those fractions containing (*R*)-alcohol of target purity were pooled, whereas fractions containing (*R*)-alcohol below target purity were recycled.

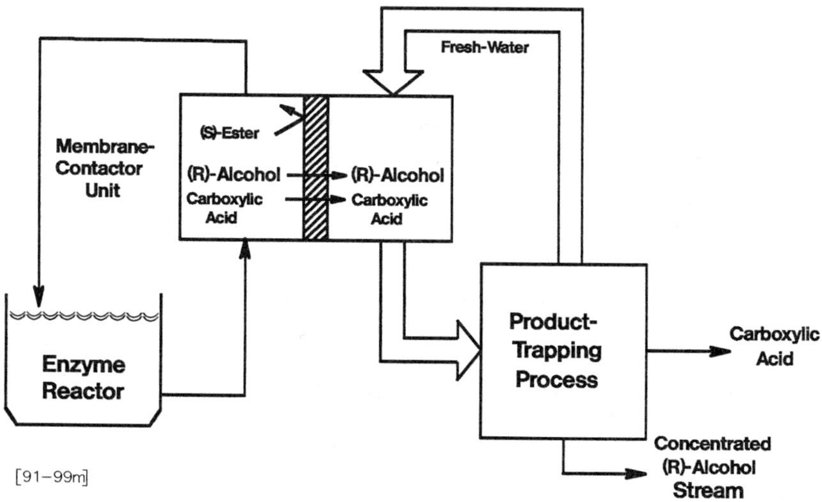

FIGURE 12. Membrane reactor for the lipase-catalyzed hydrolysis of ester highly enriched in the (*R*)-enantiomer.

CONCLUSIONS

Membrane-assisted synthesis offers a practical method to prepare chiral compounds. At Bend Research, we have demonstrated that this technology can be used to produce a range of chiral alcohols (e.g., quinuclidin-3-ol, thiolane-3-ol, unsubstituted and substituted 1-phenyl-ethanol derivatives, glycidol) of significantly higher enantiomeric and chemical purities and at higher recoveries than is possible using conventional resolution approaches. In conjunction with a pharmaceutical-company client, membrane-assisted synthesis was used to process racemic *endo*-norbornen-2-ol at a tens-of-kilograms scale to produce (*R*)- and (*S*)-*endo*-norbornen-2-ol of $>98\%$ enantiomeric purity and $>99\%$ chemical purity with recoveries as high as 85%, based on available enantiomer.

Membrane-assisted synthesis has also been shown to be useful in other applications. It has been demonstrated that membrane-assisted synthesis is a practical method to regenerate expensive reagents used in chemical and biochemical syntheses, including the expensive cofactor NADH, nicotinamide adenine dinucleotide, used in many enzyme-catalyzed oxidoreductase reactions.[15] Membrane-assisted synthesis has also been used to enhance the productivity of mammalian cells by continuously removing toxic ammonia produced during the culturing of cells.[16,17]

REFERENCES

1. SMITH, D. F., Ed. 1989. Handbook of Stereoisomers: Therapeutic Drugs. CRC Press. Boca Raton, Fla.
2. JAMALI, F., R. MEHVAR & F. M. PASUTTO. 1989. Enantioselective aspects of drug action and disposition: therapeutic pitfalls. J. Pharm. Sci. **78**(9): 695–715.
3. GROSS, M. 1989. Significance of drug stereochemistry in modern pharmaceutical research and development. Ann. Rep. Med. Chem. **25:** 323–331.
4. ARIENS, E. J. 1984. Stereochemistry, a basis for sophisticated nonsense in pharmacokinetics and clinical pharmacology. Eur. J. Clin. Pharmacol. **26**(6): 663–668.
5. ARIENS, E. J. 1986. Stereochemistry: a source of problems in medicinal chemistry. Med. Res. Rev. **6**(4): 451–466.
6. 1990. PMA Ad Hoc Committee on Racemic Mixtures. Comments on enantiomerism in the drug development process. Pharm. Tech.: 46.
7. BORMAN, S. 1990. Chirality emerges as a key issue in pharmaceutical research. Chem. Eng. News: 9.
8. DECAMP, W. H. 1989. The FDA perspective on the development of stereoisomers. Chirality **1**(1): 2–6.
9. DEUTSCH, D. H. 1991. Chiral drugs: the coming revolution. Chemtech **21**(3): 157–159.
10. LE DRIAN, C. & A. E. GREENE. 1982. Efficient stereocontrolled syntheses of racemic and natural brefeldin-A. J. Am. Chem. Soc. **104:** 5473–5483.
11. OBERHAUSER, T., M. BODENTEICH, K. FABER, G. PENN & H. GRIENGL. 1987. Enzymatic resolution of norbornane-type esters. Tetrahedron **43:** 3931–3944.
12. IRWIN, A. & B. JONES. 1976. Stereoselective horse liver alcohol dehydrogenase catalyzed oxidoreductions of some bicyclic [2.2.1] and [3.2.1] ketones and alcohols. J. Am. Chem. Soc. **98:** 8476–8482.
13. CHEN, C.-S., Y. FUJIMOTO, G. GIRDANKAS & S. SIH. 1982. Quantitative analyses of biochemical kinetic resolutions of enantiomers. J. Am. Chem. Soc. **102:** 7294–7299.
14. VAN EIKEREN, P., D. J. BROSE & R. J. RAY. 1991. Selective aqueous extraction of organics coupled with trapping by membrane separation. U.S. Patent No. 5,041,227.
15. VAN EIKEREN, P. 1989. Use of regeneration of reagents using coupled reactions and permselective barriers. European Patent Application No. EP 349204.
16. VAN EIKEREN, P. & J. M. RADOVICH. 1991. Ammonia removal from mammalian cell cultures. U.S. Patent No. 5,071,561.
17. BROSE, D. J. & P. VAN EIKEREN. 1990. A membrane-based method for removal of toxic ammonia from mammalian cell culture. Appl. Biochem. Biotechnol. **24:** 457–468.

Production of Cyclodextrins from Unliquefied Cornstarch Using Cyclodextrin Glycosyltransferase in a Membrane Bioreactor

TAE-JONG KIM, YUN-DONG LEE, AND HAK-SUNG KIM[a]

Department of Biotechnology
Korea Advanced Institute of Science and Technology
373-1, Kusung-Dong, Yusung-Ku
Taejon, Korea

INTRODUCTION

Cyclodextrins are cyclic and doughnut shaped with an interior cavity whose size and shape are determined by the number of glucose units.[1] The interior of cyclodextrins is relatively apolar compared to water, and thereby cyclodextrins have a specific function of forming an inclusion complex with many organic substances. The application of cyclodextrins has been rapidly increasing in the food, cosmetic, pharmaceutical, agrochemical, and plastics industries as emulsifiers, antioxidants, and stabilizing agents.[2]

In our previous paper,[3] cyclodextrins were efficiently produced from unliquefied corn starch without by-product in an attrition bioreactor in which hydrolysis of starch and synthesis of cyclodextrins by cyclodextrin glycosyltransferase (CGTase) occur simultaneously.

We have been investigating various approaches to increase the production yield of cyclodextrins from unliquefied corn starch. It was found that CGTase is severely inhibited by cyclodextrins.[4] The relative activity of CGTase decreased to 50% in the presence of 6.0 g/L β-cyclodextrin. Based on the above observation, a membrane bioreactor system was employed to increase the production yield of cyclodextrins by reducing product inhibition. Details are reported here.

MATERIALS AND METHODS

Materials

Cyclodextrins and cornstarch were obtained from Sigma (St. Louis, Mo.), and soluble starch was purchased from Kanto Chemical Co. (Tokyo, Japan). Yeast extract and casitone were from Difco (Detroit, Mich.). Acetonitrile and water of high-performance liquid chromatography (HPLC) grade were obtained from Burdick & Jackson (Muskegon, Mich.). All other chemicals were analytical grade.

[a] Author to whom all correspondence should be addressed.

Preparation of the Crude CGTase

The crude enzyme was prepared as described in previous papers.[3,5] Crude enzyme powder was obtained by lyophilization, and its activity was determined to be about 15 units/mg solid.

Dextrinizing Activity of CGTase

The dextrinizing activity of crude enzyme for soluble starch was measured using the method of Nakamura and Horikoshi[6] with slight modification as reported in a previous paper.[3]

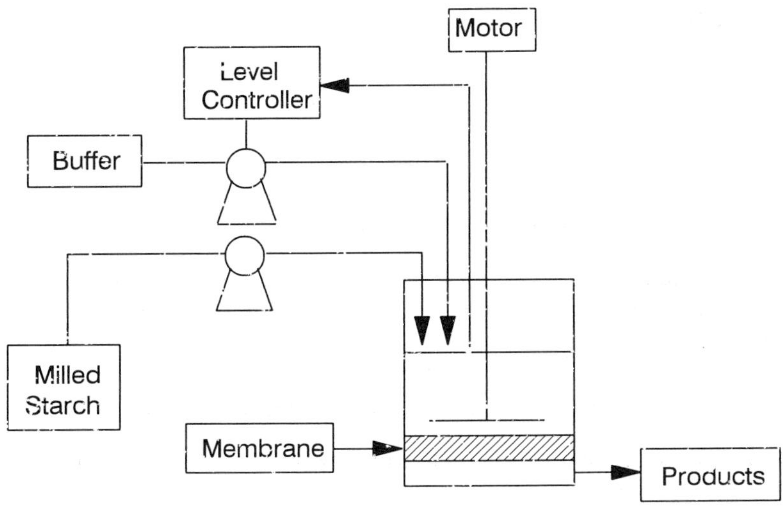

FIGURE 1. Schematic diagram of a membrane bioreactor.

Milling of Cornstarch

Cornstarch was milled for 3 hours in a 350-mL glass vessel manufactured in our laboratory.[3] Glass beads of 3 mm diameter were used as a milling medium, and agitation speed was 300 rpm. The volume of glass beads was 30% (v/v).

Membrane Bioreactor

A 200-mL ultrafiltration kit (Amicon, Danvers, Mass.) was used as a membrane bioreactor, and membrane cutoff size was 10,000 daltons. As shown in FIGURE 1, reaction mixture was continuously removed through a membrane using a pump, and the level of reaction mixture in the bioreactor was maintained constant using a level

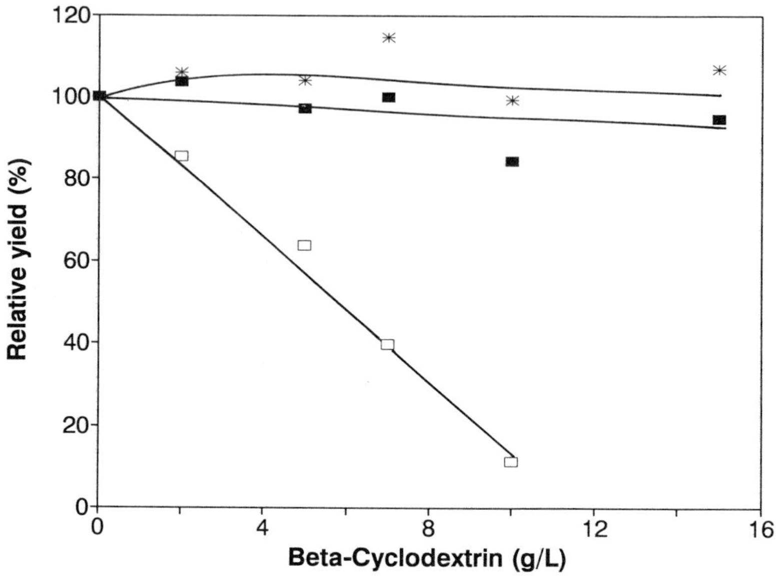

FIGURE 2. Effect of β-cyclodextrin concentration on the synthetic activity of CGTase. Symbols are α-cyclodextrin (■); β-cyclodextrin (□); γ-cyclodextrin (∗).

controller because the filtration rate varied during operation time. The filtered reaction mixture was collected using a fraction collector for analysis. The temperature was controlled at 45°C, and CGTase loading was 500 units per gram starch.

Production of Cyclodextrins in a Membrane Bioreactor

Cyclodextrins were produced with or without continuous feeding of milled starch in a membrane bioreactor. In the former case, fixed amounts of starch and CGTase were added into the membrane bioreactor, and reaction mixture was removed through a membrane using a pump. In the latter case, milled starch solution was continuously fed into the bioreactor, and reaction mixture was removed through a membrane.

Analysis

Concentration of cyclodextrins was determined by HPLC (Waters Assoc., Milford, Mass.) as reported in a previous paper.[3]

RESULTS AND DISCUSSION

Inhibition of CGTase by Cyclodextrins

As previously found,[3] the production rate of cyclodextrins decreased considerably after 10 hours of reaction. Only a 6% increase in the yield occurred during the last 14 hours even though the CGTase activity remained at a significant level. Thus, reduction in the production rate of cyclodextrins was thought to be due to the inhibition of CGTase by the cyclodextrins produced. Since CGTase prepared from *Bacillus* sp. BE101 synthesized mainly β-cyclodextrin, an inhibition study was carried out using β-cyclodextrin as an inhibitor.

As shown in FIGURE 2, CGTase was severely inhibited by the product, i.e., β-cyclodextrin. A 50% reduction in the CGTase activity was observed at 6.0 g/L of β-cyclodextrin.

Production of Cyclodextrins in a Membrane Bioreactor

As an approach to reduce the inhibition of CGTase, cyclodextrins produced were continuously removed by ultrafiltration while starch particles and CGTase were retained within the reactor. As shown in FIGURE 3, the production yield of cyclodextrins reached up to 55% in the membrane bioreactor when the starch concentration was 15%, while the yield was about 35% in the reactor without ultrafiltration. Thus, it is thought that the membrane bioreactor could be successfully applied to the production of cyclodextrins from unliquefied cornstarch.

Continuous Feeding of Substrate in a Membrane Bioreactor

In order to efficiently utilize the CGTase retained in the reactor, the substrate was continuously fed into the bioreactor. The initial concentration of starch in the

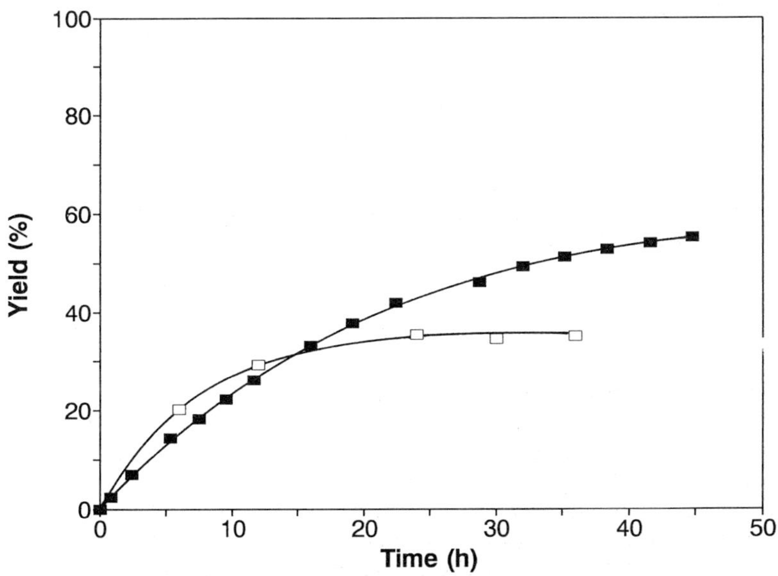

FIGURE 3. Production yield of cyclodextrins with (■) or without (□) ultrafiltration.

bioreactor was fixed at 7%, and starch concentration in the reservoir was 0.7% or 3%. The filtration rate through a membrane is severely affected by the concentration of starch in the bioreactor, which also influences the concentration of cyclodextrins in the bioreactor. Thus, the initial filtration rate was fixed at 50 mL/hour and milled starch was continuously fed until further filtration became impossible. FIGURE 4 shows the production profiles of cyclodextrins when 3% and 0.7% milled starch were continuously fed, respectively. Total amount of cyclodextrins produced per CGTase loaded in the reactor increased 3.6 and 2.4 times for 3% and 0.7% starch feeding, when compared to that without substrate feeding. Production of cyclodextrins in a membrane bioreactor with higher filtration rate is in progress in our laboratory.

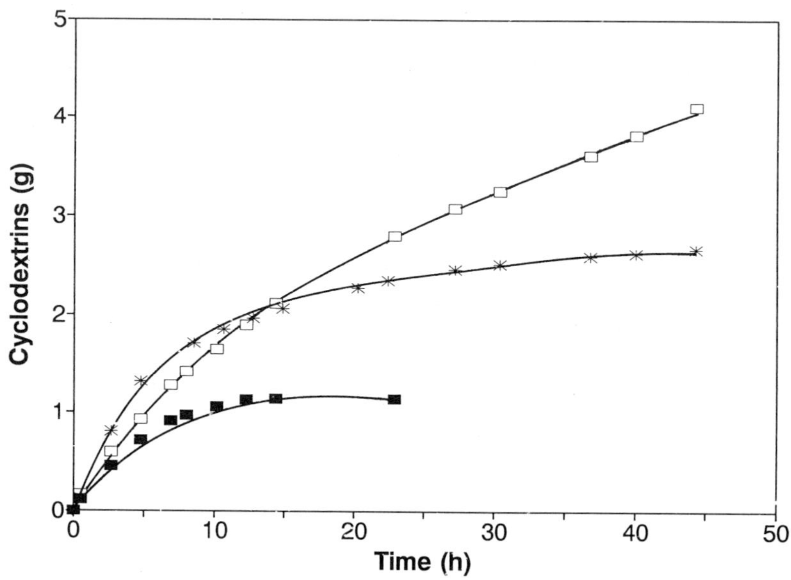

FIGURE 4. Amount of cyclodextrins produced in a membrane bioreactor with continuous feeding of starch. Concentrations of feeding starch solution were: 3% (□); 0.7% (*). Filled square indicates batch operation.

SUMMARY

Cyclodextrin glycosyltransferase (CGTase) of *Bacillus* sp. BE101 was severely inhibited by cyclodextrins. The relative activity of CGTase decreased to 50% in the presence of 6.0 g/L β-cyclodextrin. In order to increase the production yield of cyclodextrins by reducing product inhibition, cyclodextrins were produced in a membrane bioreactor. Maximum yield of cyclodextrins from 15% unliquefied corn-starch reached up to 55% (w/w) in the membrane bioreactor. The productivity of cyclodextrins per unit enzyme loaded in the reactor was significantly enhanced by continuous feeding of substrate.

REFERENCES

1. BENDER, M. L. & M. KOMIYAMA. 1978. Cyclodextrin Chemistry. Springer-Verlag. Berlin Heidelberg & New York.
2. NAGATOMO, S. 1985. Chem. Econ. Eng. Rev. **17:** 28–34.
3. LEE, Y. D. & H. S. KIM. 1991. Biotechnol. Bioeng. **37:** 795–801.
4. LEE, Y. D. & H. S. KIM. 1992. Biotechnol. Bioeng. **39:** 977–982.
5. LEE, Y. D. & H. S. KIM. 1991. Enzyme Microb. Technol. **13:** 499–503.
6. NAKAMURA, N. & K. HORIKOSHI. 1976. Agric. Biol. Chem. **40:** 755–757.

Immobilized Laccase for Must and Wine Processing[a]

A. LANTE, A. CRAPISI, G. PASINI, A. ZAMORANI, AND P. SPETTOLI

Dipartimento di Biotecnologie Agrarie
Università di Padova
via Gradenigo, 6
35131 Padova, Italy

INTRODUCTION

Whereas the use of enzyme preparations in the food industry is a well-established, rapidly expanding technology, their utilization in enology is less frequent. The "classic" wine industry is in fact still largely based on traditional methods. Moreover, musts and wines represent a hostile environment for purified enzymes, i.e., low pH, high ethanol concentration, presence of sulfur dioxide and tannins. White and "rosè" wines are known to be more sensitive than red wines to browning and to the development of madeirized flavor, which are mainly due to the presence of polyphenols. Aiming at the replacement of traditional stabilization treatments (i.e., sulfur dioxide, fining agents) with mild technologies to prevent madeirization, unconventional systems like the use of oxidative enzymes (chiefly laccase) have been proposed.[1,2] However, the residual enzyme activity in wine during storage or, in the case of musts, its inactivation by heat treatment could lead to negative effects, i.e., color browning. Therefore, the use of immobilized oxidative enzymes that can be easily recovered from the reaction mixture and reused could be exploited.

Studies have been reportedly carried out on fungal laccase immobilized by covalent attachment to CNBr-activated Sepharose 4B and by adsorption to Concanavalin A-Sepharose,[3] covalently bound to a diatomaceous silica product (R-650 Celite),[4] immobilized on porous glass beads[5,6] to detoxify the environment polluted with aromatic and phenolic compounds of agricultural or industrial origin.

Taking into account our previous trials on the approach to the use of immobilized oxidative enzymes in white wine stabilization,[7,8] in the present work we report the results on a commercial laccase (*p*-diphenol: oxygen oxidoreductase; EC 1.10.3.2) immobilized by adsorption on silica gel carrier followed by treatment with glutaraldehyde and by covalent binding on polymer carrier VA-Hydroxy Biosinth, and CNBr-activated Sepharose 4B, respectively. In addition, the possibility of an enzymatic removal of phenolics was tested in model solution, in must, and in wine.

[a] Research supported by the Consiglio Nazionale delle Ricerche.

MATERIALS AND METHODS

Enzyme

A commercial laccase from *Pyricularia oryzae* (Sigma Chemical, St. Louis, Mo.) was used without further purification.

Reagents

Silica gel GF_{254} was from Merck; CNBr-activated Sepharose 4B was purchased from Pharmacia Fine Chemicals AB, Uppsala, Sweden; VA-Hydroxy Biosinth was from Riedel de Haën; laccase from *Pyricularia oryzae,* chlorogenic acid, (+) catechin, and glutaraldehyde were obtained from Sigma Chemical, St. Louis, Mo.

Laccase Immobilization

Silica gel (250 mg) was equilibrated with 25 mg of laccase in 10 ml 0.1 M citric phosphate buffer pH 3.0 and stirred for 24 hours at 4°C. Then the mixture was centrifuged and the pellet treated with 10 ml buffered aqueous solution of glutaraldehyde (0.2% v/v). The mixture was stirred for 24 hours at 4°C, centrifuged, and washed with citric phosphate buffer to eliminate the excess of glutaradehyde.

CNBr-activated Sepharose 4B (1 g) was previously washed and swelled in 10-mM HCl, incubated with 20 mg of laccase in 10 ml 0.1-M sodium phosphate buffer pH 7.2, and stirred for 4 hours at 25°C. The preparation was incubated for 45 minutes with 1 M Tris-HCl buffer pH 8.0, and after filtration the gel was washed, three times, with coupling buffer followed by three times with 0.1-M acetate buffer at pH 4.0.

The coupling of the enzyme on VA-Hydroxy Biosinth (copolymer based on vinyl acetate and divinyl-ethylene-urea) was accomplished in a 0.1-M sodium phosphate buffer pH 7.5. The ratio between the enzyme solution and the polymer carrier was 1:10. The mixture was stirred for 12 hours at 22°C, centrifuged, and washed with a 0.1-M sodium phosphate buffer pH 7.5.

Laccase Activity

Both free and immobilized laccase activities were determined at pH 4.6 and at 25°C with 10-mM chlorogenic acid as substrate by following the O_2 uptake using a polarographic O_2 electrode. One unit (U) of activity was defined as the amount of enzyme required to consume 1.0 μmol/minute per O_2 in the assay conditions. Catechins were assayed by chemical methods.[9] The reaction in model solution was carried out for 24 hours at 25°C in a stirred batch containing 0.1-M citric phosphate buffer pH 3.0 plus 100 mg/l catechin. The reactions in must (pH 3.02) and white wine (pH 3.35) were carried out for 24 hours at 25°C in a stirred batch.

TABLE 1. Immobilization of Laccase by Adsorption followed by Treatment with Glutaraldehyde (GA) and by Covalent Bonding[a]

Carrier	EU	IY	AY
Silica gel plus GA	0.30	100	28
VA-hidroxy-biosinth	0.37	62	20
CNBr-activated Sepharose 4B	0.24	88	63

[a]EU = equilibrated enzyme units; IY = immobilization yield (IU/EU) × 100; IU = immobilized units; AY = activation yield (AU/IU) × 100; AU = active immobilized units.

RESULTS AND DISCUSSION

As shown in TABLE 1, laccase immobilized on silica gel by adsorption followed by treatment with glutaraldehyde gives an immobilization yield (100%) higher than on VA-Hydroxy or CNBr-Sepharose 4B by covalent bonding, whereas the latter system displayed the best activation yield (63%) in agreement with literature data.[3] Furthermore, the CNBr-Sepharose 4B immobilized preparation showed an operational stability which declined slowly in time to about 80% after 100 days. Also the relative activity of laccase immobilized on silica gel, after an initial decrease at about 60% (15 days), remained almost unchanged until 100 days, while that on VA-Hydroxy decreased gradually to about 50% after 35 days (FIGURE 1). These data are in accordance with the results reported by Froehner and Eriksson,[3] Leonowicz *et al.*,[5] and Spettoli *et al.*[7] FIGURE 2 shows that the enzyme immobilized on silica gel has an optimal pH of 3.5, which is a value 0.5 units less than for the free one. This effect is probably caused by an electrical field that is generated by the positively charged carrier, as observed by Moser *et al.*[10] in the immobilization of lysozyme on a silicate matrix. In contrast, Leonowicz *et al.*[5] and Rogalsky *et al.*[6] found that the optimal pH

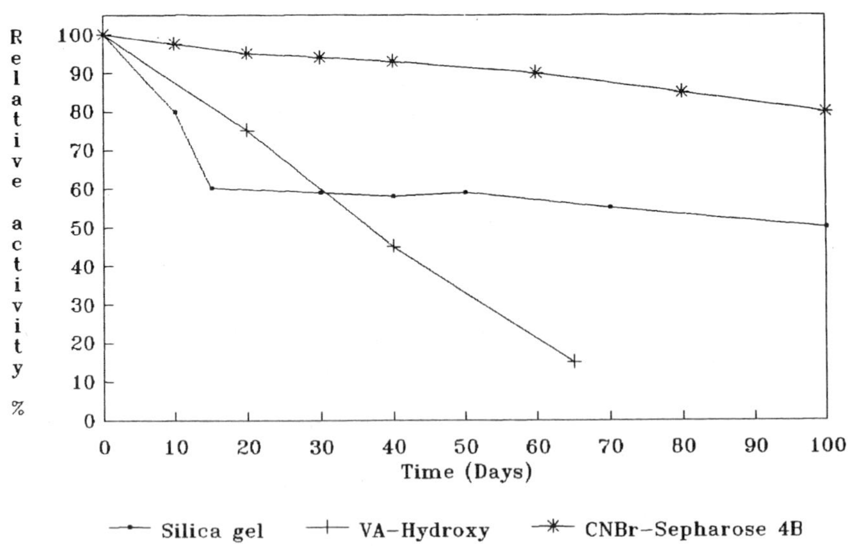

FIGURE 1. Operational stability of immobilized laccase under standard assay conditions.

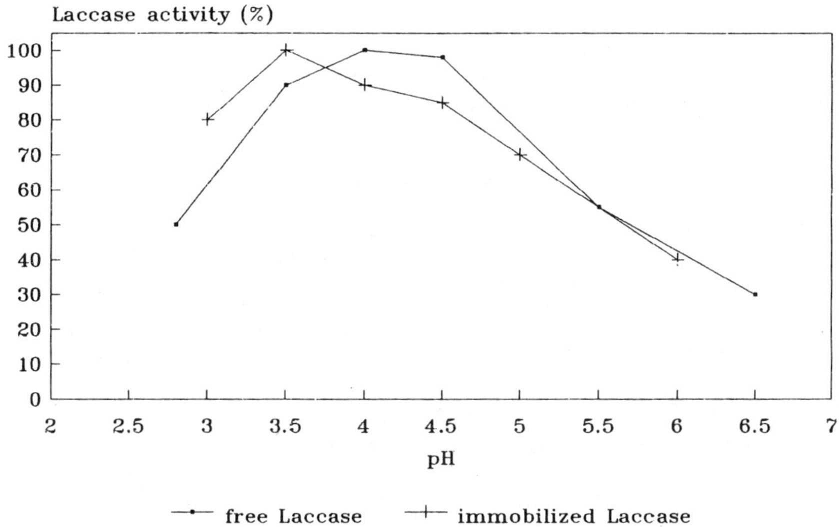

FIGURE 2. pH activity curves for free and immobilized laccase on silica gel.

for laccase immobilized on porous glass shifted slightly towards a less acidic region or showed the same pH optimum when compared to the free one, respectively. The catechin content in model solution as well as in must and wine always decreased more—about 60%, 34%, and 13%, respectively—after treatment with laccase immo-

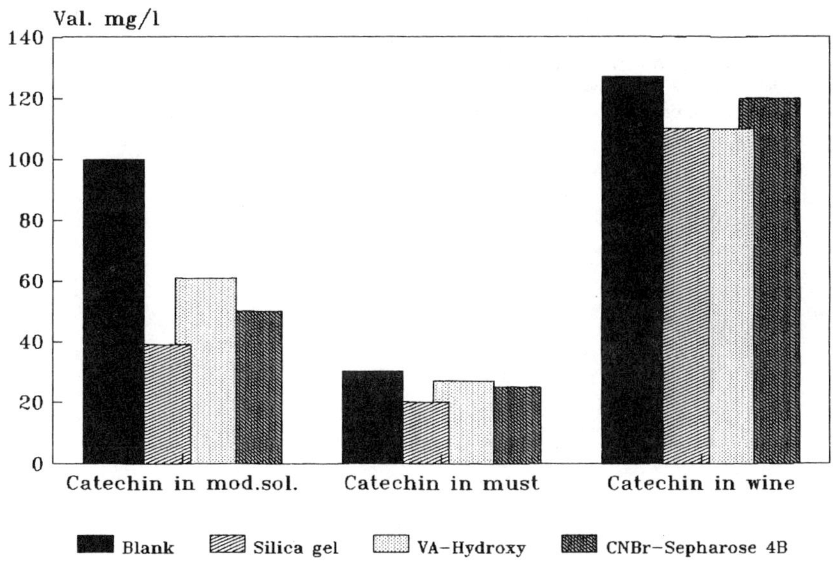

FIGURE 3. Residual of catechins after treatment with immobilized laccase in model solution, must, and wine.

bilized on silica gel (FIGURE 3). These values are in accordance with data obtained by Cantarelli[11] and with our previous experiments.[7]

Our results suggest that the use of immobilized laccase might represent a promising procedure to remove phenolics in wine-making technology.

REFERENCES

1. ZAMORANI, A. 1989. Enzymatic processing of must and wine. *In* Biotecnology Applications in Beverage Production C. Cantarelli & G. Lanzarini, Eds.: 223–246. Elsevier Applied Science. London & New York.
2. CANTARELLI, C., G. GIOVANELLI & R. PAESI. 1989. Enzymatic removal of phenolics in wine-making: four years of experimental results. *In* Innovations in Wine Technology International Symposium, Stuttgart: 134–148. Office International Vigne Vin. Paris, France.
3. FROEHNER, S. & K. E. ERIKSSON. 1975. Properties of the glycoprotein laccase immobilized by two methods. Acta Chem. Scand. **6**(B 29): 691–694.
4. SHUTTLEWORTH, K. L. & J. M. BOLLAG. 1986. Soluble and immobilized laccase catalysts for the transformation of substituted phenols. Enzyme Microb. Technol. **8:** 171–177.
5. LEONOWICZ, A., J. M. SARKAR & J. M. BOLLAG. 1988. Improvement in stability of an immobilized fungal laccase. Appl. Microbiol. Biotechnol. **29:** 129–135.
6. ROGALSKY, J., M. WOJTAS WASILEWSKA, R. APALOVIC & A. LEONOWICZ. 1991. Affinity chromatography as a rapid and convenient method for purification of fungal laccase. Biotechnol. Bioeng. **37:** 770–771.
7. SPETTOLI, P., A. LANTE, A. CRAPISI, S. GALASSI & A. ZAMORANI. 1989. An approach to the use of immobilized oxidative enzymes in white wine stabilization. International Specialized Symposium on Yeasts, XIII, Leuven: 70–71.
8. LANTE, A., A. M. PISANELLI, M. T. LOCCI & P. SPETTOLI. 1990. Activity of an immobilized fungal laccase on phenolic compounds for wine stabilization. *In* The Industrial Applications of Natural, Modified and Artificial Enzymes. S. Carrà, F. Ciardelli & J. L. Houben, Eds.: 88–89. L. Felici. Pisa, Italy.
9. MARGHERI, G. & E. FALCIERI. 1972. Importanza dell'evoluzione delle sostanze polifenoliche nei vini rossi di qualità durante l'invecchiamento. Vini d'Italia **14:** 501–511.
10. MOSER, I., P. DWORSKY & F. PITTNER. 1988. Degradation of bacterial cell walls by immobilized lysozyme. Appl. Biochem. Biotechnol. **19:** 243–249.
11. CANTARELLI, C. 1986. Trattamenti enzimatici sui costituenti fenolici dei mosti come prevenzione della maderizzazione. Vini d'Italia **28:** 87–89.

Immobilization of *Candida rugosa* Cells Having High Fumarase Activity with Polyvinyl Alcohol

LIAN-WAN YANG, XIU-YUAN WANG, AND SHAN WEI[a]

Institute of Microbiology
Academia Sinica
Beijing 100080, China

L-Malic acid is an essential compound in cellular metabolism. It is widely used as a food additive and in the pharmaceutical field. L-Malic acid is effective for the treatment of hepatic malfunctioning, especially for hyperammonemia,[2] and is also used as one of the components of amino acid infusion.

In a previous paper,[1] we reported the continuous production of L-malic acid by microbial cells having fumarase activity immobilized in polyacrylamide gel. In order to improve the immobilized cell system for production of L-malic acid, the immobilization conditions of *Candida rugosa* cells using polyvinyl alcohol (PVA), the enzymatic properties, and operational stability of the immobilized cells were further investigated. The results are presented here.

COMPARISON OF ENTRAPPING METHOD

The following methods for the preparation of immobilized cells were tested. The cells were entrapped in three kinds of gel:2.5 g of *Candida rugosa* cells were suspended in 2.5 ml of 0.9% saline solution. The suspension was added to the 12% PVA gel,[3] 15% polyacrylamide gel,[1] and 4% K-carrageenan gel[4] (final concentration), respectively.

TABLE 1. Comparison of Entrapping Method[a]

Method	Gel Strength (g/cm^2)	Fumarase Activity at 37°C (μmol/hour × g)	Activity Yield (%)
PVA-gel-entrapped cells[3]	1500	11372.4	92.3
Polyacrylamide-gel-entrapped cells[1]	850	10424.7	84.6
K-Carrageenan-gel-entrapped cells[4]	1100	7107.8	57.7

[a]Fumarase activity of 1 g wet cells: 12320.2 μmol/hour.

[a]Present address: Technical University of Berlin, Department of Chemistry, 1000 Berlin 21, Germany.

TABLE 2. Effect of PVA Concentration[a]

PVA Concentration (%)	Fumarase Activity at 37°C (μmol/hour × g)	Activity Yield (%)
5	7237	64.9
10	9347.8	83.8
12	10553.9	94.6
13	9046.2	81.1
14	8443.2	75.7
15	7237.0	64.9

[a]Fumarase activity of 1 g wet cells: 11157.1 μmol/hour.

It is clear that the most active immobilized cells were obtained by entrapping the cells in PVA gel (TABLE 1).

CONDITIONS FOR IMMOBILIZATION OF *CANDIDA RUGOSA* CELLS BY POLYVINYL ALCOHOL

PVA concentration. Candida rugosa cells were immobilized with various concentrations of PVA, to examine the effect on fumarase activity and activity yield. The results indicated that the optimal concentration of PVA was 12% (TABLE 2).

Cell concentration. TABLE 3 shows that the fumarase activity and activity yield of the immobilized cells were the highest when 2.5% of wet cells were entrapped in PVA gel.

ENZYMATIC PROPERTIES OF IMMOBILIZED *CANDIDA RUGOSA* CELLS

The basic enzymatic properties of the immobilized *Candida rugosa* cells were investigated and compared with those of native cells. Some basic parameters of the immobilized cells and native cells are listed in TABLE 4.

The most favorable preparation for the continuous production of L-malic acid was obtained when *Candida rugosa* cells were entrapped in PVA. The immobilized cell column can be used for continuous production of L-malic acid from 1-M ammonium fumarate (pH 8.5) at 30°C and showed high fumarase activity and

TABLE 3. Effect of Cell Concentration

Cell Concentration (%)	Fumarase Activity at 37°C (μmol/hours × g)	Activity Yield (%)
2.5	11372.4	95.7
5.0	9821.7	82.6
7.5	8615.5	72.5
10.0	7236.9	60.9

[a]Fumarase activity of 1 g wet cells: 11889.4 μmol/hour.

TABLE 4. Basic Enzymatic Properties of the Immobilized Cells

Optimal pH	8.5 (8.5)[a]
Optimal temperature	45°C (45°C)
pH stability	6–8 (6–8)
Heat stability	Good (poor)
Storage stability	4°C, 10 months no loss of activity

[a]Native cells in parentheses.

operational stability (about 2 months). The conversion of substrate into product is over 80%.

The main advantages of this entrapping method are as follows: (1) the entrapping method is simple; (2) the recovery of enzyme activity is high; (3) the mechanical strength of immobilized cells is better than that of the other two kinds of entrapping method; (4) the storage stability and operational stability are good; (5) the conversion of substrate into product is high; and (6) there are no by-products.

From an economic standpoint, this method gave a more satisfactory result than those of the other two immobilization methods. We have used this immobilized cell system for the industrial production of L-malic acid since 1990.

REFERENCES

1. YANG, L. W. & L. C. ZHONG. 1980. The immobilized cells of *Candida rugosa* possessing fumarase activity. Acta Microbiol. Sin. **20**(3): 296–302.
2. YAMAMOTO, K., T. TOSA, K. YAMASHITA & I. CHIBATA. 1976. Continuous production of L-malic acid by immobilized *Brevibacterium ammoniagenes* cells. Eur. J. Appl. Microbiol. **3:** 169–183.
3. YANG, L. W. 1991. Production of L-malic acid by immobilized *Candida rugosa* cells. Chinese Patent No. 91103606.7.
4. TAKATA, I., K. YAMAMOTO, T. TOSA & I. CHIBATA. 1980. Immobilization of *Brevibacterium flavum* with carrageenan and its application for continuous production of L-malic acid. Enzyme Microb. Technol. **2:** 30–36.

Enzyme Activity in Microstructured Media

JOËL CHOPINEAU, MICHEL OLLIVON,[a] AND
MARIE-DOMINIQUE LEGOY

Laboratoire de Technologie Enzymatique
URA 1442 du CNRS
BP649
60206 Compiègne, France

[a]*Equipe Physicochimie des Systèmes Polyphasés*
URA 1218 du CNRS
Université Paris-Sud
92296 Chatenay-Malabry, France

INTRODUCTION

In a previous report,[1] we have described a novel approach to the study of enzyme microenvironment interactions. Enzymatic catalysis was used to create self-evolving microstructures in which enzyme kinetics were studied. Artificial systems composed of surfactant/water/organic solvent mixtures spontaneously form different microstructured phases in which enzymes have been widely studied.[2–6] Using such systems is one way to mimic enzyme *in vivo* microenvironment. More than 40 different enzymes have been studied in micellar systems. Enzymatic activity toward a substrate is measured while microenvironment structure remains the same during the experiment. Enzymes in reverse micelles[8,9] or in liquid crystalline phases[7] obey classical Michaelis-Menten kinetics, nevertheless enzymatic activity is dependent on the ratio W_0 = [water]/[surfactant] which determines the size of the micelles or the size of the aggregates.[2,7]

Conversely in self-evolving systems, the enzyme interacts directly with the components of the system.[1] Enzyme microenvironment changes during the reaction progress. The model system octyl-β-D-glucoside/water (or glucose in water)/octanol plus the enzyme β-D-glucosidase is used to study enzyme-microenvironment dynamic interactions.[1,10] Enzyme substrates and products are the components of a classical surfactant/organic solvent/water system. The corresponding phase diagram shows different macroheterogeneous regions and microstructured domains, depending on the relative ratio of the components. β-D-Glucosidase hydrolyzes the surface-active substrate to form glucose and octanol. Starting from each homogeneous region of the phase diagram, the enzyme, due to its catalytic activity, is responsible for different phase changes. The physicochemical properties of the medium change in proportion to the progress of the enzymatic reaction. The modifications of the microenvironment are correlated with kinetic patterns specific for each phase transition. Physicochemical evolution of the reaction medium and reaction rates were followed simultaneously. Enzymatic reaction rate modifications occurring during the transitions are a function of the different microstructured phases present. Moreover, in this system octyl-β-D-glucoside is engaged in different mesophases as a monomer or under the form of aggregates. We demonstrated that octyl-β-D-glucoside monomer is the actual enzyme substrate. Then, β-D-glucosidase reaction rate modifications occurring during phase changes might also be dependent on local concentration and availability of octyl-β-D-glucoside monomer.

MATERIALS AND METHODS

Materials

Enzyme and chemicals were the same as those described in our previous report.[1] The standard buffer in all experiments was 50-mM citrate phosphate, pH 5.25, which was found to be optimum for octyl-β-D-glucoside enzymatic hydrolysis.

Phase Diagram and Phase Changes

The mixtures octyl-β-D-glucoside/water (or glucose in water)/octanol were used as media for enzymatic reactions. All experiments, whatever the technique, were carried out at 30°C and atmospheric pressure. Results are expressed in terms of total percentage mass.

Enzymatic Assay and Activity Measurement

As a reference, β-D-glucosidase was assayed on salicin in the same buffer at 37°C.[11] Enzymatic activity was found to be 11 IU (μmol glucose × $minute^{-1}$ × mg^{-1} solid). β-D-Glucosidase activity on octyl-β-D-glucoside in buffer was studied following the initial rate for liberation of glucose by high-performance liquid chromatography (HPLC). Octyl-β-D-glucoside concentration was varied within the range 3–100 mM. Enzyme solution (20 μl) was added to 10 ml of substrate solution in order that final enzyme concentration was 0.1 mg/ml. For each glucose measurement, aliquots of 350 μl were withdrawn and added to 100 μl of acetonitrile/water (40:60) solvent mixtures containing methyl-α-D-glucoside (10 mg/ml) as internal standard. A Gilson 232-401 automatic sample processor and injector was used. We have verified that the rate of the reaction was proportional to the enzyme concentration in the whole octyl-β-D-glucoside concentration range.

RESULTS AND DISCUSSION

Enzyme-Microenvironment Interactions

The pseudoternary diagram octyl-β-D-glucoside/10% glucose in water/octanol which is shown FIGURE 1A exhibits two monophasic domains and a liquid crystalline region. (a) The first domain, L_1, is an aqueous micellar solution of octyl-β-D-glucoside. (b) The second domain, L_2, located in the middle of the diagram, is a microemulsion. (c) The third region, L_α, is a liquid crystalline region that appears to have a lamellar texture. Between each two domains as L_1 and L_α, a delimited region

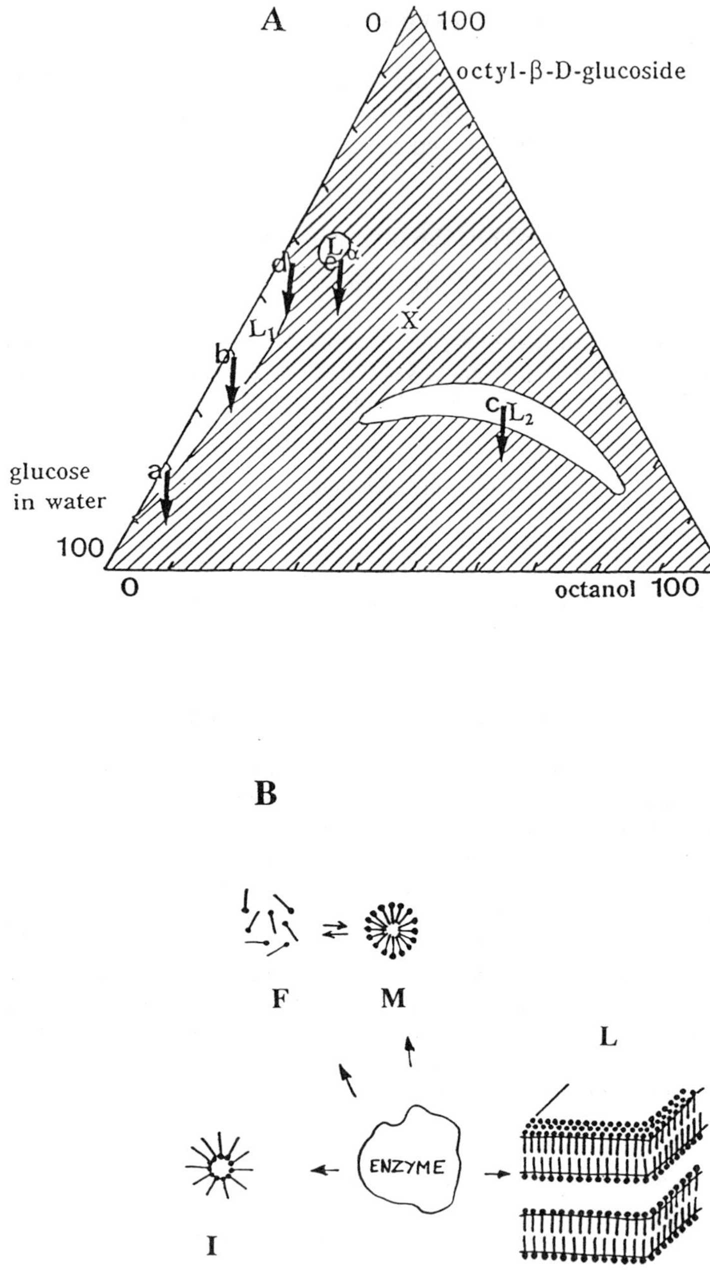

FIGURE 1. β-D-Glucosidase microenvironment interactions. (**A**) The monophasic phases and boundaries are shown for an octyl-β-D-glucoside/water/octanol phase diagram. L_1 is a micellar solution of octyl-β-D-glucoside in water, L_2 is a microemulsion, L_α is a liquid crystalline phase with a lamellar texture, and X corresponds to undetermined structures. Arrows show enzyme-catalyzed phase changes. (**B**) Schematic representation of the different physicochemical forms of octyl-β-D-glucoside in this system: free octyl-β-D-glucoside monomer (F), micelles in water (M), inverted micelles (I), lamellar phase (L).

composed of these two phases in equilibrium is present. For more details one may refer to Reference 1.

When β-D-glucosidase was added to a ternary mixture of octyl-β-D-glucoside/buffer/octanol, it hydrolyzed octyl-β-D-glucoside forming glucose and octanol. Octanol concentration increased, and the relative ratios of the components were altered such that the reaction medium changed texture. For practical purposes, phase diagram exploration (shown by arrows on FIGURE 1A) was started from each identified homogeneous region (L_1, L_2, and L_α). Inside the L_1 domain, values from 20% to 55% of octyl-β-D-glucoside were used. Only one initial composition was studied in L_2 and L_α. Experiments were done at different enzyme concentrations. After a certain period of time, the enzyme was present in a biphasic system. In most of the experiments, enzymatic activity transformed the initial single phase into a biphasic system. Starting from the monophasic domains, different possible phase changes were shown: (a) $L_1 \rightarrow L_1 + L_2$; (b) $L_1 \rightarrow L_1 + L_\alpha$; (c) $L_2 \rightarrow L_2 + X$; (d) $L_1 \rightarrow L_1 + L_\alpha$; and (e) $L_\alpha \rightarrow L_1 + L_\alpha$. β-D-Glucosidase activity has been studied for the different transitions. Glucose appearance was followed by HPLC; kinetic measurements were performed for each phase transition. These phase diagram explorations led to different β-D-glucosidase activity recordings.[1,10]

For 20 and 40%, continuous monitoring of enzymatic activity was performed by microcalorimetry, while phase changes were recorded by measuring turbidity at 400 nm.[12] Exploration patterns found for 20 and 40% were quite different and involve differences in the phase transitions but also in enzyme kinetics. In this system, microenvironment modifies enzyme activity. Changes observed may be caused by modifications of the substrate availability and more generally by restricted diffusion processes as well as direct influence of the structures involved at the boundary on the enzyme.

Enzyme Actual Substrate

In this system, octyl-β-D-glucoside, which is the enzyme substrate, can be present under different physicochemical forms as shown on FIGURE 1B. In each homogeneous region shown on the phase diagram, octyl-β-D-glucoside is engaged in different mesophases. For example, in the L_1 phase, the micellar solution of octyl-β-D-glucoside is composed of monomer molecules coexisting with micelles. Which is the physicochemical form of octyl-β-D-glucoside hydrolyzed by β-D-glucosidase: micelle, monomer, or both?

The critical micellar concentration of octyl-β-D-glucoside determined in 50-mM phosphate/citric acid buffer (pH 5.25) at 30°C was equal to 22.3 mM. A micellar solution of octyl-β-D-glucoside formed spontaneously above this value while the monomer concentration remained equal to the critical micellar concentration.[13]

Enzymatic activity measurements for octyl-β-D-glucoside hydrolysis were performed by following glucose production by HPLC, the rate of which was determined for octyl-β-D-glucoside from 3 mM to 100 mM; enzyme concentration was 0.1 mg/ml. Reaction rate versus octyl-β-D-glucoside concentration is plotted on FIGURE 2A. For octyl-β-D-glucoside concentrations from 3 mM to 20 mM, the relation between the rate of the reaction and substrate concentration followed a classical Michaelis-Menten scheme (FIGURE 2B). Apparent kinetic constants were determined from

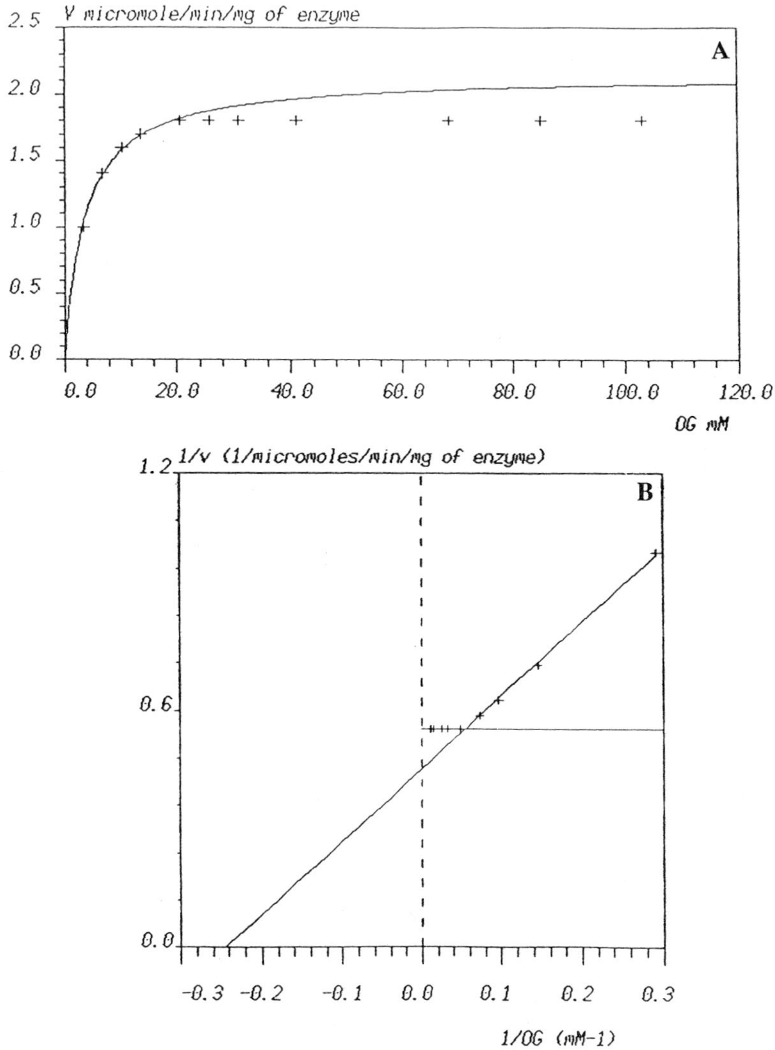

FIGURE 2. Determination of β-D-glucoside actual substrate form. (**A**) Reaction rate of glucose production was measured for different octyl-β-D-glucoside concentrations. The line represents the Michaelis-Menten curve obtained with kinetic parameters calculated from a Hanes-Woolf plot. (**B**) Lineweaver-Burk representation for octyl-β-D-glucoside hydrolysis.

Hanes-Woolf representation [S/V = f(S)]. In these experimental conditions V_{max} = 2.1 μmol × minute^{-1}/mg enzyme and K_m = 3.7 mM.

The maximum rate obtained on FIGURE 2A was 1.8 μmol × minute^{-1}/mg enzyme and was reached for 19 mM as shown by the Lineweaver-Burk representation (FIGURE 2B). Substrate availability was limited by the critical micellar concentration. It is believed that the low solubility of octanol in water leads to its preferential insertion in octyl-β-D-glucoside micelles and to the formation of mixed micelles. However, in these measurements, we neglected the influence of octanol and glucose on the critical micellar concentration. This value is in excellent agreement with the 22.3 mM value, taking into account that the critical micellar concentration of mixed micelles is lower than that of pure micelles.[14] For concentrations below the critical micellar concentration, the enzyme followed classical kinetics that can be described by the Michaelis-Menten equation. Above 19 mM, the rate was limited by the octyl-β-D-glucoside free monomer concentration so the enzyme does not follow Michaelis-Menten kinetics.

At low octyl-β-D-glucoside concentrations, the monomer is the actual enzyme substrate and micelles act as a reservoir of octyl-β-D-glucoside molecules. As soon as one molecule of octyl-β-D-glucoside was hydrolyzed, one was released by the micelles. At higher octyl-β-D-glucoside micellar concentrations, enzymatic activity is reduced (for 20 and 40% octyl-β-D-glucoside reaction rates from 0.93 down to 0.28 μmol $\times$ minute^{-1}/mg of enzyme have been found).[1,10] Viscosity of the L_1 phase, which increases with octyl-β-D-glucoside concentration, may explain this activity reduction.

CONCLUSION

The microheterogeneous (micellar and liquid crystalline) media octyl-β-D-glucoside/water (or glucose in water)/octanol plus the enzyme β-D-glucosidase is used as a model system to study enzyme-microenvironment dynamic interactions. The physicochemical properties of the medium changed in proportion to the progress of the enzymatic reaction. It was determined that octyl-β-D-glucoside free monomer was the actual enzyme substrate. The variations of the enzymatic reaction rates observed during the phase changes are a function of the different phases involved at the boundary, but they also depend on local concentration and availability of octyl-β-D-glucoside monomer in these restricted environments.

Similar microstructures and phase transitions to those reported here have been described with biological molecules.[15] For example, it has been shown that proteases behave similarly in lecithin reverse micelles as they do in AOT [sodium bis 2-(ethylhexyl sulfosuccinate)] reverse micelles.[16] Pseudohomogeneous systems such as this one, in which micellar enzymology is under study, are a novel physicochemical line of approach to problems of biology at a molecular level.

REFERENCES

1. CHOPINEAU, J., D. THOMAS & M. D. LEGOY. 1989. Dynamic interactions between enzyme activity and the microstructured environment. Eur. J. Biochem. **183:** 459–463.
2. MARTINEK, K., A. V. LEVASHOV, N. L. KLYACHKO, YU. L. KHMELNITSKY & I. V. BEREZIN. 1986. Micellar enzymology. Eur. J. Biochem. **155:** 453–468.
3. LUISI, P. L. 1985. Enzymes hosted in reversed micelles in hydrocarbon solution. Angew. Chem. Int. Ed. Engl. **24:** 439–450.
4. WAKS, M. 1986. Proteins in water restricted environment. Proteins **1:** 4–15.
5. LUISI, P. L., M. GIOMINI, M. P. PILENI & B. H. ROBINSON. 1988. Reverse micelles as hosts for proteins and small molecules. Biochim. Biophys. Acta **947:** 209–246.

6. MARTINEK, K., N. L. KLYACHKO, A. V. KABANOV, YU. L. KHMELNITSKY & A. V. LEVASHOV. 1988. Micellar enzymology: its relation to membranology. Biochim. Biophys. Acta **981:** 161–172.
7. KLYACHKO, N. L., A. V. LEVASHOV, A. V. PSHEZHETSKY, N. G. BUGDANOVA, I. V. BEREZIN & K. MARTINEK. 1986. Catalysis by enzymes entrapped into hydrated surfactant agregates having lamellar or cylindrical (hexagonal) or ball shaped (cubic) structure in organic solvents. Eur. J. Biochem. **161:** 149–154.
8. VERHAERT, R. M. D., R. HILHORST, M. VERMUE, T. J. SCHAAFSMA & C. VEEGER. 1990. Description of enzyme kinetics in reversed micelles. Eur. J. Biochem. **187:** 59–72.
9. OLDFIELD, C. 1990. Evaluation of steady state kinetic parameters for enzymes solubilized in water-in-oil microemulsion systems. Biochem. J. **272:** 15–22.
10. CHOPINEAU, J., D. THOMAS & M. D. LEGOY. 1991. Biochemical Engineering Stuttgart: 153–157. Gustav Fisher. Stuttgart, Germany.
11. LECOQ, D., J. F. HERVAGAULT, G. BROUN, G. JOLY, J. P. KERNEVEZ & D. THOMAS. 1975. The kinetic behavior of an artificial bienzyme membrane. J. Biol. Chem. **254:** 5496–5800.
12. CHOPINEAU, J. & M. OLLIVON. 1992. Enzyme activity in self-evolving microenvironment. Thermochim. Acta **204:** 35–43.
13. ROXBY, R. W. & B. P. MILLS. 1990. Micelle size distribution and free monomer concentration in aqueous solution of octyl glucoside. J. Phys. Chem. **94:** 456–459.
14. SMALL, D. M. 1986. *In* The Physical Chemistry of Lipids: from Alkanes to Phospholipids: 43–87. Plenum Press. New York & London.
15. WALDE, P., A. M. GIULIANI, C. A. BOICELLI & P. L. LUISI. 1990. Phospholipid based reversed micelles. Chem. Phys. Lipids **53:** 265–288.
16. PENG, Q. & P. L. LUISI. 1990. The behavior of proteases in lecithin reverse micelles. Eur. J. Biochem. **188:** 471–480.

A Novel Process for the Production of Dextrans of Selected Molecular Sizes

D. F. DAY AND DOMAN KIM

Audubon Sugar Institute and
Department of Microbiology
Louisiana State University
Baton Rouge, Louisiana 70803

Low molecular weight dextrans (between 4 and 70 K) have industrial uses ranging from photographic film manufacture to blood plasma extenders. We have developed a novel process that allows for direct production of specific low molecular weight dextrans at greater than 90% conversion to product. This process involves the establishment of microbial consortia of *Leuconostoc mesenteroides* with a specific strain of the yeast *Lipomyces starkeyi.* Through control of the fermentation conditions, dextrans of any desired size, from 100K to 20K, can be produced as the sole product.

INTRODUCTION

Dextran is the name given to a large class of extracellular bacterial polyglucans synthesized from sucrose by the action of the enzyme dextransucrase. They are characterized by having predominantly α 1–6 linkages. Almost all industrial dextran is produced by fermentation by the strain of *Leuconostoc mesenteroides* NRRL B-512 (F). This organism produces large amounts of polymer that contains $\sim$95% α 1–6 linkages.[1] Commercially, dextrans are produced by a two-step process. Random size polymers are produced from sucrose by *Leuconostoc mesenteroides* fermentation. These dextrans are generally of high molecular weight ($>10^6$). They are separated from the fermentation broth by alcohol precipitation and then reduced to the desired sizes by acid hydrolysis.[2] Dextrans of 150,000 daltons or less are desired for most industrial uses. A size of 10,000 daltons or less is desired for medical applications. The yields of dextran by this process are normally between 33 and 38% by weight on sucrose used. The process is limited, in part, by the viscosity of the dextran fermentation broth (practically no increase in yield can be achieved with sucrose concentrations above 15% w/v).

The key to producing marketable dextran is to produce clean preparations of controlled sizes. Efforts have been made, with limited success, to improve the existing process by separating the enzymatic synthesis of the polymer from the fermentation process.[3] We have taken a different approach, which is the subject of this report. That is to use a mixed culture fermentation where one of the strains is *L. mesenteroides* NRRL 512 (F) and the other a dextranase-producing yeast *Lipomyces starkeyi.*[4]

METHODS

Organism and Culture Conditions

The organisms used in this study are *Leuconostoc mesenteroides* NRRL B-512 (F) and *Lipomyces starkeyi* CMK-1. This *Lipomyces* is a mutant derived from *L. starkeyi* ATCC 12659 selected for growth in high sucrose concentrations. The *Lipomyces* were maintained on slants of WW agar[4] with 0.1% yeast extract, 1% starch, and the *Leuconostoc* on the fermentation media (yeast extract, phosphate, and sucrose) at room temperature.

Fermentations

Fermentations were conducted in either one-liter or 20-I volumes in New Brunswick Science Co. fermentors containing a medium composed of yeast extract (5 g/l), potassium phosphate (5 g/l), and sucrose. The initial pH was 5.7, and aeration was 0.5 v/v.

Dextran Determination

Dextran levels and polydispersity index were determined using high-pressure liquid chromatography (HPLC) on a GPC (Ultrahydrogel linear—Waters) column connected to a Dawn Model F Laser detector (Wyatt Technology, Calif.).

Process Description

The following process (Patent Pending) results in the direct production of dextran of any desired molecular size. The process requires the coinoculation of equal volumes of *Leuconostoc mesenteroides* B-512 (F) and *Lipomyces starkeyi* CMK-1. Both strains are started independently with sucrose as a carbon source. The dextran fermentation is allowed to proceed as it would normally. At the desired time (molecular size), the fermentation is harvested. The fermentation is monitored with HPLC to detect the appropriate size of product. The cells are removed by decantation, and the broth is alcohol precipitated to remove the product.

RESULTS AND DISCUSSION

This process takes advantage of some special properties of *Lipomyces starkeyi* CMK-1 that allow it to thrive in a *Leuconostoc* fermentation. *L. starkeyi* produces a dextranase that is induced when dextran is present in the media.[4] *Lipomyces starkeyi* will grow on both glucose and sucrose. These parameters allow growth in conjunction with *L. mesenteroides* in a sucrose fermentation. As dextran is produced by *Leuconostoc,* it triggers dextranase production by *Lipomyces.* The CMK-1 strain of *Lipomyces* has features that make it highly suitable for this process. It is osmotically stable, grows in high sucrose, and is constitutive for dextranase. This strain will grow and produce dextranase in 30% w/v sucrose solutions.

The dextranase that is produced during the fermentation acts upon the dextran as it is produced, resulting in shorter dextran fragments which act as acceptors for the production of more dextran. Within 24 hours, the sucrose is exhausted and theoretical yields of dextran have been reached. At this point, there are two fractions of

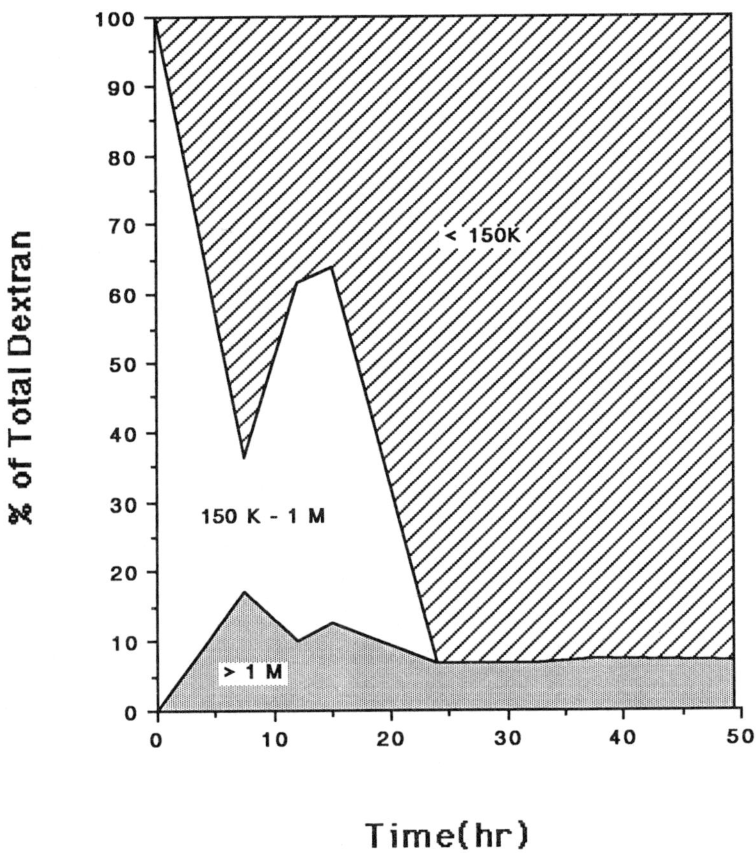

FIGURE 1. Distribution of various dextran sizes during the course of the mixed fermentation.

polymer present, a small amount of high molecular weight material (greater than 10^6 molecular size) and a fraction that is less than 150,000 molecular size. At earlier points in the fermentation, an intermediate size fraction (between 1 M and 150,000) was also present (FIGURE 1).

The fermentation is then allowed to proceed until the desired molecular size dextran is reached. Any changes take place in the 150,000 fraction. The very large

TABLE 1. Product Comparison

Sample	Yield	PDI[a]
Traditional	70–80%	~2.0
Mixed	85–95%	1.2

[a]Polydispersity index.

fraction does not appear to change size with time. We have not yet determined the composition of this fraction but think that it may be either a levan or a highly branched dextran. The fermentation is harvested when the desired size dextran is reached, and the product is separated by precipitation with 2-propanol. The yields of this process are routinely 44% by weight of the sucrose used. Theoretical conversion is 48% dextran by weight of sucrose. The product shows a cleaner HPLC profile than commercial dextran of the same size. It also does not require subsequent hydrolysis to produce lower molecular size polymers, with their associated loss of product. The yields and sizing of the dextrans produced in this manner are better than those produced in the traditional commercial production scheme and are summarized in TABLE 1.

The use of a mixed culture fermentation can produce high levels of dextran of the desired size, with a low polydispersity index, in a single-step process. This system will also allow the use of higher sucrose concentrations because of the reduced fermentation viscosities seen with the low molecular weight dextrans. The yields are consistent with the traditional methods.

REFERENCES

1. SANDFORD, P. A. & J. BAIRD. 1983. Industrial utilization of polysaccharides. *In* The Polysaccharides. G. O. Aspinall, Ed. **2:** 411–490. Academic Press. New York, N.Y.
2. BIXLER, G. H., G. E. HARRIS, R. M. MCGEE & R. A. SHURTER. 1953. A staff-industry collaborative report. Ind. Eng. Chem. **45:** 692–705.
3. JEANES, A. R. 1965. Preparation of water soluble dextran by enzymic synthesis. Methods Carbohydrate Chem. **5:** 127–132.
4. KOENIG, D. W. & D. F. DAY. 1988. Production of dextranase by *Lipomyces starkeyi.* Biotechnol. Lett. **10:** 117–122.
5. KOENIG, D. & D. DAY. 1989. The purification and characterization of a dextranase from *Lipomyces starkeyi.* Eur. J. Biochem. **183:** 161–167.

Catalytic Activity of Lipase Modified with Hydrophobic Imidoesters

M. BASRI, K. AMPON, A. B. SALLEH, W. M. Z. YUNUS, AND C. N. A. RAZAK

Enzyme and Microbial Technology Research
Fakulti Sains & Pengajian Alam Sekitar
University Pertanian Malaysia
43400 UPM Serdang, Malaysia

The alteration and improvement of the native properties of enzymes for biotechnological applications can be achieved by chemical modifications.[1,2] Modification of protein by amidination with imidoesters offers several advantages over other approaches to amino group derivatization. The reaction conditions are relatively mild, the reaction is specific for the epsilon-amino-lysil and alpha-amino residues of proteins, the amidine linkages formed are stable over a wide range of experimental conditions,[3] and the derivatized protein retains the positive charges of the amino

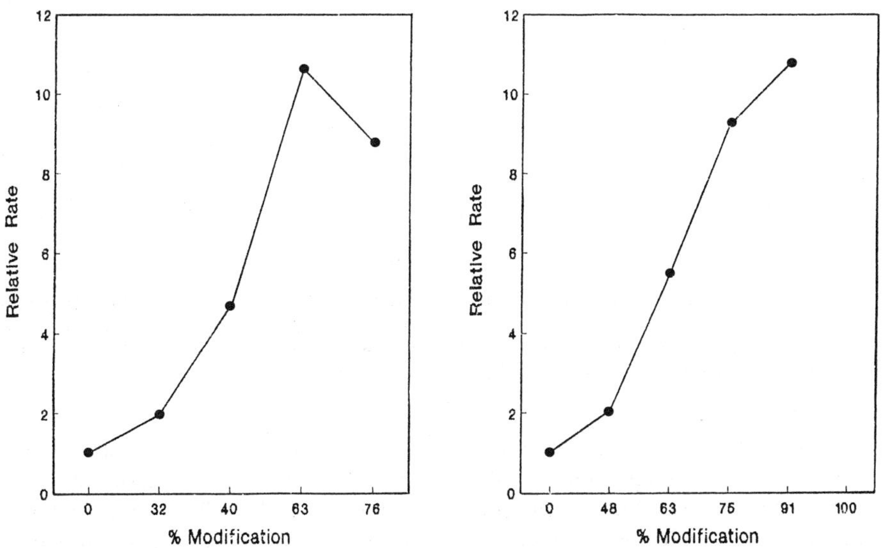

FIGURE 1. The effect of increasing degree of modification on the esterification activity of lipase derivatized with imidoester I (right) and imidoester (VI) (left). Relative rate is calculated by dividing the specific activity of the modified lipase at the indicated degree of modification with the specific activity of the unmodified lipase. The synthetic reaction system consisted of solvent (0.5 ml), alcohol (1.0 mmol), fatty acid (0.35 mmol), and unmodified or modified lipase (2–5 mg). The mixture was incubated at 37°C for 3–24 hours with continuous shaking at 150 rpm.

groups thus minimizing any changes to its physical and biological properties that are related to net charges.[4] More importantly, however, imidoesters of different hydrophobicity are easily synthesized in high yield and they are generally water soluble. The maximal activity of the modified lipase was shown to be dependent on the degree of modification by the respective imidoesters (FIGURE 1). To achieve its maximum activity, a lower degree of modification (63%) of the enzyme was observed with the more hydrophobic imidoester VI compared with imidoester I (about 90%). Similar findings had previously been reported.[5]

The half-lives of the enzyme modified with imidoesters I and VI were 5.6 and 6.3

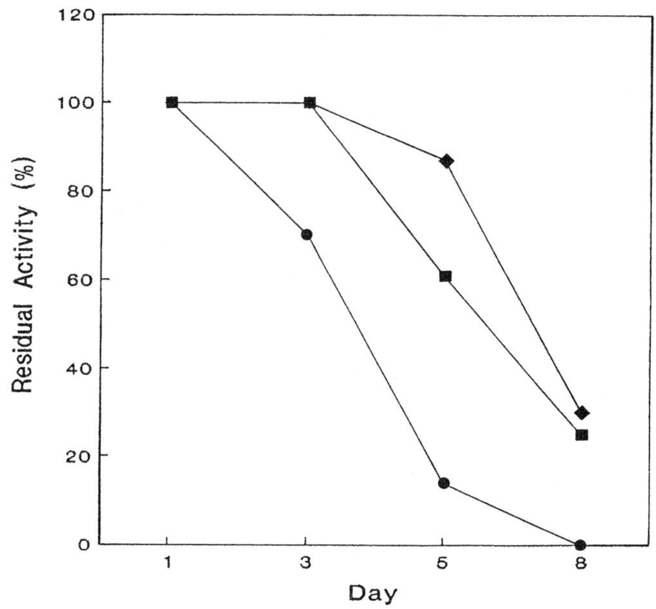

FIGURE 2. Stability of native lipase (●) and lipase modified with imidoester I (■) and imidoester VI (◆) in benzene incubated at room temperature for various time intervals. Residual activity is calculated by dividing the specific activity of the lipase at various time intervals with the specific activity of the lipase at zero time.

days, respectively, while native lipase had a half-life of 3.7 days when incubated in benzene at room temperature (FIGURE 2). In addition, modified lipase was more stable when incubated for 1 hour at a temperature from 20 to 70°C compared with the native lipase. Chemically modifying enzyme was shown to decrease the amount of heat-subjected intermolecular cross-linking caused by transamidation and intermolecular disulfide interchange, which partly caused thermal aggregation, which also correlates with the loss of enzymatic activity.[6]

Our investigation also showed that nonpolar solvent such as benzene was found to be the better solvent among all the solvents tested for synthetic reaction.

When fatty acids of different carbon chain length were used as substrates, the synthetic activity of the modified lipase was highest when myristic acid was used.

REFERENCES

1. INADA, Y., H. NISHIMURA, K. TAKAHASHI, T. YOSHIMOTO, A. R. SAHA & Y. SAITO. 1984. Ester synthesis catalyzed by polyethylene glycol–modified lipase in benzene. Biochem. Biophys. Res. Commun. **122:** 845–850.
2. KAIMAL, T. N. B. & M. SAROJA. 1989. Enhancement of catalytic activity of porcine pancreatic lipase by reductive alkylation. Biotechnol. Lett. **11**(11): 31–36.
3. HUNTER, M. J. & M. L. LUDWIG. 1962. The reaction of imidoesters with proteins and related small molecules. J. Am. Chem. Soc. **84:** 3491–3504.
4. MAKOFF, A. J. & A. D. B. MALCOLM. 1980. Properties of methyl acetimidate and its use as a protein modifying reagent. Biochem. J. **193:** 245–249.
5. AMPON, K., A. B. SALLEH, A. TEOH, W. M. Z. WAN YUNUS, C. N. A. RAZAK & M. BASRI. 1991. Sugar esterification catalysed by alkylated trypsin in dimethylformamide. Biotechnol. Lett. **13**(1): 25–30.
6. VOLKIN, D. B., A. STAUBLI, R. LANGER & A. M. KLIBANOV. 1991. Enzyme thermoinactivation in anhydrous organic solvents. Biotechnol. Bioeng. **37:** 843–853.

Mono-, Di-, and Polyaldehyde-Mediated Chemical Modification of Urease[a]

R. VENKATESH, SUDHA R. WARRIER, S. SHUBHADA, AND P. V. SUNDARAM[b]

Center for Protein Engineering and Biomedical Research
The Voluntary Health Services
Adyar, Madras—600 113, India

INTRODUCTION

ϵ-NH_2 groups, which are generally found mostly on the outside of proteins, and α-NH_2 groups are perhaps the most easily accessible to modification in proteins. They are also probably the most abundant among the various groups. Most reagents designed for reaction with proteins are targeted at the ϵ-amino groups.

In the immobilization procedures, dialdehyde-mediated cross-linking has proved to be the most popular procedure because it is simple and inexpensive. In such a procedure, among the majority of papers published so far, not a great deal of attention has been paid to designing specific conditions in the coupling reactions. Instead a large excess of glutaraldehyde, i.e., around 2.5–5%, is usually used. It is also common knowledge that the specific activity of enzymes is reduced upon immobilization, which is largely attributed to steric effects or microenvironmental condition, a term that is generally vague but probably includes more than one phenomenon.

We decided to look at this matter, that is, the effect of the reaction of mono-, di-, and polyaldehydes on enzymes in solution, in order to isolate effects due to cross-linking from the effects due to immobilization.

METHODS

Urease (35 U/mg, Sigma Chemical Co.), Dextran 70 (Pharmacia), and Ficoll 400 (Pharmacia) were periodate oxidized to produce dextran polyaldehyde and Ficoll polyaldehyde.

A 2-mg/ml urease solution made up in acetate (pH 4.3), phosphate (pH 7), or bicarbonate (pH 9) buffer was reacted with freshly made glutaraldehyde solutions at various concentrations for periods of time ranging from 30 seconds to 40 minutes and immediately assayed for activity. After the reaction with the aldehyde, the cross-linked enzyme was treated with $NaBH_4$.

Similarly, protein to aldehyde ratio (TABLE 1) was changed to follow effects on activity retained.

[a] Dedicated to Professor Fritz Eckstein on the occasion of his 60th birthday.

[b] Author to whom correspondence should be addressed.

TABLE 1. Effect of Aldehydes on Urease Activity: Effect of Reaction Time (at pH 7.0)

<table>
<tr><th rowspan="2">Aldehyde</th><th rowspan="2">Enzyme:
Aldehyde
Ratio</th><th colspan="7">Time (minutes)</th></tr>
<tr><th>1/2</th><th>1</th><th>2</th><th>3</th><th>4</th><th>5</th><th>10</th></tr>
<tr><td>Formaldehyde</td><td>1:0.1</td><td>15%
activation</td><td colspan="6">←No significant change→</td></tr>
<tr><td>Glutaraldehyde</td><td>1:0.1</td><td>13%
activation</td><td colspan="2">←→
15%
activation</td><td colspan="3">←→
22%
activation</td><td>16%
activation</td></tr>
<tr><td>Oxidized
Dextran (70)</td><td>1:17.86</td><td>32%
inhibition</td><td colspan="2">←→
29%
inhibition</td><td colspan="3">←→
40.8%
inhibition</td><td>63%
inhibition</td></tr>
<tr><td></td><td>1:0.357</td><td colspan="2">←→
2%
activation</td><td>9%
activation</td><td colspan="3">←→
No significant
change</td><td>5.1%
inhibition</td></tr>
<tr><td>Oxidized
Ficoll (400)</td><td>1:3.125</td><td colspan="2">8%
activation</td><td>6%
inhibition</td><td colspan="4">←→
No significant change</td></tr>
</table>

RESULTS

Formaldehyde, glutaraldehyde (both at 0.1:0.01 nmol/experiment), oxidized Dextran at 0.1:1.786 and 0.1:0.0357 nmol/experiment, and oxidized Ficoll at 0.1: 0.3125 nmol/experiment were reacted with urease at pH 7 for various lengths of time (TABLE 1). The mono- and dialdehydes activate the enzyme around 15–20% at this 10:1 ratio of enzyme to aldehyde.

Periodate-oxidized Dextran and Ficoll activate at low concentration when the

TABLE 2. Effect of Aldehydes on Urease Activity: Effect of pH (1-minute reactions)[a]

Aldehyde	Enzyme: Aldehyde Ratio	pH 4.4	pH 7.0	pH 9.2
Formaldehyde	1:0.1	22% activation at 1:0.1	18% activation at 1:0.1	19% activation at 1:1
	1:1000	36% inhibition	22% inhibition	40% inhibition
Glutaraldehyde	1:0.0001 to 1:100	2.6% activation at 1:0.0001	29.5% activation at 1:0.0001	No significant change at 1:0.0001
	1:1000	13% inhibition	20% inhibition	35.2% inhibition
Oxidized Dextran (70)	1:17.86	58% inhibition at 10 min	63% inhibition at 10 min	92% inhibition at 10 min
	1:0.357	6.5% inhibition at 10 min	5.1% inhibition at 10 min	0.7% inhibition at 10 min
Oxidized Ficoll (400)	1:3.125	96% inhibition at 10 min	5% inhibition at 2 min	20% inhibition at 30 sec
		No activation	8% activation at 30 sec	4% activation at 3 min

[a]Enzyme and aldehyde concentrations in nmol per experiment.

reaction time is short. At higher concentrations and longer reaction times, the enzyme is inhibited (TABLES 1 and 2).

The effect of pH on aldehyde modification of urease (1-minute reaction time) at various enzyme:aldehyde ratios yields results (TABLE 2) that, broadly speaking, indicate that the change of conformation of the protein at pH 4.4, 7.0, and 9.2 also controls the availability of ϵ-NH_2 groups. At low aldehyde concentration, the enzyme is activated; HCHO activates by about 20% at 1:0.1 ratio (0.1:0.01 nmol), and as the ratio is increased, there is no effective change in activity up to 1:100 (0.1:10 nmol), whereas at 1:1000 (0.1:100 nmol), inhibition of around 35–40% at pH 7 is seen. Glutaraldehyde activates at 0.1:0.00001 nmol/experiment (that is, 10,000:1) and inhibits at 1:1000 (0.1:100 nmol) to about 13%, 20%, and 35% at pH 4.4, 7.0, and 9.2, respectively.

DISCUSSION

Reactions of aldehydes conducted with urease solution have the following broad features:

a. At low enough aldehyde concentration, titration of an optimum number of NH_2 groups that are more easily available probably on the surface of the protein actually activates the enzyme.
b. pH of the reaction has an effect on aldehyde-mediated modifications. Often the effect of activation or inhibition is more pronounced at extremes of pH (pH 4.4 or 9.2), which suggests that the protein probably has a more open conformation than at pH 7.0. A secondary effect of pH induced inhibition at the two extremes of the pH profile curve cannot be ignored.
c. Kinetics of the reaction of aldehydes produce a classical hyperbolic curve in most cases, with initial activation reaching a peak followed by inhibition as reaction time increases. (See TABLES 1 and 2 for the effect of time.) The implication here is that, when a critical number of NH_2 groups are titrated by the aldehyde, activation reaches its peak. Beyond this point, when more NH_2 groups are attacked, enzyme inhibition results.
d. The concentration-dependent kinetics of aldehyde reaction with urease suggest that there may be two classes of NH_2 groups in urease. One is probably more on the surface and therefore more easily accessible and, when blocked, activates the enzyme. The other class of NH_2 groups are not easily accessible for modification; however, when these are blocked, inhibition results.
e. Unlike the smaller aldehydes, polymers such as Dextran and Ficoll also provide a cage of hydroxyl groups surrounding the enzyme. Whether this contributes to the modulation of enzyme activity is under investigation.
f. There are also indications that the enzyme will exhibit a similar activation pattern when the NH_2 groups are reacted with other kinds of cross-linkers that have different chemical groups. This suggests that this phenomenon may not be specific to aldehydes alone.

Immobilization of *Aspergillus ficuum* Phytase by Carbohydrate Moieties onto Cross-linked Agarose

H. CHARLES DISCHINGER, JR. AND ABUL H. J. ULLAH

Southern Regional Research Center
Agricultural Research Service
United States Department of Agriculture
1100 Robert E. Lee Boulevard
New Orleans, Louisiana 70124

INTRODUCTION

Phytase, myo-inositol hexakisphosphate phosphohydrolase, is unique among the acid phosphatases in its ability to accept phytic acid (IP6) as substrate. Some breakdown products of phytate, especially 1,3,4,5-IP4, 1,3,4,6-IP4, and 1,4,5-IP3, have been shown to mobilize intracellular calcium ions[1] and are considered to be pharmacoactive.[2–4] Attempts to produce these compounds nonenzymatically from phytate have resulted in mixtures of IP5, IP4, and IP3 isomers.[5] Purification of these isomers from the mixture may prove to be cumbersome and uneconomical. An alternative approach would be to use an immobilized-enzyme-based bioreactor to break down phytic acid, which is readily available from soybeans and other seed crops. Previous attempts to develop a bioreactor by covalent attachment of phytase via the protein backbone have resulted in severe reduction in K_{cat}.[6,7] Phytase is a heavily glycosylated protein; nearly 43% of the observed molecular weight is due to carbohydrate. In an attempt to improve the catalytic efficiency of the immobilized phytase, we took the approach of covalently attaching the molecule through its carbohydrate moieties. In this communication we present the characterization of a bioreactor produced by this technique.

MATERIALS AND METHODS

Enzyme

Aspergillus ficuum phytase (E.C.3.1.3.8) was purified according to a procedure established in this laboratory.[8]

Immobilization

Hydrazide AvidGel F (BioProbe Intl., Tustin, Calif.) was used as the support. The immobilization procedure was according to manufacturer's protocol. Phytase (9.55 mg, 400 nKat/mg) in 50-mM sodium acetate, pH 5.0 was oxidized with 10-mM

TABLE 1. Total Output of the Bioreactor versus Flow Rate

Flow Rate (cm/minute)	Temperature (°C)	Space Velocity[a]	Output (nKat)
0.25	30	0.07	9.53
0.5	30	0.15	18.46
1.0	30	0.30	30.65
2.0	30	0.60	38.08
3.0	30	0.90	41.24

[a]Space velocity = flow rate (mL/minute) divided by column volume (mL).

sodium periodate by gently shaking at room temperature for an hour. The unreacted periodate was removed by gel filtration (P2, BioRad, Richmond, Calif.). The oxidized phytase (383 nKat/mg) was reacted with the gel (3.0 ml) in the acetate buffer for 16 hours at 4°C. The gel was poured into a Kontes (Vineland, N.J.) thermally jacketed Flex column (1.0 × 10 cm) and washed extensively with the acetate buffer to remove unbound phytase.

Activity Measurement

Bioreactor activity was determined by measuring the inorganic orthophosphate (P_i) content of the column eluate according to the method of Heinonen and Lahti.[9]

RESULTS

After immobilization, 5.25 mg of phytase were covalently bound to the matrix. This represents 55% of the mass, which in the soluble form would give 2010 nKat. The bioreactor's output was asseseed by passing phytate solution through the column at various flow rates (TABLE 1). The maximum output (41.24 nKat), obtained at the highest flow rate, is only 2.05% of the total free-form activity. However, the pH and temperature profiles (FIGURES 1 and 2) of the bound enzyme mimicked those of the free form. The Michaelis constant, K_m, was estimated to be 295 μM as compared to 40 μM for the free enzyme. the K_{cat} of the immobilized phytase was estimated to be 1.2 seconds^{-1}; the reported value for the free form was 58 s^{-1}.[8] Repeated passage of the substrate through the bioreactor resulted in hydrolysis of a maximum of 31% of the available phosphate (TABLE 2).

DISCUSSION

When immobilization was achieved through the protein backbone,[6] myo-inositol di- and monophosphates were generated upon repeated substrate passage. By contrast, when immobilization was performed using carbohydrate side chains (present study), a maximum of 31% of the available phosphate was cleaved (TABLE 2), indicating the final form to be myo-inositol tetraphosphates. This may be the consequence of a difference in conformation of the catalytic active center, which resulted in differential substrate accommodation. While the immobilized enzyme is able to degrade phytic acid, the observed values for the bioreactor's output and the calculated K_{cat} indicate a diminished catalytic performance. These results are very similar to those obtained upon immobilization of phytase when the attachment was to the protein backbone,[6,7] except that in the present study, the K_m was not so severely affected as were the other kinetic parameters. These results indicate that the active catalytic center of the enzyme may be distorted by extensive cross-linking with the carbohydrate side chain of phytase. While the immobilization of this glycoprotein by its carbohydrate moieties is an appropriate approach, the resulting enzyme should not suffer steric hinderance caused by extensive cross-linking. Since the native enzyme is heavily glycosylated, prevention of cross-linking is difficult. One approach to controlling the extent of attachment is to remove, by protein engineering, those glycosylation sites that are closest to the active center. We have identified the 10 glycosylated residues of the total of 19 ASN in the molecule. Six of these are close to the active site. By using site-directed mutagenesis, it should be possible to remove enough of these to produce an enzyme that can be immobilized by this method while retaining a high level of activity.

CONCLUSIONS

Fungal phytase was immobilized onto modified cross-linked agarose by the enzyme's carbohydrate side chains.

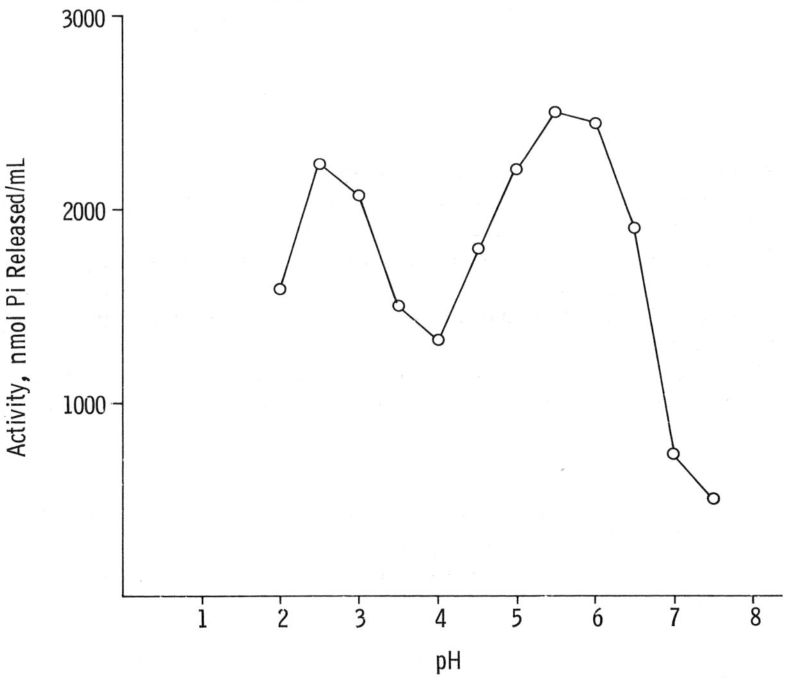

FIGURE 1. Effect of pH on the activity of the immobilized phytase.

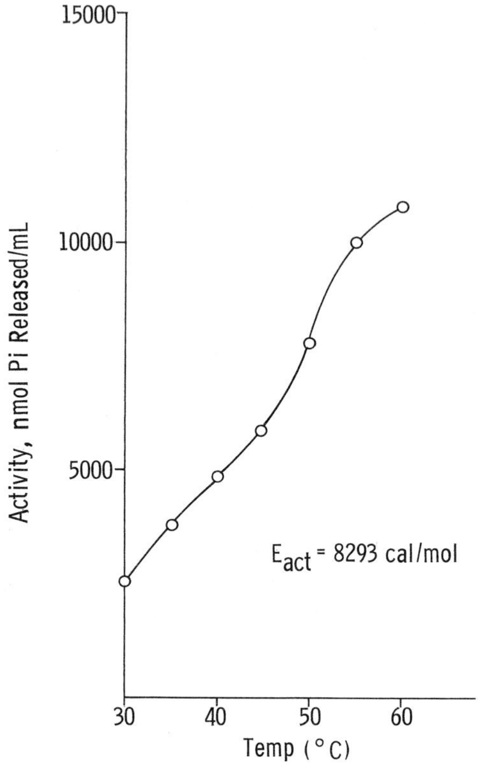

FIGURE 2. Temperature dependence of the activity of immobilized phytase.

The packed-bed bioreactor showed diminished catalytic activity by comparison with the soluble enzyme. However, the pH and temperature optima profile of the immobilized enzyme were essentially the same as those for the soluble form.

The K_m shift alone does not account for the reduction in catalytic turnover number, which indicates active center distortion induced by extensive cross-linking.

The kinetic data indicate that the final form of myo-inositol phosphate achieved by this bioreactor to be tetraphosphate.

We propose that improvement in catalytic efficiency can be brought about by reducing the extent of glycosylation. This could be realized through protein engineering.

TABLE 2. Repeated Passage of Phytate versus P_i Released from the Bioreactor

Passage No.	P_i Released (nmoles/ml)	Percent of available P_i
1	1448	24.1
2	1862	31.0
3	1859	30.9

REFERENCES

1. IRVINE, R. F., A. J. LETCHER, J. P. HELSOP & M. J. BERRIDGE. 1986. The inositol tris/tetrakisphosphate pathway-demonstration of Ins(1,4,5)P_3 3-kinase activity in animal tissues. Nature **320:** 631–634.
2. MOLLEMAN, A., B. HOITING, M. DUIN, J. VAN DEN AKKER, A. NELEMANS & A. DEN HERTOG. 1991. Potassium channels regulated by inositol 1,3,4,5-tetrakisphosphate and internal calcium in DDT1 MF-2 smooth muscle cells. J. Biol. Chem. **266:** 5658–5663.
3. IVORRA, I., R. GIGG, R. F. IRVINE & I. PARKER. 1991. Inositol 1,3,4,6-tetrakisphosphate mobilizes calcium in *Xenopus* oocytes with high potency. Biochem. J. **273:** 317–321.
4. BEZPROZVANNY, I., J. WATRAS & B. E. EHRLICH. 1991. Bell-shaped calcium-response curves of INS(1,4,5)P3- and calcium-gated channels from endoplasmic reticulum of cerebellum. Nature **351:** 751–754.
5. PHILLIPPY, B. Q., K. D. WHITE, M. R. JOHNSTON, S.-H. TAO & M. R. S. FOX. 1987. Preparation of inositol phosphates from sodium phytate by enzymatic and non-enzymatic hydrolysis. Anal. Biochem. **162:** 115–121.
6. ULLAH, A. H. J. & B. J. CUMMINS. 1987. Immobilization of *Aspergillus ficuum* extracellular phytase on modified Fractogel. Biotechnol. Appl. Biochem. **9:** 380–388.
7. ULLAH, A. H. J. & B. J. CUMMINS. 1988. *Aspergillus ficuum* extracellular phytase: immobilization on glutaraldehyde activated silicate. Ann. N.Y. Acad. Sci. **542:** 102–106.
8. ULLAH, A. H. J. & D. M. GIBSON. 1987. Extracellular phytase (E.C.3.1.3.8) from *Aspergillus ficuum* NRRL 3135: purification and characterization. Prep. Biochem. **17:** 63–91.
9. HEINONEN, J. & R. LAHTI. 1981. A new and convenient colorimetric determination of inorganic orthophosphate and its application to the assay of inorganic pyrophosphatase. Anal. Biochem. **79:** 313–317.

Optimization of the Production of β-Galactosidase by an Autolytic Strain of *Streptococcus salivarius* subsp. *thermophilus*[a]

SUSAN LINKO,[a,b] SATU ENWALD,[b]
MARJATTA VAHVASELKÄ,[c] AND
ANNIKA MÄYRÄ-MÄKINEN[c]

[b]*Laboratory of Biotechnology and Food Engineering*
Helsinki University of Technology
SF-02150 Espoo, Finland

[c]*Valio Finnish Cooperative of Dairies' Association*
Research and Development Center
Helsinki, Finland

INTRODUCTION

The technological and nutritional advantages of the enzymic hydrolysis of lactose in milk and dairy products using β-galactosidase (lactase, β-D-galactoside galactohydrolase, EC 3.2.1.23) are clearly understood.[1,2] The continuous production scale hydrolysis of cheese whey lactose by immobilized fungal β-galactosidase is an established practice since 1978.[3,4]

The predominant microbial β-galactosidase sources have been the molds *Aspergillus oryzae* and *A. niger,* and the yeast *Kluyveromyces fragilis.* An immobilized mold lactase is especially suitable for the treatment of whey, owing to the low working pH, although *A. oryzae* lactase may also be used at pH 6.7 at low temperatures.[5] The commercially available intracellular *K. fragilis* β-galactosidase has a pH optimum of about 6.5, and it has been used in appropriate dairy applications. However, the relative instability of the enzyme during immobilization and under process conditions has limited the practical use mainly to batch processing of milk with the soluble enzyme.[4,6] There is a definite need of a thermostable, food-grade β-galactosidase having a neutral pH optimum for lactose hydrolysis of milk and other dairy products, and for the production of novel food-grade oligosaccharides.[7]

Several bacteria produce extracellular β-galactosidases of a pH optimum in the neutral area, but only a few have so far been of commercial interest. The use in the food industry of the much studied *Escherichia coli* lactase is limited because *E. coli* is not a GRAS organism. The GRAS lactic acid bacteria are of great interest as potential lactase producers, but of the thermostable dairy starters, *Lactobacillus bulgaricus, L. helveticus,*[8] and *Streptococcus cremoris*[9] are relatively poor enzyme producers. However, *Streptococcus salivarius* subsp. *thermophilus* (*S. thermophilus*),

[a]The author is grateful to the Jenny and Antti Wihuri Foundation for financial support.

an organism also widely used in the production of fermented dairy foods, is a particularly promising microorganism for the production of β-galactosidase.[10,11] The intracellular *S. thermophilus* β-galactosidase is much more stable under the required process conditions than the *Kluyveromyces fragilis* β-galactosidase.[12] Of practical importance was also the discovery of an autolyzable strain *S. thermophilus* 11F,[13] which allows convenient and economic production of β-galactosidase. This enzyme has since been extensively studied and characterized.[2,14,15]

In the present paper, we report on the optimization of the *S. thermophilus* 11F β-galactosidase production in yields of commercial interest.

MATERIALS AND METHODS

Microorganism and Growth Media

The autolytic *Streptococcus thermophilus* 11F strain was obtained from the Valio Research and Development Center, Helsinki, Finland. The organism was maintained in a lyophilized state at −80°C and revived in autoclaved skim milk at 42°C for each fermentation. The inoculum was prepared by growing the bacteria (7% v/v) in 10% skim milk for ca. 4 hours at 42°C. Two different culture media were used. Two- or 10-liter batch cultures were grown on whey-permeate-based media containing (w/v) 6% dried whey UF permeate (ca. 80% lactose, 9% ash, 4% proteins, 2% organic acids, and 3% water) (Valio), and either 0.25% casein hydrolysate (Valio) and 0.25% yeast extract (Difco) (medium 1), or 1% proteose peptone (Difco), 1% yeast extract, and 0.1% (v/v) Tween 80 (medium 2). Unless otherwise explained, each component was autoclaved (15 minutes at 121°C) separately and combined aseptically before transferring into the fermentor with a peristaltic pump. In the 2-liter fed-batch fermentations, a further 300 cm^3 of fresh medium 3 (3% whey permeate, 0.5% proteose peptone, and 0.5% yeast extract) was pumped into the fermentor in the end of the exponential growth phase.

Fermentations

Fermentations were carried out in 2 dm^3 working volume Biostat® MD or 10 dm^3 working volume Biostat® E (B. Braun, Melsungen, Germany) fermentors. After the addition of the inoculum (1% v/v), the cultures were agitated at 200 rpm without aeration, and the pH was controlled at 6.5 by automatic addition of 4-M ammonia, if not otherwise stated. Varying temperatures and temperature profiles were studied to increase β-galactosidase production.

Analytical Methods

The increases in total cell mass and cell lysis were followed on the basis of turbidity at λ = 600 nm. Light microscopy was used to obtain more detailed information on autolysis (nigrosin dye) and to investigate possible contamination of the cultures (methylene blue dye). Lactose was analyzed enzymically with a Boehringer Mannheim kit no. 176 303. β-Galactosidase was assayed at 45°C with lactose as a substrate by measuring the amount of glucose released from 5.55% (w/v) lactose in 0.1-M potassium phosphate buffer (pH 7.0). The reaction was terminated by addition of perchloro acid to 0.7% final concentration. After 3 hours at room temperature, glucose was determined enzymically using a Boehringer Mannheim kit no. 124 036. One unit of β-galactosidase activity was defined as the amount of enzyme that released 1 μmol of glucose in one minute.

RESULTS AND DISCUSSION

Sterilization of the Medium

The sterilization procedure used in the preparation of the whey-based medium 1 greatly affected both the growth of *Streptococcus thermophilus* 11F and the production of β-galactosidase (TABLE 1). However, in all cases, the microorganism auto-

TABLE 1. The Effect of Sterilization of the Medium on the Growth and β-Galactosidase Production of *Streptococcus thermophilus* 11F in Batch Fermentation[a]

Sterilization Procedure[b]	Turbidity (λ = 600 nm)	β-Gal Activity (U cm^{-3})
40 min, 121°C	4.5 (6 h)	4
15 min, 121°C	ND[c]	22
15 min, 121°C *in situ* sterilization[d]	4.5 (7 h)	12
15 min, 121°C each component separately	6.4–7.1 (5–6 h)	12–38

[a]The cultures were grown at 30°C for 4 hours after which the temperature was increased to 42°C. Shaking speed was 100 rpm, and the pH 6.5 was controlled by 2-M ammonia.

[b]The medium contained 6% whey permeate, 0.25% casein hydrolyzate, and 0.25% yeast extract (medium 1).

[c]Not determined.

[d]Fermentation was carried out in a Biostat® E fermentor with a working volume of 10 dm^3.

lyzed relatively well at the end of the growth period (FIGURE 1) and β-galactosidase was released at varying activities into the medium. In these experiments, the highest activity of 38 U cm^{-3} was obtained when the medium components were autoclaved separately. The 40-minute sterilization at 121°C suggested by the fermentor manufacturer to insure the sterility of the culture vessel and the medium was too long a treatment for β-galactosidase production by *S. thermophilus* 11F. The heat treatment resulted in a Maillard browning reaction, and the medium turned dark brown. With a shorter, 15 minute sterilization time, some browning of the medium still took place, but no contaminations were observed.

Effect of Temperature on Cell Lysis

The effect of temperature on the growth of *Streptococcus thermophilus* 11F, on autolysis, and on β-galactosidase production was studied using medium 2 in Biostat®

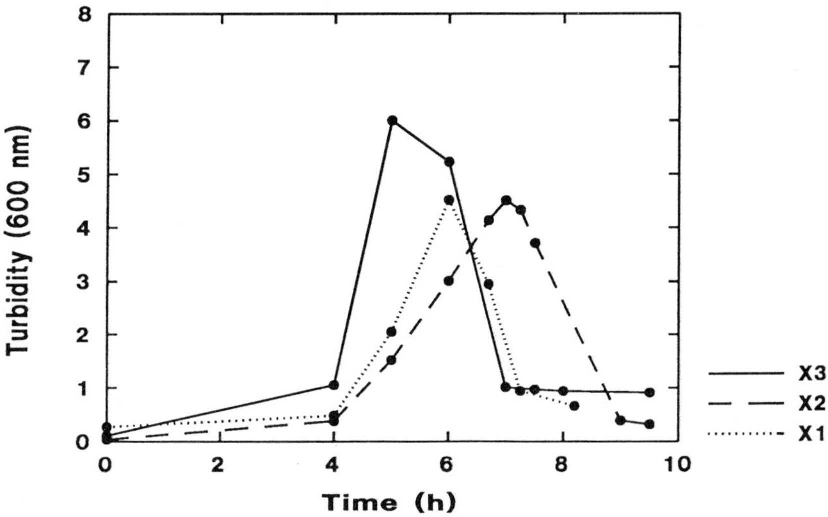

FIGURE 1 The effect of the heat treatment of the medium on the growth of *Streptococcus thermophilus* 11F; 40 minutes at 121°C (X1), 15 minutes at 121°C (X2), 15 minutes at 121°C, each component separately (X3).

MD fermentors with agitated 2 dm^3 submerged cultures under different temperatures and temperature profiles. The optimum temperature both for growth and cell lysis was about 42°C (FIGURE 2), which agrees well with the observation of Sandholm and Sarimo[13] that the maximal growth rate and lysis of *S. thermophilus* 11F are

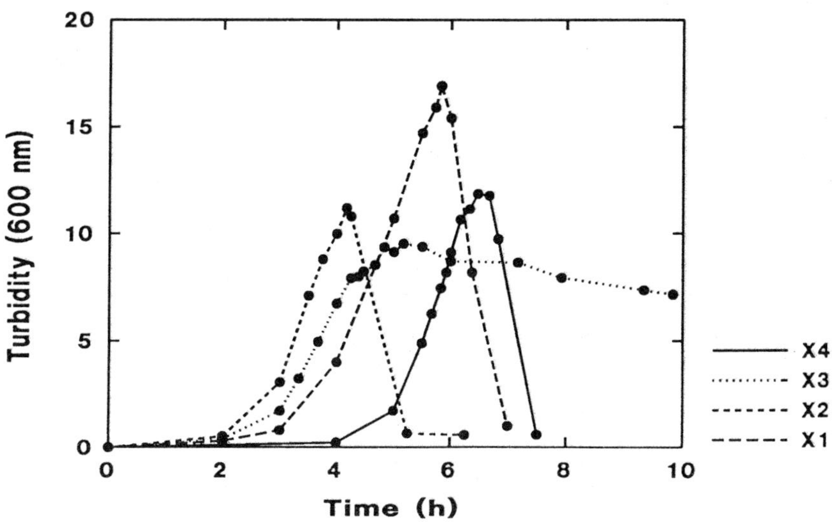

FIGURE 2 The effect of temperature on the growth of *Streptococcus thermophilus* 11F; 37°C (X1), 42°C (X2), 45°C (X3), 4 h 30°C → 42°C (X4). The medium contained 6% whey permeate, 1% proteose peptone, 1% yeast extract, and 0.1% Tween 80 (medium 2).

TABLE 2. Batch β-Galactosidase Production by Submerged *Streptococcus thermophilus* 11F Cultures

Temperature Profile	β-Galactosidase Activity ($U\ cm^{-3}$)
37°C	60 (7.5 h)
42°C	64 (6.3 h)
2 h 30°C → 42°C	58 (7.7 h)
4 h 30°C → 42°C	69 (7.5 h)
4 h 30°C → 45°C	59 (9.8 h)
6 h 30°C → 42°C	63 (10.3 h)
8 h 30°C → 42°C	47 (11.3 h)
45°C	19 (26 h)

between 40° and 45°C. At a lower temperature of 37°C, the growth rate decreased but the maximum turbidity obtained was much higher. At a constant temperature of 45°C, both the growth and the autolysis were retarded. However, if the microorganism was first grown 4 hours at 30°C followed by an upshift of the temperature to 45°C, the cell lysis was not prevented and a good yield of β-galactosidase was obtained.

Batch Fermentation

The highest β-galactosidase activity of 69 U cm^{-3} in a batch fermentation was obtained when the temperature was increased from 30° to 42°C after a 4 hour growth period (TABLE 2). In all cases, a progressive increase in the enzyme activity of the intracellular β-galactosidase released into the supernatant was observed until up to ca. 1 to 2 hours after the exponential growth phase, whereafter the activity remained quite constant. With the same strain, Smart *et al.* obtained ca. 15 U cm^{-3} of β-galactosidase on a whey-permeate-based medium in a 0.5-dm^3 working volume fermentor at constant 42°C.[14] They used a much higher agitation rate of 500 rpm, and maintained anaerobic conditions by continuous sparging with a mixture of nitrogen and carbon dioxide (95:5). Later, Smart and Richardson reported a β-galactosidase activity of ca. 20 U cm^{-3} on a complex synthetic medium.[15] Also Chang and Mahoney used a synthetic medium in the production of β-galactosidase by *S. thermophilus*

TABLE 3. Fed-Batch Fermentation of *Streptococcus thermophilus* 11F

Cultivation Temperature (°C)	Time of Nutrient Supplementation (hours)	Turbidity (600 nm)	Cell Count (cm^{-3})	β-Gal Activity ($U\ cm^{-3}$)
37[a]	5.8	19.7 (6.8 h)	ND	74
42[a]	4.9	13.1 (5.7 h)	6.3×10^9	70
37[b]	6.5	10.6 (7.4 h)	2.9×10^9	40

[a]The medium contained 6% whey permeate, 1% proteose peptone, 1% yeast extract, and 0.1% Tween 80 (medium 2).

[b]The medium contained 6% whey permeate, 0.25% casein hydrolyzate, and 0.25% yeast extract (medium 1).

11F.[2] In a 10-dm^3 culture, an enzyme activity of ca. 12 U cm^{-3} was obtained after 7 hours incubation at 42°C. In this case, the shaking speed was 150 rpm.

Under the conditions used in the present work, only about 45 to 65% of the lactose in the medium was utilized by *S. thermophilus* 11F.

Fed-Batch Fermentation

In order to increase the total cell mass for an improved β-galactosidase activity in the supernatant after the cell lysis, a fed-batch operation was investigated. In the fed-batch fermentations, the additional nutrients (medium 3) were added into the fermentor at the time turbidity had reached its maximum value. Much higher turbidities, and thus total biomass, were indeed obtained in the fed-batch fermentations with medium 2 as compared to the conventional batch fermentations. β-Galactosidase activity did not, however, increase in the same order of magnitude. The best β-galactosidase activity of 74 U cm^{-3} in the fed-batch experiments was obtained at a constant 37°C (TABLE 3). Consequently, in a fed-batch mode, equal or higher β-galactosidase activities were obtained even at constant 37°C than in a batch mode involving an upward temperature shift. The maximum activity obtained was almost four times higher than the activities reported earlier for *S. thermophilus* 11F.[2,14,15] Under the above experimental conditions at a constant 37°C, the use of the less-costly medium 1 resulted in a much slower growth and rate of autolysis, and the maximum turbidity obtained was only about one-half that obtained with the standard medium 2 used in the fed-batch fermentations (TABLE 3).

REFERENCES

1. MUSTRANTA, A., M. LINKO, M. HARJU & M. HEIKONEN. 1979. Production and use of lactase. Kem.-Kemi **6:** 306–308.
2. CHANG, B.-S. & R. R. MAHONEY. 1989. Purification and thermostability of β-galactosidase (lactase) from an autolytic strain of *Streptococcus salivarius* subsp. *thermophilus.* J. Dairy Res. **56:** 117–127.
3. FORSMAN, E.-S., M. HEIKONEN, L. KIVINIEMI, M. KREULA & P. LINKO. 1979. Kinetic investigations of the hydrolysis of milk lactose with soluble *Kluyveromyces fragilis* β-galactosidase. Milchwissenschaft **34:** 618–621.
4. HARJU, M. 1987. Lactose hydrolysis. IDF Bull. **212:** 50–55.
5. BAILEY, J. M. & M. LINKO. 1990. Production of β-galactosidase by *Aspergillus oryzae* in submerged bioreactor cultivation. J. Biotechnol. **16:** 57–66.
6. WECKSTRÖM, L., Y.-Y. LINKO & P. LINKO. 1980. Entrapment of whole cell yeast β-galactosidase in precipitated cellulose derivatives. *In* Food Process Engineering, Enzyme Engineering in Food Processing. P. Linko & J. Larinkari, Eds. **2:** 148–151. Applied Science Publishers. London, England.
7. SMART, J. B. 1991. Transferase reactions of the β-galactosidase from *Streptococcus thermophilus.* Appl. Microbiol. Biotechnol. **34:** 495–591.
8. WIERZBICKI, L. E. & F. V. KOSIKOWSKI. 1973. Lactase potential of various microorganisms grown in whey. J. Dairy Sci. **56:** 26–32.
9. RAMANA RAO, M. V. & S. M. DUTTA. 1978. Lactase activity of microorganisms. Folia Microbiol. **23:** 210.
10. RAMANA RAO, M. V. & S. M. DUTTA. 1977. Production of β-galactosidase from *Streptococcus thermophilus* grown in whey. Appl. Environ. Microbiol. **34:** 185–188.

11. GREENBERG, N. A. & R. R. MAHONEY. 1982. Production and characterization of β-galactosidase from *Streptococcus thermophilus.* J. Food Sci. **47:** 1824–1835.
12. WOYCHIK, J. H. & V. H. HOLSINGER. 1977. Use of lactase in the manufacture of dairy products. *In* Enzymes in Food and Beverage Processing. R. L. Ory & A. J. St. Angelo, Eds.: 67. American Chemical Society. Washington, D.C.
13. SANDHOLM, E. & S. S. SARIMO. 1981. Autolysis of *Streptococcus thermophilus.* FEMS Microbiol. Lett. **11:** 125–129.
14. SMART, J. B., V. L. CROW & T. D. THOMAS. 1985. Lactose hydrolysis in milk and whey using β-galactosidase from *Streptococcus thermophilus.* N.Z. Dairy Sci. Technol. **20:** 43–56.
15. SMART, J. & B. RICHARDSON. 1987. Molecular properties and sensitivity to cations of β-galactosidase from *Streptococcus thermophilus* with four enzyme substrates. Appl. Microbiol. Biotechnol. **26:** 177–185.

Optimizing Multienzyme Production

DIMITRIOS G. HATZINIKOLAOU AND BASIL J. MACRIS[a]

Biosystems Technology Laboratory
Department of Chemical Engineering, Sector IV
National Technical University of Athens
9 Iroon Polytechniou Avenue
Athens 15773, Greece

INTRODUCTION

Traditional fermentations for enzyme production are unidirected towards the enzyme of interest. However, despite optimization of process parameters towards maximum productivity of the particular enzyme, the cost of the final product is usually high.

Production of more than one enzyme in one step significantly increases the efficiency and improves the economy of the process. Very little, however, has been done in this respect. Thus, a limited number of applications exist, the most important of which refers to cellulase production, where more than one enzyme act in a concerted action for cellulose hydrolysis.[1,2] Also, glucose oxidase and catalase are commercially produced in one step, the latter enzyme being a by-product from manufacturing analytical grade glucose oxidase from *Aspergillus niger.*[3] Recently, simultaneous production of catalase and glucose oxidase from *Alternaria alternata*[4] and α- and β-amylase from *Aspergillus niger*[5] has been reported.

In this work, an attempt has been made to investigate a number of fermentation parameters affecting simultaneous production of certain enzymes. As a model system, a strain of *Aspergillus niger,* capable of simultaneously producing significant levels of cell-bound superoxide dismutase, glucose oxidase, and catalase, was used.

MATERIALS AND METHODS

Microorganism

A wild-type strain of *Aspergillus niger* isolated in our laboratory was used. The stock culture was maintained on potato-dextrose-agar at 4°C.

Media and Growth Conditions

The mineral medium used had the following composition (g/l): $(NH_4)_2HPO_4$, 0.4; KH_2PO_4, 0.2; $MgSO_4 \cdot 7H_2O$, 0.16. This medium was supplemented with the appropriate concentrations of glucose and peptone used as carbon and nitrogen sources, respectively. The initial pH of the medium was adjusted to 6.0. Sterilization

[a] Author to whom correspondence should be addressed.

was effected at 115°C for 30 minutes. Growth was carried out in 2-l Erlenmeyer flasks containing 500 ml of medium in a controlled atmosphere rotary shaker (Lab-Line, United States) at 30°C and 225 rpm.

Recovery of Enzymes

Culture samples (30–60 ml) were collected from the flasks at specific time intervals. Cells were separated on a Whatman No. 1 paper filter, washed twice with buffer (50-mM potassium phosphate, pH 7.0), and resuspended in 200 ml of the same buffer. The resulted cell suspension was introduced in a cell homogenizer (Rannie Lab Homogenizer, Model 8.30H) operating at 1000 bar. Two passes in the homogenizer were adequate for maximum release of the three cell-bound enzymes. The disrupted cell suspension was clarified by centrifugation at 15,000 × *g* for 40 minutes. Enzyme activities were determined in the clear supernatant.

Enzyme Assays

Catalase (CAT) activity was determined by following the linear rate of H_2O_2 (20 mM) consumption recorded at 240 nm at pH 7.0 and 30°C. A Hitachi 2000 spectrophotometer equipped with a thermostated cuvette holder was used. One unit (U) of CAT activity was defined as the amount of enzyme required to consume one μmole of H_2O_2 per minute, at the above-described conditions.

Glucose oxidase (GOX) was measured by the *o*-dianisidine/peroxidase method. One unit (U) of GOX activity was defined as the amount of enzyme required to oxidize one μmole of D-glucose per minute at optimum pH (5.4) and temperature (33°C).

Superoxide dismutase (SOD) was assayed by the method of alkaline dimethylsulfoxide (DMSO).[6] One unit (U) of SOD activity was defined as the amount of enzyme required for 50% inhibition of the superoxide radical mediated reduction of cytochrome c by alkaline DMSO.

RESULTS AND DISCUSSION

Definition of the Unit Value Factor

Due to large differences in the commercial value of the enzymes investigated in this work, a unit value factor (UVF), based on the mean market unit value of the crude form of each enzyme, was used. The approximate ratio of the UVFs, assigning CAT unit value as 1, was:

$$CAT/GOX/SOD = 1/7/23$$

Effect of Fermentation Time

The effect of fermentation time on enzyme production appears in FIGURE 1 and shows different optimum time for each enzyme. While SOD highest levels of activity were observed at approximately the second day of fermentation, those of GOX and CAT were observed at the fourth and seventh day, respectively. When, the maximum multienzyme yield was calculated on the basis of UVF, the total unit value (TUV) per milliliter of culture volume showed its maximum value at the sixth day of fermentation (FIGURE 1). This value was used as an alternative optimum fermentation time in the consequent experiments.

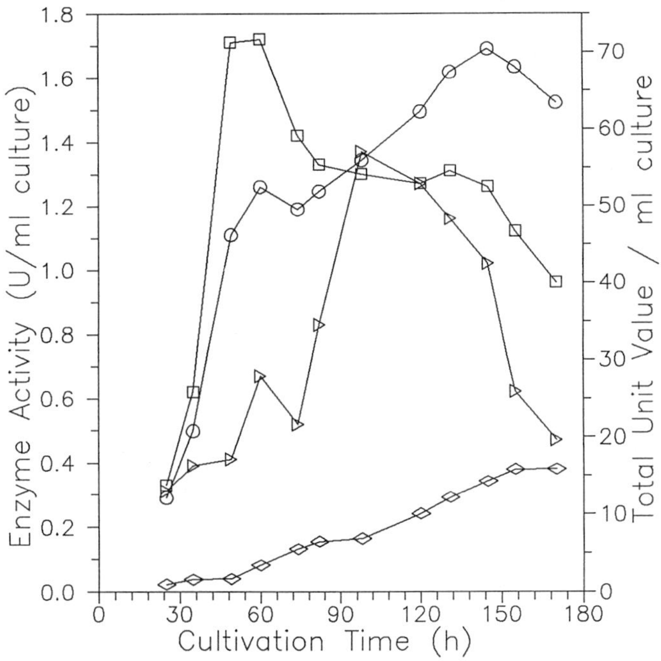

FIGURE 1. Effect of fermentation time on enzyme production and on total unit value. The mineral medium is supplemented with 25 g/l glucose and 15 g/l peptone. ◇, catalase (/100); ▷, glucose oxidase; □, superoxide dismutase; ○, total unit value.

Effect of Carbon Source Concentration

The effect of carbon source (glucose) concentration was examined in a number of experiments where the mineral medium was supplemented with 15 g/l peptone and the fermentation time was the optimum one, determined previously for each enzyme. The results appear in FIGURE 2 and show that SOD activity attained its maximum value at 40 g/l glucose, while those of CAT and GOX at 55 and 70 g/l, respectively. When, alternatively, the fermentation time was six days, different enzyme activity and TUV profiles were obtained (FIGURE 3), showing an optimum glucose concentration of 40 g/l for maximum TUV. The latter value was used as an alternative carbon source concentration in the consequent experiment.

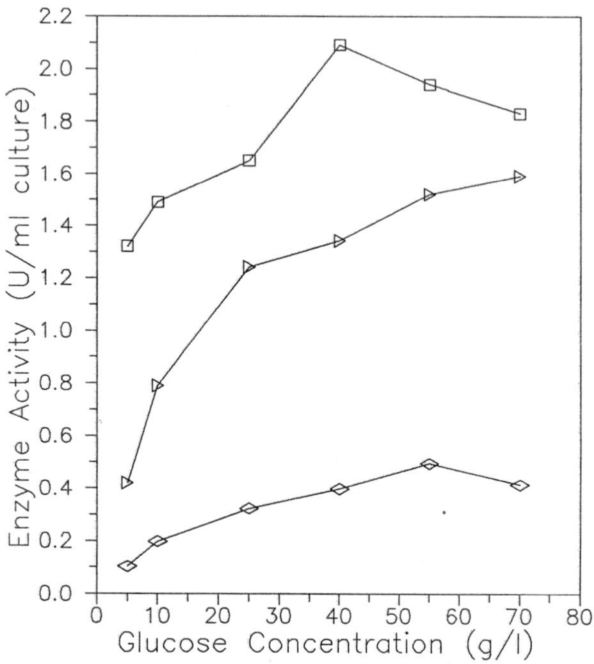

FIGURE 2. Effect of carbon source concentration on enzyme production. The mineral medium is supplemented with 15 g/l peptone. ◇, catalase (/100) at the 7th day of fermentation; ▷, glucose oxidase at the 4th day of fermentation; □, superoxide dismutase at the 2nd day of fermentation.

Effect of Nitrogen Source Concentration

Finally, the effect of nitrogen source (peptone) concentration on enzyme production and TUV was examined. The optimum peptone concentration for each enzyme was found using the optimum fermentation time and carbon source concentration determined previously. The results appear in FIGURE 4 and show different nitrogen source optima. The effect of peptone concentration on TUV and enzyme activities at the alternative optimum fermentation time and carbon source concentration appear in FIGURE 5. It is evident that the optimum TUV values were obtained at a peptone concentration different from that used for optimum enzyme activities, namely, 15 g/l.

Comparison of the Optimization Steps

The results of the three optimization steps described above are summarized in TABLE 1. In this table, both optimum growth conditions for maximum production of each enzyme and activity levels of the other two enzymes at these conditions are given. The TUV in equivalent CAT units of each optimization process has also been calculated from the UVF's ratio.

From these data, it can be seen that the optimum activity of each of the three enzymes was strongly affected by certain growth conditions, such as fermentation time and concentration of carbon and nitrogen source. Thus, we may suggest that the production profile of these enzymes can be controlled by altering these growth conditions.

A significant observation was that the optimum conditions for maximum TUV were quite different from those found for each particular enzyme. In addition, when the optimization of fermentation conditions was carried out with respect to the TUV of the enzymes and not to their optimum activity levels, the TUV attained its maximum value although none of the three enzymes was produced at its maximum level.

The results of this work showed that in multienzyme production systems, it was economically favorable to optimize the process on the basis of maximum TUV rather than on that of a particular enzyme. Further investigation should be directed towards other fermentation parameters affecting multienzyme production as well as the mathematical modeling of such a process.

CONCLUSIONS

The results of this work suggested that consecutive optimization of certain growth conditions could significantly increase the activity of catalase, glucose oxidase, and superoxide dismutase, which were simultaneously produced by the fungus

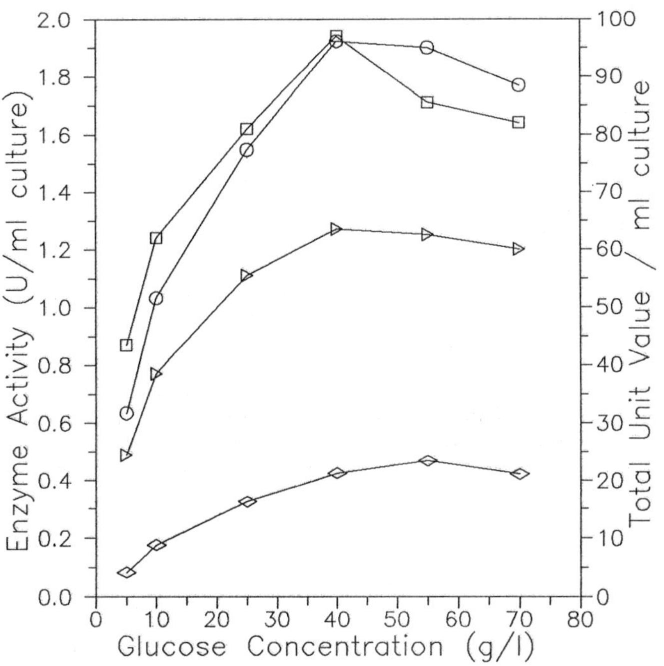

FIGURE 3. Effect of carbon source concentration on enzyme production and on total unit value. The mineral medium is supplemented with 15 g/l peptone. Fermentation time: 6 days. ◇, catalase (/100); ▷, glucose oxidase; □, superoxide dismutase; ○, total unit value.

Aspergillus niger. It was also possible to control the production profile of the three enzymes by altering certain growth conditions. In addition, the introduction of the total unit value based on the commercial value of each enzyme unit in the lowest purity form available, led to two important observations. First, the single unit value of the process was always lower than the total one, even in cases where maximum activities of the single enzyme were observed. Second, the optimum conditions for maximum total unit value of the process were different from the optimum ones for each enzyme and resulted in a significantly higher economic value of the process.

The results of this work are very promising and justify further investigation employing mathematical modeling.

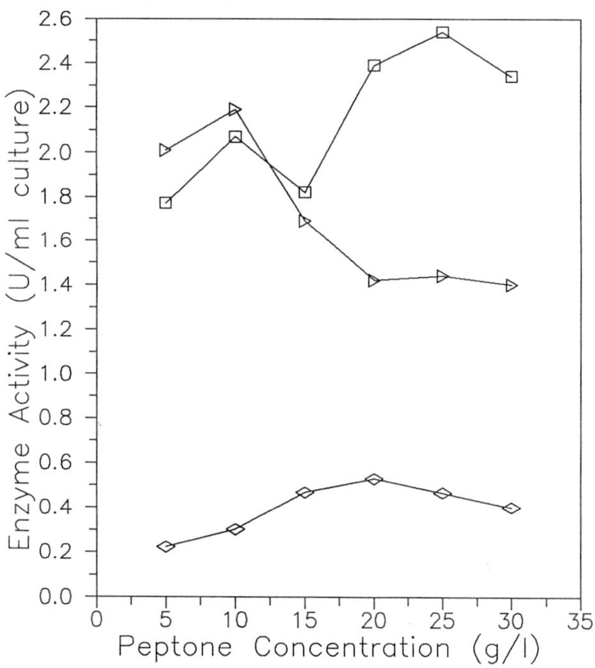

FIGURE 4. Effect of nitrogen source concentration on enzyme production. The mineral medium is supplemented with the optimum carbon source concentration for each enzyme. ◇, catalase (/100) at the 7th day of fermentation and 55 g/l glucose; ▷, glucose oxidase at the 4th day of fermentation and 70 g/l glucose; □, superoxide dismutase at the 2nd day of fermentation and 40 g/l glucose.

SUMMARY

The possibility of optimizing certain fermentation conditions towards the total unit productivity of several enzymes has been investigated.

As a model system, we used a particular strain of *Aspergillus niger* capable of simultaneously producing significant levels of superoxide dismutase, glucose oxidase,

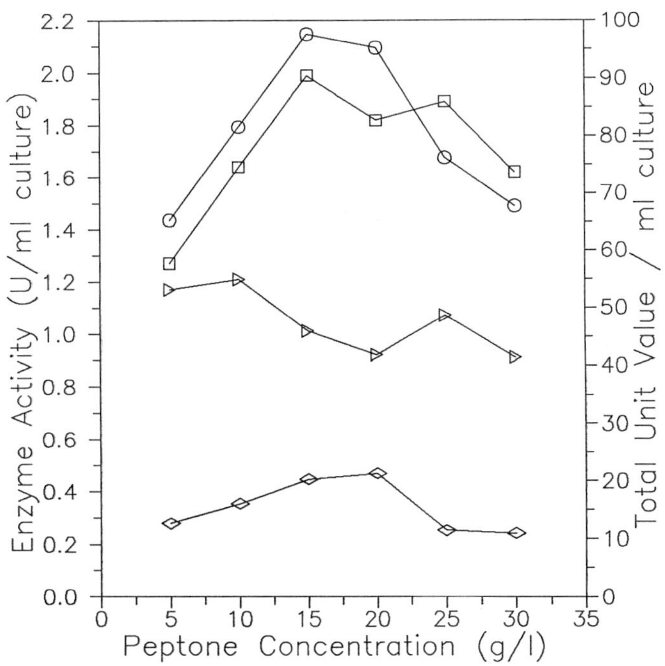

FIGURE 5. Effect of nitrogen source concentration on enzyme production and on total unit value. The mineral medium is supplemented with 40 g/l glucose. Fermentation time: 6 days. ◇, catalase (/100); ▷, glucose oxidase; □, superoxide dismutase; ○, total unit value.

and catalase. The fermentation parameters used in the optimization were the fermentation time, and the concentration of carbon and nitrogen source.

The optimum set of fermentation parameters for maximum production of each enzyme has initially been determined. The introduction of a value factor for each

TABLE 1. Comparison of Results Obtained during Consecutive Optimization of Catalase, Glucose Oxidase, and Superoxide Dismutase Produced by *A. Niger*

	Optimum Growth Conditions			Enzyme Activities at Optimum Conditions			Unit Values/ Milliliter at Optimum Conditions	
Enzyme/ Total Unit Value	Cultivation Time (days)	C Source (g/l)	N Source (g/l)	CAT (U/ml)	GOX (U/ml)	SOD (U/ml)	Single Unit Value	Total Unit Value
Catalase	7	55	20	52.9	0.60	1.34	52.9	87.92
Glucose oxidase	4	70	10	18.4	2.19	1.1	15.33	59.03
Superoxide dismutase	2	40	25	20.6	0.46	2.54	58.42	82.24
Total unit value	6	40	15	44.7	1.01	1.99	—	97.54

enzyme unit of specific purity yields the conclusion that there is a point in the parameter space, different from that of optimum production of each separate enzyme, where a maximum exists for the total unit value of the three enzymes produced. This result suggests that multienzyme optimization in fermentations may effectively reduce the cost of the production process.

REFERENCES

1. Wood, T. M. & S. I. McCrae. 1977. Carbohydr. Res. **57:** 177–133.
2. Pourquie, J. & J.-P. Desmarquest. 1989. *In* Enzyme Systems for Lignocellulose Degradation. M. P. Coughlan, Ed. **1:** 283–292. Elsevier Applied Science. London, England.
3. Frost, M. G. & D. A. Moss. 1987. *In* Biotechnology. H. J. Rehm & G. Reed, Eds. **7a:** 108–110. Verlagsgeselschaft mbH. Weinheim, Germany.
4. Caridis, K.-A., P. Christakopoulos & B. J. Macris. 1991. Appl. Microbiol. Biotechnol. **34:** 794–797.
5. Macris, B. J., D. Kekos & P. Christakopoulos. 1990. 15th Int. Carbohydr. Symp., August 12–17, Yokohama, Japan: 444.
6. Hyland, K., E. Voisin, H. Banoun & C. Auclair. 1983. Anal. Biochem. **135:** 280–287.

The Effect of Polyelectrolyte Additives on the Saccharification of Cellulase

YAN FENG, SHUGUI CAO, XIAOHONG JIN, ZHENGJI LI, AND YUHUA CHENG

The National Laboratory of Enzyme Engineering
Department of Chemistry
Jilin University
Changchun, 130023, China

INTRODUCTION

Many investigators are searching for methods to utilize plentiful cellulose effectively. One method is to convert cellulose into glucose with cellulase. At present, this process is promising but there also exist many difficulties. For example, the hydrolysis rate is too slow and the mechanisms of hydrolysis have not been sufficiently understood:[1] some reports revealed that the hydrolysis rate may be dependent not only on the intrinsic function of the enzyme, but also on the microscopic environment in the vinicity of the surface of cellulose and the active site of enzyme. In this work, we first used polyelectrolyte additives to increase the saccharification of cellulase, and the mechanism of enhancement of saccharification were elucidated from two viewpoints—the effect of polyelectrolytes on the activity of each cellulase component (C_1, C_x, β-glucosidase), and the adsorption of enzyme on the substrate.

MATERIALS AND METHODS

Enzyme Samples

Cellulase (Onozuka R-10, Yakult Honsha Co. Ltd. Japan.) was partially purified by column chromatography. Strong polycations P I [poly (*N*, *N*)-dimethyl aminopropylene bromide], P II [poly(2-hydroxyl 1-3-dimethlamino propylene chloride)], medium polyanions P III [poly (methyl acrylic acid)], and strong polyanions P IV [poly (benzethylene sulfonate)] were prepared in our lab. The other chemicals used were of reagent grade.

Saccharification

Cotton (25 mg) was hydrolyzed by 1.5 mL cellulase solution at pH 4.8 and 50°C for 24 hours in the presence of a given concentration of polyelectrolytes. The

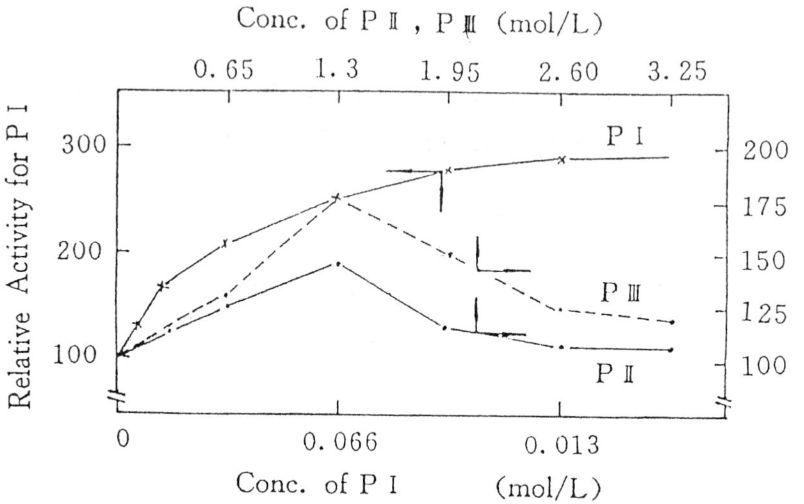

FIGURE 1. The effect of P I, P II, and P III on the saccharification of cellulase.

supernatant was supplied for determinations of reducing sugar according to Ellman's method.[2]

Activity Measurements of Endoglucanase and β-Glucosidase in the Presence of Polyelectrolytes

Carboxymethyl cellulase and salicin were hydrolyzed respectively by cellulase in the presence of polyelectrolytes for 15 minutes at pH 4.8 and 50°C. The reducing

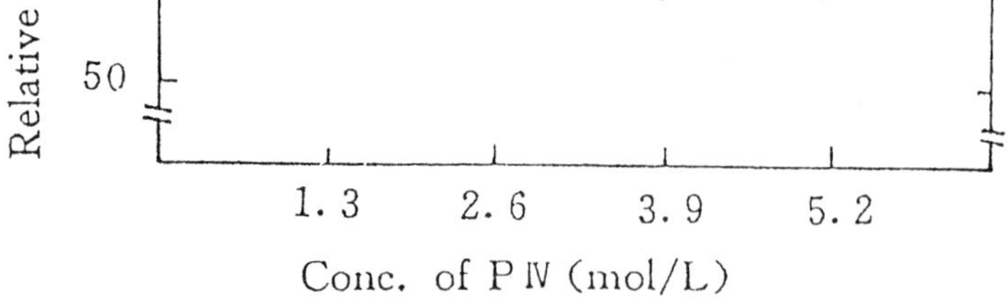

FIGURE 2. The effect of P IV on the saccharification of cellulase.

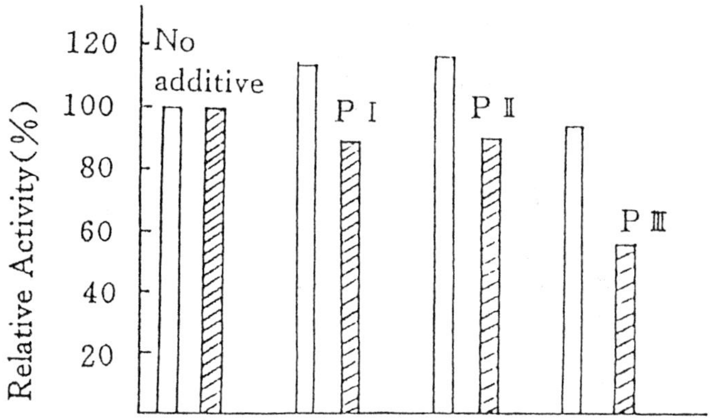

FIGURE 3. The effect of additives on the activities of C_x (white columns) and β-glucosidase (hatched columns).

suger was determined. The effect of polyelectrolytes on the activity determination was cancelled by control experiment.

The Determination of Protein

The concentration of protein was determined from absorbence at 280 nm. Bovine serum albumin was the standard.

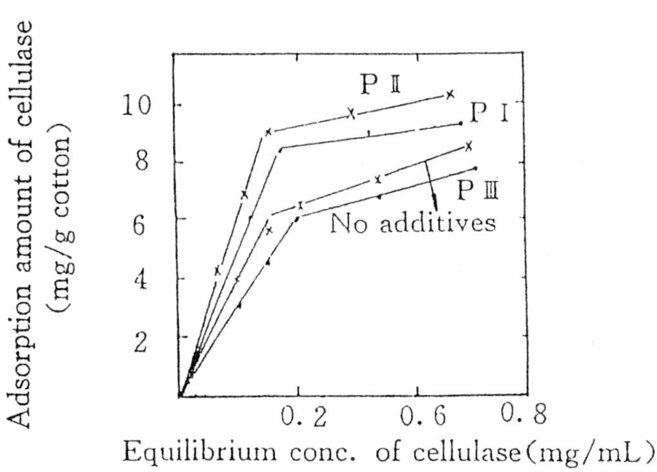

FIGURE 4. The effect of additives on the adsorption equilibrium of cellulase.

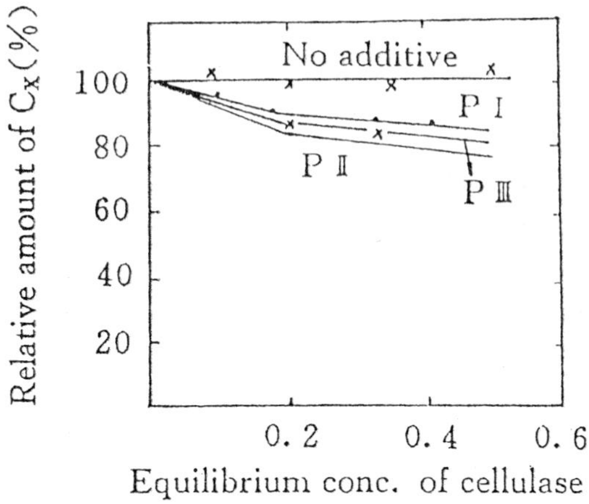

FIGURE 5. The relationship between the equilibrium concentration of cellulase and the relative amount of C_x in solution.

Adsorption of Cellulase on Cotton

The enzyme solution which contain polyelectrolytes was added to 25 mg cotton, then incubated for 30 minutes at 50°C. The concentration of protein in the supernatent was determined.

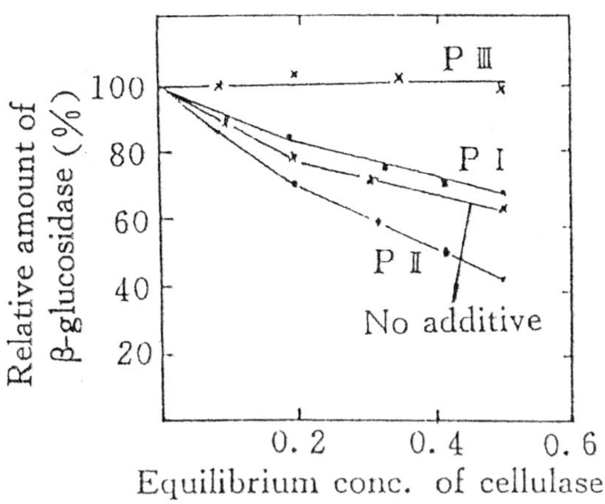

FIGURE 6. The relationship between the equilibrium concentration of cellulase and the relative amount of β-glucosidase in solution.

Determination of Concentration of β-Glucosidase and Endoglucanase in Solution during Saccharification

$$\begin{array}{c}\text{The relative amount} \\ \text{of enzyme in solution}\end{array} = \frac{\begin{array}{c}\text{The activity in solution that contains} \\ \text{polyelectrolytes after adsorption}\end{array}}{\begin{array}{c}\text{The percentage of remaining activity in the} \\ \text{presence of polyelectrolytes}\end{array}} \times \frac{1}{\begin{array}{c}\text{All the activity without adsorption} \\ \text{and polyelectrolytes}\end{array}} \times 100\%$$

RESULTS AND DISCUSSION

The Effect of Polyelectrolytes on the Saccharification of Cellulase

P I, P II, and P III enhanced the saccharification of cellulase about 190%, 45%, and 75% respectively. But P IV did not (FIGURES 1 and 2).

The Effect of Polyelectrolytes on the Activities of C_x and β-Glucosidase

Cellulase contained three components: C_1, C_x, and β-glucosidase; the process of cotton hydrolysis involved synergistic action of all enzymatic components. Since it is difficult to evaluate directly C_1 activity,[3] some effects of polyelectrolytes (P I: 6.6×10^{-2} mol/L, P II: 6.6×10^{-4} mol/L, P III: 1.3×10^{-2} mol/L) on the activities ± of C_x and β-glucosidase were studied (FIGURE 3). All polyelectrolytes inhibited β-glucosidase. P I and P II made C_x activity increase, but these increases were lower than those of saccharification. P IV made C_x activity decrease. It may be deduced that polyelectrolytes might improve C_1 activity and enhance the synergistic action by changing the activity ratio among these components.

The Adsorption of Cellulase in Cotton

Polyelectrolytes could change the adsorption amounts of anzyme (FIGURES 4, 5, and 6). Because the enzymatic components that had major disruption for cotton fiber are C_1 and C_x, it is important that polyelectrolytes increased the adsorption of C_x on the cotton.

Although polyelectrolytes with different structure did not have the same effect on the enzymatic adsorption, they increased the total adsorption of cellulase (except P III), changed the adsorption ratio of components, and enhanced the saccharification.

REFERENCES

1. OOSHIMA, H., *et al.* 1983. Biotechnol. Bioeng. **25:** 3103.
2. MILLER, G. H. 1959. Anal. Chem. **31:** 4426.
3. MANDELS, M., *et al.* 1976. Biotechnol. Bioeng. Symp. **6:** 21.

Glucose Oxidase Impette—Serum Glucose Estimation

P. V. SUNDARAM[a] AND P. G. HILL[b]

[a]*Center for Protein Engineering and Biomedical Research*
The Voluntary Health Services
Adyar, Madras—600 113, India

[b]*Clinical Chemistry Department*
Christian Medical College
Vellore, Tamil Nadu, India

INTRODUCTION

The concept of extending the use of an automatic pipette by linking a probe containing an immobilized enzyme in the form of a coiled nylon tube was first demonstrated by Sundaram and Sundaram and Jayaraman.[1,2] It was shown by these authors that such a device, called the immobilized enzyme pipette, or impette, may be used in the estimation of urea in serum samples using a probe containing immobilized urease.

This presentation is an account of the use of a glucose oxidase impette in the determination of glucose in serum. Results obtained from clinical trials conducted at a hospital at Vellore are presented.

MATERIALS AND METHODS

Glucose oxidase (100–150 U/mg) isolated from *Aspergillus niger* was obtained from Sigma Chemical Company, St. Louis, Mo. Nylon tube (1 mm i.d.) was imported from Portex Ltd., Hythe, Kent, United Kingdom. Peroxidase (2500/mg) was a Sigma Chemical Company product. All other chemicals were ultrapure grade.

Preparation of an Immobilized Enzyme Probe

The immobilized enzyme probe consists of a nylon tube (1 mm i.d.) coiled 3–4 turns, about 0.5 cm in diameter. The fully stretched length is about 10 cm. Partially hydrolyzed nylon tube was first treated with 1.25% v/v or 0.14-M glutaraldehyde made in pH 9.2 bicarbonate buffer for 40 minutes, washed with water, and filled with a 2 mg/ml solution of glucose oxidase made in a pH 7 phosphate buffer (0.05 M) and kept at 4°C overnight. The probe was then washed well with 1.0-M NaCl and kept stored, filled with the coupling buffer for storage when not in use.

FIGURE 1 shows how an impette looks when assembled.

Impette Assay Procedure

A typical impette assay consisted of serum diluted 1:40 with phosphate buffer being sucked into the impette probe by operating the automatic pipette in the normal

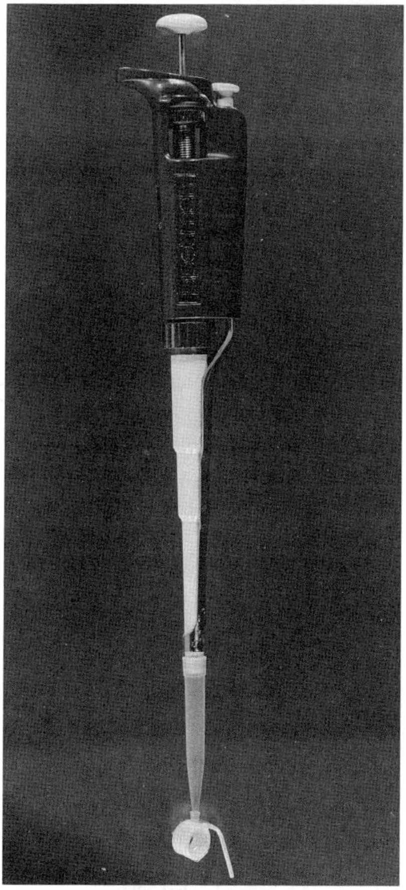

FIGURE 1

fashion. After holding the sample in the probe for 4 minutes at room temperature (34–36°C) the reacted serum was mixed with 1 ml color reagent containing 10 mg peroxidase and 15 mg *O*-dianisidine in 100 ml phosphate buffer (pH 7.0). Immediately before reading the absorbance, 1.5 ml of 5-M HCl was added and absorbance measured at 520 nm.

Caution

The pipette volume is carefully calibrated such that, when sucked in, the sample occupies the probe coil exactly without any air bubbles in it. This is a precaution necessary to exclude major errors.

Comparative Method

Continuous flow analysis using alkaline ferricyanide reduction in a Technicon AA II Auto Analyzer was the method chosen to compare with the performance of the impette method. This autoanalyzer method is based on the original method of Hoffman.[3] Serum was used for all analyses.

RESULTS

Linearity: Under the conditions described for the impette assay, linearity was achieved over a sample concentration of 0–16.7 mmol glucose per liter (0–300 mg/100 ml).

TABLE 1

	Glucose	
Serum	Target Value (mmol/liter)	Result Obtained (mmol/liter)
A	14.3	13.7
B	23.0	22.0
C	5.0	4.9

Accuracy: Assayed controls gave values close to the target value when calibration material was used, as shown in TABLE 1.

Comparison with the continuous flow ferricyanide reduction method: Fourteen samples were analyzed by both methods, with values over the range 0.72–14.1 mmol/liter. The individual results are shown in TABLE 2 and the statistical plot in FIGURE 2.

CONCLUSIONS

The preliminary conclusion was that the glucose oxidase impette can be successfully used for the determination of glucose in serum. The coiled tip containing the immobilized glucose oxidase retained activity despite regular use at room temperatures in the range 36–38°C for nearly six months.

Good correlation was observed between the impette method and the autoanalyzer ferricyanide reduction method. The method developed is a direct method.

TABLE 2

No.	Impette GOD (x mmol/liter)	Ferricyanide Reduction (y mmol/liter)
1.	8.7	7.7
2.	0.89	0.72
3.	2.8	2.7
4.	9.9	9.3
5.	3.2	4.4
6.	6.4	6.1
7.	10.8	11.7
8.	5.0	5.1
9.	4.4	4.6
10.	5.0	5.0
11.	9.8	9.1
12.	9.7	10.0
13.	15.0	13.0
14.	14.7	14.1
	$x = 0.92, y = 0.9x + 0.54$	

RELEVANCE TO CLINICS

The impette is a device that is simple to construct and operate and does not involve a vast outlay of capital. It is especially suitable for small clinics and physicians' consulting rooms. The results are quantitative and reproducible. The

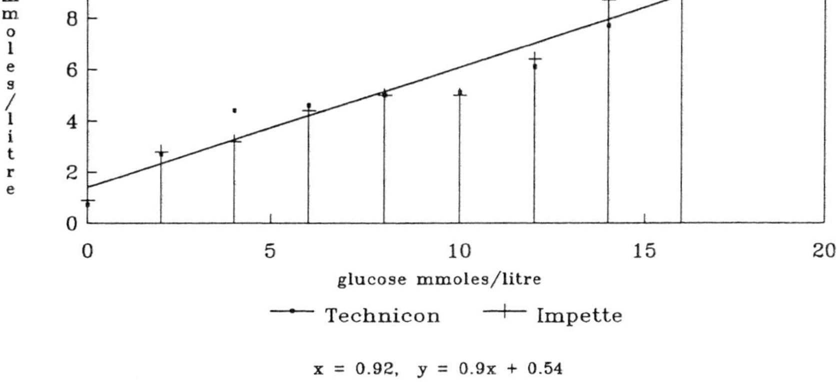

$x = 0.92, \quad y = 0.9x + 0.54$

FIGURE 2

impette probes may be reused several thousand times when used and stored with care.

Accuracy and reproducibility depend on a proper mixing of the sample with the inside walls of the probe that contains the immobilized enzyme. This can be achieved by a gentle (without causing air bubbles) and uniform suction into and pouring out of the serum sample each time.

REFERENCES

1. SUNDARAM, P. V. 1979. Immobilized-enzyme pipette. Scope and limitations of a simple device. Biochem. J. **179:** 445–447.
2. SUNDARAM, P. V. & S. JAYARAMAN. 1979. Immobilized enzyme pipette, impette. Routine estimation of serum and urine urea. Clin. Chim. Acta **94:** 309–316.
3. HOFFMAN, W. S. 1937. A rapid photometric method for the determination of glucose in blood and urine. J. Biol. Chem. **120:** 51–55.

An Affinity Ultrafiltration System for the Purification of Trypsin

HÖRDUR FILIPPUSSON AND
KRISTMUNDUR SIGMUNDSSON

Department of Biochemistry
University of Iceland
Vatnsmýrarvegur 16
Reykjavík IS-101, Iceland

INTRODUCTION

Biospecific affinity techniques offer potentially the most effective means of separating and purifying biomolecules. The combination of affinity binding with ultrafiltration is one promising way of utilizing biospecificity of binding on a large scale. Several systems that utilize affinity ultrafiltration have been reported in the literature.[1,2] Luong *et al.* reported an affinity ultrafiltration system where an acrylic polymer bearing benzamidine side chains binds trypsin specifically and retains it in an ultrafiltration device.[2] In the work reported here, we have looked at some properties of an affinity ultrafiltration system for the isolation of trypsin based on a similar polymer in order to test its suitability for purifying enzymes from animal glands.

Various glands and tissues that are by-products of meat production are sources for potentially valuable biochemicals. The work reported here is part of a project aimed at the commercial production of enzymes and other biochemicals from ovine glands. The pancreas gland is a rich source of several enzymes, and in the interest of economy, they should all be extracted and isolated from the same raw material. Thus an efficient process is needed to separate each enzyme without losing the others.

THEORETICAL

The affinity-ultrafiltration system is shown schematically in FIGURE 1. If it is assumed for a system at constant volume that one component only is specifically retained by the affinity polymer while other components pass the membrane freely, then the concentration of each of these is given by:

$$F(C_{\text{in}} - C_{\text{out}})\text{d}t = V\text{d}C_{\text{s}} \tag{1}$$

where F is the flow rate, C_{in} and C_{out} are the concentrations flowing in and out respectively, C_{s} is the concentration within the system, V is the volume of the system, and t is the time.

This does not, however, hold for an ultrafiltration system except for the smallest molecules, due to the rejection behavior of the system, which is described by the rejection coefficient:

$$\mathbf{R} = 1 - C_{\text{out}}/C_{\text{s}}. \tag{2}$$

Combined with Equation 1, this gives:

$$F[C_{\text{in}} + C_{\text{k}}(1 - R)\text{d}t = V\text{d}C_{\text{s}} \tag{3}$$

This has the general solution:

$$C_{\text{s}} = (1/(1 - R))[C_{\text{in}} + (C_0(1 - R) - C_{\text{in}})e^{-(1-R)^{\Phi}}] \tag{4}$$

If the rejection is taken into account, then the elution from the system is therefore described by the equation:

$$\text{C}_{\text{out}} = C_{\text{in}} + (C_0(1 - R) - C_{\text{in}})e^{-(1-R)^{\Phi}} \tag{5}$$

where $C_0 = C_{\text{k}}$ at $t = 0$ and $\Phi = F^*t/V$. This equation may be used to compute the changes in concentrations in the system and in the effluent.

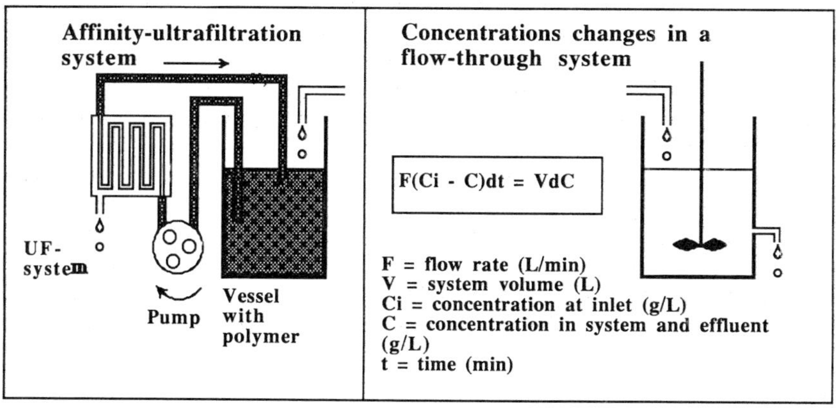

FIGURE 1. Schematic representation of the affinity-ultrafiltration system and a constant-volume flow-through vessel for comparison.

It is important that the rejection coefficient be as small as possible, in order that the elution be rapid and efficient. FIGURE 2 shows the elution profile from the ultrafilter for different values of R. Using a polysulfone membrane (Millipore), the R value for bovine trypsin (MW 24,000) was found to be 0.66 for a 100,000 MW cutoff membrane, while for a 300,000 cutoff membrane, it was found to be 0.17. Complete elution requires about half the volume using the latter as compared with the former. It is thus important to use an affinity polymer with a large MW and a membrane with a large nominal cutoff and as low as possible adsorption characteristics for proteins.

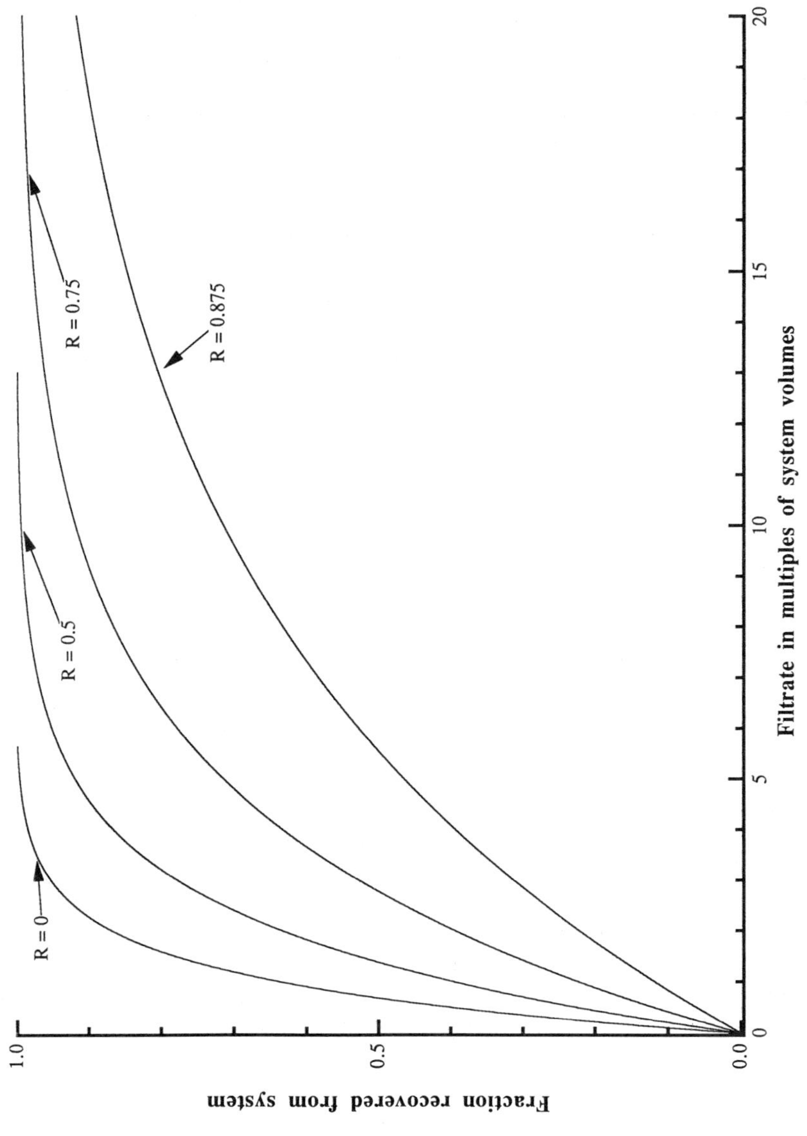

FIGURE 2. The rate of elution of a protein from an ultrafiltration system shown for different values of the rejection coefficient, *R*.

MATERIALS AND METHODS

Preparation of the Affinity Polymer

N-acryloyl-*m*-aminobenzamidine was synthesized by the reaction of acryloyl chloride and *m*-aminobenzamidine as described by Luong *et al.*[2] A solution of 0.0115 moles of this product in 20 ml of dimethyl sulfoxide was added to a solution of 0.245 moles of acrylamide in 600 ml of water and 60 μl of TEMED were added. The copolymerization was initiated by adding 80 ml of a 0.2-g/l ammonium persulfate solution. The resulting molar ratio of ammonium persulfate to the monomers was 1:3800. After polymerization, a 50-ml portion of the polymer solution was transferred to a Millipore Minitan ultrafiltration system fitted with a membrane with a MW cutoff of 300,000 and washed with large volumes of urea (8 M) and distilled water. A sample of the retentate was evaporated to dryness and weighed in order to estimate the amount of polymer that was retained by the membrane.

A sample of the retained polymer was subjected to proton NMR spectroscopy in a Bruker AC 250 spectrometer to estimate the ratio of benzamidine-carrying residues.

Enzyme Assays

Trypsin and chymotrypsin activities were measured spectrophotometrically using the substrates benzoylarginine-*p*-nitroanilide and succinyl-phenylalanyl-*p*-nitroanilide respectively in microtiter trays using a ThermoMax Microplate Reader (Molecular Devices Corp., Menlo Park, Calif.). Instrument control and data analysis were effected by means of a Macintosh computer using SOFTMax software (Molecular Devices). The assay conditions were as described by Bergmeyer.[3] Where necessary, enzyme samples were freed of benzamidine by passage through a mini–gel filtration column in a centrifuge.

Affinity-Ultrafiltration System

The system, shown schematically in FIGURE 1, consisted of a Minitan tangential flow ultrafiltration device (Millipore), a vessel for the polymer solution, and a peristaltic pump. The total volume of the polymer solution was typically 70 ml, and the pumping rate was 150 ml/minute. The system volume was maintained by adding buffer continually to replace fluid leaving the system as filtrate.

Calculations

Calculation of theoretical curves was done with the aid of a Macintosh personal computer using the Igor graphing and data analysis program (WaveMetrics, Lake Oswego, Oreg.).

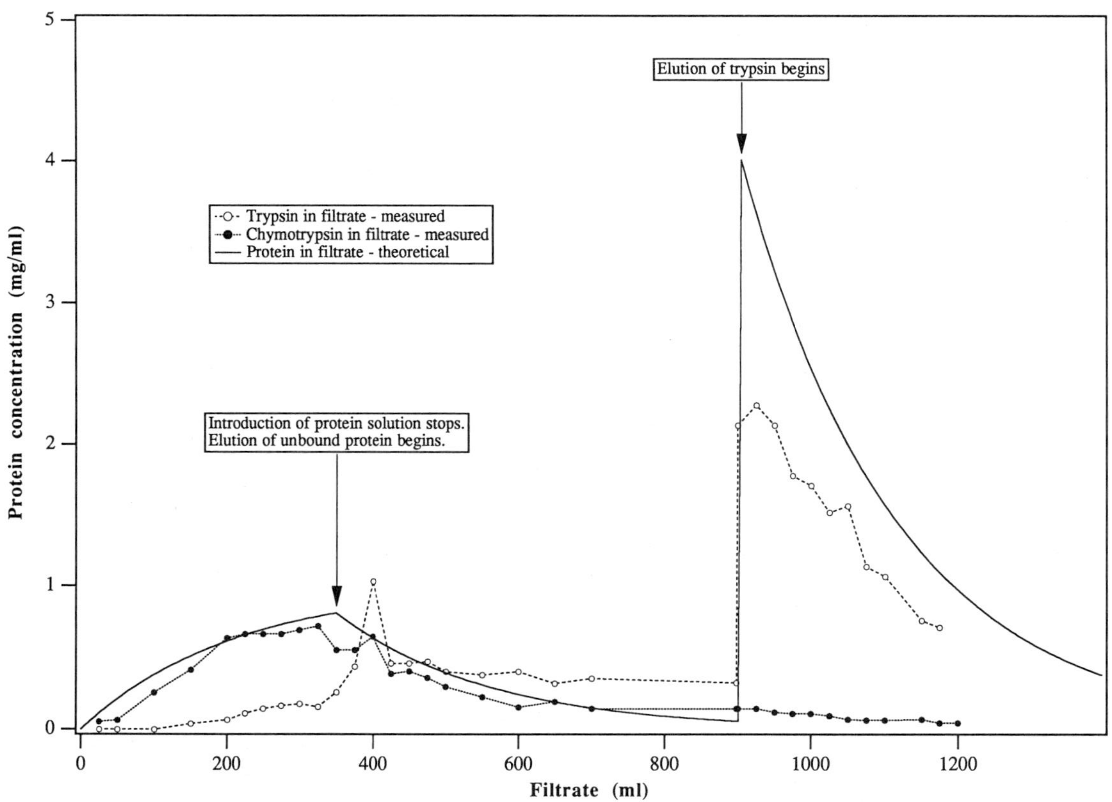

FIGURE 3. The separation of trypsin from chymotrypsin in an affinity-ultrafiltration system containing a benzamidine-containing acrylamide polymer and a 300,000 nominal cutoff membrane. Measured and theoretical curves are shown.

Preparation of Pancreas Extracts

Pancreas glands were collected from lambs at slaughter and frozen. The glands were size reduced to flakes in the frozen state by means of a domestic vegetable grater, extracted with dilute hydrochloric acid at pH 2.5 with slow stirring for 24 hours, and filtered. The extract was taken to 50% saturation with ammonium sulfate, the supernatant desalted and activated with a trace of trypsin for 6 hours at 38°C in 0.01-M calcium chloride and 0.025 mol/l tris buffer at pH 8.

RESULTS AND DISCUSSION

The theoretical considerations indicated that for such a system to be efficient in keeping working volumes low, the nominal cutoff for the membrane should be as high as possible compared with the molecules being separated. Therefore an attempt was made to prepare an affinity polymer of a high molecular weight.

By using low concentrations of initiator in the polymerization reaction mixture, the resulting polymer was large enough to be substantially retained by a 300,000 MW cutoff membrane. Analysis of UV absorption spectra of reactants and products before and after ultrafiltration, as suggested by Luong et al,[2] yielded insufficient information to estimate the composition of the polymer and the yield of polymer over MW 300,000. The composition of the polymer was therefore determined by proton NMR, and it was found that one-sixteenth of the acrylamide residues carried a benzamidine residue. The polymer yield as determined by the dry weight of the retentate was found to be approximately 80%.

Calculations based on Equation 5, taking into account the rejection coefficient of the ultrafiltration membrane, allowed the elution of the different components of a two-enzyme mixture from the system to be predicted with good accuracy.

FIGURE 3 shows the separation of a mixture of chymotrypsin (1 mg/ml) and trypsin (2 mg/ml) and the theoretical results. Trypsin was eluted from the polymer by introducing benzamidine into the system to a concentration of 50 mM.

A polymer solution of approximately 5 g/l was found to bind some 8 mg trypsin per ml, which compares favorably with most other affinity systems.

The system described was tested in the isolation of trypsin from a semicrude extract of ovine pancreas. The specific activity of the trypsin derived by affiny ultrafiltration was 36 μkat/mg, and the purification achieved over the raw tissue extract was 164-fold. These results indicate that the method is successful for the isolation of trypsin from a tissue extract. As the experimental system of 100 ml can bind some 800 mg of trypsin, this may be a practical method for enzyme recovery, if it is successfully scaled up.

REFERENCES

1. ADAMSKI-MEDDA, D., Q. T. NGUYEN & E. DELLACHERIE. 1981. J. Membr. Sci. **9:** 337–342.
2. LUONG, J. H. T., K. B. MALE & A. L. NGUYEN. 1988. Biotechnol. Bioeng. **31:** 439–446.

Transformation of *N*ε-Z-L-Lysine to Z-L-Oxylysine Using L-Amino Acid Oxidase from *Providencia alcalifaciens* and L-2-Hydroxyisocaproate Dehydrogenase from *Lactobacillus confusus*

RONALD L. HANSON, RAMESH N. PATEL, AND LASZLO J. SZARKA

Department of Microbial Technology
Bristol-Myers Squibb
New Brunswick, New Jersey 08903

Biotransformations were developed to oxidize *N*ε-Z-L-lysine and to reduce the product keto acid to L-Z-oxylysine. Lysyl oxidase from *Trichoderma viride* was relatively specific for L-lysine and had very low activity with *N*ε-substituted derivatives. L-Amino acid oxidase from *Crotalus adamanteus* venom had low activity with L-lysine but high activity with *N*ε-formyl-, *t*-BOC-, acetyl-, trifluoroacetyl-, or Z-L-lysine. L-2-Hydroxyisocaproate dehydrogenase from *Lactobacillus confusus* catalyzed the reduction by NADH of the keto acids from *N*ε-acetyl-, trifluoroacetyl-, formyl-, and Z-L-lysine but was inactive with the products from oxidation of L-lysine, L-lysine methyl ester, L-lysine ethyl ester, or *N*ε-*t*-BOC-L-lysine.

Providencia alcalifaciens (SC9036, ATCC 13159) was a good microbial substitute for the snake venom oxidase and also provided catalase. *N*ε-Z-L-Lysine was converted to Z-L-oxylysine in 95% yield with 98.5% optical purity by oxidation using *Providencia alcalifaciens* cells followed by reduction of the keto acid using L-2-hydroxyisocaproate dehydrogenase. NADH was regenerated using formate dehydrogenase from *Candida boidinii.* The *Providencia oxidase* was localized in the particulate fraction, and catalase activity was predominantly in the soluble fraction of sonicated cells. pH optima and kinetic constants were determined for the reactions.

INTRODUCTION

As requirements for optical purity of pharmaceuticals become more stringent, enzymic methods for production of chiral synthons are receiving increased attention. L-Z-oxylysine is an intermediate needed for synthesis of ceronapril, a new ACE inhibitor being developed for treatment of hypertension (FIGURE 1).[1] L-Z-Oxylysine has been prepared from L-lysine by diazotization with sodium nitrite and sulfuric acid followed by treatment of the product L-oxylysine with benzyl chloroformate.[1] As an alternative approach, we have explored the enzymic oxidation of L-lysine or

Z-L-lysine to the keto acid followed by reduction using a dehydrogenase with appropriate enantioselectivity.

MATERIALS AND METHODS

Enzyme Assays

L-Amino acid oxidase was monitored by coupling the H_2O_2 evolved in the reaction to the oxidation of *o*-dianisidine catalyzed by horseradish peroxidase.[2] The reaction solution contained in 1 ml: 50 mM potassium phosphate pH 7.4, 1-mM L-lysine or lysine derivative, 0.2-mM *o*-dianisidine, and 10 μg (2.17 U) horseradish peroxidase. The reaction was started by the addition of L-amino acid oxidase, and the increase in absorbance at 460 nm was monitored (absorbance = 11.3 mM^{-1} cm^{-1}). All continuous spectrophotometric assays were performed at 25°C.

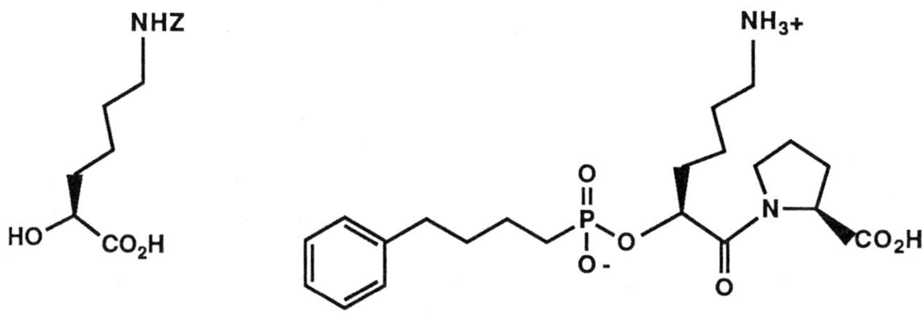

L-Z-oxylysine Ceronapril

FIGURE 1. Structures of L-Z-oxylysine and ceronapril.

The reaction solution for L-2-hydroxyisocaproate dehydrogenase (L-HIC dehydrogenase) coupled to L-amino acid oxidase contained in 1 ml: 0.1-M potassium phosphate pH 7.4, 1-mM lysine or lysine derivative, 2000 U bovine liver catalase, and 1.7 U L-HIC dehydrogenase. Reactions were started by addition of L-amino acid oxidase, and the absorbance decrease at 340 nm was monitored.

The catalase assay contained in 1.0 ml: 0.1-M potassium phosphate pH 7.4 and 0.06% H_2O_2. Absorbance decrease after addition of enzyme was monitored at 240 nm (absorbance = 0.0436 mM^{-1} cm^{-1}).

Protein was determined by the dye-binding method of Bradford,[3] using bovine serum albumin as standard.

Analytical Methods

High-performance liquid chromatography (HPLC) analysis of L-Z-lysine transformations was performed with a Hewlett-Packard hypersil C18 200 × 4.6 mm column, with 5-micron particle size. Column temperature was 40°C, mobile phase was 37%

methanol and 63% water containing 0.05% H_3PO_4, flow rate was 1 ml/minute, detection wavelength was 215 nm, and injection volume was 5 μl.

Optical purity of Z-oxylysine was determined by derivatization and separation of diastereomers by gas chromatography (GC).[4]

Microbial Growth Conditions

Providencia alcalifaciens SC9036 was obtained from the Squibb culture collection and is ATCC strain 13159. It was grown at 37°C on medium described by Szwajcer *et al.*[5] containing 1% peptone, 0.2% casein hydrolysate, 0.2% yeast extract, and 0.6% NaCl at pH 7.2 to 7.4.

Enzyme Localization

Cells (1.40 g wet weight) were sonicated in 15 ml 50-mM potassium phosphate, pH 7.4. Debris was removed by centrifugation for 10 minutes at 12,000 × *g*. The extract supernatant was centrifuged for 1 hour at 101,000 × *g* to give a supernatant and pellet fraction. The pellet was resuspended in 2 ml 50-mM potassium phosphate, pH 7.4. Oxidation of Z-lysine was measured by incubating 5-mM Z-lysine with 0.6 ml of extract or 101,000 × *g* pellet or supernatant fractions in 1.5 ml containing 0.1-M potassium phosphate pH 7.4 and 3000 U catalase for 16 hours.

Materials

L-HIC dehydrogenase was a generous gift from Dr. H. Schütte, Gesellschaft für Biotechnolische Forschung, Braunschweig, Germany. Commercial sources were: lysyl oxidase, Yamasa Shoyu Co., Ltd; L-amino acid oxidase type 1 from *Crotalus adamanteus,* Sigma Chemical Company; PEG-2000-NADH, Braunschweiger Biotechnologie.

RESULTS AND DISCUSSION

Substrate Specificities

The enzymatic approach to synthesis of Z-L-oxylysine is shown in FIGURE 2. *Crotalus adamanteus* (rattlesnake) venom L-amino acid oxidase has been reported to oxidize Z-L-lysine to the corresponding keto acid,[6] and L-HIC dehydrogenase from *Lactobacillus confusus* was found to convert the keto acid to Z-L-oxylysine (R. Robison and M. Doremus, unpublished results.) Lysyl 2-oxidase from *Trichoderma viride* was tested as a microbial alternative to the snake venom enzyme. Substrate specificity studies (TABLE 1) showed that the activity of the enzyme is greatly reduced with ε-substituted lysine derivatives. On the other hand, the snake venom oxidase was active with the ε-substitute lysine derivatives but had very little activity with L-lysine. The methyl and ethyl esters of lysine were substrates for the *T. viride* oxidase but were not utilized by the snake venom enzyme.

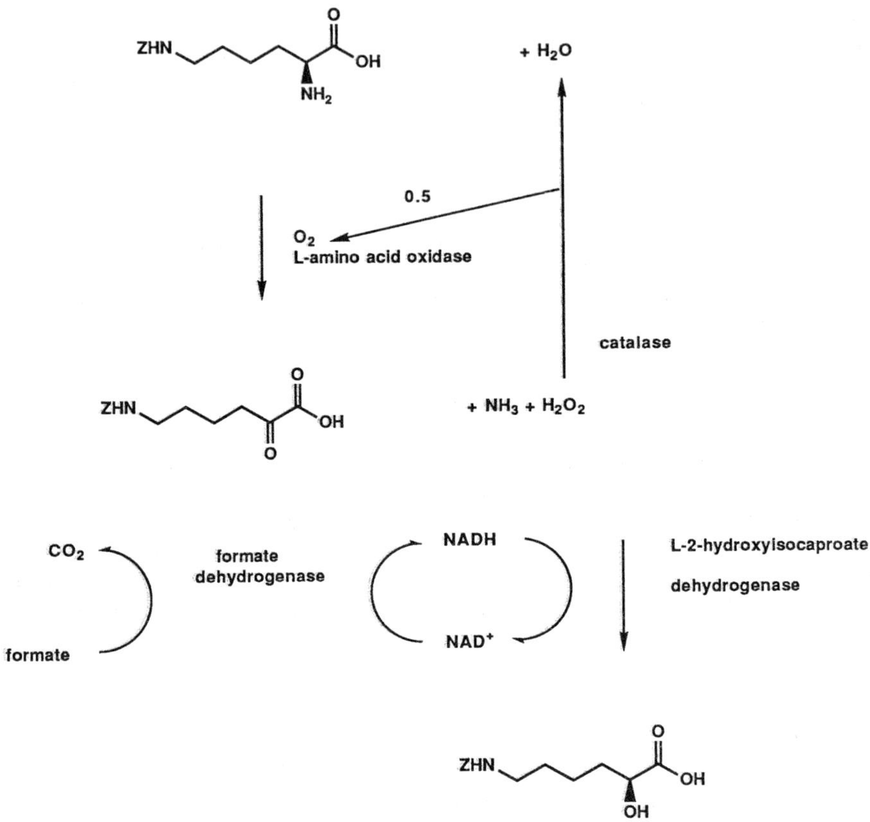

FIGURE 2. Scheme for enzymatic synthesis of L-Z-oxylysine.

The product from lysine oxidation by *T. viride* lysyl oxidase was not a substrate for L-HIC dehydrogenase (TABLE 1) or for several other dehydrogenases that were screened at pH 7.4 or 9. With snake venom L-amino acid oxidase, *N*ε-formyl-, acetyl-, Z-, and trifluoroacetyl-L-lysine were good substrates when the oxidation was coupled to L-HIC dehydrogenase, but *N*ε-t-BOC-L-lysine was not an effective substrate. The product of lysine oxidation by the *T. viride* enzyme has been shown to cyclize to a Schiff base,[7] and this may interfere with utilization by L-HIC dehydrogenase.

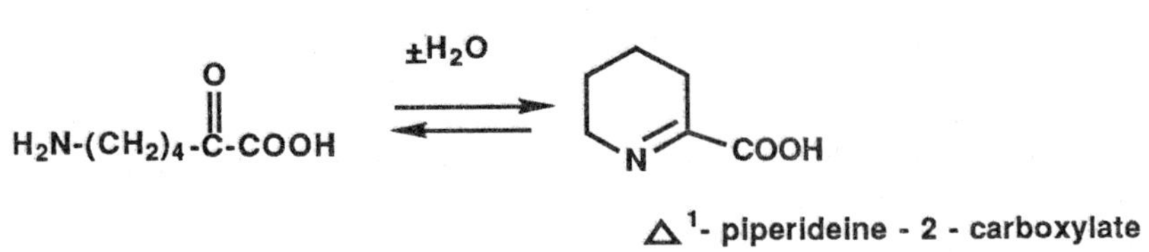

Alternatively, although L-HIC dehydrogenase shows rather broad specificity,[8] it may require a substituent on the ε-amino of lysine for activity.

In an effort to find a more active Z-L-lysine oxidase activity, several microbial strains known to possess L-amino acid oxidase were screened. Four strains of *Proteus* and *Providencia* were found to oxidize Z-L-lysine to the keto acid. *Providencia* *alcalifaciens* SC9036 (*Providencia rustiganii* ATCC 13159) gave the highest conversion to keto acid and was selected for further study.

TABLE 1. Substrate Specificities of L-Amino Acid Oxidases and L-2-Hydroxyisocaproate Dehydrogenase

Enzyme	Substrate	Specific Activity (U/mg oxidase)
T. viride lysyl oxidase	L-Lysine	1.663
	L-Oxylysine	0.000
	*N*ε-Acetyl-L-lysine	0.003
	*N*ε-Formyl-L-lysine	0.033
	*N*ε-*t*BOC-L-lysine	0.000
	*N*ε-Trifluoroacetyl-L-lysine	0.027
	*N*ε-Z Lysine (0.5 mM)	0.003
	*N*ε-Z-Lysine methyl ester	0.000
	L-Lysine methyl ester	0.847
	L-Lysine ethyl ester	1.083
	lys pro NH_2	0.000
C. adamanteus L-amino acid oxidase	L-Lysine	0.000
	L-Oxylysine	0.001
	*N*ε-Acetyl-L-lysine	0.405
	*N*ε-Formyl-L-lysine	0.338
	*N*ε-*t*BOC-L-lysine	0.254
	*N*ε-Trifluoroacetyl-L-lysine	0.406
	*N*ε-Z-Lysine	0.405
	*N*ε-Z-Lysine methyl ester	0.006
	L-Lysine methyl ester	0.001
	L-Lysine ethyl ester	0.000
	lys pro NH_2	0.001
T. viride lysyl oxidase coupled to L-2-hydroxyisocaproate dehydrogenase	L-Lysine	0.000
	L-Lysine ethyl ester	0.000
	L-Lysine methyl ester	0.000
C. adamanteus L-amino acid oxidase coupled to L-2-hydroxyisocaproate dehydrogenase	*N*ε-Acetyl-L-lysine	0.062
	*N*ε-Formyl-L-lysine	0.120
	*N*ε-*t*BOC-L-lysine	0.000
	*N*ε-Trifluoroacetyl-L-lysine	0.028
	*N*ε-Z-L-Lysine	0.120

Z-L-lysine was converted to Z-L-oxylysine in 95% yield by the following procedure: 5.6 g Z-lysine was added to 1 L of solution containing 0.1-M potassium phosphate, pH 7.4, and 10 g SC9036 cells, and the mixture was shaken at 200 rpm for 27 hours at 30°C. Z-L-Lysine is nearly insoluble at pH 7.4 but went into solution as the reaction proceeded. After 24 hours, HPLC indicated complete conversion to the keto acid.

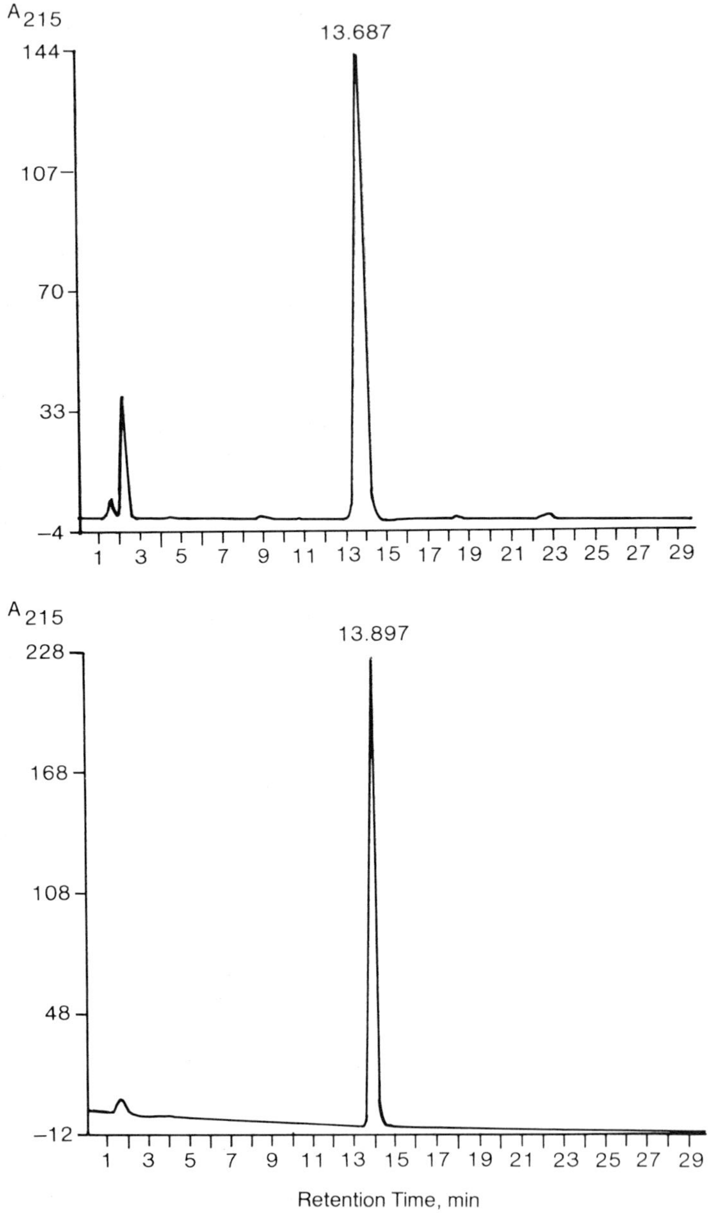

FIGURE 3. HPLC analysis of enzymatically synthesized L-*Z*-oxylysine before isolation (top) and chemically synthesized standard (bottom).

TABLE 2. Localization of Z-Lysine Oxidase and Catalase Activities in *Providencia alcalifaciens*

Fraction	Catalase (U/mg)	Oxidase[a] (% conversion)	Protein (mg/ml)
Extract	1328	97.6	4.3
101,000 × *g* pellet	318	96.8	5.1
101,000 × *g* supernatant	2197	0	2.4

[a]Percent conversion of 5-mM Z-lysine to keto acid measured by HPLC as detailed in Materials and Methods.

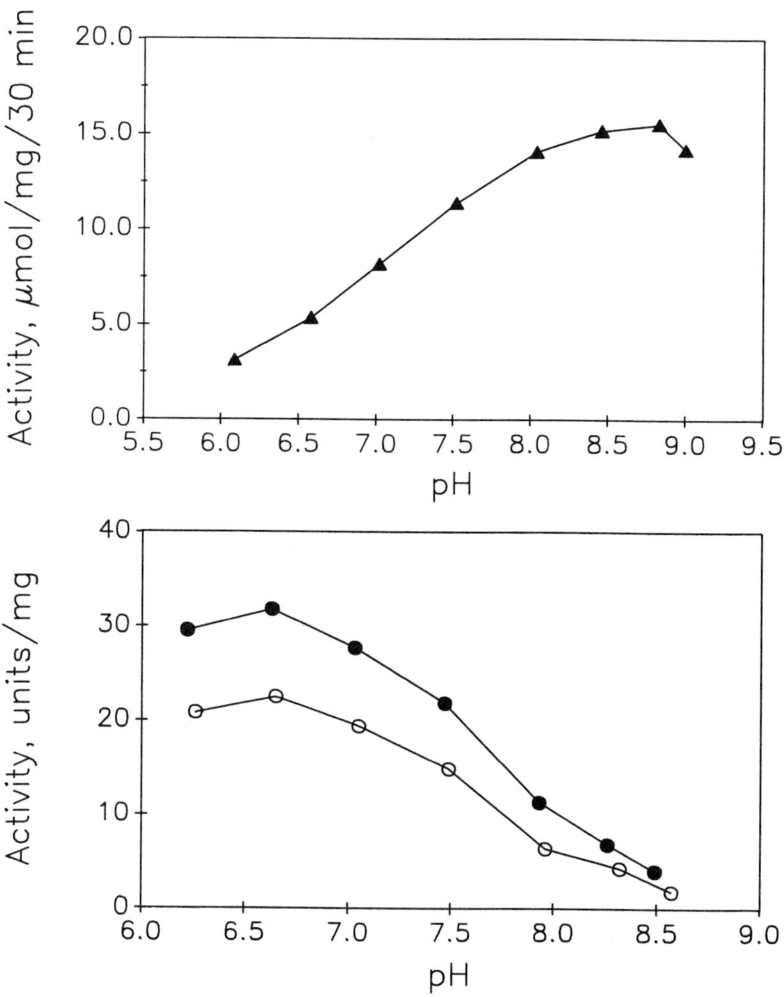

FIGURE 4. Effect of pH on biotransformations. **Top:** Oxidation of *N*ε–Z-L-lysine. Potassium phosphate buffer 0.1 M, 1 mM Z-L-lysine, and 5 mg (wet weight)/ml *P. alcalifaciens* were shaken at 28°C, 200 rpm for 30 minutes. Oxidation of Z-L-lysine was determined by HPLC. **Bottom:** Reduction of 6-Z-amino-2-oxohexanoic acid using L-HIC dehydrogenase with NADH or PEG-NADH. Assay conditions: 0.1-M potassium phosphate buffer, 2-mM 6-Z-amino-2-oxohexanoic acid, 0.3-mM pyridine nucleotide, and 1 μg L-HIC dehydrogenase in 1 ml. ●, NADH; ○, PEG-NADH.

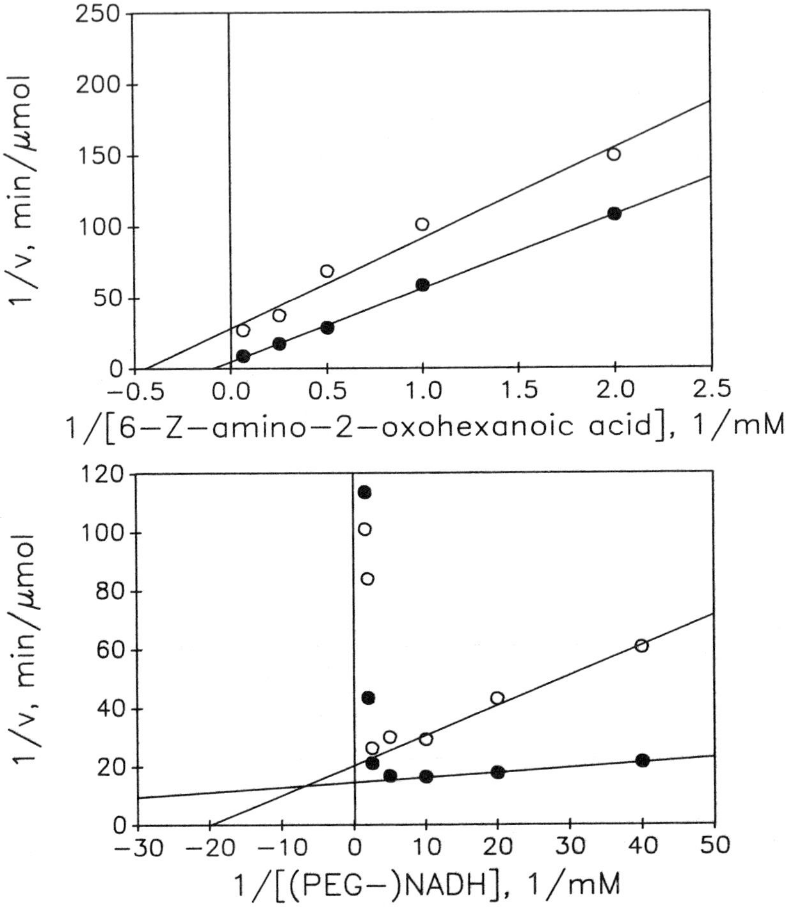

FIGURE 5. Lineweaver-Burk plots for L-HIC dehydrogenase. Assay conditions: 0.1-M potassium phosphate pH 7.4, 1 μg L-HIC dehydrogenase, and 0.2-mM reduced pyridine nucleotide (upper plot) or 4 mM 6-Z-amino-2-oxohexanoic acid (lower plot) in 1 ml. ●, NADH; ○, PEG-NADH.

After cells were removed by centrifugation, 0.2 M sodium formate, 1 mM NAD, 32 U formate dehydrogenase, and 66 U L-HIC dehydrogenase were added and the solution was incubated for 64 hours at 28°C. HPLC indicated conversion to 18.9 mM SQ30411. (FIGURE 3). The isolated product (mp 74–75°C, standard mp 76°C) had 98.4% HPLC homogeneity and 98.5% optical purity.

Enzyme Localization in Providencia alcalifaciens

L-Amino acid oxidase has been reported to be associated with a particulate fraction of the cell in several *Proteus* species.[9,10] Cells, extract, and the particulate fraction from *Providencia alcalifaciens* were effective in oxidizing Z-lysine to the keto acid, but the 101,000 × *g* supernatant had no activity (TABLE 2). No Z-lysine oxidase activity was found in the 101,000 × *g* pellet when the assay measured H_2O_2 production. When conversion to the keto acid was measured by HPLC, Z-lysine oxidase was found in the particulate fraction and was absent from the supernatant (TABLE 2). Very high catalase activity was found in both the pellet and supernatant fractions, with higher specific activity in the supernatant. The high level of catalase accounts for the inability to measure Z-lysine oxidation by H_2O_2 production.

Kinetic Parameters

The pH optimum for oxidation of Z-lysine by *Providencia alcalifaciens* cells coupled to L-HIC dehydrogenase was about 8.8 (FIGURE 4). The optimum pH for reduction of the keto acid by L-HIC dehydrogenase, using either NADH or PEG-NADH, was about 6.6 (FIGURE 4). At pH 7.4 and 0.2 mM NADH, apparent K_m for 6-Z-amino-2-oxohexanoic acid was 10.4 mM and the Lineweaver-Burk plot was linear (FIGURE 5). With 0.2 mM PEG-NADH, apparent K_m for the keto acid was 2.2 mM and apparent V_{max} was 17% of the value with NADH. At pH 7.4 and 4-mM 6-Z-amino-2-oxohexanoic acid, apparent K_m for NADH was 12 μM, and there was substrate inhibition by concentrations of NADH above 0.2 mM (FIGURE 5). Under the same conditions, apparent K_m for PEG-NADH was 50 μM and there was substrate inhibition above 0.4 mM. Apparent V_{max} under these conditions with PEG-NADH was 71% of the value with NADH. Schütte *et al.* have reported that NADH concentrations greater than 0.24 mM inhibit L-HIC dehydrogenase.[8]

REFERENCES

1. KARANEWSKY, D. S., M. C. BADIA, D. W. CUSHMAN, J. M. DEFORREST, T. DEJNEKA, M. J. LOOTS, M. G. PERRI, E. W. PETRILLO, JR. & J. R. POWELL. 1988. J. Med. Chem. **31:** 204–212.
2. BEREZOV, T. T. & E. V. LUKASHEVA. 1988. Biochem. Int. **17:** 529–534.
3. BRADFORD, M. M. 1976. Anal. Biochem. **72:** 248–254.
4. JEMAL, M. & A. I. COHEN. 1987. J. Chromatogr. **394:** 388–394.
5. SZWAJCER, E., P. BRODELIUS & K. MOSBACH. 1982. Enzyme Microbiol. Technol. **4:** 409–413.
6. MEISTER, A. 1954. J. Biol. Chem. **206:** 577–585.
7. KUSAKABE, H., K. KODAMA, A. KUNINAKA, H. YOSHINO, H. MISONO & K. SODA. 1980. J. Biol. Chem. **255:** 976–981.
8. SCHUTTE, H., W. HUMMEL & M. R. KULA. 1984. Appl. Microbiol. Biotechnol. **19:** 167–176.
9. PELMONT, J., G. ARLAUD & A. M. ROSSAT. 1972. Biochimie **54:** 1359–1374.
10. DUERRE, J. A. & S. CHAKRABARTY. 1975. J. Bacteriol. **121:** 656–663.

Coimmobilization of *Gluconobacter oxydans* and *Bacillus cereus* for Bioconversion of 2-Keto-L-gulonic Acid

ZHONG YI YUAN,[a] DONG ZHI WEI,[c] GUANG LIN YIN,[b] AND WEI KANG YUAN[c]

[a]*Shanghai Institute of Biochemistry*
[b]*Shanghai Biotechnology Research Center*
Academia Sinica
320 Yue-Yang Road
Shanghai 200031, China

[c]*East-China University of Chemical Technology*
130 Meilong Road
Shanghai 200237, China

L-Ascorbic acid or vitamin C, as an essential nutriment and an important medicine, has been and still is one of the major products to the pharmaceutical, chemical, and food industries. In addition to the conventional chemical synthesis process, the merits of fermentation by microbial cells have been emphasized in recent decades.[1] Yin *et al.* established a method to convert L-sorbose into the vitamin C precursor 2-keto-L-gulonic acid (2-KLG) by the bacterial strain N1197A, which is composed of *Gluconobacter oxydans* and *Bacillus* sp. Coexistance of both species in the fermentation broth is a prerequisite of bioconversion of 2-KLG.[2,3] An industrially successful, "two-stage fermentation process" using mixed-cell culture has been used in China during the last 15 years. Perlman and his group tried a process of 2-KLG production using a mixture of immobilized cells of *Gluconobacter melanogenus* and *Pseudomonas syringae.*[4] The results led us to consider a coimmobilized cell system as a short approach to vitamin C production.

Encouraged by recent progress in cell-immobilization techniques, we explored different immobilizing procedures to entrap both *Gluconobacter oxydans* and *Bacillus cereus* in the same gel beads, and examined the properties of immobilized cells. The preliminary results indicate that a more efficient production of 2-KLG might be achieved by coimmobilized *G. oxydans* and *B. cereus.*

METHODS

Microorganisms and Cultivation

As Yin *et al.* reported,[2] *Gluconobacter oxydans* and *Bacillus cereus* were grown in cotton-plugged Erlenmeyer flasks containing the growth medium (sorbose, 1.5%; yeast extract, 0.5%; maize, 0.3%; urea, 0.1%; K_2HPO_4, 0.2%; $CaCO_3$, 0.2%; pH 6.5–6.7) at 30°C for 24–48 hours with shaking at 220 rpm, and were harvested by centrifugation.

Coimmobilization of G. oxydans *and* B. cereus

After centrifugation, the packed cells of *G. oxydans* and *B. cereus* were suspended and entrapped under sterile conditions by polyacrylamide, CM-cellulose, alginate, chitosan, agar, and K-carrageenan, as described in Reference 5.

Culture Conditions of Immobilized and Free Cells

Immobilized cell beads (5 g) or 5 ml of the mixed cells were incubated with 25 ml of the production medium at 30°C in a 250-ml flask on a rotatory shaker at 220 rpm. The production medium for the free or immobilized mixed cell system contained sorbose, 2–8%; maize, 2%; urea, 1.2%; $CaCO_3$, 0.2%; pH 6.5–6.7.

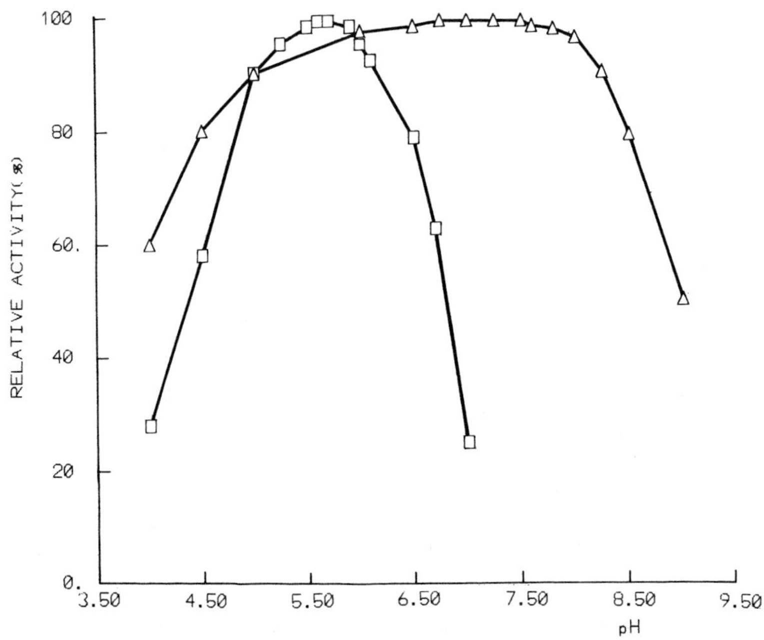

FIGURE 1. Effect of pH on 2-KLG production.

RESULTS

Coimmobilization of Cells

Both *G. oxydans* and *B. cereus* were simultaneously entrapped in different polymer gels as described above. By comparison of characteristics such as mechanical strength, biocompatibility, cell proliferation and leakage, conversion ability, and reusability of the immobilized cells, the K-carrageenan was the most efficient matrix for immobilization of these cells among other gelation materials. The carrageenan-entrapped cells were precultured in the growth medium at 30°C for 24 hours.

Morphological Observations

Immobilized cells were precultured. Distribution of cells in the gel bead was observed under the microscope and electron microscope. The cells form a dense cell layer near the surface area of the gel beads as the microphotograph shows. Transmission electron micrography indicates that similar appearance and size of cells were maintained after 2, 20, and 40 days of cultivation.

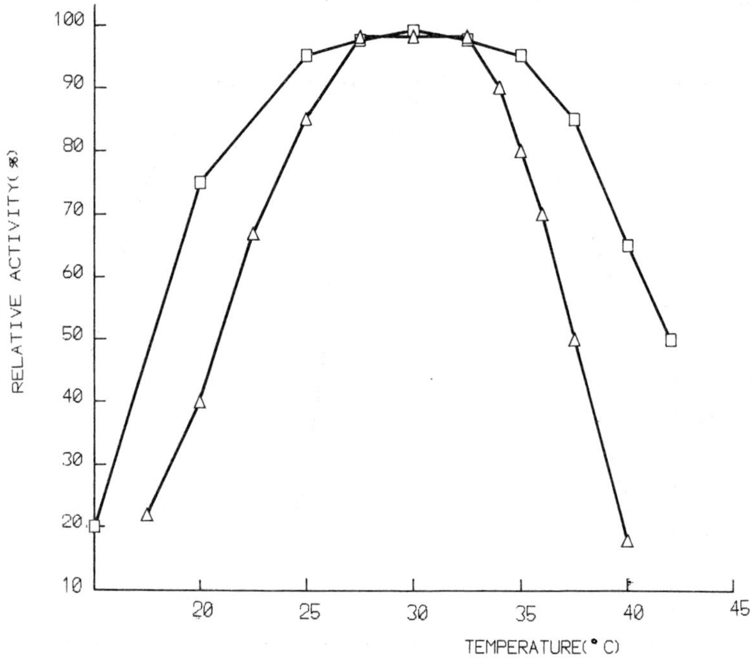

FIGURE 2. Effect of temperature on 2-KLG production.

Characteristics of the Coimmobilized Cells

The immobilization did affect the pH optimum of 2-KLG formation as shown in FIGURE 1. The optimal pH for free cells was 5–6, while the optimal pH of immobilized cells became alkali shifted and much widened. FIGURE 2 indicates a significant difference in the relationship of temperature and 2-KLG production between free and immobilized cells. After incubation of immobilized and free cells at various temperatures for 2 hours, the remaining 2-KLG bioconversion capabilities are summarized in FIGURE 3. When mixed *G. oxydans* and *B. cereus* cells were treated at temperatures between 40 and 50°C, the free cells lost activity more severely than the immobilized cells.

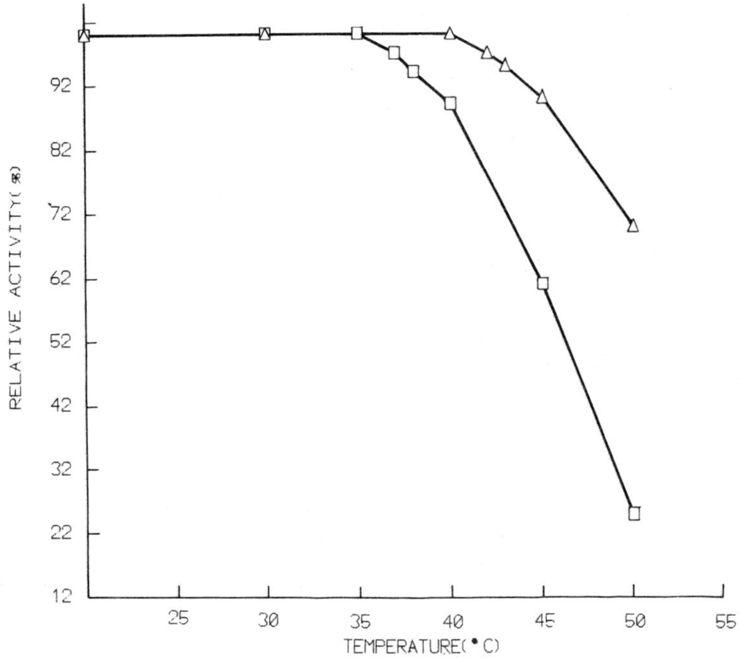

FIGURE 3. Heat stability of immobilized and free cells on capability to form 2-KLG.

Formation of 2-keto-L-Gulonic Acid

Comparison of Different Types of Immobilized Cells

G. oxydans and *B. cereus* were separately entrapped in beads. The mixed cell systems with different ratios of separately immobilized microorganisms were compared to the immobilized dual species as listed in TABLE 1. When *G. oxydans* and *B. cereus* were coimmobilized, the best conversion of L-sorbose to 2-KLG was obtained

TABLE 1. Comparison of 2-KLG Formation by Coimmobilized Cells and Mixed Immobilized Cells

Mixed Immobilized Cells		
Immobilized *G. oxydans* (g)	Immobilized *B. cereus* (g)	2-KLG Production (mg/ml)
5.0	0	3.44
4.5	0.5	1.48
4.0	1.0	1.15
3.0	2.0	0.82
2.0	3.0	0.80
1.0	4.0	0.40
0	5.0	0.00
Coimmobilized *G. oxydans* and *B. cereus* 5 g		6.06

TABLE 2. 2-KLG Production Using Coimmobilized and Free Mixed Cells after Proliferation

Cell Form	No.	2-KLG (mg/ml) 24 Hours[a]	48 Hours[a]	60 Hours[a]
Free cells	1	24.92	38.11	37.90
	2	25.37	38.38	38.41
Immobilized cells	1	42.15	42.80	42.86
	2	42.74	43.00	43.02

[a] Cultivation time.

(see the last line of TABLE 1). The 2-KLG productivity is much higher than those of immobilized *G. oxydans* and/or immobilized *B. cereus.*

2-KLG Formation of Immobilized and Free Cells

The conversion of L-sorbose to 2-KLG was carried out in the production medium. As the data listed in TABLE 2 show, the free mixed cells did not complete bioconversion until 48 hours cultivation. However, in the case of immobilized cells, 24 hours were adequate for conversion. It is considered that carrageenan maintained cells

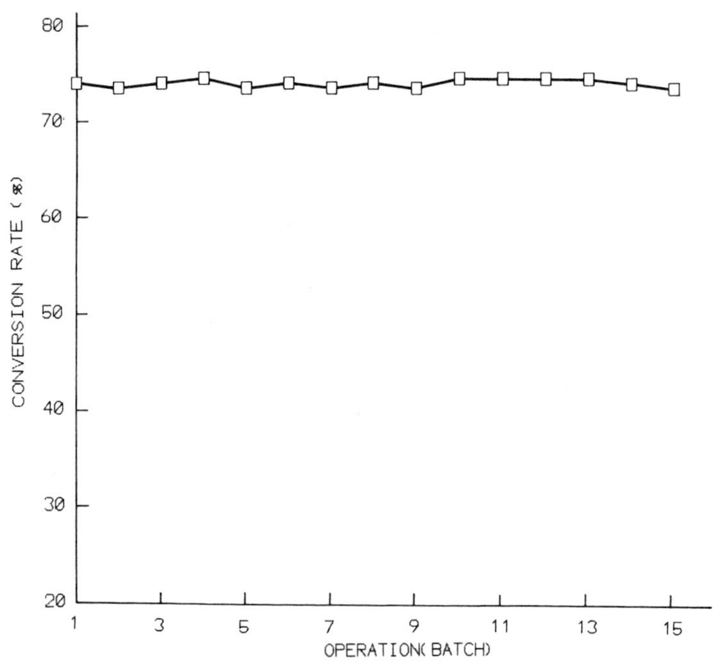

FIGURE 4. Relationship between bioconversion rate of 2-KLG and operational batches of coimmobilized cells.

inside the gel beads and provided a suitable environment for cell proliferation. In the same experimental conditions, the productivity of the immobilized cells is slightly higher than that of the free cells.

Longer Operational Stability

The stability of gel-entrapped *G. oxydans and B. cereus* cells in batch culture was investigated. About 5 ml immobilized cells or 5 ml inoculum solution was added into a 250-ml cotton-plugged flask with 25 ml production medium and incubated at 30°C on a shaker at 220 rpm. The cultivation lasted for 24 hours incubation, which is the point of maximum production rate. The results are shown in FIGURE 4. During 16 successive operations, no decay of 2-KLG production rate was found.

CONCLUSIONS

Entrapment in carrageenan gel has been successfully used for the immobilization of a *Gluconobacter oxydans* and *Bacillus cereus* cell mixture. After immobilization, the dual cell mixture improved resistance against pH and thermal inactivation. The coimmobilized cell system was an excellent approach to convert L-sorbose to 2-keto-L-gulonic acid, the precursor of L-ascorbic acid. Time was saved by using the immobilized cell system instead of the free cell culture. The gel-entrapped cells were successively utilized to produce 2-KLG over 16 batches. Therefore, we conclude that it is worth further exploration of immobilized cell processes for improving productivity, in addition to developing genetically manipulated organisms.

REFERENCES

1. JAFFE, G. M. 1984. Ascorbic acid. *In* Encyclopedia of Chemical Technology, R. E. Kirk & D. F. Othmer, Eds. **24:** 8–40. John Wiley & Sons. New York, N.Y.
2. YIN, G. L., Z. X. TAO, L. H. YU, D. S. WANG, J. L. DAN, Z. Z. YAN, W. Z. NING, C. H. WANG, S. D. WANG, H. F. JIANG, X. M. ZHANG, X. Y. FENG, Q. ZHAO & W. Q. WEI. 1980. Studies on production of vitamin C precursor 2-keto-L-gulonic acid from L-sorbose by fermentation. I. Isolation, screening and identification of 2-keto-L-gulonic acid producing bacteria. Acta Microbiol. Sinica **20:** 246.
3. YAN, Z. Z., Z. X. TAO, L. H. YU, G. L. YIN, W. Z. NING, C. H. WANG, S. D. WANG, H. F. JIANG, J. F. YU, M. S. WANG & X. J. YU. 1981. Studies on production of vitamin C precursor—2-keto-L -gulonic acid—from L-sorbose by fermentation. II. Conditions for submerged fermentation of 2-keto-L-gulonic acid. Acta Microbiol. Sinica **21:** 185.
4. MARTIN, C. K. A. & K. PERLMAN. 1976. Conversion of L-sorbose to 2-keto-L-gulonic acid by mixtures of immobilized cells of *Gluconobacter melanogenus* IFO 3293 and *Pseudomonas* species. Eur. J. Appl. Microbiol. **3:** 91.
5. TAMPION, J. & M. D. TAMPION, Eds. 1987. Immobilized Cells: Principles and Applications: 77. Cambridge University Press. Cambridge, England.

Isonitrile Derivatives of Polyacrylamide as Immobilization Matrices

LEON GOLDSTEIN AND ADINA NIV

Department of Biochemistry
The George S. Wise Faculty of Life Sciences
Tel Aviv University
Ranat Aviv 69978, Israel

INTRODUCTION

Polymeric matrices bearing isonitrile functional groups have been shown to serve as versatile supports for the immobilization of proteins and low molecular weight ligands by four-component condensation reactions (4CC) carried out in aqueous buffers at neutral pH.[1–4]

This communication describes a method for the preparation of isonitrile derivatives of polyacrylamide beads (Biogel P-100). The modified polyacrylamide beads were used as supports for immobilization of proteins and low molecular weight ligands.

RESULTS AND DISCUSSION

Preparation and Characterization of Support

Isonitrile derivatives of cross-linked polyacrylamide beads were prepared by a modification of a procedure described earlier for the preparation of isonitrile derivative of polysaccharides and other polymers bearing hydroxyl groups.[2,4,5]

The procedure consisted of:[4,5]

1. *N*-Hydroxymethylation (methylolation) of amide groups on the support by treatment with formaldehyde.[6]
2. Attachment of side chains containing isonitrile functional groups.[2,4,5] This involved (a) The use of a strong base in a polar aprotic solvent (potassium *tert*-butoxide in dimethyl sulfoxide) to ionize methylol hydroxyl groups on the support; (b) Nucleophilic attack of the polymeric alkoxide ion thus generated on an isonitrile containing a good leaving group in the omega-position; 3-tosyloxyl-1-isocyanopropane [$p\text{-}CH_3 \cdot C_6H_4 \cdot SO_2 \cdot O \cdot (CH_2)_3\text{-}NC$].

$$\vdash CONH_2 \xrightarrow[\text{pH } 9.5]{HCHO} \vdash CONHCH_2OH \quad (1)$$

$$\vdash CONHCH_2OH \xrightarrow[TosO(CH_2)_3NC]{tert-BuO^-/DMSO} \vdash CONHCH_2O(CH_2)_3NC \quad (2)$$

$$Tos = p - CH_3C_6H_4SO_2\text{-}$$

TABLE 1. Isonitrile Derivatives of Hydroxymethylated Polyacrylamide Beads[a]

	Isonitrile Derivative			Bound Trypsin[c]		
	Preparation				Active	
No.	Potass. butoxide (M)	3-Tosyloxyl-isocyano-propane	Isonitrile Content[b] (μmol/g)	Total (mg/g)	mg/g	Percent of Total
1.	0.005	0.1	5.1	3.5	3.2	91
2.	0.018	0.1	32	27	23	85
3.	0.035	0.1	71	38	31	82
4.	0.070	0.05	74	40	32	80
5.	0.070	0.1	150	49	35	71

[a]Hydroxymethyl polyacrylamide beads (Biogel P-100) of methylol groups content 3.5 mmol/g were used.

[b]Estimated from the amount of ^{35}S-methionine or tyrosine bound.

[c]Amount of total bound trypsin was estimated by a modified Lowry procedure.[2,4,5] Amount of active enzyme was determined by rate assays.[2,4,5]

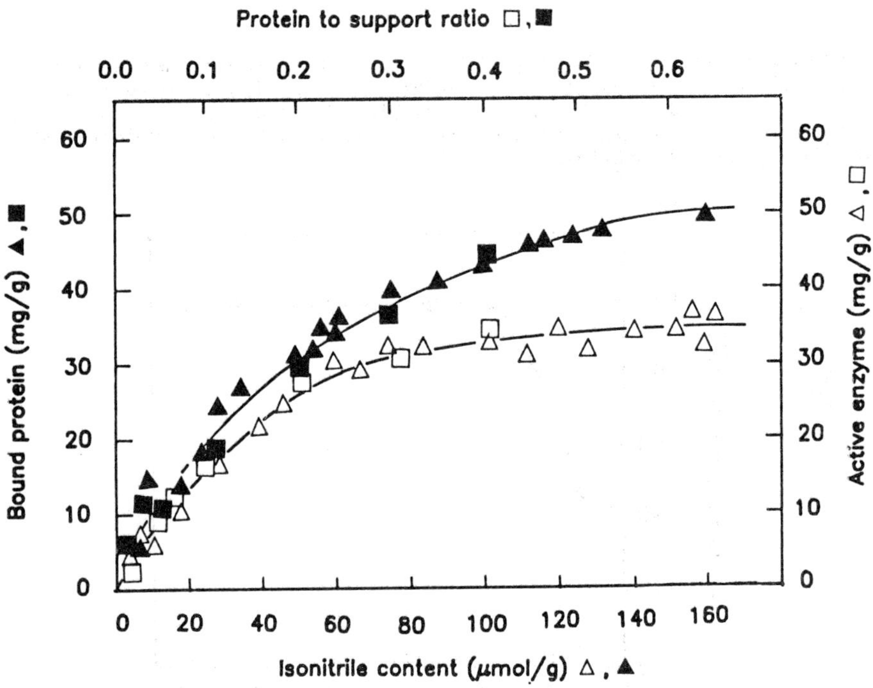

FIGURE 1. Covalent binding of trypsin to isonitrile derivatives of polyacrylamide beads. Two types of 4CC saturation binding experiments were carried out: (a) a constant amount of trypsin (20 mg) was reacted with polyacrylamide derivatives (50 mg) of varying isonitrile content (triangles) and (b) a polyacrylamide derivative of high isonitrile content (99 μmol/g) was reacted with varying amounts of trypsin (squares).[2,4,5] Filled symbols, total bound protein; open symbols, enzymically active protein.

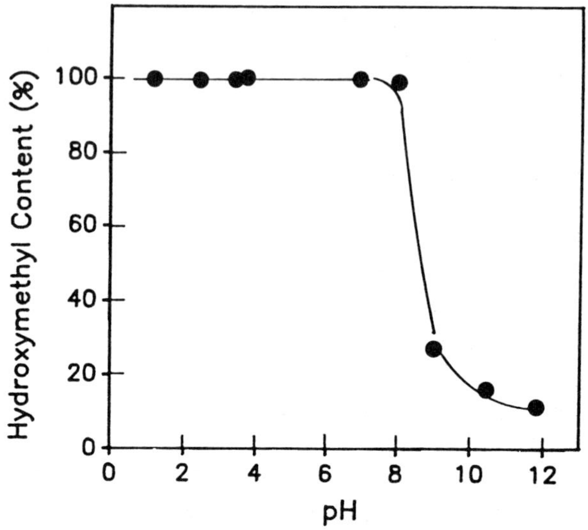

FIGURE 2. Stability of methylol groups of hydroxymethylated polyacrylamide as a function of pH; 0.5-ml aliquots of an aqueous suspension of hydroxymethylated polyacrylamide beads were added to 5 ml of a 0.1-M buffer of the specified pH and immersed in boiling water for 15 minutes. The resin was separated by centrifugation, and the methylol content determined as described by Amarant and Bohak.[6]

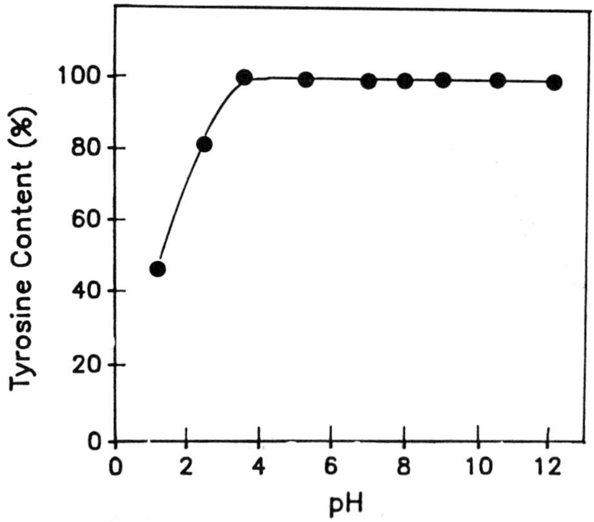

FIGURE 3. Stability of tyrosine-polyacrylamide conjugates as function of pH; 0.5-ml aliquots of an aqueous suspension of tyrosine-polyacrylamide conjugate were added to 5 ml of 0.1-M buffer of the specified pH. The reaction mixtures were immersed in boiling water for 15 minutes, and the resin separated by centrifugation. The tyrosine content of the samples was determined by a modified Lowry procedure.[2,4,5]

The sequence of reactions employed is summarized in Equations 1 and 2. The degree of hydroxymethylation of polyacrylamide beads increased hyperbolically with increasing formaldehyde concentration. Even at 4-M CH_2O, however, only about 32% of the available amide groups were hydroxymethylated in 1 hour at 50°C.[4,5] For comparison, ground particles of polyacrylamide gel (5% cross-linking, highly swollen, and permeable) and linear high molecular weight polyacrylamide in solution underwent, under similar conditions, 50% and 70% hydroxymethylation respectively. This suggests that the extent of modification of amide groups in the cross-linked polymers is probably limited by parameters related to matrix density. Modified

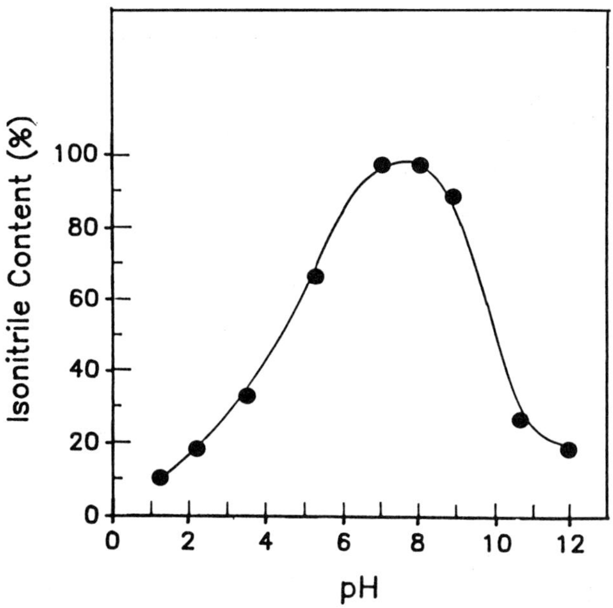

FIGURE 4. Susceptibility of the isonitrile functional group to hydrolysis as a function of pH; 30 mg of an isonitrile derivative of polyacrylamide were suspended in 5 ml of 0.1-M buffer of the specified pH and incubated at 37°C for 48 hours. The resin was removed by centrifugation and reacted with tyrosine under 4CC conditions.[4,5] The tyrosine content of the tyr-polyacrylamide conjugate was determined by a modified Lowry procedure.[2,4,5]

polyacrylamide beads of hydroxymethyl content 3–3.5 mmol/g (20 to 25% modification of amide groups) were routinely used.

The −NC content of the isonitrile derivatives of polyacrylamide, as estimated from the amount of [^{35}S]-methionine bound, varied linearly with the concentration of base or isonitrile reagent (at fixed coreagent concentration). This allowed considerable flexibility in the preparation of derivatized polyacrylamide beads of varying binding capacity as indicated by the data of TABLE 1.

The isonitrile derivatives of polyacrylamide exhibited high storage stability when kept in dry form in the cold; no significant decrease in the binding capacity of samples refrigerated at 4°C over silica gel for up to one year could be detected.

Covalent Attachment of Proteins and Low Molecular Weight Ligands

As shown in earlier publications from our laboratory, proteins and low molecular weight substances such as peptides and amino acids can be attached covalently to polymeric supports containing isonitrile functional groups by four-component condensation reactions, carried out in an aqueous buffer at neutral pH. 4CC reactions involve the simultaneous participation of amine, carboxyl, aldehyde, and isonitrile (Equation 3) and lead to the formation of an *N*-substituted peptide bond between the carboxyl (R^1-COOH) and amine (R^1-NH_2) moieties; the aldehyde and isonitrile

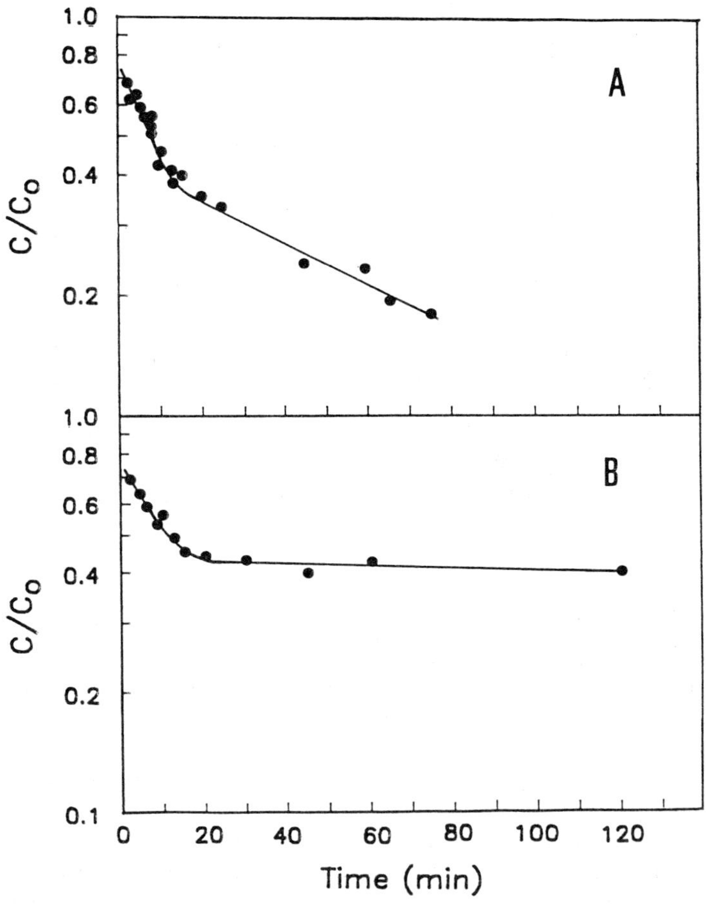

FIGURE 5. Kinetics of 4CC binding of tyrosine to an isonitrile derivative of polyacrylamide beads. (A) 4CC binding via tyrosyl amino groups. (B) 4CC binding via tyrosyl carboxyl groups. A polyacrylamide derivative of high isonitrile content (70 μmol/g) was reacted with a 10^{-3} M solution of tyrosine 1.15 M in acetaldehyde and 0.1 M in phosphate 0.5 M in acetate for binding via tyrosyl amino groups (A) or 0.1 M Tris 1.15 M in acetylaldehyde for binding via tyrosyl carboxyl groups (B).[2,4,5]

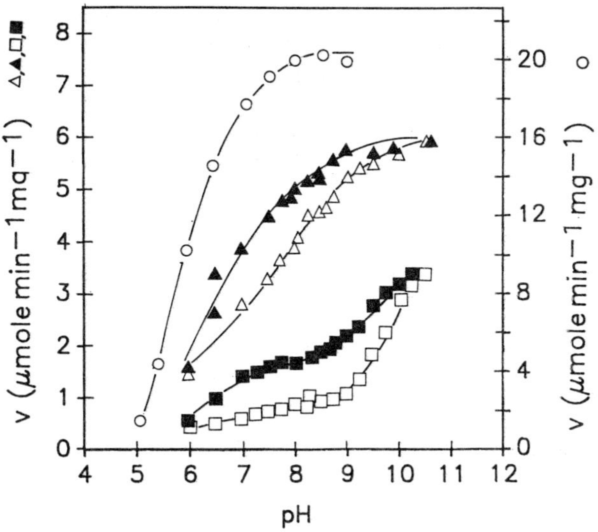

FIGURE 6. pH dependence of the specific activity of trypsin-polyacrylamide conjugates. The enzyme activity of the trypsin-polyacrylamide conjugate (batch 4, TABLE 1) was determined pH statistically[2] using 1.5×10^{-2} M *N*-benzoyl-L-arginine ethylester (10^{-2} M in KCl). Specific activities were calculated per mg immobilized protein. Open squares: intact trypsin-polyacrylamide beads, 10^{-2} M KCl (mean diameter 270 ± 80 microns); filled squares: intact beads, 10^{-2} M KCl, 10^{-3} M borate. Open triangles: ground trypsin-polyacrylamide beads (mean diameter 35 ± 10 microns), 10^{-3} M KCl. Filled triangles: ground trypsin-polyacrylamide beads, 10^{-2} M KCl, 10^{-3} M borate. Open circles: native trypsin, 10^{-2} M KCl.

components (R^3-CHO and R^4-NC) appear as the side chain attached to the amide nitrogen. The protein can serve as either amine or carboxyl donor in 4CC reactions, the actual mode of attachment to the support (R^4 in Equation 3) being determined by the overall composition of the reaction mixture. In the presence of a water-soluble aldehyde (acetaldehyde), coupling via ligand amino groups is obtained when the protein (or unprotected amino acid or peptide) is the sole amine donor, i.e., excess acetate is present in the reaction medium to provide carboxyl groups. Conversely, the presence of an amine (e.g., Tris; trishydroxymethyl aminomethane), in excess in the medium, allows the ligand to undergo covalent attachment through its carboxyl groups.[1-3]

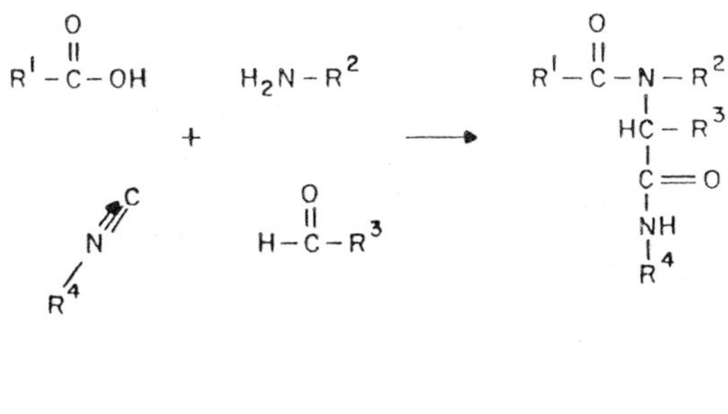

The protein-binding capacity of the polyacrylamide supports was determined under two sets of conditions (FIGURE 1). Polyacrylamide derivatives of varying isontrile content were contacted with excess trypsin under conditions of 4CC binding via protein amino groups (FIGURE 1, circles). Alternatively, a derivatized polyacrylamide preparation of high isonitrile content (99 μmol/g) was contacted with increasing amounts of trypsin (FIGURE 1, squares). The two sets of data points, which

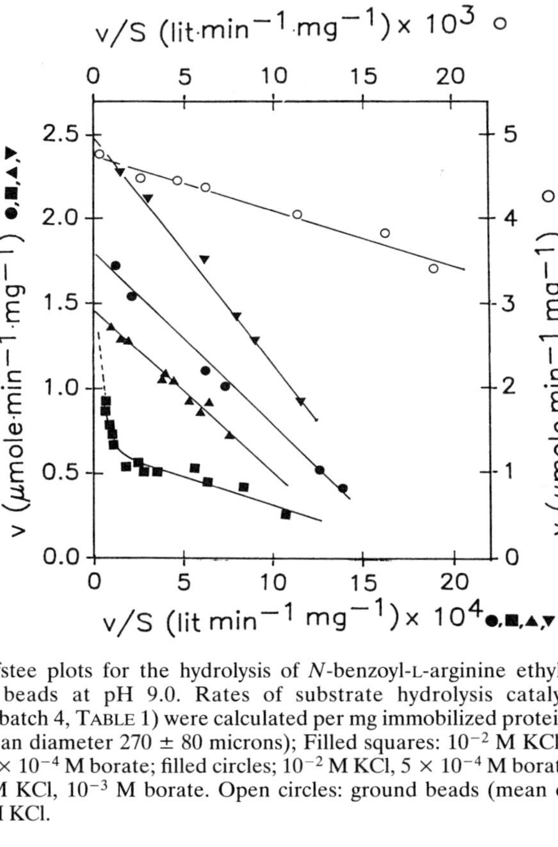

FIGURE 7. Hofstee plots for the hydrolysis of *N*-benzoyl-L-arginine ethylester by trypsin-polyacrylamide beads at pH 9.0. Rates of substrate hydrolysis catalyzed by trypsin-polyacrylamide (batch 4, TABLE 1) were calculated per mg immobilized protein. Filled symbols: intact beads (mean diameter 270 ± 80 microns); Filled squares: 10^{-2} M KCl; Filled triangles: 10^{-2} M KCl, 2.5×10^{-4} M borate; filled circles; 10^{-2} M KCl, 5×10^{-4} M borate. Filled inverted triangles, 10^{-2} M KCl, 10^{-3} M borate. Open circles: ground beads (mean diameter 35 ± 10 microns), 10^{-2} M KCl.

could be fitted on the same saturation curve, asymptoted towards 50 mg/g support and 35 mg/g support for total bound and enzymically active protein respectively. These values imply the presence of diffusional constraints on the penetration of substrate in the case of polyacrylamide beads of high protein content.

Representative data on the saturative binding of a protein, trypsin, and of two amino acids, tyrosine and ^{35}S-methionine, to different batches of modified polyacrylamide beads are summarized in TABLE 1.

STABILITY OF BONDS

In view of the fact that a number of bonds are generated in the course of the synthesis of polyacrylamide-ligand conjugates (see Equations 1–3), the susceptibility to hydrolysis of the bonds and/or groups generated at each step was investigated. To this end, residual capacities or contents were determined following exposure to buffers at different pH values.

The pH stability of the hydroxymethyl groups on methylolated polyacrylamide is shown in FIGURE 2.

The stability to pH of a tyrosine conjugate of isonitrile polyacrylamide is shown in FIGURE 3.

Comparison of the data of FIGURES 2 and 3 establishes the increased stability of the $-CONH \cdot CH_2 \cdot O \cdot (CH_2)_n$-R configuration relative to that of the hydroxymethylated amide groups ($-CONH \cdot CH_2OH$). The susceptibility of the isonitrile functional group to spontaneous hydrolysis is presented in FIGURE 4. The data show that hydrolysis of the $-NC$ group is negligible around neutral pH values.

Binding Kinetics

The kinetics of binding of low molecular weight substances to modified polyacrylamide beads by 4CC reactions was investigated using tyrosine as model compound. The biphasic first-order plots for binding of tyrosine via amino or carboxyl groups shown in FIGURE 5 conform with a two-sequential-reactions model ($k_1 = 7.7 \times 10^{-2}$ minute^{-1}; $k_2 = 0.79 \times 10^{-2}$ minute^{-1}, and $k_1 = 4.8 \times 10^{-2}$ minute^{-1}, $k_2 = 0.08$ minute^{-1} for binding via the amino and carboxyl groups of tyrosine respectively). The kinetics arguably imply the generation, with time, of diffusional constraints on ligand penetration, most probably related to changes in matrix density with the advance of the coupling reaction.

Kinetic Behavior of Polyacrylamide Trypsin Conjugates

The pH dependence of the specific activity of intact and ground polyacrylamide-trypsin beads in the absence and in the presence of buffer (10^{-3} M borate) is shown in FIGURE 6 for the hydrolysis of *N*-benzoyl-L-arginine ethylester. The pH activity profile of intact beads (mean diameter 270 ± 80 microns) at low salt concentration (10^{-3}M KCl) and in the absence of buffer, initially S shaped, levels off as the pH increases and then rises again at high pH values. The presence of 10^{-3} M borate raises the whole pH activity curve but does not significantly change its shape. Grinding of the beads (to mean diameter 35 ± 10 microns) results in additional increase in activity as well as in the abrogation of the anomalous shape of the curves. The pH activity profiles of ground beads are, however, broader than the corresponding curves of native trypsin, even in the presence of 10^{-3} M borate buffer.

This type of behavior is consistent with a "buffer-facilitated proton transport" model proposed by Engasser and Horvath[7–9] to explain the effects of buffers in proton-generating enzyme reactions controlled by diffusional constraints, viz., buffers effect protogenic catalytic events by functioning as carriers which clear the proton from the site of the reaction.[7,10]

The substrate concentration dependence of the rate of hydrolysis of *N*-benzoyl-L-arginine ethylester catalyzed by intact and ground polyacrylamide-trypsin beads at pH 9 in the absence and in the presence of borate is shown in FIGURE 7 (pK_a of boric acid is 9.2).

In the absence of buffer, intact beads exhibit large deviations from Michaelis-Menten behavior, reflected in concave rather than linear Hofstee plots. Borate even at relatively low concentrations leads to the linearization of the Hofstee plots and to increased values of V_{max} (app). Grinding the beads causes additional increase in V_{max} (app) in parallel with drastic decrease in the values of K_M (app).

The data of FIGURE 7 are consistent with the predicted behavior for a diffusionally restricted enzyme-catalyzed reaction inhibited by product (viz., protons).[11,12] In such systems, diffusional constraints would be relieved by either decreasing particle size or by substituting diffusional clearance of protons by their chemical trapping and facilitated transport through the addition of buffer.

REFERENCES

1. GOLDSTEIN, L., A. FREEMAN & M. SOKOLOVSKY. 1974. Chemically modified nylons as supports for enzyme immobilization. I. Polyisonitrile-nylon. Biochem. J. **143:** 497–509.
2. FREEMAN, A., M. SOKOLOVSKY & L. GOLDSTEIN. 1979. Isonitrile derivatives of polysaccharides as supports for the covalent fixation of proteins and other ligands. Biochim. Biophys. Acta **571:** 127–136.
3. GOLDSTEIN, L. 1981. Polymeric supports bearing isonotrile functional groups for covalent fixation of biologically active molecules (a review). J. Chromatogs. **215:** 31–43.
4. GOLDSTEIN, L. 1987. Polymers bearing isonitrile functional groups as supports for enzyme immobilization. Methods Enzymol. **135:** 90–102.
5. GOLDSTEIN, L. & A. NIV. In preparation.
6. AMARANT, T. & Z. BOHAK. 1981. Immobilization of proteins as a tool for studying primary structure around their cysteinyl residues. Appl. Biochem. Biotechnol. **6:** 237–250.
7. ENGASSER, J. M. & C. HORVATH.1974. Buffer-facilitated proton transport. pH profile of bound enzymes. Biochim. Biophys. Acta **358:** 178–192.
8. KONECNY, J. & J. STANICKA. 1975. The effects of buffer diffusion on the activity of an immobilized esterase. Biochim. Biophys. Acta **403:** 573–578.
9. KONECNY, J. & W. VOSER. 1977. Effect of carrier morphology and buffer diffusion on the expression of enzymatic activity. Biochim. Biophys. Acta **485:** 367–378.
10. GUTMAN, M. & E. NACHLIEL. 1990. The dynamic aspects of proton transfer processes. Biochim. Biophys. Acta **1015:** 391–414.
11. ENGASSER, J. M. & C. HORVATH. 1974. Inhibition of bound enzymes. Characterization of product inhibition and accumulation. Biochemistry **13:** 3849–3854.
12. GOLDSTEIN, L. 1976. Kinetic behavior of immobilized enzyme systems. Methods Enzymol. **44:** 397–443.

Continuous Production of 6-APA in an Aqueous Two-Phase System

HO NAM CHANG, YONG-HEE LEE,
AND CHEO YOUNG LEE

Bioprocess Engineering Research Center and
Department of Chemical Engineering
Korea Advanced Institute of Science and Technology
Daeduk Science Town
Taejon 305-701, Korea

INTRODUCTION

Enzyme reactions are carried out in batch, fed-batch, and immobilized cell systems. However, when a product limits the extent of conversion, it is necessary to develop a system where products are continuously removed.[1] To this end, organic and aqueous two-phase systems are extensively used to partition substrates or products separately in a phase system where enzyme catalysts are retained.[2] In the authors' laboratory, aqueous two-phase systems were used to convert acrylonitrile to acrylamide and in the production of alkaline protease.[3,4] Another advantage of aqueous two-phase systems is that cells do not have to be immobilized for reuse since the cell-containing phase can be easily recycled.

Penicillin acylase (penicillin aminohydrolase EC 3.5.1.11) hydrolyzes benzyl penicillin to 6—aminocillanic acid (6-APA) and phenoxy acetic acid (PAA). 6-APA is used as intermediate for the synthesis of a variety of semisynthetic penicillin products. Penicillin acylase is produced by many microorganisms including *Escherichia coli. E. coli* penicillin acylase is produced under glucose-limited culture conditions, and amino acid has been used as the substrate for the enzyme production.[5] Recently, a recombinant strain of *E. coli* harboring plasmid (HB101) pPAKS2 produced the enzyme using glucose as the substrate.[6]

In the aqueous reaction system, the enzyme is inhibited and deactivated by the products 6-APA and PAA. In this study, we tried to lower this inhibition and deactivation using a continuous aqueous two-phase system and could recycle whole cells containing enzymes without immobilization on solid supports.

MATERIALS AND METHODS

Enzyme Source

E. coli HB101 (pPAKS2) was used for the enzyme source. *E. coli* HB101 were transformed using the vector pKM 300 with the penicillin acylase gene from *E. coli* ATCC 11105. The production of penicillin acylase from *E. coli* HB101 (pPAKS2) is not repressed by glucose and need not be induced by phenoxyacetic acid. The strain was stored on LB agar plate at 4°C and transferred every month. Per liter of the medium consisted of glucose 10 g, yeast extract 5 g, casamino acid 5 g, K_2HPO_4 0.5 g, $MgSO_4 \cdot 7H_2O$ 0.2 g, $CaCl_2$ 0.02 g, vitamin B_1 0.001 g, and trace metal solution 0.1 ml. A one-liter trace metal solution contained the following minerals dissolved in 5-N HCl: 10 g $FeSO_4 \cdot 7H_2O$, 6 g $MnSO_4 \cdot nH_2O$, 10 g $AlCl_3 \cdot 6H_2O$, 4 g $CoCl_2$, 2 g $ZnSO_4 \cdot 7H_2O$, 2 g $Na_2MoO_4 \cdot 2H_2O$, 1 g $CuCl_2 \cdot 2H_2O$, and 0.5 g H_3BO_4. The cells were cultured for 24–30 hours at 30°C, 1 vvm of aeration, and stored in a refrigerator. The specific activity of the enzyme was 0.22–0.28 μmol of 6-APA produced/mg of cells per minute.

Analysis

The cell concentration was determined by measuring an optical density at 600 nm (Beckman spectrophotometer model 35). The activity of penicillin acylase was measured with the variation from the method of Balashingham *et al.*[7] that measures the color developed from 6-APA with a spectrophotometer. The analysis reaction

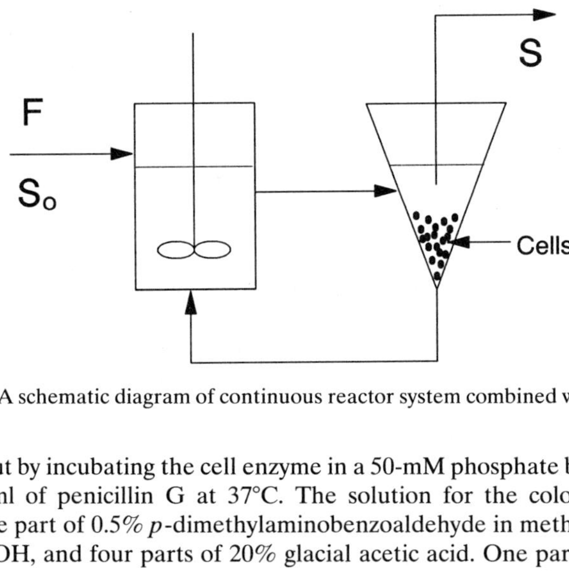

FIGURE 1. A schematic diagram of continuous reactor system combined with separator.

was carried out by incubating the cell enzyme in a 50-mM phosphate buffer of pH 7.5 with 25 mg/ml of penicillin G at 37°C. The solution for the color development consists of one part of 0.5% *p*-dimethylaminobenzoaldehyde in methanol, two parts of 0.05-N NaOH, and four parts of 20% glacial acetic acid. One part of the sample was mixed with 7 parts of the color-development solution to react for 3 minutes, and the optical density was measured at 415 nm.

The analysis of penicillin G and phenylacetic acid was done with high-performance liquid chromatography (HPLC, Water Associates model 440) with μ-Bondapak C_{18} (Water Associates). The solvent was prepared by mixing a phosphate buffer of pH 7.8 with methanol at a ratio of 7:3. The velocity of the mobile phase was 1 ml/minute. The concentration was measured at 254 nm with a UV absorption detector. The measurement of polyethylene glycol content in the aqueous two-phase system was carried out by the method of Hasko *et al.*[8] The procedure is as follows: to precipitate proteins, 1.0 ml of perchloric acid is added to 0.2 ml of a

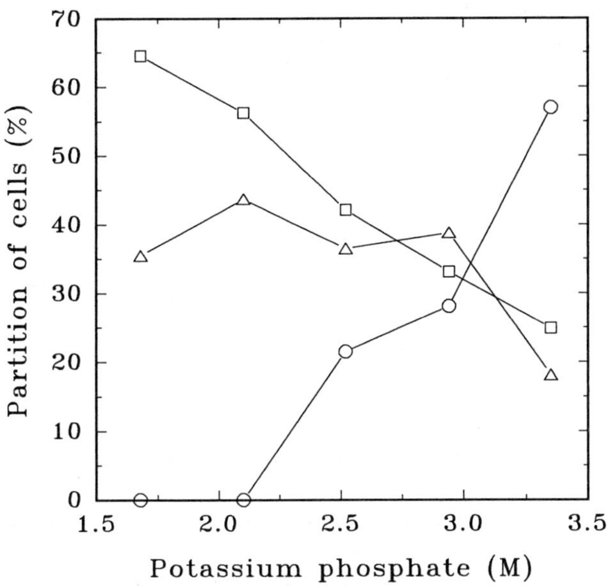

FIGURE 2. Partition of *E. coli* in PEG 4000/potassium phosphate systems (○: top phase, △: bottom phase, □: interface).

sample solution. After 15 minutes, the solution is centrifuged for 4 minutes at 10,000 × *g*. To 1 ml of the supernatant, 0.25 ml of 5% $BaCl_2$ and 0.1 ml of 0.1-M iodine solution are added and allowed to react for 15 minutes. The optical density is measured at 535 nm after the color development. The phosphate content was calculated by subtracting the weights of polyethylene glycol and water after the evaporation at 95°C in an oven.

Bioconversion Experiment

Batch and repeated batch experiments were done at 37°C in a flask (reaction volume: 40 ml) with a magnetic stirrer unless otherwise mentioned. The continuous experiment used the same reactor as in the batch experiments but with a separator where the cell-free top phase was removed and the cell-containing bottom phase was recycled back to the reactor (FIGURE 1). The volume of the separator was 35 ml.

TABLE 1. Kinetic Parameters in Buffer and Aqueous Two-Phase System of Penicillin Acylase at 37°C and pH 7.8

	Aqueous Two-Phase	Buffer System
K_i for 6-APA (mM)	95.5 mM	75.0 mM
K_i for PAA (mM)	8.5 mM	7.6 mM
K_m (mM)	1.52 mM	1.24 mM

TABLE 2. Initial Specific Productivity and Yield Obtained from the Top Phase for Each Batch Reaction in an Aqueous Two-Phase System

Batch	Initial Specific Productivity (μmol 6-APA/mg biocatalyst per minute)	Yield Obtained from the Top Phase (mol 6-APA/mol Pen. G)
1	0.96	0.80
2	0.91	0.71
3	1.00	0.89
4	1.51	0.88
5	1.18	0.68
6	0.72	0.49
7	0.65	0.57
8	0.36	0.35
9	0.45	0.60
10	0.31	0.88

RESULTS AND DISCUSSION

In aqueous two-phase bioconversion systems, it is very important to localize biocatalysts in one phase, which facilitates recycling of biocatalysts and protects the biocatalysts from deactivation. FIGURE 2 shows the distribution of the *E. coli* cells among top, interface, and bottom phases depending on the contents of potassium phosphate (20% PEG). As the content of potassium phosphate was increased, more cells partitioned in the top phase. This is due to the effects of many factors

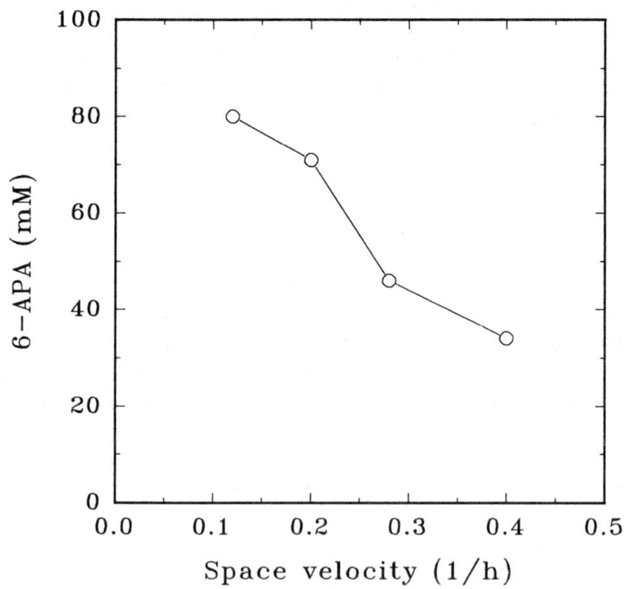

FIGURE 3. Effect of flow rate of substrate on the production of 6-APA. Measured in samples from the total phase system. Reaction volume 40 ml contained 2% (w/v) wet cells.

influencing the partition. We chose a system of 20% PEG 4000 and 8% potassium phosphate where nearly no cells partitioned in the top phase. The partition coefficients of penicillin G, 6-APA, and PAA in the top/bottom phases were 10.0, 2.52, and 1.76, respectively.

We compared the temperature and pH profiles in the two aqueous phases with respect to buffer phases and found no significant differences between them. But the inhibition by the substrate and PAA was less in the aqueous two-phase system. The kinetic constants were shown in TABLE 1.

Removing the top phase containing 6-APA and PAA and refilling it with fresh substrates could establish repeated batch experiments. Thus the whole cell enzymes localized in the bottom phase could be used over and over (TABLE 2). The enzyme activity dropped from 0.96 units/mg to 0.31 units/mg in the 10th run. FIGURE 3 shows

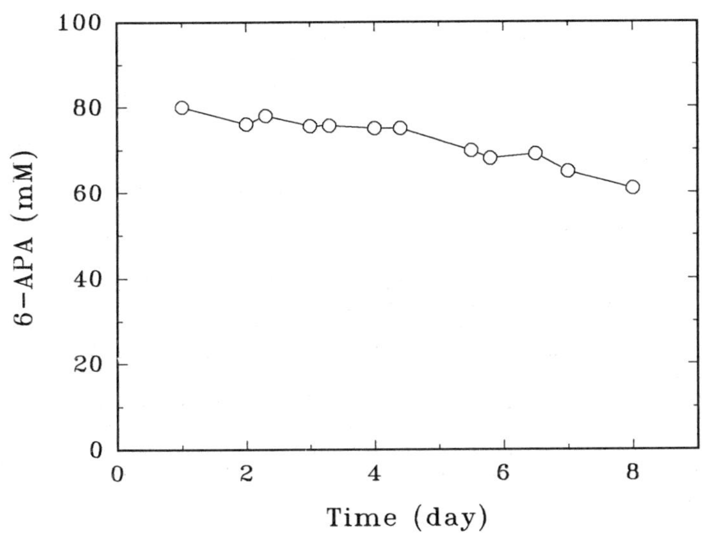

FIGURE 4. Continuous production of 6-APA using reactor-separator in aqueous two-phase system. Space velocity: 0.12 $hours^{-1}$.

the production of 6-APA depending on the space velocity in the continuous production experiment. As expected, at high space velocity, the amount of 6-APA formed decreased because of shorter residence time. Finally, the long-term production of 6-APA at a space velocity of 0.12 $hours^{-1}$ was shown (FIGURE 4). Even after the 7th day, the 6-APA production remained strong, decreasing initially only 80 mM to 65 mM. The amount of 6-APA produced per g of *E. coli* wet cells was 9.2 g in repeated batch (10 runs) and 16.9 g in the continuous experiment. The immobilized cell system suffers from activity loss due to the immobilization process and diffusion. Thus, if the enzyme does not possess long-term operational stability, the aqueous two-phase system could be an alternative to industrial use of many labile enzymes.

SUMMARY

An aqueous two-phase system of polyethylene glycol (PEG) and potassium phosphate provided a favorable environment for bioconversion of penicillin G to 6-aminopenicillanic acid (6-APA). The recombinant *E. coli* cells containing penicillin acylase were partitioned in the phosphate-rich bottom phase, and the product 6-APA in the PEG-rich top phase, which protected the enzyme from deactivation for a longer period of time. The continuous production of 6-APA by recycling the bottom phase showed a slight decrease in enzyme activity from initial 80 mM of 6-APA production to 61 mM during the eight days of operation at a space velocity of 0.12 hours^{-1}. This work showed a possibility that the aqueous two-phase whole-cell enzyme bioconversion could serve as an alternative to immobilized cells on solid matrices.

REFERENCES

1. MATTIASSON, B. 1983. Applications of aqueous-phase systems in biotechnology. Trends Biotechnol. **1:** 16–21.
2. ALBERTSSON, P. A. 1986. Partition of Cell Particles and Macromolecules. 3rd edit. John Wiley & Sons. New York, N.Y.
3. LEE, Y. H. & H. N. CHANG. 1989. Biosynthesis of acrylamide from acrylonitrile in an aqueous two-phase system. Biotechnol. Lett. **11:** 23–26.
4. LEE, Y. H. & H. N. CHANG. 1990. Production of alkaline protease by *Bacillus licheniformis* in an aqueous two-phase system. J. Ferment. Bioeng. **69:** 89–92.
5. LEE, Y. L. & H. N. CHANG. 1988. High cell density continuous culture of *E. coli* producing penicillin acylase. Biotechnol. Lett. **10:** 787–792.
6. LEE, Y. L. & H. N. CHANG. 1990. High cell density culture of a recombinant *E. coli* producing penicillin acylase in a membrane cell recycle fermenter. Biotechnol. Bioeng. **36:** 330–337.
7. BALASHINGHAM, K., D. WARBURTON & M. D. LILLY. 1972. The isolation and kinetics of penicillin amidase from *E. coli.* Biochim. Biophys. Acta **276:** 250–256.
8. HASKO, F., R. VAZILEVA & L. HALASZL. 1982. Solubility of plasma proteins in the presence of polyethylene glycol. Biotechnol. Bioeng. **24:** 1931–1939.

Molecular Weight Distribution Analysis of Water-Soluble Polymers Using Aqueous Two-Phase Systems[a]

MATS LARSSON[b] AND BO MATTIASSON[c]

[b]*Swedish Meat Research Institute*
Post Office Box 504
S-244 24 Kävlinge, Sweden

[c]*Department of Biotechnology*
Lund University
Post Office Box 124
S-221 00 Lund, Sweden

INTRODUCTION

Natural polymers, polysaccharides and proteins, are used with increasing frequency in industrial applications as well as in research. Some of these polymers have very specific activities that are used, but most often the properties from the polymer character are exploited. Precise definition of the molecular weight distribution of polymers is very important for polymer characterization since it dramatically influences the physical-chemical properties of the polymer. Polysaccharides have a broader molecular weight distribution than a specified protein. Distribution analysis of polysaccharides by classical methods is difficult and tedious as it involves fractionation by precipitation and determination of the amount and molecular weight of polymer in every fraction. The most used methods to determine the molecular weight of polymers are averaging methods, which determine the number average molecular weight (M_n) and the weight average molecular weight (M_w). M_n and M_w are defined in Equations 1 and 2.

$$M_n = \Sigma N_i M_i / \Sigma N_i \tag{1}$$

$$M_w = \Sigma N_i M_i^2 / \Sigma N_i M_i \tag{2}$$

N_i is the number of moles of molecules of molecular weight M_i in the polymer. The polydispersity index is defined as M_w/M_n.

When a biopolymer is degraded in an *in vitro* experiment, the reaction is usually quantified or characterized by measuring the amount of monomer liberated or the number of bonds broken using end-group analysis. It would, however, be more relevant to determine the molecular weight and molecular weight distribution of the remaining polymeric substrate.

We have earlier demonstrated the use of aqueous two-phase systems to character-

[a]This project was supported by the National Swedish Board for Industrial and Technical Development (NUTEC).

ize the enzymatic degradation of starch. This was accomplished by monitoring the volume of the starch-containing phase in an aqueous two-phase system.[1] The observation that phase-partitioning behavior is dependent on the molecular weight of the phase-forming polymers is supported by earlier reports.[2] In the same context, we also described a new technique to determine the binodial in-phase diagrams of aqueous two-phase systems.

This report is an extension of earlier studies, and describes the application of spectrophotometric registration of phase transitions in aqueous two-phase systems in order to determine molecular weights and polydispersity indices of proteins and polysaccharides.

MATERIALS AND METHODS

Phase Components and Chemicals

The polyethylene glycols (PEG) used were PEG 3350 (M_w = 3000–3700) and PEG 8000 (M_w = 7000–9000) obtained from Union Carbide. Dextran T10 (M_w = 9000, M_n = 4700), dextran T40 (M_w = 40,500, M_n = 25,400), dextran T70 (M_w = 70,400, M_n = 34,700), dextran T500 (M_w = 510,000, M_n = 184,000), and dextran T2000 ($M_w \approx$ 2,000,000) were purchased from Pharmacia, Sweden. The proteins used were lysosyme (L 6876), trypsin (T 8253), peroxidase (P 8250), and bovine serum albumin (BSA, A 7906) from Sigma Chemical Company. All other chemicals were of analytical grade.

Phase-Transition Experiments in Aqueous Two-Phase Systems Using Dextrans of Different Molecular Weight

Aqueous two-phase systems composed of 1% (w/w) dextran of different known molecular weights and 15% (w/w) PEG 8000 were prepared. Each phase system (5 g) was continuously stirred in a 30-ml beaker, and pumped through a flow cuvette and back to the beaker. The phase system was diluted with deionized water at constant flow rate (0.25 g/minute). The turbidity of the two-phase system was determined by continuous spectrophotometric measurement at 600 nm.[1] The transmittance was recorded. After a certain dilution, the binodial between the two-phase and one-phase area was reached and the two-phase system started to disappear. The transmittance increased rapidly until a homogeneous mixture was obtained.

Phase-Transition Experiments in Aqueous Two-Phase Systems Using Dextrans of Different Polydispersity Index

Six different two-phase systems were prepared. Each system had a total composition of 1% (w/w) dextran and 10% (w/w) PEG 8000. Dextran T10, T70, and T500 were mixed to form an average weight average molecular weight of about 70,000 Da, but with increasing polydispersity index in each system. The phase systems were diluted, and the transmittance was recorded. The polydispersity index was calculated using the molecular weight data of the fractionated dextrans supplied by the manufacturer.

Addition of Protein to a Phase System

Four enzymes or protein (50 mg) of different molecular weights were added to separate phase systems composed of 1.5% (w/w) PEG 3350 and 22.5% (w/w) potassium phosphate (64.5% K_2HPO_4 and 35.5% KH_2PO_4). The phase systems were diluted with deionized water, and the transmittance was recorded as described above.

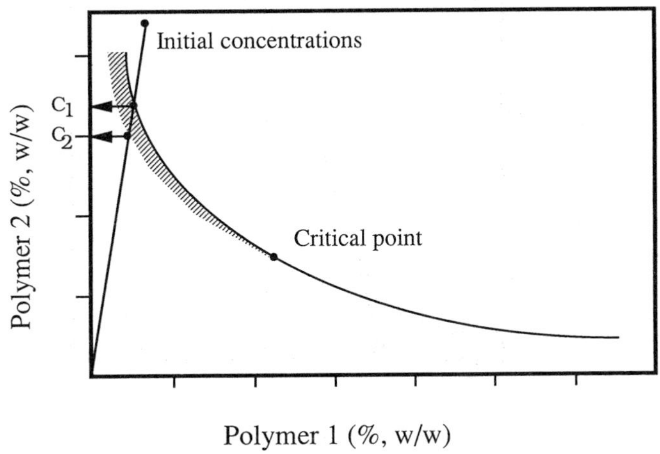

FIGURE 1. Phase diagram of an aqueous two-phase system. The shaded area represents the diffuse transition area from two phases to a homogeneous solution.

RESULTS AND DISCUSSION

Distribution analysis of polydisperse preparations of polymers is often complicated and time-consuming. However, in a recent paper, a convenient method involving the use of aqueous two-phase systems in order to characterize water-soluble polymers was reported.[3]

When two different water-soluble polymers are mixed, the solution becomes turbid under certain conditions, and after some time, phase separation takes place.[2] A typical phase diagram for an aqueous two-phase polymer system is shown in FIGURE 1. In our experiments, the two-phase systems were diluted and the stirred systems were passed through a flow cuvette in a spectrophotometer. In this way, it was possible to register the phase transition. The turbidities of the systems were recorded until the transition from a two-phase system to a homogeneous solution started. In this transition period, the light scattering dropped dramatically and when a homogeneous solution was formed, a constant value was again reached (FIGURE 2). The evaluation of the recordings was done in the sense that the points t_1 and t_2 were constructed between the intercepts of the lines drawn as tangents to the initial and the final slopes and the slope of the transition period, respectively. The concentration of the phase component (c_1 and c_2) which was not evaluated with respect to molecular weight could then be calculated using t_1 and t_2 respectively.[1] In order to study the correlation between the phase transition and the M_w of a polymer, a series of experiments was performed using a two-phase system of dextran and PEG.

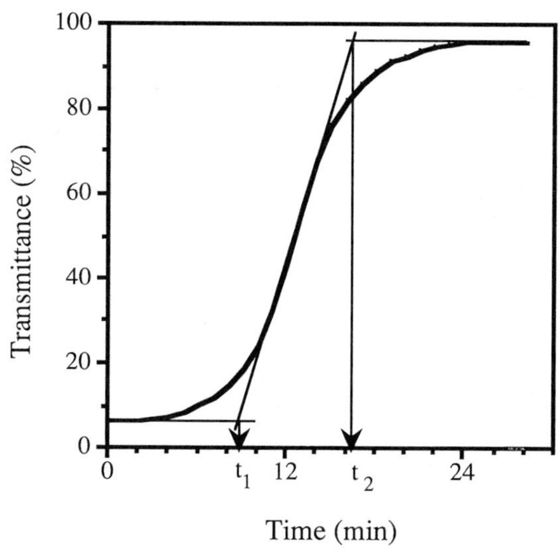

FIGURE 2. Transmittance versus time in an dilution experiment. Two points are defined, t_1 and t_2. t_1 is defined to be the time when the transition from two phases to one phase is initiated, and t_2 when a homogeneous solution appears. The concentration of phase-forming polymers that corresponds to t_1 and t_2 could then be calculated.[1]

Measurements of Molecular Weight and M_w/M_n of Biopolymers in Dextran-PEG Two-Phase Systems

When making a series of two-phase systems with fixed amounts (by weight) of PEG and dextran, interesting results were obtained by using different molecular weights of the dextran in the different systems. If the log of $(c_1 + c_2)/2$ is plotted (c_1 and c_2 are defined in FIGURE 1) versus the logarithm of the molecular weight of the dextrans, a relation as shown in FIGURE 3 is obtained. From this figure, it is obvious that there is a correlation between the phase transition and the molecular weight of the dextran used as the bottom phase component. The sharpness of the phase transition is thus correlated to the molecular weight of the phase components.

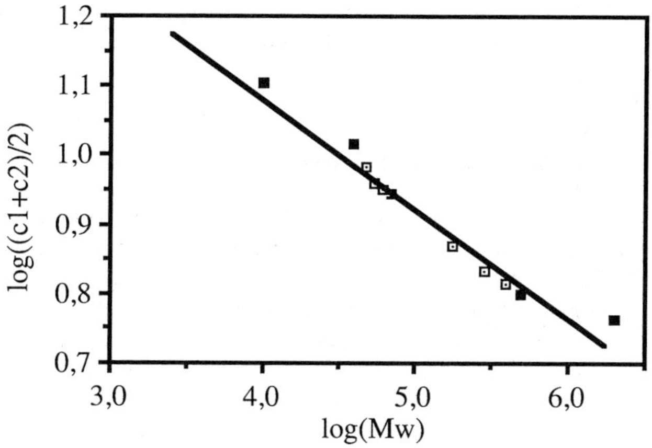

FIGURE 3. The log of $(c_1 + c_2)/2$ as a function of the logarithmic average molecular weight of dextran. Standard curve using commercial preparations (filled symbols) and mixed preparations (open symbols).

In order to study the influence of polydispersity of the polymer preparation on the phase transition, a fixed amount of dextran with different proportions of high- and low-molecular-weight dextrans was used, so that the average molecular weight was constant. In FIGURE 4, the width of the phase transition expressed as $\Delta c = (c_1 - c_2)$ is plotted versus the polydispersity index (M_w/M_n) of the dextran in the system. It is clearly seen from this curve that Δc increases with increasing polydispersity of the dextrans. This could be explained by the fact that when crossing the binodial near one of the axes, the polymers are fractionated between the phases according to their molecular weight and the transition becomes diffuse.[2] This phenomenon is accentuated when using more polydisperse polymers.

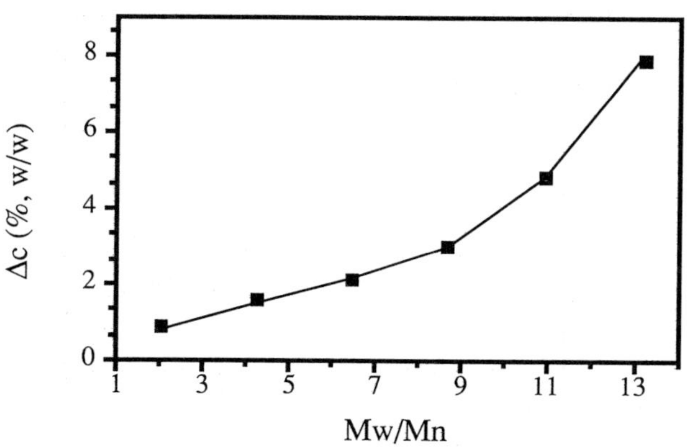

FIGURE 4. Δc as a function of polydispersity index of dextran preparations.

It is thus likely that a simple test like the one performed in these experiments can give very useful information about the molecular weight and polydispersity index of new phase components.

Molecular Weight Measurements of Proteins

Four proteins of different molecular weights were added to two-phase systems with fixed amounts (by weight) of phase components. The systems were diluted with water, and the phase transition was determined. If $(c_1 + c_2)/2$ is plotted versus the molecular weight of the proteins, a relation as shown in FIGURE 5 is obtained. From this figure, it is shown that the phase transition is affected by the molecular weight of the protein. Proteins of higher molecular weight retard the phase transition, and there seems to be a correlation between the degree of retardation and the molecular weight of the protein added.

Degradation of Dextran T2000

Dextran T2000, with an average molecular weight of about 2 million, was hydrolyzed with HCl.[3] Samples were taken out for analysis at various times, and the degradation products were analyzed both by the phase transition method and by gel permeation chromatography (GPC).

The sample to be analyzed was mixed with PEG to form an aqueous polymer two-phase system. A continuous dilution was carried out as described above. The transition values were compared with a calibration curve obtained from studying preparations of known molecular weight.

Gel permeation chromatography on Sepharose CL 4B was used as a reference method. This system was also calibrated with dextran samples of known molecular weights.

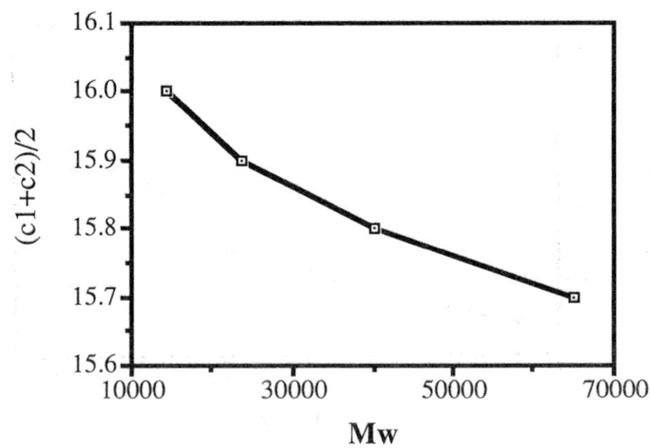

FIGURE 5. $(c_1 + c_2)/2$ as a function of the molecular weight of different proteins added (50 mg) to a system consisting of 1.5% (w/w) PEG 3350 and 22.5% (w/w) potassium phosphate.

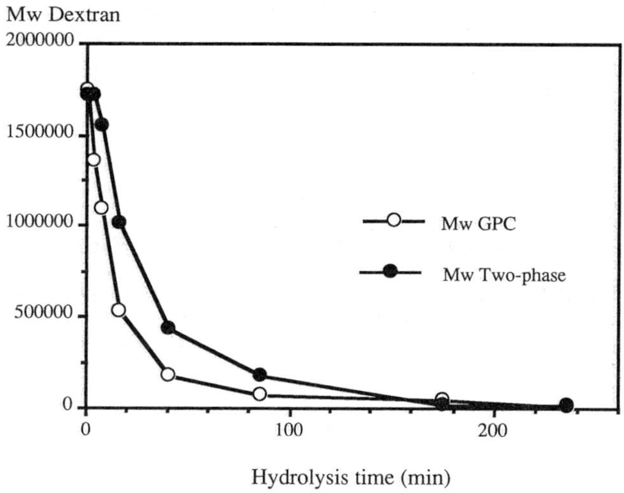

FIGURE 6. Hydrolysis of dextran T2000. Molecular weight determination using GPC (open symbols) and two-phase method (filled symbols).

Comparison of the different results from the two methods shows a significant correlation. The average molecular weights obtained for the different preparations of the hydrolytic process are plotted vs. time of hydrolysis for the two methods (FIGURE 6). Both methods give information on the change in average molecular weight during the process. It should be stressed that none of the methods are absolute. When comparing two semiquantitative methods, it is to be expected that some discrepancies appear.

CONCLUDING REMARKS

The most common method for analysis of molecular weight distribution is gel permeation chromatography. This technique gives good resolution provided components of well-defined molecular weights are studied, but when it comes to heterogeneous mixtures, broadening of the peaks will take place. Furthermore, when a number of samples must be analyzed, gel permeation chromatography is very time-consuming.

There are several publications on the gel permeation chromatography method.[4–6] Therefore, a relatively rich amount of reference material based on this technique may be found in the literature.

The gel permeation chromatography method is also sensitive to the presence of particulate matter in the material studied. Particles will cause clogging of the column. In this respect, the aqueous two-phase systems offer some unique properties in the sense that they are also useful when dealing with polymer preparations with some particulate content.

When studying degradation of polysaccharides, no effects on the phase separation behavior were observed when particulate matter was present. From this, one can draw the conclusion that the method presented here is a simple and convenient method for molecular weight distribution analysis of even crude preparations of water-soluble biopolymers.

REFERENCES

1. LARSSON, M. & B. MATTIASSON. 1988. Biotechnol. Bioeng. **31:** 979–983.
2. ALBERTSSON, P.-Å. 1971. Partition of Cell Particles and Macromolecules. Wiley. New York, N.Y.
3. LARSSON, M. & B. MATTIASSON. 1990. Biotechnol. Appl. Biochem. **12:** 405–416.
4. GRANATH, C. A. & P. FLODIN. 1961. Macromol. Chem. **48:** 160–171.
5. LAURENT, T. C. & K. A. GRANATH. 1967. Biochem. Biophys. Acta **136:** 191–198.
6. GRANATH, K. A. & B. E. KVIST. 1967. J. Chromatogr. **28:** 69–81.

Index of Contributors